Lecture Notes in Computer Science

Founding Editors

Gerhard Goos
Juris Hartmanis

Editorial Board Members

Elisa Bertino, *Purdue University, West Lafayette, IN, USA*
Wen Gao, *Peking University, Beijing, China*
Bernhard Steffen, *TU Dortmund University, Dortmund, Germany*
Moti Yung, *Columbia University, New York, NY, USA*

The series Lecture Notes in Computer Science (LNCS), including its subseries Lecture Notes in Artificial Intelligence (LNAI) and Lecture Notes in Bioinformatics (LNBI), has established itself as a medium for the publication of new developments in computer science and information technology research, teaching, and education.

LNCS enjoys close cooperation with the computer science R & D community, the series counts many renowned academics among its volume editors and paper authors, and collaborates with prestigious societies. Its mission is to serve this international community by providing an invaluable service, mainly focused on the publication of conference and workshop proceedings and postproceedings. LNCS commenced publication in 1973.

Aleš Leonardis · Elisa Ricci · Stefan Roth ·
Olga Russakovsky · Torsten Sattler · Gül Varol
Editors

Computer Vision – ECCV 2024

18th European Conference
Milan, Italy, September 29–October 4, 2024
Proceedings, Part XLIII

Editors
Aleš Leonardis
University of Birmingham
Birmingham, UK

Elisa Ricci
University of Trento
Trento, Italy

Stefan Roth
Technical University of Darmstadt
Darmstadt, Germany

Olga Russakovsky
Princeton University
Princeton, NJ, USA

Torsten Sattler
Czech Technical University in Prague
Prague, Czech Republic

Gül Varol
École des Ponts ParisTech
Marne-la-Vallée, France

ISSN 0302-9743   ISSN 1611-3349 (electronic)
Lecture Notes in Computer Science
ISBN 978-3-031-72774-0   ISBN 978-3-031-72775-7 (eBook)
https://doi.org/10.1007/978-3-031-72775-7

© The Editor(s) (if applicable) and The Author(s), under exclusive license to Springer Nature Switzerland AG 2025

This work is subject to copyright. All rights are solely and exclusively licensed by the Publisher, whether the whole or part of the material is concerned, specifically the rights of translation, reprinting, reuse of illustrations, recitation, broadcasting, reproduction on microfilms or in any other physical way, and transmission or information storage and retrieval, electronic adaptation, computer software, or by similar or dissimilar methodology now known or hereafter developed.
The use of general descriptive names, registered names, trademarks, service marks, etc. in this publication does not imply, even in the absence of a specific statement, that such names are exempt from the relevant protective laws and regulations and therefore free for general use.
The publisher, the authors and the editors are safe to assume that the advice and information in this book are believed to be true and accurate at the date of publication. Neither the publisher nor the authors or the editors give a warranty, expressed or implied, with respect to the material contained herein or for any errors or omissions that may have been made. The publisher remains neutral with regard to jurisdictional claims in published maps and institutional affiliations.

This Springer imprint is published by the registered company Springer Nature Switzerland AG
The registered company address is: Gewerbestrasse 11, 6330 Cham, Switzerland

If disposing of this product, please recycle the paper.

Foreword

Welcome to the proceedings of the European Conference on Computer Vision (ECCV) 2024, the 18th edition of the conference, which has been held biennially since its founding in France in 1990. We are holding the event in Italy for the third time, in a convention centre where we can accommodate nearly 7000 in-person attendees.

The journey to ECCV 2024 started at CVPR 2019, where the general chairs met at a sketchy bar in Long Beach to discuss ideas on how to organize an amazing future ECCV. After that, Covid came and changed conferences probably forever, but we maintained our excitement to organize an awesome in-person conference in 2024, particularly having seen the wonderful event that was ECCV 2022 in Tel Aviv. And here we are, 5 years later, delighted that we finally get to welcome all of you to Milan!

First and foremost, ECCV is about the exchange of scientific ideas, and hence the most important organizational task is the selection of the programme. To our Program Chairs, Aleš Leonardis, Elisa Ricci, Gül Varol, Olga Russakovsky, Stefan Roth, and Torsten Sattler, our entire community owe a huge debt of gratitude. As you will read in their Preface to the proceedings, they dealt with over 8,500 submissions, and their diligence, thoroughness, and sheer effort has been inspirational. From the early stages of designing the call for papers, through recruiting and selecting area chairs and reviewers, to ensuring every paper has a fair and thorough assessment, their professionalism, commitment, and attention to detail has been exemplary. From all the choices that we made to organize ECCV, choosing this amazing team of Program Chairs has been, without a doubt, our very best decision.

The Program Chairs were advised and supported by an Ethics Review Committee, Chloé Bakalar, Kate Saenko, Remi Denton, and Yisong Yue, who provided invaluable input on papers where reviewers or Area Chairs had raised ethical concerns.

The Publication Chairs, Jovita Lukasik, Michael Möller, François Brémond, and Mahmoud Ali, did great work in assembling the camera-ready papers into these Springer volumes, dealing with numerous issues that arose promptly and professionally.

With the ever-expanding importance of workshops, tutorials, and demos at our major conferences comes a corresponding increase in the challenge of organizing over 70 workshops and 9 tutorials attached to the main conference. The Workshop and Tutorial Chairs, Alessio Del Bue, Jordi Pont-Tuset, Cristian Canton, and Tatiana Tommasi, and Demo Chairs, Hyung Chang and Marco Cristani, worked tirelessly to coordinate the very varied demands and structures of these different events, yielding an extremely rich auxiliary program which will enhance the conference experience for everyone who attends.

The Industry Chairs, Shaogang Gong and Cees Snoek, working with Limor Urfaly and Lior Gelfand, managed to attract numerous companies and organizations from all around the world to the industrial Expo, making ECCV 2024 an international event that not only showcases excellent foundational research, but also presents how such research can be transformed in real products. It is notable that together with the big companies, several start-ups chose ECCV to present their business ideas and activities.

As usual in the recent main conference, we also organized Speed Mentoring and Doctoral Consortium events. While the latter is an institutional occasion allowing senior PhDs to show their work, the former is especially important to answer to the many doubts and uncertainties young scholars have while undertaking a career in Computer Vision and Artificial Intelligence. We warmly thank our Social Activities Chairs, Raffaella Lanzarotti, Simone Bianco and Giovanni Farinella, and the Doctoral Consortium Chairs, Cigdem Beyan and Or Litany, for their dedication to organize such events, so important for our young colleagues, as well as all the mentors who were available to share their experience.

As the conference becomes larger and larger, the overall budget increases to a level where uncertainties in estimates of attendance can result in significant losses, perhaps enough to significantly damage the prospects of running the conference in the future. This risk imposed the need to require authors to commit to full conference registrations, with the undesirable consequent possibility to cause hardship to authors, particularly student authors, from organizations where funding attendance is difficult. To mitigate this effect, we allocated travel grants to students who might otherwise have had difficulty in attending. Our Diversity Chairs, David Fouhey and Rita Cucchiara, put tremendous effort into a wonderfully careful and thoughtful programme, balancing diversity in all its forms with a changing budget landscape, allowing us to support the participation of many people who might not otherwise have been able to attend.

This year's conference also marks the move to a unified, multi-year conference website, based on that used for other leading computer vision and AI conferences. This was an effort that we decided to take on to make the ECCV website more familiar to the community and support the organization of the future ECCV editions. Our thanks go to Lee Campbell of Eventhosts who managed the entire design and development of the new website and provided responsive support and updates throughout the conference organization timeline.

Our Finance Chairs, Gérard Medioni and Nicole Finn, provided invaluable advice and assistance on all aspects of the financial planning of the conference, in conjunction with the professional conference organizers AIM Group, where Lavinia Ricci led a fantastic team that managed thousands of details from space planning to food to registrations. In particular, Elder Bromley and Mara Carletti were a tremendous support to the entire conference organization. The conference centre, Mico DMC, also provided an excellent service, showing considerable flexibility in adapting the venue and processes to the particular demands of an academic-focused conference.

Finally, we cannot forget to thank our Social Media Chair, Kosta Derpanis, supported by Abby Stylianou and Jia-Bin Huang, who, with their presence on social media, spread ECCV news and responded to small and big questions in a timely manner, helping the community to promptly receive information and fix their problems.

October 2024
                                                Andrew Fitzgibbon
                                                Laura Leal-Taixé
                                                  Vittorio Murino

Preface

ECCV 2024 saw a remarkable 48% increase in submissions, with 8585 valid papers submitted, compared to 5804 in ECCV 2022, 5150 in 2020, and 2439 in 2018. Of these, 2387 papers (27.9%) were accepted for publication, with 200 (2.3% overall) selected for oral presentations.

A total of 173 complete submissions were desk-rejected for various reasons. Common issues included revealing author identities in the paper or supplementary material (for example: by including the author names, referring to specific grants in the acknowledgments, including links to GitHub accounts with visible author information, etc.), exceeding the page limit or otherwise significantly altering the submission template, or posting an arXiv preprint explicitly mentioning the work as an ECCV 2024 submission. The desk-rejected submissions also include 14 papers identified as dual submissions to other concurrent conferences, such as ICML, NeurIPS, and MICCAI. Seven papers were rejected among the accepted papers due to plagiarism issues, after the Program Chairs conducted a thorough plagiarism check with the iThenticate software followed by manual inspection.

The double-blind review process was managed using the CMT system, ensuring anonymity between authors and Area Chairs/reviewers. Each paper received at least three reviews, amounting to over 26,000 reviews in total. To maintain the quality and the fairness of the reviews, we recruited 469 Area Chairs (ACs) and more than 8200 reviewers. Of the recruited reviewers, 7293 ultimately participated.

Area Chairs were chosen for their technical expertise and reputation; many of them had served in similar roles at other top conferences. We also recruited additional Area Chairs through an open call for self-nominations. Among the selected ACs, around 18% were women. Geographical distribution was taken into account during the selection process, with 31% of ACs working in Europe, 38% in the Americas, 27% in Asia, and 4% in the rest of the world. Despite our best efforts to consider diversity factors in the selection process, the gender diversity and geographic diversity of ACs is still below what we were hoping for.

To ensure a smooth decision-making process, we grouped the ACs into triplets. Upon accepting their role, ACs were requested to enter into the CMT system domain conflicts and subject areas, as well as information about their DBLP, TPMS (Toronto Paper Matching System), and OpenReview profiles. These were used to determine whether different subject areas were sufficiently covered, to detect conflicts, and to automatically form AC triplets. An AC triplet was a group of three ACs who worked together throughout the process, which helped to facilitate discussion, calibrate the decision-making process, and provide support for first-time ACs. AC triplets were manually refined by Program Chairs to ensure no triplet contained more than two major time zones to enable easier coordination for virtual AC meetings. Paper assignments were made such that no AC within the triplet was conflicted with any of the papers. Each AC was responsible for overseeing 18 papers on average.

In each triplet, we identified a Lead AC who was an experienced member of the computer vision community willing to take on additional administrative duties. The primary responsibilities of Lead ACs included coordinating the AC triplet to ensure deadlines were met and offering guidance to less experienced ACs. They also served as the main contact person with the Program Chairs and helped in handling papers if an original AC was unable to complete their tasks.

Reviewers were invited from reviewer pools of previous conferences and also selected from the pool of authors. We asked experienced ACs to recommend more potential reviewers. We further recruited reviewers through self-nominations, where researchers volunteered to review for ECCV 2024 via an open call. The self-nominations were then filtered, selecting those reviewers with at least two papers published in related conferences such as ECCV, ICCV, CVPR, ICLR, NeurIPS, etc.

Similarly to ACs, reviewers were also asked to update their CMT, Toronto Paper Matching System (TPMS), and OpenReview profiles. Reviewers had the option to classify themselves as either senior researchers or junior researchers and students, with the number of assigned papers varying accordingly (a quota of 6 for senior researchers with 3 or more times reviewing experience, and a quota of 4 otherwise). On average, each reviewer was assigned about 4 papers.

Conflicts of interest among authors, ACs, and reviewers were managed automatically via conflict domains and co-authorship information from DBLP. Paper assignments to ACs and reviewers were determined using an algorithm that combined subject-area affinity scores from CMT, TPMS, and OpenReview, along with additional criteria such as ensuring that every paper had at least one non-student reviewer. For AC assignments, we additionally incorporated a newly built affinity score based on embedding similarity with the papers on the Google Scholar profile of the ACs. Once the assignments were made, ACs and reviewers were asked to promptly report any undetected conflicts of interest or other concerns that would preclude them from handling the assigned papers to ensure immediate reassignment, if necessary.

Given the widespread use of Large Language Models (LLMs), ECCV 2024, like previous conferences, implemented a special policy regulating their use. Authors were permitted to use any tools, including LLMs, in preparing their papers, but they remained fully responsible for any misrepresentation, factual inaccuracies, or plagiarism. Reviewers were also allowed to use LLMs to refine the wording of their reviews; however, they were held accountable for the accuracy of their content and were strictly prohibited from inputting submissions into an LLM. One challenge we ran into was that tools that detect LLM-generated content were not able to distinguish between text polished vs. text written from scratch by an LLM. Thus, it became quite difficult to enforce the policy in practice.

Overall, the challenges encountered during the review process were consistent with those faced at previous computer vision conferences. For instance, a small number of reviewers ultimately did not submit their reviews or provided brief, uninformative reviews, necessitating last-minute reassignment to emergency reviewers. As the community expands and the number of submissions rises, securing enough qualified reviewers and performing a quality check of the received reviews becomes increasingly difficult.

The review process included a rebuttal phase where authors were given a week to address any concerns raised by the reviewers. Each response was limited to a single PDF page using a predefined template. Rebuttals with formatting or anonymity violations were removed by the Program Chairs. ACs then facilitated discussions with reviewers to evaluate the merits of each submission. While the goal was to reach a consensus, the final decision was left to the ACs in the triplet to ensure fairness. As is common, the ACs within each triplet were tasked with organizing meetings to discuss papers, especially borderline cases. The entire process was conducted online, with no in-person meetings. For each paper, a primary and a secondary AC were assigned from the AC triplet. The primary AC was responsible for entering the decisions and meta-reviews into the CMT system, which were then reviewed and approved by the secondary ACs. Besides making accept/reject decisions, the ACs were also required to provide recommendations regarding oral/poster presentations and award nominations.

Reviewers and ACs were asked to flag papers with potential ethical concerns. An ethics review committee, consisting of four leading researchers from both industry and academia, was appointed. The role of the committee was to evaluate papers escalated by the AC for further investigation. In total, 31 papers were flagged to the committee. The committee requested that the authors of 15 papers address specific concerns in their camera-ready submissions. The authors all complied with the requests.

Following the recent IEEE decision to ban the use of the photograph of Lena Forsén, we automatically identified all submissions that included this image. Although the ban did not officially apply to ECCV, we nevertheless shared the context with the authors of the affected papers and suggested that they remove the photograph in the camera-ready submission. All authors complied.

To help the ACs to track the overall progress of the reviewing and decision-making processes, we created AC triplet activity dashboards for each AC triplet (via Google sheets), which were updated on a regular basis. During the review process, the activity dashboard enabled ACs to track the progress of incoming reviews for each paper and promptly identify papers that necessitated attention, e.g., papers potentially requiring the invitation of emergency reviewers, or papers with suspiciously short reviews. During the decision-making process, the dashboard was also used by the Program Chairs to identify papers with incomplete review information, e.g., for which reviewers did not enter their final recommendation, the AC did not enter their final decision or meta-review, the secondary AC did not confirm the recommendation, etc.

At the end of the decision-making progress, the Program Chairs carefully examined the very small number of cases where ACs overruled a unanimous reviewer recommendation for acceptance or rejection. In such instances, it was crucial for the primary and secondary ACs to explain in detail the motivation behind the decision. To ensure a fair evaluation, an additional Area Chair, not originally part of the decision-making triplet, was appointed to provide an independent opinion and advise the Program Chairs in determining whether to uphold or reverse the overturn.

Inevitably, after decisions were released, some authors were dissatisfied. The authors were given the possibility to appeal the final decision in case of procedural issues encountered during the review process by filling out a form to directly contact the Program Chairs. Valid reasons for appealing a decision included policy errors (e.g., reviewers or ACs enforcing a non-existent policy), clerical errors (e.g., the meta-review clearly indicated an intention to accept a paper, but it was mistakenly rejected), and significant misunderstandings by the reviewers or ACs. In total, we received 59 appeals and upheld 6 of them. One was due to a clear policy error by an AC, and the paper decision changed from reject to accept following a successful appeal. The other 5 papers were already conditionally accepted – but the AC/reviewers wrongly flagged them as papers for which the associated dataset was required to be released prior to final acceptance; this condition was removed following the appeal.

After the camera-ready deadline, additional checks were conducted, particularly for papers contributing a new dataset and for those flagged for potential ethical concerns. Specifically, during the submission phase, authors were required to specify if the primary contribution of their papers was the release of a new dataset, and for these papers, a valid dataset link had to be provided by the camera-ready deadline. A total of 435 submissions included a dataset as part of their contribution, and all complied with the requirement to release the dataset by the camera-ready deadline.

We sincerely thank all the ACs and reviewers for their invaluable contributions to the review process. Their dedication and expertise were crucial in maintaining the high standards of the conference. We would also like to thank our Technical Program Chair, Sascha Hornauer, who did tremendous work behind the scenes, and to the entire CMT team for their prompt support with any technical issues.

October 2024

Aleš Leonardis
Elisa Ricci
Stefan Roth
Olga Russakovsky
Torsten Sattler
Gül Varol

Organization

General Chairs

Andrew Fitzgibbon — Graphcore, UK
Laura Leal-Taixé — NVIDIA, Italy
Vittorio Murino — University of Verona & University of Genoa, Italy

Program Chairs

Aleš Leonardis — University of Birmingham, UK
Elisa Ricci — University of Trento, Italy
Stefan Roth — Technical University of Darmstadt, Germany
Olga Russakovsky — Princeton University, USA
Torsten Sattler — Czech Technical University in Prague, Czech Republic
Gül Varol — Ecole des Ponts Paris Tech, France

Technical Program Chair

Sascha Hornauer — Mines Paris – PSL, France

Industrial Liaison Chairs

Cees Snoek — University of Amsterdam, Netherlands
Shaogang Gong — Queen Mary University of London, UK

Publication Chairs

Francois Bremond — Inria, France
Mahmoud Ali — Inria, France
Jovita Lukasik — University of Siegen, Germany
Michael Moeller — University of Siegen, Germany

Poster Chairs

Aljoša Ošep Carnegie Mellon University, USA
Zuzana Kukelova Czech Technical University in Prague, Czech Republic

Diversity Chairs

David Fouhey New York University, USA
Rita Cucchiara Università di Modena e Reggio Emilia, Italy

Local Chairs

Raffaella Lanzarotti Università degli Studi di Milano, Italy
Simone Bianco University of Milano-Bicocca, Italy

Conference Ombuds

Georgia Gkioxari California Institute of Technology, USA
Greg Mori Borealis AI & Simon Fraser University, Canada

Ethics Review Committee

Chloé Bakalar Meta, USA
Kate Saenko Boston University, USA
Remi Denton Google Research, USA
Yisong Yue Caltech, Asari AI & Latitude AI, USA

Workshop and Tutorial Chairs

Alessio Del Bue Istituto Italiano di Tecnologia, Italy
Cristian Canton Meta AI, USA
Jordi Pont-Tuset Google DeepMind, Switzerland
Tatiana Tommasi Politecnico di Torino, Italy

Finance Chairs

Gerard Medioni Amazon, USA
Nicole Finn c to c events, USA

Demo Chairs

Hyung Chang University of Birmingham, UK
Marco Cristani University of Verona, Italy

Publicity and Social Media Chairs

Konstantinos Derpanis York University & Samsung AI Centre Toronto, Canada
Jia-Bin Huang University of Maryland College Park, USA
Abby Stylianou Saint Louis University, USA

Social Activities Chairs

Giovanni Maria Farinella University of Catania, Italy
Raffaella Lanzarotti Università degli Studi di Milano, Italy
Simone Bianco University of Milano-Bicocca, Italy

Doctoral Consortium Chairs

Cigdem Beyan University of Verona, Italy
Or Litany NVIDIA & Technion, USA

Web Developer

Lee Campbell Eventhosts, USA

Area Chairs

Ehsan Adeli	Stanford University, USA
Aishwarya Agrawal	University of Montreal & Mila & DeepMind, Canada
Naveed Akhtar	University of Western Australia, Australia
Yagiz Aksoy	Simon Fraser University, Canada
Karteek Alahari	Inria, France
Jose Alvarez	NVIDIA, USA
Djamila Aouada	Interdisciplinary Centre for Security, Reliability and Trust, University of Luxembourg, Luxembourg
Andre Araujo	Google, Brazil
Pablo Arbelaez	Universidad de los Andes, Colombia
Iro Armeni	Stanford University, USA
Yuki Asano	University of Amsterdam, Netherlands
Karl Åström	Lund University, Sweden
Mathieu Aubry	Ecole des Ponts ParisTech, France
Shai Bagon	Weizmann Institute of Science, Israel
Song Bai	ByteDance, Singapore
Xiang Bai	Huazhong University of Science and Technology, China
Yang Bai	Tencent, China
Lamberto Ballan	University of Padova, Italy
Linchao Bao	University of Birmingham, UK
Ronen Basri	Weizmann Institute of Science, Israel
Sara Beery	Massachusetts Institute of Technology, USA
Vasileios Belagiannis	University of Erlangen–Nuremberg, Germany
Serge Belongie	University of Copenhagen, Denmark
Sagie Benaim	Hebrew University of Jerusalem, Israel
Rodrigo Benenson	Google, Switzerland
Cigdem Beyan	University of Bergamo, Italy
Bharat Bhatnagar	Meta Reality Labs Research, Switzerland
Tolga Birdal	Imperial College London, UK
Matthew Blaschko	KU Leuven, Belgium
Federica Bogo	Meta Reality Labs Research, Switzerland
Timo Bolkart	Google, Switzerland
Katherine Bouman	California Institute of Technology, USA
Lubomir Bourdev	Primepoint, USA
Edmond Boyer	Inria, France
Yuri Boykov	University of Waterloo, Canada
Eric Brachmann	Niantic, Germany

Wieland Brendel	University of Tübingen, Germany
Gabriel Brostow	University College London, UK
Michael Brown	York University, Canada
Andrés Bruhn	University of Stuttgart, Germany
Andrei Bursuc	valeo.ai, France
Benjamin Busam	Technical University of Munich, Germany
Holger Caesar	TU Delft, Netherlands
Jianfei Cai	Monash University, Australia
Simone Calderara	University of Modena and Reggio Emilia, Italy
Octavia Camps	Northeastern University, Boston, USA
Joao Carreira	DeepMind, UK
Silvia Cascianelli	University of Modena and Reggio Emilia, Italy
Ayan Chakrabarti	Google Research, USA
Tat-Jen Cham	Nanyang Technological University, Singapore
Antoni Chan	City University of Hong Kong, China
Manmohan Chandraker	University of California, San Diego, USA
Xiaojun Chang	University of Technology Sydney, Australia
Devendra Singh Chaplot	Mistral AI, USA
Liang-Chieh Chen	TikTok, USA
Long Chen	Hong Kong University of Science and Technology, China
Mei Chen	Microsoft, USA
Shizhe Chen	Inria, France
Shuo Chen	RIKEN, Japan
Xilin Chen	Institute of Computing Technology, Chinese Academy of Sciences, China
Anoop Cherian	Mitsubishi Electric Research Labs, USA
Tat-Jun Chin	University of Adelaide, Australia
Minsu Cho	Pohang University of Science and Technology, Korea
Hisham Cholakkal	Mohamed bin Zayed University of Artificial Intelligence, United Arab Emirates
Yung-Yu Chuang	National Taiwan University, Taiwan
Ondrej Chum	Czech Technical University in Prague, Czech Republic
Marcella Cornia	University of Modena and Reggio Emilia, Italy
Marco Cristani	University of Verona, Italy
Cristian Canton	Meta AI, USA
Zhaopeng Cui	Zhejiang University, China
Angela Dai	Technical University of Munich, Germany
Jifeng Dai	Tsinghua University, China
Yuchao Dai	Northwestern Polytechnical University, China

Dima Damen	University of Bristol, UK
Kostas Daniilidis	University of Pennsylvania, USA
Antitza Dantcheva	Inria, France
Raoul de Charette	Inria, France
Shalini De Mello	NVIDIA Research, USA
Tali Dekel	Weizmann Institute of Science, Israel
Ilke Demir	Intel Corporation, USA
Joachim Denzler	Friedrich Schiller University Jena, Germany
Konstantinos Derpanis	York University, Canada
Jose Dolz	École de technologie supérieure Montreal, Canada
Jiangxin Dong	Nanjing University of Science and Technology, China
Hazel Doughty	Leiden University, Netherlands
Iddo Drori	Boston University and Columbia University, USA
Enrique Dunn	Stevens Institute of Technology, USA
Thibaut Durand	Borealis AI, Canada
Sayna Ebrahimi	Google, USA
Mohamed Elhoseiny	King Abdullah University of Science and Technology, Saudi Arabia
Bin Fan	University of Science and Technology Beijing, China
Yi Fang	New York University, USA
Giovanni Maria Farinella	University of Catania, Italy
Ryan Farrell	Brigham Young University, USA
Alireza Fathi	Google, USA
Christoph Feichtenhofer	Meta & FAIR, USA
Rogerio Feris	MIT-IBM Watson AI Lab & IBM Research, USA
Basura Fernando	Agency for Science, Technology and Research (A*STAR), Singapore
David Fouhey	New York University, USA
Katerina Fragkiadaki	Carnegie Mellon University, USA
Friedrich Fraundorfer	Graz University of Technology, Austria
Oren Freifeld	Ben-Gurion University, Israel
Huan Fu	University of Sydney, China
Yasutaka Furukawa	Simon Fraser University, Canada
Andrea Fusiello	University of Udine, Italy
Fabio Galasso	Sapienza University, Italy
Jürgen Gall	University of Bonn, Germany
Orazio Gallo	NVIDIA Research, USA
Chuang Gan	MIT-IBM Watson AI Lab, USA
Lianli Gao	University of Electronic Science and Technology of China, China

Shenghua Gao	University of Hong Kong, China
Sourav Garg	University of Adelaide, Australia
Efstratios Gavves	University of Amsterdam, Netherlands
Peter Gehler	Zalando, Germany
Theo Gevers	University of Amsterdam, Netherlands
Golnaz Ghiasi	Google DeepMind, USA
Rohit Girdhar	Carnegie Mellon University, USA
Xavier Giro-i-Nieto	Universitat Politecnica de Catalunya, Spain
Ross Girshick	Allen Institute for Artificial Intelligence, USA
Zan Gojcic	NVIDIA, Switzerland
Vladislav Golyanik	Max Planck Institute for Informatics, Germany
Stephen Gould	Australian National University, Australia
Liangyan Gui	University of Illinois Urbana-Champaign, USA
Fatma Guney	Koc University, Turkey
Yulan Guo	Sun Yat-sen University, China
Yunhui Guo	University of Texas at Dallas, USA
Mohit Gupta	University of Wisconsin-Madison, USA
Minh Ha Quang	RIKEN Center for Advanced Intelligence Project, Japan
Bumsub Ham	Yonsei University, Korea
Bohyung Han	Seoul National University, Korea
Kai Han	University of Hong Kong, China
Xiaoguang Han	Shenzhen Research Institute of Big Data & Chinese University of Hong Kong (Shenzhen), China
Tatsuya Harada	University of Tokyo & RIKEN, Japan
Tal Hassner	Weir AI, USA
Xuming He	ShanghaiTech University, China
Felix Heide	Princeton & Algolux, USA
Anthony Hoogs	Kitware, USA
Timothy Hospedales	Edinburgh University, UK
Mahdi Hosseini	Concordia University, Canada
Peng Hu	College of Computer Science, Sichuan University, China
Gang Hua	Wormpex AI Research, USA
Jia-Bin Huang	University of Maryland College Park, USA
Qixing Huang	University of Texas at Austin, USA
Sharon Xiaolei Huang	Pennsylvania State University, USA
Junhwa Hur	Google, USA
Nazli Ikizler-Cinbis	Hacettepe University, Turkey
Eddy Ilg	Saarland University, Germany
Ahmet Iscen	Google, France

Nathan Jacobs	Washington University in St. Louis, USA
Varun Jampani	Stability AI, USA
C. V. Jawahar	International Institute of Information Technology - Hyderabad, India
Laszlo Jeni	Carnegie Mellon University, USA
Jiaya Jia	Chinese University of Hong Kong, China
Qin Jin	Renmin University of China, China
Jungseock Joo	NVIDIA & University of California, Los Angeles, USA
Fredrik Kahl	Chalmers, Sweden
Yannis Kalantidis	NAVER LABS Europe, France
Vicky Kalogeiton	Ecole Polytechnique, IP Paris, France
Evangelos Kalogerakis	University of Massachusetts Amherst, USA
Hiroshi Kawasaki	Kyushu University, Japan
Margret Keuper	University of Mannheim & Max Planck Institute for Informatics, Germany
Salman Khan	Mohamed bin Zayed University of Artificial Intelligence, United Arab Emirates
Anna Khoreva	Bosch Center for Artificial Intelligence, Germany
Gunhee Kim	Seoul National University, Korea
Junmo Kim	Korea Advanced Institute of Science and Technology, Korea
Min H. Kim	Korea Advanced Institute of Science and Technology, Korea
Seon Joo Kim	Yonsei University, Korea
Tae-Kyun Kim	Korea Advanced Institute of Science and Technology & Imperial College London, Korea
Benjamin Kimia	Brown University, USA
Hedvig Kjellström	KTH Royal Institute of Technology, Sweden
Laurent Kneip	ShanghaiTech University, China
A. Sophia Koepke	University of Tübingen, Germany
Iasonas Kokkinos	University College London, UK
Wai-Kin Adams Kong	Nanyang Technological University, Singapore
Piotr Koniusz	Data61/CSIRO & Australian National University, Australia
Jana Kosecka	George Mason University, USA
Adriana Kovashka	University of Pittsburgh, USA
Hilde Kuehne	University of Bonn, Germany
Zuzana Kukelova	Czech Technical University in Prague, Czech Republic
Ajay Kumar	Hong Kong Polytechnic University, China
Kiriakos Kutulakos	University of Toronto, Canada

Suha Kwak	Pohang University of Science and Technology, Korea
Zorah Laehner	University of Bonn, Germany
Shang-Hong Lai	National Tsing Hua University, Taiwan
Iro Laina	University of Oxford, UK
Jean-Francois Lalonde	Université Laval, Canada
Loic Landrieu	École des Ponts ParisTech, France
Diane Larlus	Naver Labs Europe, France
Viktor Larsson	Lund University, Sweden
Stéphane Lathuilière	Telecom-Paris, France
Gim Hee Lee	National University of Singapore, Singapore
Yong Jae Lee	University of Wisconsin-Madison, USA
Bastian Leibe	RWTH Aachen University, Germany
Ales Leonardis	University of Birmingham, UK
Vincent Lepetit	Ecole des Ponts ParisTech, France
Fuxin Li	Oregon State University, USA
Hongdong Li	Australian National University, Australia
Xi Li	Zhejiang University, China
Yin Li	University of Wisconsin-Madison, USA
Yu Li	International Digital Economy Academy, Singapore
Xiaodan Liang	Sun Yat-sen University, China
Shengcai Liao	Inception Institute of Artificial Intelligence, United Arab Emirates
Jongwoo Lim	Seoul National University, Korea
Ser-Nam Lim	University of Central Florida, USA
Stephen Lin	Microsoft Research, China
Tsung-Yi Lin	NVIDIA Research, USA
Yen-Yu Lin	National Yang Ming Chiao Tung University, Taiwan
Yutian Lin	Wuhan University, China
Zhe Lin	Adobe Research, USA
Haibin Ling	Stony Brook University, USA
Or Litany	NVIDIA, USA
Jiaying Liu	Peking University, China
Jun Liu	Lancaster University, UK
Miaomiao Liu	Australian National University, Australia
Si Liu	Beihang University, China
Sifei Liu	NVIDIA, USA
Wei Liu	Tencent, China
Yanxi Liu	Pennsylvania State University, USA
Yaoyao Liu	Johns Hopkins University, USA

Yebin Liu	Tsinghua University, China
Zicheng Liu	Microsoft, USA
Ziwei Liu	Nanyang Technological University, Singapore
Ismini Lourentzou	University of Illinois, Urbana-Champaign, USA
Chen Change Loy	Nanyang Technological University, Singapore
Feng Lu	Beihang University, China
Le Lu	Alibaba Group, USA
Oisin Mac Aodha	University of Edinburgh, UK
Luca Magri	Politecnico di Milano, Italy
Michael Maire	University of Chicago, USA
Subhransu Maji	University of Massachusetts, Amherst, USA
Atsuto Maki	KTH Royal Institute of Technology, Sweden
Yasushi Makihara	Osaka University, Japan
Massimiliano Mancini	University of Trento, Italy
Kevis-Kokitsi Maninis	Google Research, Switzerland
Kenneth Marino	Google DeepMind, USA
Renaud Marlet	Ecole des Ponts ParisTech & Valeo, France
Niki Martinel	University of Udine, Italy
Jiri Matas	Czech Technical University in Pragu, Czech Republic
Yasuyuki Matsushita	Osaka University, Japan
Simone Melzi	University of Milano-Bicocca, Italy
Thomas Mensink	Google Research, Netherlands
Michele Merler	IBM Research, USA
Pascal Mettes	University of Amsterdam, Netherlands
Ajmal Mian	University of Western Australia, Australia
Krystian Mikolajczyk	Imperial College London, UK
Ishan Misra	Facebook AI Research, USA
Niloy Mitra	University College London, UK
Anurag Mittal	Indian Institute of Technology Madras, India
Davide Modolo	Amazon, USA
Davide Moltisanti	University of Bath, UK
Philippos Mordohai	Stevens Institute of Technology, USA
Francesc Moreno	Institut de Robotica i Informatica Industrial, Spain
Pietro Morerio	Istituto Italiano di Tecnologia, Italy
Greg Mori	Simon Fraser University & Borealis AI, Canada
Roozbeh Mottaghi	FAIR @ Meta, USA
Yadong Mu	Peking University, China
Vineeth N. Balasubramanian	Indian Institute of Technology, India
Hajime Nagahara	Osaka University, Japan
Vinay Namboodiri	University of Bath, UK
Liangliang Nan	Delft University of Technology, Netherlands

Michele Nappi	University of Salerno, Italy
Srinivasa Narasimhan	Carnegie Mellon University, USA
P. J. Narayanan	International Institute of Information Technology, Hyderabad, India
Nassir Navab	Technical University of Munich, Germany
Ram Nevatia	University of Southern California, USA
Natalia Neverova	Facebook AI Research, France
Juan Carlos Niebles	Salesforce & Stanford University, USA
Michael Niemeyer	Google, Germany
Simon Niklaus	Adobe Research, USA
Ko Nishino	Kyoto University, Japan
Nicoletta Noceti	MaLGa-DIBRIS & Università degli Studi di Genova, Italy
Matthew O'Toole	Carnegie Mellon University, USA
Jean-Marc Odobez	Idiap Research Institute, École polytechnique fédérale de Lausanne, Switzerland
Francesca Odone	Università di Genova, Italy
Takayuki Okatani	Tohoku University & RIKEN Center for Advanced Intelligence Project, Japan
Carl Olsson	Lund University, Sweden
Vicente Ordonez	Rice University, USA
Aljosa Osep	Technical University of Munich, Germany
Martin R. Oswald	University of Amsterdam, Netherlands
Andrew Owens	University of Michigan, USA
Tomas Pajdla	Czech Technical University in Prague, Czechia
Manohar Paluri	Meta, USA
Jinshan Pan	Nanjing University of Science and Technology, China
In Kyu Park	Inha University, Korea
Jaesik Park	Seoul National University, Korea
Despoina Paschalidou	Stanford, USA
Georgios Pavlakos	University of Texas at Austin, USA
Vladimir Pavlovic	Rutgers University, USA
Marcello Pelillo	University of Venice, Italy
David Picard	École des Ponts ParisTech, France
Hamed Pirsiavash	University of California, Davis, USA
Bryan Plummer	Boston University, USA
Thomas Pock	Graz University of Technology, Austria
Fabio Poiesi	Fondazione Bruno Kessler, Italy
Marc Pollefeys	ETH Zurich & Microsoft, Switzerland
Jean Ponce	Ecole normale supérieure - PSL, France
Gerard Pons-Moll	University of Tübingen, Germany

Jordi Pont-Tuset	Google, Switzerland
Fatih Porikli	Qualcomm R&D, USA
Brian Price	Adobe, USA
Guo-Jun Qi	Westlake University, USA
Long Quan	Hong Kong University of Science and Technology, China
Venkatesh Babu Radhakrishnan	Indian Institute of Science, India
Hossein Rahmani	Lancaster University, UK
Deva Ramanan	Carnegie Mellon University, USA
Vignesh Ramanathan	Facebook, USA
Vikram V. Ramaswamy	Princeton University, USA
Nalini Ratha	University at Buffalo, State University of New York, USA
Yogesh Rawat	University of Central Florida, USA
Hamid Rezatofighi	Monash University, Australia
Helge Rhodin	University of British Columbia, Canada
Stephan Richter	Apple, Germany
Anna Rohrbach	Technical University of Darmstadt, Germany
Marcus Rohrbach	Technical University of Darmstadt, Germany
Gemma Roig	Goethe University Frankfurt, Germany
Negar Rostamzadeh	Google, Canada
Paolo Rota	University of Trento, Italy
Stefan Roth	Technical University of Darmstadt, Germany
Carsten Rother	University of Heidelberg, Germany
Subhankar Roy	University of Trento, Italy
Michael Rubinstein	Google, USA
Christian Rupprecht	University of Oxford, UK
Olga Russakovsky	Princeton University, USA
Bryan Russell	Adobe Research, USA
Michael Ryoo	Stony Brook University & Google, USA
Reza Sabzevari	Delft University of Technology, Netherlands
Mathieu Salzmann	École polytechnique fédérale de Lausanne, Switzerland
Dimitris Samaras	Stony Brook University, USA
Aswin Sankaranarayanan	Carnegie Mellon University, USA
Sudeep Sarkar	University of South Florida, USA
Yoichi Sato	University of Tokyo, Japan
Manolis Savva	Simon Fraser University, Canada
Simone Schaub-Meyer	Technical University of Darmstadt, Germany
Bernt Schiele	Max Planck Institute for Informatics, Germany
Cordelia Schmid	Inria & Google, France
Nicu Sebe	University of Trento, Italy

Laura Sevilla-Lara	University of Edinburgh, UK
Fahad Shahbaz Khan	Mohamed bin Zayed University of Artificial Intelligence, United Arab Emirates
Qi Shan	Apple, USA
Shiguang Shan	Institute of Computing Technology, Chinese Academy of Sciences, China
Viktoriia Sharmanska	University of Sussex & Imperial College London, UK
Eli Shechtman	Adobe Research, USA
Evan Shelhamer	DeepMind, UK
Lu Sheng	Beihang University, China
Boxin Shi	Peking University, China
Jianbo Shi	University of Pennsylvania, USA
Mike Zheng Shou	National University of Singapore, Singapore
Abhinav Shrivastava	University of Maryland, USA
Aliaksandr Siarohin	Snap, USA
Kaleem Siddiqi	McGill University, Canada
Leonid Sigal	University of British Columbia, Canada
Oriane Siméoni	valeo.ai, France
Krishna Kumar Singh	Adobe Research, USA
Richa Singh	Indian Institute of Technology Jodhpur, India
Yale Song	Facebook AI Research, USA
Concetto Spampinato	University of Catania, Italy
Srinath Sridhar	Brown University, USA
Bjorn Stenger	Rakuten, Japan
Abby Stylianou	Saint Louis University, USA
Akihiro Sugimoto	National Institute of Informatics, Japan
Chen Sun	Brown University, USA
Deqing Sun	Google, USA
Jian Sun	Xi'an Jiaotong University, China
Min Sun	National Tsing Hua University, Taiwan
Qianru Sun	Singapore Management University, Singapore
Yu-Wing Tai	Dartmouth College, China
Ayellet Tal	Technion, Israel
Ping Tan	Hong Kong University of Science and Technology, China
Robby Tan	National University of Singapore, Singapore
Chi-Keung Tang	Hong Kong University of Science and Technology, China
Makarand Tapaswi	International Institute of Information Technology, Hyderabad & Wadhwani AI, India
Justus Thies	Technical University of Darmstadt, Germany

Radu Timofte	University of Wurzburg & ETH Zurich, Germany
Giorgos Tolias	Czech Technical University in Prague, Czechia
Federico Tombari	Google & Technical University of Munich, Switzerland
Tatiana Tommasi	Politecnico di Torino, Italy
James Tompkin	Brown University, USA
Matthew Trager	AWS AI Labs, USA
Anh Tran	VinAI, Vietnam
Du Tran	Google, USA
Shubham Tulsiani	Carnegie Mellon University, USA
Sergey Tulyakov	Snap, USA
Dimitrios Tzionas	University of Amsterdam, Netherlands
Maria Vakalopoulou	CentraleSupelec, France
Joost van de Weijer	Computer Vision Center, Spain
Laurens van der Maaten	Meta, USA
Jan van Gemert	Delft University of Technology, Netherlands
Sebastiano Vascon	Ca' Foscari University of Venice & European Centre for Living Technology, Italy
Nuno Vasconcelos	University of California, San Diego, USA
Mayank Vatsa	Indian Institute of Technology Jodhpur, India
Javier Vazquez-Corral	Autonomous University of Barcelona, Spain
Andrea Vedaldi	Oxford University, UK
Olga Veksler	University of Waterloo, Canada
Luisa Verdoliva	University Federico II of Naples, Italy
Rene Vidal	University of Pennsylvania, USA
Bo Wang	CtrsVision, USA
Heng Wang	TikTok, USA
Jingdong Wang	Baidu, China
Jingya Wang	ShanghaiTech University, China
Ruiping Wang	Institute of Computing Technology, Chinese Academy of Sciences, China
Shenlong Wang	University of Illinois, Urbana-Champaign, USA
Ting-Chun Wang	NVIDIA, USA
Wenguan Wang	Zhejiang University, China
Xiaolong Wang	University of California, San Diego, USA
Xiaoqian Wang	Purdue University, USA
Xiaoyu Wang	Hong Kong University of Science and Technology, USA
Xin Wang	Microsoft Research, USA
Xinchao Wang	National University of Singapore, Singapore
Yang Wang	Concordia University, Canada
Yiming Wang	Fondazione Bruno Kessler, Italy

Yu-Chiang Frank Wang	National Taiwan University, Taiwan
Yu-Xiong Wang	University of Illinois, Urbana-Champaign, USA
Zhangyang Wang	University of Texas at Austin, USA
Olivia Wiles	DeepMind, UK
Christian Wolf	Naver Labs Europe, France
Michael Wray	University of Bristol, UK
Jiajun Wu	Stanford University, USA
Jianxin Wu	Nanjing University, China
Shangzhe Wu	Stanford University, USA
Ying Wu	Northwestern University, USA
Yu Wu	Wuhan University, China
Jianwen Xie	Baidu Research, USA
Saining Xie	New York University, USA
Weidi Xie	Shanghai Jiao Tong University, China
Dan Xu	Hong Kong University of Science and Technology, China
Huijuan Xu	Pennsylvania State University, USA
Xiangyu Xu	Xi'an Jiaotong University, China
Yan Yan	Illinois Institute of Technology, USA
Jiaolong Yang	Microsoft Research, China
Kaiyu Yang	California Institute of Technology, USA
Linjie Yang	ByteDance AI Lab, USA
Ming-Hsuan Yang	University of California, Merced, USA
Yanchao Yang	University of Hong Kong, China
Yi Yang	Zhejiang University, China
Angela Yao	National University of Singapore, Singapore
Mang Ye	Wuhan University, China
Kwang Moo Yi	University of British Columbia, Canada
Xi Yin	Facebook, USA
Chang D. Yoo	Korea Advanced Institute of Science and Technology, Korea
Shaodi You	University of Amsterdam, Netherlands
Jingyi Yu	Shanghai Tech University, China
Ning Yu	Netflix Eyeline Studios, USA
Qi Yu	Rochester Institute of Technology, USA
Stella Yu	University of Michigan, USA
Yizhou Yu	University of Hong Kong, China
Junsong Yuan	University at Buffalo, State University of New York, USA
Lu Yuan	Microsoft, USA
Sangdoo Yun	NAVER AI Lab, Korea
Hongbin Zha	Peking University, China

Jing Zhang — Australian National University, Australia
Lei Zhang — Hong Kong Polytechnic University, China
Ning Zhang — Facebook AI, USA
Richard Zhang — Adobe, USA
Tianzhu Zhang — University of Science and Technology of China, China
Hengshuang Zhao — University of Hong Kong, China
Qi Zhao — University of Minnesota, USA
Sicheng Zhao — Tsinghua University, China
Liang Zheng — Australian National University, Australia
Wei-Shi Zheng — Sun Yat-sen University, China
Zhun Zhong — University of Nottingham, UK
Bolei Zhou — University of California, Los Angeles, USA
Kaiyang Zhou — Hong Kong Baptist University, China
Luping Zhou — University of Sydney, Australia
S. Kevin Zhou — University of Science and Technology of China, China
Lei Zhu — Hong Kong University of Science and Technology (Guangzhou), China
Linchao Zhu — Zhejiang University, China
Andrew Zisserman — University of Oxford, UK
Daniel Zoran — DeepMind, UK
Wangmeng Zuo — Harbin Institute of Technology, China

Technical Program Committee

Sathyanarayanan N. Aakur
Andrea Abate
Wim Abbeloos
Amos L. Abbott
Rameen Abdal
Salma Abdel Magid
Rabab Abdelfattah
Ahmed Abdelkader
Eslam Mohamed Abdelrahman
Ahmed Abdelreheem
Akash Abdu Jyothi
Kfir Aberman
Shahira Abousamra
Yazan Abu Farha
Shady Abu-Hussein
Abulikemu Abuduweili
Hanno Ackermann
Chandranath Adak
Aikaterini Adam
Lukáš Adam
Kamil Adamczewski
Donald Adjeroh
Mohammed Adnan
Mahmoud Afifi
Triantafyllos Afouras
Akshay Agarwal
Madhav Agarwal
Nakul Agarwal
Sharat Agarwal
Shivang Agarwal
Vatsal Agarwal
Abhinav Agarwalla
Shivam Aggarwal
Sara Aghajanzadeh
Shashank Agnihotri
Harsh Agrawal
Antonio Agudo
Shai Aharon
Saba Ahmadi
Waqar Ahmed
Byeongjoo Ahn
Chanho Ahn
Jaewoo Ahn
Namhyuk Ahn
Seoyoung Ahn
Nilesh A. Ahuja
Yihao Ai

Abhishek Aich
Emanuele Aiello
Noam Aigerman
Kiyoharu Aizawa
Kenan Emir Ak
Reza Akbarian Bafghi
Sai Aparna Aketi
Derya Akkaynak
Ibrahim Alabdulmohsin
Yuval Alaluf
Md Ferdous Alam
Stephan Alaniz
Badour A. Sh AlBahar
Paul Albert
Isabela Albuquerque
Emanuel Aldea
Luca Alessandrini
Emma Alexander
Konstantinos P. Alexandridis
Stamatis Alexandropoulos
Saghir Alfasly
Mohsen Ali
Waqar Ali
Daniel Aliaga
Sadegh Aliakbarian
Garvita Allabadi
Thiemo Alldieck
Antonio Alliegro
Jurandy Almeida
Gal Alon
Hadi Alzayer
Ibtihel Amara
Giuseppe Amato
Sameer Ambekar
Niki Amini-Naieni
Shir Amir
Roberto Amoroso
Dongsheng An
Jie An
Junyi An
Liang An
Xiang An
Yuexuan An
Zhaochong An

Zhulin An
Saket Anand
Pavan Kumar Anasosalu Vasu
Codruta O. Ancuti
Cosmin Ancuti
Fernanda Andaló
Connor Anderson
Juan Andrade-Cetto
Bruno Andreis
Alexander Andreopoulos
Bjoern Andres
Shivangi Aneja
Dragomir Anguelov
Damith Kawshan Anhettigama
Aditya Annavajjala
Michel Antunes
Tejas Anvekar
Saeed Anwar
Evlampios Apostolidis
Srikar Appalaraju
Relja Arandjelović
Nikita Araslanov
Alexandre Araujo
Helder Araujo
Eric Arazo
Dawit Mureja Argaw
Anurag Arnab
Federica Arrigoni
Alexey Artemov
Aditya Arun
Nader Asadi
Kumar Ashutosh
Vishal Asnani
Mahmoud Assran
Amir Atapour-Abarghouei
Nikos Athanasiou
Ali Athar
ShahRukh Athar
Rowel O. Atienza
Benjamin Attal
Matan Atzmon
Nicolas Audebert
Romaric Audigier

Tristan Aumentado-Armstrong
Yannis Avrithis
Muhammad Awais
Md Rabiul Awal
Görkay Aydemir
Mehmet Aygün
Melika Ayoughi
Ali Ayub
Kumar Ayush
Mehdi Azabou
Basim Azam
Mohammad Farid Azampour
Hossein Azizpour
Vivek B. S.
Yunhao Ba
Mohammadreza Babaee
Deepika Bablani
Sudarshan Babu
Roman Bachmann
Inhwan Bae
Jeonghun Baek
Kyungjune Baek
Seungryul Baek
Piyush Nitin Bagad
Mohammadhossein Bahari
Gaetan Bahl
Sherwin Bahmani
Bing Bai
Haoyue Bai
Jiawang Bai
Jinbin Bai
Lei Bai
Qingyan Bai
Shutao Bai
Xuyang Bai
Yalong Bai
Yuanchao Bai
Yue Bai
Yunpeng Bai
Ziyi Bai
Daniele Baieri
Sungyong Baik
Ramon Baldrich

Coloma Ballester
Irene Ballester
Sonat Baltaci
Biplab Banerjee
Monami Banerjee
Sandipan Banerjee
Snehasis Banerjee
Soumya Banerjee
Sreya Banerjee
Jihwan Bang
Ankan Bansal
Nitin Bansal
Siddhant Bansal
Chong Bao
Hangbo Bao
Runxue Bao
Wentao Bao
Yiwei Bao
Zhipeng Bao
Zhongyun Bao
Amir Bar
Omer Bar Tal
Manel Baradad Jurjo
Lorenzo Baraldi
Lorenzo Baraldi
Tal Barami
Danny Barash
Daniel Barath
Adrian Barbu
Nick Barnes
Alina J. Barnett
German Barquero
Joao P. Barreto
Jonathan T. Barron
Rodrigo C. Barros
Luca Barsellotti
Myles Bartlett
Kristijan Bartol
Hritam Basak
Dina Bashkirova
Chaim Baskin
Lennart C. Bastian
Abhipsa Basu
Soumen Basu
Peyman Bateni

Anil Batra
David Bau
Zuria Bauer
Eduard Gabriel Bazavan
Federico Becattini
Nathan A. Beck
Jan Bednarik
Peter A. Beerel
Ardhendu Behera
Harkirat Singh Behl
Jens Behley
William Beksi
Soufiane Belharbi
Giovanni Bellitto
Ismail Ben Ayed
Abdessamad Ben Hamza
Yizhak Ben-Shabat
Guy Ben-Yosef
Robert Benavente
Assia Benbihi
Yasser Benigmim
Urs Bergmann
Amit H Bermano
Jesus Bermudez-Cameo
Florian Bernard
Stefano Berretti
Gabriele Berton
Petra Bevandić
Matthew Beveridge
Lalit Bhagat
Yash Bhalgat
Homanga Bharadhwaj
Skanda Bharadwaj
Ambika Saklani Bhardwaj
Goutam Bhat
Gaurav Bhatt
Neel P. Bhatt
Deblina Bhattacharjee
Anish Bhattacharya
Rajarshi Bhattacharya
Uttaran Bhattacharya
Apratim Bhattacharyya
Anand Bhattad
Binod Bhattarai
Snehal Bhayani

Brojeshwar Bhowmick
Aritra Bhowmik
Neelanjan Bhowmik
Amran Bhuiyan
Ankan Kumar Bhunia
Ayan Kumar Bhunia
Jing Bi
Qi Bi
Xiuli Bi
Michael Bi Mi
Hao Bian
Jia-Wang Bian
Shaojun Bian
Simone Bianco
Pia Bideau
Xiaoyu Bie
Roberto Bigazzi
Mario Bijelic
Bahri Batuhan Bilecen
Guillaume-Alexandre
    Bilodeau
Alexander Binder
Massimo Bini
Niccolò Bisagno
Carmen Bisogni
Anthony R. Bisulco
Sandika Biswas
Sanket Biswas
Ksenia Bittner
Kevin Blackburn-Matzen
Nathaniel Blanchard
Hunter Blanton
Emile Blettery
Hermann Blum
Deyu Bo
Janusz Bobulski
Marius Bock
Lea Bogensperger
Simion-Vlad Bogolin
Moritz Böhle
Piotr Bojanowski
Alexey Bokhovkin
Georg Bökman
Ladislau Boloni
Daniel Bolya

Lorenzo Bonicelli
Gianpaolo Bontempo
Vivek Boominathan
Adrian Bors
Damian Borth
Matteo Bortolon
Davide Boscaini
Laurie Nicholas Bose
Shirsha Bose
Mark Boss
Andrea Bottino
Larbi Boubchir
Alexandre Boulch
Terrance E. Boult
Amine Bourki
Walid Bousselham
Fadi Boutros
Nicolas C. Boutry
Thierry Bouwmans
Richard S. Bowen
Kevin W. Bowyer
Ivaylo Boyadzhiev
Aidan Boyd
Joseph Boyd
Aljaz Bozic
Behzad Bozorgtabar
Manuel Brack
Samarth Brahmbhatt
Nikolas Brasch
Cameron E. Braunstein
Toby P. Breckon
Mathieu Bredif
Romain Brégier
Carlo Bretti
Joel R. Brogan
Clifford Broni-Bediako
Andrew Brown
Ellis L. Brown
Thomas Brox
Qingwen Bu
Tong Bu
Shyamal Buch
Alvaro Budria
Ignas Budvytis
José M. Buenaposada

Kyle R. Buettner
Marcel C. Bühler
Nhat-Tan Bui
Adrian Bulat
Valay M. Bundele
Nathaniel J. Burgdorfer
Ryan D. Burgert
Evgeny Burnaev
Pietro Buzzega
Marco Buzzelli
Kwon Byung-Ki
Fabian Caba
Felipe Cadar
Martin Cadik
Bowen Cai
Changjiang Cai
Guanyu Cai
Haoming Cai
Hongrui Cai
Jiarui Cai
Jinyu Cai
Kaixin Cai
Lile Cai
Minjie Cai
Mu Cai
Pingping Cai
Rizhao Cai
Ruisi Cai
Ruojin Cai
Shen Cai
Shengqu Cai
Weidong Cai
Weiwei Cai
Wenjia Cai
Wenxiao Cai
Yancheng Cai
Yuanhao Cai
Yujun Cai
Yuxuan Cai
Zhaowei Cai
Zhipeng Cai
Zhixi Cai
Zhongang Cai
Zikui Cai
Salvatore Calcagno

Roberto Caldelli
Akin Caliskan
Lilian Calvet
Necati Cihan Camgoz
Tommaso Campari
Dylan Campbell
Guglielmo Camporese
Yigit Baran Can
Shaun Canavan
Gemma Canet Tarrés
Marco Cannici
Ang Cao
Anh-Quan Cao
Bing Cao
Bingyi Cao
Chenjie Cao
Dongliang Cao
Hu Cao
Jiahang Cao
Jiale Cao
Jie Cao
Jiezhang Cao
Jinkun Cao
Meng Cao
Mingdeng Cao
Peng Cao
Pu Cao
Qinglong Cao
Qingxing Cao
Shengcao Cao
Song Cao
Weiwei Cao
Xiangyong Cao
Xiao Cao
Xu Cao
Xu Cao
Yan-Pei Cao
Yang Cao
Yichao Cao
Ying Cao
Yuan Cao
Yuhang Cao
Yukang Cao
Yulong Cao
Yun-Hao Cao

Yunhao Cao
Yutong Cao
Zhangjie Cao
Zhiguo Cao
Zhiwen Cao
Ziang Cao
Giacomo Capitani
Luigi Capogrosso
Andrea Caraffa
Jaime S. Cardoso
Chris Careaga
Zachariah Carmichael
Giuseppe Cartella
Vincent Cartillier
Paola Cascante-Bonilla
Lucia Cascone
Pol Caselles Rico
Angela Castillo
Lluis Castrejon
Modesto
   Castrillón-Santana
Francisco M. Castro
Pedro Castro
Jacopo Cavazza
Oya Celiktutan
Luigi Celona
Jiazhong Cen
Fernando J. Cendra
Fabio Cermelli
Duygu Ceylan
Eunju Cha
Junbum Cha
Rohan Chabra
Rohan Chacko
Julia Chae
Nicolas Chahine
Junyi Chai
Lucy Chai
Tianrui Chai
Wenhao Chai
Zenghao Chai
Anish Chakrabarty
Goirik Chakrabarty
Anirban Chakraborty
Deep Chakraborty

Souradeep Chakraborty
Punarjay Chakravarty
Jacob Chalk
Loick Chambon
Chee Seng Chan
David Chan
Dorian Y. Chan
Kin Sun Chan
Matthew Chan
Adrien Chan-Hon-Tong
Sampath Chanda
Shivam Chandhok
Prashanth Chandran
Kenneth Chaney
Chirui Chang
Di Chang
Dongliang Chang
Hong Chang
Hyung Jin Chang
Ju Yong Chang
Matthew Chang
Ming-Ching Chang
MingFang Chang
Nadine Chang
Qi Chang
Seunggyu Chang
Wei-Yi Chang
Xiaojun Chang
Yakun Chang
Yi Chang
Zheng Chang
Sumohana S.
   Channappayya
Hanqing Chao
Pradyumna Chari
Korawat Charoenpitaks
Dibyadip Chatterjee
Moitreya Chatterjee
Pratik Chaudhari
Muawiz Chaudhary
Ritwick Chaudhry
Ruchika Chavhan
Sofian Chaybouti
Zhengping Che
Gullal Singh Cheema

Kunal Chelani
Rama Chellappa
Allison Y. Chen
Anpei Chen
Aozhu Chen
Bin Chen
Bin Chen
Bin Chen
Bohong Chen
Bor-Chun Chen
Bowei Chen
Brian Chen
Chang Wen Chen
Changan Chen
Changhao Chen
Changhuai Chen
Chaofan Chen
Chaofan Chen
Chaofeng Chen
Chen Chen
Cheng Chen
Chengkuan Chen
Chieh-Yun Chen
Chong Chen
Chun-Fu Richard Chen
Congliang Chen
Cuiqun Chen
Da Chen
Daoyuan Chen
Defang Chen
Dian Chen
Dian Chen
Ding-Jie Chen
Dingfan Chen
Dong Chen
Dongdong Chen
Fangyi Chen
Guang-Yong Chen
Guangyao Chen
Guangyi Chen
Guanying Chen
Guikun Chen
Guo Chen
Haibo Chen
Haiwei Chen

Hansheng Chen
Hanting Chen
Hanzhi Chen
Hao Chen
Haodong Chen
Haokun Chen
Haoran Chen
Haoxin Chen
Haoyu Chen
Haoyu Chen
Hong-You Chen
Honghua Chen
Honglin Chen
Hongming Chen
Huanran Chen
Hui Chen
Hwann-Tzong Chen
Jiacheng Chen
Jiajing Chen
Jianqiu Chen
Jiaqi Chen
Jiaxin Chen
Jiayi Chen
Jie Chen
Jie Chen
Jie Chen
Jieneng Chen
Jierun Chen
Jinghui Chen
Jingjing Chen
Jintai Chen
Jinyu Chen
Joya Chen
Jun Chen
Jun Chen
Jun Chen
Jun-Cheng Chen
Jun-Kun Chen
Junjie Chen
Kai Chen
Kai Chen
Kaifeng Chen
Kefan Chen
Kejiang Chen
Kuan-Wen Chen
Kunyao Chen
Li Chen
Liang Chen
Lianggangxu Chen
Liangyu Chen
Ling-Hao Chen
Linghao Chen
Liyan Chen
Min Chen
Min-Hung Chen
Minghao Chen
Minghui Chen
Nenglun Chen
Peihao Chen
Pengguang Chen
Ping Chen
Qi Chen
Qi Chen
Qi Chen
Qiang Chen
Qimin Chen
Qingcai Chen
Rongyu Chen
Rui Chen
Runjian Chen
Shang-Fu Chen
Shen Chen
Sherry X. Chen
Shi Chen
Shiming Chen
Shiyan Chen
Shizhe Chen
Shoufa Chen
Shu-Yu Chen
Shuai Chen
Si Chen
Siheng Chen
Simin Chen
Sixiang Chen
Sizhe Chen
Songcan Chen
Tao Chen
Tianshui Chen
Tianyi Chen
Tsai-Shien Chen
Wei Chen
Weidong Chen
Weihua Chen
Weijie Chen
Weikai Chen
Wuyang Chen
Xi Chen
Xi Chen
Xiang Chen
Xiangyu Chen
Xianyu Chen
Xiao Chen
Xiao Chen
Xiaohan Chen
Xiaokang Chen
Xiaotong Chen
Xin Chen
Xin Chen
Xin Chen
Xinghao Chen
Xingyu Chen
Xingyu Chen
Xinlei Chen
Xiuwei Chen
Xuanbai Chen
Xuanhong Chen
Xuesong Chen
Xuweiyi Chen
Xuxi Chen
Yanbei Chen
Yang Chen
Yaofo Chen
Yen-Chun Chen
Yi-Ting Chen
Yi-Wen Chen
Yilun Chen
Yinbo Chen
Yingcong Chen
Yingwen Chen
Yiting Chen
Yixin Chen
Yixin Chen
Yu Chen
Yuanhong Chen
Yuanqi Chen

Yuedong Chen
Yueting Chen
Yuhao Chen
Yuhao Chen
Yujin Chen
Yukang Chen
Yun-Chun Chen
Yunjin Chen
Yunlu Chen
Yuntao Chen
Yushuo Chen
Yuxiao Chen
Zehui Chen
Zerui Chen
Zewen Chen
Zeyuan Chen
Zeyuan Chen
Zhang Chen
Zhao-Min Chen
Zhaoliang Chen
Zhaoxi Chen
Zhaoyu Chen
Zhaozheng Chen
Zhaozheng Chen
Zhe Chen
Zhen Chen
Zhen Chen
Zheng Chen
Zhenghao Chen
Zhengyu Chen
Zhenyu Chen
Zhi Chen
Zhibo Chen
Zhimin Chen
Zhineng Chen
Zhiwei Chen
Zhixiang Chen
Zhiyang Chen
Zhiyuan Chen
Zhuangzhuang Chen
Zhuo Chen
Zhuoxiao Chen
Zihan Chen
Zijiao Chen
Ziwen Chen

Ziyang Chen
An-Chieh Cheng
Bowen Cheng
De Cheng
Erkang Cheng
Feng Cheng
Guangliang Cheng
Hao Cheng
Hao Cheng
Harry Cheng
Ho Kei Cheng
Jingchun Cheng
Jun Cheng
Ka Leong Cheng
Kelvin B. Cheng
Kevin Ho Man Cheng
Lechao Cheng
Shuo Cheng
Silin Cheng
Ta-Ying Cheng
Tianhang Cheng
Weihao Cheng
Wencan Cheng
Wensheng Cheng
Xi Cheng
Xize Cheng
Xuxin Cheng
Yen-Chi Cheng
Yi Cheng
Yihua Cheng
Yu Cheng
Zesen Cheng
Zezhou Cheng
Zhanzhan Cheng
Zhen Cheng
Zhi-Qi Cheng
Ziang Cheng
Ziheng Cheng
Soon Yau Cheong
Anoop Cherian
Daniil Cherniavskii
Ajad Chhatkuli
Cheng Chi
Haoang Chi
Hyung-gun Chi

Seunggeun Chi
Zhixiang Chi
Naoya Chiba
Julian Chibane
Aditya Chinchure
Eleni Chiou
Mia Chiquier
Chiranjeev Chiranjeev
Kashyap Chitta
Hsu-kuang Chiu
Wei-Chen Chiu
Chee Kheng Chng
Shin-Fang Chng
Donghyeon Cho
Gyusang Cho
Hanbyel Cho
Hoonhee Cho
Jae Won Cho
Jang Hyun Cho
Jungchan Cho
Junhyeong Cho
Nam Ik Cho
Seokju Cho
Seungju Cho
Suhwan Cho
Sunghyun Cho
Sungjun Cho
Sungmin Cho
Wonwoong Cho
Nathaniel E. Chodosh
Jaesung Choe
Junsuk Choe
Changwoon Choi
Chiho Choi
Ching Lam Choi
Dongyoung Choi
Dooseop Choi
Hongsuk Choi
Hyesong Choi
Jaewoong Choi
Janghoon Choi
Jin Young Choi
Jinwoo Choi
Jinyoung Choi
Jonghyun Choi

Jongwon Choi
Jooyoung Choi
Kanghyun Choi
Myungsub Choi
Seokeon Choi
Wonhyeok Choi
Bhavin Choksi
Jaegul Choo
Baptiste Chopin
Ayush Chopra
Gene Chou
Shih-Han Chou
Divya Choudhary
Siddharth Choudhary
Sunav Choudhary
Rohan Choudhury
Vasileios Choutas
Ka-Ho Chow
Pinaki Nath Chowdhury
Sanjoy Chowdhury
Sammy Christen
Fu-Jen Chu
Qi Chu
Ruihang Chu
Sanghyeok Chu
Wei-Ta Chu
Wen-Hsuan Chu
Wen-Hsuan Chu
Wei Qin Chuah
Andrei Chubarau
Ilya Chugunov
Sanghyuk Chun
Se Young Chun
Hyungjin Chung
Inseop Chung
Jaeyoung Chung
Jihoon Chung
Jiwan Chung
Joon Son Chung
Thomas A. Ciarfuglia
Umur A. Ciftci
Antonio E. Cinà
Ramazan Gokberk Cinbis
Anthony Cioppa
Javier Civera

Christopher A. Clark
James J. Clark
Ronald Clark
Brian S. Clipp
Mark Coates
Federico Cocchi
Felipe Codevilla
Cecília Coelho
Danny Cohen-Or
Forrester Cole
Julien Colin
John Collomosse
Marc Comino-Trinidad
Nicola Conci
Marcos V. Conde
Alexandru Paul
  Condurache
Runmin Cong
Tianshuo Cong
Wenyan Cong
Yuren Cong
Alessandro Conti
Andrea Conti
Enric Corona
John R. Corring
Jaime Corsetti
Jason J. Corso
Theo W. Costain
Guillaume Couairon
Robin Courant
Davide Cozzolino
Donato Crisostomi
Francesco Croce
Florinel Alin Croitoru
Ioana Croitoru
James L. Crowley
Steve Cruz
Gabriela Csurka
Alfredo Cuesta Infante
Aiyu Cui
Hainan Cui
Jiali Cui
Jiequan Cui
Justin Cui
Qiongjie Cui

Ruikai Cui
Yawen Cui
Ying Cui
Yuning Cui
Zhen Cui
Zhenyu Cui
Ziteng Cui
Benjamin J. Culpepper
Xiaodong Cun
Federico Cunico
Claudio Cusano
Sebastian Cygert
Guido Maria D'Amely di
  Melendugno
Moreno D'Incà
Feipeng Da
Mosam Dabhi
Rishabh Dabral
Konstantinos M. Dafnis
Matthew L. Daggitt
Anders B. Dahl
Bo Dai
Bo Dai
Haixing Dai
Mengyu Dai
Peng Dai
Peng Dai
Pingyang Dai
Qi Dai
Qiyu Dai
Wenrui Dai
Wenxun Dai
Zuozhuo Dai
Nicola Dall'Asen
Naser Damer
Bo Dang
Meghal Dani
Duolikun Danier
Donald G. Dansereau
Quan Dao
Son Duy Dao
Trung Tuan Dao
Mohamed Daoudi
Ahmad Darkhalil
Rangel Daroya

Abhijit Das
Abir Das
Anurag Das
Ayan Das
Nilotpal Das
Partha Das
Soumi Das
Srijan Das
Sukhendu Das
Swagatam Das
Ananyananda Dasari
Avijit Dasgupta
Gourav Datta
Achal Dave
Ishan Rajendrakumar Dave
Andrew Davison
Aram Davtyan
Axel Davy
Youssef Dawoud
Laura Daza
Francesca De Benetti
Daan de Geus
Alfredo S. De Goyeneche
Riccardo de Lutio
Maria De Marsico
Christophe De
  Vleeschouwer
Tonmoay Deb
Sayan Deb Sarkar
Dale Decatur
Thanos Delatolas
Fabien Delattre
Mauricio Delbracio
Ginger D. Delmas
Andong Deng
Bailin Deng
Bin Deng
Boyang Deng
Chaorui Deng
Cheng Deng
Congyue Deng
Jiacheng Deng
Jiajun Deng
Jiankang Deng
Jingjing Deng

Jinhong Deng
Kangle Deng
Kenan Deng
Lei Deng
Liang-Jian Deng
Shijian Deng
Weijian Deng
Xiaoming Deng
Xueqing Deng
Ye Deng
Yongjian Deng
Youming Deng
Yu Deng
Zhijie Deng
Zhongying Deng
Matteo Denitto
Mohammad Mahdi
  Derakhshani
Alakh Desai
Kevin Desai
Nishq P. Desai
Sai Vikas Desai
Jean-Emmanuel Deschaud
Frédéric Devernay
Rahul Dey
Arturo Deza
Helisa Dhamo
Amaya Dharmasiri
Ankit Dhiman
Vikas Dhiman
Yan Di
Zonglin Di
Ousmane A. Dia
Renwei Dian
Haiwen Diao
Xiaolei Diao
Gonçalo José Dias Pais
Abdallah Dib
Juan Carlos Dibene
  Simental
Sebastian Dille
Christian Diller
Mariella Dimiccoli
Anastasios Dimou
Or Dinari

Caiwen Ding
Changxing Ding
Choubo Ding
Fangqiang Ding
Guiguang Ding
Guodong Ding
Henghui Ding
Hui Ding
Jian Ding
Jian-Jiun Ding
Jianhao Ding
Li Ding
Liang Ding
Lihe Ding
Ming Ding
Runyu Ding
Shouhong Ding
Shuangrui Ding
Shuxiao Ding
Tianjiao Ding
Tianyu Ding
Wenhao Ding
Xiaohan Ding
Xinpeng Ding
Yaqing Ding
Yong Ding
Yuhang Ding
Yuqi Ding
Zheng Ding
Zhengming Ding
Zhipeng Ding
Zhuangzhuang Ding
Zihan Ding
Zihan Ding
Ziluo Ding
Anh-Dung Dinh
Markos Diomataris
Christos Diou
Alish Dipani
Alara Dirik
Ajay Divakaran
Abdelaziz Djelouah
Nemanja Djuric
Thanh-Toan Do
Tien Do

Anh-Dzung Doan
Bao Gia Doan
Khoa D. Doan
Carl Doersch
Andreea Dogaru
Nehal Doiphode
Csaba Domokos
Bowen Dong
Haoye Dong
Jiahua Dong
Jianfeng Dong
Jinpeng Dong
Junhao Dong
Junting Dong
Li Dong
Minjing Dong
Nanqing Dong
Peijie Dong
Qiaole Dong
Qihua Dong
Qiulei Dong
Runpei Dong
Shichao Dong
Shuting Dong
Sixun Dong
Siyan Dong
Tian Dong
Wei Dong
Wei Dong
Weisheng Dong
Xiaoyi Dong
Xiaoyu Dong
Xingping Dong
Yinpeng Dong
Zhenxing Dong
Jonathan C. Donnelly
Naren Doraiswamy
Gianfranco Doretto
Michael Dorkenwald
Muskan Dosi
Huanzhang Dou
Shuguang Dou
Yiming Dou
Zhiyang Frank Dou
Zi-Yi Dou

Arthur Douillard
Michail C. Doukas
Matthijs Douze
Amil Dravid
Vincent Drouard
Dawei Du
Jia-Run Du
Jinhao Du
Ruoyi Du
Weiyu Du
Xiaodan Du
Ye Du
Yingjun Du
Yong Du
Yuanqi Du
Yuntao Du
Zexing Du
Zhuoran Du
Radhika Dua
Haodong Duan
Haoran Duan
Huiyu Duan
Jiafei Duan
Jiali Duan
Peiqi Duan
Ye Duan
Yuchen Duan
Yueqi Duan
Zhihao Duan
Amanda Cardoso Duarte
Shiv Ram Dubey
Anastasia Dubrovina
Daniel Duckworth
Timothy Duff
Nicolas Dufour
Rahul Duggal
Shivam Duggal
Bardienus P. Duisterhof
Felix Dülmer
Matteo Dunnhofer
Chi Nhan Duong
Elona Dupont
Nikita Durasov
Zoran Duric
Mihai Dusmanu

Titir Dutta
Ujjal Kr Dutta
Isht Dwivedi
Sai Kumar Dwivedi
Sebastian Dziadzio
Thomas Eboli
Ramin Ebrahim Nakhli
Benjamin Eckart
Takeharu Eda
Johan Edstedt
Alexei A. Efros
Nikos Efthymiadis
Bernhard Egger
Max Ehrlich
Iván Eichhardt
Farshad Einabadi
Martin Eisemann
Peter Eisert
Mohamed El Banani
Mostafa El-Khamy
Alaa El-Nouby
Randa I. M. Elanwar
James Elder
Abdelrahman Eldesokey
Ismail Elezi
Ehsan Elhamifar
Moshe Eliasof
Sara Elkerdawy
Qing En
Ian Endres
Andreas Engelhardt
Francis Engelmann
Martin Engilberge
Kenji Enomoto
Chanho Eom
Guy Erez
Linus Ericsson
Maria C. Escobar
Marcos Escudero-Viñolo
Sedigheh Eslami
Valter Estevam
Ali Etemad
Martin Nicolas Everaert
Ivan Evtimov
Ralph Ewerth

Fevziye Irem Eyiokur
Cristobal Eyzaguirre
Gabriele Facciolo
Mohammad Fahes
Masud ANI Fahim
Fabrizio Falchi
Alex Falcon
Aoxiang Fan
Baojie Fan
Bin Fan
Chao Fan
David Fan
Haoqiang Fan
Hehe Fan
Heng Fan
Hongyi Fan
Huijie Fan
Junsong Fan
Lei Fan
Lei Fan
Lifeng Fan
Lue Fan
Mingyuan Fan
Qi Fan
Rui R. Fan
Xiran Fan
Yifei Fan
Yue Fan
Zejia Fan
Zhaoxin Fan
Zhenfeng Fan
Zhentao Fan
Zhipeng Fan
Zhiwen Fan
Zicong Fan
Sean Fanello
Chaowei Fang
Chuan Fang
Gongfan Fang
Guian Fang
Han Fang
Jiansheng Fang
Jiemin Fang
Jin Fang
Jun Fang
Kun Fang
Pengfei Fang
Rui Fang
Sheng Fang
Xiang Fang
Xiaolin Fang
Xiuwen Fang
Zhaoyuan Fang
Zhiyuan Fang
Eros Fanì
Matteo Farina
Farzan Farnia
Aiman Farooq
Azade Farshad
Mohsen Fayyaz
Arash Fayyazi
Ben Fei
Jingjing Fei
Xiaohan Fei
Zhengcong Fei
Michael Felsberg
Berthy T. Feng
Bin Feng
Brandon Y. Feng
Chao Feng
Chen Feng
Chengjian Feng
Chun-Mei Feng
Fan Feng
Guorui Feng
Hao Feng
Jianjiang Feng
Lan Feng
Lei Feng
Litong Feng
Mingtao Feng
Qianyu Feng
Ruicheng Feng
Ruili Feng
Ruoyu Feng
Ryan T. Feng
Shiwei Feng
Songhe Feng
Tuo Feng
Wei Feng
Weitao Feng
Weixi Feng
Xuelu Feng
Yanglin Feng
Yao Feng
Yue Feng
Zhanxiang Feng
Zhenhua Feng
Zunlei Feng
Clara Fernandez
Victoria Fernandez
  Abrevaya
Miguel-Ángel
  Fernández-Torres
Sira Ferradans
Claudio Ferrari
Ethan Fetaya
Mustansar Fiaz
Guénolé Fiche
Panagiotis P. Filntisis
Marco Fiorucci
Kai Fischer
Tobias Fischer
Tom Fischer
Volker Fischer
Robert B. Fisher
Alessandro Flaborea
Corneliu O. Florea
Georgios Floros
Alejandro Fontan
Tomaso Fontanini
Lin Geng Foo
Wolfgang Förstner
Niki M. Foteinopoulou
Simone Foti
Simone Foti
Victor Fragoso
Gianni Franchi
Jean-Sebastien Franco
Emanuele Frascaroli
Moti Freiman
Rafail Fridman
Sara Fridovich-Keil
Felix Friedrich
Stanislav Frolov

Andre Fu
Bin Fu
Changcheng Fu
Changqing Fu
Jianlong Fu
Jie Fu
Jingjing Fu
Keren Fu
Lan Fu
Lele Fu
Minghao Fu
Qichen Fu
Rao Fu
Taimeng Fu
Tianwen Fu
Yanwei Fu
Ying Fu
Yonggan Fu
Yun Fu
Yuqian Fu
Zehua Fu
Zhenqi Fu
Zipeng Fu
Wolfgang Fuhl
Yasuhisa Fujii
Yuki Fujimura
Katsuki Fujisawa
Kent Fujiwara
Takuya Funatomi
Christopher Funk
Antonino Furnari
Ryo Furukawa
Ryosuke Furuta
David Futschik
Akshay Gadi Patil
Siddhartha Gairola
Rinon Gal
Adrian Galdran
Silvio Galesso
Meirav Galun
Rofida Gamal
Weijie Gan
Yiming Gan
Yulu Gan
Vineet Gandhi
Kanchana Vaishnavi Gandikota
Aditya Ganeshan
Aditya Ganeshan
Swetava Ganguli
Sujoy Ganguly
Hanan Gani
Alireza Ganjdanesh
Harald Ganster
Roy Ganz
Angela F. Gao
Bin-Bin Gao
Changxin Gao
Chen Gao
Chongyang Gao
Daiheng Gao
Daoyi Gao
Difei Gao
Hang Gao
Hongchang Gao
Huiyu Gao
Junyu Gao
Junyu Gao
Kaifeng Gao
Katelyn Gao
Kuofeng Gao
Lin Gao
Ling Gao
Maolin Gao
Quankai Gao
Ruopeng Gao
Shanghua Gao
Shangqi Gao
Shangqian Gao
Shaobing Gao
Shenyuan Gao
Xiang Gao
Xiangjun Gao
Yang Gao
Yipeng Gao
Yuan Gao
Yue Gao
Yunhe Gao
Zhanning Gao
Zhi Gao
Zhihan Gao
Zhitong Gao
Zhongpai Gao
Ziteng Gao
Nicola Garau
Elena Garces
Noa Garcia
Nuno Cruz Garcia
Ricardo Garcia Pinel
Guillermo Garcia-Hernando
Saurabh Garg
Mathieu Garon
Risheek Garrepalli
Pablo Garrido
Quentin Garrido
Stefano Gasperini
Vincent Gaudilliere
Chandan Gautam
Shivam Gautam
Paul Gavrikov
Jonathan Z. Gazak
Chongjian Ge
Liming Ge
Runzhou Ge
Shiming Ge
Songwei Ge
Weifeng Ge
Wenhang Ge
Yanhao Ge
Yixiao Ge
Yunhao Ge
Yuying Ge
Zhiqi Ge
Zongyuan Ge
James Gee
Chen Geng
Daniel Geng
Xue Geng
Zhengyang Geng
Zichen Geng
Kyle Genova
Georgios Georgakis
Mariana-Iuliana Georgescu

Yiangos Georgiou
Markos Georgopoulos
Stamatios Georgoulis
Michal Geyer
Mina Ghadimi Atigh
Abhijay Ghildyal
Reza Ghoddoosian
Mohsen Gholami
Peyman Gholami
Enjie Ghorbel
Soumya Suvra Ghosal
Anindita Ghosh
Anurag Ghosh
Arnab Ghosh
Arthita Ghosh
Aurobrata Ghosh
Shreya Ghosh
Soham Ghosh
Sreyan Ghosh
Andrea Giachetti
Paris Giampouras
Alexander Gielisse
Andrew Gilbert
Guy Gilboa
Nate Gillman
Jhony H. Giraldo
Roger Girgis
Sharath Girish
Mario Valerio Giuffrida
Francesco Giuliari
Nikolaos Gkanatsios
Ioannis Gkioulekas
Alexi Gladstone
Derek Gloudemans
Aurele T. Gnanha
Hyojun Go
Akos Godo
Danilo D. Goede
Arushi Goel
Kratarth Goel
Rahul Goel
Shubham Goel
Vidit Goel
Tejas Gokhale
Vignesh Gokul
Aditya Golatkar
Micah Goldblum
S. Alireza Golestaneh
Gabriele Goletto
Lluis Gomez
Guillermo Gomez-Trenado
Alex Gomez-Villa
Nuno Gonçalves
Biao Gong
Boqing Gong
Chen Gong
Chengyue Gong
Dayoung Gong
Dong Gong
Jingyu Gong
Kaixiong Gong
Kehong Gong
Liyu Gong
Mingming Gong
Ran Gong
Rui Gong
Tao Gong
Xiaojin Gong
Xuan Gong
Xun Gong
Yifan Gong
Yu Gong
Yuanhao Gong
Yuning Gong
Yunye Gong
Cristina I. González
Adam Goodge
Nithin Gopalakrishnan
 Nair
Cameron Gordon
Dipam Goswami
Gaurav Goswami
Hanno Gottschalk
Chenhui Gou
Jianping Gou
Shreyank N. Gowda
Mohit Goyal
Julia Grabinski
Helmut Grabner
Patrick L. Grady
Alexandros Graikos
Eric Granger
Douglas R. Gray
Michael Alan Greenspan
Connor Greenwell
David Griffiths
Artur Grigorev
Alexey A. Gritsenko
Ivan Grubišić
Geonmo Gu
Jianyang Gu
Jiaqi Gu
Jiayuan Gu
Jinwei Gu
Peiyan Gu
Qiao Gu
Tianpei Gu
Xianfan Gu
Xiang Gu
Xiangming Gu
Xiao Gu
Xiuye Gu
Yanan Gu
Yiwen Gu
Yuchao Gu
Yuming Gu
Yun Gu
Zeqi Gu
Zhangxuan Gu
Banglei Guan
Junfeng Guan
Qingji Guan
Shanyan Guan
Tianrui Guan
Tongkun Guan
Ziqiao Guan
Ziyu Guan
Denis A. Gudovskiy
Antoine Guédon
Paul Guerrero
Ricardo Guerrero
Liangke Gui
Sadaf Gulshad
Manuel Günther
Chen Guo

Chenqi Guo
Chuan Guo
Guodong Guo
Haiyun Guo
Heng Guo
Hengtao Guo
Hongji Guo
Jia Guo
Jianwei Guo
Jianyuan Guo
Jiayi Guo
Jie Guo
Jie Guo
Jingcai Guo
Jingyuan Guo
Jinyang Guo
Lanqing Guo
Meng-Hao Guo
Mengxi China Guo
Minghao Guo
Pengfei Guo
Pengsheng Guo
Pinxue Guo
Qi Guo
Qin Guo
Qing Guo
Qingpei Guo
Saidi Guo
Shi Guo
Shuxuan Guo
Song Guo
Taian Guo
Tiantong Guo
Tianyu Guo
Wen Guo
Wenzhong Guo
Xianda Guo
Xiaobao Guo
Xiefan Guo
Yangyang Guo
Yanhui Guo
Yao Guo
Yichen Guo
Yong Guo
Yuan-Chen Guo
Yuliang Guo
Yuxiang Guo
Yuyu Guo
Zixian Guo
Ziyu Guo
Zujin Guo
Aarush Gupta
Akash Gupta
Akshita Gupta
Ankush Gupta
Anshul Gupta
Anubhav Gupta
Anup Kumar Gupta
Honey Gupta
Kunal Gupta
pravir singh gupta
Rohit Gupta
Saumya Gupta
Corina Gurau
Yeti Z. Gurbuz
Saket Gurukar
Siddharth Gururani
Vladimir Guzov
Matthew A. Gwilliam
Hyunho Ha
Ryo Hachiuma
Isma Hadji
Armin Hadzic
Daniel Haehn
Nicolai Haeni
Ronny Haensch
Meera Hahn
Oliver Hahn
Nguyen Hai
Yuval Haitman
Levente Hajder
Moayed Haji-Ali
Sina Hajimiri
Alexandros Haliassos
Peter M. Hall
Bumsub Ham
Ryuhei Hamaguchi
Abdullah J. Hamdi
Max Hamilton
Lars Hammarstrand
Hasan Abed Al Kader
 Hammoud
Beining Han
Boran Han
Dongyoon Han
Feng Han
Guangxing Han
Guangzeng Han
Hu Han
Jiaming Han
Jin Han
Jungong Han
Junlin Han
Kai Han
Kang Han
Kun Han
Ligong Han
Longfei Han
Mei Han
Mingfei Han
Muzhi Han
Sungwon Han
Tengda Han
Tengda Han
Wencheng Han
Xinran Han
Xintong Han
Yizeng Han
Yufei Han
Zhenjun Han
Zhi Han
Zhongyi Han
Zongbo Han
Zongyan Han
Ankur Handa
Tiankai Hang
Yucheng Hang
Asif Hanif
Param Hanji
Joëlle Hanna
Niklas Hanselmann
Nicklas A. Hansen
Fusheng Hao
Shaozhe Hao
Shengyu Hao

Shijie Hao
Tianxiang Hao
Yanbin Hao
Yu Hao
Zekun Hao
Takayuki Hara
Mehrtash Harandi
Sanjay Haresh
Haripriya Harikumar
Adam Harley
David M. Hart
Md Yousuf Harun
Md Rakibul Hasan
Connor Hashemi
Atsushi Hashimoto
Khurram Azeem Hashmi
Nozomi Hata
Ali Hatamizadeh
Ryuichiro Hataya
Timm Haucke
Joakim Bruslund Haurum
Stephen Hausler
Mohammad Havaei
Hideaki Hayashi
Zeeshan Hayder
Chengan He
Conghui He
Dailan He
Fan He
Fazhi He
Gaoqi He
Haoyu He
Hongliang He
Jiangpeng He
Jianzhong He
Jiawei He
Ju He
Jun-Yan He
Junfeng He
Lihuo He
Liqiang He
Nanjun He
Ruian He
Runze He
Ruozhen He
Shengfeng He
Shuting He
Songtao He
Tao He
Tianyu He
Tong He
Wei He
Xiang He
Xiangteng He
Xiangyu He
Xiaoxiao He
Xin He
Xingyi He
Xingzhe He
Xinlei He
Xinwei He
Yang He
Yang He
Yannan He
Yeting He
Yinan He
Ying He
Yisheng He
Yuhang He
Yuhang He
Zecheng He
Zewei He
Zexin He
Zhenyu He
Ziwen He
Ramya S. Hebbalaguppe
Peter Hedman
Deepti B. Hegde
Sindhu B. Hegde
Matthias Hein
Mattias Paul Heinrich
Aral Hekimoglu
Philipp Henzler
Byeongho Heo
Jae-Pil Heo
Miran Heo
Stephane Herbin
Alexander Hermans
Pedro Hermosilla Casajus
Jefferson E. Hernandez
Monica Hernandez
Charles Herrmann
Roei Herzig
Fabian Herzog
Georg Hess
Mauricio Hess-Flores
Robin Hesse
Chamin P. Hewa Koneputugodage
Masatoshi Hidaka
Richard E. L. Higgins
Mitchell K. Hill
Carlos Hinojosa
Tobias Hinz
Yusuke Hirota
Elad Hirsch
Shinsaku Hiura
Chih-Hui Ho
Man M. Ho
Tuan N. A. Hoang
Jennifer Hobbs
Jiun Tian Hoe
David T. Hoffmann
Derek Hoiem
Yannick Hold-Geoffroy
Lukas Höllein
Gregory I. Holste
Hiroto Honda
Cheeun Hong
Cheng-Yao Hong
Danfeng Hong
Deokki Hong
Fa-Ting Hong
Fangzhou Hong
Feng Hong
Guan Zhe Hong
Je Hyeong Hong
Ji Woo Hong
Jie Hong
Joanna Hong
Lanqing Hong
Lingyi Hong
Sangwoo Hong
Sungeun Hong
Sunghwan Hong

Susung Hong
Weixiang Hong
Xiaopeng Hong
Yan Hong
Yi Hong
Yuchen Hong
Julia Hornauer
Maxwell C. Horton
Eliahu Horwitz
Mir Rayat Imtiaz Hossain
Tonmoy Hossain
Mehdi Hosseinzadeh
Kazuhiro Hotta
Andrew Z. Hou
Fei Hou
Hao Hou
Ji Hou
Jian Hou
Jinghua Hou
Qibin Hou
Qiqi Hou
Ruibing Hou
Saihui Hou
Tingbo Hou
Xinhai Hou
Yuenan Hou
Yunzhong Hou
Zhi Hou
Lukas Hoyer
Petr Hruby
Jun-Wei Hsieh
Chih-Fan Hsu
Gee-Sern Hsu
Hung-Min Hsu
Yen-Chi Hsu
Anthony Hu
Benran Hu
Bo Hu
Dapeng Hu
Dongting Hu
Guosheng Hu
Haigen Hu
Hanjiang Hu
Hanzhe Hu
Hengtong Hu

Hezhen Hu
Jian Hu
Jian-Fang Hu
Jie Hu
Juan Hu
Junjie Hu
Kai Hu
Lianyu Hu
Lily Hu
MengShun Hu
Minghui Hu
Mu Hu
Panwen Hu
Panwen Hu
Ruizhen Hu
Shengshan Hu
Shiyu Hu
Shizhe Hu
Shu Hu
Tao Hu
Tao Hu
Vincent Tao Hu
Wenbo Hu
Xiaodan Hu
Xiaolin Hu
Xiaoling Hu
Xiaotao Hu
Xiaowei Hu
Xinting Hu
Xinting Hu
Xixi Hu
Xuefeng Hu
Yang Hu
Yaosi Hu
Yinlin Hu
Yueyu Hu
Zeyu Hu
Zhanhao Hu
zhengyu hu
Zhenzhen Hu
Zhiming Hu
Zhiming Hu
Zhongyun Hu
Zijian Hu
Andong Hua

Hang Hua
Miao Hua
Wei Hua
Mengdi Huai
Binbin Huang
Bo Huang
Buzhen Huang
Chao Huang
Chao-Tsung Huang
Chaoqin Huang
Ching-Chun Huang
Cong Huang
Danqing Huang
Di Huang
Feihu Huang
Haibin Huang
Haiyang Huang
Hao Huang
Hsin-Ping Huang
Huaibo Huang
Ian Y. Huang
Jiabo Huang
Jiahui Huang
Jiancheng Huang
Jiangyong Huang
Jiaxing Huang
Jin Huang
Jinfa Huang
Jing Huang
Jonathan Huang
Junkai Huang
Junwen Huang
Kejie Huang
Kuan-Chih Huang
Lei Huang
Libo Huang
Lin Huang
Linjiang Huang
Linyan Huang
linzhi Huang
Lun Huang
Luojie Huang
Mingzhen Huang
Qidong Huang
Qiusheng Huang

Runhui Huang
Ruqi Huang
Shaofei Huang
Sheng Huang
Sheng-Yu Huang
Shiyuan Huang
Shuaiyi Huang
Shuangping Huang
Siteng Huang
Siyu Huang
Siyuan Huang
Tao Huang
Thomas E. Huang
Tianyu Huang
Wenjian Huang
Wenke Huang
Xiaohu Huang
Xiaohua Huang
Xiaoke Huang
Xiaoshui Huang
Xiaoyang Huang
Xijie Huang
Xinyu Huang
Xuhua Huang
Yan Huang
Yaping Huang
Ye Huang
Yi Huang
Yi Huang
Yi Huang
Yi-Hua Huang
Yifei Huang
Yihao Huang
Ying Huang
Yizhan Huang
You Huang
Yue Huang
Yufei Huang
Yuge Huang
Yukun Huang
Yuyao Huang
Zaiyu Huang
Zehao Huang
Zeyi Huang
Zhe Huang
Zhewei Huang
Zhida Huang
Zhiqi Huang
Zhiyu Huang
Zhongzhan Huang
Ziling Huang
Zilong Huang
Ziqi Huang
Zixuan Huang
Ziyuan Huang
Jaesung Huh
Ka-Hei Hui
Le Hui
Tianrui Hui
Xiaofei Hui
Ahmed Imtiaz Humayun
Thomas Hummel
Jing Huo
Shuwei Huo
Yuankai Huo
Julio Hurtado
Rukhshanda Hussain
Mohamed Hussein
Noureldien Hussein
Chuong Minh Huynh
Tran Ngoc Huynh
Muhammad Huzaifa
Inwoo Hwang
Jaehui Hwang
Jenq-Neng Hwang
Sehyun Hwang
Seunghyun Hwang
Sukjun Hwang
Sunhee Hwang
Wonjun Hwang
Jeongseok Hyun
Sangeek Hyun
Ekaterina Iakovleva
Tomoki Ichikawa
A. S. M. Iftekhar
Masaaki Iiyama
Satoshi Ikehata
Sunghoon Im
Woobin Im
Nevrez Imamoglu
Abdullah Al Zubaer Imran
Tooba Imtiaz
Nakamasa Inoue
Naoto Inoue
Eldar Insafutdinov
Catalin Ionescu
Radu Tudor Ionescu
Nasar Iqbal
Umar Iqbal
Go Irie
Muhammad Zubair Irshad
Francesco Isgrò
Yasunori Ishii
Berivan Isik
Syed M. S. Islam
Mariko Isogawa
Vamsi Krishna K. Ithapu
Koichi Ito
Leonardo Iurada
Shun Iwase
Sergio Izquierdo
Sarah Jabbour
David Jacobs
David E. Jacobs
Yasamin Jafarian
Azin Jahedi
Achin Jain
Ayush Jain
Jitesh Jain
Kanishk Jain
Yash Jain
Nikita Jaipuria
Tomas Jakab
Muhammad Abdullah Jamal
Hadi Jamali-Rad
Stuart James
Nataraj Jammalamadaka
Donggon Jang
Sujin Jang
Young Kyun Jang
Youngjoon Jang
Steeven Janny
Paul Janson
Maximilian Jaritz

Ronnachai Jaroensri
Guillaume Jaume
Sajid Javed
Saqib Javed
Bhavin Jawade
Hirunima Jayasekara
Senthilnath Jayavelu
Guillaume Jeanneret
Pranav Jeevan
Rohit Jena
Tomas Jenicek
Simon Jenni
Hae-Gon Jeon
Jeimin Jeon
Seogkyu Jeon
Boseung Jeong
Jongheon Jeong
Yonghyun Jeong
Yoonwoo Jeong
Koteswar Rao Jerripothula
Artur Jesslen
Nikolay Jetchev
Ankit Jha
Sumit K. Jha
I-Hong Jhuo
Deyi Ji
Ge-Peng Ji
Hui Ji
Jianmin Ji
Jiayi Ji
Jingwei Ji
Naye Ji
Pengliang Ji
Shengpeng Ji
Wei Ji
Xiang Ji
Xiaopeng Ji
Xiaozhong Ji
Xinya Ji
Yu Ji
Yuanfeng Ji
Zhanghexuan Ji
Baoxiong Jia
Ding Jia
Jinrang Jia

Menglin Jia
Shan Jia
Shuai Jia
Tong Jia
Wenqi Jia
Xiaojun Jia
Xiaosong Jia
Xu Jia
Yiren Jian
Bo Jiang
Borui Jiang
Chen Jiang
Guangyuan Jiang
Hai Jiang
Haiyang Jiang
Haiyong Jiang
Hanwen Jiang
Hanxiao Jiang
Hao Jiang
Haobo Jiang
Huaizu Jiang
Huajie Jiang
Jianmin Jiang
Jianwen Jiang
Jiaxi Jiang
Jin Jiang
Jing Jiang
Kaixun Jiang
Kui Jiang
Li Jiang
Liming Jiang
Meirui Jiang
Ming Jiang
Ming Jiang
Peng Jiang
Peng-Tao Jiang
Runqing Jiang
Siyang Jiang
Tianjian Jiang
Tingting Jiang
Weisen Jiang
Wen Jiang
Xingyu Jiang
Xueying Jiang
Xuhao Jiang

Yanru Jiang
Yi Jiang
Yifan Jiang
Yingying Jiang
Yue Jiang
Yuming Jiang
Zeren Jiang
Zheheng Jiang
Zhenyu Jiang
Zhongyu Jiang
Zi-Hang Jiang
Zixuan Jiang
Ziyu Jiang
Zutao Jiang
Zutao Jiang
Jianbin Jiao
Jianbo Jiao
Licheng Jiao
Ruochen Jiao
Shuming Jiao
Wenpei Jiao
Zequn Jie
Dongkwon Jin
Gaojie Jin
Haian Jin
Haibo Jin
Kyong Hwan Jin
Lei Jin
Lianwen Jin
Linyi Jin
Peng Jin
Peng Jin
Sheng Jin
SouYoung Jin
Weiyang Jin
Xiao Jin
Xiaojie Jin
Xin Jin
Xin Jin
Yeying Jin
Yi Jin
Ying Jin
Zhenchao Jin
Chenchen Jing
Junpeng Jing

Mengmeng Jing
Taotao Jing
Jeonghee Jo
Yeonsik Jo
Younghyun Jo
Cameron A. Johnson
Faith M. Johnson
Maxwell Jones
Michael J. Jones
R. Kenny Jones
Ameya Joshi
Shantanu H. Joshi
Brendan Jou
Chen Ju
Wei Ju
Xuan Ju
Yan Ju
Felix Juefei-Xu
Florian Jug
Yohan Jun
Kim Jun-Seong
Masum Shah Junayed
Chanyong Jung
Claudio R. Jung
Hoin Jung
Hyunyoung Jung
Sangwon Jung
Steffen Jung
Sacha Jungerman
Hari Chandana K.
Prajwal K. R.
Berna Kabadayi
Anis Kacem
Anil Kag
Kumara Kahatapitiya
Bernhard Kainz
Ivana Kajic
Ioannis Kakogeorgiou
Niveditha Kalavakonda
Mahdi M. Kalayeh
Ajinkya Kale
Anmol Kalia
Sinan Kalkan
Jayateja Kalla
Tarun Kalluri

Uday Kamal
Sandesh Kamath
Chandra Kambhamettu
Meina Kan
Sai Srinivas Kancheti
Takuhiro Kaneko
Cuicui Kang
Dahyun Kang
Guoliang Kang
Hyolim Kang
Jaeyeon Kang
Juwon Kang
Li-Wei Kang
Mingon Kang
MinGuk Kang
Minjun Kang
Weitai Kang
Zhao Kang
Yash Mukund Kant
Yueying Kao
Saarthak Kapse
Aupendu Kar
Oğuzhan Fatih Kar
Ozgur Kara
Tamás Karácsony
Srikrishna Karanam
Neerav Karani
Mert Asim Karaoglu
Laurynas Karazija
Navid Kardan
Amirhossein Kardoost
Nour Karessli
Michelle Karg
Mohammad Reza Karimi
  Dastjerdi
Animesh Karnewar
Arjun M. Karpur
Shyamgopal Karthik
Korrawe Karunratanakul
Tejaswi Kasarla
Satyananda Kashyap
Yoni Kasten
Marc A. Kastner
Hirokatsu Kataoka
Isinsu Katircioglu

Kai Katsumata
Ilya Kaufman
Manuel Kaufmann
Chaitanya Kaul
Prannay Kaul
Prakhar Kaushik
Isaak Kavasidis
Ryo Kawahara
Yuki Kawana
Justin Kay
Evangelos Kazakos
Jingcheng Ke
Lei Ke
Tsung-Wei Ke
Wei Ke
Zhanghan Ke
Nikhil V. Keetha
Thomas Kehrenberg
Marilyn Keller
Rohit Keshari
Janis Keuper
Daniel Keysers
Hrant Khachatrian
Seyran Khademi
Wesley A. Khademi
Taras Khakhulin
Samir Khaki
Hasam Khalid
Umar Khalid
Amr Khalifa
Asif Hussain Khan
Faizan Farooq Khan
Muhammad Haris Khan
Naimul Khan
Qadeer Khan
Zeeshan Khan
Bishesh Khanal
Pulkit Khandelwal
Ishan Khatri
Muhammad Uzair Khattak
Vahid Reza Khazaie
Vaishnavi M. Khindkar
Rawal Khirodkar
Pirazh Khorramshahi
Sahil S. Khose

Kourosh Khoshelham
Sena Kiciroglu
Benjamin Kiefer
Kotaro Kikuchi
Mert Kilickaya
Benjamin Killeen
Beomyoung Kim
Boah Kim
Bumsoo Kim
Byeonghwi Kim
Byoungjip Kim
Changhoon Kim
Changick Kim
Chanho Kim
Chanyoung Kim
Dahun Kim
Dahyun Kim
Diana S. Kim
Dong-Jin Kim
Donggun Kim
Donghyun Kim
Dongkeun Kim
Dongwan Kim
Dongyoung Kim
Eun-Sol Kim
Eunji Kim
Euyoung Kim
Geeho Kim
Giseop Kim
Guisik Kim
Gwanghyun Kim
Hakyeong Kim
Hanjae Kim
Hanjung Kim
Heewon Kim
Howon Kim
Hyeongwoo Kim
Hyeongwoo Kim
Hyo Jin Kim
Hyung-Il Kim
Hyunwoo J. Kim
Insoo Kim
Jae Myung Kim
Jeongsol Kim
Jihwan Kim

Jinkyu Kim
Jinwoo Kim
Jongyoo Kim
Joonsoo Kim
Jung Uk Kim
Junho Kim
Junho Kim
Junsik Kim
Kangyeol Kim
Kunhee Kim
Kwang In Kim
Manjin Kim
Minchul Kim
Minji Kim
Minjung Kim
Namil Kim
Namyup Kim
Nayeong Kim
Sanghyun Kim
Seong Tae Kim
Seungbae Kim
Seungryong Kim
Seungwook Kim
Soo Ye Kim
Soohwan Kim
Sungnyun Kim
Sungyeon Kim
Sunnie S. Y. Kim
Tae Hyun Kim
Tae Hyung Kim
Taehoon Kim
Taehun Kim
Taehwan Kim
Taekyung Kim
Taeoh Kim
Taewoo Kim
Won Hwa Kim
Wonjae Kim
Woo Jae Kim
Young Min Kim
YoungBin Kim
Youngeun Kim
Youngseok Kim
Youngwook Kim
Akisato Kimura

Andreas Kirsch
Nikita Kister
Furkan Osman Kınlı
Marcus Klasson
Florian Kleber
Tzofi M. Klinghoffer
Jan P. Klopp
Florian Kluger
Hannah Kniesel
David M. Knigge
Byungsoo Ko
Dohwan Ko
Jongwoo Ko
Takumi Kobayashi
Muhammed Kocabas
Yeong Jun Koh
Kathlén Kohn
Subhadeep Koley
Nick Kolkin
Soheil Kolouri
Jacek Komorowski
Deying Kong
Fanjie Kong
Hanyang Kong
Hui Kong
Kyeongbo Kong
Lecheng Kong
Lingdong Kong
Linghe Kong
Lingshun Kong
Naejin Kong
Quan Kong
Shu Kong
Xianghao Kong
Xiangtao Kong
Xiangwei Kong
Xiaoyu Kong
Xin Kong
Youyong Kong
Yu Kong
Zhenglun Kong
Aishik Konwer
Nicholas C. Konz
Gwanhyeong Koo
Juil Koo

Julian F. P. Kooij
George Kopanas
Sanjeev J. Koppal
Bruno Korbar
Giorgos Kordopatis-Zilos
Dimitri Korsch
Adam Kortylewski
Divya Kothandaraman
Suraj Kothawade
Iuliia Kotseruba
Sasikanth Kotti
Alankar Kotwal
Shashank Kotyan
Alexandros Kouris
Petros Koutras
Rama Kovvuri
Dilip Krishnan
Praveen Krishnan
Ranganath Krishnan
Rohan M. Krishnan
Georg Krispel
Alexander Krull
Tianshu Kuai
Haowei Kuang
Zhengfei Kuang
Andrey Kuehlkamp
David Kügler
Arjan Kuijper
Anna Kukleva
Jonas Kulhanek
Peter Kulits
Akshay R. Kulkarni
Ashutosh C. Kulkarni
Kuldeep Kulkarni
Nilesh Kulkarni
Abhinav Kumar
Abhishek Kumar
Akash Kumar
Avinash Kumar
B. V. K. Vijaya Kumar
Chandan Kumar
Pulkit Kumar
Ratnesh Kumar
Sateesh Kumar
Satish Kumar

Suryansh Kumar
Yogesh Kumar
Nupur Kumari
Sudhakar Kumawat
Nilakshan Kunananthaseelan
Rohit Kundu
Souvik Kundu
Meng-Yu Jennifer Kuo
Weicheng Kuo
Shuhei Kurita
Yusuke Kurose
Takahiro Kushida
Uday Kusupati
Alina Kuznetsova
Jobin K. V.
Henry Kvinge
Ho Man Kwan
Hyeokjun Kweon
Donghyeon Kwon
Gihyun Kwon
Heeseung Kwon
Hyoukjun Kwon
Myung-Joon Kwon
Taein Kwon
YoungJoong Kwon
Cameron Kyle-Davidson
Christos Kyrkou
Jorma Laaksonen
Patrick Labatut
Yann Labbé
Manuel Ladron de Guevara
Florent Lafarge
Jean Lahoud
Bolin Lai
Farley Lai
Jian-Huang Lai
Shenqi Lai
Xin Lai
Yu-Kun Lai
Yung-Hsuan Lai
Zeqiang Lai
Zhengfeng Lai
Barath Lakshmanan
Rohit Lal

Rodney LaLonde
Hala Lamdouar
Meng Lan
Yushi Lan
Federico Landi
George V. Landon
Chunbo Lang
Jochen Lang
Nico Lang
Georg Langs
Raffaella Lanzarotti
Dong Lao
Yixing Lao
Yizhen Lao
Zakaria Laskar
Alexandros Lattas
Chun Pong Lau
Shlomi Laufer
Justin Lazarow
Svetlana Lazebnik
Duy Tho Le
Hieu Le
Hoang Le
Hoang Le
Thi-Thu-Huong Le
Trung Le
Trung-Nghia Le
Tung Thanh Le
Hoàng-Ân Lê
Herve Le Borgne
Guillaume Le Moing
Erik Learned-Miller
Tim Lebailly
Byeong-Uk Lee
Byung-Kwan Lee
Cheng-Han Lee
Chul Lee
Daeun Lee
Dogyoon Lee
Dong Hoon Lee
Eugene Eu Tzuan Lee
Eung-Joo Lee
Gyuseong Lee
Hsin-Ying Lee
Hwee Kuan Lee

Hyeongmin Lee	Seungyong Lee	Chao Li
Hyungtae Lee	Sohyun Lee	Chenghong Li
Jae Yong Lee	Suhyeon Lee	Chenglin Li
Jaeho Lee	Sungho Lee	Chenglong Li
Jaeseong Lee	Sungmin Lee	Chengze Li
Jaewon Lee	Suyoung Lee	Chun-Guang Li
Jangho Lee	Taehyun Lee	Daiqing Li
Jangwon Lee	Wooseok Lee	Dasong Li
Jihyun Lee	Yao-Chih Lee	Dian Li
Jiyoung Lee	Yi-Lun Lee	Dong Li
Jong-Seok Lee	Yonghyeon Lee	Fangda Li
Jongho Lee	Youngwan Lee	Feiran Li
Jongmin Lee	Leonidas Lefakis	Fenghai Li
Joo-Ho Lee	Bowen Lei	Gen Li
Joon-Young Lee	Chenyang Lei	Guanbin Li
Joonseok Lee	Chenyi Lei	Guangrui Li
Jungbeom Lee	Jiahui Lei	Guihong Li
Jungho Lee	Na Lei	Guorong Li
Jungwoo Lee	Qinqian Lei	Haifeng Li
Junha Lee	Yinjie Lei	Han Li
Junhyun Lee	Thomas Leimkuehler	Hang Li
Junyong Lee	Abe Leite	Hangyu Li
Kibok Lee	Abdelhak Lemkhenter	Hanhui Li
Kuan-Ying Lee	Jiaxu Leng	Hao Li
Kwonjoon Lee	Luziwei Leng	Hao Li
Kwot Sin Lee	Zhiying Leng	Haoang Li
Kyungmin Lee	Hendrik P. A. Lensch	Haoran Li
Kyungmoon Lee	Jan E. Lenssen	Haoxiang Li
Minhyeok Lee	Ted Lentsch	Haoxin Li
Minsik Lee	Simon Lepage	He Li
Pilhyeon Lee	Stefan Leutenegger	Heng Li
Saehyung Lee	Filippo Leveni	Hengduo Li
Sangho Lee	Axel Levy	Hongshan Li
Sanghyeok Lee	Ailin Li	Hongwei Bran Li
Sangmin Lee	Aixuan Li	Hongxiang Li
Sehun Lee	Baiang Li	Hongyang Li
Sehyung Lee	Baoxin Li	Hongyu Li
Seon-Ho Lee	Bin Li	Huafeng Li
Seong Hun Lee	Bing Li	Huan Li
Seongwon Lee	Bing Li	Hui Li
Seung Hyun Lee	Bo Li	Jiacheng Li
Seung-Ik Lee	Bowen Li	Jiahao Li
Seungho Lee	Boying Li	Jialu Li
Seunghun Lee	Changlin Li	Jiaman Li
Seungmin Lee	Changlin Li	Jiangmeng Li

Jiangtong Li
Jiangyuan Li
Jianing Li
Jianwei Li
Jianwu Li
Jiaqi Li
Jiaqi Li
Jiatong Li
Jiaxuan Li
Jiazhi Li
Jichang Li
Jie Li
Jin Li
Jinglun Li
Jingzhi Li
Jingzong Li
Jinlong Li
Jinlong Li
Jinpeng Li
Jinxing Li
Jun Li
Jun Li
Junbo Li
Juncheng Li
Junxuan Li
Junyi Li
Kai Li
Kaican Li
Kailin Li
Ke Li
Kehan Li
Keyu Li
Kun Li
Kunchang Li
Kunpeng Li
Lei Li
Lei Li
Li Li
Li Erran Li
Liang Li
Lin Li
Lincheng Li
Liulei Li
Liunian Harold Li
Lujun Li

Manyi Li
Maomao Li
Meng Li
Mengke Li
Mengtian Li
Mengtian Li
Ming Li
Ming Li
Minghan Li
Mingjie Li
Nannan Li
Nianyi Li
Peike Li
Peizhao Li
Peng Li
Pengpeng Li
Pengyu Li
Ping Li
Puhao Li
Qiang Li
Qing Li
Qingyong Li
Qiufu Li
Qizhang Li
Ren Li
Rong Li
Rongjie Li
Ru Li
Rui Li
Ruibo Li
Ruihui Li
Ruilong Li
Ruining Li
Ruixuan Li
Runze Li
Ruoteng Li
Shaohua Li
Shasha Li
Shigang Li
Shijie Li
Shikun Li
Shile Li
Shuai Li
Shuai Li
Shuang Li

Shuwei Li
Si Li
Siyao Li
Siyuan Li
Siyuan Li
Taihui Li
Tianye Li
Wanhua Li
Wanqing Li
Wei Li
Wei Li
Wei Li
Wei-Hong Li
Weihao Li
Weijia Li
Weiming Li
Wenbin Li
Wenbo Li
Wenhao Li
Wenjie Li
Wenshuo Li
Wentong Li
Wenxi Li
Xiang Li
Xiang Li
Xiang Li
Xiang Li
Xiang Li
Xiangtai Li
Xiangyang Li
Xianzhi Li
Xiao Li
Xiao Li
Xiaoguang Li
Xiaomeng Li
Xiaoming Li
Xiaoqi Li
Xiaoqiang Li
Xiaotian Li
Xiaoyu Li
Xin Li
Xin Li
Xin Li
Xinghui Li
Xingyi Li

Xingyu Li
Xinjie Li
Xinyu Li
Xiu Li
Xiujun Li
Xuan Li
Xuanlin Li
Xuelong Li
Xuelu Li
Xueqian Li
Ya-Li Li
Yanan Li
Yang Li
Yang Li
Yangyan Li
Yanjing Li
Yansheng Li
Yanwei Li
Yanyu Li
Yaohui Li
Yaowei Li
Yawei Li
Yi Li
Yi Li
Yicong Li
Yicong Li
Yifei Li
Yijin Li
Yijun Li
Yijun Li
Yikang Li
Yimeng Li
Yiming Li
Yiming Li
Yingwei Li
Yiting Li
Yixuan Li
Yize Li
Yizhuo Li
Yong Li
Yong-Lu Li
Yongjie Li
Yuanman Li
Yuanming Li
Yuelong Li

Yuexiang Li
Yuezun Li
Yuhang Li
Yuheng Li
Yulin Li
Yumeng Li
Yunfan Li
Yunheng Li
Yunqiang Li
Yunsheng Li
Yuyan Li
Yuyang Li
Zejian Li
Zekun Li
Zekun Li
Zhangheng Li
Zhangzikang Li
Zhaoshuo Li
Zhaowen Li
Zhe Li
Zhe Li
Zhen Li
Zhen Li
Zhen Li
Zheng Li
Zhengqin Li
Zhengyuan Li
Zhenyu Li
Zhichao Li
Zhihao Li
Zhihao Li
Zhiheng Li
Zhiqi Li
Zhixuan Li
Zhong Li
Zhuoling Li
Zhuowan Li
Zhuowei Li
Zhuoxiao Li
Zihan Li
Ziqiang Li
Wen Qiao Li
Dongze Lian
Long Lian
Qing Lian

Ruyi Lian
Zhouhui Lian
Jin Lianbao
Chao Liang
Chia-Kai Liang
Dingkang Liang
Feng Liang
Gongbo Liang
Hanxue Liang
Hao Liang
Hui Liang
Jiadong Liang
Jiajun Liang
Jian Liang
Jingyun Liang
Jinxiu S. Liang
Junwei Liang
Kaiqu Liang
Ke Liang
Kevin J. Liang
Luming Liang
Mingfu Liang
Pengpeng Liang
Siyuan Liang
Xiaoxiao Liang
Xinran Liang
Xiwen Liang
Yang Liang
Yixun Liang
Yongqing Liang
Youwei Liang
Yuanzhi Liang
Zhexin Liang
Zhihao Liang
Zhixuan Liang
Kang Liao
Liang Liao
Minghui Liao
Ting-Hsuan Liao
Wei Liao
Xin Liao
Yinghong Liao
Yue Liao
Zhibin Liao
Ziwei Liao

Benedetta Liberatori
Daniel J. Lichy
Maiko Lie
Qin Likun
Isaak Lim
Teck Yian Lim
Bannapol Limanond
Baijiong Lin
Beibei Lin
Cheng Lin
Chenhao Lin
Chia-Wen Lin
Chieh Hubert Lin
Chuang Lin
Chung-Ching Lin
Chunyu Lin
Ci-Siang Lin
Di Lin
Fanqing Lin
Feng Lin
Fudong Lin
Guangfeng Lin
Haotong Lin
Haozhe Lin
Hubert Lin
Hui Lin
Jason Lin
Jianxin Lin
Jiaqi Lin
Jiaying Lin
Jiehong Lin
Jierui Lin
Jintao Lin
Kai-En Lin
Ke Lin
Kevin Lin
Kevin Qinghong Lin
Kuan Heng Lin
Kun-Yu Lin
Kunyang Lin
Kwan-Yee Lin
Lijian Lin
Liqiang Lin
Liting Lin
Luojun Lin
Mingyuan Lin
Qiuxia Lin
Shaohui Lin
Shih-Yao Lin
Sihao Lin
Siyou Lin
Tiancheng Lin
Tsung-Yu Lin
Wanyu Lin
Wei Lin
Wei Lin
Wen-Yan Lin
Wenbin Lin
Xiangbo Lin
Xianhui Lin
Xiaofan Lin
Xiaofeng Lin
Xin Lin
Xudong Lin
Xue Lin
Xuxin Lin
Ya-Wei Eileen Lin
Yan-Bo Lin
Yancong Lin
Yi Lin
Yijie Lin
Yiming Lin
Yiqi Lin
Yiqun Lin
Yongliang Lin
Yu Lin
Yuanze Lin
Yuewei Lin
Zhi-Hao Lin
Zhiqiu Lin
ZhiWei Lin
Zinan Lin
Ziyi Lin
David B. Lindell
Philipp Lindenberger
Jingwang Ling
Jun Ling
Yongguo Ling
Zhan Ling
Alexander Liniger
Stefan P. Lionar
Phillip Lippe
Lahav O. Lipson
Joey Litalien
Ron Litman
Mattia Litrico
Dor Litvak
Aishan Liu
Ajian Liu
Akide L. Y. Liu
Andrew Liu
Ao Liu
Bang Liu
Benlin Liu
Bin Liu
Bin Liu
Bing Liu
Binghao Liu
Bingyuan Liu
Bo Liu
Bo Liu
Bo Liu
Boning Liu
Bowen Liu
Boxiao Liu
Chang Liu
Chang Liu
Chang Liu
Chang Liu
Chao Liu
Chengxin Liu
Chengxu Liu
Chih-Ting Liu
Chuanjian Liu
Chun-Hao Liu
Daizong Liu
Decheng Liu
Di Liu
Difan Liu
Dong Liu
Dongnan Liu
Fang Liu
Fang Liu
Fangyi Liu
Feng Liu

Fengbei Liu
Fenglin Liu
Fengqi Liu
Furui Liu
Fuxiao Liu
Haisong Liu
Han Liu
Hanwen Liu
Hanyuan Liu
Hao Liu
Haolin Liu
Haotian Liu
Haozhe Liu
Heshan Liu
Hong Liu
Hongbin Liu
Hongfu Liu
Hongyu Liu
Hsueh-Ti Derek Liu
Huidong Liu
Isabella Liu
Ji Liu
Jia-Wei Liu
Jiachen Liu
Jiaheng Liu
Jiahui Liu
Jiaming Liu
Jiancheng Liu
Jiang Liu
Jianmeng Liu
Jiashuo Liu
Jiawei Liu
Jiawei Liu
Jiayang Liu
Jiayi Liu
Jie Liu
Jie Liu
Jie Liu
Jihao Liu
Jing Liu
Jing Liu
Jing Liu
Jingyuan Liu
Jingyuan Liu
Jiuming Liu

Jiyuan Liu
Jun Liu
Kang-Jun Liu
Kangning Liu
Kenkun Liu
Kunhao Liu
Li Liu
Lijuan Liu
Lingbo Liu
Lingqiao Liu
Liu Liu
Liyang Liu
Meng Liu
Mengchen Liu
Miao Liu
Ming Liu
Minghao Liu
Minghua Liu
Mingxuan Liu
Mingyuan Liu
Nan Liu
Nian Liu
Ning Liu
Peidong Liu
Peirong Liu
Peiye Liu
Pengju Liu
Ping Liu
Qi Liu
Qiankun Liu
Qihao Liu
Qing Liu
Qingjie Liu
Richard Liu
Risheng Liu
Rui Liu
Ruicong Liu
Ruoshi Liu
Ruyu Liu
Shaohui Liu
Shaoteng Liu
Shaowei Liu
Sheng Liu
Shenglan Liu
Shikun Liu

Shilong Liu
Shuaicheng Liu
Shuaicheng Liu
Shuming Liu
Songhua Liu
Tao Liu
Tian Yu Liu
Tianci Liu
Tianshan Liu
Tongliang Liu
Tyng-Luh Liu
Wei Liu
Weifeng Liu
Weixiao Liu
Weiyu Liu
Wen Liu
Wenxi Liu
Wenyu Liu
Wenyu Liu
Wu Liu
Xian Liu
Xianglong Liu
Xianpeng Liu
Xiao Liu
Xiaohong Liu
Xiaoyu Liu
Xiaoyu Liu
Xihui Liu
Xin Liu
Xin Liu
Xinchen Liu
Xingtong Liu
Xingyu Liu
Xinhang Liu
Xinhui Liu
Xinpeng Liu
Xinwei Liu
Xinyu Liu
Xiulong Liu
Xiyao Liu
Xu Liu
Xubo Liu
Xudong Liu
Xueting Liu
Xueyi Liu

Yan Liu
Yanbin Liu
Yang Liu
Yang Liu
Yang Liu
Yang Liu
Yang Liu
Yanwei Liu
Yaojie Liu
Ye Liu
Yi Liu
Yihao Liu
Yingcheng Liu
Yingfei Liu
Yipeng Liu
Yipeng Liu
Yixin Liu
Yizhang Liu
Yong Liu
Yong Liu
Yonghuai Liu
Yongtuo Liu
Yu Liu
Yu-Lun Liu
Yu-Shen Liu
Yuan Liu
Yuang Liu
Yuanpei Liu
Yuanpeng Liu
Yuanwei Liu
Yuchen Liu
Yuchen Liu
Yuchi Liu
Yueh-Cheng Liu
Yufan Liu
Yuhao Liu
Yuliang Liu
Yun Liu
Yun Liu
Yun Liu
Yunfan Liu
Yunfei Liu
Yunze Liu
Yupei Liu
Yuqi Liu

Yuyang Liu
Yuyuan Liu
Zhaoqiang Liu
Zhe Liu
Zhe Liu
Zhen Liu
Zheng Liu
Zhenguang Liu
Zhi Liu
Zhihua Liu
Zhijian Liu
Zhili Liu
Zhuoran Liu
Ziquan Liu
Ziyi Liu
Zuxin Liu
Zuyan Liu
Josep Llados
Ling Lo
Shao-Yuan Lo
Liliana Lo Presti
Sylvain Lobry
Yaroslava Lochman
Fotios Logothetis
Suhas Lohit
Marios Loizou
Vishnu Suresh Lokhande
Cheng Long
Chengjiang Long
Fuchen Long
Guodong Long
Rujiao Long
Shangbang Long
Teng Long
Xiaoxiao Long
Zijun Long
Ivan Lopes
Vasco Lopes
Adrian Lopez-Rodriguez
Javier Lorenzo-Navarro
Yujing Lou
Brian C. Lovell
Weng Fei Low
Changsheng Lu
Chun-Shien Lu

Daohan Lu
Dongming Lu
Erika Lu
Fan Lu
Guangming Lu
Guo Lu
Hao Lu
Hao Lu
Hongtao Lu
Jiachen Lu
Jiaxin Lu
Jiwen Lu
Lewei Lu
Liying Lu
Quanfeng Lu
Shenyu Lu
Shun Lu
Tao Lu
Xiangyong Lu
Xiankai Lu
Xin Lu
Xuanchen Lu
Xuequan Lu
Yan Lu
Yang Lu
Yanye Lu
Yawen Lu
Yifan Lu
Yongchun Lu
Yongxi Lu
Yu Lu
Yu Lu
Yuzhe Lu
Zhichao Lu
Zhihe Lu
Zijia Lu
Tianyu Luan
Pauline Luc
Simon Lucey
Timo Lüddecke
Jonathon Luiten
Jovita Lukasik
Ao Luo
Cheng Luo
Chuanchen Luo

Donghao Luo
Fangzhou Luo
Gen Luo
Gongning Luo
Hongchen Luo
Jiahao Luo
Jiebo Luo
Jinqi Luo
Jinqi Luo
Jun Luo
Katie Z. Luo
Kunming Luo
Lei Luo
Mandi Luo
Mi Luo
Ruotian Luo
Sihui Luo
Tiange Luo
Wenhan Luo
Xiao Luo
Xiaotong Luo
Xiongbiao Luo
Xu Luo
Yadan Luo
Yawei Luo
Ye Luo
Yisi Luo
Yong Luo
You-Wei Luo
Yuanjing Luo
Zelun Luo
Zhengxiong Luo
Zhengyi Luo
Zhiming Luo
Zhipeng Luo
Zhongjin Luo
Zilin Luo
Ziyang Luo
Tung M. Luu
Diogo C. Luvizon
Jun Lv
Pei Lv
Yunqiu Lv
Zhaoyang Lv
Gengyu Lyu

Jiancheng Lyu
Jipeng Lyu
Junfeng Lyu
Mengyao Lyu
Mingzhi Lyu
Weimin Lyu
Xiaoyang Lyu
Xinyu Lyu
Yiwei Lyu
Youwei Lyu
Ailong Ma
Andy J. Ma
Benteng Ma
Bingpeng Ma
Chao Ma
Chuofan Ma
Cong Ma
Cuixia Ma
Fan Ma
Fangchang Ma
Fei Ma
Guozheng Ma
Haoyu Ma
Hengbo Ma
Huimin Ma
Jiahao Ma
Jianqi Ma
Jiawei Ma
Jiayi Ma
Kai Ma
Kede Ma
Lei Ma
Li Ma
Lin Ma
Liqian Ma
Lizhuang Ma
Mengmeng Ma
Ning Ma
Qianli Ma
Rui Ma
Shijie Ma
Shiqiang Ma
Shiqing Ma
Shuailei Ma
Sizhuo Ma

Tao Ma
Teli Ma
Wenxuan Ma
Wufei Ma
Xianzheng Ma
Xiaoxuan Ma
Xinyin Ma
Xinzhu Ma
Xu Ma
Yeyao Ma
Yifeng Ma
Yuexiao Ma
Yuexin Ma
Yunsheng Ma
Zhan Ma
Zhanyu Ma
Ziping Ma
Ziqiao Ma
Muhammad Maaz
Anish Madan
Neelu Madan
Spandan Madan
Sai Advaith Maddipatla
Rishi Madhok
Filippo Maggioli
Simone Magistri
Marcus Magnor
Sabarinath Mahadevan
Shweta Mahajan
Aniruddha Mahapatra
Sarthak Kumar Maharana
Behrooz Mahasseni
Upal Mahbub
Arif Mahmood
Kaleel Mahmood
Mohammed Mahmoud
Tanvir Mahmud
Jinjie Mai
Helena de Almeida Maia
Josef Maier
Shishira R. Maiya
Snehashis Majhi
Orchid Majumder
Sagnik Majumder
Ilya Makarov

Sina Malakouti
Hashmat Shadab Malik
Mateusz Malinowski
Utkarsh Mall
Srikanth Malla
Clement Mallet
Dimitrios Mallis
Abed Malti
Yunze Man
Oscar Mañas
Karttikeya Mangalam
Fabian Manhardt
Ioannis Maniadis Metaxas
Fahim Mannan
Rafal Mantiuk
Dongxing Mao
Jiageng Mao
Wei Mao
Weian Mao
Weixin Mao
Ye Mao
Yongsen Mao
Yunyao Mao
Yuxin Mao
Zhiyuan Mao
Emanuela Marasco
Matthew Marchellus
Alberto Marchisio
Diego Marcos
Alina E. Marcu
Riccardo Marin
Manuel J. Marín-Jiménez
Octave Mariotti
Dejan Markovic
Imad Eddine Marouf
Valerio Marsocci
Diego Martin Arroyo
Ricardo Martin-Brualla
Brais Martinez
Renato Martins
Damien Martins Gomes
Tetiana Martyniuk
Pierre Marza
David Masip
Carlo Masone

Timothée Masquelier
André G. Mateus
Minesh Mathew
Yusuke Matsui
Bruce A. Maxwell
Christoph Mayer
Prasanna Mayilvahanan
Amir Mazaheri
Amrita Mazumdar
Pratik Mazumder
Alessio Mazzucchelli
Amarachi B. Mbakwe
Scott McCloskey
Naga Venkata Kartheek Medathati
Henry Medeiros
Guofeng Mei
Haiyang Mei
Jie Mei
Jieru Mei
Kangfu Mei
Lingjie Mei
Xiaoguang Mei
Dennis Melamed
Luke Melas-Kyriazi
Iaroslav Melekhov
Yifang Men
Ricardo A. Mendoza-León
Depu Meng
Fanqing Meng
Jingke Meng
Lingchen Meng
Qier Meng
Qingjie Meng
Quan Meng
Yanda Meng
Zibo Meng
Otniel-Bogdan Mercea
Pablo Mesejo
Safa Messaoud
Nico Messikommer
Nando Metzger
Christopher Metzler
Vasileios Mezaris
Liang Mi

Zhenxing Mi
S. Mahdi H. Miangoleh
Bo Miao
Changtao Miao
Jiaxu Miao
Zichen Miao
Bjoern Michele
Christian Micheloni
Marko Mihajlovic
Zoltán Á. Milacski
Simone Milani
Leo Milecki
Roy Miles
Christen Millerdurai
Monica Millunzi
Chaerin Min
Cheol-Hui Min
Dongbo Min
Hyun-Seok Min
Jie Min
Juhong Min
Kyle Min
Yifei Min
Yuecong Min
Zhixiang Min
Matthias Minderer
Di Ming
Qi Ming
Xiang Ming
Riccardo Miotto
Aymen Mir
Pedro Miraldo
Parsa Mirdehghan
Seyed Ehsan Mirsadeghi
Muhammad Jehanzeb Mirza
Ashkan Mirzaei
Dmytro Mishkin
Anand Mishra
Ashish Mishra
Samarth Mishra
Shlok K. Mishra
Diganta Misra
Abhay Mittal
Gaurav Mittal

Surbhi Mittal
Trisha Mittal
Taiki Miyanishi
Daisuke Miyazaki
Hong Mo
Kaichun Mo
Sangwoo Mo
Sicheng Mo
Sicheng Mo
Zhipeng Mo
Michael Moeller
Peyman Moghadam
Hadi Mohaghegh Dolatabadi
Salman Mohamadi
Mirgahney H. Mohamed
Deen Dayal Mohan
Fnu Mohbat
Satyam Mohla
Tony C. W. Mok
Liliane Momeni
Pascal Monasse
Ajoy Mondal
Anindya Mondal
Mathew Monfort
Tom Monnier
Yusuke Monno
Eduardo F. Montesuma
Gyeongsik Moon
Taesup Moon
WonJun Moon
Dror Moran
Julie R. C. Mordacq
Deeptej S. More
Arthur Moreau
Davide Morelli
Luca Morelli
Pedro Morgado
Alexandre Morgand
Henrique Morimitsu
Matteo Moro
Lia Morra
Matteo Mosconi
Ali Mosleh

Sayed Mohammad Mostafavi Isfahani
Saman Motamed
Chong Mou
Dana Moukheiber
Pierre Moulon
Ramy A. Mounir
Théo Moutakanni
Fangzhou Mu
Jiteng Mu
Yao Mark Mu
Manasi Muglikar
Yasuhiro Mukaigawa
Amitangshu Mukherjee
Avideep Mukherjee
Prerana Mukherjee
Tanmoy Mukherjee
Anirban Mukhopadhyay
Soumik Mukhopadhyay
Yusuke Mukuta
Ravi Teja Mullapudi
Lea Müller
Norman Müller
Chaithanya Kumar Mummadi
Muhammad Akhtar Munir
Subrahmanyam Murala
Sanjeev Muralikrishnan
Ana C. Murillo
Nils Murrugarra-Llerena
Mohamed Adel Musallam
Damien Muselet
Josh David Myers-Dean
Byeonghu Na
Taeyoung Na
Muhammad Ferjad Naeem
Sauradip Nag
Pravin Nagar
Rajendra Nagar
Varun Nagaraja
Tushar Nagarajan
Seungjun Nah
Shu Nakamura
Gaku Nakano
Yuta Nakashima

Kiyohiro Nakayama
Mitsuru Nakazawa
Krishna Kanth Nakka
Yuesong Nan
Karthik Nandakumar
Paolo Napoletano
Syed S. Naqvi
Dinesh Reddy Narapureddy
Supreeth Narasimhaswamy
Kartik Narayan
Sriram Narayanan
Fabio Narducci
Erickson R. Nascimento
Muzammal Naseer
Kamal Nasrollahi
Lakshmanan Nataraj
Vishwesh Nath
Avisek Naug
Alexander Naumann
K. L. Navaneet
Pablo Navarrete Michelini
Shah Nawaz
Nazir Nayal
Niv Nayman
Amin Nejatbakhsh
Negar Nejatishahidin
Reyhaneh Neshatavar
Pedro C. Neto
Lukáš Neumann
Richard Newcombe
Alejandro Newell
Evonne Ng
Kam Woh Ng
Trung T. Ngo
Tuan Duc Ngo
Anh Nguyen
Anh Duy Nguyen
Cuong Cao Nguyen
Duc Anh Nguyen
Hoang Chuong Nguyen
Huy Hong Nguyen
Khai Nguyen
Khanh-Binh Nguyen

Khanh-Duy Nguyen
Khoi Nguyen
Khoi D. Nguyen
Kiet A. Nguyen
Ngoc Cuong Nguyen
Pha Nguyen
Phi Le Nguyen
Phong Ha Nguyen
Rang Nguyen
Tam V. Nguyen
Thao Nguyen
Thuan Hoang Nguyen
Toan Tien Nguyen
Trong-Tung Nguyen
Van Nguyen Nguyen
Van-Quang Nguyen
Thuong Nguyen Canh
Thien Trang Nguyen Vu
Haomiao Ni
Jiangqun Ni
Minheng Ni
Yao Ni
Zhen-Liang Ni
Zixuan Ni
Dong Nie
Hui Nie
Jiahao Nie
Lang Nie
Liqiang Nie
Qiang Nie
Ying Nie
Yinyu Nie
Yongwei Nie
Aditya Nigam
Kshitij N. Nikhal
Nick Nikzad
Jifeng Ning
Rui Ning
Xuefei Ning
Li Niu
Muyao Niu
Shuaicheng Niu
Wei Niu
Xuesong Niu
Yi Niu
Yulei Niu
Zhenxing Niu
Zhong-Han Niu
Shohei Nobuhara
Jongyoun Noh
Junhyug Noh
Nadhira Noor
Parsa Nooralinejad
Sotiris Nousias
Tiago Novello
Gal Novich
David Novotny
Slawomir Nowaczyk
Ewa M. Nowara
Evangelos Ntavelis
Valsamis Ntouskos
Leonardo Nunes
Oren Nuriel
Zhakshylyk Nurlanov
Simbarashe Nyatsanga
Lawrence O'Gorman
Anton Obukhov
Michael Oechsle
Ferda Ofli
Changjae Oh
Dongkeun Oh
Junghun Oh
Seoung Wug Oh
Youngtaek Oh
Hiroki Ohashi
Takehiko Ohkawa
Takeshi Oishi
Takahiro Okabe
Fumio Okura
Daniel Olmeda Reino
Suguru Onda
Trevine S. J. Oorloff
Michael Opitz
Roy Or-El
Jose Oramas
Jordi Orbay
Tribhuvanesh Orekondy
Evin Pınar Örnek
Alessandro Ortis
Magnus Oskarsson
Julian Ost
Daniil Ostashev
Mayu Otani
Naima Otberdout
Hatef Otroshi Shahreza
Yassine Ouali
Amine Ouasfi
Cheng Ouyang
Wanli Ouyang
Wenqi Ouyang
Xu Ouyang
Poojan B. Oza
Milind G. Padalkar
Johannes C. Paetzold
Gautam Pai
Anwesan Pal
Simone Palazzo
Avinash Paliwal
Cristina Palmero
Chengwei Pan
Fei Pan
Hao Pan
Jianhong Pan
Junting Pan
Liang Pan
Lili Pan
Linfei Pan
Liyuan Pan
Tai-Yu Pan
Xichen Pan
Xingjia Pan
Xinyu Pan
Yingwei Pan
Zhaoying Pan
Zhihong Pan
Zixuan Pan
Zizheng Pan
Rohit Pandey
Saurabh Pandey
Bo Pang
Guansong Pang
Lu Pang
Meng Pang
Tianyu Pang
Youwei Pang

Ziqi Pang
Omiros Pantazis
Juan J. Pantrigo
Hsing-Kuo Kenneth Pao
Marina Paolanti
Joao P. Papa
Samuele S. Papa
Dim P. Papadopoulos
Symeon Papadopoulos
George Papandreou
Toufiq Parag
Chethan Parameshwara
Foivos Paraperas
  Papantoniou
Shaifali Parashar
Alejandro Pardo
Jason R. Parham
Kranti K. Parida
Rishubh Parihar
Chunghyun Park
Daehee Park
Dongmin Park
Dongwon Park
Eunbyung Park
Eunhyeok Park
Eunil Park
Geon Yeong Park
Gyeong-Moon Park
Hyoungseob Park
Jae Sung Park
JaeYoo Park
Jin-Hwi Park
Jinhyung Park
Jinyoung Park
Jongwoo Park
JoonKyu Park
JungIn Park
Junheum Park
Kiru Park
Kwanyong Park
Seongsik Park
Seulki Park
Song Park
Sungho Park
Sungjune Park
Taesung Park
Yeachan Park
Gaurav Parmar
Paritosh Parmar
Maurizio Parton
Magdalini Paschali
Vito Paolo Pastore
Or Patashnik
Gaurav Patel
Maitreya Patel
Diego Patino
Suvam Patra
Viorica Patraucean
Badri Narayana Patro
Danda Pani Paudel
Angshuman Paul
Sneha Paul
Soumava Paul
Sudipta Paul
Sujoy Paul
Rémi Pautrat
Ioannis Pavlidis
Svetlana Pavlitska
Raju Pavuluri
Kim Steenstrup Pedersen
Marco Pedersoli
Adithya Pediredla
Pieter Peers
Jiju Peethambaran
Sen Pei
Wenjie Pei
Yuru Pei
Simone Alberto Peirone
Chantal Pellegrini
Latha Pemula
Abhirama Subramanyam
  V. B. Penamakuri
Adrian Penate-Sanchez
Baoyun Peng
Bo Peng
Can Peng
Cheng Peng
Chi-Han Peng
Chunlei Peng
Jie Peng
Jingliang Peng
Kebin Peng
Kunyu Peng
Liang Peng
Liangzu Peng
Pai Peng
Peixi Peng
Sida Peng
Songyou Peng
Wei Peng
Wen-Hsiao Peng
Xi Peng
Xiaojiang Peng
Yi-Xing Peng
Yuxin Peng
Zhiliang Peng
Ziqiao Peng
Matteo Pennisi
Or Perel
Gabriel Perez
Gustavo Perez
Juan C. Perez
Andres Felipe Perez
  Murcia
Eduardo Pérez-Pellitero
Neehar Peri
Skand Peri
Gabriel J. Perin
Federico Pernici
Chiara Pero
Elia Peruzzo
Marco Pesavento
Dmitry M. Petrov
Ilya A. Petrov
Mathis Petrovich
Vitali Petsiuk
Tomas Pevny
Shubham Milind Phal
Chau Pham
Hai X. Pham
Khoi Pham
Long Hoang Pham
Trong Thang Pham
Trung X. Pham
Tung Pham

Hoang Phan
Huy Phan
Minh Hieu Phan
Julien Philip
Stephen Phillips
Cheng Perng Phoo
Hao Phung
Shruti S. Phutke
Weiguo Pian
Yongri Piao
Luigi Piccinelli
A. J. Piergiovanni
Sara Pieri
Vipin Pillai
Wu Pingyu
Silvia L. Pintea
Francesco Pinto
Maura Pintor
Giovanni Pintore
Vittorio Pippi
Robinson Piramuthu
Fiora Pirri
Leonid Pishchulin
Francesca Pistilli
Francesco Pittaluga
Fabio Pizzati
Edward Pizzi
Benjamin Planche
Iuliia Pliushch
Chiara Plizzari
Ryan Po
GIovanni Poggi
Matteo Poggi
Kilian Pohl
Chandradeep Pokhariya
Ashwini Pokle
Matteo Polsinelli
Adrian Popescu
Teodora Popordanoska
Nikola Popović
Ronald Poppe
Samuele Poppi
Andrea Porfiri Dal Cin
Angelo Porrello
Pedro Porto Buarque de Gusmão
Rudra P. K. Poudel
Kossar Pourahmadi Meibodi
Hadi Pouransari
Ali Pourramezan Fard
Omid Poursaeed
Anish J. Prabhu
Mihir Prabhudesai
Aayush Prakash
Aditya Prakash
Shraman Pramanick
Mantini Pranav
B. H. Pawan Prasad
Meghshyam Prasad
Prateek Prasanna
Ekta Prashnani
Bardh Prenkaj
Derek S. Prijatelj
Véronique Prinet
Malte Prinzler
Victor Adrian Prisacariu
Federica Proietto Salanitri
Sergey Prokudin
Bill Psomas
Dongqi Pu
Mengyang Pu
Nan Pu
Shi Pu
Rita Pucci
Kuldeep Purohit
Senthil Purushwalkam
Waqas A. Qazi
Charles R. Qi
Chenyang Qi
Haozhi Qi
Jiaxin Qi
Lei Qi
Mengshi Qi
Peng Qi
Xianbiao Qi
Xiangyu Qi
Yuankai Qi
Zhangyang Qi
Guocheng Qian
Hangwei Qian
Jianing Qian
Qi Qian
Rui Qian
Shengsheng Qian
Shengyi Qian
Shenhan Qian
Wen Qian
Xuelin Qian
Yaguan Qian
Yijun Qian
Yiming Qian
Zhenxing Qian
Wenwen Qiang
Feng Qiao
Fengchun Qiao
Xiaotian Qiao
Yanyuan Qiao
Yi-Ling Qiao
Yu Qiao
Hangyu Qin
Haotong Qin
Jie Qin
Peiwu Qin
Siyang Qin
Wenda Qin
Xuebin Qin
Xugong Qin
Yang Qin
Yipeng Qin
Yongqiang Qin
Yuzhe Qin
Zequn Qin
Zeyu Qin
Zheng Qin
Zhenyue Qin
Ziheng Qin
Jiaxin Qing
Congpei Qiu
Haibo Qiu
Hang Qiu
Heqian Qiu
Jiayan Qiu
Jielin Qiu

Longtian Qiu
Mufan Qiu
Ri-Zhao Qiu
Weichao Qiu
Xuchong Qiu
Xuerui Qiu
Yuda Qiu
Yuheng Qiu
Zhongxi Qiu
Maan Qraitem
Chao Qu
Linhao Qu
Yanyun Qu
Kha Gia Quach
Ruijie Quan
Fabio Quattrini
Yvain Queau
Faisal Z. Qureshi
Rizwan Qureshi
Hamid R. Rabiee
Paolo Rabino
Ryan L. Rabinowitz
Petia Radeva
Bhaktipriya Radharapu
Krystian Radlak
Bodgan Raducanu
M. Usman Rafique
Francesco Ragusa
Sahar Rahimi Malakshan
Tanzila Rahman
Aashish Rai
Arushi Rai
Shyam Nandan Rai
Zobeir Raisi
Amit Raj
Kiran Raja
Sachin Raja
Deepu Rajan
Jathushan Rajasegaran
Gnana Praveen Rajasekhar
Ramanathan Rajendiran
Marie-Julie Rakotosaona
Gorthi Rama Krishna Sai Subrahmanyam
Sai Niranjan Ramachandran
Santhosh Kumar Ramakrishnan
Srikumar Ramalingam
Michaël Ramamonjisoa
Ravi Ramamoorthi
Shanmuganathan Raman
Mani Ramanagopal
Ashish Ramayee Asokan
Andrea Ramazzina
Jason Rambach
Sai Saketh Rambhatla
Sai Saketh Rambhatla
Clément Rambour
Francois Bernard Julien Rameau
Visvanathan Ramesh
Adín Ramírez Rivera
Haoxi Ran
Xuming Ran
Aakanksha Rana
Srinivas Rana
Kanchana N. Ranasinghe
Poorva G. Rane
Aneesh Rangnekar
Harsh Rangwani
Viresh Ranjan
Anyi Rao
Sukrut Rao
Yongming Rao
ZhiBo Rao
Carolina Raposo
Hanoona Abdul Rasheed
Amir Rasouli
Deevashwer Rathee
Christian Rathgeb
Avinash Ravichandran
Bharadwaj Ravichandran
Arijit Ray
Dripta S. Raychaudhuri
Sonia Raychaudhuri
Haziq Razali
Daniel Rebain
William T. Redman
Albert W. Reed
Aniket Rege
Christoph Reich
Christian Reimers
Simon Reiß
Konstantinos Rematas
Tal Remez
Davis Rempe
Bin Ren
Chao Ren
Chuan-Xian Ren
Dayong Ren
Dongwei Ren
Jiawei Ren
Jiaxiang Ren
Jing Ren
Mengwei Ren
Pengfei Ren
Pengzhen Ren
Qibing Ren
Shuhuai Ren
Sucheng Ren
Tianhe Ren
Weihong Ren
Wenqi Ren
Xuanchi Ren
Yanli Ren
Yihui Ren
Yixuan Ren
Yufan Ren
Zhenwen Ren
Zhihang Ren
Zhiyuan Ren
Zhongzheng Ren
Jose Restom
George Retsinas
Ambareesh Revanur
Ferdinand Rewicki
Manuel Rey Area
Md Alimoor Reza
Farnoush Rezaei Jafari
Hamed Rezazadegan Tavakoli
Rafael S. Rezende
Wonjong Rhee

Anthony D. Rhodes
Daniel Riccio
Alexander Richard
Christian Richardt
Luca Rigazio
Benjamin Risse
Dominik Rivoir
Luigi Riz
Mamshad Nayeem Rizve
Antonino M. Rizzo
Wes J. Robbins
Damien Robert
Jonathan Roberts
Joseph Robinson
Antonio Robles-Kelly
Mrigank Rochan
Chris Rockwell
Chris Rockwell
Ivan Rodin
Erik Rodner
Ranga Rodrigo
Andres C. Rodriguez
Cristian Rodriguez
Carlos Rodriguez-Pardo
Antonio J.
    Rodriguez-Sanchez
Barbara Roessle
Paul Roetzer
Alina Roitberg
Javier Romero
Meitar Ronen
Keran Rong
Xuejian Rong
Yu Rong
Marco Rosano
Bodo Rosenhahn
Gabriele Rosi
Candace Ross
Andreas Rössler
Giulio Rossolini
Mohammad Rostami
Edward Rosten
Daniel Roth
Karsten Roth
Mark S. Rothermel

Matthias Rottmann
Anastasios Roussos
Aniket Roy
Anirban Roy
Debaditya Roy
Shuvendu Roy
Sudipta Roy
Ahana Roy Choudhury
Amit Roy-Chowdhury
Aruni RoyChowdhury
Dávid Rozenberszki
Denys Rozumnyi
Lixiang Ru
Lingyan Ruan
Shulan Ruan
Viktor Rudnev
Daniel Rueckert
Nataniel Ruiz
Ewelina Rupnik
Evgenia Rusak
Chris Russell
Marc Rußwurm
Fiona Ryan
Dawid Damian Rymarczyk
DongHun Ryu
Sari Saba-Sadiya
Robert Sablatnig
Mohammad Sabokrou
Ragav Sachdeva
Ali Sadeghian
Arka Sadhu
Sadra Safadoust
Bardia Safaei
Ryusuke Sagawa
Avishkar Saha
Gobinda Saha
Oindrila Saha
Aditya Sahdev
Lakshmi Babu Saheer
Aadarsh Sahoo
Pritish Sahu
Aneeshan Sain
Nirat Saini
Saurabh Saini
Kuniaki Saito

Shunsuke Saito
Rahul Sajnani
Fumihiko Sakaue
Parikshit V. Sakurikar
Riccardo Salami
Soorena Salari
Mohammadreza Salehi
Leonard Salewski
Driton Salihu
Benjamin Salmon
Cristiano Saltori
Joel Saltz
Tim Salzmann
Sina Samangooei
Babak Samari
Nermin Samet
Fawaz Sammani
Leo Sampaio Ferraz
    Ribeiro
Shailaja Keyur Sampat
Alessio Sampieri
Jorge Sanchez
Pedro Sandoval-Segura
Nong Sang
Shengtian Sang
Patsorn Sangkloy
Depanshu Sani
Juan C. Sanmiguel
Hiroaki Santo
Joshua Santoso
Bikash Santra
Soubhik Sanyal
Hitesh Sapkota
Ayush Saraf
Nikolaos Sarafianos
István Sárándi
Kyle Sargent
Andranik Sargsyan
Josip Šarić
Mert Bulent Sariyildiz
Abhijit Sarkar
Anirban Sarkar
Chayan Sarkar
Michel Sarkis
Paul-Edouard Sarlin

Sara Sarto
Josua Sassen
Srikumar Sastry
Imari Sato
Takami Sato
Shin'ichi Satoh
Ravi Kumar Satzoda
Jack Saunders
Corentin Sautier
Mattia Savardi
Bogdan Savchynskyy
Mohamed Sayed
Marin Scalbert
Gianluca Scarpellini
Gerald Schaefer
Guilherme G. Schardong
David Schinagl
Phillip Schniter
Patrick Schramowski
Matthias Schubert
Peter Schüffler
Samuel Schulter
René Schuster
Klamer Schutte
Luca Scofano
Jesse Scott
Marcel Seelbach Benkner
Karthik Seemakurthy
Mattia Segù
Santi Seguí
Sinisa Segvic
Constantin Marc Seibold
Roman Seidel
Lorenzo Seidenari
Taiki Sekii
Yusuke Sekikawa
Matan Sela
Pratheba Selvaraju
Agniva Sengupta
Ahyun Seo
Jinhwan Seo
Junyoung Seo
Kwanggyoon Seo
Seonguk Seo
Seunghyeon Seo

Jinseok Seol
Hongje Seong
Ana F. Sequeira
Dario Serez
Dario Serez
David Serrano-Lozano
Pratinav Seth
Francesco Setti
Giorgos Sfikas
Mohammad Amin Shabani
Faisal Shafait
Anshul Shah
Chintan Shah
Jay Shah
Ketul Shah
Mubarak Shah
Viraj Shah
Mohamad Shahbazi
Muhammad Bilal B. Shaikh
Abdelrahman M. Shaker
Greg Shakhnarovich
Md Salman Shamil
Fahad Shamshad
Caifeng Shan
Dandan Shan
Hongming Shan
Xiaojun Shan
Chong Shang
Fanhua Shang
Jinghuan Shang
Lei Shang
Sifeng Shang
Wei Shang
Yuzhang Shang
Yuzhang Shang
Sukrit Shankar
Dian Shao
Mingwen Shao
Rui Shao
Ruizhi Shao
Shuai Shao
Shuwei Shao
Ron A. Shapira Weber
S. M. A. Sharif

Aashish Sharma
Avinash Sharma
Charu Sharma
Prafull Sharma
Prasen Kumar Sharma
Allam Shehata
Mark Sheinin
Sumit Shekhar
Oleksandr Shekhovtsov
Chuanfu Shen
Fei Shen
Fengyi Shen
Furao Shen
Hui-liang Shen
Jiajun Shen
Jianghao Shen
Jiangrong Shen
Jiayi Shen
Li Shen
Li-Yong Shen
Linlin Shen
Maying Shen
Qiu Shen
Qiuhong Shen
Shuai Shen
Shuhan Shen
Siqi Shen
Tianwei Shen
Tong Shen
Xiaolong Shen
Xiaoqian Shen
Yan Shen
Yanqing Shen
Yilin Shen
Ying Shen
Yiqing Shen
Yuan Shen
Yucong Shen
Yuhan Shen
Yunhang Shen
Zehong Shen
Zengming Shen
Zhijie Shen
Zhiqiang Shen
Hualian Sheng

Tao Sheng
Yichen Sheng
Zehua Sheng
Shivanand Venkanna Sheshappanavar
Ivaxi Sheth
Baoguang Shi
Botian Shi
Dachuan Shi
Daqian Shi
Haizhou Shi
Hengcan Shi
Jia Shi
Jing Shi
Jingang Shi
QingHongYa Shi
Ruoxi Shi
Tianyang Shi
Weishi Shi
Wu Shi
Wuxuan Shi
Xiaodan Shi
Xiaoshuang Shi
Xiaoyu Shi
Xingjian Shi
Xinyu Shi
Xuepeng Shi
Yichun Shi
Yujiao Shi
Zhenbo Shi
Zheng Shi
Zhensheng Shi
Zhenwei Shi
Zhihao Shi
Zifan Shi
Takashi Shibata
Meng-Li Shih
Yichang Shih
Dongseok Shim
Wataru Shimoda
Ilan Shimshoni
Changha Shin
Gyungin Shin
Hyungseob Shin
Inkyu Shin

Seungjoo Shin
Ukcheol Shin
Yooju Shin
Young Min Shin
Koichi Shinoda
Kaede Shiohara
Suprosanna Shit
Palaiahnakote Shivakumara
Sindi Shkodrani
Michal Shlapentokh-Rothman
Debaditya Shome
Hyounguk Shon
Sulabh Shrestha
Aman Shrivastava
Ayush Shrivastava
Gaurav Shrivastava
Aleksandar Shtedritski
Dong Wook Shu
Han Shu
Jun Shu
Xiangbo Shu
Xiujun Shu
Yang Shu
Bing Shuai
Hong-Han Shuai
Qing Shuai
Changjian Shui
Pushkar Shukla
Mustafa Shukor
Hubert P. H. Shum
Nina Shvetsova
Chenyang Si
Jianlou Si
Zilin Si
Mennatullah Siam
Sven Sickert
Désiré Sidibé
Ioannis Siglidis
Alberto Signoroni
Karan Sikka
Pedro Silva
Julio Silva-Rodríguez
Hyeonjun Sim

Jae-Young Sim
Chonghao Sima
Christian Simon
Martin Simon
Alessandro Simoni
Enis Simsar
Abhishek Singh
Apoorv Singh
Ashish Singh
Bharat Singh
Jasdeep Singh
Jaskirat Singh
Krishnakant Singh
Manish Kumar Singh
Mannat Singh
Nikhil Singh
Pravendra Singh
Rajat Vikram Singh
Simranjit Singh
Darshan Singh S.
Utkarsh Singhal
Dipika Singhania
Vasu Singla
Abhishek Kumar Sinha
Animesh Sinha
Sanjana Sinha
Saptarshi Sinha
Sudipta Sinha
Sophia A. Sirko-Galouchenko
Josef Sivic
Elena Sizikova
Geri Skenderi
Gregory Slabaugh
Habib Slim
Dmitriy Smirnov
James S. Smith
William Smith
Noah Snavely
Kihyuk Sohn
Bolivar E. Solarte
Mattia Soldan
Sobhan Soleymani
Samik Some
Nagabhushan Somraj

Jeany Son
Seung Woo Son
Byung Cheol Song
Chen Song
Guanglu Song
Jie Song
Jifei Song
Li Song
Liangchen Song
Lin Song
Luchuan Song
Mingli Song
Ran Song
Sibo Song
Sifan Song
Siyang Song
Weilian Song
Weinan Song
Wenfeng Song
Xiangchen Song
Xibin Song
Xinhang Song
Yafei Song
Yang Song
Yi-Yang Song
Yizhi Song
Yue Song
Zeen Song
Zhenbo Song
Zikai Song
Ekta Sood
Tomáš Souček
Rajiv Soundararajan
Albin Soutif-Cormerais
Jeremy Speth
Indro Spinelli
Jon Sporring
Manogna Sreenivas
Arvind Krishna Sridhar
Deepak Sridhar
Balaji Vasan Srinivasan
Pratul Srinivasan
Anuj Srivastava
Astitva Srivastava
Dhruv Srivastava

Koushik Srivatsan
Pierre-Luc St-Charles
Ioannis Stamos
Anastasis Stathopoulos
Colton Stearns
Jan Steinbrener
Jan-Martin O. Steitz
Sinisa Stekovic
Federico Stella
Michael Stengel
Alexandros Stergiou
Gleb Sterkin
Rainer Stiefelhagen
Noah Stier
Timo N. Stoffregen
Vladan Stojnić
Nick O. Stracke
Ombretta Strafforello
Julian Straub
Nicola Strisciuglio
Vitomir Struc
Yannick Strümpler
Joerg Stueckler
Chi Su
Hang Su
Hang Su
Kun Su
Rui Su
Shaolin Su
Sitong Su
Xingzhe Su
Xiu Su
Yao Su
Yiyang Su
Yongyi Su
Zhaoqi Su
Zhixun Su
Zhuo Su
Zhuo Su
Iago Suárez
Arulkumar Subramaniam
Sanjay Subramanian
A. Subramanyam
Swathikiran Sudhakaran
Yusuke Sugano

Masanori Suganuma
Yumin Suh
Mohammed Suhail
Xiuchao Sui
Yang Sui
Yao Sui
Heung-Il Suk
Pavel Suma
Baigui Sun
Baochen Sun
Bin Sun
Bo Sun
Changchang Sun
Che Sun
Cheng Sun
Chong Sun
Chunyi Sun
Gan Sun
Guofei Sun
Guoxing Sun
Haifeng Sun
Hanqing Sun
Haoliang Sun
He Sun
Heming Sun
Hongbin Sun
Huiming Sun
Jennifer J. Sun
Jian Sun
Jiande Sun
Jianhua Sun
Jiankai Sun
Jipeng Sun
Keqiang Sun
Lei Sun
Lichao Sun
Long Sun
Mingjie Sun
Peize Sun
Pengzhan Sun
Qiyue Sun
Shangquan Sun
Shanlin Sun
Shuyang Sun
Tao Sun

Tiancheng Sun
Wei Sun
Weiwei Sun
Weixuan Sun
Xianfang Sun
Xiaohang Sun
Xiaoshuai Sun
Xiaoxiao Sun
Ximeng Sun
Xuxiang Sun
Yanan Sun
Yasheng Sun
Yihong Sun
Ying Sun
Yixuan Sun
Yu Sun
Yuan Sun
Yuchong Sun
Zeren Sun
Zhanghao Sun
Zhaodong Sun
Zhaohui H. Sun
Zhicheng Sun
Zhicheng Sun
Haomiao Sun
Varun Sundar
Shobhita Sundaram
Minhyuk Sung
Kalyan Sunkavalli
Yucheng Suo
Indranil Sur
Saksham Suri
Naufal Suryanto
Vadim Sushko
David Suter
Roman Suvorov
Fnu Suya
Teppei Suzuki
Kunal Swami
Archana Swaminathan
Gurumurthy Swaminathan
Robin Swanson
Eran Swears
Alexander Swerdlow
Sirnam Swetha
Tabish A. Syed
Tanveer Syeda-Mahmood
Stanislaw K. Szymanowicz
Sethuraman T. V.
Calvin-Khang T. Ta
The-Anh Ta
Babak Taati
Samy Tafasca
Andrea Tagliasacchi
Haowei Tai
Yuan Tai
Francesco Taioli
Peng Taiying
Keita Takahashi
Naoya Takahashi
Jun Takamatsu
Nicolas Talabot
Hugues G. Talbot
Hossein Talebi
Davide Talon
Gary Tam
Toru Tamaki
Dipesh Tamboli
Andong Tan
Bin Tan
Cheng Tan
David Joseph New Tan
Fuwen Tan
Guang Tan
Jianchao Tan
Jing Tan
Jingru Tan
Lei Tan
Mingkui Tan
Mingxing Tan
Shuhan Tan
Shunquan Tan
Weimin Tan
Xin Tan
Zhentao Tan
Zhentao Tan
Masayuki Tanaka
Chen Tang
Chengzhou Tang
Chenwei Tang
Fan Tang
Feng Tang
Hao Tang
Haoran Tang
Jiajun Tang
Jiapeng Tang
Jiaxiang Tang
Jie Tang
Junshu Tang
Keke Tang
Luming Tang
Luyao Tang
Lv Tang
Ming Tang
Quan Tang
Shengji Tang
Sheyang Tang
Shitao Tang
Shixiang Tang
Tao Tang
Weixuan Tang
Xu Tang
Yang Tang
Yansong Tang
Yehui Tang
Yu-Ming Tang
Zheng Tang
Zhipeng Tang
Zitian Tang
Md Mehrab Tanjim
Julian Tanke
An Tao
Chaofan Tao
Chenxin Tao
Jiale Tao
Junli Tao
Keda Tao
Ming Tao
Ran Tao
Wenbing Tao
Xinhao Tao
Jean-Philippe G. Tarel
Laia Tarres
Laia Tarrés
Enzo Tartaglione

Keisuke Tateno
SaiKiran K. Tedla
Antonio Tejero-de-Pablos
Bugra Tekin
Purva Tendulkar
Minggui Teng
Ruwan Tennakoon
Andrew Beng Jin Teoh
Konstantinos Tertikas
Piotr Teterwak
Piotr Teterwak
Anh Thai
Kartik Thakral
Nupur Thakur
Sadbhawna Thakur
Balamurugan Thambiraja
Vikas Thamizharasan
Kevin Thandiackal
Sushil Thapa
Daksh Thapar
Jonas Theiner
Christian Theobalt
Spyridon Thermos
Fida Mohammad Thoker
Christopher L. Thomas
Diego Thomas
William Thong
Mamatha Thota
Mukund Varma
  Thottankara
Changyao Tian
Chunwei Tian
Jinyu Tian
Kai Tian
Lin Tian
Tai-Peng Tian
Xin Tian
Xinyu Tian
Yapeng Tian
Yu Tian
Yuan Tian
Yuesong Tian
Yunjie Tian
Yuxin Tian
Zhuotao Tian

Mert Tiftikci
Javier Tirado-Garín
Garvita Tiwari
Lokender Tiwari
Anastasia Tkach
Andrea Toaiari
Sinisa Todorovic
Pavel Tokmakov
Tri Ton
Adam Tonderski
Jinguang Tong
Peter Tong
Xin Tong
Zhan Tong
Francesco Tonini
Alessio Tonioni
Alessandro Torcinovich
Marwan Torki
Lorenzo Torresani
Fabio Tosi
Matteo Toso
Anh T. Tran
Hung Tran
Linh-Tam Tran
Minh-Triet Tran
Ngoc-Trung Tran
Phong Tran
Jonathan Tremblay
Alex Trevithick
Aditay Tripathi
Subarna Tripathi
Felix Tristram
Gabriele Trivigno
Emanuele Trucco
Prune Truong
Thanh-Dat Truong
Tomasz Trzcinski
Fu-Jen Tsai
Yu-Ju Tsai
Michael Tschannen
Tze Ho Elden Tse
Ethan Tseng
Yu-Chee Tseng
Shahar Tsiper
Hanzhang Tu

Rong-Cheng Tu
Yuanpeng Tu
Zhengzhong Tu
Zhigang Tu
Narek Tumanyan
Anil Osman Tur
Haithem Turki
Mehmet Ozgur Turkoglu
Daniyar Turmukhambetov
Victor G. Turrisi da Costa
Tinne Tuytelaars
Bartlomiej Twardowski
Radim Tylecek
Christos Tzelepis
Seiichi Uchida
Hideaki Uchiyama
Vishaal Udandarao
Mostofa Rafid Uddin
Kohei Uehara
Tatsumi Uezato
Nicolas Ugrinovic
Youngjung Uh
Norimichi Ukita
Amin Ullah
Markus Ulrich
Ardian Umam
Mesut Erhan Unal
Mathias Unberath
Devesh Upadhyay
Paul Upchurch
Shagun Uppal
Yoshitaka Ushiku
Anil Usumezbas
Yuzuko Utsumi
Roy Uziel
Anil Vadathya
Sharvaree Vadgama
Pratik Vaishnavi
Gregory Vaksman
Matias A. Valdenegro Toro
Lucas Valença
Eduardo Valle
Ernest Valveny
Laurens van der Maaten
Wouter Van Gansbeke

Nanne van Noord
Max W. F. van Spengler
Lorenzo Vaquero
Farshid Varno
Cristina Vasconcelos
Francisco Vasconcelos
Igor Vasiljevic
Florin-Alexandru
    Vasluianu
Subeesh Vasu
Arun Balajee Vasudevan
Vaibhav S. Vavilala
Kyle Vedder
Vijay Veerabadran
Ronny Xavier Velastegui
    Sandoval
Senem Velipasalar
Andreas Velten
Raviteja Vemulapalli
Deepika Vemuri
Edward Vendrow
Jonathan Ventura
Lucas Ventura
Jakob Verbeek
Dor Verbin
Eshan Verma
Manisha Verma
Monu Verma
Sahil Verma
Constantin Vertan
Eli Verwimp
Noranart Vesdapunt
Jordan J. Vice
Sara Vicente
Kavisha Vidanapathirana
Dat Viet Thanh Nguyen
Sudheendra
    Vijayanarasimhan
Sujal T. Vijayaraghavan
Deepak Vijaykeerthy
Elliot Vincent
Yael Vinker
Duc Minh Vo
Huy V. Vo
Khoa H. V. Vo

Romain Vo
Antonin Vobecky
Michele Volpi
Riccardo Volpi
Igor Vozniak
Nicholas Vretos
Vibashan V. S.
Ngoc-Son Vu
Tuan-Anh Vu
Khiem Vuong
Mårten Wadenbäck
Neal Wadhwa
Sophia J. Wagner
Muntasir Wahed
Nobuhiko Wakai
Devesh Walawalkar
Jacob Walker
Matthew Walmer
Matthew R. Walter
Bo Wan
Guancheng Wan
Jia Wan
Jin Wan
Jun Wan
Qiyang Wan
Renjie Wan
Wei Wan
Xingchen Wan
Yecong Wan
Zhexiong Wan
Ziyu Wan
Karan Wanchoo
Alex Jinpeng Wang
Angtian Wang
Baoyuan Wang
Benyou Wang
Biao Wang
Bin Wang
Bing Wang
Binghui Wang
Binglu Wang
Can Wang
Ce Wang
Changwei Wang
Chao Wang

Chaoyang Wang
Chen Wang
Chen Wang
Chen Wang
Chengrui Wang
Chien-Yao Wang
Chu Wang
Chuan Wang
Congli Wang
Dadong Wang
Di Wang
Dong Wang
Dong Wang
Dongdong Wang
Dongkai Wang
Dongqing Wang
Dongsheng Wang
X. Wang
Fan Wang
Fangfang Wang
Fangjinhua Wang
Fei Wang
Feng Wang
Feng Wang
Fu-Yun Wang
Gaoang Wang
Guangcong Wang
Guangming Wang
Guangrun Wang
Guangzhi Wang
Guanshuo Wang
Guo-Hua Wang
Guoqing Wang
Guoqing Wang
Haixin Wang
Haiyan Wang
Han Wang
Hanjing Wang
Hanyu Wang
Hao Wang
Hao Wang
Haobo Wang
Haochen Wang
Haochen Wang
Haohan Wang

Haoqi Wang
Haoran Wang
Haotao Wang
Haoxuan Wang
HaoYu Wang
Hengkang Wang
Hengli Wang
Hengyi Wang
Hesheng Wang
Hong Wang
Hongjun Wang
Hongxiao Wang
Hongyu Wang
Hongzhi Wang
Hua Wang
Huafeng Wang
Huan Wang
Huijie Wang
Huiyu Wang
Jiadong Wang
Jiahao Wang
Jiahao Wang
Jiahao Wang
Jiakai Wang
Jialiang Wang
Jiamian Wang
Jian Wang
Jiang Wang
Jiangliu Wang
Jianjia Wang
Jianyi Wang
Jianyuan Wang
Jiaqi Wang
Jiashun Wang
Jiayi Wang
Jiaze Wang
Jin Wang
Jinfeng Wang
Jingbo Wang
Jinghua Wang
Jingkang Wang
Jinglong Wang
Jinglu Wang
Jinpeng Wang
Jinqiao Wang

Jue Wang
Jun Wang
Jun Wang
Junjue Wang
Junke Wang
Junxiao Wang
Kai Wang
Kai Wang
Kai Wang
Kai Wang
Kaihong Wang
Kewei Wang
Keyan Wang
Kun Wang
Kup Wang
Lan Wang
Lanjun Wang
Le Wang
Lei Wang
Lei Wang
Lezi Wang
Liansheng Wang
Liao Wang
Lijuan Wang
Lijun Wang
Limin Wang
Lin Wang
Linwei Wang
Lishun Wang
Lixu Wang
Liyuan Wang
Lizhen Wang
Lizhi Wang
Longguang Wang
Luozhou Wang
Luting Wang
Mang Wang
Manning Wang
Mei Wang
Mengjiao Wang
Mengmeng Wang
Miaohui Wang
Min Wang
Naiyan Wang
Nannan Wang

Ning-Hsu Wang
Pei Wang
Peihao Wang
Peiqi Wang
Peng Wang
Pengfei Wang
Pengkun Wang
Pichao Wang
Pu Wang
Qi Wang
Qian Wang
Qiang Wang
Qiang Wang
Qiangchang Wang
Qianqian Wang
Qifei Wang
Qilong Wang
Qin Wang
Qing Wang
Qingzhong Wang
Qitong Wang
Qiufeng Wang
Ronggang Wang
Rui Wang
Rui Wang
Rui Wang
Ruibin Wang
Ruisheng Wang
Ruoyu Wang
Sai Wang
Sen Wang
Sen Wang
Shan Wang
Shaoru Wang
Sheng Wang
Sheng-Yu Wang
Shengze Wang
Shida Wang
Shijie Wang
Shipeng Wang
Shiping Wang
Shiyu Wang
Shizun Wang
Shuhui Wang
Shujun Wang

Shunli Wang
Shunxin Wang
Shuo Wang
Shuo Wang
Shuo Wang
Shuzhe Wang
Siqi Wang
Siwei Wang
Song Wang
Song Wang
Su Wang
Tan Wang
Tao Wang
Taoyue Wang
Teng Wang
Tengfei Wang
Tiancai Wang
Tianqi Wang
Tianyang Wang
Tianyu Wang
Tong Wang
Tsun-Hsuan Wang
Tuanfeng Wang
Tuanfeng Y. Wang
Wei Wang
Weihan Wang
Weikang Wang
Weimin Wang
Weiqiang Wang
Weixi Wang
Weiyao Wang
Weiyun Wang
Wen Wang
Wenbin Wang
Wenhao Wang
Wenjing Wang
Wenqian Wang
Wentao Wang
Wenxiao Wang
Wenxuan Wang
Wenzhe Wang
Xi Wang
Xi Wang
Xiang Wang
Xiao Wang

Xiao Wang
Xiao Wang
Xiaobing Wang
Xiaofeng Wang
Xiaohan Wang
Xiaosen Wang
Xiaosong Wang
Xiaoxing Wang
Xiaoyang Wang
Xijun Wang
Xijun Wang
Xinggang Wang
Xinghan Wang
Xinjiang Wang
Xinshao Wang
Xintong Wang
Xizi Wang
Xu Wang
Xuan Wang
Xuanhan Wang
Xue Wang
Xueping Wang
Xuyang Wang
Yali Wang
Yan Wang
Yan Wang
Yang Wang
Yangang Wang
Yangtao Wang
Yaohui Wang
Yaoming Wang
Yaxing Wang
Yaxiong Wang
Yi Wang
Yi Ru Wang
Yidong Wang
Yifan Wang
Yifeng Wang
Yifu Wang
Yikai Wang
Yilin Wang
Yilun Wang
Yin Wang
Yinggui Wang
Yingheng Wang

Yingqian Wang
Yipei Wang
Yiqun Wang
Yiran Wang
Yiwei Wang
Yixu Wang
Yizhi Wang
Yizhou Wang
Yizhou Wang
Yong Wang
Yu Wang
Yu-Shuen Wang
Yuan-Gen Wang
Yuchen Wang
Yude Wang
Yue Wang
Yuesong Wang
Yufei Wang
Yufu Wang
Yuguang Wang
Yuhan Wang
Yujia Wang
Yulin Wang
Yunke Wang
Yuting Wang
Yuxi Wang
YuXin Wang
Yuzheng Wang
Ze Wang
Zedong Wang
Zehan Wang
Zengmao Wang
Zeyu Wang
Zeyu Wang
Zhao Wang
Zhaokai Wang
Zhaowen Wang
Zhe Wang
Zhen Wang
Zhen Wang
Zhendong Wang
Zheng Wang
Zheng Wang
Zheng Wang
Zhengyi Wang

Zhennan Wang
Zhenting Wang
Zhenyi Wang
Zhenyu Wang
Zhenzhi Wang
Zhepeng Wang
Zhi Wang
Zhibo Wang
Zhihao Wang
Zhihui Wang
Zhijie Wang
Zhikang Wang
Zhixiang Wang
Zhiyong Wang
Zhongdao Wang
Zhonghao Wang
Zhouxia Wang
Zhu Wang
Zian Wang
Zifu Wang
Zihao Wang
Zijian Wang
Ziqiang Wang
Ziqin Wang
Ziqing Wang
Zirui Wang
Zirui Wang
Ziwei Wang
Ziyan Wang
Ziyang Wang
Ziyi Wang
Ziyun Wang
Frederik Warburg
Syed Talal Wasim
Daniel Watson
Jamie Watson
Ethan Weber
Silvan Weder
Jan Dirk Wegner
Chen Wei
Donglai Wei
Fangyin Wei
Fangyun Wei
Guoqiang Wei
Jia Wei

Jiacheng Wei
Kaixuan Wei
Kun Wei
Longhui Wei
Megan Wei
Mian Wei
Mingqiang Wei
Pengxu Wei
Ping Wei
Qiuhong Anna Wei
Shikui Wei
Tianyi Wei
Wei Wei
Wenqi Wei
Xian Wei
Xin Wei
Xing Wei
Xinyue Wei
Xiu-Shen Wei
Yi Wei
Yixuan Wei
Yunchao Wei
Yuxiang Wei
Yuxiang Wei
Zeming Wei
Zhipeng Wei
Zihao Wei
Zimian Wei
Jean-Baptiste Weibel
Luca Weihs
Martin Weinmann
Michael Weinmann
Bihan Wen
Bowen Wen
Chao Wen
Chenglu Wen
Chuan Wen
Congcong Wen
Jie Wen
Jing Wen
Qiang Wen
Rui Wen
Sijia Wen
Song Wen
Xiang Wen

Xin Wen
Yilin Wen
Youpeng Wen
Yuanbo Wen
Yuxin Wen
Chung-Yi Weng
Junwu Weng
Shuchen Weng
Wenming Weng
Yijia Weng
Zhenzhen Weng
Thomas Westfechtel
Christopher Johannes
 Wewer
Spencer Whitehead
Tobias Jan Wieczorek
Thaddäus Wiedemer
Julian Wiederer
Kevin Tirta Wijaya
Asiri Wijesinghe
Kimberly Wilber
Jeffrey R. Willette
Bryan M. Williams
Williem Williem
Christian Wilms
Benjamin Wilson
Richard Wilson
Felix Wimbauer
Vanessa Wirth
Scott Wisdom
Calden Wloka
Alex Wong
Chau-Wai Wong
Chi-Chong Wong
Ka Wai Wong
Kelvin Wong
Kok-Seng Wong
Kwan-Yee K. Wong
Yongkang Wong
Sangmin Woo
Simon S. Woo
Markus Worchel
Scott Workman
Marcel Worring
Safwan Wshah

Aming Wu
Bo Wu
Bojian Wu
Boxi Wu
Changguang Wu
Chaoyi Wu
Chen Henry Wu
Cheng-En Wu
Chenming Wu
Chenyan Wu
Chenyun Wu
Cho-Ying Wu
Chongruo Wu
Cong Wu
Dayan Wu
Di Wu
Dongming Wu
Fangzhao Wu
Fuxiang Wu
Gaojie Wu
Guanyao Wu
Guile Wu
Haiping Wu
Haiwei Wu
Haiyu Wu
Han Wu
Haoning Wu
Haoning Wu
Haotian Wu
Hefeng Wu
Huisi Wu
Jane Wu
Jay Zhangjie Wu
Jhih-Ciang Wu
Ji-Jia Wu
Jialian Wu
Jiaye Wu
Jimmy Wu
Jing Wu
Jing Wu
Jinjian Wu
Jiqing Wu
Jun Wu
Junfeng Wu
Junlin Wu
Junru Wu
Junyang Wu
Junyi Wu
Letian Wu
Lifang Wu
Lin Yuanbo Wu
Liwen Wu
Min Wu
Minye wu
Peng Wu
Penghao Wu
Qian Wu
Qiangqiang Wu
Qianyi Wu
Qingbo Wu
Rongliang Wu
Rui Wu
Rundi Wu
Shuang Wu
Shuzhe Wu
Tao Wu
Tao Wu
Tao Wu
Te-Lin Wu
Tianfu Wu
Tianhao Wu
Tianhao Wu
Ting-Wei Wu
Tong Wu
Tong Wu
Tsung-Han Wu
Tz-Ying Wu
Weibin Wu
Weijia Wu
Xian Wu
Xiao Wu
Xiaodong Wu
Xiaohe Wu
Xiaoqian Wu
Xiaoyang Wu
Xindi Wu
Xingjiao Wu
Xinxiao Wu
Xiuzhe Wu
Yang Wu
Yangzheng Wu
Yanze Wu
Yanzhao Wu
Yawen Wu
Yicheng Wu
Ying Nian Wu
Yingwen Wu
Yong Wu
Yuanwei Wu
Yue Wu
Yue Wu
Yuqun Wu
Yushu Wu
Yushuang Wu
Zhe Wu
Zheng Wu
Zhi-Fan Wu
Zhihao Wu
Zhijie Wu
Zhiliang Wu
Zhonghua Wu
Zijie Wu
Ziyi Wu
Zizhao Wu
Zongwei Wu
Zongyu Wu
Zongze Wu
Stefanie Wuhrer
Jamie M. Wynn
Monika Wysoczańska
Jianing Xi
Teng Xi
Bin Xia
Changqun Xia
Haifeng Xia
Jiaer Xia
Jiahao Xia
Kun Xia
Mingxuan Xia
Shihong Xia
Weihao Xia
Xiaobo Xia
Yan Xia
Ye Xia
Yifei Xia

Zhaoyang Xia
Zhihao Xia
Zhihua Xia
Zhuofan Xia
Zimin Xia
Chuhua Xian
Wenqi Xian
Yongqin Xian
Donglai Xiang
Jinhai Xiang
Liuyu Xiang
Tian-Zhu Xiang
Tiange Xiang
Wangmeng Xiang
Xiaoyu Xiang
Yuanbo Xiangli
Anqi Xiao
Aoran Xiao
Bei Xiao
Chunxia Xiao
Fanyi Xiao
Han Xiao
Jiancong Xiao
Jimin Xiao
Jing Xiao
Jing Xiao
Jun Xiao
Junbin Xiao
Junfei Xiao
Mingqing Xiao
Qingyang Xiao
Ruixuan Xiao
Taihong Xiao
Yang Xiao
Yanru Xiao
Yao Xiao
Yijun Xiao
Yuting Xiao
Zehao Xiao
Zeyu Xiao
Zihao Xiao
Binhui Xie
Chaohao Xie
Chi Xie
Christopher Xie
Chuanlong Xie
Fei Xie
Guo-Sen Xie
Haozhe Xie
Hongtao Xie
Jiahao Xie
Jiaxin Xie
Jin Xie
Jinheng Xie
Jiu-Cheng Xie
Jiyang Xie
Junyu Xie
Liuyue Xie
Ming-Kun Xie
Mingyang Xie
Qian Xie
Tingting Xie
Weicheng Xie
Xianghui Xie
Xiaohua Xie
Xudong Xie
Yichen Xie
Yiming Xie
You Xie
Yuan Xie
Yusheng Xie
Yutong Xie
Zeke Xie
Zhenda Xie
Zhenyu Xie
ZiYang Xie
Chaoyue Xing
Fuyong Xing
Jinbo Xing
Xiaoyan Xing
Xiaoying Xing
XiMing Xing
Xin Xing
Yazhou Xing
Yifan Xing
Yun Xing
Zhen Xing
Hongkai Xiong
Jingjing Xiong
Jinhui Xiong
Junwen Xiong
Peixi Xiong
Wei Xiong
Weihua Xiong
Yu Xiong
Yuanhao Xiong
Yuanjun Xiong
Yuwen Xiong
Zhexiao Xiong
Zhiwei Xiong
Yuliang Xiu
Alessio Xompero
An Xu
Angchi Xu
Baixin Xu
Bicheng Xu
Bo Xu
Chao Xu
Chenfeng Xu
Chenshu Xu
Chenxin Xu
Chi Xu
Dejia Xu
Dongli Xu
Feng Xu
Gangwei Xu
Haiming Xu
Haiyang Xu
Han Xu
Haofei Xu
Haohang Xu
Haoran Xu
Hongbin Xu
Hongmin Xu
Jiale Xu
Jianjin Xu
Jiaqi Xu
Jie Xu
Jilan Xu
Jinglin Xu
Jingyi Xu
Jun Xu
Kai Xu
Katherine Xu
Ke Xu

Kele Xu
Lan Xu
Lian Xu
Liang Xu
Linning Xu
Lumin Xu
Manjie Xu
Mengde Xu
Mengdi Xu
Mengmeng Frost Xu
Min Xu
Ming Xu
Mutian Xu
Peiran Xu
Peng Xu
Qi Xu
Qiang Xu
Qiangeng Xu
Qingshan Xu
Qingyang Xu
Qiuling Xu
Ran Xu
Renzhe Xu
Ruikang Xu
Runsen Xu
Runsheng Xu
Shichao Xu
Sirui Xu
Tongda Xu
Wanting Xu
Wei Xu
Weiwei Xu
Wenjia Xu
Wenju Xu
Wenqiang Xu
Xiang Xu
Xianghao Xu
Xiangyu Xu
Xiangyu Xu
Xiaogang Xu
Xiaohao Xu
Xin Xu
Xin Xu
Xin-Shun Xu
Xing Xu
Xinli Xu
Xinyu Xu
Xiuwei Xu
Xiyan Xu
Xudong Xu
Xuemiao Xu
Xun Xu
Yan Xu
Yan Xu
Yan Xu
Yangyang Xu
Yanwu Xu
Yating Xu
Yi Xu
Yi Xu
Yi Xu
Yihong Xu
YiKun Xu
Yinghao Xu
Yingyan Xu
Yinshuang Xu
Yiran Xu
Yixing Xu
Yongchao Xu
Yue Xu
Yufei Xu
Yunqiu Xu
Zexiang Xu
Zhan Xu
Zhe Xu
Zhengqin Xu
Zhenlin Xu
Zhiqiu Xu
Zhiyuan Xu
Zhongcong Xu
Zhuoer Xu
Zipeng Xu
Ziyue Xu
Zongyi Xu
Ziwei Xuan
Danna Xue
Fanglei Xue
Fei Xue
Feng Xue
Han Xue
Jianru Xue
Le Xue
Lixin Xue
Mingfu Xue
Nan Xue
Qinghan Xue
Shangjie Xue
Xiangyang Xue
Zihui Xue
Abhay Yadav
Amit Kumar Singh Yadav
Takuma Yagi
Tomas F Yago Vicente
I. Zeki Yalniz
Kota Yamaguchi
Shin'ya Yamaguchi
Burhaneddin Yaman
Toshihiko Yamasaki
Kohei Yamashita
Lee Juliette Yamin
Chaochao Yan
Hongyu Yan
Jiexi Yan
Kai Yan
Pei Yan
Qingan Yan
Qingsen Yan
Qingsong Yan
Rui Yan
Shaoqi Yan
Shi Yan
Siming Yan
Siming Yan
Siyuan Yan
Weilong Yan
Wending Yan
Xiangyi Yan
Xinchen Yan
Xingguang Yan
Xueting Yan
Yan Yan
Yichao Yan
Zhaoyi Yan
Zhiqiang Yan
Zhiyuan Yan

Zike Yan
Zizheng Yan
Keiji Yanai
Pinar Yanardag
Anqi Yang
Anqi Joyce Yang
Bangbang Yang
Baoyao Yang
Bin Yang
Binwei Yang
Bo Yang
Bo Yang
Boyu Yang
Changdi Yang
Chao Yang
Charig Yang
Cheng-Fu Yang
Cheng-Yen Yang
Chenhongyi Yang
Chuanguang Yang
De-Nian Yang
Dingcheng Yang
Dingkang Yang
Dong Yang
Erkun Yang
Fan Yang
Fan Yang
Fan Yang
Fan Yang
Fan Yang
Feng Yang
Fengting Yang
Fengxiang Yang
Fengyuan Yang
Fu-En Yang
Gang Yang
Gengshan Yang
Guandao Yang
Guanglei Yang
Haitao Yang
Hanqing Yang
Heran Yang
Honghui Yang
Huanrui Yang
Huiyuan Yang

Huizong Yang
Hunmin Yang
Jiange Yang
Jiaqi Yang
Jiawei Yang
Jiayu Yang
Jiazhi Yang
Jie Yang
Jie Yang
Jiewen Yang
Jihan Yang
Jing Yang
Jingkang Yang
Jinhui Yang
Jinlong Yang
Jinrong Yang
Jinyu Yang
Kaicheng Yang
Kailun Yang
Lan Yang
Le Yang
Lehan Yang
Lei Yang
Lei Yang
Lei Yang
Li Yang
Lihe Yang
Ling Yang
Lingxiao Yang
Linlin Yang
Lixin Yang
Longrong Yang
Lu Yang
Luwei Yang
Michael Ying Yang
Min Yang
Ming Yang
MingKun Yang
Mouxing Yang
Muli Yang
Peiyu Yang
Qi Yang
Qian Yang
Qiushi Yang
Ren Yang

Rui Yang
Ruihan Yang
Sejong Yang
Shan Yang
Shangrong Yang
Shiqi Yang
Shuai Yang
Shuai Yang
Shuang Yang
Shuo Yang
Shusheng Yang
Sibei Yang
Siwei Yang
Siyuan Yang
Siyuan Yang
Song Yang
Songlin Yang
Tianyu Yang
Tong Yang
Wankou Yang
Wenhan Yang
Wenhan Yang
Wenjie Yang
Wenqi Yang
William Yang
Xi Yang
Xi Yang
Xiangpeng Yang
Xiao Yang
Xiaofeng Yang
Xiaoshan Yang
Xin Jeremy Yang
Xingyi Yang
Xinlong Yang
Xitong Yang
Xiulong Yang
Xu Yang
Xuan Yang
Xue Yang
Xuelin Yang
Xun Yang
Yan Yang
Yan Yang
Yang Yang
Yaokun Yang

Yezhou Yang
Yiding Yang
Yijun Yang
Yijun Yang
Yin Yang
Yinfei Yang
Yixin Yang
Yongqi Yang
Yongqi Yang
Yue Yang
Yuewei Yang
Yuezhi Yang
Yujiu Yang
Yung-Hsu Yang
Yuwei Yang
Ze Yang
Ze Yang
Zetong Yang
Zhangsihao Yang
Zhaoyuan Yang
Zhen Yang
Zhenpei Yang
Zhibo Yang
Zhiwei Yang
Zhiwen Yang
Zhiyuan Yang
Zhuoqian Yang
Ziyan Yang
Ziyun Yang
Zongxin Yang
Zuhao Yang
Chengtang Yao
Cong Yao
Hantao Yao
Jiawen Yao
Lina Yao
Mingde Yao
Mingshuai Yao
Qingsong Yao
Shunyu Yao
Taiping Yao
Ting Yao
Xincheng Yao
Xinwei Yao
Xu Yao
Xufeng Yao
Yao Yao
Yazhou Yao
Yue Yao
Ziwei Yao
Sudhir Yarram
Rajeev Yasarla
Mohsen Yavartanoo
Botao Ye
Dengpan Ye
Fei Ye
Hanrong Ye
Jianglong Ye
Jiarong Ye
Jin Ye
Jingwen Ye
Jinwei Ye
Junjie Ye
Keren Ye
Maosheng Ye
Meng Ye
Meng Ye
Muchao Ye
Nanyang Ye
Peng Ye
Qi Ye
Qian Ye
Qinghao Ye
Qixiang Ye
Ruolin Ye
Shuquan Ye
Tian Ye
Vickie Ye
Wenqian Ye
Xinchen Ye
Yufei Ye
Moon Ye-Bin
Yousef Yeganeh
Chun-Hsiao Yeh
Raymond Yeh
Yu-Ying Yeh
Florence Yellin
Sriram Yenamandra
Tarun Yenamandra
Promod Yenigalla
Chandan Yeshwanth
Dong Yi
Hongwei Yi
Kai Yi
Ran Yi
Renjiao Yi
Xinyu Yi
Alper Yilmaz
Jonghwa Yim
Aoxiong Yin
Fei Yin
Fukun Yin
Jia-Li Yin
Ming Yin
Nan Yin
Ruihong Yin
Tianwei Yin
Wenzhe Yin
Xiaoqi Yin
Yingda Yin
Yu Yin
Yufeng Yin
Zhenfei Yin
Xianghua Ying
Xiaowen Ying
Naoto Yokoya
Chen YongCan
ByungIn Yoo
Innfarn Yoo
Jinsu Yoo
Sungjoo Yoo
Hee Suk Yoon
Jae Shin Yoon
Jihun Yoon
Sangwoong Yoon
Sejong Yoon
Sung Whan Yoon
Sung-Hoon Yoon
Sunjae Yoon
Youngho Yoon
Youngseok Yoon
Youngseok Yoon
Yuichi Yoshida
Ryota Yoshihashi
Yusuke Yoshiyasu

Chenyu You
Haoran You
Haoxuan You
Shan You
Yang You
Yingxuan You
Yurong You
Chan-Hyun Youn
Kim Youwang
Nikolaos-Antonios
  Ypsilantis
Baosheng Yu
Bei Yu
Bruce X. B. Yu
Chaohui Yu
Chunlin Yu
Cunjun Yu
Dahai Yu
En Yu
En Yu
Fenggen Yu
Gang Yu
Haibao Yu
Hanchao Yu
Hang Yu
Hao Yu
Hao Yu
Haojun Yu
Heng Yu
Hong-Xing Yu
Houjian Yu
Jianhui Yu
Jiashuo Yu
Jing Yu
Jiwen Yu
Jiyang Yu
Kaicheng Yu
Lei Yu
Lidong Yu
Lijun Yu
Mulin Yu
Peilin Yu
Qian Yu
Qihang Yu
Qing Yu

Rui Yu
Ruixuan Yu
Runpeng Yu
Shaozuo Yu
Shuzhi Yu
Sihyun Yu
Tan Yu
Tao Yu
Tianjiao Yu
Wei Yu
Weihao Yu
Wenwen Yu
Xi Yu
Xiaohan Yu
Xin Yu
Xin Yu
Xuehui Yu
Yingchen Yu
Yongsheng Yu
Yunlong Yu
Zehao Yu
Zhaofei Yu
Zhengdi Yu
Zhengdi Yu
Zhixuan Yu
Zhongzhi Yu
Zhuoran Yu
Zitong Yu
Chun Yuan
Chunfeng Yuan
Hangjie Yuan
Haobo Yuan
Jiakang Yuan
Jiangbo Yuan
Liangzhe Yuan
Maoxun Yuan
Shanxin Yuan
Shengming Yuan
Shuai Yuan
Shuaihang Yuan
Wentao Yuan
Xiaoding Yuan
Xiaoyun Yuan
Xin Yuan
Xin Yuan

Yixuan Yuan
Yu-Jie Yuan
Yuan Yuan
Yuan Yuan
Yuhui Yuan
Zheng Yuan
Zhuoning Yuan
Mehmet Kerim Yücel
Dongxu Yue
Haixiao Yue
Kaiyu Yue
Tao Yue
Xiangyu Yue
Zihao Yue
Zongsheng Yue
Heeseung Yun
Jooyeol Yun
Juseung Yun
Kimin Yun
Se-Young Yun
Sukwon Yun
Tian Yun
Raza Yunus
Ekim Yurtsever
Eloi Zablocki
Riccardo Zaccone
Martin Zach
Muhammad Zaigham
  Zaheer
Ilya Zakharkin
Egor Zakharov
Abhaysinh S. Zala
Pierluigi Zama Ramirez
Eduard Sebastian Zamfir
Amir Zamir
Luca Zancato
Yuan Zang
Yuhang Zang
Zelin Zang
Pietro Zanuttigh
Giacomo Zara
Samira Zare
Olga Zatsarynna
Denis Zavadski
Vitjan Zavrtanik

Jan Zdenek
Yanjie Ze
Bernhard Zeisl
John Zelek
Oliver Zendel
Ailing Zeng
Chong Zeng
Dan Zeng
Fangao Zeng
Haijin Zeng
Huimin Zeng
Jia Zeng
Jiabei Zeng
Kuo-Hao Zeng
Libing Zeng
Ling-An Zeng
Ming Zeng
Pengpeng Zeng
Runhao Zeng
Tieyong Zeng
Wei Zeng
Yan Zeng
Yanhong Zeng
Yawen Zeng
Yuyuan Zeng
Zilai Zeng
Ziyao Zeng
Kaiwen Zha
Ruyi Zha
Yaohua Zha
Bohan Zhai
Qiang Zhai
Runtian Zhai
Wei Zhai
YiKui Zhai
Yuanhao Zhai
Yunpeng Zhai
De-Chuan Zhan
Fangneng Zhan
Guanqi Zhan
Huangying Zhan
Huijing Zhan
Kun Zhan
Xueying Zhan
Aidong Zhang
Baochang Zhang
Baoheng Zhang
Baoming Zhang
Biao Zhang
Bingfeng Zhang
Binjie Zhang
Bo Zhang
Bo Zhang
Borui Zhang
Bowen Zhang
Can Zhang
Ce Zhang
Chang-Bin Zhang
Chao Zhang
Chao Zhang
Chen-Lin Zhang
Cheng Zhang
Cheng Zhang
Chenghao Zhang
Chenyangguang Zhang
Chi Zhang
Chongyang Zhang
Chris Zhang
Chuhan Zhang
Chunhui Zhang
Chuyu Zhang
Congyi Zhang
Daichi Zhang
Dan Zhang
Daoan Zhang
Daoqiang Zhang
David Junhao Zhang
Dexuan Zhang
Dingwen Zhang
Dingyuan Zhang
Dongsu Zhang
Fan Zhang
Fan Zhang
Fang-Lue Zhang
Feilong Zhang
Frederic Z. Zhang
Fuyang Zhang
Gang Zhang
Gengwei Zhang
Gengyu Zhang
Gengyuan Zhang
Gongjie Zhang
GuiXuan Zhang
Guofeng Zhang
Guozhen Zhang
Hang Zhang
Hang Zhang
Hanwang Zhang
Hao Zhang
Hao Zhang
Hao Zhang
Haokui Zhang
Haonan Zhang
Haotian Zhang
Hengrui Zhang
Hongguang Zhang
Hongrun Zhang
Hongyuan Zhang
Howard Zhang
Huaidong Zhang
Huaiwen Zhang
Hui Zhang
Hui Zhang
Jason Y. Zhang
Ji Zhang
Jiahui Zhang
Jiakai Zhang
Jiaming Zhang
Jian Zhang
Jianfu Zhang
Jiangning Zhang
Jianhua Zhang
Jianming Zhang
Jianpeng Zhang
Jianping Zhang
Jianrong Zhang
Jichao Zhang
Jie Zhang
Jie Zhang
Jie Zhang
Jimuyang Zhang
Jing Zhang
Jing Zhang
Jinghao Zhang
Jingyi Zhang

Jinlu Zhang
Jiqing Zhang
Jiyuan Zhang
Junbo Zhang
Junge Zhang
Junyi Zhang
Juyong Zhang
Kai Zhang
Kai Zhang
Kaidong Zhang
Kaihao Zhang
Kaipeng Zhang
Kaiyi Zhang
Ke Zhang
Ke Zhang
Kui Zhang
Le Zhang
Le Zhang
Lefei Zhang
Lei Zhang
Leo Yu Zhang
Li Zhang
Lianbo Zhang
Liang Zhang
Liangpei Zhang
Lin Zhang
Linfeng Zhang
Liqing Zhang
Lu Zhang
Malu Zhang
Manyuan Zhang
Mengmi Zhang
Mengqi Zhang
Mi Zhang
Min Zhang
Min-Ling Zhang
Mingda Zhang
Mingfang Zhang
Minghui Zhang
Mingjin Zhang
Mingyuan Zhang
Minjia Zhang
Ni Zhang
Pan Zhang
Peiyan Zhang

Pengze Zhang
Pingping Zhang
Qi Zhang
Qi Zhang
Qian Zhang
Qiang Zhang
Qijian Zhang
Qiming Zhang
Qing Zhang
Qing Zhang
Renrui Zhang
Rongyu Zhang
Ruida Zhang
Ruimao Zhang
Ruixin Zhang
Runze Zhang
Sanyi Zhang
Shan Zhang
Shanghang Zhang
Shaofeng Zhang
Sheng Zhang
Shengping Zhang
Shengyu Zhang
Shimian Zhang
Shiwei Zhang
Shizhou Zhang
Shu Zhang
Shuo Zhang
Siwei Zhang
Song-Hai Zhang
Tao Zhang
Tianyun Zhang
Ting Zhang
Tong Zhang
Weixia Zhang
Wendong Zhang
Wenlong Zhang
Wenqiang Zhang
Wentao Zhang
Wentian Zhang
Wenxiao Zhang
Wenxuan Zhang
Xi Zhang
Xiang Zhang
Xiang Zhang

Xianling Zhang
Xiao Zhang
Xiaohan Zhang
Xiaoming Zhang
Xiaoran Zhang
Xiaowei Zhang
Xiaoyun Zhang
Xikun Zhang
Xin Zhang
Xinfeng Zhang
Xingchen Zhang
Xingguang Zhang
Xingxuan Zhang
Xiong Zhang
Xiuming Zhang
Xu Zhang
Xuanyang Zhang
Xucong Zhang
Xuying Zhang
Yabin Zhang
Yabo Zhang
Yachao Zhang
Yahui Zhang
Yan Zhang
Yan Zhang
Yanan Zhang
Yang Zhang
Yanghao Zhang
Yawen Zhang
Yechao Zhang
Yi Zhang
Yi Zhang
Yi Zhang
Yi-Fan Zhang
Yifan Zhang
Yifei Zhang
Yifeng Zhang
Yihao Zhang
Yihua Zhang
Yimeng Zhang
Yiming Zhang
Yin Zhang
Yinan Zhang
Yinda Zhang
Ying Zhang

Yingliang Zhang
Yitian Zhang
Yixin Zhang
Yiyuan Zhang
Yongfei Zhang
Yonggang Zhang
Yonghua Zhang
Youjian Zhang
Youmin Zhang
Youshan Zhang
Yu Zhang
Yu Zhang
Yuan Zhang
Yuechen Zhang
Yuexi Zhang
Yufei Zhang
Yuhan Zhang
Yuhang Zhang
Yunchao Zhang
Yunhe Zhang
Yunhua Zhang
Yunpeng Zhang
Yunzhi Zhang
Yuxin Zhang
Yuyao Zhang
Zaixi Zhang
Zeliang Zhang
Zewei Zhang
Zeyu Zhang
Zhang Zhang
Zhao Zhang
Zhaoxiang Zhang
Zhen Zhang
Zheng Zhang
Zheng Zhang
Zhenyu Zhang
Zhenyu Zhang
Zheyuan Zhang
Zhicheng Zhang
Zhilu Zhang
Zhishuai Zhang
Zhitian Zhang
Zhiwei Zhang
Zhixing Zhang
Zhiyuan Zhang

Zhong Zhang
Zhongping Zhang
Zhongqun Zhang
Zicheng Zhang
Zicheng Zhang
Zihao Zhang
Ziming Zhang
Ziqi Zhang
Qilong Zhangli
Bin Zhao
Bingchen Zhao
Bingyin Zhao
Cairong Zhao
Can Zhao
Chen Zhao
Dong Zhao
Dongxu Zhao
Fang Zhao
Feng Zhao
Fuqiang Zhao
Gangming Zhao
Ganlong Zhao
Guiyu Zhao
Haimei Zhao
Hanbin Zhao
Handong Zhao
Jian Zhao
Jiaqi Zhao
Jie Zhao
Kai Zhao
Kaifeng Zhao
Lei Zhao
Liang Zhao
Lirui Zhao
Long Zhao
Luxi Zhao
Mingyang Zhao
Minyi Zhao
Na Zhao
Nanxuan Zhao
Pu Zhao
Qi Zhao
Qian Zhao
Qibin Zhao
Qingsong Zhao

Qingyu Zhao
Qinyu Zhao
Rongchang Zhao
Rui Zhao
Rui Zhao
Rui Zhao
Ruiqi Zhao
Shanshan Zhao
Shihao Zhao
Shiyu Zhao
Shizhen Zhao
Shuai Zhao
Siheng Zhao
Tianchen Zhao
TianHao Zhao
Tiesong Zhao
Wang Zhao
Wangbo Zhao
Weichao Zhao
Weiyue Zhao
Wenda Zhao
Wenliang Zhao
Xiangyun Zhao
Xiaoming Zhao
Xiaonan Zhao
Xiaoqi Zhao
Xin Zhao
Xin Zhao
Xingyu Zhao
Xu Zhao
Yajie Zhao
Yang Zhao
Yifan Zhao
Ying Zhao
Yiqun Zhao
Yiqun Zhao
Yizhou Zhao
Yizhou Zhao
Yucheng Zhao
Yue Zhao
Yunhan Zhao
Yuyang Zhao
Yuzhi Zhao
Zelin Zhao
Zengqun Zhao

Zhen Zhao
Zhenghao Zhao
Zhengyu Zhao
Zhou Zhao
Zibo Zhao
Zimeng Zhao
Ziwei Zhao
Ziwei Zhao
Zixiang Zhao
Jin Zhe
Anling Zheng
Ce Zheng
Chaoda Zheng
Chengwei Zheng
Chuanxia Zheng
Duo Zheng
Ervine Zheng
Guangcong Zheng
Haitian Zheng
Haiyong Zheng
Hao Zheng
Haotian Zheng
Haoxin Zheng
Huan Zheng
Huan Zheng
Jia Zheng
Jian Zheng
Jian-Qing Zheng
Jianqiao Zheng
Jianwei Zheng
Jin Zheng
Jingxiao Zheng
Kaiwen Zheng
Kecheng Zheng
Léon Zheng
Meng Zheng
Naishan Zheng
Peng Zheng
Qi Zheng
Qian Zheng
Rongkun Zheng
Shen Zheng
Shuai Zheng
Shuhong Zheng
Shunyuan Zheng

Siming Zheng
Tianhang Zheng
Wenting Zheng
Wenzhao Zheng
Xiaozheng Zheng
Xiawu Zheng
Xu Zheng
Yajing Zheng
Yalin Zheng
Yang Zheng
Ye Zheng
Yinglin Zheng
Yinqiang Zheng
Yu Zheng
Zangwei Zheng
Zehan Zheng
Zerong Zheng
Zhaoheng Zheng
Zhedong Zheng
Zhuo Zheng
Zilong Zheng
Shuaifeng Zhi
Tiancheng Zhi
Bineng Zhong
Fangcheng Zhong
Fangwei Zhong
Guoqiang Zhong
Nan Zhong
Yaoyao Zhong
Yijie Zhong
Yiqi Zhong
Yiran Zhong
Yiwu Zhong
Yunshan Zhong
Zichun Zhong
Ziming Zhong
Brady Zhou
Chong Zhou
Chu Zhou
Chunluan Zhou
Da-Wei Zhou
Dawei Zhou
Dewei Zhou
Dingfu Zhou
Donghao Zhou

Dongzhan Zhou
Fan Zhou
Hang Zhou
Hang Zhou
Hao Zhou
Hao Zhou
Haoyi Zhou
Hong-Yu Zhou
Honglu Zhou
Huayi Zhou
Jiahuan Zhou
Jiaming Zhou
Jian Zhou
Jianan Zhou
Jiantao Zhou
Jianxiong Zhou
JIngkai Zhou
Junsheng Zhou
Kailai Zhou
Kaiyang Zhou
Keyang Zhou
Kun Zhou
Lei Zhou
Mingyang Zhou
Mingyi Zhou
Mingyuan Zhou
Mo Zhou
Pan Zhou
Peng Zhou
Peng Zhou
Qianyi Zhou
Qianyu Zhou
Qihua Zhou
Qin Zhou
Qinqin Zhou
Qunjie Zhou
Sheng Zhou
Shenglong Zhou
Shijie Zhou
Shuchang Zhou
Sihang Zhou
Tao Zhou
Tianfei Zhou
Xiangdong Zhou
Xiaoqiang Zhou

Xin Zhou
Xingyi Zhou
Yan-Jie Zhou
Yang Zhou
Yang Zhou
Yanqi Zhou
Yi Zhou
Yi Zhou
Yichao Zhou
Yichen Zhou
Yin Zhou
Yiyi Zhou
Yu Zhou
Yucheng Zhou
Yufan Zhou
Yunsong Zhou
Yuqian Zhou
Yuxiao Zhou
Yuxuan Zhou
Zhenyu Zhou
Zijian Zhou
Zikun Zhou
Ziqi Zhou
Zixiang Zhou
Zongwei Zhou
Alex Z. Zhu
Benjin Zhu
Bin Zhu
Bin Zhu
Chenming Zhu
Chenyang Zhu
Deyao Zhu
Dongxiao Zhu
Fangrui Zhu
Fei Zhu
Feida Zhu
Fengqing Maggie Zhu
Guibo Zhu
Haidong Zhu
Hanwei Zhu
Hao Zhu
Hao Zhu
Heming Zhu
Jiachen Zhu
Jianke Zhu

Jiawen Zhu
Jiayin Zhu
Jinjing Zhu
Junyi Zhu
Kai Zhu
Ke Zhu
Lanyun Zhu
Lin Zhu
Linchao Zhu
Liyuan Zhu
Meilu Zhu
Muzhi Zhu
Qingtian Zhu
Ronghang Zhu
Rui Zhu
Rui Zhu
Rui-Jie Zhu
Ruizhao Zhu
Shengjie Zhu
Sijie Zhu
Siyu Zhu
Tyler Zhu
Wang Zhu
Weicheng Zhu
Wenwu Zhu
Xiangyu Zhu
Xiaofeng Zhu
Xiaoguang Zhu
Xiaosu Zhu
Xiaoyu Zhu
Xingkui Zhu
Xinxin Zhu
Xiyue Zhu
Yangguang Zhu
Yanjun Zhu
Yao Zhu
Ye Zhu
Feida Zhu
Fengqing Maggie Zhu
Guibo Zhu
Haidong Zhu
Hanwei Zhu
Hao Zhu
Hao Zhu
Heming Zhu

Jiachen Zhu
Jianke Zhu
Jiawen Zhu
Jiayin Zhu
Jinjing Zhu
Junyi Zhu
Kai Zhu
Ke Zhu
Lanyun Zhu
Lin Zhu
Linchao Zhu
Liyuan Zhu
Meilu Zhu
Muzhi Zhu
Qingtian Zhu
Ronghang Zhu
Rui Zhu
Rui Zhu
Rui-Jie Zhu
Ruizhao Zhu
Shengjie Zhu
Sijie Zhu
Siyu Zhu
Tyler Zhu
Wang Zhu
Weicheng Zhu
Wenwu Zhu
Xiangyu Zhu
Xiaofeng Zhu
Xiaoguang Zhu
Xiaosu Zhu
Xiaoyu Zhu
Xingkui zhu
Xinxin Zhu
Xiyue Zhu
Yangguang Zhu
Yanjun Zhu
Yao Zhu
Ye Zhu
Ye Zhu
Yichen Zhu
Yingying Zhu
Yousong Zhu
Yuansheng Zhu
Yurui Zhu

Zhen Zhu
Zhenwei Zhu
Zhenyao Zhu
Zhifan Zhu
Zhigang Zhu
Zhihong Zhu
Zihan Zhu
Zixin Zhu
Zunjie Zhu
Bingbing Zhuang
Jia-Xin Zhuang
Jiafan Zhuang
Peiye Zhuang
Wanyi Zhuang
Weiming Zhuang
Yihong Zhuang
Yixin Zhuang
Mingchen Zhuge
Tao Zhuo
Wei Zhuo
Yaoxin Zhuo
Bartosz Zieliński
Wojciech Zielonka
Filippo Ziliotto
Karel Zimmermann
Primo Zingaretti
Nikolaos Zioulis
Liu Ziyin
Mohammad Zohaib
Yongshuo Zong
Zhuofan Zong
Maria Zontak
Gaspard Zoss
Changqing Zou
Chuhang Zou
Danping Zou
Dongqing Zou
Haoming Zou
Longkun Zou
Shihao Zou
Xingxing Zou
Xueyan Zou
Yang Zou
Yuexian Zou
Yuli Zou
Yuliang Zou
Yunhao Zou
Zhiming Zou
Zihang Zou
Silvia Zuffi
Idil Esen Zulfikar
Maria A. Zuluaga
Ronglai Zuo
Xingxing Zuo
Xinxin Zuo
Yifan Zuo
Yiming Zuo
Reyer Zwiggelaar
Vlas Zyrianov

Contents – Part XLIII

Recursive Visual Programming 1
  Jiaxin Ge, Sanjay Subramanian, Baifeng Shi, Roei Herzig,
  and Trevor Darrell

LLaVA-Grounding: Grounded Visual Chat with Large Multimodal Models 19
  Hao Zhang, Hongyang Li, Feng Li, Tianhe Ren, Xueyan Zou,
  Shilong Liu, Shijia Huang, Jianfeng Gao, Leizhang, Chunyuan Li,
  and Jainwei Yang

Prompt-Driven Contrastive Learning for Transferable Adversarial Attacks 36
  Hunmin Yang, Jongoh Jeong, and Kuk-Jin Yoon

Learning to Adapt SAM for Segmenting Cross-Domain Point Clouds 54
  Xidong Peng, Runnan Chen, Feng Qiao, Lingdong Kong, Youquan Liu,
  Yujing Sun, Tai Wang, Xinge Zhu, and Yuexin Ma

Learning to Enhance Aperture Phasor Field for Non-Line-of-Sight Imaging 72
  In Cho, Hyunbo Shim, and Seon Joo Kim

ViewFormer: Exploring Spatiotemporal Modeling for Multi-view 3D
Occupancy Perception via View-Guided Transformers 90
  Jinke Li, Xiao He, Chonghua Zhou, Xiaoqiang Cheng, Yang Wen,
  and Dan Zhang

Fine-Grained Dynamic Network for Generic Event Boundary Detection 107
  Ziwei Zheng, Lijun He, Le Yang, and Fan Li

Take a Step Back: Rethinking the Two Stages in Visual Reasoning 124
  Mingyu Zhang, Jiting Cai, Mingyu Liu, Yue Xu, Cewu Lu, and Yong-Lu Li

AlignZeg: Mitigating Objective Misalignment for Zero-Shot Semantic
Segmentation 142
  Jiannan Ge, Lingxi Xie, Hongtao Xie, Pandeng Li, Xiaopeng Zhang,
  Yongdong Zhang, and Qi Tian

Contrastive Learning with Counterfactual Explanations for Radiology
Report Generation 162
  Mingjie Li, Haokun Lin, Liang Qiu, Xiaodan Liang, Ling Chen,
  Abdulmotaleb Elsaddik, and Xiaojun Chang

SpeedUpNet: A Plug-and-Play Adapter Network for Accelerating
Text-to-Image Diffusion Models 181
   Weilong Chai, Dandan Zheng, Jiajiong Cao, Zhiquan Chen,
   Changbao Wang, and Chenguang Ma

Reg-TTA3D: Better Regression Makes Better Test-Time Adaptive 3D
Object Detection 197
   Jiakang Yuan, Bo Zhang, Kaixiong Gong, Xiangyu Yue, Botian Shi,
   Yu Qiao, and Tao Chen

ShapeLLM: Universal 3D Object Understanding for Embodied Interaction 214
   Zekun Qi, Runpei Dong, Shaochen Zhang, Haoran Geng, Chunrui Han,
   Zheng Ge, Li Yi, and Kaisheng Ma

Content-Aware Radiance Fields: Aligning Model Complexity with Scene
Intricacy Through Learned Bitwidth Quantization 239
   Weihang Liu, Xue Xian Zheng, Jingyi Yu, and Xin Lou

Finding Visual Task Vectors 257
   Alberto Hojel, Yutong Bai, Trevor Darrell, Amir Globerson, and Amir Bar

Connecting Consistency Distillation to Score Distillation for Text-to-3D
Generation 274
   Zongrui Li, Minghui Hu, Qian Zheng, and Xudong Jiang

Event Camera Data Dense Pre-training 292
   Yan Yang, Liyuan Pan, and Liu Liu

Distractors-Immune Representation Learning with Cross-Modal
Contrastive Regularization for Change Captioning 311
   Yunbin Tu, Liang Li, Li Su, Chenggang Yan, and Qingming Huang

Rethinking Image-to-Video Adaptation: An Object-Centric Perspective 329
   Rui Qian, Shuangrui Ding, and Dahua Lin

Layer-Wise Relevance Propagation with Conservation Property for ResNet 349
   Seitaro Otsuki, Tsumugi Iida, Félix Doublet, Tsubasa Hirakawa,
   Takayoshi Yamashita, Hironobu Fujiyoshi, and Komei Sugiura

DECap: Towards Generalized Explicit Caption Editing via Diffusion
Mechanism 365
   Zhen Wang, Xinyun Jiang, Jun Xiao, Tao Chen, and Long Chen

EgoLifter: Open-World 3D Segmentation for Egocentric Perception 382
   Qiao Gu, Zhaoyang Lv, Duncan Frost, Simon Green, Julian Straub, and Chris Sweeney

MEVG: Multi-event Video Generation with Text-to-Video Models 401
   Gyeongrok Oh, Jaehwan Jeong, Sieun Kim, Wonmin Byeon, Jinkyu Kim, Sungwoong Kim, and Sangpil Kim

Open-Vocabulary SAM: Segment and Recognize Twenty-Thousand
Classes Interactively 419
   Haobo Yuan, Xiangtai Li, Chong Zhou, Yining Li, Kai Chen, and Chen Change Loy

Data-to-Model Distillation: Data-Efficient Learning Framework 438
   Ahmad Sajedi, Samir Khaki, Lucy Z. Liu, Ehsan Amjadian, Yuri A. Lawryshyn, and Konstantinos N. Plataniotis

DiffuX2CT: Diffusion Learning to Reconstruct CT Images from Biplanar
X-Rays 458
   Xuhui Liu, Zhi Qiao, Runkun Liu, Hong Li, Juan Zhang, Xiantong Zhen, Zhen Qian, and Baochang Zhang

AdaIFL: Adaptive Image Forgery Localization via a Dynamic
and Importance-Aware Transformer Network 477
   Yuxi Li, Fuyuan Cheng, Wangbo Yu, Guangshuo Wang, Guibo Luo, and Yuesheng Zhu

Author Index 495

Recursive Visual Programming

Jiaxin Ge[✉], Sanjay Subramanian, Baifeng Shi, Roei Herzig, and Trevor Darrell

UC Berkeley, Berkeley, CA, USA
gejiaxin01@gmail.com

Abstract. Visual Programming (VP) has emerged as a powerful framework for Visual Question Answering (VQA). By generating and executing bespoke code for each question, these methods show advancement in leveraging Large Language Models (LLMs) for complex problem-solving. Despite their potential, existing VP methods generate all code in a single function, which does not fully utilize LLM's reasoning capacity and the modular adaptability of code. This results in code that is suboptimal in terms of both accuracy and interpretability. Inspired by human coding practices, we propose Recursive Visual Programming (RVP), which better harnesses the reasoning capacity of LLMs, provides modular code structure between code pieces, and assigns different return types for the sub-problems elegantly. RVP approaches VQA tasks with an top-down recursive code generation approach, allowing decomposition of complicated problems into smaller parts. We show RVP's efficacy through extensive experiments on benchmarks including VSR, COVR, GQA, and NextQA, underscoring the value of adopting human-like recursive and modular programming techniques for solving VQA tasks. Our code is available at https://github.com/para-lost/RVP.

Keywords: Visual Programming · Vision-Language · VQA

1 Introduction

Visual Question Answering (VQA) lies at the intersection of computer vision and natural language processing, posing the challenge of interpreting visual data to answer questions [3,16,26]. Following the recent progress on code generation using Large Language Models (LLMs) [6,27,33], Visual Programming (VP) methods have notably advanced in this area, successfully using LLMs to generate and execute code in few-shot and zero-shot scenarios [14,30,31]. In these techniques, various vision models are provided to LLM as APIs, and the LLM is used as a planner to utilize these vision models to perform reasoning over images.

Supplementary Information The online version contains supplementary material available at https://doi.org/10.1007/978-3-031-72775-7_1.

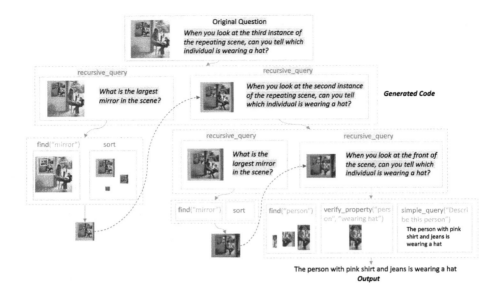

Fig. 1. A breakdown of Recursive Visual Programming for a visual question is illustrated with an image where scenes within it contain smaller versions of themselves. To locate a man wearing a hat at the third level of an image, the model first generates code that has two `recursive_query` API calls. The initial call identifies the largest mirror (representing the next image level), and the subsequent call seeks "the man with a hat" at the second level. The model iteratively generates and executes code through these calls until the final answer is produced without additional recursive calls.

Despite these advancements, such planning remains very challenging for Visual Programming. These methods require the model to generate a single code block using the predefined APIs at once. As a result, the model must handle logic, be aware of all details, and utilize the vision APIs appropriately. This has two main drawbacks. First, such approaches do not fully utilize LLM's reasoning capability, which would benefit from breaking down the problem into step-by-step sub-problems, as demonstrated in recent NLP research [4]. Secondly, they fail to utilize the full potential of code because they lack modularity between the code pieces, particularly when the code is one long piece.

We consider how humans tackle these issues. Humans tend to code in a top-down manner: first they break down complex tasks into manageable sub-problems, and then they tackle each through focused code segments. This approach allows the programmer to concentrate on the main logic flow rather than figuring out all the details at once, resulting in a more clear and elegant structure.

Inspired by this, we hypothesize that incorporating a similar recursive coding strategy would also help solve complex visual question-answering problems. To investigate whether LLMs could indeed benefit from recursive coding, we conduct preliminary experiment of the recursive coding approach on standard symbolic

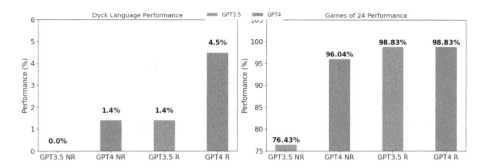

Fig. 2. Motivating Examples. Adding the prompt "solve the problem recursively" to the input significantly improves the model's performance on traditional coding tasks, Dyck language [32] and Games of 24 [38]. "NR" refers to non-recursive approach and "R" refers to recursive approach. This shows that LLMs can potentially benefit from human's recursive programming approach.

reasoning benchmarks, such as Dyck Language [32] and Games of 24 [38]. This is done by simply prompting the LLM to code recursively. We find that utilizing LLMs to code recursively leads to a notable improvement in accuracy, as shown in Fig. 2. Motivated by this result, we explore in this paper how a recursive coding methodology can be applied to VQA tasks.

Specifically, we propose a novel VP method for VQA tasks, named **R**ecursive **V**isual **P**rogramming (**RVP**). Instead of querying LLM only once to generate a single piece of code, we ask LLM to decompose the question into sub-questions and recursively query itself for each sub-question to generate modular code pieces. For each sub-question, the same process repeats, until reaching an atom level where the current question can be solved by only calling external APIs. See example in Fig. 1. In contrast to querying LLM only once, we find that using recursive queries to generate code for each sub-question can improve the readability and accuracy of the code, as shown in Fig. 3. Moreover, RVP allows assigning dynamic types to each sub-question, enabling new code pieces to return a variety of types based on the current need. Remarkably, RVP assigns return types unseen in in-context examples, such as List[str], List[ImagePatch].

Our contribution can be summarized as follows: (i) We introduce a novel recursive approach based on human programming. This approach is a top-down programming method that better harnesses LLM's reasoning capacity and enhances modularity between multiple code pieces. It is easily applicable to any existing VP model for visual question answering task. (ii) We allow dynamic return types for the sub-problems, which enhances flexibility and adaptability. (iii) We demonstrate improved performance on several standard benchmarks, such as GQA, VSR, NextQA and COVR, demonstrating the effectiveness of our approach. We also show better interpretability than non-recursive VP methods.

2 Related Work

Modular Visual Reasoning. There is a long line of work that seeks to combine modularity with the expressivity of deep neural networks. Early work on this topic converts the output of a parser [1] to a program or trains a program generator either separately from or jointly with the execution modules [2,15]. Later work finds that such models are often challenging to train on real-world datasets, particularly when interpretability of individual module outputs is expected [29]. Other work improves performance on multi-hop visual reasoning without explicitly defining a set of modules, e.g., by predicting and operating over a scene graph [18].

Visual Programming. Recent language models can accurately generate code from text descriptions when prompted with a small number of examples [8]. This advance has enabled a few-shot modular visual reasoning approach, commonly called visual programming [14], which consists of prompting a language model for a program and executing the program with pre-trained vision models. This paradigm has led to impressive results in visual question answering, video question answering, visual and tabular question answering, referring expression

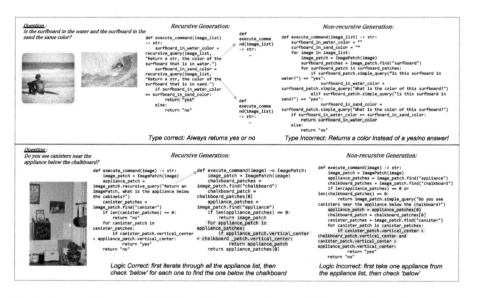

Fig. 3. Example from COVR (Upper). Current VP methods fail to provide binary 'yes' or 'no' answers, while Recursive Visual Programming method outputs the correct answer. **Example from GQA (Middle and Bottom).** Recursive outperforms current non-recursive methods by correctly addressing all details and their associated logic.

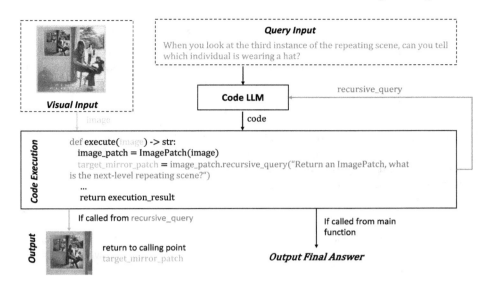

Fig. 4. An overview of our approach. When a `recursive_query` API call is occured, the model generates and executes new code based on the query, returning the output back to the original call.

comprehension, and text-based image editing [11,12,23,30,31]. Programming with LLMs has also been applied to produce plans and reward functions for robots [21,24,40]. Our work builds on these techniques by enabling the LLM to generate code in multiple steps–first planning the overall process and then filling in the details–rather than all at once.

Language Models for Reasoning Tasks. In the text domain, there is a large body of work on using language models for reasoning tasks. In particular, several papers have shown the efficacy of code generation as a scaffold for such tasks [9,13,25,34]. Aside from code generation, prior work has also shown that decomposing complex queries into simpler ones improves the performance of language models [28,42]. Our RVP approach applies this insight to improve the code generation component of visual programming.

3 Recursive Visual Programming

In this section, we introduce our method, Recursive Visual Programming (RVP). We first review the problem setup and ingredients of visual programming as established by prior work (see Sect. 3.1). We then introduce our recursive coding framework (see Sect. 3.2), which involves a multi-step prompting scheme, as well as a system for dynamic type assignment (see Sect. 3.3).

3.1 Preliminaries

In Visual Question Answering (VQA), the system f is given a visual input I (*i.e.* an image, a set of images, or a video), as well as a question Q relating to the visual content. The expected output is a textual answer A to the question:

$$A \leftarrow f(I, Q) \tag{1}$$

Visual Programming methods [14,30,31] represent f as the composition of two functions: a code generator g and a visual executor h. In the first part, the code generator generates a program P conditioned on the question, and in the second part, the executor executes the program on the visual input:

$$P \leftarrow g(Q), A \leftarrow h(P, I) \tag{2}$$

We focus on the few-shot setting in which we have only a few (at most 50) training examples–each consisting of a question, program, and answer–for a given task. Therefore, the code generator g is composed of a language model and a prompt that includes a description of the API available to the executor and in-context examples of questions along with the corresponding correct programs. We use Python as the programming language, similar to [30,31].

We evaluate our approach using the ViperGPT [31] API. For questions over images or sets of images, these functions can be called on an `ImagePatch` object, which can be any crop of the image:

1. `find(object: str)`: Returns a list of `ImagePatch` objects matching the description.
2. `exists(object: str)`: Returns a bool representing whether the described object exists.
3. `verify_property(object: str, property: str)`: Returns a bool representing whether the described object has the specified property.
4. `simple_query(question: str)`: Returns an answer to the provided question based on the given `ImagePatch`.
5. `compute_depth()`: Returns the median depth of the region in the given `ImagePatch`.
6. `crop(left: int, lower: int, right: int, upper: int)`: Crops a part of the given `ImagePatch` and returns a new `ImagePatch` representing that crop.

For questions about a video, the API includes three other functions: `trim()`, `frame_from_index()`, and `frame_iterator()`, which return an abbreviated version of the video, a specific frame, and an iterator over the frames respectively.

3.2 Recursive Visual Programming

We introduce a new method for code generation in visual programming: Recursive Visual Programming (RVP). RVP enables the language model to break the code generation process into multiple sub-questions and recursively call itself to generate a program for each sub-question. As shown in Fig. 4, we operationalize RVP by adding a new function to the API: recursive_query(image, sub-question). During execution, when recursive_query is called, it prompts the language model to generate a new code given the sub-question. This code is then executed, and its return value is returned by recursive_query. An example break down of the usage of recursive_query call is shown in Fig. 1. We set the recursive termination condition by (1) setting max depth of 10 and (2) switching to direct simple query when generated code contains a recursive call to the same input question.

To enable this process, we add examples of calling recursive_query in the prompt to show the model how to decompose the problem and utilize the answers returned by recursive_query. We provide examples of the in-context prompt of recursive_query we used in the supplementary material.

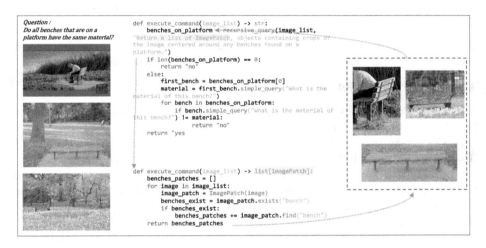

Fig. 5. Examples of dynamic type assignment in recursive_query calls. The model autonomously determines the appropriate return type in its code generation and generalize to new types unseen in in-context examples such as List[ImagePatch].

3.3 Dynamic Type Assignment

In previous visual programming frameworks, code generation mechanisms are constrained by static type assignments, *i.e.*, the return type of a given function is fixed. For instance, simple_query always returns a string. However, this rigid structure limits the model's flexibility to adapt the code to different contexts within the VQA task. RVP introduces dynamic type assignment to address this limitation. When the code generator writes a recursive_query function, it can include the expected return type as a part of the question that is passed to the recursive_query. To facilitate this, in the in-context examples that include recursive_query, we let the input 'question' specify a type. As shown in Fig. 5, we give examples of type ImagePatch, bool, float, and str being specified within the queries. When receiving this question, the model discerns the required type and generates code that declares and returns the correct type. For instance, a query like "Return a bool, is there a black cat in the picture?" will prompt the generation of code with the signature.

$$\text{execute(image)} \rightarrow \text{bool} \tag{3}$$

This signature indicates that the generated code P', when executed, will return a boolean value A_{bool} which answers the sub-question Q', illustrating the dynamic and context-sensitive nature of RVP's code generation capabilities that help to produce syntactically and semantically appropriate code for complex reasoning.

4 Experiments

In this section, we introduce the basic settings including the task formulation and parameter details. Then, we provide experiment results and in-depth analysis.

4.1 Experiment Setup

Datasets. We evaluate on four datasets spanning a wide range of visual reasoning skills and types of visual input: Visual Spatial Reasoning (VSR) [22], GQA [17], COVR [5], and NextQA [36]. For NextQA, we present results on the "Hard Temporal" subset which was curated to consist of questions that truly require temporal reasoning [7]. VSR and GQA include queries about single images, while COVR includes queries about multiple images, and NextQA includes queries about video. On each dataset, we evaluate methods by their accuracy in producing an answer that exactly matches the ground-truth answer. For NextQA, we adopt a multiple-choice setup in which we provide methods with the set of possible answer choices for each question. For VSR, a spatial statement of an image is made, and the question is framed as : "Is it true or false," followed by the statement. In this way, a VSR statement is framed as a VQA question and the model will answer either true or false.

Implementation Details. We implement RVP using the ViperGPT API and the accompanying prompt template. For each dataset, we curate a set of in-context examples which are added to the prompt. The set of question-answer pairs is shared between the recursive and non-recursive methods. We use seven in-context examples for GQA, six for NextQA, six for VSR, and eight for COVR. For code generation, use GPT-3.5-turbo via the OpenAI API[1].

---

[1] www.platform.openai.com.

4.2 Quantitative Results

Table 1 presents our results across four datasets. In all settings, we outperform ViperGPT, which is identical to RVP except that RVP includes `recursive_query` function in the API and in-context examples. This indicates that recursive programming indeed improves accuracy in visual reasoning tasks. Moreover, our method outperforms previous zero/few-shot methods in all but one of the settings. While our primary aim was not to push the state-of-the-art results, we show that integrating the clarity and elegance of recursive coding into standard VP methods not only doesn't impair performance, but can enhance it.

4.3 Qualitative Results

We provide qualitative examples in Figs. 3,6,7. In the example from COVR in Fig. 3, the non-recursive approach answers with neither "yes" nor "no," while the recursive approach does confirm to these choices. One possible explanation for this issue is that LLMs struggle to maintain coherence across long contexts [37]. RVP circumvents this issue by pushing the LLM to commit to the answer types earlier in its generation (the length from the beginning of the code to the point "return xxx" is shorter). In Fig. 6, the non-recursive approach fails to handle the logic correctly while the recursive approach correctly does. This is likely due to that RVP breaks down the code logic and each code piece only needs to focus on a single logic, while the non-recursive method attempts to handle all the logic at once. In the GQA example, a common pattern involves handling complex logical problems and multiple properties. Non-recursive methods often fell short in these cases, either overlooking critical details or processing them in a logically flawed manner. RVP manages these intricate questions by addressing each detail accurately. This difference may stem from the inherent difficulty LLMs face in processing complicated logic over extended contexts. Recursive

Table 1. Results. We report exact-match accuracy on each dataset. Best zero/few-shot result for each dataset is **bolded**. †Fully supervised results are from different models: ViLT [19] for VSR, HiTeA [39] for NextQA, VinVL-Base [41] for GQA, and VisualBERT [20] for COVR. ‡We report the results of our reproduction of ViperGPT using the officially released code and the same in-context examples used for our method (since the original in-context examples from ViperGPT were not released). We use GPT3.5-turbo while CodeVQA uses Codex, which is not available anymore.

Method	VSR		NextQA	GQA	COVR
	Random Split	Zero-shot Split	Hard Split-T	Test-dev	Test
Fully supervised†	69.3	63.0	48.6	65.1	57.9
Zero/few-shot methods					
CLIP	56.0	54.5	–	–	–
BLIPv2	–	–	–	42.31	–
CodeVQA	–	–	–	**49.0**	50.7
ViperGPT‡	61.25	61.59	47.21	44.63	51.69
RVP (ours)	63.53	**66.09**	**48.82**	45.62	**52.67**

methods simplify this by focusing on the overall logic and gradually breaking the problems down into smaller, more manageable segments. This facilitates easier and more accurate resolution.

We provide a dynamic type assignment example in Fig. 5. In this example, RVP assigns the return type List[ImagePatch] in the recursive_query call. It's worth noting that the return type List[ImagePatch] is unseen in the in-context examples. This suggests that RVP has the potential to utilize complicated and diverse data structures, highlight the effectiveness of our approach.

4.4 Ablation Study

Dynamic Type This section studies integrating dynamic type assignment within the recursive_query API. For the GQA dataset, we examined all instances with recursive patterns in the test_dev split. In COVR, where the test set is unavailable, we used the val_1000 samples from the validation set, following [30].

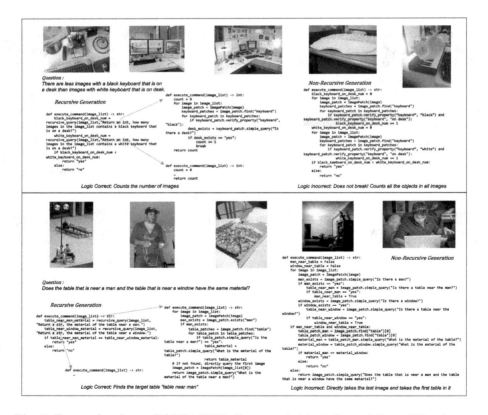

Fig. 6. Examples from COVR. RVP allows correct logic flow while traditional non-recursive VP fails to handle the logic correctly.

Fig. 7. Examples from GQA. RVP allows handling the details more elegantly and produces logically correct code.

We compare four approaches: **Non-recursive Baseline**: Standard approach without recursive features. **Fixed-Type Recursive**: The recursive_query API is constrained to return a string. **Dynamic-Implicit Recursive**: The API can return various types, but the specific type is not predefined in the query. For example, "Is there a black cat?" could yield either a boolean or a string. **Dynamic-Explicit Recursive**: Our primary approach, where the API's input query specifies the return type, like "Return a bool, is there a black cat?".

Table 2 shows our findings. The dynamic-explicit type assignment outperformed other methods, underscoring the need for clear type specification. The dynamic-implicit method showed lower performance than non-recursive due to type ambiguities, as exemplified in Fig. 8, where the first code piece assumes the return type is a bool, while the second code piece returns a str. Fixed-type assignment proves to be beneficial over non-recursive methods, indicating that recursion itself adds value, but its rigidity in type handling limits performance. This study highlights the benefits of flexible, dynamic type assignment and the crucial role of explicit type declaration for accurate and consistent responses. These insights mark a step towards more adaptable visual reasoning frameworks, emphasizing the need for precision in dynamic environments.

In-Context Examples We conduct ablation on using retrieval-based in-context example selection and using different number of in-context examples.

Retrieval-based in-context example selection. We manage to write 9 programs using the recursive method from the 50 provided by CodeVQA [30], and add our original 3 recursive programs to them. We keep the non-recursive in-context example unchanged and choose 3 recursive in-context examples from these 12 for each question. We follow [35] to use the embedding-based retrieval technique that calculates a score for each example, and then select the top 3 examples for each question. The results of GQA_val_2000 are in Table 2. We find that the recursive pattern occurance rate is significantly higher than using fixed in-context examples. This is likely because by retrieving similar recursively-decomposed questions, the model can more easily simulate their patterns, thereby attempting to solve more problems recursively. However. the overall accuracy does not improve. This suggests that although the retrieval-based method encourages more diverse and suitable patterns of recursive decomposition, there is a need for improved techniques in selecting the example pool–such as focusing on those most suitable for recursive decomposition instead of randomly selecting-along with careful tuning of the example programs.

Number of in-context examples. We randomly add one example/delete one example from the covr_1000_val, the results are shown in Table 2. The results indicate that adding the number of examples can enhance the performance, but does not have a significant impact.

Error feedback loop Recent studies have studied the debugging ability of LLMs on code [10]. We study whether modular visual code structure enables better

Table 2. Ablation study. (a) We provide ablation study results for dynamic type assignment and found that specified dynamic type assignment in RVP achieves the highest performance. (b) We provide results on using retrieval-based in-context examples and fixed examples. Retrieval-based method enables more diverse recursive patterns but requires careful choosing of the prompt pool. (c) We provide results of using different number of examples and found this doesn't influence performance significantly.

(a) Dynamic Type Ablation.

Method	GQA	COVR
Non-Recursive	55.99	60.71
Fixed-Type	65.05	60.12
Dynamic-Implicit	55.34	53.57
Dynamic-Explicit	**70.23**	**67.86**

(b) Retrieval-Based and Fixed Examples

	Fixed	Retrieval
Accuracy	54.75	53.79
Rec Rate	3.55%	14.85%
Rec Acc	71.83	57.91

(c) In-Context Example Num

Method	Accuracy
Del 1	50.65
Original	51.05
Add 1	51.35

bug identification for visual programming. We sample 20 incorrect programs for 10 questions that have type/logic errors, and use GPT3.5 to (1) identify the bug, and (2) write the correct code given the bug. For identification, the model correctly finds 7 bugs correctly for RVP and 2 for non-recursive visual programming. For correction, the model fixes 7 bugs correctly for RVP and 7 for visual programming. This suggests that clearer and more modular code structure, as presented RVP, can potentially aids bug identification.

4.5 Readability Study

This study investigates how recursive coding impacts code readability and comprehension for human programmers. Our goal is to move from complex code structures to more modular, clear, and understandable code. To evaluate this, we focus on code quality in terms of its understandability.

From the COVR dataset, we randomly select one example from each five different question types where recursive patterns occur. We present the main function of both their recursive and non-recursive code versions to 15 participants and ask them to grasp the high-level intent of the code as quickly as possible. For the non-recursive case, the main function is just the single code snippet generated by the LLM. For the recursive case, the main function consists of the code generated by the first LLM call (the high level code) but not the functions generated by the recursive queries. This helps us measure how fast users can understand the code's intention by looking at the main function, and reflects the self-explanatory nature of the code. Figure 9 shows the results. Participants consistently understand the recursive code faster than the non-recursive versions. This quicker comprehension can be attributed to the recursive code's structured and clear approach. Non-recursive code, while correct, sometimes lacks immediate clarity in conveying its primary goal.

```
def execute(image) -> str:
    image_patch = ImagePatch(image)
    is_black_skateboard=
    image_patch.recursive_query("Are there any black
    skateboards?")
    is_motorcycle = image_patch.recursive_query("Are
    there any motorcycles?")
    return bool_to_yesno(is_black_skateboard or
    is_motorcycle)

def execute(image) -> str:
    image_patch = ImagePatch(image)
    skateboard_patches = image_patch.find("skateboard")
    for skateboard_patch in skateboard_patches:
        If
        skateboard_patch.verify_property("skateboard",
        "black"):
            return "yes"
    return "no"
```

Return Type Confusion: Expect Bool but Actually Str.

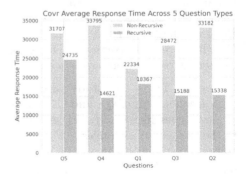

Fig. 8. Failure example of dynamic type assignment due to unspecified type, The model confuses the type and return a str instead of a bool.

Fig. 9. Comparison of average understanding time for recursive and non-recursive codes. RVP demonstrates better readability and is easy to understand.

We also calculate the average line length for all main and sub functions. For RVP, the average length of all functions is **8.47**. For non-recursive VP, the average length of the functions is **15.4**. This suggests that RVP decreases the average length the code pieces and enhances modularity.

Our findings suggest that recursive coding not only simplifies complex code but also enhances its accessibility and ease of understanding. Its orderly structure and clear logic accelerate comprehension, leading to more efficient and user-friendly programming. We provide more details of this study in the supp.

Table 3. Recursive analysis. Table (a) provides the distribution of recursive patterns by question type in GQA and COVR. We have found that recursive patterns occur commonly in question types that require reasoning. Table (b) studies different recursive query return types and their frequency in GQA and COVR. The model generalizes to unseen types. Table (c) compares accuracy across question types. Recursive methods outperform non-recursive particularly in 'Choose' and 'Query' questions.

(a) Dist. of Recursive Patterns

COVR/GQA Question Types	COVR	GQA
Compare	457	15
Mult. Ref	282	-
Quant. Attr	12	-
Spec Attr	357	-
Choose	1	25
Verify	-	68
Query	-	167
Logical	-	34

(b) Diverse Return Types

Data Type	GQA	COVR
Str	319	1828
Bool	521	592
List[str]	16	-
ImagePatch	30	-
List[ImgPatch]	1	5
Float	6	-
Int	-	914

(c) Type Accuracy

Question	Accuracy (%)	
	Non-Rec	Rec
Verify	52.94	64.71
Query	56.89	71.86
Logical	55.88	70.59
Compare	66.67	53.33
Choose	52.00	84.00

Question Type. We study which types of questions tend to use recursive and found that recursive patterns appeared in five question types within both the COVR test set and the GQA test-dev set, as shown in Table 3. The 'Verify Property' and 'Query' types in GQA, and 'Compare' and 'Multi-Reference' in COVR, were most common. This suggests recursive methods tend to occur in tasks that require comparative and referential analysis.

Dynamic Type. We study which types of dynamic return types occurs and their frequently in GQA and COVR. In Table 3, we found that RVP generalizes to new types like List[str] and List[ImagePatch], which are not included in the in-context examples. This flexibility and ability to generalize beyond in-context examples highlight the inherent adaptability of recursive methods.

Type Accuracy. We evaluate the accuracy of recursive methods across different question types. An analysis of the GQA test-dev set is presented in Table 3. It reveals that recursive methods significantly improve accuracy in 'Query', 'Logical', and 'Choose' questions. This underscores their suitability for complex reasoning and decision-making tasks.

Error Source. We sample 50 incorrect recursive examples from `GQA2000val` and manually checke the error source: 22% are caused by logic error inside subfunctions; 18% are caused by logic error between functions; 52% are caused by API model inaccuracy (i.e. detection model error); and 4% are caused by other problems like the groundtruth answer unclear.

Cost And Runtime. For the questions which require recursive query, RVP takes about 2.1 times (runtime) and about 3 times (cost) compared to ViperGPT.

5 Conclusion and Limitation

In this work, we present RVP, a visual programming method that adopts recursive coding to generate concise, elegant and easy-to-understand code, handling details and intricate logic more accurately compared with traditional visual programming methods. We conduct extensive experiments across various benchmarks and provide a comprehensive analysis of RVP. Our work is the first to explore recursive programming in the visual domain and dynamic-type code generation for visual reasoning; we believe the recursive programming concept would have a potential greater impact on other domains as well. While RVP is effective for complicated visual questions, most questions in current VQA benchmarks tend to be short and non-recursive. We believe that the benefit of our method could be demonstrated better with a new benchmark or a more complex split.

Acknowledgements. We would like to thank Ben Bogin for helping us test on the private COVR test set, and Guangyuan Jiang for helping us proof read the paper and providing valuable feedback.

References

1. Andreas, J., Rohrbach, M., Darrell, T., Klein, D.: Neural module networks. In: 2016 IEEE Conference on Computer Vision and Pattern Recognition (CVPR), pp. 39–48 (2015). https://api.semanticscholar.org/CorpusID:5276660
2. Andreas, J., Rohrbach, M., Darrell, T., Klein, D.: Learning to compose neural networks for question answering. ArXiv abs/1601.01705 (2016). https://api.semanticscholar.org/CorpusID:3130692
3. Antol, S., et al.: VQA: visual question answering. In: Proceedings of the IEEE International Conference on Computer Vision, pp. 2425–2433 (2015)
4. Besta, M., et al.: Graph of thoughts: solving elaborate problems with large language models (2024)
5. Bogin, B., Gupta, S., Gardner, M., Berant, J.: COVR: a test-bed for visually grounded compositional generalization with real images. ArXiv abs/2109.10613 (2021). https://api.semanticscholar.org/CorpusID:237592834
6. Brown, T., et al.: Language models are few-shot learners. In: Larochelle, H., Ranzato, M., Hadsell, R., Balcan, M., Lin, H. (eds.) Advances in Neural Information Processing Systems, vol. 33, pp. 1877–1901. Curran Associates, Inc. (2020). https://proceedings.neurips.cc/paper/2020/file/1457c0d6bfcb4967418bfb8ac142f64a-Paper.pdf

7. Buch, S., Eyzaguirre, C., Gaidon, A., Wu, J., Fei-Fei, L., Niebles, J.C.: Revisiting the "Video" in video-language understanding. In: Proceedings of the IEEE/CVF Conference on Computer Vision and Pattern Recognition (CVPR) (2022)
8. Chen, M., et al.: Evaluating large language models trained on code. ArXiv abs/2107.03374 (2021). https://api.semanticscholar.org/CorpusID:235755472
9. Chen, W., Ma, X., Wang, X., Cohen, W.W.: Program of thoughts prompting: disentangling computation from reasoning for numerical reasoning tasks. Trans. Mach. Learn. Res. (2023)
10. Chen, X., Lin, M., Schärli, N., Zhou, D.: Teaching large language models to self-debug. arXiv preprint arXiv:2304.05128 (2023)
11. Cheng, Z., et al.: Binding language models in symbolic languages. ArXiv abs/2210.02875 (2022). https://api.semanticscholar.org/CorpusID:252734772
12. Cho, J., Zala, A., Bansal, M.: Visual programming for text-to-image generation and evaluation. NeurIPS (2023)
13. Gao, L., et al.: PAL: program-aided language models. arXiv preprint arXiv:2211.10435 (2022)
14. Gupta, T., Kembhavi, A.: Visual programming: Compositional visual reasoning without training. In: 2023 IEEE/CVF Conference on Computer Vision and Pattern Recognition (CVPR), pp. 14953–14962 (2022). https://api.semanticscholar.org/CorpusID:253734854
15. Hu, R., Andreas, J., Rohrbach, M., Darrell, T., Saenko, K.: Learning to reason: end-to-end module networks for visual question answering. In: 2017 IEEE International Conference on Computer Vision (ICCV), pp. 804–813 (2017). https://api.semanticscholar.org/CorpusID:18682
16. Hudson, D.A., Manning, C.D.: GQA: a new dataset for real-world visual reasoning and compositional question answering. In: 2019 IEEE/CVF Conference on Computer Vision and Pattern Recognition (CVPR), pp. 6693–6702 (2019). https://api.semanticscholar.org/CorpusID:152282269
17. Hudson, D.A., Manning, C.D.: GQA: a new dataset for real-world visual reasoning and compositional question answering. In: 2019 IEEE/CVF Conference on Computer Vision and Pattern Recognition (CVPR), pp. 6693–6702 (2019). https://api.semanticscholar.org/CorpusID:152282269
18. Hudson, D.A., Manning, C.D.: Learning by abstraction: the neural state machine. In: Neural Information Processing Systems (2019). https://api.semanticscholar.org/CorpusID:195847902
19. Kim, W., Son, B., Kim, I.: ViLT: vision-and-language transformer without convolution or region supervision. In: International Conference on Machine Learning (2021). https://api.semanticscholar.org/CorpusID:231839613
20. Li, L.H., Yatskar, M., Yin, D., Hsieh, C.J., Chang, K.W.: VisualBERT: a simple and performant baseline for vision and language. ArXiv abs/1908.03557 (2019). https://api.semanticscholar.org/CorpusID:199528533
21. Liang, J., et al.: Code as policies: language model programs for embodied control. ArXiv preprint arXiv:2209.07753 (2022)
22. Liu, F., Emerson, G.E.T., Collier, N.: Visual spatial reasoning. Trans. Assoc. Comput. Linguist. **11**, 635–651 (2022). https://api.semanticscholar.org/CorpusID:248496506
23. Lu, P., et al.: Chameleon: plug-and-play compositional reasoning with large language models. arXiv preprint arXiv:2304.09842 (2023)
24. Ma, Y.J., et al.: Eureka: human-level reward design via coding large language models. arXiv preprint arXiv: Arxiv-2310.12931 (2023)

25. Madaan, A., Zhou, S., Alon, U., Yang, Y., Neubig, G.: Language models of code are few-shot commonsense learners. In: Goldberg, Y., Kozareva, Z., Zhang, Y. (eds.) Proceedings of the 2022 Conference on Empirical Methods in Natural Language Processing, pp. 1384–1403. Association for Computational Linguistics, Abu Dhabi, United Arab Emirates (2022). https://doi.org/10.18653/v1/2022.emnlp-main.90
26. Marino, K., Rastegari, M., Farhadi, A., Mottaghi, R.: OK-VQA: a visual question answering benchmark requiring external knowledge. In: 2019 IEEE/CVF Conference on Computer Vision and Pattern Recognition (CVPR), pp. 3190–3199 (2019). https://api.semanticscholar.org/CorpusID:173991173
27. OpenAI: GPT-4 technical report. ArXiv abs/2303.08774 (2023). https://api.semanticscholar.org/CorpusID:257532815
28. Press, O., Zhang, M., Min, S., Schmidt, L., Smith, N.A., Lewis, M.: Measuring and narrowing the compositionality gap in language models. ArXiv abs/2210.03350 (2022). https://api.semanticscholar.org/CorpusID:252762102
29. Subramanian, S., et al.: Obtaining faithful interpretations from compositional neural networks. In: Annual Meeting of the Association for Computational Linguistics (2020). https://api.semanticscholar.org/CorpusID:218487535
30. Subramanian, S., et al.: Modular visual question answering via code generation. In: Rogers, A., Boyd-Graber, J., Okazaki, N. (eds.) Proceedings of the 61st Annual Meeting of the Association for Computational Linguistics (Volume 2: Short Papers), pp. 747–761. Association for Computational Linguistics, Toronto, Canada (Jul 2023). https://doi.org/10.18653/v1/2023.acl-short.65, https://aclanthology.org/2023.acl-short.65
31. Sur'is, D., Menon, S., Vondrick, C.: ViperGPT: visual inference via python execution for reasoning. ArXiv abs/2303.08128 (2023). https://api.semanticscholar.org/CorpusID:257505358
32. Suzgun, M., et al.: Challenging big-bench tasks and whether chain-of-thought can solve them. arXiv preprint arXiv:2210.09261 (2022)
33. Touvron, H., et al.: LLaMA: open and efficient foundation language models. ArXiv abs/2302.13971 (2023). https://api.semanticscholar.org/CorpusID:257219404
34. Wang, X., Li, S., Ji, H.: Code4Struct: code generation for few-shot structured prediction from natural language. arXiv preprint arXiv:2210.12810 (2022)
35. Wang, Z., et al.: Language models with image descriptors are strong few-shot video-language learners. Adv. Neural. Inf. Process. Syst. **35**, 8483–8497 (2022)
36. Xiao, J., Shang, X., Yao, A., Chua, T.S.: Next-qa: next phase of question-answering to explaining temporal actions. In: Proceedings of the IEEE/CVF Conference on Computer Vision and Pattern Recognition (CVPR), pp. 9777–9786 (2021)
37. Yang, K., Klein, D., Peng, N., Tian, Y.: DOC: improving long story coherence with detailed outline control. In: Annual Meeting of the Association for Computational Linguistics (2023). https://api.semanticscholar.org/CorpusID:254877751
38. Yao, S., et al.: Tree of thoughts: deliberate problem solving with large language models. arXiv preprint arXiv:2305.10601 (2023)
39. Ye, Q., et al.: HiTeA: hierarchical temporal-aware video-language pre-training. ArXiv abs/2212.14546 (2022). https://api.semanticscholar.org/CorpusID:255340506
40. Yu, W., et al.: Language to rewards for robotic skill synthesis. Arxiv preprint arXiv:2306.08647 (2023)

41. Zhang, P., et al.: VinVL: revisiting visual representations in vision-language models. In: 2021 IEEE/CVF Conference on Computer Vision and Pattern Recognition (CVPR), pp. 5575–5584 (2021). https://api.semanticscholar.org/CorpusID:235692795
42. Zhou, D., et al.: Least-to-most prompting enables complex reasoning in large language models. ArXiv abs/2205.10625 (2022). https://api.semanticscholar.org/CorpusID:248986239

LLaVA-Grounding: Grounded Visual Chat with Large Multimodal Models

Hao Zhang[1](✉), Hongyang Li[2], Feng Li[1], Tianhe Ren[4], Xueyan Zou[5], Shilong Liu[6], Shijia Huang[7], Jianfeng Gao[3], Leizhang[4], Chunyuan Li[3], and Jainwei Yang[3]

[1] The Hong Kong University of Science and Technology, Hong Kong, China
hzhangcx@connect.ust.hk
[2] South China University of Technology, Guangzhou, China
[3] Microsoft Research, Redmond, USA
[4] International Digital Economy Academy (IDEA), Shenzhen, China
[5] University of Wisconsin-Madison, Madison, USA
[6] Tsinghua University, Beijing, China
[7] The Chinese University of Hong Kong, Hong Kong SAR, China

Abstract. With the recent significant advancements in large multimodal models (LMMs), the importance of their grounding capability in visual chat is increasingly recognized. Despite recent efforts to enable LMMs to support grounding, their capabilities for grounding and chat are usually separate, and their chat performance drops dramatically when asked to ground. The problem is the lack of a dataset for grounded visual chat (GVC). Existing grounding datasets only contain short captions. To address this issue, we have created GVC data that allows for the combination of grounding and chat capabilities. To better evaluate the GVC capabilities, we have introduced a benchmark called Grounding-Bench. Additionally, we have proposed a model design that can support GVC and various types of visual prompts by connecting segmentation models with language models. Experimental results demonstrate that our model outperforms other LMMs on Grounding-Bench. Furthermore, our model achieves competitive performance on classic grounding benchmarks like RefCOCO/+/g and Flickr30K Entities.

1 Introduction

With the success of large language models (LLMs) like GPT-4 [19] and the open-sourced substitutes LLaMA [24], researchers are eager to leverage their strong

---

H. Zhang, H. Li—Equal Contribution.
C. Li, J. Yang—Directional Lead.
J. Gao, Leizhang—Equal Advisory.

Supplementary Information The online version contains supplementary material available at https://doi.org/10.1007/978-3-031-72775-7_2.

language capabilities in the field of vision. This enthusiasm has led to a surge in the development of large multimodal models (LLMs). Previous LMMs, such as LLaVA [14] and miniGPT-4 [36], have demonstrated exceptional visual chat abilities by generating plausible responses based on images and user instructions. However, they often encounter challenges in providing responses that exhibit a fine-grained understanding of images, including specific regions and alignment with related image regions-this is often referred to as visual grounding.

Recognizing the significance of visual grounding for LMMs, recent research efforts have focused on developing grounding and referring capabilities for LMMs [2,3,7,25,30]. While these models have achieved performance comparable to specialized models [15,17] on classic grounding benchmarks such as RefCOCO [5] and Flickr30K [23], they often treat grounding as a distinct task that requires customized prompts to initiate. Consequently, their text responses undergo significant changes when tasked with grounding. Most models, such as MiniGPT-v2 [2] and CogVLM-Grounding [25], can only generate short captions when performing grounding, as they are primarily trained on grounding caption data like Flickr30K. As illustrated in Fig. 1(a), these earlier models struggle to excel simultaneously in both chat and grounding tasks. BuboGPT [35] maintains chat capability by leveraging an external grounding model for grounding, but this approach can be constrained by the performance of the language encoder in the grounding model. Shikra [3] engages in referential dialog, which includes grounded chat, but its performance is limited due to the scarcity of available data. All existing LMMs [2,3,25,30] only support outputting coordinates as text, which restricts localization performance, and they do not support pixel-wise grounding and referring. In summary, previous LMMs struggle to perform grounded visual chat effectively due to the scarcity of grounded visual chat data and suboptimal model designs. Furthermore, they lack the capability for pixel-wise grounding and referring.

To address these challenges, we contribute to grounded visual chat in three key areas: data creation, network architecture, and benchmarking. When annotating grounding data, previous methods such as Kosmos-2 [22] and GPT4ROI [34] rely on pretrained grounding models or detection models to predict bounding boxes based on existing captions. In contrast, we label grounded visual chat data using human-labeled object detection data [11].

Our data creation process begins by leveraging GPT-4 [19], following the data creation method used in LLaVA [14]. We provide GPT-4 with chat data and ground-truth instances, instructing it to match instances with noun phrases in the chat data. This approach benefits from the high quality of human-labeled instances and chat data generated by GPT-4, ensuring minimal noise in the data annotation pipeline. In total, we annotated $150K$ grounded visual chat data.

In terms of network architecture, we propose connecting the output features of the Language Model (LLM) with a grounding model to handle grounding tasks, relieving the language model from the burden of vision localization tasks. For this purpose, we use the open-set segmentation and detection model OpenSeeD [32] as the grounding model, enabling both box and pixel-level grounding simultaneously.

To evaluate the capability of grounded visual chat, we introduce the Grounding Bench, a benchmark that assesses grounding and chat performances concurrently. Built upon the foundation of LLaVA bench, our benchmark evaluates chat and phrase grounding in three contexts: conversation, detailed description, and complex reasoning. Additionally, recognizing that grounded detailed description is the most challenging aspect of grounded visual chat, we propose grounded recall and precision metrics. Grounded recall measures the proportion of ground-truth instances correctly mentioned and grounded, while grounded precision measures the accuracy of groundings or predicted boxes. We also calculate the F_1 score, a combination of precision and recall. To evaluate the correctness of semantic matching since the models generate free-form phrases, we rely on GPT-4.

In summary, our contributions are as follows:

1. We introduce a data annotation pipeline to label high-quality Grounded Visual Chat (GVC) data. Leveraging human-labeled object detection data [11] and harnessing the robust matching capability of GPT-4 [20], we have successfully labeled 150K GVC instances using the LLaVA instruction tuning dataset.
2. We present an end-to-end model, named LLaVA-Grounding (LLaVA-G for brevity), which connects a Large Multimodal Model (LMM) with a grounding model to facilitate grounded visual chat. Our model supports both object and pixel-level grounding, accommodating various visual prompts such as mark, click, box, and scribble. Table 1 demonstrates that our model offers a broader range of input and output prompt types compared to other LMMs.
3. We establish the Grounding-Benchbenchmark for evaluating grounded visual chat and propose an auto-evaluation pipeline aided by GPT-4. This benchmark assesses grounded visual chat capabilities and provides performance metrics for other state-of-the-art methods.
4. Through extensive experiments, we demonstrate that our model surpasses other grounding LMMs in terms of performance on Grounding-Bench, while also achieving competitive results on classic grounding benchmarks like RefCOCO/+/g and Flickr30K.

2 Method

2.1 Overview

To advance the development of grounded visual chat for Large Multimodal Models (LMMs), we introduce a comprehensive pipeline for labeling grounded visual chat data, a tailored modeling approach designed for the grounded visual chat task, and a benchmark for evaluating grounded visual chat performance, as illustrated in Fig. 1(b). We will provide further details on these three components in the following subsections.

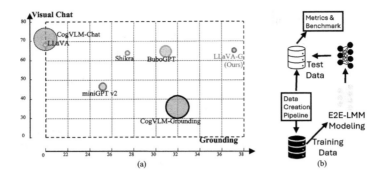

Fig. 1. (a) A comparison on the integrated ability of visual grounding and visual chat of open-source LMMs on Grounding-Bench. LLaVA-G achieves a good trade-off on both abilities simultaneously. For CogVLM [25], two different model checkpoints are released: CogVLM-Grounding is the grounding model and CogVLM-Chat is the chat model. Grounding and Visual Chat scores represent the F_1 score and Chat scores of detailed descriptions in Table 3, respectively. Circle size indicates the model size. (b) An overview of our main contributions. We use the data creation pipeline to create training and test data. The training data is used to train our LLaVA-G. The test data is used to build our Grounding-Bench.

2.2 Grounded Visual Chat Data Creation

To perform grounded visual chat (GVC) effectively, it is crucial to have high-quality data that encompasses both meaningful conversations and accurate grounding. We have constructed our dataset based on LLaVA instruction tuning data for two primary reasons. Firstly, the conversations within this dataset are generated by GPT-4, known for its high linguistic quality. Secondly, the images used are sourced from COCO, which contains human-annotated grounding box instances.

Our data annotation process aims to associate phrases from conversations with specific instances. To achieve this, we leverage the capabilities of GPT-4. As illustrated in Table 2, we provide GPT-4 with ground-truth (GT) boxes containing class labels and a sentence from the conversation. We task GPT-4 with matching noun phrases from the sentence to the GT instances. Once noun phrases are successfully grounded by GPT-4, we mark them with special start tokens, $\langle g_s \rangle$ and $\langle g_e \rangle$, followed by a token, $\langle seg \rangle$, which corresponds to the output feature used by the grounding model to segment the grounded region. An example of a question and its answer in the dataset is as follows:

Q: What is the man doing? A: $\langle g_s \rangle$ The man $\langle g_e \rangle$ $\langle seg \rangle$ is using $\langle g_s \rangle$ a clothing iron $\langle g_e \rangle$ $\langle seg \rangle$ on the back of $\langle g_s \rangle$ a yellow taxi $\langle g_e \rangle$ $\langle seg \rangle$.

For each $\langle seg \rangle$, we have a corresponding segmentation mask. This annotated data forms the basis of our Grounded Visual Chat (GVC) dataset. To support visual prompts in user instructions, we apply a similar annotation process to instances in the question itself. The resulting data appears as follows:

> **Context type 1: Boxes (for data annotation)**
> 1.person: [0.681, 0.242, 0.774, 0.694], 2.person: [0.63, 0.222, 0.686, 0.516],
> 3.person: [0.444, 0.233, 0.487, 0.34], 4.backpack: [0.384, 0.696, 0.485, 0.914],
> 5.backpack: [0.755, 0.413, 0.846, 0.692], 6.suitcase: [0.758, 0.413, 0.845, 0.69],
> 7.suitcase: [0.1, 0.497, 0.173, 0.579], 8.bicycle: [0.282, 0.363, 0.327, 0.442],
> 9.car: [0.786, 0.25, 0.848, 0.322], 10.car: [0.783, 0.27, 0.827, 0.335],
> 11.car: [0.86, 0.254, 0.891, 0.3], 12.car: [0.261, 0.101, 0.787, 0.626]
>
> **Context type 2: user responses (for data annotation)**
> The image is an underground parking area with a black sport utility vehicle (SUV) parked. There are three people in the scene, with one person standing closer to the left side of the vehicle, another person in the middle, and the third person on the right side. They are all working together to pack their luggage into the SUV for a trip.
>
> **Response: grounded responses (for data annotation)**
> The image is an underground parking area with a (black sport utility vehicle) [10.car] (SUV) parked. There are (three people) [1.person, 2.person, 3.person] in the scene, with (one person) [3.person] standing closer to the left side of the vehicle, (another person) [2.person] in the middle, and (the third person) [1.person] on the right side. They are all working together to pack (their luggage) [4.backpack, 5.backpack, 6.suitcase, 7.suitcase] into the SUV for a trip.

> **Context type 3: predicted grounded responses (for evaluation)**
> The depiction is of a below-ground parking facility, where a sleek, black vehicle [9.car] is situated. In the vicinity of this SUV, a trio of individuals [1.person, 2.person, 3.person] is engaged in an activity: the first person [1.person] is adjacent to the left side of the vehicle, the second [2.person] is situated centrally, and the third [3.person] is near the right side. They are collaboratively arranging their travel bags in the SUV, signaling the onset of an impending journey.
>
> **Response:** TP_{pred} and TP_{gt} **(for evaluation)**
> "a sleek, black vehicle" [9.car] - Incorrectly referred.
> "a trio of individuals" [1.person, 2.person, 3.person] - 3 Correctly referred.
> "the first person" [1.person] - Incorrectly referred.
> "the second" [2.person] - Correctly referred.
> "the third" [3.person] - Incorrectly referred.
>
> There are 4 correct references (TP_{pred}) and 3 correctly referred entities (TP_{gt}).

Fig. 2. Illustrate the data annotation (top) and the evaluation (bottom) with language GPT4. The top table shows the contexts and responses for data annotation. The bottom table shows the contexts and responses for evaluating the recall and precision of grounded description. Note that the Context 1 for evaluation is same as that for data annotation. Note that the visual image is not used to prompt GPT4, we only show it here as a reference.

Q: What is the object $\langle obj \rangle$ doing? A: $\langle g_s \rangle$ The man $\langle g_e \rangle$ $\langle seg \rangle$ is using $\langle g_s \rangle$ a clothing iron $\langle g_e \rangle$ $\langle seg \rangle$ on the back of $\langle g_s \rangle$ a yellow taxi $\langle g_e \rangle$ $\langle seg \rangle$.

It's important to note that we modify "the man" to "the object" in cases where the model might disregard the visual prompts. For each $\langle obj \rangle$ in the question, we provide a corresponding segmentation mask. This dataset is referred to as GVC-R (Grounded Visual Chat with Referring). To examine the quality, we randomly sampled 100 images to do a human study. It turned out the overall grounding precision of GPT-4 is up to **95.4%**, with some occasional failures when multiple people are grounded separately.

2.3 Network Architectures

Since our network architecture is nearly identical to LLaVA, with the exception of the additional prompt encoder and grounding model, we will only introduce these two parts in this section. For the other components of our architecture, please refer to LLaVA [14].

Prompt Encoder. For an input image $\mathbf{X}_v$ and a visual prompt $\mathbf{X}_p$, we employ the pre-trained Semantic-SAM as the prompt encoder. This encoder extracts visual features based on the input image and visual prompts, denoted as $\mathbf{X}_p =$

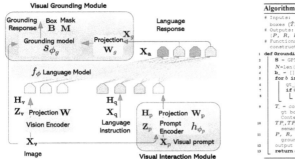

Fig. 3. (left) Network architecture of our LLaVA-Grounding contains a CLIP vision encoder, a LLM, a prompt encoder, a grounding model and the corresponding projection layers. LLaVA-Grounding expands LLaVA with two additional modules highlighted in blue blocks: the visual interaction module that accepts user drawing and visual grounding module that outputs object masks/boxes. The yellow tokens represents the visual prompt feature aligned to language embedding space. The light green output tokens represent the grounding features which are the last-layer hidden feature of the language model corresponding to ⟨seg⟩ tokens. (right) The pseudo code to evaluate our Grounding Bench.

$h(\mathbf{X_v}, \mathbf{X_p})$. To convert these prompt features into language embedding tokens $\mathbf{H_p}$ of the same dimensionality as the word embedding space in the language model, we use a simple linear layer with a trainable projection matrix $\mathbf{W_p}$:

$$\mathbf{H_p} = \mathbf{W_p} \cdot \mathbf{X_p}, \text{ where } \mathbf{X_p} = h\left(\mathbf{X_v}, \mathbf{X_p}\right) \quad (1)$$

This results in a sequence of visual tokens $\mathbf{H_p}$. It's worth noting that there are special tokens ⟨obj⟩ in $\mathbf{X_q}$ with word embeddings as placeholders, and visual tokens in $\mathbf{H_p}$ replace the word embeddings of ⟨obj⟩ in $\mathbf{H_q}$.

Grounding Model. In addition to the language response $\mathbf{X_a}$, our model also produces features $\mathbf{X_g}$ for grounding. These features correspond to the last layer hidden features of the language model that align with the ⟨seg⟩ tokens. We initially map these features to a grounding space using a trainable projection matrix $\mathbf{W_g}$. Subsequently, we employ a pretrained OpenSeeD model as the grounding model to generate bounding boxes $\mathbf{B}$ and masks $\mathbf{M}$. This process can be defined as follows:

$$\mathbf{B}, \mathbf{M} = s\left(\mathbf{X_v}, \mathbf{W_g} \cdot \mathbf{X_g}\right) \quad (2)$$

Here, $s(\cdot, \cdot)$ represents the grounding model, which takes the image $\mathbf{X_v}$ and the grounding features as input.

2.4 Training

We propose a three-stage training strategy, as illustrated in Table 2. These stages are **pretraining for alignment, instruction tuning for grounded**

visual chat, and **extension to visual prompt**. A unified representation of our instruction-following data is presented as follows:

$$\begin{aligned} \texttt{Human}: \mathbf{X_v} <\texttt{\textbackslash n}> \mathbf{X_q}(\mathbf{X_p})\texttt{<STOP>} \\ \texttt{Assistant}: \mathbf{X_a}(\mathbf{X_g})\texttt{<STOP>}\texttt{\textbackslash n} \end{aligned} \quad (3)$$

In this representation, $\mathbf{X_p}$ and $\mathbf{X_g}$ are enclosed in brackets, indicating that they are optional. During training, the model is trained to predict the assistant's answers, including the grounded instances and where to stop. Consequently, only the green sequence/tokens are used to compute the loss in the auto-regressive model.

Stage 1: Pretraining for Alignment. Stage 1 focuses on feature alignment for the visual encoder and granularity alignment for the grounding model.

Feature Alignment for Vision Encoder. As shown in Table 2, we utilize the RefCOCO/+/g, COCO 2017train, Visual Genome, LLaVA 585K image caption, and Flickr30K Entities datasets for Stage 1. Both LLaVA 585K and Flickr30K Entities datasets consist of image caption pairs and are used to train the projection layer $\mathbf{W}$ for feature alignment in the vision encoder. The conversation construction approach aligns with that of LLaVA, where a question is randomly selected from Appendix: Table 13 as $\mathbf{X_q}$, and the original caption is used as $\mathbf{X_a}$. The learnable parameter for this part is denoted as $\theta = \{\mathbf{W}\}$.

Feature and Granularity Alignment for Grounding Model. To facilitate grounding, we need to align the features $\mathbf{X_g}$ output by the language model with the vocabulary space of the grounding model. For this purpose, we train on the RefCOCO/+/g, COCO 2017train, Visual Genome, and Flickr30K Entities datasets. The approach to construct instruction-following data is as follows:

1. For RefCOCO/+/g and Visual Genome, the user instruction $\mathbf{X_q}$ is randomly selected from Appendix: Table 12, and $\mathbf{X_a}$ consists only of the special token $\langle seg \rangle$. COCO 2017train follows the same approach as RefCOCO/+/g, but with a distinction: the class name of an instance serves as its referring text.
2. In contrast, the Flickr30K Entities dataset differs from the image caption data mentioned earlier. Here, the user instruction is followed by a suffix randomly chosen from Appendix: Table 14. This suffix signals the model to produce a response in grounding format, as described in Sect. 2.2. The response $\mathbf{X_a}$ is then converted into the grounding format by inserting special tokens $\langle g_s \rangle$, $\langle g_e \rangle$, and $\langle seg \rangle$ into $\mathbf{X_a}$ to mark noun phrases.

Given the instruction-following data, the last-layer hidden features of the language model corresponding to $\langle seg \rangle$ tokens $\mathbf{X_g}$ are mapped to the grounding vocabulary space by multiplying them with $\mathbf{W_g}$. Additionally, since our grounding model is pretrained on COCO and Object365, which have different granularities compared to the Visual Genome and Flickr30K grounding data, we also train the grounding model to align these granularities.

In summary, the learnable parameters for Stage 1 are denoted as $\theta = \{\mathbf{W}, \mathbf{W_g}, \phi_g\}$.

Table 1. A comparison of input referring and output grounding format of LMMs. Our model is the only one that supports all the input and output formats in the table.

	input				output			
	text	click	box	mark	text	box	mask	mark
LLaVA [9]	✓				✓			
MiniGPT-4 [36]	✓				✓			
GPT4ROI [34]	✓		✓		✓			
Shikra [3]	✓				✓			
Ferret [30]	✓	✓			✓	✓		
MiniGPTv2 [2]	✓				✓	✓		
LLaVA1.5 [13]	✓				✓	✓		
CogVLM-Grounding [25]	✓				✓	✓		
LLaVA-G (Ours)	✓	✓	✓	✓	✓	✓	✓	✓

Table 2. Blue, green and red means the training data and tasks in the 1st, 2nd, and 3rd stages, respectively. "Grounding" means only predict boxes and "Grounding Seg" means predict masks. For Flickr30K, we use SAM to label pseudo GT masks. "Chat with VP" means chat with visual prompts.

	Grounding	Grounding Seg	Visual Chat	Chat with VP
RefCOCO/+/g [5,31]	✓	✓		✓
Visual Genome [6]	✓			✓
COCO train2017 [11]	✓	✓		
LLaVA 585K [14]			✓	
Flickr30K [23]	✓	✓	✓	
LLaVA 150K [14]			✓	
GVC 2.2	✓	✓		
GVC-R 2.2				✓

Stage 2: Instruction Tuning for Grounded Visual Chat. In the second training stage, we leverage the Grounded Visual Chat (GVC) data, excluding visual prompts, for instruction tuning. To also support chat without grounding, we incorporate LLaVA 158K instruction-following data. During this stage, we freeze the CLIP vision encoder and focus on fine-tuning the other components of the model. The learnable parameters in this stage are denoted as $\theta = \{\mathbf{W}, \mathbf{W_g}, \phi, \phi_g\}$.

The data format consists of instruction data containing $\langle seg \rangle$ tokens in the answer, accompanied by several grounding annotations. The number of grounding annotations corresponds to the number of $\langle seg \rangle$ tokens present. In this stage, we calculate both language loss and grounding losses. The language loss is computed in the same manner as in LLaVA for the answer tokens and "STOP" tokens. The grounding losses encompass box, mask, and matching losses. Box and mask losses are utilized solely for training the grounding model, while the matching loss is propagated to the language model.

Stage 3: Extension to Visual Prompt. In the third stage, we introduce support for visual prompts as an additional component by training only h_{ϕ_p} and the projection layer $\mathbf{W_p}$. As detailed in Table 2, the training data includes RefCOCO/+/g, Visual Genome, and GVC-R. In contrast to Stage 1, for RefCOCO/+/g and Visual Genome, we provide visual prompts for the ground truth (GT) instances and instruct the model to predict captions. The text instruction $\mathbf{X_p}$ is randomly selected from Table 15, where $\langle obj \rangle$ tokens serve as placeholders, and their input embeddings will be replaced by prompt features. The text answer $\mathbf{X_a}$ comprises the original referring expressions.

In this stage, the learnable parameters are represented as $\theta = \{\phi_p, \mathbf{W_p}\}$, where ϕ_p is trained to output boxes and masks corresponding to visual prompts, and $\mathbf{W_p}$ is trained to align visual prompt features with the language embedding space.

Set-of-Mark (SoM) Prompts. (Optional) In addition to visual prompts (such as clicks and boxes) that can be handled through the prompt encoder, our model also supports marks as visual prompts, similar to the approach presented in [27]. These marks consist of alphanumerics and masks that are directly overlaid on the image. To illustrate, consider the data sample in Sec.2.2. Let's assume we overlay marks labeled as ⟨1⟩, ⟨2⟩, and ⟨3⟩ on the "man", "iron", and "taxi" in the input image. This results in the Grounded and Referring Visual Chat (GRVC) data taking the form:

Q: What is the object ⟨1⟩ doing? A: The man ⟨1⟩ is using a clothing iron ⟨2⟩ on the back of a yellow taxi ⟨3⟩.

It's important to note that both the question and answer consist of text only. Therefore, in order to support marks as visual prompts, we specifically fine-tune the language part of the model.

2.5 Grounding-Bench

Benchmark Creation. We introduce a benchmark named Grounding-Bench to assess a model's grounded visual chat capability. To evaluate both grounding and chat abilities concurrently, we build this benchmark on top of LLaVA Bench (COCO), which comprises chat data generated by GPT4 and instance annotations from MSCOCO. To enhance the robustness of Grounding-Bench, we expand our test dataset to include 1000 images with 7000 entities, all sourced from the MSCOCO 2014val split. These images are converted into grounded visual chat data using our data creation pipeline, forming the basis of our test dataset.

Task Definition. Grounded visual chat tasks involve taking an image X_V and a user instruction I as input and generating a caption T accompanied by bounding boxes **b**, with each bounding box corresponding to a specific phrase.

Evaluate Chat Scores. Our benchmark evaluation encompasses two main aspects: chat scores and grounded response scores. We outline the evaluation process for Grounding-Bench in Algorithm 1 in Fig. 3 (right). Chat scores are akin to those used in LLaVA Bench. However, in contrast, we instruct the model to produce grounded responses. Subsequently, we process the output to remove special tokens and boxes, yielding the pure-text response for evaluation.

Evaluate Grounded Response Scores. For grounded responses, we specifically evaluate the grounded detailed description task. Our evaluation includes metrics such as recall (R) for completeness, precision (P) for hallucination, and the F_1 score (F_1) to combine both aspects. R measures the proportion of entities correctly mentioned and grounded in the description, while P assesses the proportion of correctly predicted groundings. A grounding is deemed correct only when the box matches a ground truth (GT) box with an IoU greater than 0.5, and their semantics are accurately matched. To determine TP_{pred} and TP_{gt} for GPT4, we provide Context types 1 and 3, as shown in the bottom block in

Fig. 2. For example, in the provided example, $N_{pred} = 7$ and $N_{gt} = 12$. Based on GPT4's response, we calculate $TP_{pred} = 4$ and $TP_{gt} = 3$. Consequently, we obtain $P = 0.57$, $R = 0.25$, and $F_1 = 0.35$. We examine the agreement rate between humans and GPT-4 evaluator and find it is almost **100%** for pairs labeled as positive by humans and **86%** for negative pairs because GPT-4 sometimes struggles to distinguish objects of the same category.

3 Experiments

In this section, we will first introduce our experimental settings. Then, we will compare our model with other state-of-the-art models on our benchmark, Grounding-Bench. Next, we will evaluate our model against other grounding models on challenging Referring Expression Comprehension (REC) and Referring Expression Segmentation (RES) tasks on RefCOCO, RefCOCO+, and RefCOCOg. The results will demonstrate that our model outperforms other grounding LLMs with the same number of parameters on both REC and RES tasks, and ours is the only model capable of handling both REC and RES effectively. Afterward, we will conduct an evaluation of the support for various types of visual prompts. Finally, we will perform ablation studies on our modeling and data creation processes to validate our method.

Table 3. A comparison on our Grounding-Bench. For each model, we use the prompt template recommended by the paper. The results in grounded response scores are two parts in each grid where the left one is evaluated on the 1000 images of our Grounding-Bench and the right one is on the 30 images in LLaVA Bench (COCO). * denotes Shikra with a special prompt for grounded description recommended by the paper. We make GPT4-V+SoM grey because it uses external model to label marks.

Model	#Vision params(B)	Grounded Response Scores			Chat Scores				Phrase grounding
		Recall	Precision	F_1	Detail desc.	Conv.	Reasoning	All	
LLaVA [14]	0.30	-	-	-	69.1	82.0	92.6	81.2	-
Bubo-GPT [35]	2.00	26.2|25.7	37.2|31.3	30.7|28.2	65.0	75.9	93.4	78.2	-
Shikra [3]	0.30	21.1|21.6	39.8|38.4	27.6|27.7	64.7	75.4	86.4	75.5	64.29
Shikra*	0.30	22.0|28.7	44.6|48.6	29.4|36.1	41.8	-	-	-	-
miniGPT v2 [2]	1.00	20.6|25.3	33.6|39.1	25.6|30.7	48.0	51.0	38.7	45.8	-
CogVLM-Grounding [25]	10.0	22.3|27.5	56.3|62.5	32.0|38.2	35.8	47.8	22.2	34.9	-
CogVLM-Chat	10.0	-	-	-	73.1	86.9	92.1	84.2	-
LLaVA-G (Ours)	0.35	28.6|36.3	52.7|53.4	37.1|43.2	67.2	78.7	91.1	79.3	81.6

3.1 Experimental Settings

To facilitate result reproduction, we provide detailed settings. Our language model is initialized from a pretrained Vicuna-7b v1.3, the grounding model is initialized from the vision part of an OpenSeeD Tiny model pretrained on COCO and Object365, and the interactive encoder is initialized from a Semantic-SAM Tiny model pretrained on COCO with three granularities.

In the first training stage, we freeze the language model and train the grounding model, prompt encoder, and projection layers with a learning rate of 1×10^{-4}. For the second stage, we train the language model and projection layers with a learning rate of 2×10^{-5}, while training the grounding model with a learning rate of 1×10^{-4} while freezing the CLIP vision encoder and the prompt encoder.

3.2 Grounding-Bench

To demonstrate the effectiveness of our method in Grounded Visual Chat (GVC), we compare our method with other strong LMMs that support visual grounding on our benchmark. As shown in Table 3, the results in grounded response scores are presented in two parts for each grid. The left one is evaluated on the 1000 images of our Grounding-Bench, and the right one is on the 30 images in LLaVA Bench (COCO). All the numbers for grounding LMMs are evaluated using their official prompt templates for grounding to ensure the best performance. The results show that our method outperforms all open-source methods in both grounded response scores and chat scores on grounded responses, except for CogVLM-Chat and LLaVA, which are chat models. GPT4-V achieves the best performance on grounded detailed description with the help of SoM, but it is a combination of two models. Among open-source methods, GogVLM is second only to ours in terms of the F_1 score for grounded detailed description, but it has the lowest GPT evaluated scores. Shikra's chat scores are second only to ours. We also annotated 30 images in LLaVA Bench (COCO) as grounded detailed description and reported phrase grounding performance of our model and Shikra for reference (Table 4).

Table 4. Performance comparison on the referring expression comprehension (REC) referring expression segmentation (RES) and phrase grounding tasks. We mark the best results with bold. * denotes the zero-shot results are reported. Since CogVLM-Grounding is a larger model with 4B vision model and 6B connection module, we make it grey.

Models	RefCOCO			RefCOCO+			RefCOCOg			Flickr30k Entities	
	REC	RES		REC	RES		REC	RES			
	ACC@0.5	mIoU	cIoU	ACC@0.5	mIoU	cIoU	ACC@0.5	mIoU	cIoU	val	test
ReLA [12]	–	–	73.80	–	–	66.00	–	–	65.00	–	–
PolyFormer-L[15]	–	76.94	75.96	–	72.15	69.33	–	71.15	69.20	–	–
UniTAB [28]	86.32	–	–	78.70	–	–	79.96	–	–	78.76	79.58
MDETR [4]	86.75	–	–	79.52	–	–	81.64	–	–	82.3	83.8
GLIP-T* [10]	50.42	–	–	49.50	–	–	66.09	–	–	–	–
GDINO-T [17]	**89.19**	–	–	81.09	–	–	84.15	–	–	–	–
Kosmos-2* [22]	52.32	–	–	45.48	–	–	60.57	–	–	77.80	78.70
LISA-7B [3]	–	–	74.9	–	–	65.1	–	–	67.9	–	–
MiniGPT v2-7B [3]	88.06	–	–	79.58	–	–	84.19	–	–	–	–
Shikra-7B [3]	87.01	–	–	81.60	–	–	82.27	–	–	75.84	76.54
Ferret-7B [30]	87.49	–	–	80.78	–	–	83.93	–	–	80.39	82.21
LLaVA-G-7B (Ours)	89.16	**79.68**	**77.13**	**81.68**	**72.92**	**68.79**	**84.82**	**74.39**	**71.54**	**83.03**	**83.62**

3.3 Traditional Grounding Benchmarks

We also evaluate our model on classic grounding benchmarks, including RefCOCO/+/g for Referring Expression Comprehension (REC) and Referring Expression Segmentation (RES), and Flickr30K Entities for Phrase Grounding. For this experiment, we use the 7B language model with the grounding model using the Swin-Tiny backbone. Our model is trained for the first stage with RefCOCO/+/g, Visual Genome, and Flickr30K Entities. Our model stands out as the only LMM that can excel in both REC and RES tasks. On the REC task, our model outperforms all LMMs, except for CogVLM-Grounding, which utilizes a 4B vision model and a 6B connection module. On RES and Phrase grounding tasks, our model surpasses all LMMs. One advantage of our model is its ability to be trained on both box and mask data, allowing us to leverage Visual Genome to enhance our RES performance.

3.4 Visual Prompts

Table 5. The top1 accuracy of phrase grounding on Flickr30K. The first row is our original pipeline with grounding model to do phrase grounding.

Model	Ground type	α Mark	Size Mark	val	test
Ours	-	-	-	83.0	83.6
Ours	Mark	0.4 / 0.4	20	72.1	73.7
Ours	Mark	0.4 / 0.2	30	75.1	75.4
Ours	Mark	0.2 / 0.2	30	76.6	77.9

Table 6. The comparison on COCO object classification accuracy. Numbers except for our method are from PVIT [1] paper. They evaluate llava by cropping the regions of GT boxes.

LLava	Shikra	GPT4ROI	PVIT	Ours-T click	Ours-T box
40	53.9	64	64.5	70.8	71.5

We demonstrate our support for various types of visual prompts, including marks, clicks, and boxes.

Support for Marks as Visual Prompts. In Table 5, we showcase our phrase grounding performance on Flikr30K when using marks as visual prompts. It's important to note that for this experiment, we overlay GT masks and marks onto the input image, positioning the marks at the centers of the masks. The table presents results for different α values of masks and sizes of marks. We also include the performance of Shikra-7B and Ours-7B with the grounding model for comparison. The results clearly indicate that our model effectively supports the use of marks for grounding.

Support for Clicks and Boxes as Visual Prompts. To further demonstrate our support for visual prompts, we compare object classification accuracy on COCO val2017 in Table 6. We report our performances achieved using clicks and boxes as visual prompts. Additionally, we include numbers from the PVIT paper for reference. The results illustrate that our referring accuracy surpasses that of previous works, with boxes yielding better accuracy than clicks.

Table 7. An example of grounded detailed descriptions generated by LLaVA-G and other LMMs. We use the best prompt recommended by the paper for each model.

Grounded description example 1:	
User	Describe the following image with grounding.
LLaVA-G	The image depicts a group of people sitting at a large, white table, engaged in activities. There are four people in the scene, with three of them sitting close to each other and one person sitting further to the right. They are all using laptop computers, likely working or collaborating on a project together. The table is well organized, featuring multiple laptops, a large TV screen on the left side, and a clock on the wall. There are also several books and a cell phone placed on the table, indicating a casual and comfortable environment. A chair is positioned near the left side of the table, and there are two bottles placed nearby, possibly containing drinks for the people working at the table.
User	Can you provide a description of the image and include the coordinates [[x0,y0,x1,y1]] for each mentioned object?
CogVLM-Grounding	A group of people are sitting at a table with laptops.
User	Provide a comprehensive description of the image <image> and specify the positions of any mentioned objects in square brackets.
Shikra	A group of people are in a meeting , some with laptops, some with laptops, and some without .
User	[grounding] Describe the scene in detail.
MiniGPTv2	A group of people are sitting around a large table in a meeting room , with laptops and phones on the table

3.5 Visualizations

Compare with Other Models. We provide visual comparisons between our model and other grounding LMMs, including CogVLM-Grounding, Shikra, and MiniGPTv2, in Table 7 and Appendix: Table 16, and 17. These visualizations highlight that our model produces longer descriptions, detects more objects, and maintains high precision. Furthermore, our model demonstrates the ability to ground open-set concepts, such as "dragons," which do not exist in our grounded visual chat data. These visual comparisons underscore the strengths and capabilities of our model in the context of grounded visual chat.

Other functionalities. We show other functionalities of our model such as **visual chat with visual prompts, marks as visual prompts, and potential hallucination removal** with visualizations in Appendix.

3.6 Ablation Studies

Table 8. A comparison on LLaVA-Bench. "GVC" is "No" means it outputs pure-text response without grounding.

	GVC	LLaVA-Bench (COCO)				LLaVA-Bench (In-the-Wild)			
		Conv.	Detail	Reasoning	All	Conv.	Detail	Reasoning	All
LLaVA		82.0	69.1	92.6	81.2	42.6	51.9	68.9	57.1
LLaVA-G	Yes	74.8	68.5	95.3	79.7	38.5	40.1	75.1	55.8
LLaVA-G	No	79.3	71.2	92.8	81.2	47.7	44.6	70.0	57.2

Table 9. The comparison when using different number of queries in the grounding model. "#Q" is the number of queries.

	RefCOCO			RefCOCO+			RefCOCOg		
#Q	ACC	cIoU	mIoU	ACC	cIoU	mIoU	ACC	cIoU	mIoU
50	86.71	74.77	77.6	77.91	64.97	69.68	82.37	68.46	72.43
100	86.58	74.70	77.40	77.23	64.08	69.02	81.99	68.02	72.06
300	86.35	74.26	77.19	77.78	64.68	69.54	81.92	67.89	71.85

In this section, we provide insights into our visual chat capability and the design of the grounding model through various ablation studies.

Maintaining Visual Chat Capability. We demonstrate that our model retains strong visual chat capabilities by comparing it with LLaVA on LLaVA Bench (Table 8). The results indicate that our model's visual chat performance is comparable to LLaVA, whether responding with or without grounding.

Number of Queries in Grounding Model. Table 9 presents our model's performance on Referring Expression Comprehension (REC) and Referring Expression Segmentation (RES) tasks with different numbers of queries. The results reveal that using 50 queries is sufficient for both tasks and achieves optimal performance. This finding highlights the efficiency of our approach in handling these tasks.

Detaching the Grounding Model. We investigate the impact of detaching the grounding model on both chat and grounding performance. Detaching the grounding model means stopping gradients from propagating from the grounding model to the Language Model (LLM). Table 10 compares the detached model's performance with the original model. The results demonstrate that detaching the grounding model leads to slightly improved chat performance but significantly compromises the grounding performance. This indicates the importance of the grounding model in maintaining high-quality visual chat with grounding capabilities.

Table 10. Ablations on our benchmark. "Detach GD" means stop gradient from the grounding model to language model.

Model	Detach GD	Grounded detail description			Chat scores			
		Recall	Precision	F_1	Detail desc.	Conv.	Reasoning	All
Ours	✓	25.1	58.2	35.1	61.6	86.3	94.9	81.2
Ours		36.3	53.4	43.2	67.2	78.7	91.1	79.3

4 Related Work

Large Multi-modal Models. The expansion of Large Language Models (LLMs) has led to the development of Large Multi-modal Models (LMMs), notably LLaVA [9] and MiniGPT-4 [36], which have enriched models with visual instruction capabilities through GPT-4 and tailored prompts. Other key contributions, such as mPLUG-DocOwl [29], Otter [8], LLaMa-Adaptor [33], and InternGPT [18], have advanced LMMs by introducing varied strategies. Vision-LLM [26], GPT4RoI [34], and PVIT [1] exemplify the push towards detailed multimodal understanding through innovative feature extraction methods.

Grounding Large Multi-modal Models. Based on their architectural characteristics and functionalities, Grounding LMMs can be classified into three distinct categories. **The first category** of grounding models predicts textual box coordinates, exemplified by Kosmos-2 [22], Shikra [3], MiniGPT v2 [36], Ferret [30], and CogVLM [25]. Kosmos-2 is noted for its zero-shot grounding using a specialized dataset, while Shikra enhances referral dialogues with box inputs/outputs. MiniGPT v2 and CogVLM demonstrate versatility and high performance in vision-language tasks, respectively, leveraging task-specific tokens and a robust vision model. Despite some relying on smaller or less refined datasets, these models have significantly advanced visual grounding capabilities. **The second category** involves models that employ a separate grounding model for grounded chat, exemplified by BuboGPT [35] and LLaVA-PLUS [16]. However, these models often face performance limitations at the language encoder of the grounding model. **The third category** adopts an approach where the output of a language model is fed into a grounding model to decode masks and boxes. LISA [7] is a representative model in this category, with a primary focus on various segmentation tasks rather than chat interactions. In many previous works, there has been a trade-off between grounding and chat abilities, with data and evaluation metrics typically emphasizing one of these aspects. In contrast, our dataset and benchmark prioritize assessing the compositional abilities of both grounding and chat interactions, providing a unique perspective in this field.

5 Conclusion

This paper introduced LLaVA-Grounding, an AI assistant that combines visual chat and grounding capabilities. We began by creating a grounded visual chat dataset using a novel data creation pipeline. Subsequently, we proposed an end-to-end model architecture that integrates a grounding model with a Language Model (LM) for effective grounding. Additionally, we introduced Grounding-Bench as a comprehensive benchmark for evaluating grounded visual chat performance, covering both chat and grounding aspects. Our experiments demonstrated that LLaVA-Grounding consistently outperforms other open-source LM models in both chat and grounding tasks, showcasing its effectiveness. Furthermore, LLaVA-Grounding excelled in traditional grounding benchmarks, highlighting its versatility. However, we acknowledge that LLaVA-Grounding has limitations in terms of semantic scope, and future work could explore extending the dataset and data labeling methods to open-vocabulary settings.

References

1. Chen, C., et al.: Position-enhanced visual instruction tuning for multimodal large language models (2023)
2. Chen, J., et al.: MiniGPT-v2: large language model as a unified interface for vision-language multi-task learning. arXiv:2310.09478 (2023)
3. Chen, K., Zhang, Z., Zeng, W., Zhang, R., Zhu, F., Zhao, R.: Shikra: unleashing multimodal LLM's referential dialogue magic. arXiv preprint arXiv:2306.15195 (2023)
4. Kamath, A., Singh, M., LeCun, Y., Synnaeve, G., Misra, I., Carion, N.: MDETR – modulated detection for end-to-end multi-modal understanding (2021)
5. Kazemzadeh, S., Ordonez, V., Matten, M., Berg, T.: ReferitGame: referring to objects in photographs of natural scenes. In: Proceedings of the 2014 Conference on Empirical Methods in Natural Language Processing (EMNLP), pp. 787–798 (2014)
6. Krishna, R., et al.: Visual Genome: connecting language and vision using crowd-sourced dense image annotations (2016)
7. Lai, X., et al.: LISA: reasoning segmentation via large language model. arXiv preprint arXiv:2308.00692 (2023)
8. Li, B., Zhang, Y., Chen, L., Wang, J., Yang, J., Liu, Z.: Otter: a multi-modal model with in-context instruction tuning. arXiv preprint arXiv:2305.03726 (2023)
9. Li, C., et al.: LLaVA-Med: training a large language-and-vision assistant for biomedicine in one day. arXiv preprint arXiv:2306.00890 (2023)
10. Li, L.H., et al.: Grounded language-image pre-training. In: CVPR (2022)
11. Lin, T.Y., et al.: Microsoft COCO: common objects in context. In: ECCV (2014)
12. Liu, C., Ding, H., Jiang, X.: GRES: generalized referring expression segmentation. In: Proceedings of the IEEE/CVF Conference on Computer Vision and Pattern Recognition, pp. 23592–23601 (2023)
13. Liu, H., Li, C., Li, Y., Lee, Y.J.: Improved baselines with visual instruction tuning (2023)
14. Liu, H., Li, C., Wu, Q., Lee, Y.J.: Visual instruction tuning. arXiv preprint arXiv:2304.08485 (2023)
15. Liu, J., et al.: PolyFormer: referring image segmentation as sequential polygon generation (2023)
16. Liu, S., et al.: LLaVA-Plus: learning to use tools for creating multimodal agents (2023)
17. Liu, S., et al.: Grounding DINO: marrying DINO with grounded pre-training for open-set object detection. arXiv preprint arXiv:2303.05499 (2023)
18. Liu, Z., et al.: InternGPT: solving vision-centric tasks by interacting with ChatGPT beyond language (2023)
19. OpenAI: GPT-4 technical report (2023)
20. OpenAI: GPT-4 technical report (2023)
21. OpenAI: GPT-4v(ision) system card (2023). https://cdn.openai.com/papers/GPTV_System_Card.pdf
22. Peng, Z., et al.: Kosmos-2: grounding multimodal large language models to the world. arXiv preprint arXiv:2306.14824 (2023)

23. Plummer, B.A., Wang, L., Cervantes, C.M., Caicedo, J.C., Hockenmaier, J., Lazebnik, S.: Flickr30k Entities: collecting region-to-phrase correspondences for richer image-to-sentence models. In: ICCV (2015)
24. Touvron, H., et al.: LLaMA: open and efficient foundation language models. arXiv preprint arXiv:2302.13971 (2023)
25. Wang, W., et al.: CogVLM: visual expert for pretrained language models (2023)
26. Wang, W., et al.: VisionLLM: large language model is also an open-ended decoder for vision-centric tasks (2023)
27. Yang, J., Zhang, H., Li, F., Zou, X., Li, C., Gao, J.: Set-of-mark prompting unleashes extraordinary visual grounding in GPT-4V (2023)
28. Yang, Z., et al.: UniTAB: unifying text and box outputs for grounded vision-language modeling (2022)
29. Ye, J., et al.: mPLUG-DocOwl: modularized multimodal large language model for document understanding. arXiv preprint arXiv:2307.02499 (2023)
30. You, H., et al.: Ferret: refer and ground anything anywhere at any granularity (2023)
31. Yu, L., Poirson, P., Yang, S., Berg, A.C., Berg, T.L.: Modeling context in referring expressions (2016)
32. Zhang, H., et al.: A simple framework for open-vocabulary segmentation and detection. arXiv preprint arXiv:2303.08131 (2023)
33. Zhang, R., et al.: LLaMA-Adapter: efficient fine-tuning of language models with zero-init attention (2023)
34. Zhang, S., et al.: GPT4RoI: instruction tuning large language model on region-of-interest. arXiv preprint arXiv:2307.03601 (2023)
35. Zhao, Y., Lin, Z., Zhou, D., Huang, Z., Feng, J., Kang, B.: BuboGPT: enabling visual grounding in multi-modal LLMs. arXiv preprint arXiv:2307.08581 (2023)
36. Zhu, D., Chen, J., Shen, X., Li, X., Elhoseiny, M.: MiniGPT-4: enhancing vision-language understanding with advanced large language models. arXiv preprint arXiv:2304.10592 (2023)

Prompt-Driven Contrastive Learning for Transferable Adversarial Attacks

Hunmin Yang[1,2], Jongoh Jeong[1], and Kuk-Jin Yoon[1(✉)]

[1] KAIST, Daejeon, South Korea
kjyoon@kaist.ac.kr
[2] ADD, Daejeon, South Korea
https://PDCL-Attack.github.io

Abstract. Recent vision-language foundation models, such as CLIP, have demonstrated superior capabilities in learning representations that can be transferable across diverse range of downstream tasks and domains. With the emergence of such powerful models, it has become crucial to effectively leverage their capabilities in tackling challenging vision tasks. On the other hand, only a few works have focused on devising adversarial examples that transfer well to both unknown domains and model architectures. In this paper, we propose a novel transfer attack method called **PDCL-Attack**, which leverages the CLIP model to enhance the transferability of adversarial perturbations generated by a generative model-based attack framework. Specifically, we formulate an effective prompt-driven feature guidance by harnessing the semantic representation power of text, particularly from the ground-truth class labels of input images. To the best of our knowledge, we are the first to introduce prompt learning to enhance the transferable generative attacks. Extensive experiments conducted across various cross-domain and cross-model settings empirically validate our approach, demonstrating its superiority over state-of-the-art methods.

Keywords: Adversarial Attack · Transferability · Foundation Model · Vision-Language Model · Prompt Learning · Contrastive Learning

1 Introduction

Regarding the training of deep neural networks, it is typically assumed that the training and testing data are independent and identically distributed. This common principle may impair the generalization of trained models in situations involving domain shifts, which are frequently encountered in real-world deployments. To address this issue, numerous approaches have been proposed in the fields of domain adaptation and generalization. They fundamentally share a common objective: to extract domain-invariant semantic features that prove to be effective for target domains and downstream tasks.

Supplementary Information The online version contains supplementary material available at https://doi.org/10.1007/978-3-031-72775-7_3.

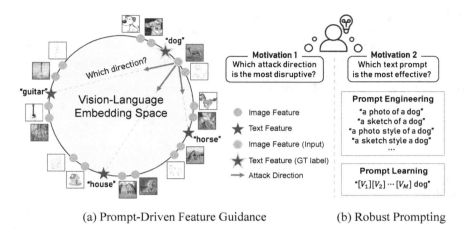

Fig. 1. Our motivation. In a joint vision-language space, a single text can encapsulate core semantics that align with numerous images from diverse domains. Leveraging this principle, our approach utilizes representative prompt-driven text features to enhance the transferable adversarial attacks. On the adversary's side, two clear challenges arise: (a) Generating effective prompt-driven feature guidance, and (b) Identifying robust prompts which maximize the effectiveness.

Another related but distinct area of research delves into the transferability aspect of adversarial examples. While these examples inherently possess somewhat transferable characteristics, naïvely crafted adversarial examples are often limited in their transferability, particularly in strict black-box settings. To address this challenge, researchers have explored transfer-based attacks with the aim of enhancing the transferability of adversarial examples. Notably, recent generative model-based attacks [2,31,32,36,40,49,54] showcase superior adversarial transferability compared to iterative approaches [3,4,8,9,14,25–29,33–35,43,48]. Furthermore, generative attacks possess a distinct advantage in inference time complexity over instance-specific iterative attacks, highlighting their practical utility in real-world scenarios. GAP [36] initially established a generative attack framework, and CDA [31] introduced domain-agnostic generation of adversarial examples by employing relativistic cross-entropy loss. LTP [40] and BIA [54] utilized the mid-layer features of surrogate models, which contain shared information across different architectures, making them effective for targeting various victim models. FACL-Attack [49] enhanced the transferability by employing a feature contrastive approach in the frequency domain. Several other studies have employed vision-language models [2] and object-centric features [1] to acquire additional feature guidance for multi-object scene attacks. As outlined in Table 1, attack strategies have evolved as adversaries leverage advanced open-source models, resulting in more effective feature guidance.

In pursuit of acquiring highly generalizable representations, vision-language foundation models [22,37,50–53] have emerged. These models have undergone

Table 1. Evolving attack strategies. This summarizes characteristics of generative model-based transfer attacks in a chronological order. Adversaries are equipped with more powerful tools such as mid-layer features of surrogate model [40,54] and vision-language foundation model [2]. Our work further enhances the attack effectiveness by designing a novel prompt-driven attack loss (**Ours†**) and employing a learnable prompt (**Ours**). Please refer to Table 4 for comparative evaluation results.

Attack	Surrogate Model			Vision-Language Foundation Model				Prompt Type	
	Output Logits		Features	Image Features		Text Features			
	Absolute	Relative	Mid-layer	Clean	Adv.	GT	Adv.	Heuristic	Learnable
GAP [36]	✓	–	–	–	–	–	–	–	–
CDA [31]	✓	✓	–	–	–	–	–	–	–
LTP [40]	–	–	✓	–	–	–	–	–	–
BIA [54]	–	–	✓	–	–	–	–	–	–
GAMA [2]	–	–	✓	✓	–	–	✓	✓	–
Ours†	–	–	✓	✓	✓	✓	✓	✓	–
Ours	–	–	✓	✓	✓	✓	✓	–	✓

large-scale training on vast amounts of web-scale data, yielding versatile features that serve as powerful priors for various downstream tasks. In this context, we posited that generative attacks could also harness the capabilities of these potent models as illustrated in Fig. 1. Our work draws inspiration from recent studies [6, 11,12,21] that employ CLIP [37] as a tool for text-driven feature manipulation and domain extension. These works have validated the superior representation power of CLIP, emphasizing a key insight that text can function as a prototype for various styles of images within a joint vision-language space. PromptStyler [6] further employs prompt learning [55,56] to improve the prompt-driven domain generalization performance.

Building upon this insight, we introduce a novel generative attack method that leverages CLIP [37] and prompt learning [56] to effectively train the perturbation generator. Our method, dubbed **PDCL-Attack**, integrates both pre-trained CLIP image and text encoders into the generative attack framework, further enhancing the effectiveness by employing a learnable prompt. Our work is differentiated from GAMA [2] in that we have re-designed the contrastive loss by separating the heterogeneous surrogate and CLIP features, and employing text features extracted from the ground-truth (GT) class labels of input images. Our method incorporates perturbed image features from CLIP, which can provide effective loss gradients for training the robust generator. To fully unleash the power of prompt-driven features, we pre-train the context words of the prompt to enhance the robustness of CLIP to distribution shifts. In comparison to GAMA [2], our method with enhanced attack loss design (**Ours†**, 1.87%p ↑) and incorporating prompt learning (**Ours**, 4.65%p ↑) demonstrates significant improvements. Our contributions are summarized as follows:

- We propose a novel prompt-driven contrastive learning (PDCL) method to enhance the training of the perturbation generator by leveraging the semantic representation power of text features from class labels of input images.
- This work is the first attempt to introduce prompt learning with a vision-language model to enhance the generative model-based transfer attacks.
- PDCL-Attack achieves state-of-the-art attack transferability across various domains and model architectures, surpassing previous approaches.

2 Related Work

Generative Model-Based Attack. Generative attacks [1,2,31,32,36,40,49,54] leverage adversarial training [13], using a pre-trained surrogate model as a discriminator to generate adversarial perturbations effective across entire data distributions. This approach is computationally efficient and advantageous, as it can simultaneously generate diverse forms of perturbations across multiple images. CDA [31] seeks to enhance the training of the generator by incorporating both the cross-entropy (CE) and the relativistic CE loss. The work initially introduced domain-agnostic perturbations as well as model-agnostic ones. LTP [40] and BIA [54] utilize features extracted from the mid-level layers of the surrogate model, which have been examined to contain a higher degree of shared information among various model architectures. FACL-Attack [49] exploits frequency domain manipulations to boost the attack transferability. Several studies have utilized vision-language models [2] and object-centric features [1] to acquire effective features for attacking multi-object scenes. Our work further enhances the training by employing a vision-language model and prompt learning.

Vision-Language Foundation Model. Recent advancements in large vision-language models (VLM) [22,37,50–53] have demonstrated superior capabilities in learning generic representations. This significant progress has been made possible by leveraging enormous web-scale training datasets. In particular, Contrastive Language Image Pre-training (CLIP) [37] employs 400 million image-text pairs for contrastive pre-training of image and text encoders within a joint vision-language embedding space. The zero-shot CLIP model of ViT-L/14 exhibits impressive image classification performance, achieving a top-1 accuracy of 76.2% on ImageNet-1K [39], on par with that of a fully-supervised ResNet101 [16]. CLIP demonstrates superior domain generalization capability compared to supervised trained models. For ImageNet-Sketch [47], zero-shot CLIP achieves 60.2%, while the fully-supervised ResNet101 only attains 25.2%. Furthermore, context optimization [55,56] has further improved the few-shot performance of the employed CLIP model. Interestingly, this trained context enhances CLIP's robustness to distribution shifts, as highlighted in [56]. Inspired by this line of research, our work incorporates CLIP and prompt learning into the generative attack framework to enable more effective transfer attacks.

Text-Driven Feature Guidance. As both image and text features are mapped into a joint vision-language embedding space as in CLIP [37], various vision tasks focused on generalization or adaptation have benefited from

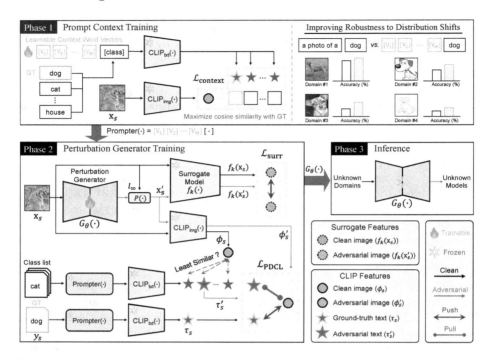

Fig. 2. Overview of PDCL-Attack. For effective transfer attacks leveraging CLIP [37], our proposed pipeline consists of three serial stages; Phase 1 and 2 are the training stage, and Phase 3 is the inference stage. The goal of Phase 1 is to pre-train Prompter($\cdot$), optimizing the context words $[\mathbf{V}_1][\mathbf{V}_2]\cdots[\mathbf{V}_M]$ to yield generalizable text features in Phase 2. In Phase 1, only the learnable context word vectors are updated, while the weights of the CLIP image encoder $\text{CLIP}_{\text{img}}(\cdot)$ and text encoder $\text{CLIP}_{\text{txt}}(\cdot)$ remain fixed. In Phase 2, we train a generator model $G_\theta(\cdot)$ which crafts adversarial perturbations for encouraging a surrogate model $f_k(\cdot)$ to produce mispredictions for input images $\mathbf{x}_s$. The generator $G_\theta(\cdot)$ crafts the ℓ_∞-budget bounded adversarial image $\mathbf{x}'_s$ via a perturbation projector $P(\cdot)$. In Phase 3, we employ the trained generator $G_\theta(\cdot)$ to yield transferable adversarial examples on unknown domains and victim models.

text-driven feature guidance. LADS [11] introduced a method for extending the trained model to new domains based solely on language descriptions of distribution shifts. PODA [12] introduced a straightforward feature augmentation method for zero-shot domain adaptation, guided by a single textual description of the target domain. PromptStyler [6] proposed a text-driven style generation method to enhance the domain generalization performance. From the perspective of utilizing generic text features to guide model training, our work aligns with these efforts. We aim to explore an effective text-driven feature contrastive method to enhance the training of the perturbation generator towards a more robust regime.

3 Proposed Approach: PDCL-Attack

Problem Definition. Transfer attacks aim to craft adversarial examples which are transferable in black-box settings. This task aims to create adversarial examples that are not just effective against a single surrogate model but also capable of deceiving other victim models, even those with different architectures or trained on different data distributions. To that end, crafting transferable adversarial examples involves identifying vulnerabilities that are shared among diverse models and domains, effectively addressing the challenge of generalization. Our goal is to devise an effective transfer attack method that can train a robust perturbation generator model, denoted as $G_\theta(\cdot)$, to generate adversarial examples which are transferable to arbitrary domains and model architectures.
Overview of PDCL-Attack. Overall framework of our proposed method is shown in Fig. 2. PDCL-Attack introduces a novel method for training a perturbation generator by leveraging CLIP [37] and prompt learning [56]. Our attack pipeline consists of three sequential stages: Phases 1 and 2 constitute the training stage, whereas Phase 3 corresponds to the inference stage. The objective of Phase 1 is to pre-train a Prompter($\cdot$) which can extract generalizable text features. In Phase 1, learnable context word vectors of Prompter($\cdot$) are trained through few-shot learning by minimizing the cross-entropy loss on pairs of input images and their corresponding class labels. In Phase 2, we train a perturbation generator $G_\theta(\cdot)$ which can craft adversarial perturbations that cause mispredictions of the surrogate model $f_k(\cdot)$. By inputting the clean images $\mathbf{x}_s$, the generator $G_\theta(\cdot)$ crafts the ℓ_∞-bounded adversarial images $\mathbf{x}'_s$ via a perturbation projector $P(\cdot)$. In Phase 3, we use the trained generator $G_\theta(\cdot)$ to craft transferable adversarial examples in black-box scenarios.

3.1 Phase 1: Prompt Context Training

The objective of Phase 1 is to train learnable context words used in Prompter($\cdot$), facilitating the extraction of generalizable text features in Phase 2. To fully harness the capabilities of CLIP [37] model, recent studies [55,56] have shown the effectiveness of leveraging context optimization. These methods have demonstrated enhanced performance of the CLIP model across various distribution-shifting scenarios. We adhere to the recent protocol outlined in [55,56] and integrate it to enhance our attack framework. We utilize the same training dataset (*i.e.*, ImageNet-1K [39] which contains diverse distribution shifts) as in Phase 2.
Optimal Context Training. We model the context words used in Prompter($\cdot$) via learnable continuous vectors $[\mathbf{V}_1][\mathbf{V}_2]\cdots[\mathbf{V}_M]$ as follows:

$$\text{Prompter}(\cdot) = [\mathbf{V}_1][\mathbf{V}_2]\cdots[\mathbf{V}_M][\ \cdot\ ], \qquad (1)$$

where each $[\mathbf{V}_m]$ ($m \in \{1, ..., M\}$) is a trainable 512-dimensional float vector with the same word embedding size of CLIP [37], and M is the number of context words. The training is exclusively focused on these word vectors, whereas the weights of the CLIP image encoder $\text{CLIP}_{\text{img}}(\cdot)$ and text encoder $\text{CLIP}_{\text{txt}}(\cdot)$

Algorithm 1. Perturbation Generator Training (Phase 2)

Require: Source domain data distribution $\mathcal{X}_s$, perturbation generator $G_\theta(\cdot)$, perturbation projector $P(\cdot)$, perturbation budget ϵ, surrogate model $f_k(\cdot)$, CLIP image encoder $\text{CLIP}_{\text{img}}(\cdot)$, CLIP text encoder $\text{CLIP}_{\text{txt}}(\cdot)$, `Prompter`$(\cdot)$
Ensure: Freeze the pre-trained $f_k(\cdot)$, $\text{CLIP}_{\text{img}}(\cdot)$, $\text{CLIP}_{\text{txt}}(\cdot)$, `Prompter`$(\cdot)$
1: Randomly initialize $G_\theta(\cdot)$
2: **repeat**
3: Randomly sample a mini-batch $\mathbf{x}_s \sim \mathcal{X}_s$ with batch size n
4: Forward pass $\mathbf{x}_s$ through $G_\theta(\cdot)$ and generate unbounded adversarial examples $\tilde{\mathbf{x}}'_s = G_\theta(\mathbf{x}_s)$
5: Bound the adversarial examples using $P(\cdot)$ such that:
$$\|P(\tilde{\mathbf{x}}'_s) - \mathbf{x}_s\|_\infty \leq \epsilon$$
6: Forward pass $\mathbf{x}'_s = P(\tilde{\mathbf{x}}'_s)$ and $\mathbf{x}_s$ through $f_k(\cdot)$
7: Compute the surrogate model loss $\mathcal{L}_{\text{surr}}$ using Eq. (4)
8: Forward pass $\mathbf{x}'_s$ through $\text{CLIP}_{\text{img}}(\cdot)$
9: Forward pass class labels through $\text{CLIP}_{\text{txt}}(\cdot)$ via `Prompter`$(\cdot)$
10: Compute the prompt-driven contrastive loss $\mathcal{L}_{\text{PDCL}}$ using Eq. (8)
11: Compute the total loss $\mathcal{L}$ using Eq. (9)
$$\mathcal{L} = \mathcal{L}_{\text{surr}} + \mathcal{L}_{\text{PDCL}}$$
12: Backward pass and update $G_\theta(\cdot)$
13: **until** $G_\theta(\cdot)$ converges

remain unchanged. The training objective is to optimize $[\mathbf{V}_1][\mathbf{V}_2]\cdots[\mathbf{V}_M]$ by minimizing the cross-entropy loss computed using pairs of input images and GT class labels through few-shot learning. When the input images $\mathbf{x}_s$ are passed through $\text{CLIP}_{\text{img}}(\cdot)$, the produced image features can be represented as ϕ_s. With K class labels, we can obtain text features $\{\tau_i\}_{i=1}^K$ by feeding the class labels into $\text{CLIP}_{\text{txt}}(\cdot)$. Then, the prediction probability can be computed as:

$$p(y = i|\mathbf{x}_s) = \frac{\exp(\cos(\tau_i, \phi_s)/\lambda)}{\sum_{j=1}^K \exp(\cos(\tau_j, \phi_s)/\lambda)}, \quad (2)$$

where λ is the temperature parameter learned by CLIP [37], and $\cos(\cdot, \cdot)$ denotes the standard cosine similarity.

Using the prediction probability and a one-hot encoded label vector l, the context training loss can be computed as follows:

$$\mathcal{L}_{\text{context}} = -\sum_{j=1}^K l_j \cdot \log p(y = j), \quad (3)$$

where the trained `Prompter`$(\cdot)$ along with $\text{CLIP}_{\text{txt}}(\cdot)$ is utilized in Phase 2.

3.2 Phase 2: Perturbation Generator Training

This phase aims to train a perturbation generator model $G_\theta(\cdot)$ capable of crafting transferable perturbations as described in Algorithm 1. We randomly initialize the generator $G_\theta(\cdot)$. Given a batch of clean images $\mathbf{x}_s$ with a batch size of n, $G_\theta(\cdot)$ generates unbounded adversarial examples. Subsequently, these are constrained by the perturbation projector $P(\cdot)$ within a predefined budget of ϵ.

Surrogate Model Loss $\mathcal{L}_{\text{surr}}$. The generated adversarial images $\mathbf{x}'_s$ and clean images $\mathbf{x}_s$ are passed through the surrogate model $f_k(\cdot)$, where we extract k-th mid-layer features. We define the surrogate model loss $\mathcal{L}_{\text{surr}}$ as follows:

$$\mathcal{L}_{\text{surr}} = \cos(f_k(\mathbf{x}'_s), f_k(\mathbf{x}_s)), \tag{4}$$

where $\cos(\cdot,\cdot)$ denotes the standard cosine similarity.

Prompt-Driven Contrastive Loss $\mathcal{L}_{\text{PDCL}}$. As CLIP [37] employs contrastive learning to learn representations in a joint vision-language embedding space, we design a contrastive learning [7,15] based attack loss for effectively harnessing CLIP features. Our loss design differs from GAMA [2] since we separate heterogeneous surrogate and CLIP features, and utilizes representative text features extracted from ground-truth (GT) class labels of input images. Additionally, our method leverages CLIP's perturbed image features, which can provide effective loss gradients. Given clean images $\mathbf{x}_s$ of batch size n, we first feed $\mathbf{x}_s$ into the CLIP image encoder $\text{CLIP}_{\text{img}}(\cdot)$ for acquiring clean image features ϕ_s. We input $\mathbf{x}_s$ into $G_\theta(\cdot)$ followed by $P(\cdot)$ and craft adversarial images by $\mathbf{x}'_s = P(G_\theta(\mathbf{x}_s))$. We then pass $\mathbf{x}'_s$ through $\text{CLIP}_{\text{img}}(\cdot)$, and CLIP's adversarial image features ϕ'_s can be obtained as follows:

$$\phi'_s = \text{CLIP}_{\text{img}}(\mathbf{x}'_s). \tag{5}$$

Since we utilize a large-scale ImageNet-1K [39] dataset for training, we have corresponding GT class labels y_s for each $\mathbf{x}_s$. If there are K classes in the training dataset (e.g., $K = 1,000$ for ImageNet-1K), we can obtain the corresponding set of K text features using the CLIP text encoder $\text{CLIP}_{\text{txt}}(\cdot)$. In contrast to GAMA [2], we pass y_s as input to the pre-trained $\text{Prompter}(\cdot)$, which yields more effective text-driven prompt inputs for $\text{CLIP}_{\text{txt}}(\cdot)$. Then, text features τ_s extracted from the GT class labels of input images can be computed as follows:

$$\tau_s = \text{CLIP}_{\text{txt}}(\text{Prompter}(y_s)). \tag{6}$$

With adversarial image features ϕ'_s as the anchor point, we employ adversarial text features τ'_s as the positives, which are computed by identifying the least similar text features compared to the input image features ϕ_s. Specifically, we compute a set of text features denoted as $\mathcal{T} = [\mathcal{T}_1, \mathcal{T}_2, \cdots, \mathcal{T}_K]$ using the class names of all K classes, where each $\mathcal{T}_i$ corresponds to $\text{CLIP}_{\text{txt}}(\text{Prompter}(y_i))$. For each batch, we randomly select $n = 16$ candidates from $\mathcal{T}$ for computational efficiency and identify the least similar candidate, using its label as y'_s. We define the adversarial text features τ'_s as follows:

$$\tau'_s = \text{CLIP}_{\text{txt}}(\text{Prompter}(y'_s)). \tag{7}$$

We use ϕ'_s from Eq. (5) as the anchor, τ_s from Eq. (6) as the negatives, and τ'_s from Eq. (7) as the positives to constitute our contrastive loss. Finally, our prompt-driven contrastive loss can be formulated as follows:

$$\mathcal{L}_{\text{PDCL}} = \|\phi'_s - \tau'_s\|_2^2 + \max(0, \alpha - \|\phi'_s - \tau_s\|_2)^2, \qquad (8)$$

where α denotes the desired margin between ϕ'_s and τ_s. All the features are ℓ_2-normalized before the loss calculation. As a result, $\mathcal{L}_{\text{PDCL}}$ facilitates more robust training of the generator $G_\theta(\cdot)$ by leveraging the prototypical semantic characteristics of text-driven features, boosted by the frozen robust Prompter$(\cdot)$.

Total Loss $\mathcal{L}$. Our generator $G_\theta(\cdot)$ is trained by minimizing $\mathcal{L}$ as follows:

$$\mathcal{L} = \mathcal{L}_{\text{surr}} + \mathcal{L}_{\text{PDCL}}, \qquad (9)$$

where the total loss $\mathcal{L}$ facilitates $G_\theta(\cdot)$ to be trained towards more robust regime via effective prompt-driven feature guidance.

3.3 Phase 3: Inference Using the Frozen Generator

Since our perturbation generator $G_\theta(\cdot)$ has been trained robustly, we freeze it for the final inference stage of crafting adversarial examples. With the frozen $G_\theta(\cdot)$, we can generate adversarial examples by applying it to arbitrary image inputs, even if they belong to target data distributions different from the source domain. We assess the performance of the trained $G_\theta(\cdot)$ in black-box scenarios across diverse domains and model architectures. The generated adversarial examples are expected to exhibit superior cross-domain and cross-model transferability.

4 Experiments

4.1 Experimental Setting

Datasets and Attack Scenarios. Since our goal is to generate adversarial examples which show high transferability across various domains and models, we carry out experiments in challenging black-box scenarios, namely *cross-domain* and *cross-model* settings. Building upon a recent work [54] that demonstrated remarkable attack transferability by employing a large-scale training dataset (*i.e.*, ImageNet-1K [39]), we also leverage ImageNet-1K to train our perturbation generator and prompter. The efficacy of the trained generator is assessed by conducting evaluations on three additional datasets: CUB-201-2011 [46], Stanford Cars [24], and FGVC Aircraft [30]. Specifically for the *cross-domain* setting, we evaluate our method on unknown target domains (*i.e.*, CUB-201-2011, Stanford Cars, FGVC Aircraft) and victim models distinct from both the source domain (*i.e.*, ImageNet-1K) and the surrogate model. For the *cross-model* setting, we evaluate our method against black-box models with varying architectures, maintaining a white-box domain setup using ImageNet-1K.

Surrogate and Victim Models. The perturbation generator is trained on ImageNet-1K against a pre-trained surrogate model of VGG-16 [41]. For the *cross-domain* setting, we employ fine-grained classification victim models trained by DCL framework [5]. These models are based on three different backbones: ResNet50 (Res-50) [16], SENet154, and SE-ResNet101 (SE-Res101) [19]. For the *cross-model* setting, we explore various different model architectures such as Res-50, ResNet152 (Res-152) [16], DenseNet121 (Dense-121), DenseNet169 (Dense-169) [20], Inception-v3 (Inc-v3) [42], MNasNet [44], and ViT [10].

Implementation Details. We adhere closely to the implementations employed in recent generative model-based attacks [2,54] to ensure a fair comparison. The mid-layer from which we extract features from the surrogate model (*i.e.*, VGG-16 [41]) is *Maxpool.3*. For the CLIP [37] model, we use ViT-B/16 [10] for the image encoder and Transformer [45] for the text encoder, consistent with the configurations used in GAMA [2]. We train the perturbation generator using Adam optimizer [23] with $\beta_1 = 0.5$ and $\beta_2 = 0.999$. The learning rate is set to 0.0002, and we use a batch size of 16 for a single epoch training. The perturbation budget for crafting the adversarial image is constrained to $\ell_\infty \leq 10$. We use a contrastive loss margin of $\alpha = 1$. Regarding the prompt context training of Phase 1, we follow the standard protocol [37,56]. Learnable word vectors are randomly initialized by a zero-mean Gaussian distribution with standard deviation of 0.02. We use a SGD optimizer with a learning rate of 0.002, and train the word vectors with a maximum epoch of 50 by 16-shot learning on ImageNet-1K [39]. The number of context words M is set to 16 and related analyses are shown in Table 4. We also employ distribution-shifted versions of ImageNet (-V2 [38], -Sketch [47], -A [18], -R [17]) to identify the robustness of the learned context words. More details are provided in the Supplementary Material.

Evaluation Metric and Competitors. Our primary evaluation metric for assessing the attack effectiveness is the top-1 classification accuracy. The competitors include state-of-the-art generative model-based attacks, such as GAP [36], CDA [31], LTP [40], BIA [54], and GAMA [2]. We train all the baselines [2,31,36,40,54] on the same ImageNet-1K dataset for fair comparison.

4.2 Main Results

Cross-Domain Evaluation Results. We compare our method with the state-of-the-art generative model-based attacks [2,31,36,40,54] on various black-box domains with black-box models. In the training stage, we utilize ImageNet-1K [39] as the source domain to train a perturbation generator model against a pre-trained surrogate model of VGG-16 [41]. In the inference stage, we evaluate the trained perturbation generator on various unknown target domains, namely CUB-200-2011 [46], Stanford Cars [24], and FGVC Aircraft [30], using different victim model architectures. Specifically, we employ several fine-grained classification models that have been trained using the DCL framework [5]. These victim models are based on three different backbones: Res-50 [16], SENet154 and SE-ResNet101 (SE-Res101) [19]. We assess the effectiveness of our trained perturbation generator in this challenging black-box attack scenario.

As shown in Table 2, our method exhibits superior attack effectiveness with significant margins on most cross-domain benchmarks, which are also cross-model. This highlights the robust and potent transferability of our crafted adversarial examples, enabled by prompt-driven feature guidance and prompt learning to maximize its effectiveness. We conjecture that the remarkable generalization ability of PDCL-Attack might be attributed to the synergy between our two proposed methods: harnessing the superior representation power of CLIP's text features and improving it further with a prompt learning method to produce generalizable text features. In essence, our approach indeed enhances the perturbation generator's capability to generalize across various black-box domains and state-of-the-art model architectures.

Table 2. Cross-domain evaluation results. The perturbation generator is trained on ImageNet-1K [39] with VGG-16 [41] as the surrogate model and evaluated on black-box domains with models. We compare the top-1 classification accuracy after attacks (↓ is better) with the perturbation budget of $\ell_\infty \leq 10$. **Best** and second best.

Method	CUB-200-2011			Stanford Cars			FGVC Aircraft			AVG.
	Res-50	SENet154	SE-Res101	Res-50	SENet154	SE-Res101	Res-50	SENet154	SE-Res101	
Clean	87.33	86.81	86.59	94.25	93.35	92.96	92.14	92.05	91.84	90.81
GAP [36]	68.85	74.11	72.73	85.64	84.34	87.84	81.40	81.88	76.90	79.30
CDA [31]	69.69	62.51	71.00	75.94	72.45	84.64	71.53	58.33	63.39	69.94
LTP [40]	30.86	52.50	62.86	34.54	65.53	73.88	15.90	60.37	52.75	49.91
BIA [54]	32.74	52.99	58.04	39.61	69.90	70.17	28.92	60.31	46.92	51.07
GAMA [2]	34.47	54.02	57.66	30.16	69.80	63.82	25.29	58.42	**43.41**	48.56
Ours	**22.97**	**49.19**	**54.92**	**22.58**	**64.95**	**63.70**	**15.81**	**53.83**	47.25	**43.91**

Cross-Model Evaluation Results. While our method demonstrates the effectiveness in enhancing the attack transferability in the strict black-box scenario as shown in Table 2, we further conducted investigations in a controlled white-box domain scenario, specifically within ImageNet-1K [39]. We train a generator against a surrogate model of VGG-16 [41], and subsequently assess its effectiveness on victim models with various architectures such as ResNet50 (Res-50), ResNet152 (Res-152) [16], DenseNet121 (Dense-121), DenseNet169 (Dense-169) [20], Inception-v3 (Inc-v3) [42], MNasNet [44], and ViT [10].

As shown in Table 3, our method mostly achieves competitive performance even in the white-box domain scenario. We conjecture that our CLIP-driven guidance, coupled with optimized prompt context, can improve the training of the generator in crafting more resilient perturbations, ultimately showcasing enhanced generalization capabilities in unknown feature spaces of victim models. It is noteworthy that incorporating the pre-trained CLIP ViT-B/16 [10] image encoder alongside the surrogate model might be one of the factors which could enhance the ViT-based transferability. Nonetheless, compared to GAMA [2] which also utilizes the same CLIP encoder backbone, our method surpasses it particularly on ViT-based model transferability. We posit that incorporating text feature guidance from GT labels and adversarial image-driven loss gradients further improves the training of the perturbation generator.

Table 3. Cross-model evaluation results. The perturbation generator is trained on ImageNet-1K [39] with VGG-16 [41] as the surrogate model and evaluated on black-box models. We compare the top-1 classification accuracy after attacks (↓ is better) with the perturbation budget of $\ell_\infty \leq 10$. **Best** and second best.

Method	Res-50	Res-152	Dense-121	Dense-169	Inc-v3	MNasNet	ViT-B/16	ViT-L/16	AVG.
Clean	74.61	77.34	74.22	75.75	76.19	66.49	79.56	80.86	75.63
GAP [36]	57.87	65.50	57.94	61.37	63.30	42.47	72.89	76.69	54.34
CDA [31]	36.27	51.05	38.89	42.67	54.02	33.10	68.73	74.22	53.24
LTP [40]	<u>21.70</u>	<u>39.88</u>	<u>23.42</u>	**25.46**	41.27	45.28	72.44	76.75	43.28
BIA [54]	25.36	42.98	26.97	32.35	41.20	34.31	<u>67.05</u>	73.23	42.93
GAMA [2]	24.82	43.22	24.84	30.81	<u>35.10</u>	**27.96**	67.33	<u>73.16</u>	<u>40.91</u>
Ours	**20.87**	**38.62**	**21.26**	<u>29.01</u>	**32.99**	<u>28.00</u>	**65.53**	**72.52**	**38.60**

Table 4. Ablation study on our proposed losses. With the same surrogate model loss $\mathcal{L}_\text{surr}$, **Ours**† using a hand-crafted prompt (*i.e.*, "a photo of a [class]") outperforms $\mathcal{L}_\text{GAMA}$ [2] by our improved CLIP-guidance loss $\mathcal{L}_\text{PDCL}$. **Ours** using a learned prompt (*i.e.*, "[V$_1$][V$_2$]···[V$_{16}$] [class]") further enhances the attack effectiveness by our context training loss $\mathcal{L}_\text{context}$. **Best** and second best.

Method	$\mathcal{L}_\text{surr}$	$\mathcal{L}_\text{GAMA}$	$\mathcal{L}_\text{PDCL}$	$\mathcal{L}_\text{context}$	Cross-Domain	Cross-Model
Clean	–	–	–	–	90.85	75.63
BIA [54]	✓	–	–	–	51.07	42.93
GAMA [2]	✓	✓	–	–	48.56	40.91
Ours†	✓	–	✓	–	<u>46.69</u>	<u>40.35</u>
Ours	✓	–	✓	✓	**43.91**	**38.60**

4.3 More Analyses

Ablation Study on Our Proposed Losses. In transfer-based attack scenarios, the adversary leverages surrogate models to simulate potential unknown victim models and craft adversarial examples. In this work, we additionally leverage a vision-language foundation model (*i.e.*, CLIP [37]) and formulate effective loss functions to enhance the training. In Table 4, we evaluate the effect of our proposed losses used for improving the transferability of adversarial examples in both cross-domain and cross-model scenarios. Remarkably, in **Ours**†, our approach achieves state-of-the-art results even without employing prompt learning. This highlights the effectiveness of our proposed CLIP-driven attack loss $\mathcal{L}_\text{PDCL}$ compared to GAMA [2]. This demonstrates the effectiveness of the text-driven semantic feature guidance from GT labels and back-propagated loss gradients from adversarial CLIP image features. Moreover, incorporating prompt learning $\mathcal{L}_\text{context}$ into the framework further boosts the transferability as demonstrated in **Ours**. When employing Prompter(·) trained with $\mathcal{L}_\text{context}$, the text-driven guidance facilitated by $\mathcal{L}_{\mathrm{PDCL}}$ is strengthened by the extraction of more generalizable features. Compared to the baselines, our method consistently improves both cross-domain and cross-model attack transferability.

Table 5. Effect of learnable context words. Learnable context words outperform hand-crafted heuristic ones, and increasing their capacity further improves the attack effectiveness. **Best** and second best.

Type	# of words	Text Prompt	Accuracy ($\downarrow$)
Heuristic	$M = 4$	"a photo of a [class]"	46.69
		"a sketch of a [class]"	47.02
	$M = 5$	"a photo style of a [class]"	46.14
		"a sketch style of a [class]"	47.70
		"a [$\mathbf{V}_{\mathrm{rand}}$] style of a [class]"	47.81
Learnable	$M = 4$	"[$\mathbf{V}_1$][$\mathbf{V}_2$][$\mathbf{V}_3$][$\mathbf{V}_4$] [class]"	<u>45.44</u>
	$M = 16$	"[$\mathbf{V}_1$][$\mathbf{V}_2$]$\cdots$[$\mathbf{V}_{16}$] [class]"	**43.91**

Effect of Learnable Context Words. In Table 5, we compare our prompt learning method with various hand-crafted text prompts in a cross-domain attack scenario. Using domain-specific heuristic prompts, such as "a [domain] of a [class]" or "a [domain] style of a [class]", CLIP's textual guidance might fall into sub-optimal regime due to the domain-specific overfitting. Considering this, we investigated a domain randomization approach by randomly initializing a word vector, i.e., "a [$\mathbf{V}_{\mathrm{rand}}$] style of a [class]". In this trial, we randomly initialize [$\mathbf{V}_{\mathrm{rand}}$] for each training iteration. However, the effectiveness is comparable to that of the naïve heuristic prompts, indicating the necessity for a meticulously crafted prompt to attain better performance. We conjecture that our superior results might be attributed to the enhanced representational power of generalizable text features attained through learned context words. This also aligns with recent findings [55,56] that prompt learning not only enhances the downstream task performance, it can even improve the robustness of the trained model in distribution-shifted scenarios. Remarkably, increasing the capacity of learnable context words ($M = 16$) further enhances the attack effectiveness.

Attack Effectiveness and Perceptibility. While our work primarily focuses on crafting more effective perturbations, it is also crucial to carefully examine the image quality of the adversarial examples for real-world deployment. Therefore, we investigate how the perturbation budget affects the transferability of cross-domain attacks in Table 6. We compare the top-1 classification accuracy after attacks using the generator trained with the perturbation budget of $\ell_\infty \leq 10$. Across each test-time perturbation budget, our method consistently demonstrates superior attack performance. In other words, ours can achieve higher attack transferability with lower perturbation power and better image

Table 6. Cross-domain attack effectiveness w.r.t. perturbation budget. Given the same test-time perturbation budget, our method consistently demonstrates superior attack effectiveness. **Best** and second best.

$\ell_\infty \leq$	6	7	8	9	10
BIA [54]	78.76	72.32	65.03	57.75	51.07
GAMA [2]	<u>75.23</u>	<u>68.41</u>	<u>61.52</u>	<u>54.66</u>	<u>48.56</u>
Ours	**72.07**	**64.67**	**57.04**	**50.23**	**43.91**

quality, which are significant advantages in real-world scenarios. PDCL-Attack can generate effective and high-quality adversarial images as shown in Fig. 3.

5 Limitations and Broader Impacts

Limitations. Since our method employs a vision-language foundation model, it entails several limitations. Firstly, the attack effectiveness depends on the quality of the pre-trained vision-language model. This could be alleviated with the progress of vision-language foundation models. Secondly, our approach leveraging CLIP [37] incurs higher training computational costs. For instance, our method takes longer training time than BIA [54] (∼33 vs. 12 h). Thirdly, our

Fig. 3. Qualitative results. PDCL-Attack successfully fools the classifier, causing it to predict the clean image labels (in black) as the mispredicted class labels shown at the bottom (in red). From top to bottom: clean images, unbounded adversarial images, and bounded ($\ell_\infty \leq 10$) adversarial images which are *actual inputs* to the classifier. (Color figure online)

method necessitates the availability of ground-truth class labels for text-driven feature guidance. Lastly, we limited our evaluation to an image classification task. We posit that CLIP's generalizable text features could also offer benefits across various tasks, and we defer this exploration to future work.

Broader Impacts. Our research community is facing remarkable progress regarding large language models (LLMs) and multi-modal foundation models. It is crucial to note that adversaries also have access to these powerful open-source models, and potential threats are escalating with their advancement. This work aims to raise awareness of such risks and encourage research into developing methodologies for training more robust models.

6 Conclusion

In this paper, we have introduced a novel generator-based transfer attack method that harnesses the capabilities of a vision-language foundation model and prompt learning. We design an effective attack loss by incorporating image and text features extracted from the CLIP model and integrating these with a surrogate model-based loss. We find that text feature guidance from ground-truth labels and adversarial image-driven loss gradients enhance the training. Extensive evaluation results validate the effectiveness of our approach in black-box scenarios with unknown distribution shifts and variations in model architectures.

Acknowledgements. This work was partially supported by the Agency for Defense Development grant funded by the Korean Government. We thank Junhyeong Cho for his insightful discussions and valuable comments.

References

1. Aich, A., Li, S., Song, C., Asif, M.S., Krishnamurthy, S.V., Roy-Chowdhury, A.K.: Leveraging local patch differences in multi-object scenes for generative adversarial attacks. In: Proceedings of the IEEE/CVF Winter Conference on Applications of Computer Vision, pp. 1308–1318 (2023)
2. Aich, A., et al.: GAMA: generative adversarial multi-object scene attacks. Adv. Neural. Inf. Process. Syst. **35**, 36914–36930 (2022)
3. Carlini, N., Wagner, D.: Towards evaluating the robustness of neural networks. In: 2017 IEEE Symposium on Security and Privacy (SP), pp. 39–57. IEEE (2017)
4. Chen, P.Y., Sharma, Y., Zhang, H., Yi, J., Hsieh, C.J.: EAD: elastic-net attacks to deep neural networks via adversarial examples. In: Proceedings of the AAAI Conference on Artificial Intelligence (2018)
5. Chen, Y., Bai, Y., Zhang, W., Mei, T.: Destruction and construction learning for fine-grained image recognition. In: CVPR (2019)
6. Cho, J., Nam, G., Kim, S., Yang, H., Kwak, S.: PromptStyler: prompt-driven style generation for source-free domain generalization. In: Proceedings of the IEEE/CVF International Conference on Computer Vision (ICCV) (2023)
7. Chuang, C.Y., Robinson, J., Lin, Y.C., Torralba, A., Jegelka, S.: Debiased contrastive learning. Adv. Neural. Inf. Process. Syst. **33**, 8765–8775 (2020)

8. Croce, F., Hein, M.: Reliable evaluation of adversarial robustness with an ensemble of diverse parameter-free attacks. In: International Conference on Machine Learning, pp. 2206–2216. PMLR (2020)
9. Dong, Y., Liao, F., Pang, T., Su, H., Zhu, J., Hu, X., Li, J.: Boosting adversarial attacks with momentum. In: Proceedings of the IEEE Conference on Computer Vision and Pattern Recognition, pp. 9185–9193 (2018)
10. Dosovitskiy, A., et al.: An image is worth 16×16 words: transformers for image recognition at scale. ICLR (2021)
11. Dunlap, L., et al.: Using language to extend to unseen domains. In: The Eleventh International Conference on Learning Representations (2022)
12. Fahes, M., Vu, T.H., Bursuc, A., Pérez, P., de Charette, R.: Pøda: prompt-driven zero-shot domain adaptation. In: ICCV (2023)
13. Goodfellow, I., et al.: Generative adversarial networks. Commun. ACM **63**(11), 139–144 (2020). https://doi.org/10.1145/3422622
14. Goodfellow, I.J., Shlens, J., Szegedy, C.: Explaining and harnessing adversarial examples. arXiv preprint arXiv:1412.6572 (2014)
15. Hadsell, R., Chopra, S., LeCun, Y.: Dimensionality reduction by learning an invariant mapping. In: 2006 IEEE Computer Society Conference on Computer Vision and Pattern Recognition (CVPR'06), vol. 2, pp. 1735–1742. IEEE (2006)
16. He, K., Zhang, X., Ren, S., Sun, J.: Deep residual learning for image recognition. In: CVPR (2016)
17. Hendrycks, D., et al.: The many faces of robustness: a critical analysis of out-of-distribution generalization. In: Proceedings of the IEEE/CVF International Conference on Computer Vision, pp. 8340–8349 (2021)
18. Hendrycks, D., Zhao, K., Basart, S., Steinhardt, J., Song, D.: Natural adversarial examples. In: Proceedings of the IEEE/CVF Conference on Computer Vision and Pattern Recognition, pp. 15262–15271 (2021)
19. Hu, J., Shen, L., Sun, G.: Squeeze-and-excitation networks. In: Proceedings of the IEEE Conference on Computer Vision and Pattern Recognition, pp. 7132–7141 (2018)
20. Huang, G., Liu, Z., van der Maaten, L., Weinberger, K.Q.: Densely connected convolutional networks. In: CVPR (2017)
21. Huang, Z., Zhou, A., Ling, Z., Cai, M., Wang, H., Lee, Y.J.: A sentence speaks a thousand images: domain generalization through distilling clip with language guidance. In: Proceedings of the IEEE/CVF International Conference on Computer Vision, pp. 11685–11695 (2023)
22. Jia, C., et al.: Scaling up visual and vision-language representation learning with noisy text supervision. In: International Conference on Machine Learning, pp. 4904–4916. PMLR (2021)
23. Kingma, D.P., Ba, J.: Adam: a method for stochastic optimization. In: ICLR (2015)
24. Krause, J., Stark, M., Deng, J., Fei-Fei, L.: 3D object representations for fine-grained categorization. In: 2013 IEEE International Conference on Computer Vision Workshops, pp. 554–561 (2013). https://doi.org/10.1109/ICCVW.2013.77
25. Kurakin, A., Goodfellow, I.J., Bengio, S.: Adversarial examples in the physical world. In: Artificial Intelligence Safety and Security, pp. 99–112. Chapman and Hall/CRC (2018)
26. Long, Y., Zhang, Q., Zeng, B., Gao, L., Liu, X., Zhang, J., Song, J.: Frequency domain model augmentation for adversarial attack. In: Computer Vision–ECCV 2022: 17th European Conference, Tel Aviv, Israel, October 23–27, 2022, Proceedings, Part IV, pp. 549–566. Springer (2022). https://doi.org/10.1007/978-3-031-19772-7_32

27. Lorenz, P., Harder, P., Straßel, D., Keuper, M., Keuper, J.: Detecting AutoAttack perturbations in the frequency domain. arXiv preprint arXiv:2111.08785 (2021)
28. Lu, Y., et al.: Enhancing cross-task black-box transferability of adversarial examples with dispersion reduction. In: CVPR (2020)
29. Madry, A., Makelov, A., Schmidt, L., Tsipras, D., Vladu, A.: Towards deep learning models resistant to adversarial attacks. arXiv preprint arXiv:1706.06083 (2017)
30. Maji, S., Rahtu, E., Kannala, J., Blaschko, M.B., Vedaldi, A.: Fine-grained visual classification of aircraft. ArXiv **abs/1306.5151** (2013), https://api.semanticscholar.org/CorpusID:2118703
31. Naseer, M.M., Khan, S.H., Khan, M.H., Shahbaz Khan, F., Porikli, F.: Cross-domain transferability of adversarial perturbations. In: Advances in Neural Information Processing Systems, vol. 32 (2019)
32. Naseer, M., Khan, S., Hayat, M., Khan, F.S., Porikli, F.: On generating transferable targeted perturbations. In: Proceedings of the IEEE/CVF International Conference on Computer Vision, pp. 7708–7717 (2021)
33. Naseer, M., Khan, S.H., Hayat, M., Khan, F.S., Porikli, F.: A self-supervised approach for adversarial robustness. In: CVPR (2020)
34. Nguyen, A., Yosinski, J., Clune, J.: Deep neural networks are easily fooled: high confidence predictions for unrecognizable images. In: Proceedings of the IEEE Conference on Computer Vision and Pattern Recognition, pp. 427–436 (2015)
35. Papernot, N., McDaniel, P., Jha, S., Fredrikson, M., Celik, Z.B., Swami, A.: The limitations of deep learning in adversarial settings. In: 2016 IEEE European Symposium on Security and Privacy (EuroS&P), pp. 372–387. IEEE (2016)
36. Poursaeed, O., Katsman, I., Gao, B., Belongie, S.: Generative adversarial perturbations. In: Proceedings of the IEEE Conference on Computer Vision and Pattern Recognition, pp. 4422–4431 (2018)
37. Radford, A., et al.: Learning transferable visual models from natural language supervision. In: International Conference on Machine Learning, pp. 8748–8763. PMLR (2021)
38. Recht, B., Roelofs, R., Schmidt, L., Shankar, V.: Do ImageNet classifiers generalize to imagenet? In: International Conference on Machine Learning, pp. 5389–5400. PMLR (2019)
39. Russakovsky, O., et al.: ImageNet large scale visual recognition challenge. Int. J. Comput. Vis. **115**(3), 211–252 (2015). https://doi.org/10.1007/s11263-015-0816-y
40. Salzmann, M., et al.: Learning transferable adversarial perturbations. Adv. Neural. Inf. Process. Syst. **34**, 13950–13962 (2021)
41. Simonyan, K., Zisserman, A.: Very deep convolutional networks for large-scale image recognition. In: Bengio, Y., LeCun, Y. (eds.) ICLR (2015)
42. Szegedy, C., Vanhoucke, V., Ioffe, S., Shlens, J., Wojna, Z.: Rethinking the inception architecture for computer vision. In: CVPR (2016)
43. Szegedy, C., et al.: Intriguing properties of neural networks. arXiv preprint arXiv:1312.6199 (2013)
44. Tan, M., et al.: MnasNet: platform-aware neural architecture search for mobile. In: Proceedings of the IEEE/CVF Conference on Computer Vision and Pattern Recognition, pp. 2820–2828 (2019)
45. Vaswani, A., et al.: Attention is all you need. In: Advances in Neural Information Processing Systems, vol. 30 (2017)
46. Wah, C., Branson, S., Welinder, P., Perona, P., Belongie, S.: The Caltech-UCSD Birds-200-2011 Dataset. Tech. Rep., California Institute of Technology (2011)

47. Wang, H., Ge, S., Lipton, Z., Xing, E.P.: Learning robust global representations by penalizing local predictive power. In: Advances in Neural Information Processing Systems, vol. 32 (2019)
48. Xie, C., et al.: Improving transferability of adversarial examples with input diversity. In: CVPR (2019)
49. Yang, H., Jeong, J., Yoon, K.J.: FACL-Attack: frequency-aware contrastive learning for transferable adversarial attacks. In: Proceedings of the AAAI Conference on Artificial Intelligence, vol. 38, pp. 6494–6502 (2024)
50. Yang, J., et al.: Unified contrastive learning in image-text-label space. In: Proceedings of the IEEE/CVF Conference on Computer Vision and Pattern Recognition, pp. 19163–19173 (2022)
51. Yang, J., et al.: Vision-language pre-training with triple contrastive learning. In: Proceedings of the IEEE/CVF Conference on Computer Vision and Pattern Recognition, pp. 15671–15680 (2022)
52. Yao, L., et al.: FILIP: fine-grained interactive language-image pre-training. arXiv preprint arXiv:2111.07783 (2021)
53. You, H., et al.: Learning visual representation from modality-shared contrastive language-image pre-training. In: Avidan, S., Brostow, G., Cissé, M., Farinella, G.M., Hassner, T. (eds.) Computer Vision – ECCV 2022: 17th European Conference, Tel Aviv, Israel, October 23–27, 2022, Proceedings, Part XXVII, pp. 69–87. Springer Nature Switzerland, Cham (2022). https://doi.org/10.1007/978-3-031-19812-0_5
54. Zhang, Q., Li, X., Chen, Y., Song, J., Gao, L., He, Y., Xue, H.: Beyond ImageNet Attack: Towards crafting adversarial examples for black-box domains. arXiv preprint arXiv:2201.11528 (2022)
55. Zhou, K., Yang, J., Loy, C.C., Liu, Z.: Conditional prompt learning for vision-language models. In: IEEE/CVF Conference on Computer Vision and Pattern Recognition (CVPR) (2022)
56. Zhou, K., Yang, J., Loy, C.C., Liu, Z.: Learning to prompt for vision-language models. Int. J. Comput. Vis. (IJCV) (2022). https://doi.org/10.1007/s11263-022-01653-1

Learning to Adapt SAM for Segmenting Cross-Domain Point Clouds

Xidong Peng[1], Runnan Chen[2], Feng Qiao[3], Lingdong Kong[4], Youquan Liu[5], Yujing Sun[2], Tai Wang[6], Xinge Zhu[7]([✉]), and Yuexin Ma[1]([✉])

[1] ShanghaiTech University, Shanghai, China
{pengxd,mayuexin}@shanghaitech.edu.cn
[2] The University of Hong Kong, Hong Kong, China
[3] RWTH Aachen University, Aachen, Germany
[4] National University of Singapore, Singapore, Singapore
[5] Hochschule Bremerhaven, Bremerhaven, Germany
[6] Shanghai AI Laboratory, Shanghai, China
[7] The Chinese University of Hong Kong, Hong Kong, China

Abstract. Unsupervised domain adaptation (UDA) in 3D segmentation tasks presents a formidable challenge, primarily steming from the sparse and unordered nature of point clouds. Especially for LiDAR point clouds, the domain discrepancy becomes obvious across varying capture scenes, fluctuating weather conditions, and the diverse array of LiDAR devices in use. Inspired by the remarkable generalization capabilities exhibited by the vision foundation model, SAM, in the realm of image segmentation, our approach leverages the wealth of general knowledge embedded within SAM to unify feature representations across diverse 3D domains and further solves the 3D domain adaptation problem. Specifically, we harness the corresponding images associated with point clouds to facilitate knowledge transfer and propose an innovative hybrid feature augmentation methodology, which enhances the alignment between the 3D feature space and SAM's feature space, operating at both the scene and instance levels. Our method is evaluated on many widely-recognized datasets and achieves state-of-the-art performance.

Keywords: Unsupervised Domain Adaptation · 3D Segmentation · Feature Alignment · Vision Foundation Model

---

X. Zhu and Y. Ma—This work was supported by NSFC (No.62206173), Natural Science Foundation of Shanghai (No.22dz1201900), Shanghai Sailing Program (No.22YF1428700), MoE Key Laboratory of Intelligent Perception and Human-Machine Collaboration (ShanghaiTech University), Shanghai Frontiers Science Center of Human-centered Artificial Intelligence (ShangHAI).

Supplementary Information The online version contains supplementary material available at https://doi.org/10.1007/978-3-031-72775-7_4.

1 Introduction

3D scene understanding is fundamental for many real-world applications, such as autonomous driving, robotics, smart cities, etc. Based on the point cloud, 3D segmentation is a critical task for scene understanding, which requires assigning semantic labels for each point. Current deep learning-based solutions [44,48] rely heavily on massive annotated data, which are high-cost and lack generalization capability for handling domain shifts. Unsupervised domain adaptation is significant for alleviating data dependency. However, unlike images with dense and regular representation, point clouds, especially LiDAR point clouds of large scenes, are unstructured and sparse, and have overt differences in patterns for various capture devices. Although some studies [38,39,46] have extended 2D techniques to solve the 3D UDA problem, the performance is still limited due to the essential defect of point cloud representation.

Considering that RGB cameras yield dense, color-rich, and structured data, and more importantly, they represent minor discrepancies across various devices, certain 3D UDA methods [4,5,19] utilize the synergy of LiDAR and camera capabilities to achieve more comprehensive and precise perception, and further enhance adaptation capabilities for 3D segmentation tasks. However, these methods usually train 2D and 3D networks simultaneously, demanding substantial online computing resources. Vision foundation models (VFMs), such as the Segment Anything Model (SAM) [23], have garnered significant attention due to their remarkable performance in addressing open-world vision tasks. Such models are trained on massive image data with tremendous parameters. Compared with a common model trained on limited data, VFMs have more general knowledge and much stronger generalization capability. Many works such as [7,8] have emerged recently to transfer the general 2D vision knowledge of VFMs to 3D and have achieved promising performance.

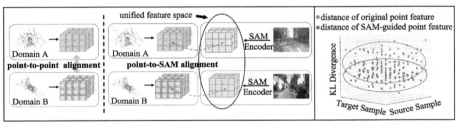

Fig. 1. (a) Comparison of 3D UDA paradigms. Different from aligning two point feature domains directly, our method makes both the source domain and target domain align with the SAM feature space. (b) Visualization of the feature distance across different datasets, where smaller values indicate a more similar distribution. It is obvious that after mapping to SAM feature space, point feature distributions from disparate domains become much more aligned.

Based on SAM, focusing on image segmentation, we propose a novel paradigm for 3D UDA segmentation. As shown in Fig. 1(a), different from previous UDA approaches that strive to align the target domain to the source domain so that the model trained on labeled source data can also work on target data without annotation, our method makes both the source domain and target domain align with the SAM feature space. SAM feature space contains more general knowledge, which provides a friendly space to unify the feature representation from different domains. We utilize RGB images to assist point clouds in our framework. However, unlike the methods mentioned above only using images to provide auxiliary information, we take images as a bridge to align diverse 3D feature spaces to the SAM feature space, so we do not need to train extra 2D networks and we can process the image offline for less computing resources. Moreover, considering that the 3D feature space created by the source-domain data and the target-domain data is still much smaller than the SAM feature space, we propose a hybrid feature augmentation method at both scene and instance levels to generate more 3D data with diverse feature patterns in a broader data domain, which can further benefit the 3D-to-SAM feature alignment. In particular, we make full use of the masks generated by SAM to mix instance-level point clouds with the other domains. This technique can maintain the geometric completeness of instances, which is beneficial for semantic recognition.

To verify that our idea of SAM-guided UDA is reliable, we randomly choose data from the source and target datasets and calculate the differences of feature distributions from source and target domain by KL divergence. The qualitative results are shown in Fig. 1(b), where the feature distribution differences after mapping to SAM feature space truly become much smaller. Moreover, to verify the effectiveness of our method, we compare it with current SOTA works on extensive 3D UDA segmentation settings and our method outperforms others by a large margin, improving about 14% mIoU for VirtualKITTI-to-SematicKITTI, about 15% mIoU for Waymo-to-nuScenes, and about 20% mIoU for nuScenes-to-SemanticKITTI domain adaptation. Surprisingly, our unsupervised method achieves comparable performance with the supervised method for city-changing and light-changing settings on the nuScenes dataset. We also test our method on more challenging tasks, such as panoptic segmentation and domain generalization, showing that our method is robust and has good generalization capability.

In summary, our contributions are as follows:

- We propose a novel unsupervised domain adaptation approach for 3D segmentation, leveraging the foundational model SAM to guide the alignment of features from diverse 3D data domains into a unified domain.
- We introduce a hybrid feature augmentation strategy at both scene and instance levels, generating more distinct feature patterns across a broader data domain for better feature alignment.
- We conduct extensive experiments on large-scale datasets and achieve SOTA performance.

2 Related Work

2.1 Point Cloud Semantic Segmentation

Point cloud semantic segmentation [17,48] is a rapidly evolving field, and numerous research works have contributed to advancements in this area. The pioneering approach PointNet [34] directly processes point clouds without voxelization and revolutionizes 3D segmentation by providing a novel perspective on point cloud analysis. Further, PointNet++ [35] extends PointNet with hierarchical feature learning through partitioning point clouds into local regions. To handle sparse point cloud data efficiently within large-scale scenes, a framework called SparseConvNet [16] has been specifically crafted. It excels in processing sparse 3D data and has been effectively utilized in various applications, including 3D semantic segmentation. MinkUNet [9] represents a significant advancement in point cloud semantic segmentation. Employing multi-scale interaction networks, MinkUNet enhances the segmentation of point clouds, effectively addressing the challenges posed by 3D spatial data. Our 3D segmentation networks are the popular SparseConvNet and MinkUNet. Due to the sparse characteristics of point cloud data, many current methods [18,25,45] add corresponding dense image information to facilitate point cloud segmentation tasks. Our method also uses image features to assist point cloud segmentation, and additionally, we utilize the 2D segmentation foundation model to achieve effective knowledge transfer.

2.2 Domain Adaptation for 3D Segmentation

Unsupervised Domain Adaptation (UDA) aims at transferring knowledge learned from a source annotated domain to a target unlabelled domain, and there are already several UDA methods proposed for 2D segmentation [6,22,47,50]. In recent years, domain adaptation techniques have gained increasing traction in the context of 3D segmentation tasks. [46] leverage a"Complete and Label" strategy to enhance semantic segmentation of LiDAR point clouds by recovering underlying surfaces and facilitating the transfer of semantic labels across varying LiDAR sensor domains. CosMix [38] introduces a sample mixing approach for UDA in 3D segmentation, which stands as the pioneering UDA approach utilizing sample mixing to alleviate domain shift. It generates two new intermediate domains of composite point clouds through a novel mixing strategy applied at the input level, mitigating domain discrepancies. However, due to the sparsity and irregularity of the point cloud, the disparity across different point cloud data domains is larger compared to that across 2D image domains, which makes it difficult to mitigate the variation across domains.

With the development of multi-modal perception [1,10] in autonomous driving, prevalent 3D datasets [2,13,15,29] include both 3D point clouds and corresponding 2D images, making leveraging multi-modality for addressing domain shift challenges in point clouds convenient. xMUDA [19,20] shows the power of combining 2D and 3D networks within a single framework, which achieves outstanding performance by aggregating the scores from these two branches.

This achievement is attributed to the complementary nature resulting from the diverse modalities processed by each branch. [32] introduce Dynamic Sparse-to-Dense Cross-Modal Learning (DsCML) to enhance the interaction of multi-modality information, ultimately boosting domain adaptation sufficiency, while [5] elucidate this complementarity of image and point cloud through an intuitive explanation centered on the effective receptive field, and proposes to feed both modalities to both branches. However, in practice, training two networks with distinct architectures is difficult to converge and demands substantial computing resources due to increased memory. Our method uses the pre-trained foundation model to process the image data, guaranteeing the quality of the image features and enabling the training process to focus on the 3D model.

2.3 Vision Foundation Models

The rise of foundation models [12,21,40] has garnered significant attention which are trained on extensive datasets, consequently demonstrating exceptional performance. Foundation models [42,49] have seen significant advancements in the realm of 2D vision, and several research studies extend these foundation models to comprehend 3D information. Representative works CLIP [36] leverage contrastive learning techniques to train both text and image encoders. CLIP2Scene [8] extends the capabilities of CLIP by incorporating a 2D-3D calibration matrix to facilitate a deeper comprehension of 3D scenes and OpenScene [33] focuses on zero-shot learning for 3D scenes through aligning point features in CLIP feature space to enable open vocabulary queries for 3D points. The Meta Research team recently launched the 'Segment Anything Model' [23], trained on an extensive dataset of over 1 billion masks from 11 million images. Utilizing efficient prompting, SAM can generate high-quality masks for image instance segmentation. The integration of flexible prompting and ambiguity awareness enables SAM with robust generalization capabilities for various downstream segmentation challenges. Many methods [7,8,28] take it as an off-the-shelf tool and distillate the knowledge to solve 3D problems by 2D-3D feature alignment. In our work of tackling the UDA of 3D segmentation, we utilize SAM to provide 2D prior knowledge for 3D feature alignment in a wider data domain.

3 Method

3.1 Problem Statement

We explore UDA for 3D segmentation, in which we have the source domain, denoted as $D_S = \{P_S, I_S, Y_S\}$ with paired input, namely point cloud P_S and image I_S, as well as annotated labels Y_S for each point, and the target domain denoted as $D_T = \{P_T, I_T\}$ without any annotation. Using these data, we train a 3D segmentation model that can generalize well to the target domain. 3D data from various domains have obvious differences in distribution and patterns, leading to over-fitting problems when models trained in one domain try to analyze data from another. The main solution is to align different features despite domain differences to achieve the generalization capability of the model. Different from previous works, our novel paradigm is to map data from distinct domains into a unified feature space, ensuring the model performs consistently across domains.

3.2 Framework Overview

The vision foundation model, SAM, is trained by massive image data, which contains general vision knowledge and provides a friendly feature space to unify diverse feature representations. Taking 2D images as the bridge, the 3D feature space of different domains can be indirectly unified by bringing them closer to the SAM feature space based on 2D-to-3D knowledge distillation. Based on this, we design a novel SAM-guided UDA method for 3D segmentation, as Fig. 2 shows. Specifically, given a point cloud input P, the point encoder M generates a point embedding $F_{point} \in \mathbb{R}^{n \times d}$ in the d-dimensional latent feature space. Concurrently, the corresponding image input I is passed through the SAM encoder for a c-channels image embedding $F_{image} \in \mathbb{R}^{h \times w \times c}$. Utilizing the correspondence between the point cloud and image, we acquire SAM-guided point embedding $\hat{F}_{point} \in \mathbb{R}^{n \times d}$ to compute the alignment loss L_{align} with the original point embedding F_{point}, serving the purpose of using SAM as a bridge to integrate the features of diverse data domains into a unified feature space. Notably, during training, the input for feature alignment consists of data from both source

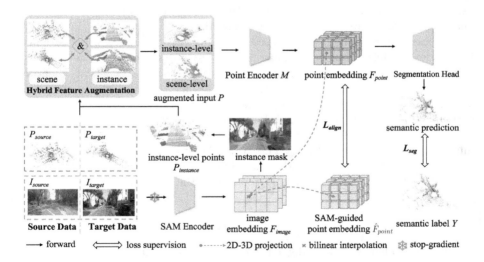

Fig. 2. Pipeline of our method. The point cloud is fed into the point encoder for point embeddings at the top, and the corresponding images are passed through the SAM encoder for image embeddings at the bottom, from which we obtain SAM-guided point embedding with the 2D-3D projection. Alignment loss L_{align} is calculated based on the SAM-guided features and original features. Furthermore, augmented inputs provide diverse feature patterns boosting the 3D-to-SAM feature alignment.

and target domains. We named this process as **SAM-guided Feature Alignment**. At the same time, as for labeled data Y, segmentation loss L_{seg} is also calculated as semantic supervision. During model training, only the point cloud branch of the whole pipeline is trained, and the gradient is not calculated in the image branch, which makes our method more lightweight. Furthermore, a **Scene-Instance Hybrid Feature Augmentation** is designed, which consists of scene-level and instance-level mix-up strategy. These mix-up strategies boost the variance of training data and generalize the network capability under the convex combination of the source domain and target domain data. Notably, the instance-level feature augmentation could maintain the local geometric relationship between two domains and make the subsequent alignment efficient.

3.3 SAM-Guided 3D Feature Alignment

Previous UDA methods usually align the feature space of the target domain to that of the source domain so that the model trained on the source domain with labeled data can also recognize the data from the novel domain. However, the distributions and patterns of 3D point clouds in various datasets have substantial differences, making the alignment very difficult. SAM [23], a 2D foundation model, is trained with a huge dataset of 11M images, granting it robust generalization capabilities to address downstream segmentation challenges effectively. If we can align features extracted from various data domains into the unified feature space represented by SAM, the model trained on the source domain can effectively handle the target data with the assistance of the universal vision knowledge existing in the SAM feature space.

We focus on training a point-based 3D segmentation model, while SAM is a foundation model trained on 2D images, which presents a fundamental challenge: how to bridge the semantic information captured in 2D images with the features extracted from 3D points. Most outdoor large-scale datasets with point clouds and images provide calibration information to project the 3D points into the corresponding images. With this information, we can easily translate the coordination of points P from the 3D LiDAR coordinate system P_{lidar} to the 2D image coordinate system P_{image}. This transformation can be formally expressed as Eq. 1, where rotation R_{ext} and translation T_{ext} represent the extrinsic parameters of the camera, and matrix K represents the intrinsic parameters of the camera.

$$zP_{image} = K(R_{ext}P_{lidar} + T_{ext}) \tag{1}$$

Once we calculate the projected 2D positions of points in the image coordinate system, we can determine their corresponding positions in the SAM-guided image embedding F_{image}, which is generated from the image by the SAM feature extractor. As the positions of points in the image embedding typically are not integer values, we perform bilinear interpolation based on the surrounding semantic features in the image embedding corresponding to the point, which to

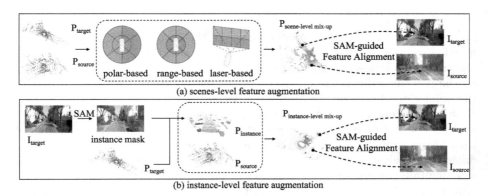

Fig. 3. Hybrid feature augmentation by data mixing for better 3D-to-SAM feature alignment. Part(a) illustrate all the scene-level approaches including polar-based, range-based, and laser-based point mix-up, where different color represents points from distinct domain. Part(b) shows the data flow of mixing the point data with instance-level data from another domain with an instance mask, where we take source data as an example for instance-level point generation and vice versa.

some extent alleviates the effect of calibration errors and allows us to derive the SAM-guided feature of each point, denoted as Eq. 2.

$$\hat{F}_{point} = \textbf{Bilinear}(F_{image}, P_{image}) \qquad (2)$$

Then, the original point embedding F_{point} from both source and target domains are all required to align with their corresponding SAM-guided features $\hat{F}_{point}$. Specifically, we utilize the cosine function to measure the similarity of F_{point} and $\hat{F}_{point}$, employing it as the alignment loss L_{align} during training. With the supervision of L_{align}, features obtained by the point encoder M will gradually converge towards the feature space represented by SAM, achieving the purpose of extracting features within a unified feature space from the input of different domains. The formulation of the loss function for feature alignment is shown as Eq. 3.

$$L_{align} = 1 - \textbf{cos}(\hat{F}_{point}, F_{point}) \qquad (3)$$

3.4 Scene-Instance Hybrid Feature Augmentation

3D point features of the source-domain data and the target-domain data only cover subsets of the 3D feature space, which are limited to align with the whole SAM feature space with more universal knowledge. Therefore, more 3D data with diverse feature patterns in a broader data domain is needed to achieve more effective 3D-to-SAM feature alignment.

Previous works [24,43] usually focus on synthesizing data by combining data in the source domain and target domain at the scene level, including polar-based, range-based, and laser-based, as shown in Fig. 3(a). Polar-based point mix-up selects semi-circular point cloud data from two different domains based on the polar coordinates of the point cloud. Range-based point mix-up divides the point cloud by its distance from the center, synthesizing circular point data close to the center and ring point data farther away from the center. Laser-based point mix-up determines the part of point clouds based on the number of laser beams, combining points with positive and negative laser pitch angles from different domains for synthesis. These ways of scene-level feature augmentation can maintain the general pattern of LiDAR point clouds as much as possible and improve the data diversity. Moreover, they are simple to process without any requirement for additional annotations such as real or pseudo-semantic labels. We adopt these three kinds of scene-level data augmentation in our method.

However, scene-level data augmentation will, to some extent, destroy the completeness of the point cloud of instances in the stitching areas and affect the exploitation of local geometric characteristics of point clouds. To further increase the data diversity and meanwhile keep the instance feature patterns of LiDAR point clouds for better semantic recognition, we propose an instance-level augmentation method. Benefiting from the instance mask output from SAM, we can thoroughly exploit the instance-level geometric features. Compared with pre-trained 3D segmentation models, SAM provides more accurate and robust instance masks and enables us to avoid extra warm-up for a pre-train model, simplifying the whole training process. Therefore, we perform instance-level data synthesis as Fig. 3(b) shows. Specifically, we begin by employing SAM to generate instance masks for input images from either the source or target domain (We take target data as the example in the figure). Next, we use the calibration matrix to project the corresponding point cloud into the image. The instance information of each point is determined according to whether the projection position of the point cloud falls within a specific instance mask, and then we randomly select some points with $20 \sim 30$ specific instances, mixed with the point cloud from the other domain by direct concatenation to achieve point augmentation at the instance level.

In practice, we combine all the ways of feature augmentation at both scene level and instance level with a random-selection strategy for a more comprehensive feature augmentation, which generates a more diverse set of point cloud data with varied feature patterns. Then, the augmented points are fed into the point encoder M to obtain the point embedding F_{point} with distinct feature patterns in a broader data domain beyond the source domain and target domain for more effective SAM-guided feature alignment. Notably, to maintain the consistency of the point cloud and the image, we extract SAM-guided point embedding based on the corresponding original image embedding.

4 Experiment

We first introduce datasets and implementation details. After that, we explore several domain shift scenarios and conduct comparisons for 3D segmentation. Then, we conduct extensive ablation studies to assess submodules of our method. Finally, we extend our method to more challenging tasks to show its generalization capability.

4.1 Dataset Setup

We first follow the benchmark introduced in xMUDA [20] to evaluate our method, comprehending four domain shift scenarios, including (1) USA-to-Singapore, (2) Day-to-Night, (3) VirtualKITTI-to-SemanticKITTI and (4) A2D2-to-Semantic-KITTI. The first two leverage nuScenes [3] as their dataset, consisting of 1000 driving scenes in total with 40k annotated point-wise frames. Specifically, the former differs in the layout and infrastructure while the latter exhibits severe illumination changes between the source and the target domain. The third is more challenging since it is the adaptation from synthetic to real data, implemented by adapting from VirtualKITTI [14] to SemanticKITTI [2] while the fourth involves A2D2 [15] and SemanticKITTI as different data domains, where the domain discrepancy lies in the distinct density and arrangement of 3D point clouds captured by different devices since the A2D2 is captured by 16-beam LiDAR and the SemanticKITTI uses 64-beam LiDAR. For the above settings, noted that only 3D points visible from the camera are used for training and testing, specifically, only one image and corresponding points for each sample are used for training.

Since we only use the image combined with SAM as offline assistance for the training of a 3D segmentation network instead of training a new 2D segmentation network, we focus on comparing the performance of the 3D segmentation network and enable model training with the whole point cloud sample because of less computational cost, even if some part of it is not visible in the images. For the part of the point cloud that cannot be covered by the image, alignment loss is not calculated, only segmentation loss is calculated. Thus, we also compare our method with others trained with the whole 360° view of the point cloud, in which three datasets are involved including nuScenes, SemanticKITTI, and Waymo [29]. In these settings, we use 6 images in nuScenes covering 360° view, 1 image in SemanticKITTI covering 120° view, and 5 images in Waymo covering 252° view. More information is introduced in the Appendix. For metric, We compute the Intersection over the Union per class and report the mean Intersection over the Union (mIoU).

4.2 Implementation Details

We make source and target labels compatible across these experiments. For all benchmarks in prior multi-modal UDA methods, we strictly follow class mapping like xMUDA for a fair comparison, while we map the labels of the dataset in other experiments into 10 segmentation classes in common. Our method is implemented by using the public PyTorch [31] repository MMDetection3D [11] and all the models are trained on a single 24GB GeForce RTX 3090 GPU. To compare fairly, we use SparseConvNet [16] with U-Net architecture as the 3D backbone network when following the benchmark introduced in xMUDA to evaluate our method and use MinkUNet32 [9] as the 3D backbone network when following the setup of taking the whole 360° point cloud as input, which is also the backbone of the state-of-the-art uni-modal method CosMix. For the image branch, the ViT-h variant SAM model is utilized to generate image embedding for SAM-guided feature alignment and instance masks for hybrid feature augmentation in an offline manner. We keep the proportion of mixed data and normal data from the source and target domain the same during model training. Before the data is fed into the 3D network, data augmentation such as vertical axis flipping, random scaling, and random 3D rotations are widely used like all the compared methods. For the model training strategies, we choose a batch size of 8 for both source data and target data, then mix the data batch for training at each iteration. Besides, we adopt AdamW as the model optimizer and One Cycle Policy as the learning-rate scheduler.

4.3 Experimental Results and Comparison

Tables 1 and 2 show the experimental results and performance comparison with previous UDA methods for 3D segmentation under the setup introduced in Sect. 4.1. Each experiment contains two reference methods in common, a baseline model named **Source only** trained only on the source domain and an upper-bound model named **Oracle** trained only on the target data with annotations. Table 1 focuses on four domain shift scenarios introduced by xMUDA [20] and comparison with these multi-modal methods based on xMUDA such as DsCML [32] and MM2D3D [5]. Among them, MM2D3D fully exploits the complementarity of image and point cloud and proposes to feed two modalities to both branches, achieving better performance. Our method outperforms it by +6.8% (USA → Singapore), +0.3% (Day → Night), +14.6% (v.KITTI →

Table 1. Results under four domain shift scenarios introduced by xMUDA. We report all the 3D network performance of compared multi-modal UDA methods in terms of mIoU. Note that the 3D backbone in these experiments is SparseConvNet [16].

Method	USA → Singapore		Day → Night		v.KITTI → Sem.KITTI		A2D2 → Sem.KITTI	
Source only	62.8	+0.0	68.8	+0.0	42.0	+0.0	35.9	+0.0
xMUDA [20]	63.2	+0.4	69.2	+0.4	46.7	+4.7	46.0	+10.1
DsCML [32]	52.3	−10.5	61.4	−7.4	32.8	−9.2	32.6	−3.3
MM2D3D [5]	66.8	+4.0	70.2	+1.4	50.3	+8.3	46.1	+10.2
Ours	**73.6**	**+10.8**	**70.5**	**+1.7**	**64.9**	**+22.9**	**52.1**	**+16.2**
Oracle	76.0	−	69.2	−	78.4	−	71.9	−

Table 2. Results under four domain shift scenarios with 360° point cloud, where not all the points are visible in the images. We report the 3D network performance in terms of mIoU. Note that the 3D backbone in these experiments is MinkUNet32 [9].

Method	nuScenes → Sem.KITTI		Sem.KITTI → nuScenes		nuScenes → Waymo		Waymo → nuScenes	
Source only	27.7	+0.0	28.1	+0.0	29.4	+0.0	21.8	+0.0
PL [30]	30.0	+2.3	29.0	+0.9	31.9	+2.5	22.3	+0.5
CoSMix [37]	30.6	+2.9	29.7	+1.6	31.5	+2.1	30.0	+8.2
MM2D3D [5]	30.4	+2.7	31.9	+3.8	31.3	+1.9	33.5	+11.7
MM2D3D*	32.9	+5.2	33.7	+5.6	34.1	+4.7	37.5	+15.7
Ours	**48.5**	**+20.8**	**42.9**	**+14.8**	**44.9**	**+15.5**	**48.2**	**+26.4**
Oracle	70.3	–	78.3	–	79.9	–	78.3	–

Sem.KITTI), +6.0% (A2D2 → Sem.KITTI) respectively, because our method aligns all the features into a unified feature space with the guidance of SAM instead of simply aligning features from image and point cloud in 2D and 3D network. Table 2 focuses on the scenarios where not all the point clouds are visible in the images and we re-implement three methods by their official codes. PL [30] uses the prediction from the pre-trained model as pseudo labels for unlabelled data to retrain this model, which is widely used in UDA methods. CosMix [37] trains a 3D network with only the utilization of a point cloud, which generates new intermediate domains through a mixing scene-level strategy to mitigate domain discrepancies. MM2D3d is the SOTA multi-modal method as described above, but it needs all the points visible in the image for the best performance. Our method surpasses them by at least +17.9% (nuScenes → Sem.KITTI), +11.0% (Sem.KITTI → nuScenes), +13.0% (nuScenes → Waymo), +14.7% (Waymo → nuScenes) respectively by a large margin, since hybrid feature augmentation can provide more intermediate domains and SAM-guided feature alignment can help map the whole point cloud into the unified feature space.

According to the results above, our method outperforms others by a large margin, attributed to the full utilization of the general knowledge provided by SAM. To further prove that the achievement is due to not only SAM but also our novel paradigm, we also use SAM for current SOTA multi-modal UDA work and get results in the row of "MM2D3D*". Because it trains both 2D and 3D networks simultaneously, under its original framework, we can only refine the supervision signals of 2D network using the instance mask output of SAM. As seen from the results, SAM can improve its performance but very limited. Our method not only uses instance masks but also makes full use of the general features extracted by SAM, ensuring the superiority of our method. Qualitative results are shown in Fig. 4, where predictions in the ellipses demonstrate that source-only and MM2D3D models often infer wrong and mingling results, especially for the person category, while our method can provide correct and more fine-grained segmentation. More qualitative results are in the Appendix.

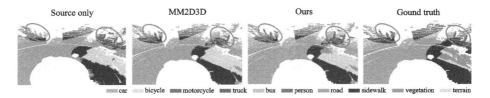

Fig. 4. Visualization of the domain adaptation from nuScenes to SemanticKITTI.

Table 3. Ablation study. Baseline means the result of the source-only model indicating the lower-bound and Pseudo Label means re-training the model with pseudo labels.

Setting	Baseline	SAM-guided Feature Alignment	Hybrid Feature Augmentation		Pseudo Label	mIoU
			Scene-level	Instance-level		
(1)	✓					27.7
(2)	✓	✓				34.0
(3)	✓		✓	✓		28.6
(4)	✓	✓	✓	✓		40.1
(5)	✓	✓		✓		39.0
(6)	✓	✓	✓	✓		44.0
(7)	✓	✓	✓	✓	✓	**48.5**

4.4 Ablation Study

To show the effectiveness of each module of our method, we conduct ablation studies on nuScenes-to-SemanticKITTI UDA. We also analyze and show the effect of other vision foundation models on our method.

Effectiveness of Model Components We first analyze the effects of all the submodules in our method in Table 3, containing SAM-guided Feature Alignment, Hybrid Feature Augmentation, and Pseudo Label. SAM-guided Feature Alignment aligns all the point features with the corresponding feature embeddings output by SAM, guiding the 3D network map point cloud into the unified feature space represented by SAM while Hybrid Feature Augmentation generates additional point cloud data of the intermediate domain for feature extraction to maximize the effect of feature alignment. Setting (1), (2), (3), and (6) in the table shows that combining the two submodules improves performance by a large margin. Besides, re-training the model with pseudo labels is a strategy widely used in UDA tasks and it also improves the performance.

Effectiveness of Hybrid Feature Augmentation For the detailed ablation of feature alignment, we adopt hybrid strategies for diverse data with distinct feature patterns, which not only mix up points at the scene level in polar-based, range-based, and laser-based ways but also at the instance level with the help of instance mask output by SAM. Random selection in all these point mix-up ways forms this feature augmentation. Setting (4), (5), (6) in Table 3 shows that both mix-up methods can help feature alignment with more distinct features but the hybrid strategy raises the best performance. More ablations are in the Appendix. Moreover, since masks generated by SAM do not contain semantic labels, we conduct an additional experiment to prove the validity of instance-level augmentation by replacing the generated SAM instance masks with the ground truth semantic masks under the setting (6). Compared with the original result (mIoU=44.0), the performance of using ground truth semantic masks is mIoU=42.7, demonstrating that although masks generated by SAM cannot be identical to the ground truth, the contained semantic information is consistent, and the randomness of our augmentation further improve the performance.

Effect of Different Point-to-Pixel Coverage for Alignment. For point clouds not covered by images, we do not calculate feature alignment loss and solely calculate segmentation loss with ground truth or pseudo label. When conducting the experiments, we used all available images to ensure that as many point cloud data as possible could find corresponding features on the pictures for SAM-guided feature alignment. We also conduct additional experiments to demonstrate the impact of the number of available images for feature alignment, as shown in Table 4. The results indicate that a greater number of available images correlates with improved experimental outcomes. Importantly, as long as SAM-guided feature alignment is achievable, the performance does not significantly degrade even with limited coverage.

Table 4. The effect of image-point coverage. Different numbers represent the number of used pictures of nuScenes.

Covered Images for Alignment	baseline	2	4	6
nuScenes → Sem.KITTI	27.7	43.9	45.6	**48.5**
Sem.KITTI → nuScenes	28.1	38.4	40.6	**42.9**

Effect of Vision Foundation Model. We also extend our method to other vision foundation models, such as InternImage [41], serving for image-based tasks. Specifically, we replace the SAM-based image encoder with InternImage to guide the feature alignment in a similar manner. Compared with the baseline (mIoU=27.7), the performance of using InternImage is mIoU=36.9. A consistent performance gain can be obtained, which also verifies and validates our insight, i.e., the generic feature space of the VFM can ease the feature alignment.

4.5 More Challenging Tasks

Since we achieve the purpose of mapping data from different domains into a unified feature space, the extracted feature can be used for some more challenging tasks. We show some extension results of our method in Table 5. The left subtable shows the results of UDA for panoptic segmentation, a more challenging task requiring instance-level predictions. With more accurate and fine-grained semantic prediction, our method achieves promising results. The right subtable shows the results of domain generalization, in which target data only can be used for testing. In this subtable, models are trained with nuScenes and SemanticKITTI and then evaluated with A2D2 dataset. With the ability of stronger data-to-feature mapping, our method outperforms the current SOTA method [26]. In the future, we seek to explore the potential of our method on more tasks, such as 3D detection.

Table 5. Extension on more challenging tasks, such as UDA for Panoptic Segmentation(left) and Domain Generalization(right), where N, S, A represent nuScenes, SemanticKITTI and A2D2 dataset.

Task	Method	PQ	PQ†	RQ	SQ	mIoU
nuScenes → Sem.KITTI	Source only	14.0	21.6	19.9	55.8	27.7
	PL	15.9	22.7	22.2	58.1	29.7
	Ours	**34.3**	**38.4**	**42.6**	**55.9**	**48.5**
	Oracle	50.5	52.2	57.8	77.2	70.3
Sem.KITTI → nuScenes	Source only	15.6	22.1	20.7	52.7	28.2
	PL	16.8	23.0	21.7	48.3	29.0
	Ours	**24.6**	**30.7**	**30.8**	**60.0**	**42.9**
	Oracle	40.7	44.9	47.2	83.8	78.3

Method	N,S → A
Baseline	45.0
xMUDA [20]	44.9
Dual-Cross [27]	41.3
BEV-DG [26]	55.1
Ours	**57.2**

5 Conclusion

In this paper, we acknowledge the limitations of existing UDA methods in handling the domain discrepancy present in 3D point cloud data and propose a novel paradigm to unify feature representations across diverse 3D domains by leveraging the powerful generalization capabilities of the vision foundation model, significantly enhancing the adaptability of 3D segmentation models. Hybrid feature augmentation strategy is also proposed for better 3D-SAM feature alignment. Extensive experiments show that our method surpasses all compared SOTA methods by a large margin.

References

1. Bai, X., et al.: TransFusion: robust lidar-camera fusion for 3D object detection with transformers. In: Proceedings of the IEEE/CVF Conference on Computer Vision and Pattern Recognition, pp. 1090–1099 (2022)
2. Behley, J., et al.: SemanticKITTI: a dataset for semantic scene understanding of lidar sequences. In: Proceedings of the IEEE/CVF International Conference on Computer Vision, pp. 9297–9307 (2019)
3. Caesar, H., et al.: nuScenes: a multimodal dataset for autonomous driving. In: Proceedings of the IEEE/CVF Conference on Computer Vision and Pattern Recognition, pp. 11621–11631 (2020)

4. Cao, H., Xu, Y., Yang, J., Yin, P., Yuan, S., Xie, L.: MoPA: multi-modal prior aided domain adaptation for 3D semantic segmentation. arXiv preprint arXiv:2309.11839 (2023)
5. Cardace, A., Ramirez, P.Z., Salti, S., Di Stefano, L.: Exploiting the complementarity of 2D and 3D networks to address domain-shift in 3D semantic segmentation. In: Proceedings of the IEEE/CVF Conference on Computer Vision and Pattern Recognition, pp. 98–109 (2023)
6. Chang, W.L., Wang, H.P., Peng, W.H., Chiu, W.C.: All About Structure: adapting structural information across domains for boosting semantic segmentation. In: Proceedings of the IEEE/CVF Conference on Computer Vision and Pattern Recognition, pp. 1900–1909 (2019)
7. Chen, R., et al.: Towards label-free scene understanding by vision foundation models. In: Advances in Neural Information Processing Systems (2023)
8. Chen, R., et al.: Clip2scene: towards label-efficient 3D scene understanding by clip. In: Proceedings of the IEEE/CVF Conference on Computer Vision and Pattern Recognition, pp. 7020–7030 (2023)
9. Choy, C., Gwak, J., Savarese, S.: 4D Spatio-Temporal ConvNets: Minkowski convolutional neural networks. In: Proceedings of the IEEE/CVF Conference on Computer Vision and Pattern Recognition, pp. 3075–3084 (2019)
10. Cong, P., et al.: Weakly supervised 3D multi-person pose estimation for large-scale scenes based on monocular camera and single lidar. In: Proceedings of the AAAI Conference on Artificial Intelligence, vol. 37, pp. 461–469 (2023)
11. Contributors, M.: MMDetection3D: OpenMMLab next-generation platform for general 3D object detection (2020). https://github.com/open-mmlab/mmdetection3d
12. Devlin, J., Chang, M.W., Lee, K., Toutanova, K.: BERT: pre-training of deep bidirectional transformers for language understanding. arXiv preprint arXiv:1810.04805 (2018)
13. Fong, W.K., et al.: Panoptic nuScenes: a large-scale benchmark for lidar panoptic segmentation and tracking. IEEE Robot. Autom. Lett. **7**(2), 3795–3802 (2022)
14. Gaidon, A., Wang, Q., Cabon, Y., Vig, E.: Virtual worlds as proxy for multi-object tracking analysis. In: Proceedings of the IEEE/CVF Conference on Computer Vision and Pattern Recognition, pp. 4340–4349 (2016)
15. Geyer, J., et al.: A2D2: audi autonomous driving dataset. arXiv preprint arXiv:2004.06320 (2020)
16. Graham, B., Engelcke, M., Van Der Maaten, L.: 3D semantic segmentation with submanifold sparse convolutional networks. In: Proceedings of the IEEE/CVF Conference on Computer Vision and Pattern Recognition, pp. 9224–9232 (2018)
17. Guo, Y., Wang, H., Hu, Q., Liu, H., Liu, L., Bennamoun, M.: Deep learning for 3D point clouds: a survey. IEEE Trans. Pattern Anal. Mach. Intell. **43**(12), 4338–4364 (2020)
18. He, D., Abid, F., Kim, J.H.: Multimodal fusion and data augmentation for 3D semantic segmentation. In: IEEE International Conference on Control, Automation and Systems, pp. 1143–1148 (2022)
19. Jaritz, M., Vu, T.H., Charette, R.d., Wirbel, E., Pérez, P.: xMUDA: cross-modal unsupervised domain adaptation for 3D semantic segmentation. In: Proceedings of the IEEE/CVF Conference on Computer Vision and Pattern Recognition, pp. 12605–12614 (2020)
20. Jaritz, M., Vu, T.H., De Charette, R., Wirbel, É., Pérez, P.: Cross-modal learning for domain adaptation in 3D semantic segmentation. IEEE Trans. Pattern Anal. Mach. Intell. **45**(2), 1533–1544 (2022)

21. Jia, C., et al.: Scaling up visual and vision-language representation learning with noisy text supervision. In: International Conference on Machine Learning, pp. 4904–4916. PMLR (2021)
22. Kim, M., Byun, H.: Learning texture invariant representation for domain adaptation of semantic segmentation. In: Proceedings of the IEEE/CVF Conference on Computer Vision and Pattern Recognition, pp. 12975–12984 (2020)
23. Kirillov, A., et al.: Segment anything. In: Proceedings of the IEEE/CVF International Conference on Computer Vision (2023)
24. Kong, L., Ren, J., Pan, L., Liu, Z.: Lasermix for semi-supervised lidar semantic segmentation. In: Proceedings of the IEEE/CVF Conference on Computer Vision and Pattern Recognition, pp. 21705–21715 (2023)
25. Krispel, G., Opitz, M., Waltner, G., Possegger, H., Bischof, H.: Fuseseg: Lidar point cloud segmentation fusing multi-modal data. In: Proceedings of the IEEE/CVF Winter Conference on Applications of Computer Vision, pp. 1874–1883 (2020)
26. Li, M., Zhang, Y., Ma, X., Qu, Y., Fu, Y.: BEV-DG: cross-modal learning under Bird's-eye view for domain generalization of 3D semantic segmentation. arXiv preprint arXiv:2308.06530 (2023)
27. Li, M., et al.: Cross-domain and cross-modal knowledge distillation in domain adaptation for 3D semantic segmentation. In: Proceedings of the ACM International Conference on Multimedia, pp. 3829–3837 (2022)
28. Liu, Y., et al.: Segment any point cloud sequences by distilling vision foundation models. arXiv preprint arXiv:2306.09347 (2023)
29. Mei, J., et al.: Waymo Open Dataset: Panoramic video panoptic segmentation. In: European Conference on Computer Vision, pp. 53–72. Springer (2022). https://doi.org/10.1007/978-3-031-19818-2_4
30. Morerio, P., Cavazza, J., Murino, V.: Minimal-entropy correlation alignment for unsupervised deep domain adaptation. arXiv preprint arXiv:1711.10288 (2017)
31. Paszke, A., Gross, S., Massa, e.: Pytorch: an imperative style, high-performance deep learning library. In: Advances in Neural Information Processing Systems, vol. 32 (2019)
32. Peng, D., Lei, Y., Li, W., Zhang, P., Guo, Y.: Sparse-to-dense feature matching: intra and inter domain cross-modal learning in domain adaptation for 3D semantic segmentation. In: Proceedings of the IEEE/CVF International Conference on Computer Vision, pp. 7108–7117 (2021)
33. Peng, S., Genova, K., Jiang, C., Tagliasacchi, A., Pollefeys, M., Funkhouser, T., et al.: OpenScene: 3D scene understanding with open vocabularies. In: Proceedings of the IEEE/CVF Conference on Computer Vision and Pattern Recognition, pp. 815–824 (2023)
34. Qi, C.R., Su, H., Mo, K., Guibas, L.J.: PointNet: deep learning on point sets for 3D classification and segmentation. In: Proceedings of the IEEE/CVF Conference on Computer Vision and Pattern Recognition, pp. 652–660 (2017)
35. Qi, C.R., Yi, L., Su, H., Guibas, L.J.: PointNet++: deep hierarchical feature learning on point sets in a metric space. In: Advances in Neural Information Processing Systems, vol. 30 (2017)
36. Radford, A., et al.: Learning transferable visual models from natural language supervision. In: International Conference on Machine Learning, pp. 8748–8763. PMLR (2021)
37. Saltori, C., Galasso, F., Fiameni, G., Sebe, N., Poiesi, F., Ricci, E.: Compositional semantic mix for domain adaptation in point cloud segmentation. IEEE Trans. Pattern Anal. Mach. Intell. (2023)

38. Saltori, C., Galasso, F., Fiameni, G., Sebe, N., Ricci, E., Poiesi, F.: CoSMix: compositional semantic mix for domain adaptation in 3D lidar segmentation. In: European Conference on Computer Vision, pp. 586–602 (2022)
39. Shaban, A., Lee, J., Jung, S., Meng, X., Boots, B.: LiDAR-UDA: self-ensembling through time for unsupervised lidar domain adaptation (2023)
40. Touvron, H., et al.: LLaMA: open and efficient foundation language models. arXiv preprint arXiv:2302.13971 (2023)
41. Wang, W., et al.: InternImage: exploring large-scale vision foundation models with deformable convolutions. In: Proceedings of the IEEE/CVF Conference on Computer Vision and Pattern Recognition, pp. 14408–14419 (2023)
42. Wang, X., Zhang, X., Cao, Y., Wang, W., Shen, C., Huang, T.: SegGPT: segmenting everything in context. arXiv preprint arXiv:2304.03284 (2023)
43. Xiao, A., Huang, J., Guan, D., Cui, K., Lu, S., Shao, L.: PolarMix: general data augmentation technique for lidar point clouds. Adv. Neural. Inf. Process. Syst. **35**, 11035–11048 (2022)
44. Xu, Y., et al.: Human-centric scene understanding for 3D large-scale scenarios. arXiv preprint arXiv:2307.14392 (2023)
45. Yan, X., et al.: 2DPASS: 2D priors assisted semantic segmentation on LiDAR point clouds. In: Avidan, S., Brostow, G., Cissé, M., Farinella, G.M., Hassner, T. (eds.) Computer Vision – ECCV 2022: 17th European Conference, Tel Aviv, Israel, October 23–27, 2022, Proceedings, Part XXVIII, pp. 677–695. Springer Nature Switzerland, Cham (2022). https://doi.org/10.1007/978-3-031-19815-1_39
46. Yi, L., Gong, B., Funkhouser, T.: Complete & Label: a domain adaptation approach to semantic segmentation of lidar point clouds. In: Proceedings of the IEEE/CVF Conference on Computer Vision and Pattern Recognition, pp. 15363–15373 (2021)
47. Zhang, Y., Wang, Z.: Joint adversarial learning for domain adaptation in semantic segmentation. In: Proceedings of the AAAI Conference on Artificial Intelligence, vol. 34, pp. 6877–6884 (2020)
48. Zhu, X., et al.: Cylindrical and asymmetrical 3D convolution networks for lidar segmentation. In: Proceedings of the IEEE/CVF Conference on Computer Vision and Pattern Recognition, pp. 9939–9948 (2021)
49. Zou, X., et al.: Segment everything everywhere all at once. arXiv preprint arXiv:2304.06718 (2023)
50. Zou, Y., Yu, Z., Kumar, B., Wang, J.: Unsupervised domain adaptation for semantic segmentation via class-balanced self-training. In: European Conference on Computer Vision, pp. 289–305 (2018)

Learning to Enhance Aperture Phasor Field for Non-Line-of-Sight Imaging

In Cho(✉)⬤, Hyunbo Shim⬤, and Seon Joo Kim⬤

Yonsei University, Seoul, South Korea
join@yonsei.ac.kr

Abstract. This paper aims to facilitate more practical NLOS imaging by reducing the number of samplings and scan areas. To this end, we introduce a phasor-based enhancement network that is capable of predicting clean and full measurements from noisy partial observations. We leverage a denoising autoencoder scheme to acquire rich and noise-robust representations in the measurement space. Through this pipeline, our enhancement network is trained to accurately reconstruct complete measurements from their corrupted and partial counterparts. However, we observe that the naïve application of denoising often yields degraded and over-smoothed results, caused by unnecessary and spurious frequency signals present in measurements. To address this issue, we introduce a phasor-based pipeline designed to limit the spectrum of our network to the frequency range of interests, where the majority of informative signals are detected. The phasor wavefronts at the aperture, which are band-limited signals, are employed as inputs and outputs of the network, guiding our network to learn from the frequency range of interests and discard unnecessary information. The experimental results in more practical acquisition scenarios demonstrate that we can look around the corners with 16× or 64× fewer samplings and 4× smaller apertures. Our code is available at https://github.com/join16/LEAP.

Keywords: Non-line-of-sight imaging · Deep learning

1 Introduction

Non-line-of-sight (NLOS) imaging aims to reconstruct scenes that are hidden in direct line-of-sight systems, with a laser illuminating a relay wall, and a time-resolved detector recording the returning photons. The ability to perceive occluded objects has captivated many researchers due to its wide range of future applications, such as medical imaging, rescue operations, and autonomous driving. Representative NLOS imaging methods [20–22, 27, 45] have shown that the hidden scenes can be reconstructed in high-quality if measurements are captured with sufficient sampling points, acquisition time, and scanning areas.

Supplementary Information The online version contains supplementary material available at https://doi.org/10.1007/978-3-031-72775-7_5.

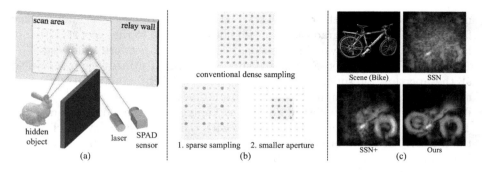

Fig. 1. (a) A typical NLOS imaging system. (b) More practical acquisition scenarios of NLOS imaging: sparse sampling and scanning with smaller apertures. (c) Results on confocal 16 × 16 measurements of Bike [20]. Our method exhibits high-quality results with 16× fewer sampling points and a shorter acquisition time, whereas previous signal recovery network (SSN) [41] and simple addition of the denoising criterion to SSN (SSN+) fail to correctly reconstruct the hidden objects.

Beyond the recent advances, we extend our focus to more practical acquisition scenarios relevant to real-world applications, most of which do not offer sufficient scanning time and areas. Reducing the scan areas, the number of samplings, and thereby the scanning time yields much degraded results in previous methods, due to their theoretical resolution limits and increased effects of noise. To mitigate this issue, recent studies leverage custom-designed arrays of single-photon avalanche diode (SPAD) sensors [24,25,28] or optimization-based methods [23,43], incurring additional expenses for hardware or computations.

In this paper, we introduce a phasor-based enhancement network that leverages learned priors and phasor-based frequency filtering to facilitate NLOS imaging under more practical acquisition setups (Fig. 1 (b)), where measurements are acquired with fewer samplings, smaller scan areas, and reduced acquisition time. We begin by tailoring a denoising autoencoder scheme [40], where we place the denoising criterion on top of the missing signal recovery problem [41]. This enables our model to attain rich and noise robust representations in the measurement space. Specifically, our network processes partial measurements corrupted by Poisson noise [6,10,26], and is trained to accurately predict the optimal measurements, containing sufficient scanning points and clean signals. After training, our method enables high-quality NLOS reconstruction in partial sampling scenarios, which is achieved by applying inverse NLOS methods to the predictions.

While this straightforward application of the denoising criterion alleviates the effects of noise, this training scheme leads the enhancement network to be parameterized across the entire frequency spectrum, often producing degraded and over-smoothed results. Conversely, in measurements of NLOS imaging, the majority of informative signals are concentrated within a specific frequency range while signals in other frequency ranges mostly contain coarse structures and noise. We observe that these unnecessary and spurious frequency signals are key factors causing such degradation, but do not contribute to the reconstruction of hidden volumes. Thus, we aim to prune them from our network's interests.

To this end, we propose a phasor-based scheme that utilizes the phasor field at the aperture for supervision. In phasor field NLOS methods [21,22], aperture wavefronts are computed by the convolution of the measurements and the illumination function, typically defined as Gaussian in the frequency domain. The aperture wavefronts are thus signals with a limited frequency band, where the majority of the informative signals are observed. By leveraging these band-limited signals as inputs and outputs of the network, we constrain the network operations within the frequency range of interests. This guides our network to discard unnecessary signals, achieving substantial improvement in reconstruction quality and better generalization capability to real-world measurements.

Coupling the phasor-based network with the denoising autoencoding scheme, we name our method as Learning to Enhance Aperture Phasor field (LEAP). We validate our model in sparse sampling and smaller aperture scenarios, on both confocal and non-confocal measurements. The experimental results showcase the effectiveness of our phasor-based enhancement network, demonstrating that we can look around the corners with 16× or 64× fewer samplings, and 4× smaller apertures, all without incurring additional costs.

2 Related Work

NLOS Imaging and Methods. The concept of NLOS imaging was originally proposed in [15] and experimentally validated in [39]. It has been further developed in subsequent research involving SPAD sensors. Methods for NLOS imaging can be broadly divided into two streams: active [12,13,20,27,37] methods with a controllable light source and passive methods relying on the indirect light [3,4,31,32,36,44]. We follow the line of the active methods using a SPAD sensor and a laser, which usually offer a wider range of reconstructable objects.

Early methods of active NLOS imaging propose back-projection based solutions [1,2,16,38] with $O(N^5)$ computational complexity. Successive researches [20,21,27,45] alleviate such computational costs by fast Fourier transform (FFT) based inverse methods with $O(N^3 log N)$ complexity. These include Light Cone Transform (LCT) [27] based on 3D convolution, DLCT with a vector deconvolution [45], and the wave-based solution [20] using Stolt's method.

Recently proposed phasor field NLOS methods [21,22] formulate NLOS imaging as a wave propagation problem. These methods show that NLOS imaging can be solved with well-established line-of-sight propagation operators such as the Rayleigh-Sommerfeld diffraction (RSD) integral. The faster RSD algorithm with ring and radius-based samplings [14] further boosts up the efficiency. These methods, including the phasor field and other inverse methods, exhibit remarkable results on measurements captured with sufficient samplings and scan areas. Our goal is to extend these recent advances to more practical scanning setups.

NLOS Imaging in Practical Scenarios. Several attempts [12,19,23,25,28, 42,43] have been made to address NLOS imaging with reduced acquisition time.

One promising approach is increasing the number of pixels with the arrays of high-end SPAD sensors [25,28]. Despite clear advantages, SPAD arrays also introduce additional hardware costs depending on the number of sensing pixels. Another stream of works explores optimization-based algorithms for sparse sampling scenarios [23,43], which suffer from huge computational costs of the optimization. Apart from these works, our method achieves more practical NLOS imaging with negligible computational costs and without additional hardware.

Deep Learning and NLOS. A number of recent methods employ neural networks for NLOS imaging [6,7,17,24,29,33,47]. Due to the generalization issue of the first learning-based method [7], successive studies [6,17,24,46] employ the physics-based models after the lightweight convolution layers, and focus on refining propagated spatial feature volumes with learned priors. These models rely on the employed propagators and have insufficient capacity to extract meaningful representations in partial sampling scenarios. Recently, to address NLOS imaging with fewer samplings, Signal Super-resolution Network (SSN) [41] performs super-resolution on sparsely sampled measurements, and Li et al. extend LFE [6] by employing signal recovery network before propagating feature volumes. However, solitary learning of the signal recovery problem leads SSN to be vulnerable to noise. We also illustrate that the naïve application of the denoising criterion or its incorporation with volume refinement [18] often fail to bring meaningful improvement, highlighting the necessity of adequate frequency management.

3 Proposed Method

3.1 Preliminary of Phasor Field NLOS Imaging

The goal of NLOS imaging is to reconstruct hidden scenes from measurements of indirect multi-bounce light reflections. Short laser pulses illuminate a set of points $\mathbf{x_p}$ on a relay wall P. The light scatters towards the hidden object and some photons hit the object and return back to the relay wall. Scanning a set of points $\mathbf{x_c}$ on a relay wall C produces the impulse response $H(\mathbf{x_p} \to \mathbf{x_c}, t)$.

Recent phasor field NLOS methods have demonstrated that NLOS imaging can be viewed as a diffractive wave propagation problem with a virtual camera, which can be solved with line-of-sight diffraction operators [21,22]. The phasor wavefront at the virtual aperture can be computed from $H(\mathbf{x_p} \to \mathbf{x_c}, t)$:

$$\mathcal{P}(\mathbf{x_c}, t) = \int_P [\mathcal{P}(\mathbf{x_p}, t) * H(\mathbf{x_p} \to \mathbf{x_c}, t)]\, d\mathbf{x_p}, \qquad (1)$$

where $\mathcal{P}(\mathbf{x_p}, t)$ is the wavefront of the virtual illumination source and $*$ is the convolution operator in time. The hidden scenes can be reconstructed from $\mathcal{P}(\mathbf{x_c}, t)$ using the wave propagation operator $\Phi(\cdot)$:

$$I(\mathbf{x_v}) = \Phi(\mathcal{P}(\mathbf{x_c}, t)), \qquad (2)$$

where $\mathbf{x_v}$ is a point in the hidden scenes being imaged. The propagation operator $\Phi(\cdot)$ is commonly formulated using the Rayleigh-Sommerfeld Diffraction (RSD) integral [14,21]. Despite their remarkable results, the reconstruction quality of diffraction-based NLOS methods depends on the quality of the measurements.

Resolution Limit. The spatial resolution of the phasor camera is determined as $0.61\lambda L/d$, where λ is the wavelength, L is the imaging distance and d is the diameter of the virtual aperture [22]. Since the minimum achievable wavelength is determined by the sampling distance Δ_p ($\lambda > 2\Delta_p$), increasing Δ_p or reducing d theoretically limits the spatial resolution of the systems. We aim to increase the achievable resolution of the imaging systems by exploiting learned priors, to recover full measurements from partial inputs.

Illumination Phasor Field. The phasor field NLOS imaging is mostly implemented as a virtual transient camera with a short Gaussian shape flash [21,22]. The corresponding illumination function is $\mathcal{P}(\mathbf{x_p}, t) = \delta(\mathbf{x_p} - \mathbf{x_{ls}})(e^{i\Omega_c t}e^{-\frac{t^2}{2\sigma^2}})$, and its Fourier domain representation can be expressed as

$$\mathcal{P}_\mathcal{F}(\mathbf{x_p}, \Omega) = \delta(\mathbf{x_p} - \mathbf{x_{ls}}) \left(2\pi\delta(\Omega - \Omega_C) * \sigma\sqrt{2\pi}e^{-\frac{\sigma^2\Omega^2}{2}}\right), \quad (3)$$

where $\mathbf{x_{ls}}$ is the virtual light source position and Ω_C is the central frequency determined by the wavelength λ. The illumination phasor field in the frequency domain $\mathcal{P}_\mathcal{F}(\mathbf{x_p}, \Omega)$ is defined as Gaussian (Fig. 2, top-left), which works as a band-pass filter. The computed phasor wavefront at the aperture is thus band-limited signals, indicating that signals only in a certain frequency range are necessary for reconstructing hidden scenes.

3.2 Denoising and Frequency of Interests

NLOS imaging suffers from measurements with an extremely low signal-to-noise ratio (SNR). Reducing the number of scan points, and thereby reducing the number of total detected photons, amplifies the effects of noise. Since sensor noise is commonly modeled as Poisson distribution [10,26], we propose to apply the denoising criterion for Poisson noise on top of the signal recovery problem [41]. Unfortunately, such a training scheme guides the network to recover the entire frequency components, due to the effects of the sensor noise across the entire spectrum [10]. The network trained with this scheme often yields degraded and over-smoothed results with missing fine details.

To examine the effects of noise on frequency components, we visualize the frequency components of the rendered measurements of Stanford Bunny, both with and without noise. For better understanding, we also visualize reconstructed scenes using FK [20] and a band-pass filter, retaining frequency components within a given range and discards others (details in Supplement A). As shown in Fig. 2, informative signals are mostly observed around the central frequency of the illumination function (range B). By adding Poisson noise, some artifacts appear in a lower frequency range and higher frequency components become indistinguishable from noise. On the other hand, signals near the central frequency still contain clearly visible shapes of objects, are more robust to the noise, and thus are easier to recover. This motivates us to restrict our network's spectrum, by utilizing the aperture phasor field as band-limited inputs and outputs.

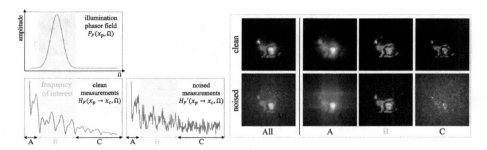

Fig. 2. (**left**) The illumination function in the frequency domain (top), and amplitudes of measurements of the Stanford bunny, both clean and with Poisson noise (bottom). Signals at the center pixel are visualized. (**right**) Reconstruction results of FK [20] on frequency-filtered measurements. Informative signals are mostly observed in a certain frequency range, whereas the Poisson noise affects across the entire spectrum [10].

3.3 Learning to Enhance Aperture Phasor Field

Based on the above observations, we propose the phasor-based neural network, coined as Learning to Enhance Aperture Phasor field (LEAP), which can predict clean and full measurements from noisy partial observations. We assume a single virtual illumination point $\mathbf{x_{ls}}$ for simplicity, which removes the integral over P.

Figure 3 depicts the overview of our proposed method. We begin by sampling partial inputs from full measurements $H(\mathbf{x_p} \rightarrow \mathbf{x_c}, t)$ and corrupting them with Poisson noise. Then the enhancement network takes these noisy partial inputs and predicts the optimal phasor field at the aperture, containing full scans and clean signals, in the frequency domain. We train our network by minimizing the L1 distance between the predicted and the optimal phasor field at the aperture. After training, hidden scenes are reconstructed by propagating the predicted phasor field using the RSD algorithm. We describe details on each part below.

Sensor Noise Simulation. To simulate the strong effects of the noise in NLOS imaging, we follow the computational model of SPAD [6,26,31] and utilize Poisson distribution to model the sensor noise. Considering the cumulative photon counting procedure, we model the sensor noise with multiple exposure levels as

$$\begin{aligned} X &= (\eta \tilde{H}(\mathbf{x_p} \rightarrow \mathbf{x_c}) * g) + d, \\ H'(\mathbf{x_p} \rightarrow \mathbf{x_c}, t) &\sim Poisson(c \cdot X), \end{aligned} \quad (4)$$

where H' is noised measurements, $\tilde{H}$ is partial measurements subsampled from H. η is the photon detection efficiency and g models the time jitter [26]. c controls the exposure time and d models the background noise, including both ambient light and dark counts. Once the measurements are partially sampled and corrupted with noise, they are taken to the network as inputs.

Input Phasor Field Convolution. The noise-augmented partial inputs are then convolved with multiple illumination functions. We employ a set of illumination wavefronts with multiple wavelengths, of which frequency ranges are chosen

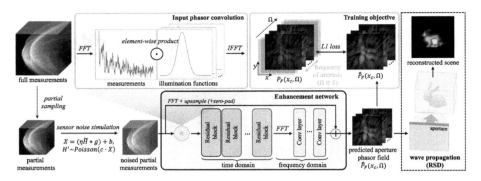

Fig. 3. The overview of the proposed LEAP. Our model takes noisy partial measurements and learns to predict clean and complete phasor wavefronts at the aperture. Hidden scenes are reconstructed by propagating the predicted phasor field with RSD [21].

to be near the target frequency range. The convolved outputs $F = \{f_1, f_2, ..., f_i\}$ with the multiple wavelengths $\{\lambda_1, \lambda_2, ..., \lambda_i\}$ are computed in the frequency domain using the convolution theorem, which can be described as

$$f_i(\mathbf{x_c}, t) = \mathcal{F}^{-1}(\mathcal{F}(H'(\mathbf{x_p} \to \mathbf{x_c}), t) \cdot \mathcal{P}_\mathcal{F}^i(\mathbf{x_p}, \Omega)), \tag{5}$$

where $\mathcal{P}_\mathcal{F}^i(\mathbf{x_p}, \Omega)$ is the illumination phasor field in the frequency domain with a wavelength λ_i. Both real and imaginary components of F are concatenated and passed to the enhancement network for feature extraction.

Enhancement Network. We employ a 3D residual convolutional neural network (CNN) as the enhancement network. Our network extracts feature volumes from F with several 3D residual blocks, and then transforms them into the frequency domain. Then 3 convolution layers further extract features from these frequency volumes and predict the residuals, both real and imaginary parts in the frequency domain. The residuals are then added with the upsampled (and zero-padded in smaller aperture cases) inputs to predict the clean and full measurements $\hat{H}(\mathbf{x_p} \to \mathbf{x_c}, \Omega)$. Our model finally computes the aperture phasor field:

$$\hat{\mathcal{P}}_\mathcal{F}(\mathbf{x_c}, \Omega) = \hat{H}(\mathbf{x_p} \to \mathbf{x_c}, \Omega) \cdot \mathcal{P}_\mathcal{F}(\mathbf{x_p}, \Omega), \tag{6}$$

where $\mathcal{P}_\mathcal{F}(\mathbf{x_p}, \Omega)$ is the target illumination phasor wavefront with the wavelength λ_T, which will also be used to compute the ground truth phasor field. We provide details of the network architecture in Supplement B.2.

Compared with SSN [41], our network only consists of 3D convolution blocks and does not include computationally expensive attention branches, resulting in the improved efficiency of our model (see Sect. 5, runtime analysis).

Training Objective and Reconstruction. To train the network, we minimize the L1 distance between the predicted phasor field $\hat{\mathcal{P}}_\mathcal{F}(\mathbf{x_c}, \Omega)$ and the target phasor field $\mathcal{P}_\mathcal{F}(\mathbf{x_c}, \Omega)$ at the aperture. The target aperture wavefront is computed by convolve the optimal measurements with the target illumination function $\mathcal{P}_\mathcal{F}(\mathbf{x_p}, t)$. Since the informative frequency components needed for reconstructing hidden scenes are determined by λ_T, we only minimize the loss for such components. The training objective of our method can be described as

$$\mathcal{L} = \sum_{\Omega' \in S} ||\hat{\mathcal{P}}_\mathcal{F}(\mathbf{x_c}, \Omega') - \mathcal{P}_\mathcal{F}(\mathbf{x_c}, \Omega')||_1, \qquad (7)$$

where $S = [\Omega_C - \Delta\Omega, \Omega_C + \Delta\Omega]$ is a range where the coefficients of input wavefront is larger than a peak ratio γ, and the central frequency Ω_C is determined by the target wavelength λ_T. Supervising the network with the aperture wavefront restricts the frequency spectrum of the training objective, confining the operations of our network to the frequency range of interests.

Once we predict the enhanced phasor field at the aperture, the hidden scenes can be reconstructed using the existing wave propagation operators. We utilize the 2D fast Fourier transform (FFT) based RSD algorithm [21] in this work.

4 Experiment

To demonstrate the effectiveness of the proposed method, we conduct the experiments in two practical acquisition scenarios: (1) sparse sampling and (2) scanning with a smaller aperture, of which scanning patterns are depicted in Fig. 1 (b). Details of the experimental setup can be found in Supplement C, D.

Evaluation Scenarios. The acquisition scenarios are more specifically divided into 4 setups: confocal sparse scanning with 16 × 16 samplings (denoted as Conf-16), confocal sparse scanning with 8 × 8 points (Conf-8), confocally scanning the smaller area with size 1 m × 1 m and 16 × 16 samplings (Conf-small), and non-confocal sparse scanning of 16 × 16 points (Non-16).

Following previous works [18,23,41,43], we assess the performance in partial sampling scenarios with the subset of the measurements, of which scanning points are evenly sampled from full measurements with appropriate spatial strides (and center-crop for the smaller aperture case). We aim to recover target measurements with 2 m × 2 m apertures, 32 ps bin resolution and 64 × 64 samplings, which are then used to reconstruct 64 × 64 × 64 hidden volumes.

Baselines. We compare our method with several representative baselines: FK [20], LCT [27], RSD [21] with nearest and trilinear interpolation methods, SSCR [23] as an optimization-based few-shot NLOS method, and LFE [6], USM [18], SSN [41] as learning-based baselines. SSN and USM are designed to solve NLOS imaging with partial measurements: SSN recovers missing signals from partial measurements, while USM extends the architecture of LFE, consisting of the signal recovery network, the feature propagator, and the volume refinement module.

We follow the original paper to reproduce SSN [41] as their codes are not available. Results of the learning-based methods are reproduced with our synthetic dataset. For a fair comparison, LFE and USM are modified to employ RSD as a propagator, are trained using our noise augmentation and supervised with 2D labels generated by projecting the outputs of RSD using the optimal measurements. We refer to Supplement E for more baselines, details, and results.

Implementation Detail. Our model is implemented with PyTorch and trained 160 epochs using a single RTX A5000 GPU, which takes less than a day in our environment. We employ 7 wavelengths for the input phasor field convolution, $\gamma = 0.1$, and the target wavelength $\lambda_T = 9.375$ cm. We deliver 2D projected results of all methods, our model and baselines, obtained with maximum intensity projection. More implementation details are described in Supplement D.

Table 1. Quantitative results on the synthetic dataset. RMSE values of USM [18] are omitted as its original version does not include depth map reconstruction.

Method	Conf-16			Conf-8			Conf-small			Non-16		
	PSNR↑	SSIM↑	RMSE↓	PSNR↑	SSIM↑	RMSE↓	PSNR↑	SSIM↑	RMSE↓	PSNR↑	SSIM↑	RMSE↓
RSD$_{Nearest}$	14.85	0.1515	0.8232	12.65	0.0855	0.8919	19.73	0.3743	0.3073	19.67	0.3218	0.5020
RSD$_{Linear}$	14.62	0.1631	0.7536	11.28	0.0760	0.8884	19.90	0.4664	0.2046	19.29	0.3224	0.4856
LFE [6]	23.05	0.6729	0.2247	17.45	0.4098	0.3282	22.92	0.6826	0.2852	29.47	0.8077	0.2898
USM [18]	29.99	**0.8994**	-	25.75	0.8235	-	26.94	0.8519	-	35.14	0.9313	-
SSN [41]	23.27	0.4506	0.2699	21.49	0.4426	0.2259	21.98	0.4629	0.2036	29.45	0.6367	0.1798
Ours	**32.02**	0.8949	**0.0892**	**28.07**	**0.8472**	**0.0962**	**28.31**	**0.8556**	**0.0969**	**37.45**	**0.9625**	**0.1414**

4.1 Synthetic Dataset Evaluation

To train and validate our model, we generate a synthetic NLOS dataset from ShapeNet [5] using the NLOS renderer of [6]. We use 15,000 objects from all categories for generation, 11,000 objects for training and 4,000 objects for validation. The generated synthetic dataset consists of measurements with 2 m × 2 m scan area, 64 × 64 sampling points and 32 ps bin resolution with time jitter.

For quantitative comparisons, We measure peak-signal-to-noise ratio (PSNR) and structural similarity index (SSIM) for the visual quality, and root-mean-square error (RMSE) for the accuracy of the reconstructed geometry. The 2D projected results of RSD with the optimal measurements are served as the ground truth intensity images. Due to the variance of reconstructed albedo values, we compare with methods based on RSD: RSD with nearest (RSD$_{Nearest}$) and trilinear interpolations (RSD$_{Linear}$), and learning-based methods. Measurements with the sensor noise model in Sect. 3.3 are used for the evaluation.

Results. Table 1 reports the quantitative results on the synthetic dataset in all 4 evaluation scenarios. The proposed model outperforms all other methods in both terms of visual quality and geometry. RSD with both interpolation methods produce worst results. SSN [41] fails to learn robust representations from noisy measurements and yields inaccurate results. LFE also fails to deliver meaningful results, indicating difficulties in exploiting learned priors from inaccurately propagated feature volumes (see results of RSD). By exploiting the signal recovery network on top of LFE, USM achieves improvement compared to LFE and

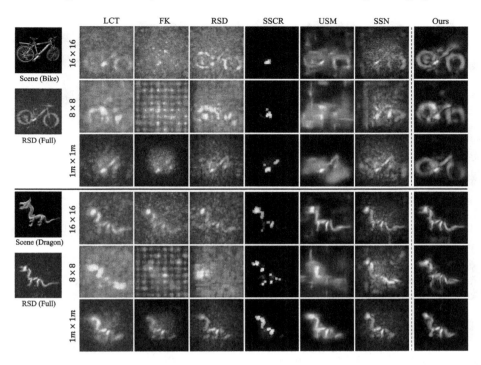

Fig. 4. Qualitative results on Bike, Dragon from the Stanford real-world dataset [20]. We report the results of FK, LCT, and RSD with the nearest interpolation. Evaluation scenarios involve 16 × 16 and 8 × 8 sparse samplings with 2 m × 2 m apertures, and the 1 m × 1 m smaller aperture with 16 × 16 samplings.

delivers the second best results. Nevertheless, the performance gap between USM and our model indicates that our phasor-based scheme guides our model to effectively extract informative signals from noisy partial inputs. For more results on the synthetic dataset, please refer to Supplement F.1.

4.2 Confocal Real-World Evaluation

To evaluate the generalization capability to real-world measurements, we compare the results on the Stanford confocal real-world dataset [20]. We choose Bike and Dragon instances for the evaluation, which have lower photon counts and SNR compared to other instances, making them more challenging to reconstruct. The original measurements of Stanford dataset are captured with 512 × 512 samplings, 2 m × 2 m apertures and 32 ps bin resolution. We first downsample the measurements by 2×, which results in slight increase of the exposure time per pixel. Then we sub-sample the partial measurements with spatial strides and center crop. We use measurements with approximately 55 ms exposure per pixel, which corresponds to the 60 min exposure time of the original ones. Results of

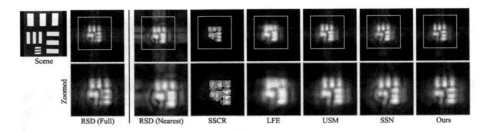

Fig. 5. Qualitative results of non-confocal 16 × 16 sparse samplings, on Resolution of the real-world dataset [21]. All methods employing signal recovery networks produce plausible results with sufficiently long exposure time and white diffuse objects.

the trilinear interpolation and LFE are omitted in this section due to the space constraints, and they can be found in Supplement E.3.

Sparse Sampling. Results in the 16 × 16 sparse sampling scenario are reported in Fig. 4 (first, fourth rows). Our model outperforms all other baselines, producing clean results with details. The inverse NLOS methods with the nearest interpolation only reconstruct coarse shapes of the objects, most of which are hardly identifiable due to the artifacts. SSCR only reconstructs some parts of the objects. SSN fails to learn noise-robust representations due to the absence of denoising, leading to outputs with severe artifacts. USM produces plausible outputs in general, yet some artifacts can be observed in its results. In contrast, our model successfully reconstructs clean shapes of the objects with fine details, e.g., the rear wheel of Bike and the head of Dragon.

The efficacy of our method is more highlighted with 8 × 8 sparse sampling, having 4× shorter scanning time than 16 × 16. As shown in the second, fifth rows of Fig. 4, none of the baselines deliver plausible results. On the other hand, our method delivers the results where several parts of the objects, e.g. the wheels of Bike and the legs of Dragon, are clearly visible. Our results in sparse sampling scenarios showcase the effectiveness of our phasor-based network under real-world noise, leading to significant improvement compared to other baselines.

Smaller Aperture. Our model also achieves high-quality results in the smaller aperture case, as shown in Fig. 4 (third, sixth rows). Results of the nearest interpolation only contain coarse shapes of the objects near the aperture with severe artifacts. Again, SSCR misses many parts of the objects, USM produces distorted shapes of the objects, and SSN suffer from the effects of the noise. The results of our method, in which many parts of the objects placed out of the aperture are reconstructed (the wheels of Bike), demonstrate that the proposed model can realize applications with limited apertures.

4.3 Non-confocal Real-World Evaluation

Finally, we evaluate our method in the non-confocal 16 × 16 sparse sampling scenario. We compare the results on the measurements of Resolution provided

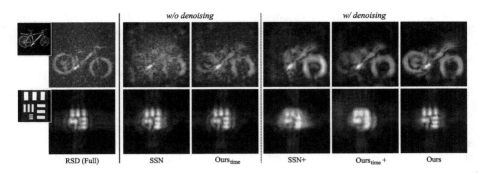

Fig. 6. Qualitative ablation results on the denoising criterion. Results in 16 × 16 sparse sampling scenarios under both confocal (top) and non-confocal (bottom) setups are reported. While the naïve addition of denoising effectively reduces background noise, it often yields over-smoothed and degraded results with missing details.

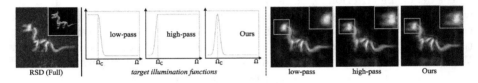

Fig. 7. Qualitative ablation results on the frequency filtering. Left plots indicate the modified target illumination function used for the models. Results on 16 × 16 sparsely sampled confocal measurements of Dragon are reported.

in [21], captured with 1.8 m × 1.3 m apertures, 1 cm sampling distance, 4 ps bin resolution and 1 s exposure per pixel. We first apply spatial zero-pad and temporal averaging to the measurements to make 2 m × 2 m apertures and 32 ps bin resolution. Then the partial inputs are sub-sampled with spatial strides.

As in Fig. 5, our model exhibits much improved performance, showing the results closes to the results of RSD with full measurements. On the other hand, results of RSD with the interpolation methods are blurry and contain some artifacts. LFE only reconstructs coarse shapes of the object. Interestingly, all learning-based methods employing signal recovery networks (USM, SSN, Ours) seem to produce compelling results under the low-noise condition, where the exposure time per pixel is sufficient (about 20× longer than the Stanford dataset) and the target objects with high-reflectivity. We further probe effects of the denoising criterion and noise robustness of the models in the following sections, and provide more results in the sparse non-confocal scenario in Supplement F.6.

4.4 Ablation Study

To verify the effectiveness of the proposed concepts, we conduct the ablations in both confocal and non-confocal 16 × 16 sparse sampling scenarios.

Table 2. Ablation results on the denoising criterion. '+' indicates that the models (SSN+, Ours$_{time}$+) are trained with the denoising criterion.

Method	Conf-16		Non-16	
	PSNR↑	RMSE↓	PSNR↑	RMSE↓
SSN [41]	23.27	0.2699	29.45	0.1798
Ours$_{time}$	23.85	0.2352	29.89	0.1736
SSN+	29.55	0.0949	35.47	0.1435
Ours$_{time}$+	30.69	0.0924	36.36	0.1423
Ours	**32.02**	**0.0892**	**37.45**	**0.1414**

Table 3. Ablation results on the phasor-based frequency filtering. The "low-pass" model passes frequencies lower than the central frequency Ω_C, and the "high-pass" model passes frequencies higher than Ω_C.

Method	Conf-16		Non-16	
	PSNR↑	RMSE↓	PSNR↑	RMSE↓
low-pass	31.23	0.0903	36.72	0.1420
high-pass	31.54	0.0909	37.09	0.1422
Ours	**32.02**	**0.0892**	**37.45**	**0.1414**

Denoising and the Phasor-based Network. We first conduct the ablation on the denoising criterion for the signal recovery problem. We compare our model with several variants of signal recovery networks: SSN [41] and the same network trained with the denoising criterion (SSN+), our enhancement network in the time domain (w/o phasor-based scheme, Ours$_{time}$) with and without denoising. We report the results on Bike and Resolution for the real-world evaluations.

The ablation results on the synthetic dataset and the real-world measurements are reported in Table 2 and Fig. 6. Adding denoising criterion greatly helps the network to remove background noise and learn more robust representations. However, as shown in Fig. 6, applying denoising criterion often results in over-smoothed outputs (Resolution) and missing details of the objects (the rear wheel of Bike). By applying the phasor-based scheme, our model achieves both quantitative and qualitative improvements, producing clean results with fine details of the objects. Notably, attention branches employed in SSN make no meaningful improvement but rather results in degraded outputs and expensive computations compared to our model. Based on these results, we conclude that the naïve application of the denoising often leads to the unfavorable results, highlighting the effectiveness of our phasor-based network.

Frequency Filtering. Next, we explore the effects of frequency filtering by modifying the target illumination function. We compare with two illumination functions: (1) passing all frequencies smaller than the central frequency Ω_C (low-pass), (2) passing all frequencies higher than Ω_C (high-pass). Shapes of each illumination function are shown in Fig. 7, with their frequency ranges closely related to those in Fig. 2. These models also employ additional wavelengths in the input phasor convolution to sufficiently cover the target frequency ranges.

As reported in Table 3 and Fig. 7, both low-pass and high-pass models deliver degraded results compared to our model. Interestingly, the low-pass model delivers worse results than the high-pass model, failing to reconstruct details of the objects in the real-world (see Fig. 7, the head of Dragon). This also corresponds to the observations made in prior works [8,30], which discovered the spectral bias of neural networks towards lower-frequency signals. Such results demonstrate that employing phasor wavefronts as band-limited signals for the supervision guides our network to effectively avoid the spectral bias.

5 Analysis and Discussion

Runtime Analysis. Our network takes 20.7 ms for processing measurements with 16×16 samplings, and the GPU version of RSD takes 3.9 ms to reconstruct $64 \times 64 \times 64$ volumes. On the other hand, SSN employs computationally expensive attention branches and thus takes 72.9 ms to process 16×16 inputs. As a result, our entire pipeline takes less than 25 ms, highlighting the suitability of our method for real-time applications. The latency is measured using a single RTX 3090 GPU. Please refer to Supplement E.3 for comparisons with more baselines.

Noise Robustness of Learning-based Methods. To examine the noise robustness of learning-based methods, we report the results on Bike with shorter and longer exposure times per pixel. As shown in Fig. 8, all methods produce plausible results with sufficiently long exposure time. While the denoising criterion seems less effective in such low-noise conditions, its effects are clearly highlighted when the exposure time (and thereby the total scanning time) is reduced. Simple addition of denoising to SSN (SSN+) or incorporating with the volume refinement network (USM) fail to reveal some details of the object, e.g. the wheels of Bike. Our model consistently deliver high-quality results and presents its noise robustness with a shorter exposure time and a shorter scanning time.

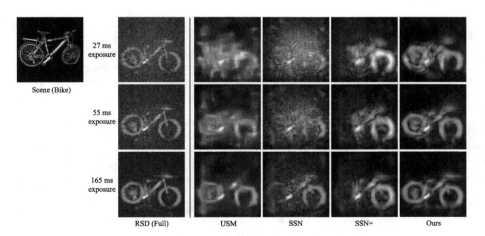

Fig. 8. Results on sparse 16×16 confocal measurements of Bike [20], with shorter (27 ms per pixel) and longer (165 ms per pixel) exposure time. The denoising criterion becomes evidently effective as the exposure time is reduced, but it fails to reveal some details of the object. Our method consistently presents high-quality results, exhibiting the noise robustness of the proposed phasor-based scheme.

Incorporating with Other NLOS Methods. While our model predicts the optimal phasor wavefronts at the aperture in the frequency domain, we can still incorporate with other inverse NLOS methods with slight modification. We refer to Supplement F.3 for more information and results with other NLOS methods.

Limitation and Future Works. Since the proposed LEAP exploits learned priors from the synthetic dataset, the performance of our model is affected by the quality of the generated samples. This can be further improved by using more complicated objects and precise NLOS renderer [9]. In addition, incorporating recent advances in generative diffusion models [11,34,35] would also be an interesting direction for the future works.

6 Conclusion

We proposed the learning-based method LEAP, which learns to enhance noisy partial measurements and enables NLOS imaging with fewer samplings and smaller apertures. Our enhancement network with the phasor-based scheme presented its effectiveness in various scanning scenarios, showing high-fidelity results while being robust to the noise. We believe our method can serve as an effective solution to address the exhaustive scanning procedures in NLOS imaging.

Acknowledgements. This work was supported by the Samsung Research Funding Center (SRFC-IT2001-04), Artificial Intelligence Innovation Hub under Grant RS-2021-II212068, and an Institute of Information & communications Technology Planning & Evaluation (IITP) grant funded by the Korean Government (MSIT) (No. RS-2020-II201361, Artificial Intelligence Graduate School Program (Yonsei University)).

References

1. Ahn, B., Dave, A., Veeraraghavan, A., Gkioulekas, I., Sankaranarayanan, A.C.: Convolutional approximations to the general non-line-of-sight imaging operator. In: Proceedings of the IEEE/CVF International Conference on Computer Vision (ICCV), pp. 7889–7899 (2019)
2. Arellano, V., Gutierrez, D., Jarabo, A.: Fast back-projection for non-line of sight reconstruction. Opt. Express **25**(10), 11574–11583 (2017)
3. Batarseh, M., Sukhov, S., Shen, Z., Gemar, H., Rezvani, R., Dogariu, A.: Passive sensing around the corner using spatial coherence. Nat. Commun. **9**(1), 1–6 (2018)
4. Boger-Lombard, J., Katz, O.: Passive optical time-of-flight for non line-of-sight localization. Nat. Commun. **10**(1), 1–9 (2019)
5. Chang, A.X., et al.: ShapeNet: an information-rich 3D model repository. Tech. Rep. arXiv:1512.03012 [cs.GR], Stanford University — Princeton University — Toyota Technological Institute at Chicago (2015)
6. Chen, W., Wei, F., Kutulakos, K.N., Rusinkiewicz, S., Heide, F.: Learned feature embeddings for non-line-of-sight imaging and recognition. ACM Trans. Graph. (TOG) **39**(6), 1–18 (2020)
7. Chopite, J.G., Hullin, M.B., Wand, M., Iseringhausen, J.: Deep non-line-of-sight reconstruction. In: Proceedings of the IEEE/CVF Conference on Computer Vision and Pattern Recognition (CVPR), pp. 960–969 (2020)

8. Fridovich-Keil, S., Gontijo Lopes, R., Roelofs, R.: Spectral bias in practice: the role of function frequency in generalization. Adv. Neural Inf. Process. Syst. (NeurIPS) **35**, 7368–7382 (2022)
9. Galindo, M., Marco, J., O'Toole, M., Wetzstein, G., Gutierrez, D., Jarabo, A.: A dataset for benchmarking time-resolved non-line-of-sight imaging (2019). https://graphics.unizar.es/nlos
10. Hernandez, Q., Gutierrez, D., Jarabo, A.: A computational model of a single-photon avalanche diode sensor for transient imaging. arXiv preprint arXiv:1703.02635 (2017)
11. Ho, J., Jain, A., Abbeel, P.: Denoising diffusion probabilistic models. Adv. Neural Inf. Process. Syst. (NeurIPS) **33**, 6840–6851 (2020)
12. Isogawa, M., Chan, D., Yuan, Y., Kitani, K., O'Toole, M.: Efficient non-line-of-sight imaging from transient Sinograms. In: Vedaldi, A., Bischof, H., Brox, T., Frahm, J.-M. (eds.) ECCV 2020. LNCS, vol. 12352, pp. 193–208. Springer, Cham (2020). https://doi.org/10.1007/978-3-030-58571-6_12
13. Isogawa, M., Yuan, Y., O'Toole, M., Kitani, K.M.: Optical non-line-of-sight physics-based 3D human pose estimation. In: Proceedings of the IEEE/CVF Conference on Computer Vision and Pattern Recognition (CVPR), pp. 7013–7022 (2020)
14. Jiang, D., Liu, X., Luo, J., Liao, Z., Velten, A., Lou, X.: Ring and radius sampling based phasor field diffraction algorithm for non-line-of-sight reconstruction. IEEE Trans. Pattern Anal. Mach. Intell. (TPAMI) **44**(11), 7841–7853 (2021)
15. Kirmani, A., Hutchison, T., Davis, J., Raskar, R.: Looking around the corner using transient imaging. In: Proceedings of the IEEE/CVF International Conference on Computer Vision (ICCV), pp. 159–166. IEEE (2009)
16. Laurenzis, M., Velten, A.: Nonline-of-sight laser gated viewing of scattered photons. Opt. Eng. **53**(2), 023102 (2014)
17. Li, Y., Peng, J., Ye, J., Zhang, Y., Xu, F., Xiong, Z.: NLOST: Non-line-of-sight imaging with transformer. In: Proceedings of the IEEE/CVF Conference on Computer Vision and Pattern Recognition (CVPR), pp. 13313–13322 (2023)
18. Li, Y., Zhang, Y., Ye, J., Xu, F., Xiong, Z.: Deep non-line-of-sight imaging from under-scanning measurements. In: Advances in Neural Information Processing Systems (NeurIPS), vol. 36 (2023)
19. Liao, Z., Jiang, D., Liu, X., Velten, A., Ha, Y., Lou, X.: FPGA accelerator for real-time non-line-of-sight imaging. IEEE Trans. Circuits Syst. I Regul. Pap. **69**(2), 721–734 (2021)
20. Lindell, D.B., Wetzstein, G., O'Toole, M.: Wave-based non-line-of-sight imaging using fast f-k migration. ACM Trans. Graph. (TOG) **38**(4), 1–13 (2019)
21. Liu, X., Bauer, S., Velten, A.: Phasor field diffraction based reconstruction for fast non-line-of-sight imaging systems. Nat. Commun. **11**(1), 1–13 (2020)
22. Liu, X., et al.: Non-line-of-sight imaging using phasor-field virtual wave optics. Nature **572**(7771), 620–623 (2019)
23. Liu, X., Wang, J., Xiao, L., Fu, X., Qiu, L., Shi, Z.: Few-shot non-line-of-sight imaging with signal-surface collaborative regularization. In: Proceedings of the IEEE/CVF Conference on Computer Vision and Pattern Recognition (CVPR), pp. 13303–13312 (2023)
24. Mu, F., et al.: Physics to the rescue: deep non-line-of-sight reconstruction for high-speed imaging. IEEE Trans. Pattern Anal. Mach. Intell. (TPAMI) (2022)
25. Nam, J.H., et al.: Low-latency time-of-flight non-line-of-sight imaging at 5 frames per second. Nat. Commun. **12**(1), 6526 (2021)

26. O'Toole, M., Heide, F., Lindell, D.B., Zang, K., Diamond, S., Wetzstein, G.: Reconstructing transient images from single-photon sensors. In: Proceedings of the IEEE/CVF Conference on Computer Vision and Pattern Recognition (CVPR), pp. 1539–1547 (2017)
27. O'Toole, M., Lindell, D.B., Wetzstein, G.: Confocal non-line-of-sight imaging based on the light-cone transform. Nature **555**(7696), 338–341 (2018)
28. Pei, C., et al.: Dynamic non-line-of-sight imaging system based on the optimization of point spread functions. Opt. Express **29**(20), 32349–32364 (2021)
29. Plack, M., Callenberg, C., Schneider, M., Hullin, M.B.: Fast differentiable transient rendering for non-line-of-sight reconstruction. In: Proceedings of the IEEE/CVF Winter Conference on Applications of Computer Vision (WACV), pp. 3067–3076 (2023)
30. Rahaman, N., et al.: On the spectral bias of neural networks. In: International conference on machine learning (ICML), pp. 5301–5310. PMLR (2019)
31. Saunders, C., Murray-Bruce, J., Goyal, V.K.: Computational periscopy with an ordinary digital camera. Nature **565**(7740), 472–475 (2019)
32. Seidel, S.W., Murray-Bruce, J., Ma, Y., Yu, C., Freeman, W.T., Goyal, V.K.: Two-dimensional non-line-of-sight scene estimation from a single edge occluder. IEEE Trans. Comput. Imaging **7**, 58–72 (2020)
33. Shen, S., et al.: Non-line-of-sight imaging via neural transient fields. IEEE Trans. Pattern Anal. Mach. Intell. (TPAMI) **43**(7), 2257–2268 (2021)
34. Song, Y., Ermon, S.: Generative modeling by estimating gradients of the data distribution. In: Advances in Neural Information Processing Systems (NeurIPS), vol. 32 (2019)
35. Song, Y., Sohl-Dickstein, J., Kingma, D.P., Kumar, A., Ermon, S., Poole, B.: Score-based generative modeling through stochastic differential equations. In: International Conference on Learning Representations (ICLR) (2020)
36. Tanaka, K., Mukaigawa, Y., Kadambi, A.: Polarized non-line-of-sight imaging. In: Proceedings of the IEEE/CVF Conference on Computer Vision and Pattern Recognition (CVPR), pp. 2136–2145 (2020)
37. Tsai, C.Y., Kutulakos, K.N., Narasimhan, S.G., Sankaranarayanan, A.C.: The geometry of first-returning photons for non-line-of-sight imaging. In: Proceedings of the IEEE/CVF Conference on Computer Vision and Pattern Recognition (CVPR), pp. 7216–7224 (2017)
38. Velten, A., Willwacher, T., Gupta, O., Veeraraghavan, A., Bawendi, M.G., Raskar, R.: Recovering three-dimensional shape around a corner using ultrafast time-of-flight imaging. Nat. Commun. **3**(1), 1–8 (2012)
39. Velten, A., et al.: Femto-photography: capturing and visualizing the propagation of light. ACM Trans. Graph. (TOG) **32**(4), 1–8 (2013)
40. Vincent, P., Larochelle, H., Bengio, Y., Manzagol, P.A.: Extracting and composing robust features with denoising autoencoders. In: International Conference on Machine Learning (ICML), pp. 1096–1103 (2008)
41. Wang, J., Liu, X., Xiao, L., Shi, Z., Qiu, L., Fu, X.: Non-line-of-sight imaging with signal superresolution network. In: Proceedings of the IEEE/CVF Conference on Computer Vision and Pattern Recognition (CVPR) (2023)
42. Willomitzer, F., Rangarajan, P.V., Li, F., Balaji, M.M., Christensen, M.P., Cossairt, O.: Fast non-line-of-sight imaging with high-resolution and wide field of view using synthetic wavelength holography. Nat. Commun. **12**(1), 6647 (2021)
43. Ye, J.T., Huang, X., Li, Z.P., Xu, F.: Compressed sensing for active non-line-of-sight imaging. Opt. Express **29**(2), 1749–1763 (2021)

44. Yedidia, A.B., Baradad, M., Thrampoulidis, C., Freeman, W.T., Wornell, G.W.: Using unknown occluders to recover hidden scenes. In: Proceedings of the IEEE/CVF Conference on Computer Vision and Pattern Recognition (CVPR), pp. 12231–12239 (2019)
45. Young, S.I., Lindell, D.B., Girod, B., Taubman, D., Wetzstein, G.: Non-line-of-sight surface reconstruction using the directional light-cone transform. In: Proceedings of the IEEE/CVF Conference on Computer Vision and Pattern Recognition (CVPR), pp. 1407–1416 (2020)
46. Yu, Y., et al.: Enhancing non-line-of-sight imaging via learnable inverse kernel and attention mechanisms. In: Proceedings of the IEEE/CVF Conference on Computer Vision and Pattern Recognition (CVPR), pp. 10563–10573 (2023)
47. Zhu, D., Cai, W.: Fast non-line-of-sight imaging with two-step deep remapping. ACS Photonics **9**(6), 2046–2055 (2022)

ViewFormer: Exploring Spatiotemporal Modeling for Multi-view 3D Occupancy Perception via View-Guided Transformers

Jinke Li[1]( ), Xiao He[1], Chonghua Zhou[2], Xiaoqiang Cheng[1], Yang Wen[1], and Dan Zhang[1]

[1] Uisee Foundation Research and Development, Beijing, China
jinke.li@uisee.com
[2] University of Science and Technology of China, Hefei, China

Abstract. 3D occupancy, an advanced perception technology for driving scenarios, represents the entire scene without distinguishing between foreground and background by quantifying the physical space into a grid map. The widely adopted projection-first deformable attention, efficient in transforming image features into 3D representations, encounters challenges in aggregating multi-view features due to sensor deployment constraints. To address this issue, we propose our learning-first view attention mechanism for effective multi-view feature aggregation. Moreover, we showcase the scalability of our view attention across diverse multi-view 3D tasks, including map construction and 3D object detection. Leveraging the proposed view attention as well as an additional multi-frame streaming temporal attention, we introduce ViewFormer, a vision-centric transformer-based framework for spatiotemporal feature aggregation. To further explore occupancy-level flow representation, we present FlowOcc3D, a benchmark built on top of existing high-quality datasets. Qualitative and quantitative analyses on this benchmark reveal the potential to represent fine-grained dynamic scenes. Extensive experiments show that our approach significantly outperforms prior state-of-the-art methods. The codes are available at https://github.com/ViewFormerOcc/ViewFormer-Occ.

Keywords: 3D Occupancy · Occupancy flow · Multi-view Interaction · Spatiotemporal Modeling · Streaming video pipeline

1 Introduction

The vision-centric autonomous driving (AD) systems are attracting extensive attention in recent years, promoting the research domain perceiving the real 3D world from 2D images. 3D object detection is a traditional task, representing

Supplementary Information The online version contains supplementary material available at https://doi.org/10.1007/978-3-031-72775-7_6.

foreground objects with limited categories. As illustrated in Fig. 1(a), the pedestrian is detected as a 3D bounding box, while the uncommon suitcase is hardly defined in driving scenarios. Once such a suitcase appears on the road, an AD system relying only on object detection is unable to secure the driving safety. In contrast, the 3D occupancy representation, unifying the concept of foreground and background, and quantifying the entire 3D space into voxel-wise cells with semantic labels, shows superior performance, where objects like suitcases can be effectively defined as category-agnostic occupancies as shown in Fig. 1(b). Beyond static occupancy, occupancy flow is crucial for representing dynamic scenes as shown in Fig. 1(d). In this paper, we introduce *ViewFormer*, a transformer-based framework designed to predict 3D occupancy and occupancy flow with multi-camera images as input.

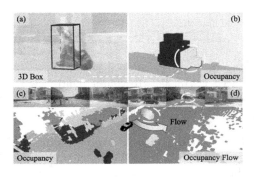

Fig. 1. Treating objects simply as 3D boxes lacks the sense of the background, such as the suitcase in (**a**). Defining 3D space as occupancies (**b**) and (**c**) is more effective in representing objects. Beyond static occupancy, occupancy flow is crucial to perceive dynamic scenes. In the case of a turning car in (**d**), different flow directions of occupancies can be clearly observed.

Typical vision-centric 3D perception frameworks can be decomposed into spatial interaction and temporal interaction components. The former is responsible for transforming image features into 3D space, while the latter, on one hand, enhances the 3D features, and on the other hand, models dynamic scenes to reason velocity information. Regarding spatial interaction, to reduce computational costs, sparse methods have been explored to transform multi-view image features into 3D space [18,27,39]. For instance, deformable attention [42], originally designed for monocular images, is extended to the multi-view field by BEVFormer [18] through projecting 3D reference points onto multiple images, which is referred to as the projection-first method in this paper. This method is widely used to aggregate multi-view features and predict 3D occupancy [13,19,31]. Nevertheless, we identify two predominant issues of this approach.

To gain a better understanding, let's review the projection-first method illustrated in Fig. 2(a), which projects the 3D reference point of a voxel query onto images first and then performs deformable attention [42]. However, a critical limitation arises when a 3D reference point, fixed during training, is projected outside the image size for a specific camera, the projection-first method no longer applies deformable attention to extracting features for this reference point. As a consequence, features from this camera are masked out throughout the entire dataset. Notably, this issue is widespread, as seen in scenarios like nuScenes [5,9], where numerous 3D reference points can only be projected onto a single camera due to sensor deployment. Additionally, as the deformable attention utilized

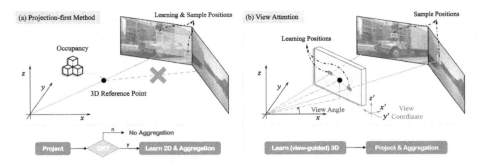

Fig. 2. Constrained by fixed reference points, the projection-first method (**a**) introduced in [18] fails to collect multi-view features. In contrast, our learning-first view attention (**b**) gathers features from multiple cameras more adequately.

in the method learns sample points on the image plane, the 2D sample area corresponding to a 3D object undergoes rapid variations with changes in depth due to perspective transformation, such scale inconsistency poses a convergence challenge to 3D perception tasks.

To address the aforementioned issues, we introduce our learning-first *view attention*, facilitating multi-view feature aggregation. As shown in Fig. 2(b), to overcome the constraint imposed by fixed reference points on feature collection, we adopt a strategy to learn local regions in 3D space for a given query. The corresponding 3D points of these regions are then projected onto multiple images for feature aggregation. Consequently, the extraction of features across cameras becomes a data-driven process. For higher efficiency and faster convergence, we define the learned 3D regions in a local view coordinate (VC) system, providing view guidance. These regions remain invariant as the query's view angle changes, introducing effective rotational invariance in perception around the vehicle. Furthermore, learning regions directly in 3D space preserves a consistent 3D spatial scale, avoiding the challenges correlated to perspective transformation.

Temporal modeling in computer vision often emphasizes the efficiency of leveraging video data [4,24,28]. The popular multi-camera temporal method BEVFormer [18] achieves temporal interaction solely with a single historical frame, leading to limited performance. Besides, in BEVFormer, employing a sliding window approach during training and switching to online video during inference brings inconsistency, and the use of windowed data also increases training time due to redundant inference of many frames. Therefore, we propose our streaming temporal attention mechanism with multi-frame interaction, in which the utilization of a streaming memory mechanism [26,35] significantly reduces training time without additional inference latency.

In aspect of dataset, although prior work [31] delves into object-level occupancy flow by assigning the center velocity of an object to all its internal occupancies, the exploration of occupancy-level flow representation remains limited. As shown in Fig. 1(d), fine-grained occupancy flow provides more detailed information such as motion directions for different parts of a turning car. Furthermore, occupancy-level flow has the potential to represent objects whose shapes vary in motion for future research. Hence, we create our *FlowOcc3D* dataset, building upon nuScenes [5,9] and Occ3D [34] datasets, featuring fine-grained occupancy-level flow annotations.

In summary, our contributions are four main aspects: 1) We identify limitations of the widely used projection-first method and propose our view attention to more effectively transform multi-view features into 3D space. We demonstrate its scalability across various multi-view 3D tasks, which can be new baselines for future multi-view 3D perception research. 2) We introduce ViewFormer, a vision-centric transformer-based framework that incorporates the novel view attention and multi-frame streaming temporal attention, enhancing spatiotemporal modeling for multi-view 3D perception. 3) We create FlowOcc3D, a high-quality occupancy-level flow benchmark. Qualitative and quantitative analyses are conducted to compare models trained on occupancy-flow and object-flow, revealing the potential of fine-grained representation in dynamic scenes. 4) ViewFormer achieves state-of-the-art performance across diverse benchmarks, surpassing previous methods by substantial margins.

2 Related Work

2D Image to 3D Space Transformation. To transform 2D image features to 3D space, bottom-up methods [17,27,29] rely on pixel-wise depth prediction, where the limited reconstruction density due to corresponding image resolution poses a challenge for the demanding dense 3D occupancy representation. In contrast, top-down transformer-based methods [18,23,39] are more flexible by allowing arbitrary predefined resolutions for 3D space and implementing feature transformation through query-to-feature interactions. For transformer-based approaches, sparse deformable attention [18] is particularly suitable to handle a large number of occupancy instances with low computational overhead. However, we find that extending the monocular deformable attention to multi-view tasks through the projection-first method [18], which is also widely adopted in [13,19,20,31], presents shortcomings, motivating us to explore our view attention to adequately collect image features for multi-view AD systems.

3D Scene Reconstruction. The occupancy network, introduced by Tesla [1], brings the concept of the occupancy grid map, which has long been utilized in the domains of robotic mapping and planning [8,30], into the AD field. Semantic scene completion [32] is similar to the goals of occupancy tasks discussed in this paper. MonoScene [6] leverages a U-Net architecture to infer dense 3D occupancy from a single image. VoxFormer [16] introduces depth estimation to guide voxel queries, and OccDepth [25] adopts stereo depth information to improve occupancy prediction. TPVFormer [13] proposes a tri-perspective view representation to aggregate image features. FB-OCC [19] constructs 3D features via forward-backward transformations. In addition, [31,38] contribute high-quality occupancy benchmarks to facilitate the research community. Although

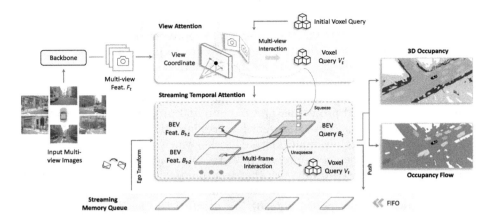

Fig. 3. ViewFormer pipeline. In our ViewFormer, the multi-view features F_t are first extracted from the multiple images via a backbone. Then we introduce the view attention specific for addressing the limitations of the existing projection-first method, allowing us to aggregate multi-view features for voxels V'_t more adequately. In our streaming temporal attention, we squeeze the voxel queries V'_t into the BEV queries B_t with concern of the computing complexity. Each BEV cell of B_t interacts with historical multi-frame BEV features stored in the streaming memory queue, where we utilize ego transformation to compensate ego motion. The voxels V_t obtained from unsqueezing the updated BEV features are subsequently fed into 3D occupancy and occupancy flow prediction. We push the updated BEV queries into the memory queue for subsequent temporal interaction in the video stream pipeline.

[31] assigns coarse object-level velocity directly to object occupancies, fine-grained occupancy-level flow representation still remain unexplored, by which we are motivated to present our occupancy-level flow dataset FlowOcc3D, as well as a feasible framework based on our view attention and streaming temporal attention for motion 3D occupancy prediction.

3 Methodology

In this paper, we propose a general framework, named ViewFormer, gathering spatiotemporal features from multi-view images adequately for unified prediction of 3D occupancy and occupancy flow, as shown in Fig. 3. Our approach stands out due to two key components: the view attention and streaming temporal attention, both of which compose a transformer encoder, wherein queries circulate in the form of voxels and BEV cells, ensuring efficient computation while extracting fine-grained representative 3D features.

Spatial Interaction. Following bird's-eye-view (BEV) perception methods [18, 23,39], we extract multi-view image features F_t via an image backbone and then update queries through spatial interaction. The difference is that we directly use voxel queries V'_t instead of BEV queries to collect finer-grained 3D occupancy features, where a relatively small number of query channels is set to reduce computational effort for a large number of voxel queries. Serving as the core of spatial interaction, our view attention is introduced specific for transforming multi-view image features into 3D space, exceeding a simple extension of the 2D deformable attention commonly used in [13,18,19,31].

Temporal Modeling. Inspired by the concept of streaming video methods [26, 35], we construct a streaming memory queue to dynamically store historical features spanning N frames during both the training and inference phases, which follows the first-in, first-out (FIFO) principle [35] for entry and exit. Considering the increased burden of storage and computation, we proceed with temporal modeling at the 2D BEV level. Specifically, the voxel queries V'_t are squeezed into BEV-level queries B_t along the z-axis, each BEV cell of B_t interacts with the historical multi-frame BEV features stored in the memory queue, where we utilize ego transformation to compensate ego motion. The voxels V_t obtained from unsqueezing the updated BEV queries are then fed into 3D occupancy and occupancy flow prediction. Meanwhile, we push the updated BEV queries into the memory queue for subsequent temporal interaction in the video stream pipeline.

3.1 View Attention

Faced with the challenge of 3D scenarios, we employ dense voxel queries ($V_t \in \mathbb{R}^{Z \times H \times W \times C_{\text{Voxel}}}$) as containers to facilitate fine-grained voxel feature extraction throughout the entire pipeline. Here Z, H, and W represent the 3D dimensions of the space, and C_{Voxel} denotes the feature channels. To effectively construct spatial 3D features with reasonable receptive fields in a vision-centric AD system, the keystone is to aggregate coherent 3D features from discrete multi-view images, for which we introduce our learning-first view attention mechanism to address the limitations of the existing projection-first method, as discussed Sect. 1.

Our view attention is implemented by abstracting an occupancy-related view coordinate (VC) system $\mathcal{T}$ as depicted in Fig. 2(b). Given multi-view features F_t, a voxel query $q \in \mathbb{R}^{C_{\text{Voxel}}}$ and its corresponding fixed 3D reference point p, view attention can be formulated as:

$$\text{ViewAttn}(q, p, F_t) = \sum_{m=1}^{M} W_m \sum_{k=1}^{K} \sum_{j=1}^{J} A_{mkj} W'_m F_t(p + \Delta p^{\mathcal{T}}_{mk}), \quad (1)$$

where M, K and J are the number of attention heads, sample points and cameras respectively, $W_m \in \mathbb{R}^{C_{\text{Voxel}} \times C_v}$, $W'_m \in \mathbb{R}^{C_v \times C_{\text{Voxel}}}$ are the learning weights ($C_v = C_{\text{Voxel}}/M$ by default), A_{mkj} is the normalized attention weight, $\Delta p^{\mathcal{T}}_{mk}$ denotes a learnable sample point associated with query q in the VC system $\mathcal{T}$, and $F_t(\cdot)$ represents extracting features from F_t by projecting the learnable sample points onto multi-view images.

Now, let's delve into how we represent learnable sample points in the view-guided VC system $\mathcal{T}$. Given the coordinate (x, y, z) of a reference point $\boldsymbol{p}$ for a query $\boldsymbol{q}$, we define its view angle θ as the angle between its projection line on the x-y plane and the x-axis, as illustrated in Fig. 2(b). The VC system $\mathcal{T}$ of this query can be obtained by translating the origin of the ego-centric perception coordinate system to the reference point and rotating it around the z-axis by the view angle. The corresponding rotation matrix $\boldsymbol{R}(\theta)$ are calculated as:

$$\theta = \arctan2(y, x),$$
$$\boldsymbol{R}(\theta) = \begin{bmatrix} \cos\theta & -\sin\theta & 0 \\ \sin\theta & \cos\theta & 0 \\ 0 & 0 & 1 \end{bmatrix}. \quad (2)$$

Through a transformation process, a learnable sample point $\Delta \boldsymbol{p}^{\mathcal{T}}$, initially defined in the local VC system $\mathcal{T}$, can be converted to be a 3D sample point $\boldsymbol{p}_s$ in the perception coordinate system, which is denoted as:

$$\boldsymbol{p}_s = \boldsymbol{p} + \boldsymbol{R}(\theta) \cdot \Delta \boldsymbol{p}^{\mathcal{T}}. \quad (3)$$

Subsequently, we project all the 3D sample points associated with a query onto the multi-view images via camera projection matrices for multi-view spatial interaction, aggregating features from F_t into voxel queries V'_t.

3.2 Streaming Temporal Attention

Streaming Memory Queue. Drawing inspiration from the efficient 3D point cloud perception [14,15], where it is observed that fine-grained 3D feature representation is not necessarily required for 3D space, we recognize that combining the voxel level and the BEV level can significantly enhance computation efficiency and alleviate storage consumption. Therefore, we establish our BEV-level streaming memory queue $\boldsymbol{B} = \{B_i \in \mathbb{R}^{H \times W \times C_{\text{BEV}}}, i = t-1, ..., t-N\}$ with $C_{\text{BEV}} > C_{\text{Voxel}}$, to dynamically store historical features spanning N frames during both the training and inference phases. The latest BEV features are pushed into the memory queue per frame for later BEV-level temporal interaction.

Multi-frame Temporal Interaction. We perform a multi-frame streaming temporal interaction. We observe that, compared to voxel queries V'_t directly interacting with all historical BEV features in the memory queue, implementing BEV-level temporal interaction not only reduces computational overhead but also improves accuracy. Specifically, we first compress the voxel queries V'_t into BEV queries $B_t \in \mathbb{R}^{H \times W \times C_{\text{BEV}}}$ along the z-axis, of which each BEV cell interacts with all the memory BEV features by leveraging the deformable attention [42], and the updated BEV queries are then recovered to be the voxel queries V_t through a feed-forward function for final predictions. We attribute this improvement to the presence of a substantial number of empty voxels, the compressed BEV queries B_t lead to more pure information and thus enhance the temporal interaction.

To compensate ego motion, we apply feature warping by transforming all memory BEV features to the current frame as in [11,17,26]. Taking the last frame $t-1$ as an example, given the ego pose matrices of the last frame T_{t-1} and the current frame T_t, the transformation matrix T^t_{t-1} between two frames is calculated as follow:

$$T^t_{t-1} = T^{inv}_t \cdot T_{t-1}. \qquad (4)$$

We align the last BEV features B_{t-1} to the current frame:

$$\tilde{B}_t = T^t_{t-1} \cdot B_{t-1}, \qquad (5)$$

where $\tilde{B}_t$ is the aligned BEV features in the local coordinate system of the current ego pose. Then, we adopt deformable attention [42] to achieve interaction between the current BEV queries B_t and the aligned multi-frame BEV features $\tilde{B}_t$ for the final updated BEV queries. The multi-frame aggregation mechanism follows a multi-scale aggregation [42].

4 Optimization

4.1 Occupancy Flow Generation

Although OpenOcc [31] has already generated object-level flow annotations by assigning the center velocity of an object to all its internal occupancies, the occupancy-level flow representation still remains unexplored. As illustrated in Fig. 4, fine-grained occupancy flow is able to capture more accurate flow vectors for different parts of a turning vehicle, offering the potential for a more precise representation of dynamic scenes. This capability is beneficial for decision-making in an AD system. Hence, we build our FlowOcc3D dataset with occupancy-level flow annotations. Figure 4(e) illustrates the generation process of flow annotations. As nuScenes [5,9] does not provide explicit temporal association for occupancies themselves, we leverage the temporal association of objects to track their internal occupancies. Given the pose matrices O_t and O_{t-1}

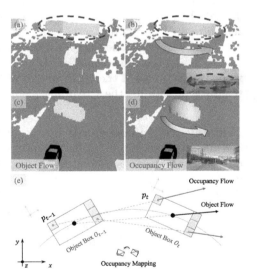

Fig. 4. Occupancy flow vs. object flow. Object flow assigns only a single flow vector to the entire object as in (**a**) and (**c**), while occupancy flow provides finer-grained flow vectors for all occupancy grids as in (**b**) and (**d**), where the color and brightness represent the flow direction and magnitude respectively.

of an object in two frames, along with an occupancy center p_t within the object box at O_t, where O_t, O_{t-1} and p_t are all defined in the global coordinate system, by taking the box as a rigid body, we map p_t to obtain its historical position p_{t-1} in the temporal frame $t - 1$, via $p_{t-1} = O_{t-1} \cdot O_t^{inv} \cdot p_t$. Then, the corresponding flow vector can be computed by $f = (p_t - p_{t-1})/\Delta t$, where Δt denotes the time interval.

4.2 Loss Function

Our final loss function is comprised of four parts, the focal loss [21] $\mathcal{L}_{focal}$ for supervising occupancy state, the cross-entropy loss $\mathcal{L}_{ce}$ as well as the Lovasz softmax loss [3] $\mathcal{L}_{ls}$ for semantic classification, and the L1 loss $\mathcal{L}_{l1}$ with λ to adjust loss weight for flow regression, which can be formulated as:

$$\mathcal{L} = \mathcal{L}_{focal} + \mathcal{L}_{ce} + \mathcal{L}_{ls} + \lambda \mathcal{L}_{l1}. \tag{6}$$

4.3 Implementation Details

Network. Following the experimental setup [18,31], we employ two backbones: ResNet50 (Res50) [10] initialized from ImageNet [7], and ResNet101 (Res101) [10] initialized from FCOS3D [36] for experiments in Table 1 and Table 2. In respect of experiments in Table 3 and the ablation section, we utilize InternImage-Tiny (InterT) [37] initialized from COCO [22] as image backbone. Our transformer encoder has 4 layers. For the detection space of the voxel queries V_t, we define the dimensions as: $H = 100$, $W = 100$, $Z = 8$, and $C_{\text{Voxel}} = 72$, while the channels of the BEV queries B_t are set to $C_{\text{BEV}} = 126$. We use $N = 4$ temporal frames for the streaming memory queue.

Training and Inference. By default, we train our ViewFormer with a streaming video approach [26,35] for 24 epochs. The learning rate is set to 2×10^{-4}, and the image size is 256 × 704. We also apply interpolation method on predictions to align with ground truth whose resolutions are inconsistent with our detection space. For the ease of comparison among different methods, we predict and evaluate BEV-level flow, and for visualization, we map the BEV-level flow to each voxel cell to recover occupancy flow. Inference FPS is measured with a single RTX 3090 GPU, while the training time is recorded using 8 A100 GPUs.

5 Evaluation

To evaluate the performance of our method comprehensively, we utilize the high-quality 3D occupancy benchmarks Occ3D [34] and OpenOcc [31]. Moreover, based on the nuScenes [5,9] and Occ3D [34] datasets, we build our FlowOcc3D with occupancy-level flow annotations to study the potential of fine-grained occupancy flow representation, on which we further conduct extensive experiments.

5.1 Datasets

Both of the Occ3D [34] and OpenOcc [31] occupancy datasets are derived from nuScenes [5,9], a substantial driving dataset that comprises videos of 1000 scenes equipped with multi-view cameras. It includes 34,149 annotated frames for all 700 training scenes and 150 validation scenes.

Benchmark Details. Occ3D [34] defines the detection space $\mathcal{V} = [-40\,\text{m}, 40\,\text{m}] \times [-40\,\text{m}, 40\,\text{m}] \times [-1\,\text{m}, 5.4\,\text{m}]$, in the ego vehicle coordinate system to generate occupancy data. The space $\mathcal{V}$ is voxelized with a resolution of $\Delta s = 0.4\,\text{m}$, producing $200 \times 200 \times 16$ voxel grids to represent the 3D environment. Furthermore, the dataset includes occupancy visibility masks for various sensors, facilitating performance improvement and evaluation in diverse tasks. To create our **FlowOcc3D** dataset, we attach the occupancy-level flow information to the occupancy cells of Occ3D [34] for each annotated frame. **OpenOcc** [31] defines the detection space $\mathcal{V} = [-50\,\text{m}, 50\,\text{m}] \times [-50\,\text{m}, 50\,\text{m}] \times [-5\,\text{m}, 3\,\text{m}]$ in LiDAR coordinate system, and produces $200 \times 200 \times 16$ voxel grids by voxelizing $\mathcal{V}$ with a resolution of $\Delta s = 0.5\,\text{m}$.

Evaluation Metrics. Occ3D utilizes the mean Intersection-over-Union (mIoU) across categories. OpenOcc uses both the mIoU and the single-class IoU_{geo} as metrics to evaluate the occupancy state. Besides these two metrics, we

Table 1. 3D Occupancy Prediction on Occ3D benchmark. Our ViewFormer outperforms previous SOTAs by significant margins.

Method	Backbone	mIoU	others	barrier	bicycle	bus	car	const. veh.	motorcycle	pedestrian	traffic cone	trailer	truck	driv. surf.	other flat	sidewalk	terrain	manmade	vegetation
MonoScene [6]	Res101	6.06	1.75	7.23	4.26	4.93	9.38	5.67	3.98	3.01	5.90	4.45	7.17	14.91	6.32	7.92	7.43	1.01	7.65
TPVFormer [13]	Res101	27.83	7.22	38.90	13.67	40.78	45.90	17.23	19.99	18.85	14.30	26.69	34.17	55.65	35.47	37.55	30.70	19.40	16.78
OccFormer [40]	Res101	21.93	5.94	30.29	12.32	34.40	39.17	14.44	16.45	17.22	9.27	13.90	26.36	50.99	30.96	34.66	22.73	6.76	6.97
BEVFormer [18]	Res101	26.88	5.85	37.83	17.87	40.44	42.43	7.36	23.88	21.81	20.98	22.38	30.70	55.35	28.36	36.00	28.06	20.04	17.69
CTF-Occ [34]	Res101	28.53	8.09	39.33	20.56	38.29	42.24	16.93	24.52	22.72	21.05	22.98	31.11	53.33	33.84	37.98	33.23	20.79	18.00
FB-OCC [19]	Res50	39.11	**13.57**	44.74	27.01	**45.41**	49.10	25.15	26.33	27.86	27.79	32.28	36.75	80.07	42.76	51.18	55.13	42.19	37.53
ViewFormer (ours)	Res50	**41.85**	12.94	**50.11**	**27.97**	44.61	**52.85**	**22.38**	**29.62**	**28.01**	**29.28**	**35.18**	**39.40**	**84.71**	**49.36**	**57.44**	**59.69**	**47.37**	**40.56**

Table 2. 3D Occupancy Prediction on OpenOcc benchmark. Our ViewFormer demonstrates significant performance improvements over previous SOTAs in terms of both the mIoU and IoU_{geo}.

Method	Backbone	mIoU	IoU$_{geo}$	barrier	bicycle	bus	car	const. veh.	motorcycle	pedestrian	traffic cone	trailer	truck	driv. surf.	other flat	sidewalk	terrain	manmade	vegetation
TPVFormer [13]	Res101	23.67	37.47	27.95	12.75	33.24	**38.70**	12.41	17.84	11.65	8.49	16.42	26.47	47.88	25.43	30.62	30.18	15.51	23.12
OccNet [31]	Res101	26.98	41.08	**29.77**	16.89	34.16	37.35	15.58	**21.92**	**21.29**	**16.75**	16.37	26.23	50.74	27.93	31.98	33.24	20.80	30.68
ViewFormer (ours)	Res101	**27.37**	**41.86**	28.18	**17.96**	**34.49**	37.03	**16.00**	21.53	19.50	13.56	15.59	**26.52**	**51.48**	**31.81**	**34.73**	**34.77**	**22.20**	**32.62**
BEVDet4D [11]	Res50	9.85	18.27	13.56	0.00	13.04	26.98	0.61	1.20	6.76	0.93	1.93	12.63	27.23	11.09	13.64	12.04	6.42	9.56
BEVDepth [17]	Res50	11.88	23.45	15.15	0.02	20.75	27.05	1.10	2.01	9.69	1.45	1.91	14.31	31.92	7.88	17.08	16.27	8.76	14.75
BEVDet [12]	Res50	12.49	27.46	16.06	0.11	18.27	21.09	2.62	1.42	7.78	1.08	3.4	13.76	33.89	10.84	17.55	22.03	11.72	18.15
OccNet [31]	Res50	19.48	37.69	20.63	5.52	24.16	27.72	9.79	7.73	13.38	7.18	10.68	18.00	46.13	20.6	26.75	29.37	16.90	27.21
ViewFormer (ours)	Res50	**23.84**	**40.40**	**24.35**	**12.55**	**28.87**	**31.87**	**15.35**	**17.89**	**14.76**	**8.53**	**14.18**	**23.54**	**49.02**	**29.05**	**32.35**	**32.16**	**17.93**	**29.00**

additionally compute the mean absolute velocity error (mAVE) across categories to evaluate occupancy flow on our FlowOcc3D benchmark.

5.2 Main Results

3D Occupancy on Occ3D. We compare our ViewFormer with previous state-of-the-art methods on the 3D occupancy task in Table 1, where the baselines BEVFormer [18], TPVFormer [13], CTF-Occ [40] and FB-OCC [19] adopt the projection-first spatial interaction method as analyzed in Sect. 1. In terms of training setup, we follow FB-OCC [19], the winner of 3D occupancy challenge of CVPR 2023 [31,34], which includes 90-epoch training, the same pretrained weights, and depth supervision for image backbone. As shown in Table 1, in contrast to the strong baseline FB-OCC [19], which utilizes longer video sequences, our model achieves 2.74 mIoU improvement, proving its superiority.

3D Occupancy on OpenOcc. We conduct experiments on OpenOcc in Table 2, where we adopt the same configuration as OccNet [31], with an image size of 450 × 800. Our approach demonstrates significant performance improvements over the baselines. Especially, we surpass the state-of-the-art OccNet [31] by 4.36 on mIoU and 2.71 on IoU_{geo} when utilizing Res50 as the backbone.

Occupancy and Flow on FlowOcc3D. We validate our ViewFormer against previous state-of-the-art methods on the 3D occupancy and flow tasks with our FlowOcc3D in Table 3, training both the occupancy and flow heads simultaneously. The two selected baselines, BEVFormer [18] and FB-OCC [19], employ two typical temporal modeling approaches that can directly affect the flow prediction. The former utilizes a transformer-based single-frame interaction trained in a sliding-window fashion, while the latter adopts a CNN-based streaming video interaction from [26]. We add a flow head to the two methods and retrain them on our FlowOcc3D for 24 epochs with an image size of 256 × 704, matching our ViewFormer setup. The results in Table 3 show that our method surpasses the two baselines in both the occupancy and flow prediction. It can be observed that, under the 24-epoch setup, our model outperforms FB-OCC by 5.2 mIoU,

Table 3. 3D Occupancy and Occupancy Flow Prediction on FlowOcc3D. Our ViewFormer achieves the best result among previous SOTAs. *: we add a flow head for flow prediction to BEVFormer [18] and FB-OCC [19] and retrain them on FlowOcc3D.

Method	Backbone	Flow Head	mIoU	IoU_{geo}	mAVE↓	others	barrier	bicycle	bus	car	const. veh.	motorcycle	pedestrian	traffic cone	trailer	truck	driv. surf.	other flat	sidewalk	terrain	manmade	vegetation
BEVFormer* [18]	InternT	✓	33.61	67.41	0.695	8.44	40.80	12.57	37.70	45.76	18.44	12.14	22.52	21.25	25.97	29.40	80.92	38.19	48.56	52.58	40.98	35.07
FB-OCC* [19]	InternT	✓	37.36	69.73	0.433	10.95	40.08	23.14	42.87	47.17	21.43	23.84	27.24	24.50	31.54	37.36	80.31	42.26	50.14	55.44	39.98	36.84
ViewFormer (ours)	InternT	✓	**42.54**	**72.36**	**0.412**	**13.63**	**49.35**	**28.74**	**46.63**	**52.71**	21.04	**29.63**	**30.34**	**30.53**	**34.01**	**41.04**	**85.23**	**50.63**	**58.68**	**61.63**	**47.72**	**41.56**
BEVFormer [18]	InternT	✗	36.93	68.49	-	8.56	42.94	19.34	47.02	49.59	20.36	22.62	24.69	19.77	30.11	35.24	82.11	40.86	50.62	54.81	42.56	36.55
FB-OCC [19]	InternT	✗	38.69	69.95	-	11.4	41.42	24.27	46.01	49.38	24.56	27.06	28.09	25.61	32.23	38.46	80.97	42.99	50.95	56.15	40.55	37.61
ViewFormer (ours)	InternT	✗	**43.61**	**72.46**	-	**13.82**	**50.32**	**29.49**	**49.24**	**54.52**	24.34	**32.72**	**31.09**	**31.49**	**34.44**	**41.62**	**85.47**	**51.27**	**59.03**	**62.15**	**48.33**	**42.06**

2.63 IoU$_{geo}$ and 2.1% mAVE, highlighting the rapid convergence of our method. We also report the results without the flow head under the same settings.

5.3 Ablations and Analysis

View Attention. To validate the effectiveness of our view attention, we conduct experiments as presented in Table 4. Since this module serves as a spatial interaction and is not involved in temporal modeling, we temporarily disable the flow head. The findings are as below. 1) We replace our view attention module with the projection-first deformable attention as BEVformer [18], the results show that our learning-first view attention leads to an improvement of 1.22 mIoU and 0.43 IoU$_{geo}$ under approximate computational complexity, revealing the limitation of the commonly used projection-first method for multi-view feature aggregation. 2) Learning sample points directly in the ego-centric perception coordinate system instead of the query-specific view coordinate (VC) system results in performance degradation of -0.75 mIoU, which validates our motivation mentioned in Sect. 1 that the rotation invariance introduced by our view attention effectively reduces learning complexity, accelerates convergence, and enhances accuracy. Detailed convergence experiments can be found in supplementary materials.

Table 4. Ablation study for the view attention. "Projection-first" denotes the extension of deformable attention used in the multi-view field as [18]. "w/o VC trans." denotes learning sample points directly in the ego-centric perception coordinate system without view angle rotation.

Method	mIoU	IoU$_{geo}$	car	truck	vege.
View Attention	**43.61**	**72.46**	**54.52**	**41.62**	**42.06**
Projection-first	42.39	72.03	53.16	40.58	40.99
w/o VC trans.	42.86	72.26	53.30	40.70	41.93

Visualization Analysis. We visualize our view attention in Fig. 5. The green cross mark denotes a query's reference point, which can only be projected onto the left image. The learned 3D points are projected onto multi-view images represented by filled circles, with colors to indicate attention weights. In such scenarios, the projection-first method, like BEVFormer [18] that only gathers features

Fig. 5. Visualization on ViewAttn. (Color figure online)

Table 5. Results on multi-view map construction. We apply our view attention to MapTR [20].

Method	mAP	AP_{ped}	AP_{div}	AP_{bound}
MapTR	44.73	39.69	44.59	49.91
w/ViewAttn	**50.55**	**45.85**	**52.82**	**52.98**

Table 6. Results on multi-view 3D object detection. We apply our view attention to DETR3D [39].

Method	mAP↑	NDS↑	mATE↓	mAVE↓
DETR3D	0.347	0.422	0.765	0.876
w/ViewAttn	**0.388**	**0.441**	**0.712**	**0.874**

from the image where the reference point can be projected, fails to collect features from the right image, as discussed in Sect. 1. In contrast, our method is competent to gather features from both images, allowing for constructing more semantically informative and robust features for this query.

Application of View Attention. We also apply our view attention to other tasks to assess its scalability, including MapTR [20] for HD map construction and DETR3D [39] for 3D object detection, both of them collect multi-view image features in a similar way to the projection-first method analyzed in Sect. 1. We replace their corresponding feature collection module with our view attention. The evaluation metrics are available in their respective literatures. We train MapTR for 24 epochs with an image size of 324×576, and Res50 as the backbone. Table 5 indicates our ViewAttn-MapTR achieves a substantial 5.82% improvement on mAP, demonstrating a remarkable increase in convergence speed. For training DETR3D, we use Res101-DCN as the backbone without CBGS [41]. Table 6 shows our ViewAttn-DETR3D also results in 4.1% and 1.9% improvements on mAP and NDS respectively. Unlike a considerable amount of recent researches mainly focusing on temporal modeling, and lacking the study of the fundamental spatial interaction for multi-camera 3D perception, our work reveals the limitations of the widely used projection-first method and highlights substantial research opportunities that remain unexplored.

Table 7. Ablation study for the streaming temporal attention. "w/o TempAttn." denotes disabling the streaming temporal attention module.

Method	mIoU	IoU_{geo}	mAVE↓	FPS↑	car	truck	vege.
w/o TempAttn.	39.28	68.38	1.125	**4.2**	49.97	37.38	37.27
Voxel to BEV	41.62	71.53	1.104	3.9	52.16	40.59	40.68
BEV to BEV	**42.54**	**72.36**	**0.412**	4.1	**52.71**	41.04	**41.56**
Queue $N=1$	41.67	71.18	0.426	4.1	52.56	40.36	40.28
Queue $N=2$	42.28	71.95	0.421	4.1	52.53	40.75	41.11
Queue $N=3$	42.42	72.18	0.415	4.1	52.70	**41.20**	41.48

Streaming Temporal Attention. We train both the occupancy head and the flow head simultaneously to evaluate our temporal modeling in Table 7. Overall, our streaming temporal attention improves the mIoU by 3.26 and produces more reasonable flow prediction. Notably, the "BEV-to-BEV" interaction method of our streaming temporal attention as depicted in Sect. 3.2 outperforms the "Voxel-to-BEV" interaction method by 0.92 mIoU, with less computational effort by reducing the number of interaction queries. Due to the presence of a substantial number of empty voxels, compressing voxel queries into BEV queries leads to more pure information and thus enhances the temporal interaction. Additionally, high mAVE indicates that the "Voxel-to-BEV" interaction method faces convergence issues in the flow task.

Length of the Memory Queue. We also study the influence of the queue length in Table 7, which shows that the performance improves as the queue length increases, proving the effectiveness of our multi-frame interaction mechanism compared to the single-frame interaction mechanism with $N = 1$ as BEV-Former [18]. The performance starts to plateau after $N = 3$ as in [35]. We use $N = 4$ in our framework, greater than which the improvement becomes ignorable. As presented in Table 7 under the "FPS" column, thanks to the efficient streaming memory mechanism, the additional time cost brought by multi-frame temporal interaction is negligible.

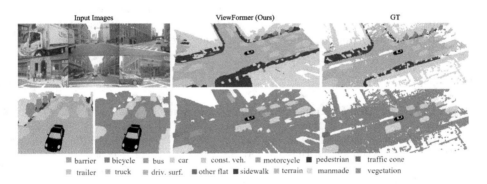

Fig. 6. Qualitative results of 3D occupancy and occupancy flow prediction. 3D occupancies are color-coded according to semantic categories. For flow visualization, we utilize color and brightness to represent the flow direction and magnitude respectively, following the convention of the optical flow fields [2,33].

5.4 Qualitative Results

We present qualitative results of our ViewFormer in Fig. 6. An additional animated video is also provided in the supplementary material. Moreover, we demonstrate a qualitative comparison for flow predictions supervised by object-level flow (a) and occupancy-level flow (b) in Fig. 7, the model trained with our occupancy-level FlowOcc3D dataset achieves more reasonable results for a turning car, providing fine-grained motion information for AD systems.

Fig. 7. Visualization on flow predictions. (a) Flow predictions supervised by object-level flow. (b) Flow predictions supervised by our FlowOcc3D.

6 Conclusion

We present the ViewFormer framework for 3D occupancy and occupancy flow prediction, featuring our proposed view attention that addresses the limitations of the existing projection-first spatial interaction method, as well as the streaming temporal attention designed for multi-frame temporal interaction. Furthermore, we build a novel occupancy-level flow benchmark FlowOcc3D to explore the potential of occupancy flow representation for dynamic scenes, which we also contribute to the research community. Our approach demonstrates significant advancements over previous state-of-the-art methods.

References

1. Tesla AI Day (2021). https://www.youtube.com/watch?v=j0z4FweCy4M
2. Baker, S., Scharstein, D., Lewis, J.P., Roth, S., Black, M.J., Szeliski, R.: A database and evaluation methodology for optical flow. In: ICCV, pp. 1–8 (2007)
3. Berman, M., Triki, A.R., Blaschko, M.B.: The Lovász-Softmax loss: a tractable surrogate for the optimization of the intersection-over-union measure in neural networks. In: CVPR, pp. 4413–4421 (2018)
4. Brazil, G., Pons-Moll, G., Liu, X., Schiele, B.: Kinematic 3D object detection in monocular video. In: Vedaldi, A., Bischof, H., Brox, T., Frahm, J.-M. (eds.) ECCV 2020. LNCS, vol. 12368, pp. 135–152. Springer, Cham (2020). https://doi.org/10.1007/978-3-030-58592-1_9
5. Caesar, H., et al.: nuScenes: a multimodal dataset for autonomous driving. In: CVPR, pp. 11618–11628 (2020)
6. Cao, A.Q., de Charette, R.: MonoScene: monocular 3D semantic scene completion. In: CVPR, pp. 3991–4001 (2022)
7. Deng, J., Dong, W., Socher, R., Li, L.J., Li, K., Fei-Fei, L.: ImageNet: a large-scale hierarchical image database. In: CVPR, pp. 248–255 (2009)

8. Dezert, J., Moras, J., Pannetier, B.: Environment perception using grid occupancy estimation with belief functions. In: FUSION, pp. 1070–1077 (2015)
9. Fong, W.K., et al.: Panoptic nuScenes: a large-scale benchmark for LiDAR panoptic segmentation and tracking, pp. 3795–3802 (2022)
10. He, K., Zhang, X., Ren, S., Sun, J.: Deep residual learning for image recognition. In: CVPR, pp. 770–778 (2016)
11. Huang, J., Huang, G.: BEVDet4D: exploit temporal cues in multi-camera 3D object detection. arXiv preprint arXiv:2203.17054 (2022)
12. Huang, J., Huang, G., Zhu, Z., Yun, Y., Du, D.: BEVDet: high-performance multi-camera 3D object detection in bird-eye-view. arXiv preprint arXiv:2112.11790 (2021)
13. Huang, Y., Zheng, W., Zhang, Y., Zhou, J., Lu, J.: Tri-perspective view for vision-based 3D semantic occupancy prediction. In: CVPR, pp. 9223–9232 (2023)
14. Lang, A.H., Vora, S., Caesar, H., Zhou, L., Yang, J., Beijbom, O.: PointPillars: fast encoders for object detection from point clouds. In: CVPR, pp. 12697–12705 (2019)
15. Li, J., He, X., Wen, Y., Gao, Y., Cheng, X., Zhang, D.: Panoptic-PHNet: towards real-time and high-precision LiDAR panoptic segmentation via clustering pseudo heatmap. In: CVPR, pp. 11799–11808 (2022)
16. Li, Y., et al.: VoxFormer: sparse voxel transformer for camera-based 3D semantic scene completion. arXiv preprint arXiv:2302.12251 (2023)
17. Li, Y., et al.: BEVDepth: acquisition of reliable depth for multi-view 3D object detection (2022). arXiv preprint arXiv:2206.10092
18. Li, Z., et al.: BEVFormer: learning bird's-eye-view representation from multi-camera images via spatiotemporal transformers. In: Avidan, S., Brostow, G., Cisse, M., Farinella, G.M., Hassner, T. (eds.) Computer Vision – ECCV 2022. LNCS, vol. 13669, pp. 1–18. Springer, Cham (2022). https://doi.org/10.1007/978-3-031-20077-9_1
19. Li, Z., Yu, Z., Wang, W., Anandkumar, A., Lu, T., Alvarez, J.M.: FB-BEV: BEV representation from forward-backward view transformations. In: ICCV, pp. 6919–6928 (2023)
20. Liao, B., et al.: MapTR: structured modeling and learning for online vectorized HD map construction. In: ICLR (2023)
21. Lin, T., Goyal, P., Girshick, R.B., He, K., Dollár, P.: Focal loss for dense object detection. In: ICCV, pp. 2999–3007 (2017)
22. Lin, T.-Y., et al.: Microsoft COCO: common objects in context. In: Fleet, D., Pajdla, T., Schiele, B., Tuytelaars, T. (eds.) ECCV 2014. LNCS, vol. 8693, pp. 740–755. Springer, Cham (2014). https://doi.org/10.1007/978-3-319-10602-1_48
23. Liu, Y., Wang, T., Zhang, X., Sun, J.: PETR: position embedding transformation for multi-view 3D object detection. In: Avidan, S., Brostow, G., Cisse, M., Farinella, G.M., Hassner, T. (eds.) Computer Vision – ECCV 2022. LNCS, vol. 13687, pp. 531–548. Springer, Cham (2022). https://doi.org/10.1007/978-3-031-19812-0_31
24. Luo, W., Yang, B., Urtasun, R.: Fast and furious: real time end-to-end 3D detection, tracking and motion forecasting with a single convolutional net. In: CVPR, pp. 3569–3577 (2018)
25. Miao, R., et al.: OccDepth: a depth-aware method for 3D semantic scene completion. arXiv:2302.13540 (2023)
26. Park, J., et al.: Time will tell: new outlooks and a baseline for temporal multi-view 3D object detection. In: ICLR (2023)

27. Philion, J., Fidler, S.: Lift, Splat, Shoot: encoding images from arbitrary camera rigs by implicitly unprojecting to 3D. In: Vedaldi, A., Bischof, H., Brox, T., Frahm, J.-M. (eds.) ECCV 2020. LNCS, vol. 12359, pp. 194–210. Springer, Cham (2020). https://doi.org/10.1007/978-3-030-58568-6_12
28. Qi, C.R., et al.: Offboard 3D object detection from point cloud sequences. In: CVPR, pp. 6134–6144 (2021)
29. Reading, C., Harakeh, A., Chae, J., Waslander, S.L.: Categorical depth distribution network for monocular 3D object detection. In: CVPR, pp. 8555–8564 (2021)
30. Schreiber, M., Belagiannis, V., Gläser, C., Dietmayer, K.: Dynamic occupancy grid mapping with recurrent neural networks. In: ICRA, pp. 6717–6724 (2021)
31. Sima, C., et al.: Scene as occupancy. In: ICCV, pp. 8406–8415 (2023)
32. Song, S., Yu, F., Zeng, A., Chang, A.X., Savva, M., Funkhouser, T.: Semantic scene completion from a single depth image. In: CVPR, pp. 1746–1754 (2017)
33. Teed, Z., Deng, J.: RAFT: recurrent all-pairs field transforms for optical flow. In: Vedaldi, A., Bischof, H., Brox, T., Frahm, J.-M. (eds.) ECCV 2020. LNCS, vol. 12347, pp. 402–419. Springer, Cham (2020). https://doi.org/10.1007/978-3-030-58536-5_24
34. Tian, X., Jiang, T., Yun, L., Wang, Y., Wang, Y., Zhao, H.: Occ3D: a large-scale 3D occupancy prediction benchmark for autonomous driving. arXiv preprint arXiv:2304.14365 (2023)
35. Wang, S., Liu, Y., Wang, T., Li, Y., Zhang, X.: Exploring object-centric temporal modeling for efficient multi-view 3D object detection. In: ICCV, pp. 11618–11628 (2023)
36. Wang, T., Zhu, X., Pang, J., Lin, D.: FCOS3D: fully convolutional one-stage monocular 3D object detection. In: ICCV, pp. 913–922 (2021)
37. Wang, W., et al.: InternImage: exploring large-scale vision foundation models with deformable convolutions. In: Proceedings of the IEEE/CVF Conference on Computer Vision and Pattern Recognition (CVPR), pp. 14408–14419, June 2023
38. Wang, X., et al.: OpenOccupancy: a large scale benchmark for surrounding semantic occupancy perception. arXiv preprint arXiv:2303.03991 (2023)
39. Wang, Y., Guizilini, V., Zhang, T., Wang, Y., Zhao, H., Solomon, J.: DETR3D: 3D object detection from multi-view images via 3D-to-2D queries. In: CoRL, pp. 180–191 (2021)
40. Zhang, Y., Zhu, Z., Du, D.: OccFormer: dual-path transformer for vision-based 3D semantic occupancy prediction. arXiv preprint arXiv:2304.05316 (2023)
41. Zhu, B., Jiang, Z., Zhou, X., Li, Z., Yu, G.: Class-balanced grouping and sampling for point cloud 3D object detection. arXiv preprint arXiv:1908.09492 (2019)
42. Zhu, X., Su, W., Lu, L., Li, B., Wang, X., Dai, J.: Deformable DETR: deformable transformers for end-to-end object detection. In: ICLR (2021)

Fine-Grained Dynamic Network for Generic Event Boundary Detection

Ziwei Zheng, Lijun He, Le Yang, and Fan Li(✉)

Xi'an Jiaotong University, Xi'an, China
ziwei.zheng@stu.xjtu.edu.cn, {lijunhe,lifan}@mail.xjtu.edu.cn,
yangle15@xjtu.edu.cn

Abstract. Generic event boundary detection (GEBD) aims at pinpointing event boundaries naturally perceived by humans, playing a crucial role in understanding long-form videos. Given the diverse nature of generic boundaries, spanning different video appearances, objects, and actions, this task remains challenging. Existing methods usually detect various boundaries by the same protocol, regardless of their distinctive characteristics and detection difficulties, resulting in suboptimal performance. Intuitively, a more intelligent and reasonable way is to adaptively detect boundaries by considering their special properties. In light of this, we propose a novel dynamic pipeline for generic event boundaries named DyBDet. By introducing a multi-exit network architecture, DyBDet automatically learns the subnet allocation to different video snippets, enabling fine-grained detection for various boundaries. Besides, a multi-order difference detector is also proposed to ensure generic boundaries can be effectively identified and adaptively processed. Extensive experiments on the challenging Kinetics-GEBD and TAPOS datasets demonstrate that adopting the dynamic strategy significantly benefits GEBD tasks, leading to obvious improvements in both performance and efficiency compared to the current state-of-the-art. The code is available at https://github.com/Ziwei-Zheng/DyBDet.

Keywords: Generic event boundary detection · Dynamic network · Long-form video understanding

1 Introduction

Video comprehension has gained significant traction due to the rapid growth of online resources. According to cognitive science [32], humans naturally tend to parse long videos into meaningful segments. In light of this, Generic Event Boundary Detection [35] (GEBD) has been introduced to localize distinctive event boundaries, aiming to advance the field of long-form video understanding.

Generic event boundaries are taxonomy-free and encompass various low-level and high-level semantic changes. Modeling diverse boundaries of different semantic levels in the same way could entangle the feature extraction and result in inferior performance. For instance, the shot change in Fig. 1 (a) can be efficiently and

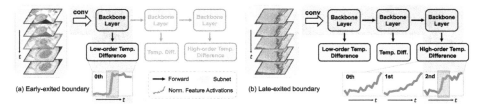

Fig. 1. Adaptive inference for video snippets within the dynamic architecture. We use temporal differences with different orders to capture the distinctive boundary features and plot the normalized activations for better visualization. The ground-truth boundaries are highlighted with red lines. (a): The boundary of the shot change can be obviously identified with low-level appearance features w/o temporal difference and can exit early to save computations. (b): The action change relies on features of high-level semantics and high-order temporal differences to reveal the boundary. (Color figure online)

effectively detected with appearance-level information only. In contrast, exploring high-level semantics is crucial for detecting the sub-action change in Fig. 1 (b). However, applying deep models to video snippets that contain shot changes may vanish the coarse-level but distinctive appearance features, as well as unnecessary computational waste. Therefore, detecting generic event boundaries with the same protocol regardless of their distinctive characteristics and detection difficulties can lead to suboptimal performance and efficiency.

To address this issue, we propose a dynamic pipeline capable of sample-dependent detections. A multi-exit network is proposed in DyBDet, allowing adaptive inference for various boundaries by automatically learning subnet allocation. The boundaries corresponding to low-level semantic information, such as shot changes, are only passed through the subnet with shallow layers to capture appearance information and simple temporal dependencies. While boundaries needing high-level semantics for boundary identification will traverse the whole network for in-depth spatiotemporal modeling. By inferring different inputs in a fine-grained manner, the processing of generic event boundaries can be specialized to enhance the performance and efficiency of the overall detection system.

To further ensure various boundaries can be effectively detected with specialized strategies, we propose a multi-order difference detector to explore the most distinctive localized change pattern of each candidate, which is crucial for boundary identification. Specifically, a multi-order difference encoder and a pair-wise contrast module are proposed to compose the boundary detector. Detectors with different order combinations are attached at varying depths of the backbone to form the overall dynamic architecture. The temporal difference describes the "change" within adjacent frames, which is crucial for event boundary detection [38]. Figure 1 (a) shows that the shot change has a clear boundary itself based on features from shallow layers of the backbone. However, sub-action changes in Fig. 1 (b) are much more complex in time, needing extra high-order temporal differences to reveal the distinctive boundary information, which is usually overlooked by previous methods. Therefore, by introducing multi-order temporal differences to the dynamic model, various boundaries can be easily distinguished and adaptively processed, leading to substantial improvement.

To the best of our knowledge, we are the first to introduce dynamic processing for generic event boundary detection. Extensive experiments and studies on Kinetics-GEBD [35] and TAPOS [33] dataset demonstrate the effectiveness and efficiency of adaptive inference to achieve the new state-of-the-art.

2 Related Work

Temporal Detection Tasks and GEBD. Temporal detection tasks are proposed to detect clip-level instances in untrimmed videos almost at a similar difficulty level, including shot boundary detection [9,36,39], temporal action segmentation [1,2,8,24] and localization [4,31,34,50]. [35] first proposed generic event boundary detection (GEBD) to localize the taxonomy-free moments that humans naturally perceive event boundaries. The boundaries can be used for further long-form video comprehension. Recent works either follow a similar fashion of [35] to slice the long video into adjacent overlapped snippets as independent samples [13,20,38], or take the whole video as input with continuous predictions [18,25]. Specifically, [21] proposes a recursive parsing algorithm based on the temporal self-similarity matrix to enhance local modeling. [38] aims to characterize the motion pattern with dense difference maps. Unlike these methods process all boundaries with the same protocol, we propose a dynamic pipeline for specialized processing with higher performance and efficiency.

Dynamic Neural Networks. Due to the favourability of efficiency while maintaining high performance, dynamic neural networks are attracting more attention. Different from the conventional methods that handle various inputs with a fixed network, dynamic networks [10,17,27,43,48,51] are capable of sample-wise adaptive inference. Each sample can receive customized processing with adapted subnets within the dynamic architecture [3,11,44,45,47]. When adopted to downstream tasks [5,30,42,46,49], the subnets in dynamic networks usually still share most of the parameters for the consideration of efficiency, and the distinctive characteristics for inputs are not fully explored. Different from them, the dynamic network in our method excels in fine-grained detection, leveraging the special properties inherent in each sample.

3 Method

3.1 Overview

Figure 2 illustrates the overall architecture. Given a video sequence with T frames, we aim to adaptively allocate subnets for different types of event boundaries. DyBDet is mainly composed of two key designs: the **multi-exit network** to enable adaptive inference at the video snippet level and the **multi-order difference detector** to achieve the dynamism by capturing the most distinctive motion patterns for boundary detection.

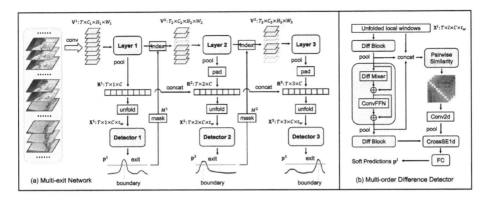

Fig. 2. Overview of the proposed DyBDet. Boundaries are highlighted with red lines. (a): the **multi-exit network** to enable frame-level adaptive inference, (b): the **multi-order difference detector** to distinguish boundaries with various characteristics. (Color figure online)

3.2 Multi-exit Network

To achieve adaptive inference, we design a dynamic network with L different boundary detectors, where these intermediate detectors are attached at varying depths of the image backbone (shown in the left plane of Fig. 2).

Shared Backbone and Separated Detectors. Given the frame $\mathbf{I}_t$ at the timestamp t and the centered local video snippet $\mathbf{V} = [\mathbf{I}_{t-k}, \ldots, \mathbf{I}_{t+k}]$, the output of the l-th detector ($l \in [1, \ldots, L]$) can be represented by:

$$\mathbf{p}_t^l = h_l(f_l([\mathbf{I}_{t-k}, \ldots, \mathbf{I}_{t+k}], \theta_l), \phi_l), \tag{1}$$

where f_l is the subnet of backbone and θ_l denotes the partially shared parameters, h_l is the specific boundary detector with parameter ϕ_l. The combination of a shared backbone and separated detectors allows more targeted processing to various boundaries while still maintaining the efficiency property of multi-exit networks. $\mathbf{p}_t^l \in (0, 1)$ denotes the prediction score after the sigmoid function.

Multi-scale Spatio-temporal Features. Following the common paradigm [21,25], T frames are first sampled from the raw video, and sent into the image backbone (ResNet50 [12] by default). Spatial features from different layers $[\mathbf{V}^1, \ldots, \mathbf{V}^L]$ are iteratively extracted during adaptive inference, severing as representations of different levels of semantics. For each scale, spatial pooling is then performed for each frame to save computations and obtain the 1d temporal feature. All temporal features from previous scales (if exist) are collected and concatenated to the current scale l to obtain enriched feature $\mathbf{R}^l \in \mathbb{R}^{T \times l \times C}$.

Continuous Prediction. Due to the ambiguous annotations for generic boundaries, we formulate GEBD as a global continuous prediction task supervised by soft labels [25]. A much longer video sequence is sent to the network directly so that the huge repetitive computations of the backbone among adjacent overlapped windows can be saved. Due to the local property [25,37] of event boundaries, we unfold the sequence into local windows for precise detection but on features after being extracted from the backbone. Specifically, k frames before and after each candidate position in $\mathbf{R}^l$ are collected to form the local centered window of t_w frames, where $t_w = 2k + 1$. The overall partitioned feature is $\mathbf{X}^l \in \mathbb{R}^{T \times l \times C \times t_w}$, which corresponds to the input of l-th attached boundary detector. Each local window corresponds to each timestamp and can produce a confidence score to represent the probability of being the boundary. Since the predictions are supervised by soft labels, the change of confidence scores between adjacent frames is smooth, we can then perform peak estimation to get boundaries $\mathbf{b}$:

$$\mathbf{b} = \{t \mid \mathbf{p}_t > \mathbf{p}_{t-1}, \mathbf{p}_t > \mathbf{p}_{t+1}, \mathbf{p}_t > \epsilon\}, \qquad (2)$$

where each timestamp corresponds to the local maximum that satisfies the classification threshold ϵ (e.g. 0.5) is considered as the boundary.

Partial Exit Criterion. We can easily determine whether a boundary candidate should exit at the l-th detector based on whether its prediction score exceeds a pre-defined threshold ϵ^l. However, since the long video sequence we send into the network usually contains multiple boundaries with various characteristics, we can not simply follow existing multi-exit networks [17,47] to have the entire sequence exit at once. Therefore, we design a partial exit criterion from the global perspective on the whole sequence. Specifically, due to the locality of event boundaries, we only allow features around the detected point to exit and generate a mask $\mathbf{M}^l \in \{0,1\}^T$ indicating whether to drop or keep for features at each timestamp:

$$\mathbf{M}_t^l = \mathbb{I}(t \notin [b - t_\mu^l, b + t_\mu^l]), \quad b \in \mathbf{b}, \quad t \in [1, \ldots, T], \qquad (3)$$

where t_μ^l determines the amount of exited features around the detected boundaries $\mathbf{b}$. Then we use the mask $\mathbf{M}^l$ to index which parts of the backbone features are kept:

$$\mathbf{V}^{l+1} = \{f_l(\mathbf{V}_t) \mid \mathbf{M}_t^l = 1\}. \qquad (4)$$

Since t_μ is adjustable, it controls the allocation of subnets to different samples as well as the trade-off between performance and computational costs. After partial exit, the remaining frame features are reorganized into a new sequence and fed into deeper backbone layers to obtain high-level semantics. Since the simply-merged features exhibit significant differences at the points where frames were previously exited and result in false boundaries, we pad the exited positions with repetitions of the nearby frame features to preserve the local information, and the sequence is padded back to the initial length T before being sent to corresponding detectors. Furthermore, the boundaries detected within the padded frame features are also undesirable and will not be recorded.

Considering that the computational cost of the backbone network f is significantly larger than that of the detector h (over 10× in FLOPs), the efficiency gained from allowing more frames to exit early far outweighs the additional burden caused by hard-to-detect samples traversing all detectors.

3.3 Multi-order Difference Detector

The generic event boundary types span simple shot changes to complex sub-action changes. Besides the overall dynamic architecture, we further propose the multi-order difference detector utilizing coarse-to-fine features from the backbone and ensuring these various boundaries can be detected with specialized strategies. The detectors should explore distinctive representations among various boundaries and distinguish them to ensure the effectiveness of dynamism. The detector based on multi-order temporal differences captures and amplifies the most significant motion pattern for boundary identification by modeling localized changes. From the right plane of Fig. 2, the detector is composed of a multi-order difference encoder (MDE) and a pairwise contrast module (PCM) for boundary identification and localization.

Multi-order Difference Encoder. A transformer-liked architecture is proposed to capture multi-order temporal dependencies, which is composed of n blocks with pooling layers among them. The localized instantaneous change (i.e., the gradient of the time series) is crucial for boundary detection tasks [38]. From Fig. 1, the shot change has a clear discrete change point of its feature activation curve. For the sub-action change, the boundary information can only be distinguished after the 2nd temporal difference. In order to model various localized relationships, we propose the difference mixer (Diff Mixer in Fig. 2) with different choices of orders to capture the most distinctive motion patterns for various boundaries. Following ConvMixer [41], we first use small kernel convolutions to enable local interactions. Given a input tensor $\mathbf{X} \in \mathbb{R}^{T \times l \times C \times t_w}$, the mixer g is:

$$g(\mathbf{X}) = \text{BN}(\sigma(\text{DWConv}(\mathbf{X}))), \quad (5)$$

where BN is batch normalization [19], σ is the non-linear activation function and DWConv is the depth-wise temporal convolution [14]. To model the localized change pattern, we compute the first and second-order temporal differences to approximate multi-order derivatives. The final output of the proposed multi-order difference mixer can be written as:

$$\begin{aligned} \mathbf{X}_d &= \mathbf{X}_d^0 + \mathbf{X}_d^1 + \mathbf{X}_d^2 \\ &= g(\mathbf{X}) + g(\Delta \mathbf{X}) + g(\Delta g(\Delta \mathbf{X})), \end{aligned} \quad (6)$$

where $\Delta[\cdot]$ denotes the temporal difference for the input sequence. We decompose the complex instantaneous motion representation into its multi-order derivatives, revealing the crucial change pattern for better boundary identification. Based on different order choices and combinations in Eq. 6, the boundary information of candidate snippets with various temporal dependencies can be easily captured.

The vanilla FFN is replaced with ConvFFN to improve robustness [29]. We chose the max pooling with stride 1 as the connection among blocks. Adjacent features are merged gradually to capture deep action-level representations without losing discriminativeness. The features from different blocks are concatenated as $\mathbf{D} \in \mathbb{R}^{T \times nl \times C \times t_w}$, which is the input of the following pairwise contrast module.

Pairwise Contrast Module. Although the complex temporal dependencies have been decomposed into different levels through multi-order differences, a boundary can only be identified within the given video snippet, and it is still ambiguous to localize the specific timestamp where the change occurs. The representation of a video snippet centered by the candidate frame remains under-explored. We found that frames preceding and following the central frame should exhibit dissimilarity between groups but similarity within each group. In order to enhance the localization ability, we proposed a pairwise contrast module by utilizing the relationship of frames inside the given snippet. Specifically, we calculate the frame-level pairwise similarity as below:

$$\mathbf{S}_t(i,j) = \mathtt{Sim}(\mathbf{D}_t^i, \mathbf{D}_t^j), \quad i,j \in [1,\ldots,t_w], \quad t \in [1,\ldots,T], \qquad (7)$$

where $\mathbf{S}_t \in \mathbb{R}^{nl \times t_w \times t_w}$ is the similarity map of n features from different blocks in MDE, $\mathbf{D}_i^t$ and $\mathbf{D}_j^t$ are features inside the given snippet and $\mathtt{Sim}$ is the cosine similarity function. Since the crucial boundary information relies on the local diagonal patterns in $\mathbf{S}$ [21], the obtained similarity map is then sent into a light-weighted encoder composed of a few layers of convolutions for feature refinement.

Lastly, we fuse the outputs of two branches in the proposed detector through cross SE (squeeze-and-excitation) [15] modules to re-weighting and balancing the features. The fused features are passed through MLP-based classifiers to generate the final predictions.

3.4 Training Details

Gaussian Smoothing. As mentioned before, we follow the continuous paradigm that each frame-wise prediction is merged to obtain the scores $\mathbf{p}$ with length T. Since the annotations for GEBD are subjective and ambiguous, directly using these hard labels to optimize the network could lead to poor generalization ability. Therefore, we smoothed the one-hot labels with a Gaussian kernel to generate soft labels $\widetilde{\mathbf{y}}$. The window size of the Gaussian kernel remains the same as the length of the video snippet and $\sigma = 1$ in all experiments.

Loss Function. After obtaining the predictions from detectors, we calculate the binary cross entropy loss for each one. The total training loss can be written as:

$$\mathcal{L} = \sum_l \alpha_l [-\widetilde{\mathbf{y}}_l \log(\mathbf{p}_l) - (1 - \widetilde{\mathbf{y}}_l) \log(1 - \mathbf{p}_l)], \qquad (8)$$

where losses of L detectors are summed together by weight α to balance the optimization of different subnets.

4 Experiment

4.1 Experimental Settings

Datasets. We perform evaluations on Kinetics-GEBD [35] and TAPOS [33] datasets following others. Kinetics-GEBD is a large-scale dataset comprising 1,290,000 temporal boundaries, designed to annotate generic event boundaries in the wild. Each video is annotated with an average of 4.77 boundaries. The TAPOS dataset contains 21 different actions in Olympic sports videos, which are more challenging to detect. There are 13,094 action instances for training and 1,790 instances for validation. Following [35], we trim each action instance with its action label hidden and conducting experiments on them.

Evaluation Protocol. We employ the F1 score as the measurement for GEBD. As mentioned in [35], the Relative Distance (Rel.Dis.) represents the error between the detected and ground truth timestamps divided by the length of the corresponding action instance. Each detection result is compared with the annotations from each rater, and the highest F1 score is considered the final outcome. We present F1 scores across various Rel.Dis. thresholds ranging from 0.05 to 0.5, with an increment of 0.05. We mainly compare F1@0.05 with others, which keep the same evaluation protocol as the official CVPR'23 LOVEU challenge.

Implementation Details. In practise, we choose ResNet50 [12] pretained on ImageNet [6] as the backbone for a fair comparison. Since the video duration and fps are similar in Kinetics-GEBD, we uniformly sample 100 frames (i.e., $T = 100$) from each video as the inputs. The TAPOS has a large variety of duration over instances from a few seconds to around 5 min. Following [37], we split the instances without overlapping and sample 100 frames by keeping a similar fps to Kinetics-GEBD. The scores of sub-instances are merged together to generate the final prediction. The k is set to 8, and the length of the local window t_w is 17. We choose the last 3 layers of ResNet50 for adaptive inference, and the number of blocks n in the transformer-liked encoder is 3. The cosine function is chosen as the measurement of the pairwise similarity. The whole model is trained end-to-end for 20 epochs with Adam [22] and a base learning rate of 1e−2.

4.2 Main Results

Kinetics-GEBD. We evaluated the proposed DyBDet on the validation set of Kinetics-GEBD [35] in Table 1. From the results, our approach significantly outperforms all counterparts, particularly under the most strict Rel.Dis. constraint. Specifically, our method achieves improvements of 1.9% and 0.9% in F1@0.05 and average compared to the current sota, respectively. Moreover, when equipping with a much more powerful CSN [40] backbone, we can achieve extra improvement significantly, demonstrating DyBDet is not limited by the backbone choices. We also provide the results of computational costs (measured in FLOPs) in Fig. 3.

Table 1. Comparisons in terms o F1 score (%) on Kinetics-GEBD. †: CSN backbone.

Method	F1 @ Rel. Dis.									
	0.05	0.1	0.15	0.2	0.25	0.3	0.35	0.4	0.45	0.5	avg
BMN [28]	18.6	20.4	21.3	22.0	22.6	23.0	23.3	23.7	23.9	24.1	22.3
BMN-StartEnd [28]	49.1	58.9	62.7	64.8	66.0	66.8	67.4	67.8	68.1	68.3	64.0
TCN-TAPOS [23]	46.4	56.0	60.2	62.8	64.5	65.9	66.9	67.6	68.2	68.7	62.7
TCN [23]	58.8	65.7	67.9	69.1	69.8	70.3	70.6	70.8	71.0	71.2	68.5
PC [35]	62.5	75.8	80.4	82.9	84.4	85.3	85.9	86.4	86.7	87.0	81.7
SBoCo-Res50 [21]	73.2	-	-	-	-	-	-	-	-	-	86.6
Temporal Perceiver [37]	74.8	82.8	85.2	86.6	87.4	87.9	88.3	88.7	89.0	89.2	86.0
CVRL [26]	74.3	83.0	85.7	87.2	88.0	88.6	89.0	89.3	89.6	89.8	86.5
DDM-Net [38]	76.4	84.3	86.6	88.0	88.7	89.2	89.5	89.8	90.0	90.2	87.3
SC-Transformer [25]	77.7	84.9	87.3	88.6	89.5	90.0	90.4	90.7	90.9	91.1	88.1
DyBDet (ours)	**79.6**	**85.8**	**88.0**	**89.3**	**90.1**	**90.7**	**91.1**	**91.5**	**91.7**	**91.9**	**89.0**
DyBDet (ours)†	**83.1**	**88.4**	**90.4**	**91.3**	**92.0**	**92.5**	**92.7**	**93.0**	**93.2**	**93.3**	**91.0**

Table 2. Comparison with others in terms of F1 score (%) on TAPOS.

Method	F1 @ Rel. Dis.									
	0.05	0.1	0.15	0.2	0.25	0.3	0.35	0.4	0.45	0.5	avg
ISBA [7]	10.6	17.0	22.7	26.5	29.8	32.6	34.8	36.9	38.2	39.6	30.2
TCN [23]	23.7	31.2	33.1	33.9	34.2	34.4	34.7	34.8	34.8	34.8	64.0
CTM [16]	24.4	31.2	33.6	35.1	36.1	36.9	37.4	38.1	38.3	38.5	35.0
TransParser [33]	28.9	38.1	43.5	47.5	50.0	51.4	52.7	53.4	54.0	54.5	47.4
PC [35]	52.2	59.5	62.8	64.6	65.9	66.5	67.1	67.6	67.9	68.3	64.2
Temporal Perceiver [37]	55.2	66.3	71.3	73.8	75.7	76.5	77.4	77.9	78.4	78.8	73.2
DDM-Net [38]	60.4	68.1	71.5	73.5	74.7	75.3	75.7	76.0	76.3	76.7	72.8
SC-Transformer [25]	61.8	69.4	72.8	74.9	76.1	76.7	77.1	77.4	77.7	78.0	74.2
DyBDet (ours)	**62.5**	**70.1**	**73.4**	**75.6**	**76.7**	**77.2**	**77.5**	**77.9**	**78.1**	**78.4**	**74.7**

Since our method is **dynamic**, we can adjust the amount of exited features t_μ to control the computational budgets and obtain a continuous curve standing for adaptive inference, indicating the flexibility of the proposed method in practical deployment with different computational constraints. As the frames are only inferred once within the backbone, we performed a 3.2% improvement with 11.3× fewer FLOPs compared to DDM-Net [38]. Notably, our proposed DyBDet achieved a significant improvement of 4.7% compared to

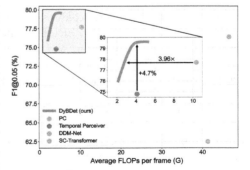

Fig. 3. Comparisons of F1@0.05 v.s. FLOPs on Kinetics-GEBD with others. We report the average FLOPs per frame through the whole inference pipeline.

Table 3. Study on modules in the detector.

PCM	MDE	Fuse	0.05	avg.
			52.6	65.5
	✓		73.2	85.0
✓			76.3	87.2
✓	✓		78.5	88.4
✓	✓	✓	**79.6**	**89.0**

Table 4. Study on design variations in MDE.

Mixer	#blk	Param.	0.05	avg.
SA	3	27.3	77.6	88.0
Conv1d	3	29.7	78.2	88.3
Conv1d	6	37.0	78.5	88.4
M-Diff	6	27.6	78.2	88.4
M-Diff	3	22.5	**79.6**	**89.0**

Table 5. Composition of difference order choices.

Det.1	Det.2	Det.3	0.05	avg.
0	*0*	*0*	76.3	87.2
+1	*+1*	*+1*	78.5	88.5
+1+2	*+1+2*	*+1+2*	78.9	88.7
learned			78.2	88.3
0	*+1*	*+1+2*	**79.6**	**89.0**

Temporal Perceriver [37] under the same computational budget, and 3.96× fewer FLOPs compared to SC-Transformer [25] under the same detection performance. We also found that if all frame features are exited at the first detector, the detection performance is already comparable while requiring much fewer computations, highlighting the effectiveness of the proposed method.

TAPOS. We also conduct experiments on TAPOS [33] in Table 2. DyBDet also achieves state-of-the-art and increases the performance by 0.7% and 0.5% in F1@0.05 and average, respectively. Since TAPOS contains more fine-grained action instances than Kinetics-GEBD and with a much larger variety of video durations, the results also demonstrate the scalability of the proposed DyBDet.

4.3 Ablation Study

Multi-order Difference Detector. To achieve adaptive processing in the dynamic architecture, we need a detector that can distinguish and capture the most representative features of various types of boundaries. The proposed multi-order difference detector is composed of the multi-order difference encoder (MDE) and the pairwise contrast module (PCM). The former adaptively models temporal differences and the latter encompasses pairwise similarities to enhance the boundary representation. From the ablation results in Table 3, the similarity map captured from the pairwise contrast module (PCM) greatly improves the detection performance, which is similar to the conclusions from [21,38]. The localized change pattern along the temporal dimension from PCM is regarded as a significant boundary representation, which is intuitive and effective. Considering the variations of temporal dependencies of different types of boundaries, we further enhance the boundary features with the multi-order difference encoder (MDE). The naive per-frame features are replaced by the multi-order temporal differences, enabling the capability of modeling complex changes as well as the distinguishment for different kinds of boundaries, leading to further performance gains of the dynamic architecture from 76.3% to 78.5%. Since the proposed MDE is composed of several transformer-liked layers and gradually merges boundary features, the output can also contribute to boundary detection. Simply fusing the outcomes from two modules can obtain the best result.

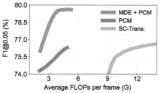
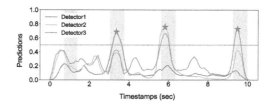

Fig. 4. Dynamic networks with partial exit composed by different detectors. MDE+PCM indicates the detector in DyBDet.

Fig. 5. The predictions of different detectors w/o partial exit. The red line represents the threshold ϵ, and the gray area indicates the ground-truth labels w.r.t F1@0.05. Stars are detected boundaries. (Color figure online)

Temporal Differences. To further verify the performance is not gained from in-depth temporal feature extraction in MDE, we replace the multi-order temporal differences (M-Diff.) in the token mixer with more parameterized operations and also increase the number of blocks. From Table 4, simply adopting networks with more parameters for temporal modeling does not yield higher performance, indicating the importance of multi-order temporal difference for boundary detection. Moreover, computing differences in deep temporal features can also be harmful since the features of adjacent frames are becoming similar [34].

Affinity with Dynamic Architecture. Besides the capability of boundary detection, the proposed multi-order difference detector is designed to ensure various boundaries can be effectively detected with different strategies. To study the affinity with the dynamic architecture, we replace the detectors attached at different depths of the backbone with other alternatives and perform adaptive inference in Fig. 4. Although the detector in SC-Transformer [25] can achieve high performance alone, it suffers from poor generalizations to the dynamic network due to its heaviness. The performance even drops 1.9% under the same computational budget. We also examine the detector composed of PCM alone, which can achieve comparable performance as discussed in Table 3 while being extremely lightweight. However, we can not witness performance saturation when given more computational budgets, indicating that almost all boundaries favor the deepest subnet. As PCM fails to effectively distinguish various boundaries, the adaptive inference of the corresponding dynamic network loses its intended significance. Different from them, since the proposed detector can adaptively explore the most distinctive features of boundaries through multi-order differences, DyBDet can effectively detect various boundaries with different subnets and already achieve the best performance without traversing the whole network for all candidates. We also plot the predictions w.r.t subnets corresponding to different detectors in Fig. 5, and each subnet can partially find specific boundaries. If these boundaries are aggregated by the multi-exit network, we can obtain the best overall performance, indicating the effectiveness of adaptive inference.

Table 6. Study on the partial exit criterion.

Strategy	F1@0.05	GFLOPs
Post Combine	78.1	5.25
Exit w/o Pad	78.4	4.03
Exit LinearPad	79.0	4.27
Exit RepeatPad	**79.6**	4.15

Table 7. Comparisons on inference latency (ms).

Method	F1@0.05	latency
PC [35]	62.5	5.32
DDM-Net [38]	76.4	25.75
SC-Trans. [25]	77.7	1.03
DyBDet	**79.6**	**0.84**

Table 8. Study on different local window sizes.

#win t_w	F1@0.05	F1@avg
9	78.0	88.2
13	78.6	88.6
17	**79.6**	**89.0**
21	79.5	89.0

Order Choices. To further study how the proposed detector achieves adaptive inference, we provide the results of different choices of temporal difference orders for the three detectors (Det.1/2/3) attached at varying depths of the backbone in Table 5. We found that although high-order differences can achieve performance gain, adding it for all detectors equally is worse than the choices of different order combinations. Intuitively, boundaries like shot changes can be easily detected through pairwise contrast in PCM, adding additional temporal differences could introduce noises and vanish the most distinctive boundary representations. We also provide the results that the computed orders are summed together through learnable parameters, yet the performance drops by 1.4%. Since the image backbone is trained together with detectors, letting samples with complex temporal dependencies exit early could do harm to the training of the backbone as early layers lose sensitivities to capture low-level appearance information, which could be essential. The findings demonstrate the separated order choices in DyBDet can benefit the joint training of the supernet and boost the overall performance.

Partial Exit Criterion. We propose the partial exit criterion to enable dynamism, and we study its contribution in Table 6. We first replace the partial exit strategy by simply combining the detection results from different detectors (Post Combine). However, this leads to a 1.5% decrease in performance along with additional computational overhead. The partial exit criterion makes the detection of different boundaries distinguishable, indicating the effectiveness of dynamic detection for GEBD. We also study different padding methods as discussed in Sect. 3.2. Padding the previously exited timestamps can mitigate their influence on subsequent detections. Moreover, using repetitions of nearby remaining frame features for padding outperforms linear interpolation.

Inference Latency. In addition to the efficiency results w.r.t FLOPs, we also provide the practical inference latency on the GeForce RTX 4090 GPU. From Table 7, the proposed dynamic method can achieve high performance with practical efficiency without customized CUDA kernels.

Window Size. We also conduct ablations on different lengths of the local window size for detectors in Table 8. Given an input sequence of length T, the larger window size t_w means a larger temporal respective field for boundary detection. However, due to the locality of boundaries, increasing the window size does not necessarily bring performance growth while with huge extra computations.

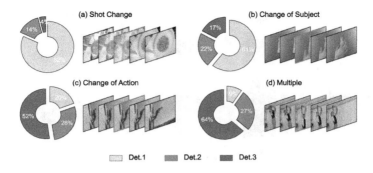

Fig. 6. Class-specific distribution of True Positive samples on different detectors.

4.4 Visualization and Discussion

Class-Specific Statistics. We analyze the class-specific statistics in Fig. 6 to understand how DyBDet achieves fine-grained detection for generic boundaries. Note that although class-specific annotations (like shot changes) are available in Kinetics-GEBD, we treat them as binary labels during training and evaluation. We choose the top four categories with the highest quantity and report the distribution of detection results on different detectors for each class. Since only binary labels are available in predictions, we are unable to compute a confusion matrix for each category, so we count the number of True Positive (TP) samples as an alternative metric. For cases with large appearance-level shifts like shot changes and dominate subject changes, they favor shallow subnets and tend to exit early. On the other hand, identifying action changes and multiple movements requires deeper and more complex spatiotemporal modeling. As a result, generic boundaries with various characteristics can be automatically detected with the most suitable subnet, achieving high overall performance and efficiency.

Pairwise Similarity Map. We present examples with t_w frames in Fig. 7 and visualize their pairwise similarity maps of $t_w \times t_w$ captured by detectors with different orders of temporal difference. The red lines present the ground-truth boundaries. The pairwise similarity maps are then sent to the contrast module to amplify the discriminatives. Therefore, the diagonal pattern (similarity within the same side of frame groups and dissimilarity between groups) is essential to maximizing the boundary information [21]. From the top panel of Fig. 7, samples like shot changes have distinctive background changes rather than the main objects. These low appearance-level features are only preserved in the shallow layers of the backbone. As a result, we can only distinguish a clear boundary based on the features from *layer1*. Moreover, these characteristics also have fewer temporal dependencies, mainly captured from the zero-order difference of Det.1. On the other hand, samples like sub-action changes on the bottom panel require high semantic features from deeper layers of *layer3*, as well as complex temporal change pattern modeling to identify the ambiguous boundary. The crucial

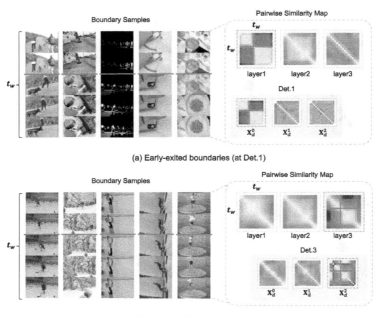

Fig. 7. Visualizations of the pairwise similarity maps vary detectors and computed orders. Red lines indicate the boundaries. (a): the early-exited samples have clear boundary information only based on the lowest level features from *layer1* and the zero-order difference $\mathbf{X}_d^0$. (b): the late-exited samples have more discriminate characteristics based on high-level features from *layer3* and the second-order difference $\mathbf{X}_d^2$. (Color figure online)

diagonal patterns are found mainly in the second-order difference of Det.3, indicating high-level features with high-order temporal differences are needed. Overall, the visualizations demonstrate the demands of adaptive processing for GEBD.

5 Conclusion

In this paper, we present a dynamic model for generic event boundary detection (GEBD) named DyBDet. Unlike existing methods that detect various boundaries with the same protocol, we aim to dynamically allocate subnets to different video snippets based on their distinctive features, outperforming the previous methods by a large margin. As for **future works**, we plan to evaluate the generalizability of our method on more tasks with temporal localization.

Acknowledgement. This work was supported in part by the National Science and Technology Major Project under Grant 2022ZD0115803, the Natural Science Basic Research Plan in Shaanxi Province of China under Grant 2023-JC-JQ-51, and the National Natural Science Foundation of China under Grants 62206215.

References

1. Aakur, S.N., Sarkar, S.: A perceptual prediction framework for self supervised event segmentation. In: CVPR, pp. 1197–1206 (2019)
2. Alayrac, J.B., Laptev, I., Sivic, J., Lacoste-Julien, S.: Joint discovery of object states and manipulation actions. In: ICCV, pp. 2127–2136 (2017)
3. Chen, Z., Li, Y., Bengio, S., Si, S.: You Look Twice: GaterNet for dynamic filter selection in CNNs. In: CVPR, pp. 9172–9180 (2019)
4. Cheng, F., Bertasius, G.: TallFormer: temporal action localization with a long-memory transformer. In: Avidan, S., Brostow, G., Cisse, M., Farinella, G.M., Hassner, T. (eds.) Computer Vision – ECCV 2022. ECCV 2022. LNCS, vol. 13694, pp. 503–521. Springer, Cham (2022). https://doi.org/10.1007/978-3-031-19830-4_29
5. Dai, X., et al.: Dynamic head: unifying object detection heads with attentions. In: CVPR, pp. 7373–7382 (2021)
6. Deng, J., Dong, W., Socher, R., Li, L.J., Li, K., Fei-Fei, L.: ImageNet: a large-scale hierarchical image database. In: CVPR, pp. 248–255. IEEE (2009)
7. Ding, L., Xu, C.: Weakly-supervised action segmentation with iterative soft boundary assignment. In: CVPR, pp. 6508–6516 (2018)
8. Farha, Y.A., Gall, J.: MS-TCN: multi-stage temporal convolutional network for action segmentation. In: CVPR, pp. 3575–3584 (2019)
9. Gygli, M.: Ridiculously fast shot boundary detection with fully convolutional neural networks. In: 2018 International Conference on Content-Based Multimedia Indexing (CBMI), pp. 1–4. IEEE (2018)
10. Han, Y., Huang, G., Song, S., Yang, L., Wang, H., Wang, Y.: Dynamic neural networks: a survey. IEEE TPAMI **44**(11), 7436–7456 (2021)
11. Han, Y., et al.: Latency-aware unified dynamic networks for efficient image recognition. IEEE TPAMI (2024)
12. He, K., Zhang, X., Ren, S., Sun, J.: Deep residual learning for image recognition. In: CVPR, pp. 770–778 (2016)
13. Hong, D., Li, C., Wen, L., Wang, X., Zhang, L.: Generic event boundary detection challenge at CVPR 2021 technical report: cascaded temporal attention network (CASTANET). arXiv preprint arXiv:2107.00239 (2021)
14. Howard, A.G., et al.: MobileNets: efficient convolutional neural networks for mobile vision applications. arXiv preprint arXiv:1704.04861 (2017)
15. Hu, J., Shen, L., Sun, G.: Squeeze-and-excitation networks. In: CVPR, pp. 7132–7141 (2018)
16. Huang, D.-A., Fei-Fei, L., Niebles, J.C.: Connectionist temporal modeling for weakly supervised action labeling. In: Leibe, B., Matas, J., Sebe, N., Welling, M. (eds.) ECCV 2016. LNCS, vol. 9908, pp. 137–153. Springer, Cham (2016). https://doi.org/10.1007/978-3-319-46493-0_9
17. Huang, G., Chen, D., Li, T., Wu, F., van der Maaten, L., Weinberger, K.: Multi-scale dense networks for resource efficient image classification. In: ICLR (2018)
18. Huynh, V.T., Yang, H.J., Lee, G.S., Kim, S.H.: Generic event boundary detection in video with pyramid features. arXiv preprint arXiv:2301.04288 (2023)
19. Ioffe, S., Szegedy, C.: Batch normalization: accelerating deep network training by reducing internal covariate shift. In: ICML, pp. 448–456. PMLR (2015)
20. Kang, H., Kim, J., Kim, K., Kim, T., Kim, S.J.: Winning the CVPR'2021 kinetics-GEBD challenge: contrastive learning approach. arXiv preprint arXiv:2106.11549 (2021)

21. Kang, H., Kim, J., Kim, T., Kim, S.J.: UBoCo: unsupervised boundary contrastive learning for generic event boundary detection. In: CVPR, pp. 20073–20082 (2022)
22. Kingma, D.P., Ba, J.: Adam: a method for stochastic optimization. arXiv preprint arXiv:1412.6980 (2014)
23. Lea, C., Reiter, A., Vidal, R., Hager, G.D.: Segmental spatiotemporal CNNs for fine-grained action segmentation. In: Leibe, B., Matas, J., Sebe, N., Welling, M. (eds.) ECCV 2016. LNCS, vol. 9907, pp. 36–52. Springer, Cham (2016). https://doi.org/10.1007/978-3-319-46487-9_3
24. Lei, P., Todorovic, S.: Temporal deformable residual networks for action segmentation in videos. In: CVPR, pp. 6742–6751 (2018)
25. Li, C., et al.: Structured context transformer for generic event boundary detection. arXiv preprint arXiv:2206.02985 (2022)
26. Li, C., Wang, X., Wen, L., Hong, D., Luo, T., Zhang, L.: End-to-end compressed video representation learning for generic event boundary detection. In: CVPR, pp. 13967–13976 (2022)
27. Li, Y., et al.: Learning dynamic routing for semantic segmentation. In: CVPR, pp. 8553–8562 (2020)
28. Lin, T., Liu, X., Li, X., Ding, E., Wen, S.: BMN: boundary-matching network for temporal action proposal generation. In: ICCV, pp. 3889–3898 (2019)
29. Mao, X., et al.: Towards robust vision transformer. In: CVPR, pp. 12042–12051 (2022)
30. Ming, Q., Zhou, Z., Miao, L., Zhang, H., Li, L.: Dynamic anchor learning for arbitrary-oriented object detection. In: AAAI, vol. 35, pp. 2355–2363 (2021)
31. Nag, S., Zhu, X., Song, Y.Z., Xiang, T.: Post-processing temporal action detection. In: CVPR, pp. 18837–18845 (2023)
32. Radvansky, G.A., Zacks, J.M.: Event perception. Wiley Interdisc. Rev. Cogn. Sci. **2**(6), 608–620 (2011)
33. Shao, D., Zhao, Y., Dai, B., Lin, D.: Intra-and inter-action understanding via temporal action parsing. In: CVPR, pp. 730–739 (2020)
34. Shi, D., Zhong, Y., Cao, Q., Ma, L., Li, J., Tao, D.: TriDet: temporal action detection with relative boundary modeling. In: CVPR, pp. 18857–18866 (2023)
35. Shou, M.Z., Lei, S.W., Wang, W., Ghadiyaram, D., Feiszli, M.: Generic event boundary detection: a benchmark for event segmentation. In: ICCV, pp. 8075–8084 (2021)
36. Souček, T., Moravec, J., Lokoč, J.: TransNet: a deep network for fast detection of common shot transitions. arXiv preprint arXiv:1906.03363 (2019)
37. Tan, J., Wang, Y., Wu, G., Wang, L.: Temporal perceiver: a general architecture for arbitrary boundary detection. IEEE TPAMI **45**(10), 12506–12520 (2023)
38. Tang, J., Liu, Z., Qian, C., Wu, W., Wang, L.: Progressive attention on multilevel dense difference maps for generic event boundary detection. In: CVPR, pp. 3355–3364 (2022)
39. Tang, S., Feng, L., Kuang, Z., Chen, Y., Zhang, W.: Fast video shot transition localization with deep structured models. In: Jawahar, C.V., Li, H., Mori, G., Schindler, K. (eds.) ACCV 2018. LNCS, vol. 11361, pp. 577–592. Springer, Cham (2019). https://doi.org/10.1007/978-3-030-20887-5_36
40. Tran, D., Wang, H., Torresani, L., Feiszli, M.: Video classification with channel-separated convolutional networks. In: Proceedings of the IEEE/CVF International Conference on Computer Vision, pp. 5552–5561 (2019)
41. Trockman, A., Kolter, J.Z.: Patches are all you need? arXiv preprint arXiv:2201.09792 (2022)

42. Wang, J., Li, F., An, Y., Zhang, X., Sun, H.: Towards robust LiDAR-camera fusion in BEV space via mutual deformable attention and temporal aggregation. IEEE TCSVT **34**(7), 5753–5764 (2024)
43. Wang, X., Yu, F., Dou, Z.Y., Darrell, T., Gonzalez, J.E.: SkipNet: learning dynamic routing in convolutional networks. In: ECCV, pp. 409–424 (2018)
44. Wang, Y., Chen, Z., Jiang, H., Song, S., Han, Y., Huang, G.: Adaptive focus for efficient video recognition. In: ICCV, pp. 16249–16258 (2021)
45. Wang, Y., et al.: AdaFocus V2: end-to-end training of spatial dynamic networks for video recognition. In: CVPR, pp. 20030–20040. IEEE (2022)
46. Xie, Z., Zhang, Z., Zhu, X., Huang, G., Lin, S.: Spatially adaptive inference with stochastic feature sampling and interpolation. In: Vedaldi, A., Bischof, H., Brox, T., Frahm, J.-M. (eds.) ECCV 2020. LNCS, vol. 12346, pp. 531–548. Springer, Cham (2020). https://doi.org/10.1007/978-3-030-58452-8_31
47. Yang, L., Han, Y., Chen, X., Song, S., Dai, J., Huang, G.: Resolution adaptive networks for efficient inference. In: CVPR, pp. 2369–2378 (2020)
48. Yang, L., et al.: CondenseNet V2: sparse feature reactivation for deep networks. In: CVPR, pp. 3569–3578 (2021)
49. Yang, L., Zheng, Z., Wang, J., Song, S., Huang, G., Li, F.: AdaDet: an adaptive object detection system based on early-exit neural networks. IEEE Trans. Cogn. Dev. Syst. **16**(1), 332–345 (2023)
50. Zhang, C.L., Wu, J., Li, Y.: ActionFormer: localizing moments of actions with transformers. In: Avidan, S., Brostow, G., Cisse, M., Farinella, G.M., Hassner, T. (eds.) Computer Vision – ECCV 2022. ECCV 2022. LNCS, vol. 13664, pp. 492–510. Springer, Cham (2022). https://doi.org/10.1007/978-3-031-19772-7_29
51. Zheng, Z., et al.: Dynamic spatial focus for efficient compressed video action recognition. IEEE TCSVT **34**(2), 695–708 (2024)

Take a Step Back: Rethinking the Two Stages in Visual Reasoning

Mingyu Zhang[1], Jiting Cai[1], Mingyu Liu[2], Yue Xu[1], Cewu Lu[1], and Yong-Lu Li[1(✉)]

[1] Shanghai Jiao Tong University, Shanghai, China
{sjtuzmy2003,Caijiting,silicxuyue,lucewu,yonglu_li}@sjtu.edu.cn
[2] Zhejiang University, Hangzhou, China
mingyuliu@zju.edu.cn

Abstract. As a prominent research area, visual reasoning plays a crucial role in AI by facilitating concept formation and interaction with the world. However, current works are usually carried out separately on small datasets thus lacking generalization ability. Through rigorous evaluation of diverse benchmarks, we demonstrate the shortcomings of existing ad-hoc methods in achieving cross-domain reasoning and their tendency to data bias fitting. In this paper, we revisit visual reasoning with a two-stage perspective: (1) symbolization and (2) logical reasoning given symbols or their representations. We find that the reasoning stage is better at generalization than symbolization. Thus, it is more efficient to implement symbolization via **separated** encoders for different data domains while using a **shared** reasoner. Given our findings, we establish design principles for visual reasoning frameworks following the separated symbolization and shared reasoning. The proposed two-stage framework achieves impressive generalization ability on various visual reasoning tasks, including puzzles, physical prediction, and visual question answering (VQA), encompassing 2D and 3D modalities. We believe our insights will pave the way for generalizable visual reasoning. **Our code is publicly available at** https://mybearyzhang.github.io/projects/TwoStageReason.

Keywords: Visual Reasoning · Symbolization and Reasoning · Generalization Ability

1 Introduction

Reasoning ability [38] is a concentrated embodiment of human intelligence, serving as the foundation for concept formation, cognitive understanding of the world, and interaction with the environment. Specifically, visual reasoning, being

---

M. Zhang and J. Cai—Equal contribution.

Supplementary Information The online version contains supplementary material available at https://doi.org/10.1007/978-3-031-72775-7_8.

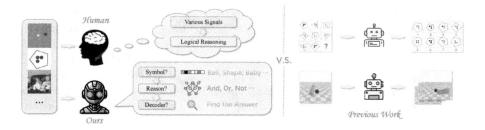

Fig. 1. Comparison between end-to-end model, human, and our framework. Previous works usually use a specific end-to-end model for each task, while our framework shares a logical reasoner similar to human intelligence.

one of the primary modes through which humans acquire information and understanding, has been the focus of extensive research. In recent years, with the advancements in deep learning, numerous works [15,19,34,35,46,52] on visual reasoning have been proposed. Additionally, various datasets [3,4,15,20,21,25,57] have also emerged to evaluate the reasoning models.

However, a notable limitation of existing visual reasoning works lies in their direct reliance on entangling the recognition and reasoning phrases via end-to-end deep learning models, *e.g.*, recognizing the concepts in images while answering logical questions [57]. However, this paradigm has obvious limitations as follows: 1) Reasoning annotation (rule, relation) is much more costly and difficult than symbol annotation (triangle, cube, apple), thus current rigorous visual reasoning datasets are usually small. Therefore, current approaches tend to be heavily task-specific on small datasets [29,56,57], hindering their generalization potential. 2) Pursuing a versatile model for symbol recognition and logical reasoning simultaneously may be inefficient and challenging. Even recent large language models (LLM) struggle with diverse visual reasoning tasks [59].

In this paper, we argue that visual reasoning is rooted in first grounding symbols derived from visual signals, followed by logical reasoning, as shown in Fig. 1. Thus, a question arises: **should the two stages be entangled or disentangled?** Reasoning naturally has a generalization property than symbolization. For example, we use *similar* logic to analyze the rules of different tasks (*e.g.*, play go, do the math, and discover anomaly) but very *different* knowledge to recognize the alphabet and objects. Thus, we assume that disentangled symbolization and reasoning would be a more wise choice. The recent success of Large Language Models (LLM) on text-based reasoning tasks [44] also validates this point, as LLMs directly leverage the abstract symbols (language) derived from human observation and focus on high-level linguistic tasks. Relatively, Multi-Modal Large Language Models (MLLM) still struggle with visual reasoning [13] even with more parameters. Recently, another related research trend is the neuro-symbolic method. The neuro-symbolic approach transforms raw inputs into explicit symbols for subsequent reasoning and analysis [54]. However, neuro-symbolic methods often remain confined to a single dataset [2,16,43], making it challenging to achieve generalization across different tasks.

We conduct comprehensive experiments on various tasks of multiple benchmarks with significant domain gaps to verify our assumption. We formulate the symbolization stage as the representation extraction with Deep Neural Networks (DNN) and implement the logical reasoner with various architectures (MLP, CNN, GNN, Transformer, Neuro-Symbolic models, LLM, *etc.*). We mainly investigate **two key questions**: (1) Within a trained DNN, where does the symbolization phrase conclude? That is, identifying the suitable symbol (representations) for reasoning, such as the depth of the model, feature characteristics, *etc.* (2) Given the abstracted symbols, what types of models and training strategies are most suitable for reasoning and endowing the generalization capability?

For the **first** problem, we find that different tasks and domains need very different scales of parameters or model depths to achieve a good symbolization. Thus, for a specific domain, a small **separated** in-domain encoder is enough to extract symbols from the data for the subsequent reasoning phrase. Though a general and large foundation model like CLIP [39] can do well on some tasks, it still struggles on tasks with a huge domain gap with its training data [32,39]. For the **second** problem, our results reveal that existing methods struggle to execute cross-domain reasoning, instead fitting biases aligned with training data. Thus, maybe we should or can only achieve a generalizable **shared** reasoner via training it on various reasoning tasks (puzzles, physical prediction, VQA) and data domains (2D, 3D, text), *i.e.*, "**approximation principle**".

Building upon our findings, we build a concise framework with separated encoders to achieve the optimal symbolization for different data domains and a shared reasoner following the "approximation principle". Our method performs exceptionally well with fewer parameters across cross-domain benchmarks.

In general, our contributions are: (1) We conclude an efficient two-stage perspective for visual reasoning drawing inspiration from previous visual reasoning networks. (2) We investigate the optimal design principles of symbolization and logical reasoning for visual reasoning. (3) Accordingly, we introduce a concise framework with decent performance on multiple datasets with domain gaps.

2 Related Work

Visual Reasoning. Visual reasoning is a subfield of computer vision and artificial intelligence that aims to enable machines to reason about visual information in a human-like manner. It involves using machine learning techniques to analyze images or videos and making decisions based on that analysis. Several recent studies have explored various architectures for deep neural networks designed specifically for visual reasoning tasks such as the Visual Question Answering (VQA) [15,20], video causal inference [4,31], dynamic prediction for 3D scenes [21], *etc.* However, most of them are domain-specific and lack generalization ability. These architectures have typically combined convolutional neural networks for visual feature extraction with recurrent neural networks or LSTM for language processing [52]. Another area of research in visual reasoning is the development of knowledge-based systems [46], which rely on structured representations of knowledge about the world to reason about visual information. For example, Hong et al. [19] built a system that could reason about the layout of objects in a scene or the relationships between different objects in an image. However, the essence of reasoning itself and the certain stages of visual reasoning, which is a basic problem in this domain, remains unclear.

Neuro-symbolic Learning. Neuro-symbolic learning [11,12,16,30,43,48,53,54,58] combines symbolic reasoning with neural networks to address complex problems in domains like computer vision [27,30,55], natural language processing [17,33], and knowledge inference [6,28]. Neuro-symbolic methods are categorized into three types [54]: 1) *Learning for reasoning* uses neural systems to enhance symbolic reasoning; 2) *Reasoning for learning* uses symbolic systems to aid neural learning and improve interpretability; and 3) *Learning-reasoning* involves bidirectional interaction between neural and symbolic systems. Wu et al. [34] and Gupta et al. [16] have proposed notable neuro-symbolic approaches for image feature extraction and problem-solving. Despite its promise, neuro-symbolic integration has limitations, particularly in generalizing reasoning abilities across tasks [54]. Our framework differs from neuro-symbolic methods in two key ways: (1) It includes both *end-to-end* and *neuro-symbolic models*, with the latter as a subset; (2) It can explain the generalization of the reasoner, which neuro-symbolic methods cannot.

Visual Reasoning Benchmarks. Researchers have developed various cross-modal visual reasoning benchmarks to evaluate reasoning models, including 2D puzzles, 3D physical prediction, Visual Question Answering (VQA), *etc.* 2D puzzle datasets explore relationships among visual elements. Notable datasets include RAVEN [57], SVRT [29], CVR [56], Bongard-HOI [23], and Bongard-LOGO [37]. Intuitive physics datasets, such as CoPhy [4], Filtered-CoPhy [21], Space [9], and Space++ [8], focus on how humans perceive and reason about the physical world. VQA requires machines to understand both natural language questions and images [47], with datasets including VQAv1 [3], VQAv2 [15], CLEVR [25], and GQA [20].

3 Preliminary

In this section, we first formulate visual reasoning following the proposed two-stage view.

3.1 Two Stages

As described above, visual reasoning can be divided into two stages: the symbolization stage extracts symbolic representations of the underlying data, and the reasoning stage performs logical reasoning.

For humans, different modalities of visual and auditory information collected from our sensors are converted into electrical signals through different pathways and then sent to the cerebellar cortex to perform logical reasoning [10]. Analogously, separated task-specific symbolizers and a shared domain-independent reasoner would be a reasonable choice for a general visual reasoning machine. Besides, the reasoner should be capable of performing unified reasoning on input information from various modalities. In other words, *the essence of reasoning lies in its generalization ability.*

Symbolization Stage. During the stage of symbolization, we implement various task-oriented feature extraction networks. These networks employ symbol encoders tailored for each task, transforming multi-modal inputs (text, image, video) into symbol representations. Formally, suppose we have n tasks. For the i-th task, we have the input data $\boldsymbol{x}^i$ and the task t^i, and the task-oriented encoder E^i. Then we get the symbol representation set $\boldsymbol{f}^i$ via:

$$\boldsymbol{f}^i = E_i(\boldsymbol{x}^i \mid t^i). \tag{1}$$

Reasoning Stage. The reasoner is fed by symbolic representations for each specific task, in a bid to capture a deeper and more comprehensive understanding of the underlying patterns and relationships embedded within the data. For symbol representation sets $\{\boldsymbol{f}^{(i)}\}_{i=1}^n$ of all tasks, we send them into the reasoner R, and get its reasoning result set $\{\boldsymbol{c}^i\}_{i=1}^n$ after the logic processing to facilitate problem-solving across various modalities:

$$\{\boldsymbol{c}^i\}_{i=1}^n = R(\{\boldsymbol{f}^i\}_{i=1}^n). \tag{2}$$

Task-Specific Heads. The final part of our framework is the task-specific heads, which take the reasoning results from the reasoner as input and generate task-specific answers. For different tasks, we need to construct task-specific classification or regression heads H_i to get the final output s^i. That says:

$$s^i = H_i(\boldsymbol{c}^i \mid t^i). \tag{3}$$

Next, we can compare the output with the ground truth to compute gradients for training the entire framework.

4 Symbolization-Reasoning Framework

4.1 Entanglement v.s. Disentanglement

Given the two stages, a natural question arises: should the symbol encoder (symbolization) and reasoner (reasoning) be shared or separated for tasks?

To validate our *shared reasoner only* assumption, we conduct a comparison between different designs (Fig. 2): 1) **Both-Separated**: the symbol encoder and logical reasoner are all separated (ad-hoc models for each task); 2) **Both-Shared**: both the encoder and reasoner are shared. 3) **Shared-Encoder-Only**:

only the symbol encoder is shared; 4) **Shared-Reasoner-Only**: only the reasoner is shared.

We compare the above four designs on several multi-modal visual reasoning benchmarks [23,29,50,56,57]. For the shared encoder/reasoner, we adopt more parameters to balance them with the sum of parameters of the separated encoders/reasoners. In the experiments of Sect. 5, we find that the Shared-Reasoner-Only (Type 4) outperforms Shared-Encoder-Only (Type 3) and Both-Shared (Type 1 and 2) a lot on all benchmarks. Besides, the Shared-Reasoner-Only even beats the ad-hoc Both-Separated on some benchmarks, validating its superiority and generalization ability.

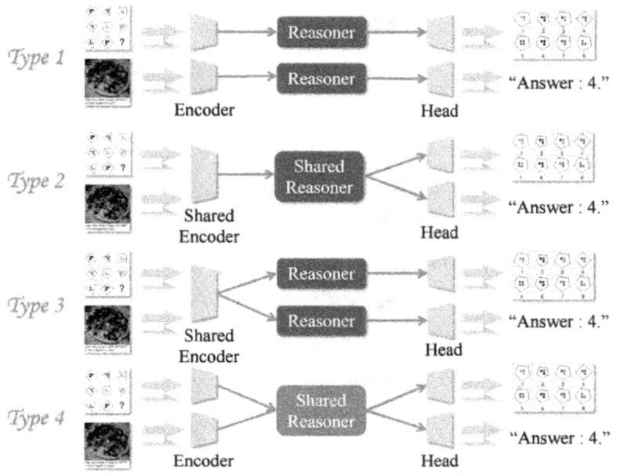

Fig. 2. Entanglement v.s. Disentanglement. Type 1: the symbol encoder and reasoner are all separated; Type 2: both the encoder and reasoner are shared; Type 3: only the encoder is shared; Type 4: only the reasoner is shared.

4.2 Symbolization Depth

Next, we probe the appropriate depth of the symbol encoder for different tasks. The symbolization stage involves processing inputs from different domains and mapping them to the conceptual level, *i.e.*, symbols. Though we can use binary or index-like (one-hot) representations for symbols, in the context of deep learning, we choose the more representative way, *i.e.*, high-dimensional features extracted from DNN. Intuitively, different tasks may need different levels of symbolization, Then, the next question is how to determine the level of abstraction for each task.

To answer this question, we design experiments to quantitatively probe the degree of symbolization in the process. To control variables, we employ the same feature extraction network (ResNet [18]) for cross-domain tasks while continuously adjusting the network depth. At different depths, we connect the output of the symbolization network to the same reasoner and measure the *accuracy* as an indicator of symbolization completion.

We hypothesize that when symbolization is complete, the network depth-accuracy curve will exhibit a distinct inflection point. By monitoring the occurrence of this inflection point, we can select the appropriate depth for the reasoner for different tasks, as shown in Fig. 3. In experiments, we find the result consistent with our common sense: both excessively shallow and deep networks are detrimental to reasoning tasks. On one hand, if the encoder is too shallow, symbolization will not be fully achieved before being inputted into the reasoner, while the reasoner would have to conduct part of the symbolization work, thus impacting the performance of reasoning. On the other hand, deeper networks tend to overfit to one single task, thus weakening the performance of the shared reasoner which aims for generalization. The symbolization depths also vary for tasks in different domains. Improperly setting the depth leads to conflicts in the shared parameters, resulting in the poor performance of deeper layers.

4.3 Reasoner Architecture

Next, we want to figure out which architecture is more suitable for the reasoner, which is a problem of a long history. Many works have been proposed and achieved improvements on various visual reasoning benchmarks [15,24,57]. Here, each task is handled by its respective encoder and task head, designed to adapt to the inherent characteristics of its data. We use a *shared reasoner* for all tasks and take into the symbol representations following Eq. 2.

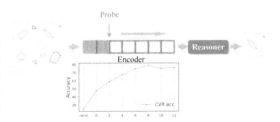

Fig. 3. Probing process of symbolization. We vary the depths of the symbol encoder (ResNet) and train the framework while recording the accuracy at each encoder depth. An inflection point occurs in the curve at moderate depths.

We choose a line of architectures as the reasoner candidates: Multilayer Perceptron (MLP), Convolutional Neural Networks (CNN), and Transformer that achieve great success across numerous tasks. We also explore a hybrid neuro-symbolic model [34] combining the representational capacity of neural networks with the interpretability of symbolic systems. Furthermore, we adopt the popular graphical and autoregressive models: Graph Convolutional Networks (GCN) [14] and MiniGPT-4 [5,60]. The above models afford us a comprehensive and diverse range of methodologies. If a robust and consistent performance is witnessed across various domain-diverse datasets, it leads us to hypothesize the presence of a certain type of architecture that excels specifically at logical reasoning.

4.4 Generalization of Reasoner

Last but not least, we aim to verify our "approximation principle", *i.e.*, a good reasoner affording generalizable logical reasoning ability can be approached by training it with diverse tasks and data from diverse domains. We believe that reasoning connotes universality and generalization. Therefore, we initially train a complete two-stage model on one task, and then directly take its reasoner and pair it with another symbol encoder of another task. If the reasoner has generalization ability, it should suit the encoder of the other tasks well. However, in our test, a reasoner trained with only one task/domain usually generalizes not well. Thus, we next verify whether the generalization ability of the reasoner is better given the training on more tasks/domains, as shown in Fig. 4.

We find that, as more and more data from different tasks and domains are involved, the overall task becomes increasingly challenging. However, the reasoner will concentrate on "pure" reasoning instead of task/domain-specific solving thus endowing better generalization ability (detailed in Sect. 5). That says, our "approximation principle" is reasonable. Thus, we can predict that the reasoner should perform better on out-of-domain tasks as the training data and task increase. Moreover, a shared and cross-task trained reasoner makes the whole visual reasoning framework lighter and more efficient.

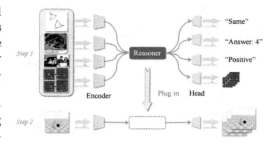

Fig. 4. "Approximation principle" verification with a shared reasoner. In step 1, the process entails the selection of 1–4 datasets, namely SVRT, Bongard-HOI, CoPhy-Balls, and VQAv2, to train the reasoner. This combination offers a total of 15 possible permutations. In step 2, the proficiently trained reasoner is subjected to rigorous testing on the CoPhy-Collision dataset for evaluation and validation purposes.

5 Experiments

In this section, we present our experimental settings and empirical results to support our assumptions. Our experiments are designed to answer the following questions:
(1) Are there any differences in transferability between the symbolization and reasoning stages?
(2) Given the presence of two stages, what is the impact of different levels of symbolization depth, and can we identify a clear indicator for the termination of symbolization?
(3) Which architectures suit the reasoning best and can the "approximation principle" guide the training of the reasoner to achieve better generalization?

5.1 Dataset and Setting

2D Puzzles. We use RAVEN [57], CVR [56], SVRT [50], Bongard-LOGO [37], and Bongard-HOI [23] as our 2D datasets to assess the reasoning ability of neural networks. CVR and SVRT focus on the attributes like shape, color, size, *etc.*, of geometric objects. RAVEN, beyond that, needs to deal with positional and logical relationships such as AND, OR, XOR, *etc.* Bongard-LOGO and Bongard-HOI follow the rules of the Bongard problem, aiming to find the deep common features and concepts among contrastive samples. More details of these datasets will be elaborated in the Supplementary.

2D VQA. We choose VQAv2 [15], which is a bias-free and balanced dataset. Built upon COCO, VQAv2 [15] carefully selects questions and images to ensure

Table 1. Entanglement v.s. Disentanglement. We demonstrate the impact of using shared modules in the two-stage framework, *i.e.*, both separated encoder and reasoner, both shared, shared encoder and shared reasoner as shown in Fig. 2.

	RAVEN	CVR	SVRT	Bongard-HOI	Bongard-LOGO
Both-Separated (ad-hoc)	53.40	74.04	86.20	61.64	72.16
Both-Shared	35.86	55.48	51.58	52.65	60.83
Shared-Encoder-Only	42.06	54.82	58.24	52.65	63.17
Shared-Reasoner-Only	**55.72**	**72.40**	**86.64**	**57.65**	**69.64**

an equal distribution of answers for each question type. It consists of 1,000 natural language descriptions, each question having ten options from which the model needs to make a selection.

3D Intuitive Physics. We select Filtered-CoPhy [21]. It involves three tasks: block tower, reasoning about the stability of stacked blocks; balls, reasoning about the motion rules of spherical objects; and collision, reasoning about the physical phenomena during object collisions. As we focus on the reasoning more instead of frame prediction, here we only utilize the differences between scene keypoints as the measurement.

Implementation Details. All the networks are trained for within 100 epochs using the Adam optimizer [26]. The learning rate and weight decay are finetuned with the assistance of Optuna [1]. For our **lite** reasoner, we choose modules from MLP, CNN, Transformer, *etc.*, with parameters *no more than 100M*. To test the generalization of the reasoner, we freeze the reasoner network and train the encoder and head modules. All experiments were conducted on 4 Titan XP GPUs except for the experiment with LLM performed on 1 V100-32 GB GPU. More details of network structure and hyperparameter configuration of each setting and model will be elaborated in the Supplementary.

5.2 Entanglement v.s. Disentanglement Analysis

To compare the three types in Fig. 2, we train the models using five datasets: RAVEN [57], CVR [56], SVRT [29], Bongard-HOI [23], and Bongard-LOGO [37]. To control variables and facilitate model sharing, for all types above we adopt ResNet-18 [18] as the encoder and an MLP as the reasoner.

The results are depicted in Table 1, we find that the performance of the Shared-Reasoner-Only is comparable with the ad-hoc Both-Separated on all five datasets and even higher on RAVEN and SVRT. Besides, Shared-Encoder-Only and Both-Shared show obvious inferior performance on all datasets. This reflects the validity of our design in employing task-specific symbolizers and shared reasoners across multiple tasks.

5.3 Optimal Symbolization Depth

Next, we aim to identify the boundary between the two stages by probing the depth of the symbol encoder as shown in Fig. 3. The shared reasoner we use is

Table 2. Detailed results of the impact of varying encoder depths in the symbolization stage. Within a certain range, different tasks exhibit an improvement in scores as the symbol encoder deepens, however, each task displays a different inflection point.

	Random	ResBlock 0	ResBlock 2	ResBlock 4	ResBlock 6	ResBlock 8	ResBlock 10	ResBlock 12
RAVEN	12.5	**53.09**	53.64	56.49	56.04	54.50	39.12	40.80
CVR	25	47.56	58.73	67.56	**74.65**	79.20	75.42	76.22
SVRT	50	51.96	51.84	53.49	66.33	**87.33**	88.92	90.22
Bongard-HOI	50	55.51	56.46	55.93	**61.97**	61.23	60.48	61.02
Bongard-LOGO	50	56.16	62.50	66.50	**69.48**	68.74	68.02	65.58

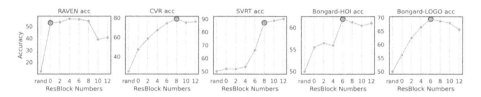

Fig. 5. Performance curve of varying encoder depth in symbolization stage. We present the results across RAVEN, CVR, SVRT, Bongard-LOGO, and Bongard-HOI. The highlight points refer to the distinct inflection points.

an MLP. By observing the changes of the *accuracy*, we hope to find the distinct **inflection point** and termination of the symbolization grounding stage. To ensure fairness, we employ the ResNet18 [18] encoder across the following 2D datasets: RAVEN [57], CVR [56], SVRT [29], Bongard-LOGO [37], Bongard-HOI [23]. For each benchmark, we train individual models to their optimal performance and then interrupt the trained networks at various depths to probe the symbolization termination points. We connect the output of the separated symbol encoders to a shared reasoner and record the accuracy at different interruption points, which is treated as evidence of symbolization termination. The results are illustrated in Table 2, which presents outcomes of various datasets following the deepening of the symbolization encoder. We also visualize the results in Fig. 5.

As observed in Fig. 5 and Table 2, for each benchmark, the network depth exhibits an initial increase followed by a plateau. The inflection point positions vary across different tasks due to their varying levels of difficulty and the requisite degree of symbol abstraction. The inflection point of Bongard-HOI is much deeper than RAVEN, showing that the former is more difficult to symbolize and needs a deeper symbolization network to acquire a sophisticated high-dimensional feature. These results verify the necessity of employing symbolization networks with varying depths for datasets of different complexities and illustrate the reasonable boundary between the two stages.

5.4 One-for-All Reasoner Architecture

Next, we aim to figure out the suitable architecture for the reasoner and test its effect in the Shared-Reason-Only design. We choose 9 cross-domain datasets,

Table 3. Impact of shared reasoner implemented with different architectures in the Shared-Reason-Only type.

	RAVEN	CVR	SVRT	Bongard-HOI	Bongard-LOGO	CoPhyBall	CoPhyBlocktower	CoPhyCollision	VQA
Both Separated (ad-hoc)	53.40	78.30	92.50	66.43	72.30	6.53	0.39	2.91	54.08
MLP	55.72	72.40	86.64	57.65	69.64	**8.01**	**0.84**	**2.98**	**50.06**
CNN	**58.47**	74.04	86.20	58.47	62.16	8.87	1.01	3.04	41.47
Transformer	49.36	78.48	84.20	**62.50**	69.32	9.34	1.80	3.74	32.64
GCN	52.72	**78.75**	**88.10**	61.36	**71.27**	8.17	0.95	3.31	42.19
Neuro-Symbolic	27.89	69.69	77.72	55.21	59.68	11.23	3.04	5.31	36.62

Table 4. Comparison between SOTA baselines and our One-for-All model. The "–" means the original papers do not provide the results on the corresponding dataset.

	RAVEN	CVR	SVRT	Bongard-HOI	Bongard-LOGO	VQA
Ad-hoc SOTA (Lite)	53.4 [57]	78.3 [56]	92.5 [36]	57.7 [41]	65.4 [37]	54.1 [3]
	(ResNet)	(ResNet-50 SSL)	(ResNet-50)	(CoCoOp)	(ProtoNet)	(LSTM+CNN)
One-for-All (MLP Reasoner)	55.7	72.4	86.6	57.7	69.6	50.1
Ad-hoc SOTA (Heavy)	94.1 [42]	–	96.9 [36]	66.4 [40]	75.3 [40]	76.4 [24]
	(Rel-AIR)		(CorNet-S)	(TPT)	(SVM+MIMIC)	(ROUGE_L)
One-for-All (MiniGPT-4 Reasoner)	21.6	58.3	97.8	53.8	54.8	63.6

RAVEN [57], CVR [56], SVRT [29], Bongard-HOI [23], Bongard-LOGO [37], Filtered-CoPhy [21], VQAv2 [15] to conduct our experiment, as solving various reasoning problems with different domains can better demonstrate the model's reasoning abilities.

We design task-specific encoders and heads based on the requirements of different tasks. As for reasoner, we test CNN, MLP, Transformer [45], GCN [14], hybrid Neuro-Symbolic model [34], and MiniGPT-4 [5,60]. We first individually train each dataset to obtain the best results with the separated encoders and separated heads. Then we perform joint training on multiple datasets using one shared reasoner.

In Table 3, we conclude that within all the architectures, MLP surprisingly performs the best on four datasets and is comparable in the other five datasets. Besides, the GCN performs well on three datasets following the previous experience in reasoning works [14]. However, other architectures that are usually thought to be more advanced like Transformer do not show obvious advantages. Thus, we choose the MLP as a lite reasoner in our One-for-All model.

In Table 4, our One-for-All model demonstrates impressive performance even compared to the ad-hoc state-of-the-art (SOTA) across a majority of tasks. We divide the SOTA according to their complexity into lite and heavy, as depicted in Table 4. From the result, One-for-All is comparable with the lite ad-hoc SOTA and even outperforms the lite SOTA on some datasets like RAVEN. This experiment demonstrates that the reasoning stage has quite a different performance-parameter relationship with recognition tasks. A lite reasoner may also perform well on reasoning if trained on multi-domain tasks.

Since reasoning ability cannot be solely measured by accuracy, we also evaluate reasoning capabilities using *reasoning consistency*. For each task, we use the

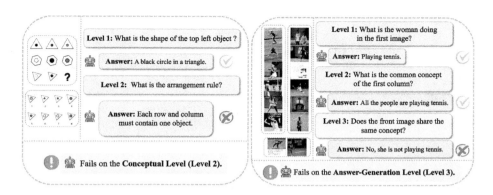

Fig. 6. Failure case analysis of LLM-based model. On RAVEN, it fails on the Symbolization level; while on Bongard-HOI, it fails on the Answer Generation level.

Table 5. Comparison of accuracy and reasoning consistency between the ad-hoc methods and our one-for-all model. Our one-for-all model demonstrates decent performance.

Metric	Model	RAVEN	CVR	SVRT	Bongard-HOI
Accuracy	Both-Separated	53.40	**74.04**	86.20	**61.64**
	One-for-All	**55.72**	72.40	**86.64**	57.65
Consistency	Both-Separated	76.94	52.35	99.42	98.93
	One-for-All	**77.89**	**53.51**	**99.73**	**99.55**

same encoder and reasoner parameters with two questioning methods: *"What is the answer to this question?"* and *"Is a certain option correct?"*. A model with good reasoning ability should yield consistent results across both methods, unlike a random model which may be inconsistent. We use the F1 score to measure the consistency between these methods, as shown in Table 5. Our One-for-All model, trained jointly on multiple datasets, shows higher consistency compared to models trained individually, demonstrating its potential in genuine reasoning.

To further evaluate the performance of LLM, we adopt MiniGPT-4 [60] as the shared reasoner. Our One-for-All model also shows superiority given a similar model scale. Surprisingly, our lite One-for-All reasoner surpasses the MiniGPT-4 on specific tasks, *e.g.*, RAVEN, and Bongard-HOI, providing compelling evidence that there is no absolute positive correlation between the number of model parameters and the model's reasoning capabilities.

To analyze the performance of LLM-based models, we probe the task according to our two-stage framework design and examine them separately: (1) *Symbolization*: whether LLM-based models can recognize the elements of the problem. (2) *Conceptual*: whether LLM-based models can learn specific concepts behind the tasks and reason about them. (3) *Answer Generation*: whether LLM-based models can utilize the concepts it learns to solve problems. Using MiniGPT-4 [60] as a representative, we summarize the typical responses of LLM-based models to three-level problems in RAVEN and Bongard in Fig. 6.

We find that LLMs may encounter certain hallucination circumstances [22,49,51] while solving visual reasoning tasks. As shown in Fig. 6, for the RAVEN problem, MiniGPT-4 succeeds in the first level of identifying the object while failing in the second stage of reasoning with the arrangement rule. For RAVEN problems, MiniGPT-4 fails to accurately identify the logical patterns. For the Bongard problem, MiniGPT-4 succeeds in the first level of recognizing human activity and the second level of grasping reasoning logically, yet it fails at the answer generation level and gets lost when utilizing rules to answer questions. Given the above cases, we can gain an understanding of the shortcomings of the LLM-based models in reasoning tasks, namely its good concept comprehension ability but insufficient performance in logical reasoning and answer generation. In future work, we plan to train our framework with a larger LLM on more visual reasoning datasets in different domains. We believe our work would inspire both visual and general reasoning studies.

5.5 Approximation Principle Verification

Next, we verify that training the reasoner with data from multiple domains can guarantee a reasoner with better generalization ability. We conduct experiments on SVRT [29], Bongard-HOI [23], Balls task from Filtered-CoPhy [21], Collision task from Filtered-CoPhy [21], and VQAv2 [15]. These datasets encompass 2D puzzles, 3D video, and VQA tasks, offering diverse and multi-modal data. We utilize the Filtered-CoPhy Collision task as the benchmark for testing.

We involve more and more cross-domain datasets to train the reasoner and pair it with the separated encoder of the target test dataset. We report the comparison in Table 6. Given the inherent gaps between the various datasets, we introduce a highly lightweight, MLP-based adapter into the reasoner prior. To equalize the contribution of each dataset to the reasoning engine, we adjust the sample sizes used for training across datasets. Specifically, we use sample sizes of 1,000 and 3,000.

Table 6 reveals a gradual improvement of the reasoner with an increasing number of training datasets. Although handling more datasets from diverse

Table 6. Validation of "approximation principle" using the Collision task in Filtered-CoPhy. We train the reasoner using data from more and more domains and then transfer the reasoner to the target Filtered-CoPhy in inference. We use the keypoint error as the metric of Filtered-CoPhy (the smaller the better). We set two tracks, i.e., training with 1,000 or 3,000 samples from each training dataset.

	SVRT	Bongard-HOI	CoPhyBall	VQAv2	1,000 Samples↓	3,000 Samples↓
$n=1$	✓				13.75	13.09
		✓			14.40	9.90
			✓		18.04	24.80
				✓	14.00	14.85
$n=2$	✓	✓			11.34	12.76
	✓		✓		10.13	7.41
	✓			✓	11.24	10.13
		✓	✓		9.99	9.53
		✓		✓	13.72	5.56
			✓	✓	9.34	6.29
$n=3$	✓	✓	✓		8.06	7.39
	✓	✓		✓	8.61	6.75
	✓		✓	✓	5.54	6.46
		✓	✓	✓	10.28	6.18
$n=4$	✓	✓	✓	✓	5.85	4.88

Table 7. Symbol encoders trained from scratch and pre-trained on ImageNet.

	RAVEN	CVR	SVRT
Trained from Scratch	**53.56**	72.40	**87.84**
ImageNet Pretrained	52.21	**74.82**	85.53

Table 8. Comparison between CLIP as a general encoder and our best One-for-All model. Our One-for-All model surpasses using CLIP as a general encoder in most tasks.

	RAVEN	CVR	SVRT	Bongard-HOI	Bongard-LOGO
One-for-All (shared reasoner)	**55.72**	72.40	**86.64**	**57.65**	**69.64**
CLIP (both shared)	23.59	**74.60**	65.64	51.10	58.09

domains significantly enhances the complexity, the trained reasoner performs well on the out-of-domain Filtered-CoPhy. This shows that the reasoner would concentrate on task-agnostic pure reasoning as the domains of the training dataset increase, which validates our "approximation principle".

5.6 Additional Ablation Study

This section mainly involves experiments in pretraining the encoder using the large-scale image dataset and using the CLIP model as a general encoder. For more experiments please refer to the supplementary material.

Pre-trained Model. We present the ablation on using the pre-trained model or not for symbol encoder in Table 7. We select RAVEN [57], CVR [56] and SVRT [29] dataset, and we employ ImageNet [7] as the pretrain dataset. We can find that the results are very close. The possible reason is that there is an obvious domain gap between ImageNet and the three reasoning datasets.

CLIP as General Encoder. We test whether CLIP [39], a general and large foundation model can do well acting as the general symbol encoder. We utilize CLIP as the visual encoder for multi-modal datasets, followed by an MLP as the reasoner, and employ task-specific head networks. As shown in Table 8, we find that using CLIP yields inferior outcomes to the best One-for-All approach, even after fine-tuning. This validates that even large models such as CLIP are unable to accomplish the symbolization of different datasets, thereby affirming the rationale behind our separated encoder, shared reasoner framework design.

6 Conclusion

In this work, we conclude a two-stage perspective for visual reasoning: the symbolization transforms data into symbolic representation, while the reasoning performs logical reasoning. We show that compared with symbolization, reasoning is more task-agnostic and can be shared by cross-domain tasks. Thus we introduce a concise framework consisting of separated symbol encoders and a shared reasoner. It is crucial to select the complexity of the symbolization and use multi-domain data to train the reasoner to pursue generalization. Our framework achieves decent performance on multiple datasets with domain gaps. We believe our work would pave the way for generalizable visual reasoning systems.

Acknowledgments. This work is supported in part by the National Natural Science Foundation of China under Grants No. 62306175.

References

1. Akiba, T., Sano, S., Yanase, T., Ohta, T., Koyama, M.: Optuna: a next-generation hyperparameter optimization framework. In: Proceedings of the 25th ACM SIGKDD International Conference on Knowledge Discovery & Data Mining, pp. 2623–2631 (2019)
2. Amizadeh, S., Palangi, H., Polozov, A., Huang, Y., Koishida, K.: Neuro-symbolic visual reasoning: disentangling. In: ICML, pp. 279–290. PMLR (2020)
3. Antol, S., et al.: VQA: visual question answering. In: ICCV, December 2015
4. Baradel, F., Neverova, N., Mille, J., Mori, G., Wolf, C.: CoPhy: counterfactual learning of physical dynamics. arXiv preprint arXiv:1909.12000 (2019)
5. Chen, J., et al.: MiniGPT-v2: large language model as a unified interface for vision-language multi-task learning. arXiv preprint arXiv:2310.09478 (2023)
6. Cornelio, C., Stuehmer, J., Hu, S.X., Hospedales, T.: Learning where and when to reason in neuro-symbolic inference. In: ICLR (2022)
7. Deng, J., Dong, W., Socher, R., Li, L.J., Li, K., Fei-Fei, L.: ImageNet: a large-scale hierarchical image database. In: 2009 CVPR, pp. 248–255. IEEE (2009)
8. Duan, J., Yu, S., Poria, S., Wen, B., Tan, C.: PIP: physical interaction prediction via mental simulation with span selection. In: Avidan, S., Brostow, G., Cisse, M., Farinella, G.M., Hassner, T. (eds.) Computer Vision – ECCV 2022. ECCV 2022. LNCS, vol. 13695, pp. 405–421. Springer, Cham (2022). https://doi.org/10.1007/978-3-031-19833-5_24
9. Duan, J., Yu, S., Tan, C.: Space: a simulator for physical interactions and causal learning in 3D environments. In: ICCV, pp. 2058–2063 (2021)
10. Funamizu, A., Kuhn, B., Doya, K.: Neural substrate of dynamic Bayesian inference in the cerebral cortex. Nat. Neurosci. **19**(12), 1682–1689 (2016)
11. Garcez, A.D., et al.: Neural-symbolic learning and reasoning: contributions and challenges. In: 2015 AAAI (2015)
12. Garcez, A.D., et al.: Neural-symbolic learning and reasoning: a survey and interpretation. Neuro-Symbolic Artif. Intell. State Art **342**(1), 327 (2022)
13. Gong, T., et al.: Multimodal-GPT: a vision and language model for dialogue with humans. arXiv preprint arXiv:2305.04790 (2023)
14. Gori, M., Monfardini, G., Scarselli, F.: A new model for learning in graph domains. In: Proceedings of 2005 IEEE International Joint Conference on Neural Networks, vol. 2, pp. 729–734. IEEE (2005)
15. Goyal, Y., Khot, T., Summers-Stay, D., Batra, D., Parikh, D.: Making the V in VQA matter: elevating the role of image understanding in visual question answering. In: CVPR, July 2017

16. Gupta, T., Kembhavi, A.: Visual programming: compositional visual reasoning without training. In: CVPR, pp. 14953–14962 (2023)
17. Hamilton, K., Nayak, A., Božić, B., Longo, L.: Is neuro-symbolic AI meeting its promises in natural language processing? A structured review. Semant. Web (Preprint), 1–42 (2022)
18. He, K., Zhang, X., Ren, S., Sun, J.: Deep residual learning for image recognition. In: CVPR, pp. 770–778 (2016)
19. Hong, Y., Yi, L., Tenenbaum, J., Torralba, A., Gan, C.: PTR: a benchmark for part-based conceptual, relational, and physical reasoning. NeurIPS **34**, 17427–17440 (2021)
20. Hudson, D.A., Manning, C.D.: GQA: a new dataset for real-world visual reasoning and compositional question answering. In: CVPR, June 2019
21. Janny, S., Baradel, F., Neverova, N., Nadri, M., Mori, G., Wolf, C.: Filtered-CoPhy: unsupervised learning of counterfactual physics in pixel space. In: ICLR (2022)
22. Ji, Z., Tiezheng, Y., Xu, Y., Lee, N., Ishii, E., Fung, P.: Towards mitigating LLM hallucination via self reflection. In: The 2023 Conference on Empirical Methods in Natural Language Processing (2023)
23. Jiang, H., Ma, X., Nie, W., Yu, Z., Zhu, Y., Anandkumar, A.: Bongard-HOI: benchmarking few-shot visual reasoning for human-object interactions. In: CVPR, pp. 19056–19065 (2022)
24. Jiang, H., Misra, I., Rohrbach, M., Learned-Miller, E., Chen, X.: In defense of grid features for visual question answering. In: CVPR (2020)
25. Johnson, J., Hariharan, B., van der Maaten, L., Fei-Fei, L., Lawrence Zitnick, C., Girshick, R.: CLEVR: a diagnostic dataset for compositional language and elementary visual reasoning. In: CVPR, July 2017
26. Kingma, D.P., Ba, J.: Adam: a method for stochastic optimization. arXiv preprint arXiv:1412.6980 (2014)
27. Kroshchanka, A., Golovko, V., Mikhno, E., Kovalev, M., Zahariev, V., Zagorskij, A.: A neural-symbolic approach to computer vision. In: Golenkov, V., Krasnoproshin, V., Golovko, V., Shunkevich, D. (eds.) Open Semantic Technologies for Intelligent Systems, OSTIS 2021. CCIS, vol. 1625, pp. 282–309. Springer, Cham (2021). https://doi.org/10.1007/978-3-031-15882-7_15
28. Lemos, H., Avelar, P., Prates, M., Garcez, A., Lamb, L.: Neural-symbolic relational reasoning on graph models: effective link inference and computation from knowledge bases. In: Farkaš, I., Masulli, P., Wermter, S. (eds.) ICANN 2020. LNCS, vol. 12396, pp. 647–659. Springer, Cham (2020). https://doi.org/10.1007/978-3-030-61609-0_51
29. Li, T., Dubout, C., Wampler, E.K., Yantis, S., Geman, D., et al.: Comparing machines and humans on a visual categorization test (2011)
30. Li, Y.L., et al.: HAKE: a knowledge engine foundation for human activity understanding. TPAMI **45**(7), 8494–8506 (2022)
31. Li, Y.L., et al.: Beyond object recognition: a new benchmark towards object concept learning. In: ICCV (2023)
32. Liu, H., Li, C., Li, Y., Lee, Y.J.: Improved baselines with visual instruction tuning. arXiv preprint arXiv:2310.03744 (2023)
33. Liu, Z., Wang, Z., Lin, Y., Li, H.: A neural-symbolic approach to natural language understanding. arXiv preprint arXiv:2203.10557 (2022)
34. Mao, J., Gan, C., Kohli, P., Tenenbaum, J.B., Wu, J.: The neuro-symbolic concept learner: interpreting scenes, words, and sentences from natural supervision. In: ICLR (2019). https://openreview.net/forum?id=rJgMlhRctm

35. McDuff, D., et al.: CausalCity: complex simulations with agency for causal discovery and reasoning. In: Conference on Causal Learning and Reasoning, pp. 559–575. PMLR (2022)
36. Messina, N., Amato, G., Carrara, F., Gennaro, C., Falchi, F.: Recurrent vision transformer for solving visual reasoning problems. In: Sclaroff, S., Distante, C., Leo, M., Farinella, G.M., Tombari, F. (eds.) Image Analysis and Processing – ICIAP 2022. LNCS, vol. 13233, pp. 50–61. Springer, Cham (2022). https://doi.org/10.1007/978-3-031-06433-3_5
37. Nie, W., Yu, Z., Mao, L., Patel, A.B., Zhu, Y., Anandkumar, A.: BONGARD-LOGO: a new benchmark for human-level concept learning and reasoning. In: NeurIPS (2020)
38. Pearl, J., Mackenzie, D.: The Book of Why: The New Science of Cause and Effect. Basic Books, New York (2018)
39. Radford, A., et al.: Learning transferable visual models from natural language supervision. In: ICML, pp. 8748–8763. PMLR (2021)
40. Raghuraman, N., Harley, A.W., Guibas, L.: Cross-image context matters for Bongard problems (2023)
41. Shu, M., et al.: Test-time prompt tuning for zero-shot generalization in vision-language models. NeurIPS **35**, 14274–14289 (2022)
42. Spratley, S., Ehinger, K., Miller, T.: A closer look at generalisation in RAVEN. In: Vedaldi, A., Bischof, H., Brox, T., Frahm, J.M. (eds.) Computer Vision–ECCV 2020: 16th European Conference, Glasgow, UK, 23–28 August 2020, Proceedings, Part XXVII 16, pp. 601–616. Springer, Cham (2020). https://doi.org/10.1007/978-3-030-58583-9_36
43. Surís, D., Menon, S., Vondrick, C.: ViperGPT: visual inference via python execution for reasoning. arXiv preprint arXiv:2303.08128 (2023)
44. Tsai, C.F., Zhou, X., Liu, S.S., Li, J., Yu, M., Mei, H.: Can large language models play text games well? Current state-of-the-art and open questions. arXiv preprint arXiv:2304.02868 (2023)
45. Vaswani, A., et al.: Attention is all you need. In: NeurIPS, vol. 30 (2017)
46. Wen, Z., Peng, Y.: Multi-level knowledge injecting for visual commonsense reasoning. IEEE Trans. Circuits Syst. Video Technol. **31**(3), 1042–1054 (2020)
47. Wu, Q., Teney, D., Wang, P., Shen, C., Dick, A., van den Hengel, A.: Visual question answering: a survey of methods and datasets. Comput. Vis. Image Underst. **163**, 21–40 (2017). https://doi.org/10.1016/j.cviu.2017.05.001, Language in Vision
48. Wu, X., Li, Y.L., Sun, J., Lu, C.: Symbol-LLM: leverage language models for symbolic system in visual human activity reasoning. In: NeurIPS (2023)
49. Xu, Z., Jain, S., Kankanhalli, M.: Hallucination is inevitable: an innate limitation of large language models. arXiv preprint arXiv:2401.11817 (2024)
50. Yang, L., et al.: Neural prediction errors enable analogical visual reasoning in human standard intelligence tests (2023)
51. Yao, J.Y., Ning, K.P., Liu, Z.H., Ning, M.N., Yuan, L.: LLM lies: hallucinations are not bugs, but features as adversarial examples. arXiv preprint arXiv:2310.01469 (2023)
52. Yi, K., et al.: CLEVRER: collision events for video representation and reasoning. arXiv preprint arXiv:1910.01442 (2019)
53. Yi, K., Wu, J., Gan, C., Torralba, A., Kohli, P., Tenenbaum, J.: Neural-symbolic VQA: disentangling reasoning from vision and language understanding. In: NeurIPS, vol. 31 (2018)
54. Yu, D., Yang, B., Liu, D., Wang, H., Pan, S.: A survey on neural-symbolic learning systems. Neural Networks **166**, 105–126 (2023)

55. Yu, D., Yang, B., Wei, Q., Li, A., Pan, S.: A probabilistic graphical model based on neural-symbolic reasoning for visual relationship detection. In: CVPR, pp. 10609–10618 (2022)
56. Zerroug, A., Vaishnav, M., Colin, J., Musslick, S., Serre, T.: A benchmark for compositional visual reasoning. arXiv preprint arXiv:2206.05379 (2022)
57. Zhang, C., Gao, F., Jia, B., Zhu, Y., Zhu, S.C.: RAVEN: a dataset for relational and analogical visual reasoning. In: CVPR (2019)
58. Zhang, J., Chen, B., Zhang, L., Ke, X., Ding, H.: Neural, symbolic and neural-symbolic reasoning on knowledge graphs. AI Open **2**, 14–35 (2021)
59. Zhao, H., et al.: MMICL: empowering vision-language model with multi-modal in-context learning. arXiv preprint arXiv:2309.07915 (2023)
60. Zhu, D., Chen, J., Shen, X., Li, X., Elhoseiny, M.: MiniGPT-4: enhancing vision-language understanding with advanced large language models. arXiv preprint arXiv:2304.10592 (2023)

AlignZeg: Mitigating Objective Misalignment for Zero-Shot Semantic Segmentation

Jiannan Ge[1], Lingxi Xie[2], Hongtao Xie[1(✉)], Pandeng Li[1], Xiaopeng Zhang[2], Yongdong Zhang[1], and Qi Tian[2]

[1] University of Science and Technology of China, Hefei, China
{gejn,lpd}@mail.ustc.edu.cn, {htxie,zhyd73}@ustc.edu.cn
[2] Huawei Inc., Shenzhen, China
tian.qi1@huawei.com

Abstract. A serious issue that harms the performance of zero-shot visual recognition is named **objective misalignment**, *i.e.*, the learning objective prioritizes improving the recognition accuracy of seen classes rather than unseen classes, while the latter is the true target to pursue. This issue becomes more significant in zero-shot image segmentation because the stronger (*i.e.*, pixel-level) supervision brings a larger gap between seen and unseen classes. To mitigate it, we propose a novel architecture named **AlignZeg**, which embodies a comprehensive improvement of the segmentation pipeline, including proposal extraction, classification, and correction, to better fit the goal of zero-shot segmentation. **(1) Mutually-Refined Proposal Extraction.** AlignZeg harnesses a mutual interaction between mask queries and visual features, facilitating detailed class-agnostic mask proposal extraction. **(2) Generalization-Enhanced Proposal Classification.** AlignZeg introduces synthetic data and incorporates multiple background prototypes to allocate a more generalizable feature space. **(3) Predictive Bias Correction.** During the inference stage, AlignZeg uses a class indicator to find potential unseen class proposals followed by a prediction postprocess to correct the prediction bias. Experiments demonstrate that AlignZeg markedly enhances zero-shot semantic segmentation, as shown by an average 3.8% increase in hIoU, primarily attributed to a 7.1% improvement in identifying unseen classes, and we further validate that the improvement comes from alleviating the objective misalignment issue.

Keywords: Zero-shot learning · Semantic segmentation

1 Introduction

Semantic segmentation [5,7,15,27,36,49,53,61] is a fundamental task in computer vision [8,41,42,55] that is critical for image understanding. Traditional

Supplementary Information The online version contains supplementary material available at https://doi.org/10.1007/978-3-031-72775-7_9.

segmentation [12,33,45] approaches depend on large amounts of annotated data to perform effectively. Given the vast and diverse nature of real-world data, it is not feasible to annotate every new image. Zero-shot semantic segmentation [3] has emerged to address this challenge, leveraging the model trained on a limited dataset to segment and recognize classes not seen during training.

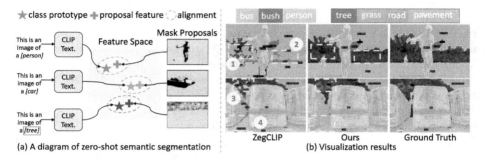

Fig. 1. (a) Proposal features are aligned with class prototypes. "CLIP Text." represents CLIP Text Encoder. (b) Red boxes select the unseen classes. Yellow dashed boxes select the misclassified areas, *e.g.*, "*tree*" (unseen) → "*bush*" (seen) ②, "*road*" (unseen) → "*pavement*" (unseen) ④. These errors show the objective misalignment issue. (Color figure online)

Recently, the introduction of large-scale vision-language models, such as CLIP [54], has advanced this task further. As shown in Fig. 1 (a), the visual features of seen classes are aligned with the semantic features, *i.e.*, class prototypes, which enables the model to detect unseen classes. Models such as SimBaseline [64], Zegformer [18], and OVSeg [44] have proposed a two-stage scheme that first generates class-agnostic mask proposals and then extracts corresponding mask visual features for zero-shot classification. This scheme effectively extends CLIP's zero-shot abilities to the pixel level. However, these methods often necessitate multiple forward passes per image, leading to inefficiency. In response, models like ZegCLIP [72], DeOP [23], and SAN [63] construct their networks based on the CLIP image encoder for a streamlined process. However, these methods suffer a phenomenon which we refer to as **"objective misalignment"**, *i.e.*, they focus on improving the accuracy on seen classes but ignore the unseen classes. In the semantic segmentation task, characterized by a strong supervision signal, this misalignment issue becomes more pronounced. It not only constrains the full potential of CLIP's zero-shot performance but also introduces biases into the model's predictions. As illustrated in Fig. 1 (b), this misalignment tends to cause inaccurate segmentation results on unseen classes. For example, the model might mistakenly segment "*tree*" class as semantically similar categories like "*grass*" ① and "*bush*" ②.

To address the issue mentioned above, we propose a novel framework named AlignZeg, which aligns the training and inference objectives of the model with zero-shot tasks. Initially, we introduce the **Mutually-Refined Proposal Extraction**, which employs mask queries and visual features to mutually refine each other, leading to the extraction of more detailed mask proposals. These high-quality class-agnostic proposals can generalize effectively to unseen classes, thereby reducing the model's sensitivity to seen classes. Subsequently, we propose the **Generalization-Enhanced Proposal Classification**, an innovative approach that integrates generality constraints within the feature space, enhancing its generalizability to unseen classes. This includes integrating synthetic features to prevent over-specialization towards seen classes, and employing a multi-background prototype strategy for diverse background representations. While these developments enhance zero-shot semantic segmentation by optimizing towards the zero-shot task objective, the exclusive use of seen class data in training inevitably leads to prediction bias. Therefore, we propose **Predictive Bias Correction** to filter potential unseen class proposals and adjust the corresponding prediction scores, thereby explicitly alleviating the prediction bias. Notably, our approach assists in judging potential unseen classes at the proposal level, first introducing a new effective step to the segmentation pipeline.

We evaluate AlignZeg in the Generalized Zero-Shot Semantic Segmentation (GZS3) setting and achieve an average improvement of 3.8% in hIoU, with a notable average increase of 7.1% in mIoU($\mathcal{U}$) for unseen classes. Additionally, in the strict Zero-Shot Semantic Segmentation (ZS3) setting, which only considers unseen classes during testing, AlignZeg also achieves state-of-the-art (SOTA) performance in both pAcc and mIoU. These experiments demonstrate that our model effectively mitigates the issue of objective misalignment.

2 Related Works

Zero-Shot Visual Recognition. Zero-shot visual recognition, aimed at identifying unseen categories, typically employs semantic descriptors like attributes [29] and word vectors [57] to bridge seen and unseen categories. Classical methods [10,11,20,24,58,66,67] create a shared space between features and descriptors but often focus more on seen classes, leading to an **objective misalignment** issue. Recent works [9,21,28,46,48,65] address this by extracting attribute-related regions. However, the reliance on manually annotated attributes limits generalizability across diverse scenarios. Recently, large-scale vision-language models [30,39,40,54,59] like CLIP [54] have advanced by training on numerous image-text pairs. While CLIP's broad semantic coverage offers good zero-shot capabilities, its focus on image level limits its efficacy in pixel-level recognition.

Zero-Shot Semantic Segmentation. Zero-shot semantic segmentation [3,17, 25,26,35,38,43,47,69,70] is an emerging research task. Earlier methods [2,14,22] focused on linking visual content with class descriptions by a shared space. Recently, the advent of CLIP [54] has shifted focus to pixel-level zero-shot recognition using vision-language models. Recent methods like SimBaseline [64] and

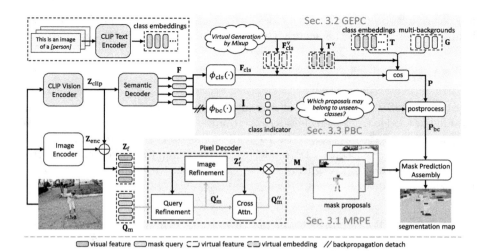

Fig. 2. Overall framework. Our method mitigates the objective misalignment between semantic segmentation and zero-shot task through three main components: Mutually-Refined Proposal Extraction (MRPE), Generalization-Enhanced Proposal Classification (GEPC), and Proposal-based Bias Correction (PBC), the latter of which is applied during the inference process. Parameters in gray are fixed.

Zegformer [18] have introduced a two-stage zero-shot segmentation approach: class-agnostic mask proposals extraction followed by their zero-shot classification. While effectively scaling down CLIP's capabilities to pixel level, challenges persist in integrating cropped images with CLIP. OVSeg [44] bridged this gap with mask prompt tuning. However, the two-stage process is inefficient due to multiple CLIP encoder iterations. Models like ZegCLIP [72], DeOP [23], and SAN [63] overcome this by utilizing the CLIP encoder directly. They not only simplified the process but also increased computational efficiency by necessitating just one forward pass. However, these approaches primarily focus on adapting CLIP for pixel-level tasks with classification loss on seen classes. This objective misaligns with the intrinsic nature of zero-shot tasks and often results in overfitting. This **objective misalignment** issue is more pronounced in the context of segmentation tasks due to their stronger supervision.

Prediction Bias. To reduce objective misalignment in zero-shot visual recognition, addressing the classifier's prediction bias towards seen classes is a pivotal way. Researchers have explored strategies like advanced domain detector design [1], entropy-based score adjustments [51], distance-based gating networks [37], and generated-based distance analysis [67], to distinguish between seen and unseen categories. However, adapting these methods to pixel-level tasks remains challenging. [69] extended these ideas to zero-shot segmentation with an unknown prototype for unknown mask extraction. However, its final performance is notably sensitive to the accuracy of the unknown mask extraction.

3 Method

Problem Definition. Zero-shot semantic segmentation aims to extract transferable knowledge from seen classes $\mathcal{C}_{\text{seen}}$ to subsequently recognize unseen classes $\mathcal{C}_{\text{unseen}}$, where $\mathcal{C}_{\text{seen}} \cap \mathcal{C}_{\text{unseen}} = \varnothing$. To achieve this objective, during the training process, we leverage the CLIP model to derive semantic representations from class labels within $\mathcal{C}_{\text{seen}}$. During the testing phase, we evaluate the model on both seen and unseen classes, $i.e.$, $\mathcal{C}_{\text{test}} = \mathcal{C}_{\text{unseen}} \cup \mathcal{C}_{\text{seen}}$.

Overview. Most existing methods prioritize improving recognition performance on seen classes, which deviates from the objectives of zero-shot tasks. In zero-shot semantic segmentation, this deviation, termed "objective misalignment", is exacerbated due to the strong supervision signals inherent in segmentation tasks. To realign the training process with zero-shot objectives, we introduce two key components: Mutually-Refined Proposal Extraction (MRPE) and Generalization-Enhanced Proposal Classification (GEPC), as detailed in Sect. 3.1 and Sect. 3.2, respectively. MRPE refines mask proposals through interactions between queries and features, while GEPC enhances the generalizability of proposal features for each mask proposal, countering seen class dominance with virtual features and diverse backgrounds. We also implement Predictive Bias Correction (PBC), outlined in Sect. 3.3, to explicitly reduce the model's bias towards seen classes by filtering potential unseen class proposals and adjusting their prediction scores.

3.1 Mutually-Refined Proposal Extraction

An intuitive idea to get the mask proposals is to use a simple transformer decoder like [13,18,64] to decode masks from visual features via mask queries. However, mask queries of this technique lack sufficient interaction with visual features. This limitation can lead to challenges in reliably extracting class-agnostic masks, especially from samples that are complex and varied in nature. Therefore, we propose an enhanced strategy in the pixel decoder, where mask queries and visual features are mutually refined, which enables the extraction of high-quality, class-agnostic masks that are more effectively generalized to unseen classes.

Directly training a feature extraction network may over-fit the training set, while directly using CLIP features cannot guarantee good segmentation. Therefore, we extract complementary visual features $\mathbf{Z}_{\text{enc}}$ and $\mathbf{Z}_{\text{clip}}$ via a learnable image encoder and a fixed CLIP image encoder, respectively. Then we fuse them by $\mathbf{Z}_{\text{f}} = \mathbf{Z}_{\text{enc}} + \mathbf{Z}_{\text{clip}} \in \mathbb{R}^{(\frac{H}{16} \times \frac{W}{16}) \times D}$ for following pixel decoder, where H and W are the input image's dimensions. We follow [72] to use prompt learning [31] to fine-tune the CLIP features to further adapt CLIP to the current dataset.

Then, we employ the queries to decode masks from the visual features. We first conduct query refinement shown in Fig. 2. Specifically, we apply a cross-attention mechanism [6] to adjust the mask queries $\mathbf{Q}_m \in \mathbb{R}^{N \times D}$ by

$$\mathbf{Q}'_m = \text{softmax}\left(\frac{\mathbf{Q}_m \mathbf{W}_Q (\mathbf{Z}_f \mathbf{W}_K)^\top}{\sqrt{D}}\right) \mathbf{Z}_f \mathbf{W}_V, \tag{1}$$

where $\mathbf{W}_Q, \mathbf{W}_K, \mathbf{W}_V$ are the mapping functions for the queries, keys and values, and D is their feature dimension. $\mathbf{Q}_m$ is randomly initialized and is learnable. N denotes the total number of queries, which corresponds to the extraction of N distinct mask proposals. Then we use residual connections and an MLP layer to prompt the queries by $\mathbf{Q}'_m = \text{MLP}(\mathbf{Q}_m + \mathbf{Q}'_m)$. We also conduct image refinement to adjust the visual information. This process first includes a cross-attention mechanism to adapt the visual features and get $\mathbf{Z}'_f$ via Eq. (1). The only difference is that we obtain queries from $\mathbf{Z}_f$, and get keys and values from $\mathbf{Q}'_m$. Then we get prompted visual features by $\mathbf{Z}'_f = \text{MLP}(\mathbf{Z}_f + \mathbf{Z}'_f)$.

Finally, we apply the mutually prompted mask queries and visual features to get final queries $\mathbf{Q}''_m$, which can also refer to the process expressed by Eq. (1). Then we generate final mask proposals by $\mathbf{M} = \mathbf{Q}''_m \cdot \mathbf{Z}'^{\top}_f$.

3.2 Generalization-Enhanced Proposal Classification

After generating mask proposals, we extract the corresponding proposal features and use CLIP for zero-shot classification. However, traditional methods mainly use classification loss on seen classes, which does not align with the objectives of zero-shot tasks. This approach is likely to result in a wide-span occupancy of the feature space by seen classes, as illustrated in Fig. 3 (left), ultimately impacting model generalization negatively. To address this, we introduce two straightforward yet effective strategies: **feature expansion strategy** and **background diversity strategy**. The former generates features outside the distribution of seen classes, while the latter focuses on learning diversified background prototypes. Collectively, these strategies assist in expanding the feature space available for unseen classes, thereby aiding in their better recognition.

Class Embedding and Proposal Feature Extraction. We first utilize the CLIP text encoder to extract class embeddings $\mathbf{T} \in \mathbb{R}^{C \times D}$ as the class prototypes, where C denotes the number of classes. To account for categories beyond the predefined set, we concurrently learn a background token $\mathbf{g} \in \mathbb{R}^{1 \times D}$. To maximize the generalization capacity of CLIP's visual information, we follow the approach in [63] to reuse the shadow [CLS] token of the CLIP image encoder as the prediction queries and employ the last layers of the CLIP image encoder as the semantic decoder. More detailed descriptions can be found in [63]. The output proposal features are defined as $\mathbf{F} \in \mathbb{R}^{N \times D}$. Subsequently, we utilize a feed-forward network $\phi_{\text{cls}}(\cdot)$ to refine the features and obtain $\mathbf{F}_{\text{cls}} = \phi_{\text{cls}}(\mathbf{F}) \in \mathbb{R}^{N \times D}$. Thus, the prediction scores can be derived as $\mathbf{P} = [\mathbf{F}_{\text{cls}} \cdot \mathbf{T}^\top, \mathbf{F}_{\text{cls}} \cdot \mathbf{g}^\top] \in \mathbb{R}^{N \times (C+1)}$.

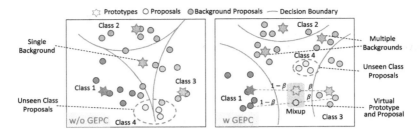

Fig. 3. Without GEPC, driven mainly by classification loss, the model focuses on seen classes, allowing them to dominate the feature space. GEPC introduces synthetic features and diverse backgrounds, helping provide a more generalizable feature space.

Feature Expansion Strategy. Our objective is to enhance the feature space originating from seen classes by synthesizing features beyond those classes. Mixup techniques [16,34,56] like vanilla mixup [68] and manifold mixup [60] provide valuable insights. However, vanilla mixup's image-level interpolation falls short for our segmentation model, whereas manifold mixup's feature-level interpolation better meets our requirements. Moreover, diverging from the manifold mixup's strategy of blending one-hot class labels, we additionally mix class embeddings while interpolating visual features. This allows us to maintain alignment between visual features and class embeddings even after interpolation. Specifically, we first collect seen class proposal features $\mathbf{F}_{cls}^{b,s} \in \mathbb{R}^{N_{b,s} \times D}$ and corresponding class embeddings $\mathbf{T}^{b,s} \in \mathbb{R}^{N_{b,s} \times D}$ from a mini-batch. Both sets are then shuffled to produce $\dot{\mathbf{F}}_{cls}^{b,s}$ and $\dot{\mathbf{T}}^{b,s}$, respectively. Following this, we blend the original and shuffled features to synthesize the mixed features shown in Fig. 3 (right):

$$\mathbf{F}_{cls}^{v} = \beta \cdot \mathbf{F}_{cls}^{b,s} + (1-\beta) \cdot \dot{\mathbf{F}}_{cls}^{b,s},$$
$$\mathbf{T}^{v} = \beta \cdot \mathbf{T}^{b,s} + (1-\beta) \cdot \dot{\mathbf{T}}^{b,s}, \quad (2)$$

where $\beta \in [0,1]$ is sampled from a Beta distribution $\text{Beta}(\alpha, \alpha)$, $\mathbf{F}_{cls}^{v}$ and $\mathbf{T}^{v}$ are the virtual proposal features and the corresponding virtual prototypes, respectively. We set $\alpha = 3$ to encourage the virtual samples to deviate significantly from the seen classes. To align the virtual features with virtual prototypes while separating them from seen classes, we calculate the logits by

$$\mathbf{P}^{v}[i] = [\mathbf{F}_{cls}^{v}[i] \cdot \mathbf{T}^{\top}, \mathbf{F}_{cls}^{v}[i] \cdot \mathbf{T}^{v}[i]^{\top}], i = 0, 1, \cdots, N_{b,s}, \quad (3)$$

where $\mathbf{P}^{v} \in \mathbb{R}^{N_{b,s} \times (C+1)}$. The labels $\mathbf{Y}^{v}$ of these generated features are all set to class $C+1$. Finally, we employ a cross-entropy loss $\mathcal{L}_{vir} = \text{CE}(\mathbf{P}^{v}, \mathbf{Y}^{v})$ to optimize. By generating virtual features that lie outside the distribution of seen classes, this strategy tends to reserve more feature space for categories outside the seen class, thereby facilitating better generalization to unseen classes.

Background Diversity Strategy. Another drawback of recent methods lies in the insufficient representation of complex and varied background categories using a single fixed prototype. To address this issue, we introduce multiple background prototypes to preserve the diversity of backgrounds. Specifically, we set M background prototypes $\mathbf{G} = [\mathbf{g}_1; \mathbf{g}_2; \cdots ; \mathbf{g}_M] \in \mathbb{R}^{M \times D}$. For each proposal feature $\mathbf{f}_{\text{cls}} \in \mathbb{R}^{1 \times D}$, the multiple background logits are calculated by $\mathbf{p}_{\text{mg}} = \mathbf{f}_{\text{cls}} \cdot \mathbf{G}^\top \in \mathbb{R}^{1 \times M}$. We then obtain the final background score by weighted summation by $p_g = \text{sum}(\text{softmax}(\mathbf{p}_{\text{mg}}) \odot \mathbf{p}_{\text{mg}})$, where $\odot$ is the element-wise product. Then the final logits are $\mathbf{P} = [\mathbf{F}_{\text{cls}} \cdot \mathbf{T}^\top, \mathbf{p}_g] \in \mathbb{R}^{N \times (C+1)}$. To maintain the diverse distribution of the backgrounds, we constrain their distances by

$$\mathcal{L}_{\text{reg}} = \frac{2}{M(M-1)} \sum_{i=1}^{M-1} \sum_{j=i+1}^{M} \mathbf{g}_i \cdot \mathbf{g}_j^\top. \qquad (4)$$

As shown in Fig. 3 and Fig. 8, our approach tends to offer a more expansive feature space, enhancing the generalization ability for unseen classes.

3.3 Predictive Bias Correction

While the aforementioned methods optimize the model towards the objectives of zero-shot tasks, an inevitable prediction bias towards seen classes arises due to the model being trained on seen categories. To address this, we propose a simple yet effective **Predictive Bias Correction** to assist in identifying whether each proposal contains potentially unseen categories. Specifically, we employ $\phi_{\text{bc}}(\cdot)$, a binary classification model comprising two fully connected layers followed by a sigmoid layer, to learn a class indicator for each proposal:

$$\mathbf{I} = \phi_{\text{bc}}(\mathbf{F}) \in \mathbb{R}^{N \times 1}, \qquad (5)$$

where each element of $\mathbf{I}$ lies in the range $[0,1]$. A value closer to 1 means that the seen class will not be included in the proposal. To train $\phi_{\text{bc}}(\cdot)$, it is imperative to identify the positive and negative proposals and allocate the appropriate ground truth labels accordingly:

- **Positive proposals.** During training, proposals extracted by the model can be divided into seen class proposals and background proposals. We select the seen class proposals as the positive proposals and assign the label 0 to them.
- **Negative proposals.** We select negative proposals from background proposals, particularly excluding the ones overlapping with seen class regions. Specifically, we evaluate each proposal against seen classes present in the image by calculating a loss matrix $\mathcal{L}_{\text{mask}}^i(\mathbf{M}, \mathbf{M}_i^{\text{gt}})$, comparing mask proposals $\mathbf{M}$ against the ground truth mask $\mathbf{M}_i^{\text{gt}}$ for the i-th seen class in the image. A higher value in $\mathcal{L}_{\text{mask}}^i$ suggests the proposal is less likely to contain the i-th seen class in the image. As shown in Fig. 4, only proposals consistently ranking within the top-K values across all $\mathcal{L}_{\text{mask}}^i$ are considered as negative. This criterion ensures that our negative proposals are less likely to contain segments of any seen classes, thus more likely to represent pure background or undefined classes. We assign label 1 to these proposals.

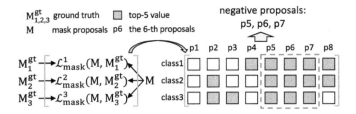

Fig. 4. Example of negative proposal selection. In an image with 3 seen classes, p5, p6, and p7 are selected as negative proposals since they are in the top-K (K = 5) based on loss for all seen classes contained in the image.

- **Others.** Proposals that are neither selected as positive nor negative are excluded from optimization due to their low quality.

After selecting the positive and negative proposals, we can optimize $\phi_{bc}(\cdot)$ using a binary cross-entropy (BCE) loss:

$$\mathcal{L}_{bc} = -\frac{1}{N_{bc}} \sum_{i=1}^{N_{bc}} [y_i \log(I) + (1 - y_i) \log(1 - I)], \quad (6)$$

where N_{bc} is the number of selected proposals, and I is the class indicator of the current proposal. In the inference phase, the trained $\phi_{bc}(\cdot)$ can be used to obtain $\mathbf{I}$, enabling the use of a threshold γ to identify possible unseen class proposals and suppress their seen class scores:

$$\mathbf{P}_{bc}[i,j] = \begin{cases} v_{\min}, & \text{if } \mathbf{I}[i] > \gamma \text{ and } \mathcal{C}_{test}[j] \in \mathcal{S}, \\ \mathbf{P}[i,j], & \text{otherwise}, \end{cases} \quad (7)$$

where $\mathcal{S}$ is the set of seen classes, $v_{\min}$ is the minimum value in $\mathbf{P}$, and i represents the i-th proposal. $\mathbf{I}[i] > \gamma$ selects the potential proposals containing unseen classes, and $\mathcal{C}_{test}[j] \in \mathcal{S}$ selects the seen class scores that need to be suppressed. In this way, possible unseen class proposals can be well screened and their prediction bias for seen classes can be well mitigated.

3.4 Optimization

We follow Mask2former [12] and optimize our model using a classification loss $\mathcal{L}_{ce}$ and a mask loss $\mathcal{L}_{mask}$, which includes a dice loss and a binary cross-entropy loss. The overall loss is

$$\mathcal{L} = \mathcal{L}_{bc} + \lambda_1 \cdot \mathcal{L}_{ce} + \lambda_2 \cdot \mathcal{L}_{mask} + \lambda_3 \cdot \mathcal{L}_{vir} + \lambda_4 \cdot \mathcal{L}_{reg}, \quad (8)$$

where λ_1, λ_2, λ_3 and λ_4 are coefficients.

4 Experiments

4.1 Datasets and Evaluation Metrics

Datasets. We evaluate our method on three widely used datasets: **PASCAL VOC 2012** [19], **COCO-Stuff 164K** [4] and **PASCAL Context** [52]. PASCAL VOC 2012 includes 10,582 images of 15 seen classes for training, and 1,449 images of 15 seen classes and 5 unseen classes for testing. COCO-Stuff 164K consists of 118,287 training images and 5,000 testing images. The classes are split into 156 seen and 15 unseen classes. PASCAL Context contains 59 semantic classes, of which 49 are seen and 10 are unseen. Its training and test sets contain 4,996 and 5,104 images, respectively. Since the background category has no explicit semantics, they are ignored. We also evaluate the generalization ability to other datasets. Following [44,63,64], we train the model on COCO-Stuff 164K with all 171 classes, and evaluate on five datasets: **A-150** [71], **A-847** [71], **VOC** [19], **P-59** [52] and **P-459** [52]. A-150 has 2K test images of 150 classes, and A-847 has the same images but is labeled with 847 classes. VOC indicates PASCAL VOC 2012, and has 1449 images of 20 classes. P-59 and P-459 are both from PASCAL Context, but contain 59 and 459 categories, respectively.

Evaluation Metrics. Following [18,64,72], we calculate the mean of class-wise intersection over union on both seen and unseen categories, resulting in mIoU($\mathcal{S}$) and mIoU($\mathcal{U}$), respectively. Simultaneously, we compute their harmonic mean, denoted as hIoU $= \frac{2 \times \text{mIoU}(\mathcal{U}) \times \text{mIoU}(\mathcal{S})}{\text{mIoU}(\mathcal{U}) + \text{mIoU}(\mathcal{S})}$, as a comprehensive evaluation metric. We also measure pixel-wise classification accuracy (pAcc) for reference assessment.

Table 1. GZS3 results on three benchmarks. The highest scores are emphasized in bold. m($\mathcal{U}, \mathcal{S}$) represents mIoU($\mathcal{U}, \mathcal{S}$), and h is the abbreviation of hIoU.

Methods	Venue	PASCAL VOC 2012				COCO-Stuff 164K				PASCAL Context			
		pAcc	m($\mathcal{S}$)	m($\mathcal{U}$)	h	pAcc	m($\mathcal{S}$)	m($\mathcal{U}$)	h	pAcc	m($\mathcal{S}$)	m($\mathcal{U}$)	h
SPNet [62]	CVPR'19	–	78	15.6	26.1	–	35.2	8.7	14	–	–	–	–
ZS3 [3]	NeurIPS'19	–	77.3	17.7	28.7	–	34.7	9.5	15	52.8	20.8	12.7	15.8
CaGNet [22]	ACM MM'20	80.7	78.4	26.6	39.7	56.6	33.5	12.2	18.2	–	24.1	18.5	21.2
SIGN [14]	ICCV'21	–	75.4	28.9	41.7	–	32.3	15.5	20.9	–	–	–	–
Joint [2]	ICCV'21	–	77.7	32.5	45.9	–	–	–	–	–	33	14.9	20.5
ZegFormer [18]	CVPR'22	–	86.4	63.6	73.3	–	36.6	33.2	34.8	–	–	–	–
SimBaseline [64]	ECCV'22	90	83.5	72.5	77.5	60.3	39.3	36.3	37.8	–	–	–	–
SAN [63]	CVPR'23	94.7	92.3	78.9	85.1	58.4	40.1	39.2	39.6	81.5	53.5	56.6	55
ZegCLIP [72]	CVPR'23	94.6	91.9	77.8	84.3	62	40.2	41.4	40.8	76.2	46	54.6	49.9
DeOP [23]	ICCV'23	92.5	88.2	74.6	80.8	62.2	38.0	38.4	38.2	–	–	–	–
MAFT [32]	NeurIPS'23	–	91.5	80.7	85.7	–	**40.6**	40.1	40.3	–	–	–	–
Ours	–	**96.6**	**93.9**	**88.2**	**91.0**	**64.4**	40.2	**50.0**	**44.6**	**82.2**	**53.7**	**61.7**	**57.4**

4.2 Implementation Details

Our image encoder is built upon an 8-layer Transformer architecture, with each layer having a dimension of 240 and a patch size of 16. Consistent with the model used in [72], we employ the CLIP ViT-B/16 as our base model. Following [63], the CLIP vision encoder and semantic decoder comprise the first 6 and last 3 layers of CLIP ViT-B/16, respectively. Further details of the Semantic Decoder can be found in **the supplementary materials** and [63]. The input sizes for the CLIP Vision Encoder and Image Encoder are 320^2 and 640^2, respectively, to cater to their individual focuses on classification and segmentation. We set the batch size to 16. Our exploration is confined to inductive zero-shot learning, deliberately excluding the unseen pseudo labels generation and self-training process, as these do not align with real-world scenarios. The number of training iterations on PASCAL VOC, COCO-Stuff, and PASCAL Context datasets is 20K, 80K, and 40K, respectively. We follow [12] to set $\lambda_1 = 2$, $\lambda_2 = 5$. The parameter λ_3 and λ_4 are empirically set to 10^{-2} and 10^{-4}, respectively. K is set to 50. N is set to 100, and M is set to 20 for COCO-Stuff and 10 for others.

4.3 Comparison with State-of-the-Art Methods

Results of Generalized Zero-Shot Semantic Segmentation. In Table 1, we compare with prior approaches. Our method achieves superior performance across nearly all evaluated metrics, particularly in the pivotal hIoU metric where we surpass the SOTA performance on PASCAL VOC 2012, COCO-Stuff 164K, and PASCAL Context by margins of 5.3%, 3.8%, and 2.4%, respectively. These gains are largely due to our improved recognition of unseen classes, as shown by our mIoU($\mathcal{U}$) metric outperformance by 7.5%, 8.6%, and 5.1% on these datasets. This indicates that our design for proposal extraction, classification, and correction in the segmentation pipeline can effectively alleviate the objective misalignment issue. Moreover, our model also sets new benchmarks on the pAcc, further demonstrating its exceptional segmentation capabilities. We also present results

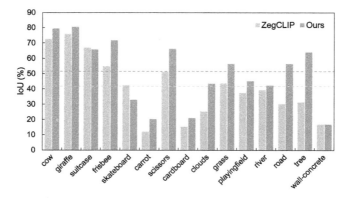

Fig. 5. The IoU results for unseen classes on the COCO-Stuff. The dashed line represents the average result, *i.e.*, mIoU($\mathcal{U}$).

Table 2. Results on ZS3 with unseen categories only. No prediction bias with this setting.

Methods	VOC		COCO	
	mIoU	pAcc	mIoU	pAcc
ZegFormer [18]	81.3	91.4	64.9	81.3
SimBaseline [64]	86.7	94.7	67.1	82.9
SAN [63]	94.1	97.8	76.9	84.8
ZegCLIP [72]	86.3	93.9	70.8	77.7
Ours	94.6	97.9	82.9	91.2

Table 3. Results of generalization ability to other datasets.

Methods	A-847	P-459	A-150	P-59	VOC
ZegFormer [18]	–	–	–	36.1	85.6
SimBaseline [64]	7	8.7	20.5	47.7	88.4
OVSeg [44]	7.1	11	24.8	53.3	92.6
SAN [63]	10.1	12.6	27.5	53.8	94
ZegCLIP [72]	–	–	–	41.2	93.6
Ours	10.4	16.2	28.1	54.3	94.7

for each unseen class on the COCO dataset as shown in Fig. 5. Compared to the SOTA method, ZegCLIP, our approach shows improvements in almost all unseen categories. This also indicates that our method can effectively alleviate the objective misalignment issue.

Results of zero-shot semantic segmentation. We also extend our evaluations to the zero-shot semantic segmentation setting, *i.e.*, $\mathcal{C}_{test} = \mathcal{C}_{unseen}$, which inherently avoids the bias of classifying unseen classes as seen by only considering unseen categories. The results of other methods, as presented in Tab. 2, were obtained using their provided codes. Table 2 reveals that our model consistently achieves optimal performance on both metrics across the two datasets. Notably, on the COCO dataset, our model achieves a relative improvement of 12.1% and 6.0% in mIoU compared to ZegCLIP and SAN, respectively. This enhancement in performance underscores our model's ability to effectively generalize knowledge from seen to unseen categories.

Results of Generalization Ability to Other Datasets. We also assess our model's ability to generalize across various datasets, a challenge due to varying data distributions as highlighted in recent studies [23,63,64]. In this scenario, unseen classes are more prevalent, thus the problem of prediction bias-which our PBC is adept at alleviating-is less prominent. Despite this, Table 3 demonstrates that our method maintains commendable performance, specifically when compared with the GZS3 SOTA method, ZegCLIP, with improvements of 13.1% and 1.1% on the P-59 and VOC datasets, respectively. This indicates that our approach can uphold its effectiveness in generalizing to new datasets.

Table 4. The impact of different components. PBC represents Predictive Bias Correction. MRPE indicates Mutually-Refined Proposal Extraction. FES (Feature Expansion Strategy) and BDS (Background Diversity Strategy) are key components of the GEPC (Generalization-Enhanced Proposal Classification).

Methods	PASCAL VOC 2012				COCO-Stuff 164K			
	pAcc	mIoU($\mathcal{S}$)	mIoU($\mathcal{U}$)	hIoU	pAcc	mIoU($\mathcal{S}$)	mIoU($\mathcal{U}$)	hIoU
Baseline	94.1	91.8	73.1	81.4	57.4	38.0	38.5	38.3
+PBC	95.1	92.9	75.4	83.2	60.5	37.9	44.1	40.8
+PBC+MRPE	95.5	92.5	82.1	87.1	62.1	40.2	43.0	41.6
+PBC+MRPE+FES	96.3	93.3	85.4	89.2	63.5	40.1	46.7	43.2
+PBC+MRPE+FES+BDS	96.6	93.9	88.2	91.0	64.4	40.2	50.0	44.6

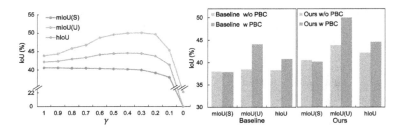

Fig. 6. Left: Parameter γ experiments in PBC. Right: Illustration of the effect of PBC.

4.4 Ablative Studies

Effect of Different Components. Our ablation study results are in Table 4. We establish our baseline model based on the framework depicted in Fig. 2, where the pixel decoder is implemented using a straightforward transformer decoder, akin to Maskformer, and is optimized solely with $\mathcal{L}_{ce}$ and $\mathcal{L}_{mask}$. The initial incorporation of Predictive Bias Correction (PBC) resulted in an average 2.15% increase in the hIoU across two datasets. This shows PBC effectively reassigns proposals. Subsequent integration of MRPE (Mutually-Refined Proposal Extraction) yielded more accurate class-agnostic masks, enhancing both pAcc and hIoU. Table 5 further demonstrates this point. Additionally, we implement Feature Expansion Strategy (FES) and Background Diversity Strategy (BDS) to impose semantic classification constraints from two different perspectives. FES, by introducing virtual features, effectively mitigates the feature space's excessive occupancy by seen classes, leading to an average increase of 1.1% in pAcc and 1.85% in hIoU. BDS further enhances feature space generalization by promoting background class diversity during training, which ultimately propels the model to achieve its optimal results. Moreover, Fig. 6 (right) shows that PBC further reduces prediction bias, even with the effective model "Ours w/o PBC". This suggests that our training optimizations, such as FES and BDS, also enhance the feature space for PBC, as seen in Fig. 3 (right).

Table 5. Comparison with Mask-Former's pixel decoder. "Ours w M-PDec" indicates that we replace our pixel decoder with the pixel decoder from MaskFormer.

Methods	mIoU($\mathcal{S}$)	mIoU($\mathcal{U}$)	hIoU
Ours w M-PDec	93.6	85.9	89.6
Ours	93.9	88.2	91.0

Table 6. Validation of the quality of class-agnostic mask proposals. We assign all the proposals with ground-truth class labels.

Methods	mIoU($\mathcal{S}$)	mIoU($\mathcal{U}$)	hIoU
Baseline	95.0	89.4	92.1
Ours	95.7	92.0	93.8

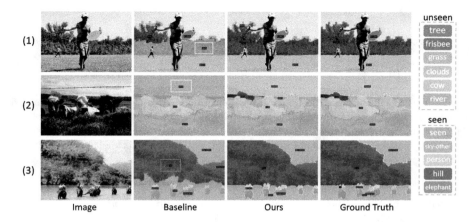

Fig. 7. Visualization of segmentation on the COCO-Stuff 164K. Rows (1) and (2) depict the unseen classes segmentation results, marking misclassified regions as "seen" label ▬. Row (3) illustrates the results for both seen and unseen categories. The four columns respectively display the original image, the segmentation results from the Baseline, the segmentation results from our method, and the ground truth.

Evaluation of Class-Agnostic Mask Quality. We assign correct class labels to mask results to assess class-agnostic mask quality only, shown in Table 6. Key observations include: (1) The Baseline achieves high accuracy, indicating satisfactory existing mask performance; (2) Our model surpasses Baseline, validating MRPE's effectiveness in enhancing class-agnostic mask extraction; (3) Referencing Table 6 and Fig. 5, we find classification to be the primary obstacle to improving zero-shot performance, seconded by mask quality, echoing findings from [44]. Our method has made advancements in both aspects.

Effect of γ. As shown in Fig. 6 (left), as γ decreases, an increasing number of proposals are categorized as unseen classes. This shift, while causing only minor fluctuations in mIoU($\mathcal{S}$), contributes to a substantial increase in mIoU($\mathcal{U}$), resulting in the optimal hIoU at $\gamma = 0.3$. This phenomenon indicates that our model can effectively uncover potential unseen class proposals, thereby aligning the inference objective more closely with the zero-shot learning goal.

Visualization. Segmentation results, shown in Fig. 7, reveal model performance. In the first (1) and second (2) rows, the Baseline tends to erroneously classify unseen categories, such as trees and clouds, as the "seen" category. In contrast, our method can mitigate this bias, resulting in substantially improved results. The third row (3) highlights our method's superiority, showcasing its capability to generate more accurate segmentation masks (mask of the river) while concurrently making precise predictions for both seen and unseen categories. More visualization results can be found in **the supplementary materials**.

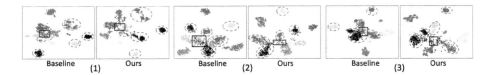

Fig. 8. T-SNE [50] visualization of proposal features for five categories each from 156 seen and 15 unseen classes on COCO-Stuff 164K, presented in three random samplings (1)–(3) to mitigate the impact of outliers. Unseen categories are marked by red dashed boxes and feature entanglements are highlighted with black rectangles. (Color figure online)

T-SNE Plot. We visualize proposal features using our method versus the Baseline, as shown in Fig. 8. Our method's features display greater dispersion and clearer category margins, showing the effectiveness of Generalization-Enhanced Proposal Classification (GEPC) in enhancing feature distinctiveness. For example, Fig. 8 (3) illustrates that while the Baseline model's features for several unseen categories and one seen category are entangled, leading to misclassification, our approach effectively separates these features, improving distinguishability.

5 Conclusion

In this paper, we introduce AlignZeg, an innovative architecture tackling the objective misalignment issue in zero-shot image segmentation by enhancing the entire segmentation pipeline. It integrates Mutually-Refined Proposal Extraction, Generalization-Enhanced Proposal Classification, and Predictive Bias Correction, enhancing unseen class segmentation while ensuring seen class performance. Notably, the Correction component aids in judging potential unseen classes at the proposal level, adding a novel step to the segmentation pipeline. Empirical results across various benchmarks demonstrate AlignZeg's superior performance in both Generalized and Strict Zero-Shot Semantic Segmentation. Our approach could also play a pivotal role in cross-modal pre-training, utilizing large-scale data for more effective zero-shot learning.

Acknowledgments. This work is supported by the National Key Research and Development Program of China (2022YFB3104700), the National Nature Science Foundation of China (62121002, U23B2028, 62232006), CAS Project for Young Scientists in Basic Research (Grant No. YSBR-067). We acknowledge the support of GPU cluster built by MCC Lab of Information Science and Technology Institution, USTC. We also thank the USTC supercomputing center for providing computational resources for this project.

References

1. Atzmon, Y., Chechik, G.: Adaptive confidence smoothing for generalized zero-shot learning. In: Proceedings of the IEEE/CVF Conference on Computer Vision and Pattern Recognition, pp. 11671–11680 (2019)
2. Baek, D., Oh, Y., Ham, B.: Exploiting a joint embedding space for generalized zero-shot semantic segmentation. In: Proceedings of the IEEE/CVF International Conference on Computer Vision, pp. 9536–9545 (2021)
3. Bucher, M., Vu, T.H., Cord, M., Pérez, P.: Zero-shot semantic segmentation. In: Advances in Neural Information Processing Systems, vol. 32 (2019)
4. Caesar, H., Uijlings, J., Ferrari, V.: COCO-Stuff: thing and stuff classes in context. In: Proceedings of the IEEE Conference on Computer Vision and Pattern Recognition, pp. 1209–1218 (2018)
5. Cai, K., et al.: MixReorg: cross-modal mixed patch reorganization is a good mask learner for open-world semantic segmentation. In: Proceedings of the IEEE/CVF International Conference on Computer Vision, pp. 1196–1205 (2023)
6. Carion, N., Massa, F., Synnaeve, G., Usunier, N., Kirillov, A., Zagoruyko, S.: End-to-end object detection with transformers. In: Vedaldi, A., Bischof, H., Brox, T., Frahm, J.-M. (eds.) ECCV 2020. LNCS, vol. 12346, pp. 213–229. Springer, Cham (2020). https://doi.org/10.1007/978-3-030-58452-8_13
7. Chen, J., et al.: Exploring open-vocabulary semantic segmentation from clip vision encoder distillation only. In: Proceedings of the IEEE/CVF International Conference on Computer Vision, pp. 699–710 (2023)
8. Chen, P., et al.: Point-to-box network for accurate object detection via single point supervision. In: Avidan, S., Brostow, G., Cissé, M., Farinella, G.M., Hassner, T. (eds.) Computer Vision – ECCV 2022. LNCS, vol. 13669, pp. 51–67. Springer, Cham (2022). https://doi.org/10.1007/978-3-031-20077-9_4
9. Chen, S., et al.: TransZero: attribute-guided transformer for zero-shot learning. In: Proceedings of the AAAI Conference on Artificial Intelligence, vol. 36, pp. 330–338 (2022)
10. Chen, S., et al.: Free: feature refinement for generalized zero-shot learning. In: Proceedings of the IEEE/CVF International Conference on Computer Vision, pp. 122–131 (2021)
11. Chen, Z., et al.: Semantics disentangling for generalized zero-shot learning. In: Proceedings of the IEEE/CVF International Conference on Computer Vision, pp. 8712–8720 (2021)
12. Cheng, B., Misra, I., Schwing, A.G., Kirillov, A., Girdhar, R.: Masked-attention mask transformer for universal image segmentation. In: Proceedings of the IEEE/CVF Conference on Computer Vision and Pattern Recognition, pp. 1290–1299 (2022)
13. Cheng, B., Schwing, A., Kirillov, A.: Per-pixel classification is not all you need for semantic segmentation. Adv. Neural. Inf. Process. Syst. **34**, 17864–17875 (2021)
14. Cheng, J., Nandi, S., Natarajan, P., Abd-Almageed, W.: SIGN: spatial-information incorporated generative network for generalized zero-shot semantic segmentation. In: Proceedings of the IEEE/CVF International Conference on Computer Vision, pp. 9556–9566 (2021)
15. Cho, S., et al.: CAT-Seg: cost aggregation for open-vocabulary semantic segmentation. arXiv preprint arXiv:2303.11797 (2023)
16. Chou, Y.Y., Lin, H.T., Liu, T.L.: Adaptive and generative zero-shot learning. In: International Conference on Learning Representations (2020)

17. Deng, R., et al.: Segment Anything Model (SAM) for digital pathology: assess zero-shot segmentation on whole slide imaging. arXiv preprint arXiv:2304.04155 (2023)
18. Ding, J., Xue, N., Xia, G.S., Dai, D.: Decoupling zero-shot semantic segmentation. In: Proceedings of the IEEE/CVF Conference on Computer Vision and Pattern Recognition, pp. 11583–11592 (2022)
19. Everingham, M., Winn, J.: The PASCAL visual object classes challenge 2012 (VOC2012) development kit. Pattern Anal. Stat. Model. Comput. Learn., Tech. Rep **2007**(1–45), 5 (2012)
20. Ge, J., Xie, H., Li, P., Xie, L., Min, S., Zhang, Y.: Towards discriminative feature generation for generalized zero-shot learning. IEEE Trans. Multimedia (2024)
21. Ge, J., Xie, H., Min, S., Li, P., Zhang, Y.: Dual part discovery network for zero-shot learning. In: Proceedings of the 30th ACM International Conference on Multimedia, pp. 3244–3252 (2022)
22. Gu, Z., Zhou, S., Niu, L., Zhao, Z., Zhang, L.: Context-aware feature generation for zero-shot semantic segmentation. In: Proceedings of the 28th ACM International Conference on Multimedia, pp. 1921–1929 (2020)
23. Han, C., Zhong, Y., Li, D., Han, K., Ma, L.: Open-vocabulary semantic segmentation with decoupled one-pass network. In: Proceedings of the IEEE/CVF International Conference on Computer Vision, pp. 1086–1096 (2023)
24. Han, Z., Fu, Z., Chen, S., Yang, J.: Contrastive embedding for generalized zero-shot learning. In: Proceedings of the IEEE/CVF Conference on Computer Vision and Pattern Recognition, pp. 2371–2381 (2021)
25. He, S., Ding, H., Jiang, W.: Primitive generation and semantic-related alignment for universal zero-shot segmentation. In: Proceedings of the IEEE/CVF Conference on Computer Vision and Pattern Recognition, pp. 11238–11247 (2023)
26. He, S., Ding, H., Jiang, W.: Semantic-promoted debiasing and background disambiguation for zero-shot instance segmentation. In: Proceedings of the IEEE/CVF Conference on Computer Vision and Pattern Recognition, pp. 19498–19507 (2023)
27. Huo, X., Xie, L., Hu, H., Zhou, W., Li, H., Tian, Q.: Domain-agnostic prior for transfer semantic segmentation. In: Proceedings of the IEEE/CVF conference on Computer Vision and Pattern Recognition, pp. 7075–7085 (2022)
28. Huynh, D., Elhamifar, E.: Fine-grained generalized zero-shot learning via dense attribute-based attention. In: Proceedings of the IEEE/CVF Conference on Computer Vision and Pattern Recognition, pp. 4483–4493 (2020)
29. Jayaraman, D., Grauman, K.: Zero-shot recognition with unreliable attributes. In: Advances in Neural Information Processing Systems, vol. 27 (2014)
30. Jia, C., et al.: Scaling up visual and vision-language representation learning with noisy text supervision. In: International Conference on Machine Learning, pp. 4904–4916. PMLR (2021)
31. Jia, M., et al.: Visual prompt tuning. In: Avidan, S., Brostow, G., Cisse, M., Farinella, G.M., Hassner, T. (eds.) Computer Vision – ECCV 2022. LNCS, vol. 13693, pp. 709–727. Springer, Cham (2022). https://doi.org/10.1007/978-3-031-19827-4_41
32. Jiao, S., Wei, Y., Wang, Y., Zhao, Y., Shi, H.: Learning mask-aware clip representations for zero-shot segmentation. In: Thirty-Seventh Conference on Neural Information Processing Systems (2023)
33. Jin, Z., et al.: Mining contextual information beyond image for semantic segmentation. In: Proceedings of the IEEE/CVF International Conference on Computer Vision, pp. 7231–7241 (2021)

34. Kalantidis, Y., Sariyildiz, M.B., Pion, N., Weinzaepfel, P., Larlus, D.: Hard negative mixing for contrastive learning. Adv. Neural. Inf. Process. Syst. **33**, 21798–21809 (2020)
35. Karazija, L., Laina, I., Vedaldi, A., Rupprecht, C.: Diffusion models for zero-shot open-vocabulary segmentation. arXiv preprint arXiv:2306.09316 (2023)
36. Kirillov, A., et al.: Segment anything. arXiv:2304.02643 (2023)
37. Kwon, G., Al Regib, G.: A gating model for bias calibration in generalized zero-shot learning. IEEE Trans. Image Process. (2022)
38. Li, J., Chen, P., Qian, S., Jia, J.: TagClip: improving discrimination ability of open-vocabulary semantic segmentation. arXiv preprint arXiv:2304.07547 (2023)
39. Li, L.H., Yatskar, M., Yin, D., Hsieh, C.J., Chang, K.W.: VisualBERT: a simple and performant baseline for vision and language. arXiv preprint arXiv:1908.03557 (2019)
40. Li, P., et al.: MomentDiff: generative video moment retrieval from random to real. In: Advances in Neural Information Processing Systems, pp. 65948–65966 (2023)
41. Li, P., et al.: Progressive spatio-temporal prototype matching for text-video retrieval. In: Proceedings of the IEEE/CVF International Conference on Computer Vision, pp. 4100–4110 (2023)
42. Li, P., Xie, H., Min, S., Ge, J., Chen, X., Zhang, Y.: Deep Fourier ranking quantization for semi-supervised image retrieval. Trans. Image Process. **31**, 5909–5922 (2022)
43. Li, P., Wei, Y., Yang, Y.: Consistent structural relation learning for zero-shot segmentation. Adv. Neural. Inf. Process. Syst. **33**, 10317–10327 (2020)
44. Liang, F., et al.: Open-vocabulary semantic segmentation with mask-adapted clip. In: Proceedings of the IEEE/CVF Conference on Computer Vision and Pattern Recognition, pp. 7061–7070 (2023)
45. Liu, J., Bao, Y., Xie, G.S., Xiong, H., Sonke, J.J., Gavves, E.: Dynamic prototype convolution network for few-shot semantic segmentation. In: Proceedings of the IEEE/CVF Conference on Computer Vision and Pattern Recognition, pp. 11553–11562 (2022)
46. Liu, M., Li, F., Zhang, C., Wei, Y., Bai, H., Zhao, Y.: Progressive semantic-visual mutual adaption for generalized zero-shot learning. In: Proceedings of the IEEE/CVF Conference on Computer Vision and Pattern Recognition, pp. 15337–15346 (2023)
47. Liu, X., et al.: Delving into shape-aware zero-shot semantic segmentation. In: Proceedings of the IEEE/CVF Conference on Computer Vision and Pattern Recognition, pp. 2999–3009 (2023)
48. Liu, Y., et al.: Goal-oriented gaze estimation for zero-shot learning. In: Proceedings of the IEEE/CVF Conference on Computer Vision and Pattern Recognition, pp. 3794–3803 (2021)
49. Luo, H., Bao, J., Wu, Y., He, X., Li, T.: SegClip: patch aggregation with learnable centers for open-vocabulary semantic segmenaztion. In: International Conference on Machine Learning, pp. 23033–23044. PMLR (2023)
50. Van der Maaten, L., Hinton, G.: Visualizing data using t-SNE. J. Mach. Learn. Res. **9**(11) (2008)
51. Min, S., Yao, H., Xie, H., Wang, C., Zha, Z.J., Zhang, Y.: Domain-aware visual bias eliminating for generalized zero-shot learning. In: Proceedings of the IEEE/CVF Conference on Computer Vision and Pattern Recognition, pp. 12664–12673 (2020)
52. Mottaghi, R., et al.: The role of context for object detection and semantic segmentation in the wild. In: Proceedings of the IEEE Conference on Computer Vision and Pattern Recognition, pp. 891–898 (2014)

53. Pastore, G., Cermelli, F., Xian, Y., Mancini, M., Akata, Z., Caputo, B.: A closer look at self-training for zero-label semantic segmentation. In: Proceedings of the IEEE/CVF Conference on Computer Vision and Pattern Recognition, pp. 2693–2702 (2021)
54. Radford, A., et al.: Learning transferable visual models from natural language supervision. In: International Conference on Machine Learning, pp. 8748–8763. PMLR (2021)
55. Rao, Y., Chen, G., Lu, J., Zhou, J.: Counterfactual attention learning for fine-grained visual categorization and re-identification. In: Proceedings of the IEEE/CVF International Conference on Computer Vision, pp. 1025–1034 (2021)
56. Roy, A., Shah, A., Shah, K., Dhar, P., Cherian, A., Chellappa, R.: FeLMi: few shot learning with hard Mixup. Adv. Neural. Inf. Process. Syst. **35**, 24474–24486 (2022)
57. Socher, R., Ganjoo, M., Manning, C.D., Ng, A.: Zero-shot learning through cross-modal transfer. In: Advances in Neural Information Processing Systems, vol. 26 (2013)
58. Su, H., Li, J., Chen, Z., Zhu, L., Lu, K.: Distinguishing unseen from seen for generalized zero-shot learning. In: Proceedings of the IEEE/CVF Conference on Computer Vision and Pattern Recognition, pp. 7885–7894 (2022)
59. Su, W., et al.: VL-BERT: pre-training of generic visual-linguistic representations. arXiv preprint arXiv:1908.08530 (2019)
60. Verma, V., et al.: Manifold mixup: better representations by interpolating hidden states. In: International Conference on Machine Learning, pp. 6438–6447. PMLR (2019)
61. Wu, W., Zhao, Y., Shou, M.Z., Zhou, H., Shen, C.: DiffuMask: synthesizing images with pixel-level annotations for semantic segmentation using diffusion models. arXiv preprint arXiv:2303.11681 (2023)
62. Xian, Y., Choudhury, S., He, Y., Schiele, B., Akata, Z.: Semantic projection network for zero-and few-label semantic segmentation. In: Proceedings of the IEEE/CVF Conference on Computer Vision and Pattern Recognition, pp. 8256–8265 (2019)
63. Xu, M., Zhang, Z., Wei, F., Hu, H., Bai, X.: Side adapter network for open-vocabulary semantic segmentation. In: Proceedings of the IEEE/CVF Conference on Computer Vision and Pattern Recognition, pp. 2945–2954 (2023)
64. Xu, M., et al.: A simple baseline for open-vocabulary semantic segmentation with pre-trained vision-language model. In: Avidan, S., Brostow, G., Cisse, M., Farinella, G.M., Hassner, T. (eds.) Computer Vision – ECCV 2022. ECCV 2022. LNCS, vol. 13689, pp. 736–753. Springer, Cham (2022). https://doi.org/10.1007/978-3-031-19818-2_42
65. Xu, W., Xian, Y., Wang, J., Schiele, B., Akata, Z.: Attribute prototype network for zero-shot learning. Adv. Neural. Inf. Process. Syst. **33**, 21969–21980 (2020)
66. Xu, W., Xian, Y., Wang, J., Schiele, B., Akata, Z.: VGSE: visually-grounded semantic embeddings for zero-shot learning. In: Proceedings of the IEEE/CVF Conference on Computer Vision and Pattern Recognition, pp. 9316–9325 (2022)
67. Yue, Z., Wang, T., Sun, Q., Hua, X.S., Zhang, H.: Counterfactual zero-shot and open-set visual recognition. In: Proceedings of the IEEE/CVF Conference on Computer Vision and Pattern Recognition, pp. 15404–15414 (2021)
68. Zhang, H., Cisse, M., Dauphin, Y.N., Lopez-Paz, D.: Mixup: beyond empirical risk minimization. arXiv preprint arXiv:1710.09412 (2017)
69. Zhang, H., Ding, H.: Prototypical matching and open set rejection for zero-shot semantic segmentation. In: Proceedings of the IEEE/CVF International Conference on Computer Vision, pp. 6974–6983 (2021)

70. Zheng, Y., Wu, J., Qin, Y., Zhang, F., Cui, L.: Zero-shot instance segmentation. In: Proceedings of the IEEE/CVF Conference on Computer Vision and Pattern Recognition, pp. 2593–2602 (2021)
71. Zhou, B., Zhao, H., Puig, X., Fidler, S., Barriuso, A., Torralba, A.: Scene parsing through ADE20K dataset. In: Proceedings of the IEEE Conference on Computer Vision and Pattern Recognition, pp. 633–641 (2017)
72. Zhou, Z., Lei, Y., Zhang, B., Liu, L., Liu, Y.: ZegCLIP: towards adapting clip for zero-shot semantic segmentation. In: Proceedings of the IEEE/CVF Conference on Computer Vision and Pattern Recognition, pp. 11175–11185 (2023)

Contrastive Learning with Counterfactual Explanations for Radiology Report Generation

Mingjie Li[1], Haokun Lin[2], Liang Qiu[1], Xiaodan Liang[2(✉)], Ling Chen[3], Abdulmotaleb Elsaddik[2], and Xiaojun Chang[2,4]

[1] Stanford University, Palo Alto, CA 94305, USA
{lmj695,qiuliang}@stanford.edu
[2] MBZUAI, Abu Dhabi, UAE
{haokun.lin,xiaodan.liang,a.elsaddik,xiaojun.chang}@mbzuai.ac.ae
[3] University of Technology Sydney, Ultimo, NSW 2007, Australia
ling.chen@uts.edu.au
[4] School of Information Science and Technology, University of Science and Technology of China, Hefei, China

Abstract. Due to the common content of anatomy, radiology images with their corresponding reports exhibit high similarity. Such inherent data bias can predispose automatic report generation models to learn entangled and spurious representations resulting in misdiagnostic reports. To tackle these, we propose a novel **C**ounter**F**actual **E**xplanations-based framework (CoFE) for radiology report generation. Counterfactual explanations serve as a potent tool for understanding how decisions made by algorithms can be changed by asking "what if" scenarios. By leveraging this concept, CoFE can learn non-spurious visual representations by contrasting the representations between factual and counterfactual images. Specifically, we derive counterfactual images by swapping a patch between positive and negative samples until a predicted diagnosis shift occurs. Here, positive and negative samples are the most semantically similar but have different diagnosis labels. Additionally, CoFE employs a learnable prompt to efficiently fine-tune the pre-trained large language model, encapsulating both factual and counterfactual content to provide a more generalizable prompt representation. Extensive experiments on two benchmarks demonstrate that leveraging the counterfactual explanations enables CoFE to generate semantically coherent and factually complete reports and outperform in terms of language generation and clinical efficacy metrics.

Keywords: Medical report generation · Counterfactual explanation · Contrastive learning

---

Code is available at: https://github.com/mlii0117/CoFE.

© The Author(s), under exclusive license to Springer Nature Switzerland AG 2025
A. Leonardis et al. (Eds.): ECCV 2024, LNCS 15101, pp. 162–180, 2025.
https://doi.org/10.1007/978-3-031-72775-7_10

1 Introduction

Automatically generating reports can reduce the load on radiologists and potentially increase the accuracy and consistency of interpretations. This is achieved by translating intricate radiology images into semantically coherent and clinically reliable free texts. However, in comparison to generic captioning tasks, Radiology Report Generation (RRG) presents a significant challenge, often yielding unsatisfactory performance when employing direct captioning methods [30,42] in the field of radiology. The difficulty arises due to the severe data bias within the limited image-report pair data available, a challenge that has been extensively acknowledged and discussed [4,16,23,27,36,45].

Given the shared anatomical content, radiology images tend to display significant similarity to one another, with abnormal or lesioned areas typically occupying minor portions of the images [24,44]. This similarity also extends to the accompanying reports, where several sentences often describe normal tissues. However, the clinical usefulness of radiology reports hinges on the accurate depiction of abnormalities. This intrinsic data bias tends to lead models to learn spurious and intertwined visual features, resulting in the generation of inaccurate diagnostic reports. To mitigate data bias, various successful concepts have been proposed by existing methods to enhance learning representations, such as employing contrastive learning [23,28], incorporating medical knowledge [27,49], and implementing relational memory [4] etc.

Recently, Tanida et al. [36] achieved promising performances by detecting abnormal regions using a pre-trained detector with pseudo labels. They then utilized these features to guide a pre-trained large language model (LLM) in generating reports. Identifying critical regions that cover abnormalities or lesions not only enhances non-spurious visual representation but also improves the explainability of RRG models. Ideally, the method would employ golden annotations to train a lesion detector capable of accurately localizing abnormal regions or lesions. However, existing RRG benchmarks lack such annotations. Relying on weakly supervised signals from pseudo labels [36,46] can result in misalignment. Furthermore, the limited size of available medical data may prevent the full unleashing of the potential of LLMs.

To address these challenges, we introduce a novel concept: counterfactual explanation (CE). The concept of CE [12,43] has surfaced in machine learning as an insightful tool to comprehend how models' decisions can be changed. CE offers a hypothetical alternative to the observed data, allowing for the assessment and comprehension of models through 'what if' scenarios. This technique has been integrated into diagnostic models to not only improve diagnostic accuracy but also enhance explainability [5,37]. For example, Tanyel et al. [37] proposed a CE to identify the minimal feature change and effectively demonstrated which features are more informative in differentiating two tumor types from MRI brains. Inspired by this, we aim to make this progress further interactive and explainable by explaining the global feature change in specific local regions. In particular, we propose a CE as *'what if we exchange the patch between two images, will the diagnosis shift?'* to identify critical regions within

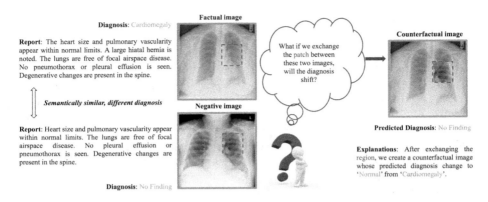

Fig. 1. A conceptual overview of our proposed counterfactual explanations is presented. Such CEs help to construct a counterfactual image by iteratively exchanging a patch between factual (positive) and negative images until the predicted diagnosis shift occurs. In this instance, the box in red covering the heart is identified as the critical region that causes the diagnosis shift. (Color figure online)

images that may cover abnormalities or lesions, providing insights into the diagnosis process. For instance, as illustrated in Fig. 1, we generate a counterfactual image by iteratively swapping a patch between semantically similar images with different diagnosis labels until a shift in predicted diagnosis is achieved. Due to aforementioned similarities, exchanging a patch in the same position between two radiology images, particularly those that are semantically similar but carry different labeled diagnoses, does not disrupt the anatomical content. Notably, previous methods only integrate the CE into the decision-making process, lacking the ability to convey factual or counterfactual information effectively. In contrast, we translate our CE into a prompt that can present the key concept of CE and encapsulate the factual and counterfactual content. This prompt can yield more comprehensive instructions to LLMs and facilitate the elicitation of their knowledge.

In this paper, we propose a **Co**unter**F**actual **E**xplanations-based framework (CoFE) for radiology report generation. CoFE is capable of learning non-spurious visual representations and effectively leverage the capabilities of LLMs. First, we introduce a novel type of CEs for RRG tasks and propose a counterfactual generation process through contrastive learning, which constructs a counterfactual image and a learnable prompt. Specifically, we adopt a negative sampling strategy to discover the most semantically similar negative sample from the data bank, based on text similarity and diagnosis label. By iteratively exchanging patches between factual (positive) and negative samples until a predicted diagnosis change occurs, we pinpoint the critical region and create a counterfactual image. We then employ contrastive learning within a joint optimization framework to differentiate representations between factual and counterfactual samples, enabling the model to learn non-spurious visual representations. Subsequently, we employ a pretrained LLM, GPT-2 Medium [34], as a decoder to generate reports. To fine-tune the LLM efficiently, we propose a learnable prompt that encapsulates both factual and counterfactual content. This prompt can elicit the embedded knowledge within the LLM, which is helpful to generate semantically coherent and factually complete reports.

We evaluate our proposed method on two benchmarks, IU-Xray [6] and MIMIC-CXR [18] respectively. Extensive experiments demonstrate that our approach can outperform previous competitive methods in metrics that measure descriptive accuracy and clinical correctness. It indicates that leveraging CEs to learn non-spurious representations and prompt the generation process can improve the quality of predicted reports.

2 Related Work

2.1 Medical Report Generation

The pursuit of automating medical report generation through machine learning aims to alleviate the workload on radiologists. Numerous effective concepts exist to learn non-spurious representations to mitigate inherent data bias. Relation memory [3,4] can prompt enhancement by boosting interaction efficiency of cross-modal memory network. Integrating medical knowledge is another solution, researchers utilize graph structural data [23,48] or medical tags [17,22] to incorporate prior knowledge into image encoding. Additional models [47,49] also enhance performance by integrating knowledge information, with strategies including multi-modal semantic alignment and multi-label classification pre-training. To identify the abnormalities, PPKED [27] employs a unique architecture to mimic human learning processes. Tanida *et al.* [36] utilize a lesion detector pre-trained by pseudo labels to attain the non-spurious features and lead a pretrained GPT-2 [34]. Due to the lack of annotations, weakly supervised signals from pseudo labels may lead to misalignment. Although, large pretrained models showcase the adaptability in learning medical representations [31], the scarcity of data may limit the potential of LLMs. In this paper, our method concentrates on learning non-spurious representations by identifying critical regions and employing a robust prompt to fine-tune the LLM efficiently.

2.2 Counterfactual Explanations Reasoning

The advent of counterfactual explanations (CEs) construction has driven significant innovations, particularly in computer vision applications, enhancing both accuracy and interpretability. CEs have the potential to relieve existing methodologies from the reliance on extensive training data and meticulous annotations, by asking 'what if' scenario to explore self-supervised signals. Fang *et al.* [8] and Kim *et al.* [19] have introduced systems and frameworks, such as Counterfactual Generative Networks (CGN), designed to augment interpretability and resilience to CEs inputs without compromising accuracy. Similarly, CPL [12] proficiently generates counterfactual features and has exhibited remarkable efficacy in tasks like image-text matching and visual question answering. Ji *et al.* [15] specifically target video relationship detection, constructing CEs to elucidate their influence on factual scenario predictions. Further, the studies by Yang et al. [50] on PubMedQA highlight the crucial role of CEs, generated via ChatGPT, in learning causal features in counterfactual classifiers, demonstrating the versatility and broad applicability of counterfactual methods across various domains. In this paper, we employ CEs to enhance the RRG models, especially where acquiring a substantial amount of golden annotations is prohibitively expensive.

2.3 Prompt Tuning

Prompt tuning is a method in natural language processing (NLP) used to efficiently modify the behavior of a pre-trained LLM, based on specific prompts or trigger phrases. This approach involves fine-tuning the model on a set of prompts or queries and their corresponding responses, allowing the model to respond more accurately or appropriately to those or similar prompts [35,53]. For example, Guo et al. [11] applied Q-Learning to optimize soft prompts, while PTuning v2 [29] demonstrated that continuous prompt tuning could match the performance of fine-tuning in various scenarios. This technique has also garnered significant interest in the field of computer vision. CoOp [33] introduced a strategy for continuous prompt optimization to negate the need for prompt design, and CoCoOp [52] expanded upon this by learning an instance-conditional network to generate a unique input-conditional token for each image. Fischer et al. [9] also prove the adaptability of prompt tuning in medical image segmentation tasks. However, the reliance on empirical risk minimization presents challenges, necessitating advancements to avoid spurious representations. In this paper, we aim to propose a generalizable prompt incorporating factual and counterfactual content to efficiently refine the medical LLMs.

3 Methodology

In this section, we introduce the detailed implementations of our proposed **Coun**-ter**F**actual **E**xplanations-based framework (CoFE). As shown in Fig. 2, our CoFE mainly consists of two uni-modal encoders, one cross-modal encoder, a language decoder, and a counterfactual generation module with four training objectives. We first introduce the backbone of CoFE and then describe the counterfactual generation process in detail.

3.1 Set Up

In this work, we aim to integrate counterfactual explanations into report generation models to learn non-spurious visual representations and efficiently generate high-quality reports. Radiology report generation tasks require a model to translate a complex radiology image I into a generic report $T = \{y_1, y_2, \ldots, y_n\}$. We denote the target report by $\hat{T} = \{\hat{y}_1, \hat{y}_2, \ldots, \hat{y}_{\hat{n}}\}$. n and $\hat{n}$ represent the number of tokens in a report. In addition to corresponding reports, we also utilize the

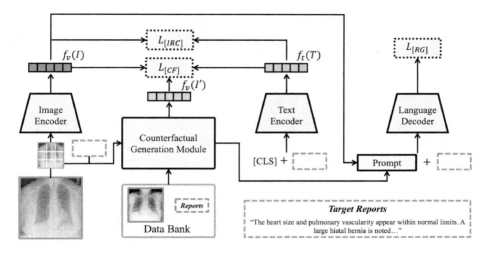

Fig. 2. Illustration of our proposed **C**ounter**F**actual **E**xplanations-based framework (CoFE). CoFE consists of two unimodal encoders, one cross-modal encoder, one language decoder, and our proposed counterfactual generation module that can construct a counterfactual image and a learnable prompt, respectively. The entire framework is trained through joint optimization, mainly employing contrastive learning paradigms for radiology report generation.

diagnosis label C for each examination in this work. Since not all existing benchmarks provide such annotations, we use the CheXPert labeling tool [13] to label ground truth reports with 14 different medical terminologies. Notably, we assign the label "No Finding" when CheXPert does not extract any terminologies. For instances with multiple labels, we adopt a strategy during training where only one label is randomly retained for each case.

Automatic report generation systems are typically based on encoder-decoder frameworks. The encoder generally aims to convert the given image I into dense visual vectors $f_v(I)$. The decoder is usually a sequence processing network, which translates $f_v(I)$ to a report T. To generate the CE and calculate the image report contrastive loss, we involve an additional text encoder following the BLIP [21] architecture, drawing inspiration from the successful concepts found in [23, 28].

3.2 Encoders

Image Encoder. Different from prior works employing CNNs, we use a pre-trained ViT [7]-S as the image encoder $f_v(\cdot)$. ViT enables finer semantic feature extraction by dividing images into more patches, whose resolution is 16×16, compared to conventional CNNs 7×7. A [CLS] token is also prepended before input to the encoder layers. The encoder layer process, $\mathbf{f}_e(\cdot)$, is defined as:

$$\mathbf{f}_e(x) = \mathrm{LN}(\mathrm{FFN}(e_{attn}) + e_{attn}), \tag{1}$$

$$e_{attn} = \mathrm{LN}(\mathrm{MHA}(x) + x), \tag{2}$$

where FFN and LN represent Feed Forward Network [40] and Layer Normalization operation [1], respectively. MHA [40] (multi-head attention) splits attention into n heads, with each head, Att($\cdot$), defined as:

$$\text{Att}(x) = \text{softmax}(\frac{\mathbf{Q}^x(\mathbf{K}^x)^\top}{\sqrt{d}})\mathbf{V}^x. \quad (3)$$

with $d = 384$ being the embedding space dimension, and $\{\mathbf{Q}, \mathbf{K}^*, \mathbf{V}^*\}$ representing the corresponding *Query, Key, Value* vectors. The resulting output, the encoded visual vectors $\mathbf{f}_I$, will be used for report generation.

Text Encoder. We employ a PubMedBERT [10], pre-trained with abstracts and full texts from PubMed[1], as our text encoder $f_t(\cdot)$. It extracts textual representations $f_t(T)$ from positive and negative reports, which will be utilized for calculating the image-report contrastive (IRC) loss to facilitate learning robust and generalizable medical visual and textual representations. We utilize momentum image and text encoders to extract positive and negative data representations in a batch. Then we first calculate the softmax-normalized image-to-report similarity $f_m^{i2t}(I)$ and the report-to-image similarity $f_m^{t2i}(T)$ for the image I and its paired report T by $f_m^{i2r}(I) = \frac{\exp s(I, T_m)/\tau}{\sum_{m=1}^M \exp s(I, T_m)/\tau}$, with τ as a learnable temperature parameter. The IRC loss can be written as:

$$\mathcal{L}_{\text{IRC}} = \frac{1}{2}(\mathcal{L}_{\text{ce}}(g^{t2i}(T), f^{t2i}(T)) + \mathcal{L}_{\text{ce}}(g^{i2t}(I), f^{i2t}(I))). \quad (4)$$

where $g(\cdot)$ denotes the ground truth of similarity.

3.3 Decoding Process

Recognizing the advanced capabilities of Large Language Models (LLMs) in various language generation tasks, we utilize a GPT-2 Medium model, which is pre-trained on the PubMed dataset, as our language decoder. In contrast to R2GenGPT, which employs a frozen LLaMa-7B [38], our choice of GPT-2, with its 355 million parameters, allows for full fine-tuning. This adaptability enables GPT-2 to more effectively cater to the nuances of report generation tasks. GPT-2, an auto-regressive model leveraging self-attention, conditions each output token in a sequence on its previous tokens for report generation. The entire process can be represented as:

$$p(T|I) = \prod_{t=1}^{n} p(y_t|y_1, \ldots, y_{t-1}, I). \quad (5)$$

Here, y_t is the input token at time step t. The typical objective for report generation is minimizing cross-entropy loss between the predicted and ground

---
[1] https://pubmed.ncbi.nlm.nih.gov.

truth token sequences. With ground truth report $\hat{R}$, all modules are optimized to maximize $p(\mathbf{y}|I)$ by minimizing:

$$\mathcal{L}_{\text{RG}} = -\sum_{t=1}^{\hat{n}} \log p(\hat{y}_t|\hat{y}_1, \cdots, \hat{y}_{t-1}, I). \tag{6}$$

3.4 Counterfactual Generation

In this section, we will explain how to generate counterfactual features, encompassing a counterfactual image and a learnable prompt in detail. Counterfactual images are pivotal, allowing the model to discern non-spurious features through contrasting representations between factual and counterfactual images. The learnable prompt encapsulates both factual and counterfactual contents and then efficiently refines the pre-trained LLM.

Negative Sampling Strategy. Counterfactual features combine features derived from both factual (positive) and negative data. We first propose a negative sampling strategy to select the negative data from a data bank. Such negative data should have different labels and are difficult to distinguish from factual data. To implement this, we first construct a data bank, denoted by D, containing candidate data, each instance d_i is annotated with {Image I_i, Report T_i, Label C_i}, maintaining the balanced distribution of diagnostic labels within the data bank. Next, we select the negative data from the data bank, as $d^- = \arg\max_i \text{BLEUScore}(T, T_i)$ and $C \neq C_i$. The BLEUScore function calculates the BLEU [32] score, setting the factual report as a reference and the negative data as a candidate. The so-selected $d^- = I^-, T^-, C^-$ is earmarked as negative data, exhibiting textual semantics similar to the original data but possessing distinct labels, emphasizing their inherent dissimilarity. The entire procedure is visually depicted in Fig. 3.

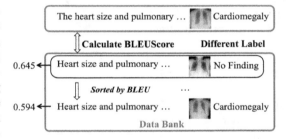

Fig. 3. Illustration of negative sampling strategy. The objective is to select a negative sample that is mostly similar in semantics but carries a different diagnostic label from the data bank.

Counterfactual Image. After selecting the negative data from candidates, we proceed to generate counterfactual images combining factual and negative images, thereby enhancing non-spurious representations through contrastive learning. As shown in Fig. 4, a factual image, I, is presented in the form of n patches: $I = p_1, p_2, ..., p_n$; its corresponding negative image is represented as

$I^- = p_1^-, p_2^-, ..., p_n^-$. From the 1st to the n-th patch, each patch of the negative image replaces the patch of the factual image at the corresponding position. The modified image is denoted by $I' = (1-u)*I + u*I^-$, where u is a one-hot vector to present the index of the replaced patch. Subsequent to each replacement, the modified image is fed into a pre-trained and frozen discriminator composed of the image encoder and a Multilayer Perceptron (MLP) to predict the logits for the diagnostic label C'. The replacement process ceases once $C' \neq C$, culminating in the acquisition of the counterfactual image I'. This methodology enables the identification of critical regions that prompt models to alter the predicted diagnosis. In essence, such regions contain pivotal information pertinent to the examination. It helps to mitigate inherent data bias and facilitates the model's focus on these critical regions, learning non-spurious and robust visual representations.

Learnable Prompt. Another key component of our counterfactual features is a learnable prompt, designed to elicit knowledge and leash the potential of pre-trained LLMs. Frequently used prompts in caption tasks, such as "the caption is..." or "describe [visual tokens]", clarify the task but often lack comprehensive instructions. To rectify this deficiency, we embed both factual and counterfactual content within the learnable prompt to attain more generalizable representations. As suggested by [39], our prompt incorporates detailed instructions and is formulated by concatenating the factual visual tokens, factual label, counterfactual label, and the index of the patch with supplementary text. The training prompt is articulated as "The u patch of image contains critical features for diagnosing C. Generate a diagnostic report for the image by describing critical entities including tubes, pneumothorax, pleural effusion, lung opacity, cardiac silhouette, hilar enlargement, and mediastinum."

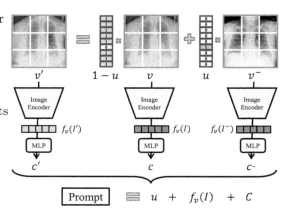

Fig. 4. Illustration of the counterfactual generation process, including a counterfactual image and a learnable prompt.

3.5 Joint Optimization

In addition to the image-report contrastive loss, and report generation loss, we introduce a novel contrastive loss aimed at amplifying the proficiency of visual representation learning. Specifically, the factual image feature $f_v(I)$, text feature $f_t(T)$, and counterfactual image feature $f_v(I')$ are employed to compute the counterfactual loss, $\mathbf{L}_{cf}$, thereby extending the divergence between the counterfactual features and the features of the original data. This can be represented as:

$$\mathcal{L}_{CF} = -\log \frac{e^{\frac{f_v(I)@f_t(T)}{\tau}}}{e^{\frac{f_v(I)@f_t(T)}{\tau}} + e^{\frac{f_v(I')@f_t(T)}{\tau}}} \quad (7)$$

Here, @ denotes cosine similarity. The total training loss is written as:

$$\mathcal{L} = \mathcal{L}_{IRC} + \lambda_{rg}\mathcal{L}_{RG} + \lambda_{cf}\mathcal{L}_{CF} \quad (8)$$

where λ_{cf} and λ_{rg} denote the loss weights, we assign a value of 1.2 to λ_{cf} and 1.5 to λ_{rg} based on performance on the validation set.

3.6 Inference

Our inference methodology is streamlined, as the counterfactual generation module is operational exclusively during the training phase, thereby enhancing the learning of both visual and textual representations. Given an image, the hidden embeddings $f_V(I)$ are seamlessly concatenated with the prompt and fed into the LLM to generate diagnostic reports. In a similar vein, the prompt is refined to "Generate a diagnostic report for the image by describing critical entities including tubes, pneumothorax, pleural effusion, lung opacity, cardiac silhouette, hilar enlargement, and mediastinum."

4 Experiments

4.1 Datasets, Evaluation Metrics and Settings

Datasets. We validate the efficacy of our proposed CoFE using the **IU-Xray** [6] and **MIMIC-CXR** [18] benchmarks. The settings adopted by [4] are utilized to uniformly split and preprocess the datasets and reports, ensuring a fair comparison. IU-Xray [6], a prevalent benchmark for evaluating RRG systems, comprises 3,955 reports and 7,470 images. After excluding cases with only one image as per [4,20], 2069/296/590 cases are allocated for training/validation/testing respectively. We utilize CheXPert to extract terminologies from reports and assign labels to each examination. MIMIC-CXR [18], the most extensive radiology dataset publicly available, includes 368,960 images and 222,758 reports. It has officially segmented subsets and has spurred the development of structurally explorative child datasets like RadGraph [14].

Metrics. We employ two types of metrics to evaluate the quality of our predicted reports. First, **natural language generation** (NLG) are employed to assess the descriptive precision of the predicted reports, with CIDEr [41] and BLEU [32] being primary. BLEU is primarily designed for machine translation, evaluating word n-gram overlap between reference and candidate, repeating frequent sentences can also achieve high scores. Conversely, CIDEr, developed for captioning systems, rewards topic terms and penalizes frequent ones, thus is more fitting for evaluating reports in RRG tasks. Additionally, ROUGE-L [25] and METEOR [2] are also considered for comprehensive comparison. Lastly, **clinical efficacy** metrics, a more recent innovation, ascertain the clinical accuracy of reports by using the CheXPert labeling tool to annotate predicted reports. Subsequent classification measurements like F1-Score, Precision, and Recall assess the aptness of the generated reports in describing abnormalities.

Table 1. The performance in NLG metrics of our proposed method compared to other competitive methods on the IU-Xray and MIMIC-CXR datasets. The highest figures in each column are highlighted in bold.

IU-Xray					MIMIC-CXR				
Methods	CIDEr	BLEU-4	ROUGE-L	METEOR	Methods	CIDEr	BLEU-4	ROUGE-L	METEOR
R2Gen	0.398	0.165	0.371	0.187	R2Gen	0.253	0.103	0.277	0.142
KERP	0.280	0.162	0.339	–	CMN	–	0.106	0.278	0.142
HRGP	0.343	0.151	0.322	–	TopDown	0.073	0.092	0.267	0.129
MKG	0.304	0.147	0.367	–	PPKED	0.237	0.106	0.284	0.149
PPKED	0.351	0.168	0.376	0.190	RGRG	**0.495**	**0.126**	0.264	0.168
MGSK	0.382	**0.178**	0.381	–	MGSK	0.203	0.115	0.284	–
CMCL	–	0.162	0.378	0.186	CMCL	–	0.097	0.281	0.133
DCL	0.586	0.163	0.383	0.193	DCL	0.281	0.109	0.284	0.150
CoFE	**0.731**	0.175	**0.438**	**0.202**	**CoFE**	0.453	0.125	**0.304**	**0.176**

Experimental Settings. For both datasets, we only utilize the front view examinations. We first pre-train the ViT-S for 10 epochs using diagnosis labels. Given the distinct domain difference between medical and general texts, a pre-trained PubMedBert [10] is utilized as both a tokenizer and a text encoder. The training is conducted on 4 NVIDIA 2080 Ti GPUs, spanning 50 epochs with batch sizes of 8. The model checkpoint achieving the highest CIEDr metric is selected for testing. An Adam optimizer, with a learning rate of 1e−4 and a weight decay of 0.02, is applied. The learning rate drops 10 every 2 epochs and stops at 1e−6. We set the size of data bank to 1,380. Note that all encoded vectors are projected by a linear transformation layer into a dimension of $d = 384$.

4.2 Main Results

Descriptive Accuracy. We compare our CoFE with several competitive RRG methods on two benchmarks. R2Gen [4] and CMN [3] are two widely-used baseline models implementing relation memory. KERP [20], PPKED [27], MKG [51] and MGSK [48] are proposed to integrate medical knowledge with typical RRG backbones. CMCL [26] and DCL [23] employ contrastive learning to further improve performance. As presented in Table 1, our method notably outperforms all competing approaches, attaining the highest figures across almost all the metrics, with a CIDEr score of 0.731 and a BLEU-4 score of 0.175 on IU-xray. Similarly, our method demonstrates competitive performance on the MIMIC-CXR dataset, achieving the highest ROUGE-L score of 0.304 and METEOR score of 0.176. These results showcase the superior capability of our method in generating matched and semantically similar reports.

Clinical Correctness. We also evaluate our method by Clinical Efficacy (CE) metrics on the MIMIC-CXR dataset to evaluate the clinical correctness of our predicted reports. In Table 2, we compare the performance against several baseline models, DCL, R2Gen, and MKSG, respectively. Most notably, our CoFE achieves the SOTA performance across all the clinical efficacy metrics, with a Precision of 0.489, Recall of 0.370, and F1-score of 0.405. This performance-boosting underscores the effectiveness of integrating counterfactual explanations, enabling the model to generate more clinically correct and relevant reports.

4.3 Analysis

In this section, we conduct ablation studies and a case study on IU-Xray and MIMIC-CXR datasets to investigate the proficiency of each key component in CoFE. Specifically, Table 3 presents the quantitative analysis of CoFE on IU-Xray measuring descriptive accuracy. In addition, clinical correctness evaluation is reported in Table 2. We employ a BLIP without the cross-modal encoder as our base model. We found that abandoning this module can save computation sources and achieve similar performances.

Table 2. The comparison of the clinical efficacy metrics on the MIMIC-CXR dataset.

Methods	Precision	Recall	F1-score
DCL	0.471	0.352	0.373
R2Gen	0.333	0.273	0.276
MKSG	0.458	0.348	0.371
Base	0.325	0.271	0.273
+ LLMs	0.396	0.323	0.317
+ prompt	0.463	0.355	0.366
+ $\mathcal{L}_{CF}$ (full)	**0.489**	**0.370**	**0.405**

Effect of Pre-trained LLMs. Compared with the base model in setting (a), illustrated in Table 3, where we utilize a pre-trained PubMedBert and a 355M-parameter GPT-2 Medium as the text encoder and language decoder, there is a significant enhancement in all metrics, with CIDEr improving from 0.363 to 0.510, emphasizing the impactful role of LLMs in enhancing the report generation performance. Specifically, PubMedBert can encode the reports into better textual representations, while GPT-2 has the capability to generate semantically coherent and logically consistent reports.

Non-spurious Representation Learning. The primary motivation for integrating counterfactual explanations is to enhance non-spurious visual representations by contrasting the representations between factual and counterfactual images. When comparing setting (c) to Setting (a) and the full model to setting (b), a significant performance boost is observable across all metrics. For instance, CIDEr elevates from 0.510 to 0.678, and from 0.698 to 0.731, respectively. Additionally, the final BLEU-4 metrics reach 0.175, achieving the SOTA performances. These notable elevations highlight the importance of non-spurious representation learning capabilities in radiology report generation tasks.

Effect of Prompt Tuning. To fully elicit pre-trained knowledge and unleash the potential of LLMs, we propose a learnable prompt that encapsulates both factual and counterfactual content to refine the LLMs. Observing setting (a) *vs* (b) and (c) *vs* the full model, it is evident that our proposed prompt can further augment performance, especially evident in ROUGE-L, which elevates from 0.355 to 0.373 and from 0.381 to 0.428, respectively. This increment underscores the effectiveness of our prompt in refining the model's natural language generation capability. Furthermore, as shown in Table 2, this prompt can also increase the clinical correctness of the predicted reports.

Fig. 5. Illustration of reports generated by R2Gen, DCL and our CoFE. The text in blue demonstrates the ground truth diagnosis labels. The red text represents the accurately matched abnormalities. (Color figure online)

Negative Sampling Strategy. The key point to constructing a counterfactual image is selecting negative data which have different labels and are difficult to be distinguished from the factual data. To verify this, we employ a random sampling strategy in which candidate data are indiscriminately selected as the negative sample. The incorporation of this random sampling strategy in settings (c) and (d) results in a discernible degeneration in the model's capability to generate high-quality reports. For instance, the CIDEr metrics drop from 0.678 to 0.653, while Bleu-4 scores decrease to 0.156. This slight decline across almost all performance metrics elucidates the influential role of our negative sampling strategy in pinpointing the most suitable negative data.

Table 3. Quantitative analysis of our proposed method on the IU-Xray dataset. We employ a vanilla BLIP without loading pre-trained parameters as the base model.

Settings	LLMs	Prompt	$\mathcal{L}_{CF}$	Random Sampling	CIDEr	BLEU-4	ROUGE-L	METEOR
Base					0.363	0.132	0.248	0.154
(a)	✓				0.510	0.151	0.355	0.180
(b)	✓	✓			0.698	0.160	0.373	0.183
(c)	✓		✓		0.678	0.164	0.381	0.191
(d)	✓		✓	✓	0.653	0.156	0.392	0.186
(e)	✓	✓	✓	✓	0.706	0.166	0.407	0.196
CoFE	✓	✓	✓		**0.731**	**0.175**	**0.428**	**0.202**

Qualitative Analysis. In Fig. 5, we present two samples from MIMIC-CXR and corresponding reports generated by R2Gen, DCL and our CoFE. R2Gen seems to lack specificity and detailed insights, providing a more generalized statement about the conditions and missing several key abnormalities mentioned in the ground truth, such as pulmonary nodules and pleural effusion. The DCL model is somewhat more aligned with the ground truth, acknowledging the unchanged appearance of the cardiac silhouette and the presence of extensive bilateral parenchymal opacities. However, it fails to mention the presence of pulmonary nodules and the pleural effusion in the right middle fissure specifically. In contrast, CoFE addresses pleural effusion, atelectasis, and the absence of pneumonia and pneumothorax, making it more in alignment with certain elements of the ground truth. These observations prove that our CoFE can generate factual complete and consistent reports.

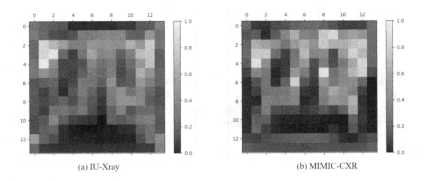

Fig. 6. Heatmaps that illustrate the frequency at which individual patches were replaced during the training process of two distinct medical imaging datasets: IU-Xray on the left and MIMIC-CXR on the right.

Frequency of Replaced Patches: In Fig. 6, we present two heatmaps that illustrate the frequency at which specific patches were replaced to construct counterfactual images during the training process on the IU-Xray and MIMIC-CXR datasets. The color gradient in each patch ranges from dark to light, with the lightest shade denoting the highest replacement frequency. The observed patterns across the heatmaps indicate that patches associated with critical anatomical regions, such as the heart, lungs, and air spaces, are replaced more frequently. This suggests that during training, these areas are particularly targeted as they likely contain critical diagnostic information (counterfactual contexts), underlining their clinical significance in disease detection or condition assessment. The distribution of frequency substantiates our methodology for constructing counterfactual images, which effectively identifies and emphasizes the most pertinent regions, thereby potentially enhancing the representation learning process.

5 Conclusion

In this paper, we present a novel framework, Counterfactual Explanations-based Framework (CoFE), designed for radiology report generation. To address the inherent data bias, we introduce a novel counterfactual concept, allowing CoFE to identify critical regions and construct a counterfactual image during training. By contrasting the representations between factual and counterfactual features, CoFE is adept at learning non-spurious visual representations. Subsequently, we summarize the counterfactual generation process into a learnable prompt, enabling the efficient fine-tuning of a pre-trained LLM. Experiments on two widely-recognized benchmarks verify the efficacy of our approach in generating factual, comprehensive, and coherent reports.

Acknowledge. This work is supported by ARC DP210101347.

References

1. Ba, J.L., Kiros, J.R., Hinton, G.E.: Layer normalization. arXiv preprint arXiv:1607.06450 (2016)
2. Banerjee, S., Lavie, A.: Meteor: an automatic metric for MT evaluation with improved correlation with human judgments. In: Proceedings of the ACL Workshop on Intrinsic and Extrinsic Evaluation Measures for Machine Translation and/or Summarization, pp. 65–72 (2005)
3. Chen, Z., Shen, Y., Song, Y., Wan, X.: Cross-modal memory networks for radiology report generation. In: Proceedings of the 59th Annual Meeting of the Association for Computational Linguistics and the 11th International Joint Conference on Natural Language Processing (Volume 1: Long Papers), pp. 5904–5914 (2021)
4. Chen, Z., Song, Y., Chang, T., Wan, X.: Generating radiology reports via memory-driven transformer. In: Proceedings of the Conference on Empirical Methods in Natural Language Processing (2020)

5. Dai, X., Keane, M.T., Shalloo, L., Ruelle, E., Byrne, R.M.J.: Counterfactual explanations for prediction and diagnosis in XAI. In: Conitzer, V., Tasioulas, J., Scheutz, M., Calo, R., Mara, M., Zimmermann, A. (eds.) AIES 2022: AAAI/ACM Conference on AI, Ethics, and Society, Oxford, United Kingdom, 19–21 May 2021, pp. 215–226. ACM (2022). https://doi.org/10.1145/3514094.3534144
6. Demner-Fushman, D., et al.: Preparing a collection of radiology examinations for distribution and retrieval. J. Am. Med. Inform. Assoc. **23**(2), 304–310 (2016)
7. Dosovitskiy, A., et al.: An image is worth 16 × 16 words: transformers for image recognition at scale. In: International Conference on Learning Representations (2020)
8. Fang, Z., Kong, S., Fowlkes, C., Yang, Y.: Modularized textual grounding for counterfactual resilience. In: Proceedings of the IEEE/CVF Conference on Computer Vision and Pattern Recognition, pp. 6378–6388 (2019)
9. Fischer, M., Bartler, A., Yang, B.: Prompt tuning for parameter-efficient medical image segmentation. CoRR abs/2211.09233 (2022). https://doi.org/10.48550/arXiv.2211.09233
10. Gu, Y., et al.: Domain-specific language model pretraining for biomedical natural language processing (2020)
11. Guo, H., Tan, B., Liu, Z., Xing, E.P., Hu, Z.: Efficient (soft) Q-learning for text generation with limited good data. In: Goldberg, Y., Kozareva, Z., Zhang, Y. (eds.) Findings of the Association for Computational Linguistics: EMNLP 2022, Abu Dhabi, United Arab Emirates, 7–11 December 2022, pp. 6969–6991. Association for Computational Linguistics (2022). https://doi.org/10.18653/v1/2022.findings-emnlp.518
12. He, X., et al.: CPL: counterfactual prompt learning for vision and language models. In: Proceedings of the 2022 Conference on Empirical Methods in Natural Language Processing, pp. 3407–3418 (2022)
13. Irvin, J., et al.: CheXpert: a large chest radiograph dataset with uncertainty labels and expert comparison. In: Proceedings of the AAAI Conference on Artificial Intelligence, vol. 33, pp. 590–597 (2019)
14. Jain, S., et al.: RadGraph: extracting clinical entities and relations from radiology reports. In: Thirty-Fifth Conference on Neural Information Processing Systems Datasets and Benchmarks Track (Round 1) (2021)
15. Ji, X., Chen, J., Wu, X.: Counterfactual inference for visual relationship detection in videos. In: 2023 IEEE International Conference on Multimedia and Expo (ICME), pp. 162–167. IEEE (2023)
16. Jin, H., Che, H., Lin, Y., Chen, H.: PromptMRG: diagnosis-driven prompts for medical report generation. In: Proceedings of the AAAI Conference on Artificial Intelligence, vol. 3, pp. 2607–2615 (2024)
17. Jing, B., Xie, P., Xing, E.: On the automatic generation of medical imaging reports. In: Proceedings of the 56th Annual Meeting of the Association for Computational Linguistics (Volume 1: Long Papers), pp. 2577–2586 (2018)
18. Johnson, A.E.W., et al.: MIMIC-CXR: a large publicly available database of labeled chest radiographs. arXiv preprint arXiv:1901.07042 (2019)
19. Kim, J., Kim, M., Ro, Y.M.: Interpretation of lesional detection via counterfactual generation. In: 2021 IEEE International Conference on Image Processing (ICIP), pp. 96–100. IEEE (2021)
20. Li, C.Y., Liang, X., Hu, Z., Xing, E.P.: Knowledge-driven encode, retrieve, paraphrase for medical image report generation. In: Proceedings of the AAAI Conference on Artificial Intelligence, vol. 33, pp. 6666–6673 (2019)

21. Li, J., Li, D., Xiong, C., Hoi, S.C.H.: BLIP: bootstrapping language-image pre-training for unified vision-language understanding and generation. In: Chaudhuri, K., Jegelka, S., Song, L., Szepesvári, C., Niu, G., Sabato, S. (eds.) International Conference on Machine Learning, ICML 2022, 17–23 July 2022, Baltimore, Maryland, USA. Proceedings of Machine Learning Research, vol. 162, pp. 12888–12900. PMLR (2022). https://proceedings.mlr.press/v162/li22n.html
22. Li, M., Cai, W., Verspoor, K., Pan, S., Liang, X., Chang, X.: Cross-modal clinical graph transformer for ophthalmic report generation. In: Proceedings of the IEEE/CVF Conference on Computer Vision and Pattern Recognition, pp. 20656–20665 (2022)
23. Li, M., Lin, B., Chen, Z., Lin, H., Liang, X., Chang, X.: Dynamic graph enhanced contrastive learning for chest X-ray report generation. In: IEEE/CVF Conference on Computer Vision and Pattern Recognition, CVPR 2023, Vancouver, BC, Canada, 17–24 June 2023, pp. 3334–3343. IEEE (2023). https://doi.org/10.1109/CVPR52729.2023.00325
24. Li, M., Liu, R., Wang, F., Chang, X., Liang, X.: Auxiliary signal-guided knowledge encoder-decoder for medical report generation. In: World Wide Web, pp. 1–18 (2022)
25. Lin, C.Y.: ROUGE: a package for automatic evaluation of summaries. In: Text Summarization Branches Out. Association for Computational Linguistics, July 2004
26. Liu, F., Ge, S., Wu, X.: Competence-based multimodal curriculum learning for medical report generation. arXiv preprint arXiv:2206.14579 (2022)
27. Liu, F., Wu, X., Ge, S., Fan, W., Zou, Y.: Exploring and distilling posterior and prior knowledge for radiology report generation. In: Proceedings of the IEEE/CVF Conference on Computer Vision and Pattern Recognition, pp. 13753–13762 (2021)
28. Liu, F., Yin, C., Wu, X., Ge, S., Zhang, P., Sun, X.: Contrastive attention for automatic chest X-ray report generation. In: Findings of the Association for Computational Linguistics, pp. 269–280 (2021)
29. Liu, X., Ji, K., Fu, Y., Du, Z., Yang, Z., Tang, J.: P-Tuning v2: prompt tuning can be comparable to fine-tuning universally across scales and tasks. CoRR abs/2110.07602 (2021). https://arxiv.org/abs/2110.07602
30. Lu, J., Xiong, C., Parikh, D., Socher, R.: Knowing when to look: adaptive attention via a visual sentinel for image captioning. In: 2017 IEEE Conference on Computer Vision and Pattern Recognition, CVPR 2017, Honolulu, HI, USA, 21–26 July 2017, pp. 3242–3250. IEEE Computer Society (2017). https://doi.org/10.1109/CVPR.2017.345
31. Mohsan, M.M., Akram, M.U., Rasool, G., Alghamdi, N.S., Abdullah-Al-Wadud, M., Abbas, M.: Vision transformer and language model based radiology report generation. IEEE Access **11**, 1814–1824 (2023). https://doi.org/10.1109/ACCESS.2022.3232719
32. Papineni, K., Roukos, S., Ward, T., Zhu, W.J.: BLEU: a method for automatic evaluation of machine translation. In: Proceedings of the 40th Annual Meeting of the Association for Computational Linguistics, July 2002
33. Peng, Z., Hui, K.M., Liu, C., Zhou, B.: Learning to simulate self-driven particles system with coordinated policy optimization. In: Advances in Neural Information Processing Systems, vol. 34 (2021)
34. Radford, A., et al.: Language models are unsupervised multitask learners. OpenAI Blog **1**(8), 9 (2019)

35. Shin, T., Razeghi, Y., Logan IV, R.L., Wallace, E., Singh, S.: AutoPrompt: eliciting knowledge from language models with automatically generated prompts. arXiv preprint arXiv:2010.15980 (2020)
36. Tanida, T., Müller, P., Kaissis, G., Rueckert, D.: Interactive and explainable region-guided radiology report generation. In: IEEE/CVF Conference on Computer Vision and Pattern Recognition, CVPR 2023, Vancouver, BC, Canada, 17–24 June 2023, pp. 7433–7442. IEEE (2023). https://doi.org/10.1109/CVPR52729.2023.00718
37. Tanyel, T., Ayvaz, S., Keserci, B.: Beyond known reality: exploiting counterfactual explanations for medical research. CoRR abs/2307.02131 (2023). https://doi.org/10.48550/arXiv.2307.02131
38. Touvron, H., et al.: LLaMA: open and efficient foundation language models. arXiv preprint arXiv:2302.13971 (2023)
39. Tu, T., et al.: Towards generalist biomedical AI. CoRR abs/2307.14334 (2023). https://doi.org/10.48550/arXiv.2307.14334
40. Vaswani, A., et al.: Attention is all you need. In: Advances in Neural Information Processing Systems, vol. 30 (2017)
41. Vedantam, R., Lawrence Zitnick, C., Parikh, D.: CIDEr: consensus-based image description evaluation. In: Proceedings of the IEEE Conference on Computer Vision and Pattern Recognition (2015)
42. Vinyals, O., Toshev, A., Bengio, S., Erhan, D.: Show and tell: a neural image caption generator. In: IEEE Conference on Computer Vision and Pattern Recognition, CVPR 2015, Boston, MA, USA, 7–12 June 2015, pp. 3156–3164. IEEE Computer Society (2015). https://doi.org/10.1109/CVPR.2015.7298935
43. Virgolin, M., Fracaros, S.: On the robustness of sparse counterfactual explanations to adverse perturbations. Artif. Intell. **316**, 103840 (2023). https://doi.org/10.1016/j.artint.2022.103840
44. Voutharoja, B.P., Wang, L., Zhou, L.: Automatic radiology report generation by learning with increasingly hard negatives. CoRR abs/2305.07176 (2023). https://doi.org/10.48550/arXiv.2305.07176
45. Wang, Z., Liu, L., Wang, L., Zhou, L.: METransformer: radiology report generation by transformer with multiple learnable expert tokens. In: IEEE/CVF Conference on Computer Vision and Pattern Recognition, CVPR 2023, Vancouver, BC, Canada, 17–24 June 2023, pp. 11558–11567. IEEE (2023). https://doi.org/10.1109/CVPR52729.2023.01112
46. Wang, Z., Zhou, L., Wang, L., Li, X.: A self-boosting framework for automated radiographic report generation. In: IEEE Conference on Computer Vision and Pattern Recognition, CVPR 2021, Virtual, 19–25 June 2021, pp. 2433–2442. Computer Vision Foundation/IEEE (2021). https://doi.org/10.1109/CVPR46437.2021.00246, https://openaccess.thecvf.com/content/CVPR2021/html/Wang_A_Self-Boosting_Framework_for_Automated_Radiographic_Report_Generation_CVPR_2021_paper.html
47. Xu, D., et al.: Vision-knowledge fusion model for multi-domain medical report generation. Inf. Fusion **97**, 101817 (2023). https://doi.org/10.1016/j.inffus.2023.101817
48. Yang, S., Wu, X., Ge, S., Zhou, S., Xiao, L.: Knowledge matters: chest radiology report generation with general and specific knowledge. Med. Image Anal. **80**, 102510 (2022)
49. Yang, S., Wu, X., Ge, S., Zheng, Z., Zhou, S.K., Xiao, L.: Radiology report generation with a learned knowledge base and multi-modal alignment. Med. Image Anal. **86**, 102798 (2023). https://doi.org/10.1016/j.media.2023.102798

50. Yang, Z., Liu, Y., Ouyang, C., Ren, L., Wen, W.: Counterfactual can be strong in medical question and answering. Inf. Process. Manag. **60**(4), 103408 (2023). https://doi.org/10.1016/j.ipm.2023.103408
51. Zhang, Y., Wang, X., Xu, Z., Yu, Q., Yuille, A., Xu, D.: When radiology report generation meets knowledge graph. In: Proceedings of the AAAI Conference on Artificial Intelligence, vol. 34, pp. 12910–12917 (2020)
52. Zhou, K., Yang, J., Loy, C.C., Liu, Z.: Conditional prompt learning for vision-language models. In: Proceedings of the IEEE/CVF Conference on Computer Vision and Pattern Recognition, pp. 16816–16825 (2022)
53. Zhou, K., Yang, J., Loy, C.C., Liu, Z.: Learning to prompt for vision-language models. Int. J. Comput. Vision **130**(9), 2337–2348 (2022)

SpeedUpNet: A Plug-and-Play Adapter Network for Accelerating Text-to-Image Diffusion Models

Weilong Chai[✉][iD], Dandan Zheng[iD], Jiajiong Cao[iD], Zhiquan Chen[iD], Changbao Wang[iD], and Chenguang Ma[iD]

Ant Group, Beijing, China
{weilong.cwl,yuandan.zdd,jiajiong.caojiajio,
zhiquan.zhiquanche,changbao.wcb,chenguang.mcg}@antgroup.com

Abstract. Text-to-image diffusion models (SD) exhibit significant advancements while requiring extensive computational resources. Existing acceleration methods usually require extensive training and are not universally applicable. LCM-LoRA, trainable once for diverse models, offers universality but rarely considers ensuring the consistency of generated content before and after acceleration. This paper proposes SpeedUpNet (SUN), an innovative acceleration module, to address the challenges of universality and consistency. Exploiting the role of cross-attention layers in U-Net for SD models, we introduce an adapter specifically designed for these layers, quantifying the offset in image generation caused by negative prompts relative to positive prompts. This learned offset demonstrates stability across a range of models, enhancing SUN's universality. To improve output consistency, we propose a Multi-Step Consistency (MSC) loss, which stabilizes the offset and ensures fidelity in accelerated content. Experiments on SD v1.5 show that SUN leads to an overall speedup of more than **10** times compared to the baseline 25-step DPM-solver++, and offers two extra advantages: (1) training-free integration into various fine-tuned Stable-Diffusion models and (2) state-of-the-art FIDs of the generated data set before and after acceleration guided by random combinations of positive and negative prompts. Code is available (Project: https://williechai.github.io/speedup-plugin-for-stable-diffusions.github.io).

Keywords: Diffusion models · Acceleration · Adapter network

W. Chai and D. Zheng—Equal contribution.

1 Introduction

In recent years, significant advancements have been made in the field of generative models, particularly in text-to-image generation, with Denoising Diffusion Probabilistic Models (DDPMs) [3] playing a crucial role. To further enhance the generation quality of text-to-image diffusion models, classifier-free guidance (CFG) [4] is widely used in large-scale generative frameworks [14,19–21]. However, the iterative sampling procedure for diffusion models costs extensive computational resource, and CFG doubles the inference latency because it demands one diffusion process for the positive prompt and another for the negative.

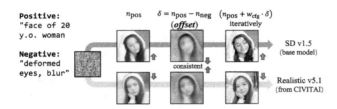

Fig. 1. Visualization of offset between positive and negative guidances. While finetuned SD can generate images of very different styles, the substraction of predictions guided by positive and negative text (*offset*) is relatively **consistent** in different SDs.

Based on the above problems, many efforts have been made on the topic of fast sampling and distillation of diffusion models. Advanced sampling strategies [7,8,25] significantly decrease the diffusion steps from several hundreds to 25 without training. Structural pruning [6] proposes that a smaller "student" model can be trained to mimic the output of the "teacher" model. To reduce the inference steps of the diffusion model, Progressive distillation [22] and Consistency Models [26] learn to iteratively reduce the sampling steps. Guided-Distill [12] and Latent Consistency Models (LCM) [9,10] augment above methods to text-to-image diffusion models, where CFG process is particularly considered in their distillation processes.

These methods have shown producing high-quality images in less than 4 sampling steps, but they still have several limitations in practicality. First, existing distillation methods require fine-tuning of the entire diffusion network and corresponding training data, which makes them difficult to apply to a new pre-trained model. Second, efficient finetuning methods that use LoRA [10] may result in significant visual differences between the images generated before and after acceleration. This is also accompanied by the inaccuracy of the indicators, when the selected dataset (LAION5B or MCOCO) for FID and CLIP-score can be very different in terms of data distribution from the dataset used for training the stylized SD model. Additionally, input from negative prompts is often simplified or discarded during accelerations, which weakens the adjustability of the accelerated models.

To address these limitations, we propose a novel and universal acceleration adapter called SpeedUpNet (SUN). Once trained on a base Stable Diffusion (SD) [20] model, SUN can be easily plugged into various fine-tuned SD models (such as different stylized models) to significantly improve inference efficiency while maintaining content consistency and negative prompt control. In particular, SUN is implemented in a teacher-student distillation framework, where the student has the same architecture with the teacher model except for an additional adapter network. During the training, only the adapter of the student, which consists of several cross-attention layers, are optimized with the other parameters frozen. The adapter network takes the negative prompt embedding as an extra input to the diffusion model, allowing for CFG-like effects in one inference. SUN adapter consists of several cross-attention operations to calculate the offset of the negative prompt relative to the positive prompt on each of the attention layers in the U-Net. As depicted in Fig. 1, the offset δ between negative and positive text embeddings, which is usually utilized to improve image quality, is noted to be a variable associated with text inputs and is unrelated to the model's style. As a result, the trained adapter network can be generalized to other stylized T2I diffusion models.

Additionally, SUN introduces a Multi-Step Consistency (MSC) loss to ensure a harmonious balance between reducing inference steps and maintaining consistency in the generated output. Different from the existing method that gradually change the inverse diffusion trajectory to a new one for one-step generation such as LCM [9] and Guided-Distill [12], MSC divides the original dense trajectory into a few (e.g. 4) stages, with each stage being approached by an accelerated inference. By mapping the output of each stage to the point of the original trajectory, this method avoids cumulative errors during acceleration, thus maintaining the consistency of the output image. Consequently, SUN significant reduces in the number of inference steps to just 4 steps and eliminates the need for CFG, which leads to an overall speedup of more than 10 times for SD models compared to the 25-step dpm-solver++. To sum up, our contributions are as follows:

- First, we propose a novel and universal acceleration module called SpeedUpNet (SUN), which can be seamlessly integrated into different fine-tuned SD models without training, once it is trained on a base SD model.
- Second, we propose a method that supports classifier-free guidance distillation with controllable negative prompts and utilizes Multi-Step Consistency (MSC) loss to enhance content consistency between the generated outputs before and after acceleration.
- Third, experimental results demonstrate SUN achieves a remarkable speedup of over 10 times on diffusion models. SUN fits various style models as well as generation tasks (including Inpainting [11], Image-to-Image and ControlNet [29]) without extra training, and achieve better results compared to existing SOTA methods.

2 Related Work

2.1 Diffusion Models and Classifier-Free Guidance

Diffusion Models have achieved great success in image generation [3,15,18,25]. Classifier-free guidance (CFG [4]) is an technique for improving the sample quality of text-to-image diffusion models, which has been applied in models such as GLIDE [14], Stable Diffusion [20] and DALL·E 2 [19]. It incorporates a guidance weight that balances the trade-off between sample quality and diversity during the generation process. However, it should be noted that this approach increases the computational load due to the requirement of evaluating both conditional and unconditional (positive and negative prompts) models at each sampling step, thus necessitating optimization strategies to improve speed.

2.2 Accelerating Diffusion Models

The advanced sampling strategies (including DDIM [25], DPM-Solver [7] and DPM-Solver++ [8]) significantly decrease the number of diffusion steps from several hundreds to around 25. On structural pruning, BK-SDM [6] introduces different types of efficient diffusion models and propose various distillation strategies. On step distillation, Progressive Distillation (PD) [22] and Guided-Distill [12] propose progressive distillation methods, where a student model can generate high-quality images with only 2 diffusion steps. Additionally, Consistency Models (CM) [26] generate image in a single step by utilizing consistency mapping derived from ODE trajectories. Based on CM, Latent Consistency Models (LCM) [9] is proposed for accelerating text-to-image synthesis tasks. Some recent studies, such as UFOGen [27] and ADD [23] use adversarial techniques to obtain high-quality images in fewer steps. By incorporating LoRA into the distillation process of LCM, without fine-tuning the entire network, LCM-LoRA [10] achieves a reduction in the memory overhead of distillation, as well as the ability for accelerating diverse models and tasks.

2.3 HyperNetwork and Adapter Methods

These approaches primarily focus on fine-tuning pre-existing models for specific tasks without extensive retraining. HyperNetworks [2], with the aim of training a small recurrent neural network to influence the weights of a larger one, have found their way into adjusting the behavior of GANs and diffusion models. To retrofit existing models with new capabilities, adapters have been shown effective in vision-language tasks and text-to-image generation. ControlNet [29] tailors SD output by conditioning. T2I-adapters [13] offers fine-tuned control over attributes such as color and style. IP-Adapter [28], which is an efficient and lightweight adapter, enables image prompt capability for pretrained text-to-image diffusion models.

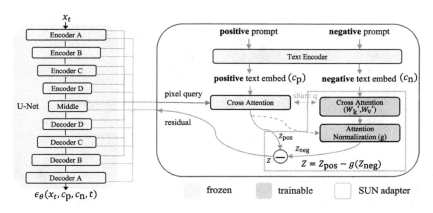

Fig. 2. The overall framework of the proposed SUN. SUN adapter is introduced to process and understand the negative prompt, which consists of several cross attention (CA) blocks. Each CA of SUN is placed side by side on each block of the original U-Net. Each block introduces a new K matrix and a V matrix, while sharing the Q with the original U-Net. Attention Normalization technique is proposed for stabilizability.

3 Method

3.1 Problem Formulation

The latent diffusion process can be inferred by optimizing the subsequent equation:

$$\mathcal{L} = \mathbb{E}_{x,\epsilon,p,t}[||\epsilon_\theta(x_t, \mathrm{E}(p), t) - \epsilon||_2^2], \tag{1}$$

where x symbolizes the noisy latent representation of an image, p is the corresponding prompt, E represents the text encoder transforming p to a conditional embedding, and t symbolizes a time step, sampled from a uniform distribution $t \sim \mathrm{Uniform}(0, 1)$. The noise ϵ adheres to a standard Gaussian distribution, i.e., $\epsilon \sim N(0, I)$. During the inference process, two texts, a positive prompt p_p and a negative prompt p_n, are applied as conditions of two independent diffusion steps:

$$\begin{aligned}\epsilon_p &= \epsilon_\theta(x_t, \mathrm{E}(p_p), t), \\ \epsilon_n &= \epsilon_\theta(x_t, \mathrm{E}(p_n), t), \\ \hat{\epsilon} &= w\epsilon_p + (1-w)\epsilon_n,\end{aligned} \tag{2}$$

where ϵ_p, ϵ_n, and $\hat{\epsilon}$ represent the positive noise, negative noise, and the final noise, respectively. This process requires two forwards of the model in order to compute the final noise, leading to potential computational inefficiency. In this study, we propose a strategy to predict the final noise in a single forward pass.

$$\hat{\epsilon} = \epsilon_\theta(x_t, \mathrm{E}(p_p), \phi(\mathrm{E}(p_n)), t), \tag{3}$$

where ϕ is a decoupled network with ϵ_θ and E, which can be optimized independently.

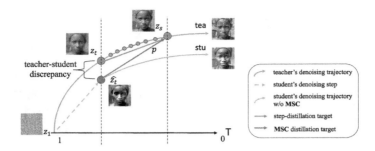

Fig. 3. An illustration of Multi-step Consistency (MSC). When distilling a faster student model, teacher-student discrepancy exists and gradually accumulates, causing the content generated by the student to be inconsistent with the teacher (from the same noise). Based on the step distillation method, MSC is used to train the student to approach the teacher's trajectory even when error occurs, thus ensuring consistency in multi-step samplings.

3.2 Negative-Positive Offset Learning

The overall framework and details of our method is illustrated in Fig. 2. During training, the negative text embedding is fed into the SUN adapter, which consists of several cross-attention blocks. And the SUN adapter interacts with the original U-Net through the cross-attention blocks. During inference, the SUN adapter can be directly plugged into any fine-tuned SDs.

To embody the interaction amongst ϵ_θ, E and the proposed decoupled network ϕ, a cross-attention mechanism is integrated. The original interaction between ϵ_θ and E can be expressed as follows:

$$Z_p = \text{MHSA}(Q, K, V), \tag{4}$$

where MHSA refers to the multi-head self-attention operation, $Q = ZW_q$, $K = E(p_p)W_k$, $V = E(p_p)W_v$ are the query, key, and value matrices of the attention operation. Respectively, W_q, W_k, W_v represent the weight matrices of the trainable linear projection layers. To insert negative text embedding, a new attention operation is added for the decoupled network ϕ:

$$Z_n = \text{MHSA}(Q, K', V'), \tag{5}$$

where Q is shared from Eq. 4, and $K' = \text{E}(p_n)W'_k$, $V' = \text{E}(p_n)W'_v$ represent the key and value of the negative text embedding.

The computation of the final feature, denoted as Z, is critical to the overall interaction between ϵ_θ, E, and ϕ as it encapsulates the negative impact of the text p_n. This is demonstrated through a subtraction operation:

$$Z = Z_p - g(Z_n), \tag{6}$$

where g is a function called **Attention Normalization** which lies in the necessity to balance the contributions from the positive and negative text prompts

---
Algorithm 1: Training of SpeedUpNet

Input: image-caption dataset $\mathcal{X}$, negative prompt dataset $\mathcal{Y}$, stable diffusion base model ϵ_θ, SUN adapter ϕ with parameter θ_ϕ.

1 **while** *not converged* **do**
2   Sample $(x, \boldsymbol{p}_p) \sim p(X)$, $\boldsymbol{p}_n \sim p(Y)$;
3   Forward $\boldsymbol{c}_p \leftarrow \mathrm{E}(\boldsymbol{p}_p)$, $\boldsymbol{c}_n \leftarrow \mathrm{E}(\boldsymbol{p}_n)$, $o \leftarrow \epsilon_\theta(\boldsymbol{x}_t, \boldsymbol{c}_p, \boldsymbol{c}_n, t)$;
4   Calculate $\mathcal{L}_\text{cls}$ via Eq. 8;
5   Calculate $\tilde{\epsilon}_\text{step}$ via Eq. 10;
6   Calculate $\mathcal{L}_\text{msc}$ on o and $\tilde{\epsilon}_\text{step}$ via Eq. 11 and Eq. 12;
7   Calculate $\mathcal{L}$ via Eq. 14;
8   Update θ_ϕ with gradient $\nabla_{\theta_\phi}(\mathcal{L})$;
9 **end**
Output: θ_ϕ

and to regulate the scale of the feature vectors. It is defined as:

$$g(Z_n) = \alpha Z_n \times \mathrm{norm}(Z_p)/\mathrm{norm}(Z_n) + \beta, \qquad (7)$$

where norm is a function that computes the magnitude of a vector, providing an objective measure of the contribution from each feature. The parameters α and β are learnable weights that allow the model to adaptively control the strength of influence of the negative prompt. Attention Normalization g helps to improve the generalization of SUN, which we describe in the experiments chapter.

3.3 Multi-Step Consistency (MSC) Distillation

Vanilla CFG Distillation. To imitate the behavior of classifier-free diffusion model, one of the objective is to encourage the output of the student to resemble the prediction by classifier-free guidance:

$$\mathcal{L}_\text{cfg} = \mathbb{E}_{\boldsymbol{x}_0, \boldsymbol{c}_p, \boldsymbol{c}_n, t} \left\| \hat{\boldsymbol{\epsilon}} - \boldsymbol{\epsilon}_\theta(\boldsymbol{x}_t, \boldsymbol{c}_p, \boldsymbol{c}_n, t) \right\|^2, \qquad (8)$$

where $\boldsymbol{c}_p = \mathrm{E}(\boldsymbol{p}_p)$ is the conditional embedding of the positive prompt, $\boldsymbol{c}_n = \mathrm{E}(\boldsymbol{p}_n)$ represents the conditional embedding of the negative prompt, and $\hat{\boldsymbol{\epsilon}} = w\boldsymbol{\epsilon}_p + (1-w)\boldsymbol{\epsilon}_n$ is the final noise given by teacher's CFG. It is worth to notice that there are two differences from the original CFG-Distill [12] method. First, the original SD model θ is frozen and only the parameters of the adapter network ϕ are optimized. Second, various negative prompts are used in training instead of a fixed empty prompt to ensure that the model is still controlled by negative prompts when producing content. These changes make the optimization goal closer to the inference procedure, and make the adapter more versatile.

Multi-step Consistency Loss. In order to further improve the model to sample high-quality and consistent images in fewer steps, we use optimized step-distillation and add MSC loss on this basis to reduce the gap between the student and the teacher. As the SUN adapter has already accepted negative prompts as input, there is no need for pre-distillation to remove CFG as done in Guided-Distill [12]. To maintain a stable teacher, we also choose not to progressively distill it multiple times like PD [22]. Given the noisy input x_t at time t and the teacher's sampling process from time t to s by N steps in continuous time space, the objective for the student network is to obtain the same diffusion state $\tilde{x}_s$ at time s in one step. To perform the sampling process in the continuous time space, we divide the time t to s into N segments, and get $\Delta = (t-s)/N$ as the time interval for each inference process. For t' in $[t, t-\Delta, t-2\Delta, ..., s+\Delta]$ and let $t'' = t' - \Delta$, the ideal noisy sample $\tilde{x}_s$ at time s can be obtained by iteratively inferencing the teacher network via classifier-free guidance:

$$x_{t''} = \alpha_{t''} \frac{x_{t'} - \sigma_{t'} \hat{\epsilon}_{\theta t'}(x_{t'})}{\alpha_{t'}} + \sigma_{t''} \hat{\epsilon}_{\theta'}(x_{t'}). \tag{9}$$

In order for the model to generate $\tilde{x}_s$ from x_t in one step, the network should predict $\tilde{\epsilon}_{\text{step}}$ approximately. According to DDIM updating rule, we have

$$\tilde{\epsilon}_{\text{step}} = \frac{(\tilde{x}_s - \frac{\alpha_s x_t}{\alpha_t})}{\sigma_s - \frac{\alpha_s \sigma_t}{\alpha_t}}. \tag{10}$$

The corresponding step-distillation loss is calculated as

$$\mathcal{L}_{\text{step}}(x_t) = \mathbb{E}_{x_0, c_p, c_n, t} \| \tilde{\epsilon}_{\text{step}} - \epsilon_\theta(x_t, c_p, c_n, t) \|^2. \tag{11}$$

It is important to note that there is a discrepancy between the output of the student network and the teacher network, and this discrepancy will accumulate with iterative sampling, leading to inaccurate results. To address the issue, we introduce the MSC loss to rectify the step-distillation loss (as shown in Fig. 3). When selecting the value of x_t, we randomly replace it with the student's output from the previous moment $\hat{x}_t$ with a probability of p. Regardless of whether the input is sampled from the teacher's sampling process or the student's, the student is forced to generate $\tilde{x}_s$ to ensure that the next moment follows the original trajectory without deviation:

$$\mathcal{L}_{\text{msc}} = \mathcal{L}_{\text{step}}(i x_t + (1-i) \hat{x}_t), \tag{12}$$

where

$$P(i=0) = p,\ P(i=1) = 1 - p. \tag{13}$$

Considering with CFG-distill loss, the overall optimization target is

$$\mathcal{L} = \mathcal{L}_{\text{cfg}} + \lambda \mathcal{L}_{\text{msc}}. \tag{14}$$

The entire training process proceeds as shown in Algorithm 1.

4 Experiments

4.1 Details of Implementation

Dataset for Training. To train the proposed network, we use LAION-Aesthetics-6+ which is a sub set of LAION-5B [24] containing 12M text-image pairs with predicted aesthetics scores higher than 6. Each sample from the original dataset includes one prompt (deemed a positive prompt) and a corresponding image. Subsequently, we leveraged two distinct strategies to collect negative prompts: (1) extracting negative prompts from AIGC websites such as PromptHero [17]; (2) utilizing large language models to generate a negative counterpart for the positive prompt. We then split every negative prompt into phrases with comma, resulting in a total of 832 distinct phrases. During the training, in order to generate a negative prompt, we uniformly sample 0 to 100 phrases from all the phrases, and then join them into a complete prompt (e.g. "watermark, blurry, ugly, bad anatomy, bad hands, error, missing fingers").

Configuration for Experiments. We use the widely-used Stable Diffusion v1.5 [20] for the base model. In our SUN adapter, we incorporate 16 cross-attention modules that is trainable during the distillation, resulting in a total parameter count of 18.5 M. Our method is implemented based on the Diffusers library [5] and PyTorch [16]. The training is launched on a single machine with 4 A100 GPUs for approximately 5k steps using the batch size of 32. Utilizing acceleration libraries allows the model to be trained on a single machine with 8 V100 GPUs for around 20k steps with a batch size of 8. The results derived from both machine configurations are competitive. We utilize the AdamW optimizer, maintaining a constant learning rate of 0.0001 and a weight decay of 0.01. The training process involves resizing the image's shortest side to 512, followed by a 512×512 center crop. For MSC loss, λ is set to 1.0, Δ is set 0.25, p is set 0.1.

Baselines and Evaluation. For training-free methods, we use DDIM [25], DPM-Solver [7], and DPM-Solver++ [8] schedulers. For training-requiring methods, we compare with Guided-Distill [12] and LCM [9]. Since there has been no open-sourced training-required method before, we reproduce Guided-Distill following the paper, on our dataset configurations. Since our method also belongs to adaptating-free methods that require training only once and can used with other pre-trained models, and we compare it to LCM-LoRA [10]. Following previous works, we test on LAION-Aesthetics-6+ dataset. We use FID and CLIP scores to evaluate the performances, where we generate 30K images using 10K text prompts of test set with 3 random seeds.

4.2 Qualitative Results

Without further training, we insert SUN pre-trained based on SD v1.5 into different popular diffusion models from CIVITAI [1], including Anything v5, Realistic

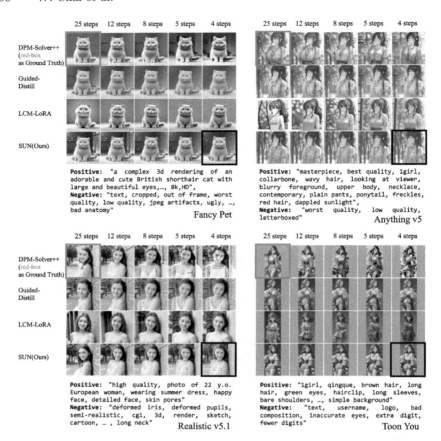

Fig. 4. Generation comparisons with different SOTA methods on different numbers of diffusion steps. The proposed SUN can produce high-quality images with only a few steps. In addition, the proposed SUN achieves the highest consistency to the ground truth with only 4 steps.

Vision v5.1, Toon You, and Fancy Pet. These models all use the same network structure and noise-prediction as SD v1.5. To display the result, we mainly compare our method with LCM-LoRA, which is also a training-free acceleration method. Additionally, we compare with Guided-Distill, which requires training, to perform further training on each model for comparisons. We regard the results of DPM-solver ++ (25 steps) as ground truths. By reducing the number of inference steps of each method, we compare the difference between its generated results and the ground truth.

As shown in Fig. 4, DPM-solver++ produce significantly poorer quality images when using a smaller number of steps (e.g. 4, 8 steps). LCM-LoRA can enhance the quality of images in the above situation without training, but the generated images may vary significantly from the ground truth. In contrast, SUN not only produces high-quality images but also generates consistent results with the ground truth at different choices of sampling steps. This reflects that SUN as a universal acceleration module is more versatile when being plugged into new models. Compared with the training-hungry method, SUN also has advantages in content consistency, which shows that MSC objective plays a role in reducing student-teacher discrepancies. At the same time, since SUN only trains the adapter parameters, it maintains well the image style and quality from the original model.

Table 1. Quantitative results on LAION-Aesthetic-6+ dataset. With training only a few parameters on cross-attention, SUN achieves the best FID/CLIP scores above the existing adapting-free methods, the results are also competitive with SOTA methods that require finetuning the entire diffusion model. Guidance scale is 8.0, resolution is 512×512.

Method	Params	Adapting free	FID ↓			CLIP score ↑		
			4-step	8-step	12-step	4-step	8-step	12-step
DDIM [25]	0	✓	22.38	13.83	12.97	0.258	0.292	0.315
DPM++ [8]	0	✓	18.43	12.20	12.03	0.266	0.295	**0.336**
Guided-Distill [12]	860M	✗	15.12	13.89	12.44	0.272	0.281	0.314
LCM [9]	860M	✗	**11.10**	**11.84**	12.02	0.286	0.288	0.320
LCM-Lora [10]	67.5M	✓	16.83	14.30	13.11	0.271	0.277	0.319
SUN (Ours)	18.5M	✓	13.23	12.08	**11.98**	**0.288**	**0.297**	0.328

4.3 Quantitative Evaluation

We first use SD v1.5 to test because all distillation-based acceleration methods are trained on SD v1.5. As shown in Table 1, SUN contains the smallest number of parameters among all distillation methods, making it more efficient in training and better to reduce the risk of overfitting. The quality of the generated images is evaluated mainly by using a standard test set as the reference. SUN is a competitive method in distribution difference (FID) and semantic consistency (CLIP score), and it achieves the best results when compared to other methods with the same parameter magnitude.

Furthermore, we evaluate the quantitative result of SUN as a universal acceleration add-on and compare it with existing techniques. We tested three different models that have been already fine-tuned on specialized datasets. As the styles of the new models are diverse, there is no standard reference set, such as LAION-5B or MSCOCO, to evaluate FIDs. To better reflect the consistency of the generated images before and after acceleration, for testing pretrained diffusion model, we use the 25-step DPM-Solver++ to generate 30k samples using the same prompts

Table 2. Quantitative results on knowledge distillation FID with various pretrained diffusion models. Each generated samples set is compared to the corresponding ground truth set generated by 25-step-DPMSolver++ scheduler (using prompts from LAION-Aesthetic-6+). SUN significantly surpasses baselines in 4, 8, and 12 steps, demonstrating its ability to seamlessly switch to other diffusion models without any training. Guidance scale is 8.0, resolution is 512 × 512.

Method	Params	Training free	Rea v5.1 ↓		RevA ↓		Any v5 ↓	
			4-step	8-step	4-step	8-step	4-step	8-step
DDIM [25]	0	✓	25.32	21.94	27.22	22.46	29.88	23.39
DPM-Solver++ [8]	0	✓	24.01	21.12	26.02	21.38	29.06	22.75
Guided-Distill [12]	860M	✗	20.31	16.33	22.40	17.23	25.57	18.49
LCM-Lora [10]	67.5M	✓	21.88	17.42	23.44	18.11	26.34	19.77
SUN (Ours)	18.5M	✓	**19.60**	**15.73**	**20.27**	**16.00**	**22.52**	**16.17**

in Sect. 4.1, and then take them as reference for computing FID. As shown in Fig. 2, SUN is demonstrated to surpass other acceleration methods on all models, therefore being a preferable acceleration method.

Table 3. Time consumption for an 512 × 512 image (seconds) using Diffusers Pipeline. *Non batch parallel* puts positive prompts and negative prompts into two batches for the inference process.

Method (steps)	V100 (FP32)		M1Pro (FP16)	
	pipeline	unet	pipeline	unet
DPM-Solver++ (25)	3.42	3.16	21.24	20.09
DPM-Solver++ (25) (*non batch parallel*)	3.67	3.42	22.21	21.07
DPM-Solver++ (4)	0.684	0.420	3.97	2.94
Guided-Distill (4)	0.459	0.243	2.42	1.55
LCM-LoRA (4)	0.521	0.317	2.56	1.69
SUN (Ours) (4)	**0.485**	**0.274**	**2.50**	**1.62**

As an important supplement, we test the time consumption of each acceleration method on different hardware platforms (Table 3). SUN is faster than baseline (DPM-solver++ 25 steps) by more than 10x in terms of U-Net time consumption, and is faster than LCM-LoRA due to the advantage of parameter quantity.

Table 4. Ablative study of hyperparameter p in Multi-step Consistency loss. Evaluated by FID, using a excessively large value makes training difficult.

p(MSC)	Rea v5.1(4)	(8)	Any v5(4)	(8)
0.0	22.41	18.77	25.62	20.98
0.1	**19.60**	**15.73**	**22.52**	**16.17**
0.25	20.13	17.22	24.01	16.69

Fig. 5. The proposed SUN maintains the controllability of negative prompts when eliminating the need for CFG.

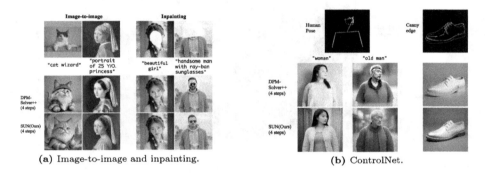

(a) Image-to-image and inpainting.   (b) ControlNet.

Fig. 6. Without extra training, SUN can accelerate other image-generation tasks, such as inpainting and image-to-image generation. SUN is also compatible with ControlNet.

4.4 Other Results

Alter the Negative Prompt. Since the ability to modify negative prompts is essential for creators in image creation, we further use different negative prompts for one positive prompt. The experimental results (Fig. 5) showed that SUN effectively learned the content of the negative prompt rather than fitting a specific style, achieving the same effect as CFG.

Image-to-Image and Inpainting. Besides text-to-image generation, SUN can also be used as a plug-in to accelerate image-to-image as well as inpaining diffusion models. As shown in the Fig. 6a, without any training on the target model, SUN is able to generate results comparable to the original model with only 4 steps.

ControlNets. Additional structure control is a popular application for text-to-image diffusion models. As our SUN does not change the original network

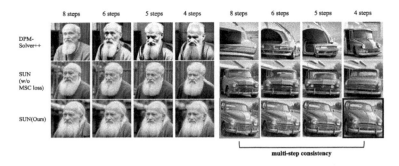

Fig. 7. Ablation on our proposed Multi Step Consistency loss. The addition of MSC allows for the generation of samples with consistent content in 4 to 8 or more steps.

Fig. 8. Ablation on our proposed Attention Normalization. It enables SUN as a pluggable module to have stable generation capabilities on different pre-trained diffusion models (Realistic Vision V5.1 and Rev Animated).

Fig. 9. Ablative study of Δ in the training strategy (4 steps). 0.25 achieves better performance in quality and consistency.

structure, it is fully compatible with existing controllable tools (as shown in Fig. 6b).

4.5 Ablation Study

In our research, we carried out ablation studies to evaluate two key methodological contributions of Sect. 3. Figure 7 demonstrates that MSC is crucial in ensuring that the model generates consistent content, whether in very few or multiple steps. As shown in Fig. 8, Attention Normalization further reduces the fitting degree of SUN to the base model and helps achieve high-quality generation capabilities on different pre-trained models. Additionally, we do ablation studies (Fig. 9 and Table 4) to assess the impact of training strategy parameters on the results.

5 Conclusion

In this work, we introduced SpeedUpNet (SUN), a novel and universal Stable-Diffusion acceleration module that can be seamlessly integrated into different fine-tuned Stable-Diffusion models without further training, once it is trained on a base Stable-Diffusion model. SUN proposes a method that utilizes an adapter for the cross-attention layers in U-Net, along with a Multi-Step Consistency (MSC) loss. This approach is specifically designed to quantify and stabilize the offset in image generation caused by negative prompts relative to positive prompts. Our empirical evaluations demonstrate that SUN significant reduces in the number of inference steps to just 4 steps and eliminates the need for classifier free guidance, which leads to a speedup of over 10 times compared to the baseline 25-step DPM-solver++, while preserving both the quality and generation consistency during the acceleration. Moreover, SUN is compatible with other generation tasks such as Inpainting [11] and Image-to-Image generation, enabling the use of controllable tools like ControlNet [29].

References

1. Civitai: Civitai website (2023). https://civitai.com/
2. Heathen: Hypernetwork style training, a tiny guide, stable-diffusion-webui (2022)
3. Ho, J., Jain, A., Abbeel, P.: Denoising diffusion probabilistic models. In: Advances in Neural Information Processing Systems, vol. 33, pp. 6840–6851 (2020)
4. Ho, J., Salimans, T.: Classifier-free diffusion guidance. arXiv preprint arXiv:2207.12598 (2022)
5. Hugging face Inc.: Diffusers (2023). https://huggingface.co/docs/diffusers/index
6. Kim, B.K., Song, H.K., Castells, T., Choi, S.: On architectural compression of text-to-image diffusion models. arXiv preprint arXiv:2305.15798 (2023)
7. Lu, C., Zhou, Y., Bao, F., Chen, J., Li, C., Zhu, J.: DPM-Solver: a fast ode solver for diffusion probabilistic model sampling in around 10 steps. arXiv preprint arXiv:2206.00927 (2022)
8. Lu, C., Zhou, Y., Bao, F., Chen, J., Li, C., Zhu, J.: DPM-Solver++: fast solver for guided sampling of diffusion probabilistic models. arXiv preprint arXiv:2211.01095 (2022)
9. Luo, S., Tan, Y., Huang, L., Li, J., Zhao, H.: Latent consistency models: synthesizing high-resolution images with few-step inference. arXiv preprint arXiv:2310.04378 (2023)
10. Luo, S., et al.: LCM-LoRA: a universal stable-diffusion acceleration module. arXiv preprint arXiv:2311.05556 (2023)

11. Meng, C., et al.: SDEdit: guided image synthesis and editing with stochastic differential equations. arXiv preprint arXiv:2108.01073 (2021)
12. Meng, C., et al.: On distillation of guided diffusion models. In: Proceedings of the IEEE/CVF Conference on Computer Vision and Pattern Recognition, pp. 14297–14306 (2023)
13. Mou, C., et al.: T2I-Adapter: learning adapters to dig out more controllable ability for text-to-image diffusion models. arXiv preprint arXiv:2302.08453 (2023)
14. Nichol, A., et al.: GLIDE: towards photorealistic image generation and editing with text-guided diffusion models. arXiv preprint arXiv:2112.10741 (2021)
15. Nichol, A.Q., Dhariwal, P.: Improved denoising diffusion probabilistic models. In: International Conference on Machine Learning, pp. 8162–8171. PMLR (2021)
16. Paszke, A., et al.: PyTorch: an imperative style, high-performance deep learning library. In: Advances in Neural Information Processing Systems, vol. 32 (2019)
17. PromptHero: Prompthero (2023). https://prompthero.com/featured
18. Ramesh, A., Dhariwal, P., Nichol, A., Chu, C., Chen, M.: Hierarchical text-conditional image generation with CLIP Latents 1(2), 3. arXiv preprint arXiv:2204.06125 (2022)
19. Ramesh, A., et al.: Zero-shot text-to-image generation. In: International Conference on Machine Learning, pp. 8821–8831. PMLR (2021)
20. Rombach, R., Blattmann, A., Lorenz, D., Esser, P., Ommer, B.: High-resolution image synthesis with latent diffusion models. In: Proceedings of the IEEE/CVF Conference on Computer Vision and Pattern Recognition, pp. 10684–10695 (2022)
21. Saharia, C., et al.: Photorealistic text-to-image diffusion models with deep language understanding. In: Advances in Neural Information Processing Systems, vol. 35, pp. 36479–36494 (2022)
22. Salimans, T., Ho, J.: Progressive distillation for fast sampling of diffusion models. arXiv preprint arXiv:2202.00512 (2022)
23. Sauer, A., Lorenz, D., Blattmann, A., Rombach, R.: Adversarial diffusion distillation. arXiv preprint arXiv:2311.17042 (2023)
24. Schuhmann, C., et al.: LAION-5B: an open large-scale dataset for training next generation image-text models (2022)
25. Song, J., Meng, C., Ermon, S.: Denoising diffusion implicit models. arXiv preprint arXiv:2010.02502 (2020)
26. Song, Y., Dhariwal, P., Chen, M., Sutskever, I.: Consistency models (2023)
27. Xu, Y., Zhao, Y., Xiao, Z., Hou, T.: UFOGen: you forward once large scale text-to-image generation via diffusion GANs. arXiv preprint arXiv:2311.09257 (2023)
28. Ye, H., Zhang, J., Liu, S., Han, X., Yang, W.: IP-Adapter: text compatible image prompt adapter for text-to-image diffusion models. arXiv preprint arXiv:2308.06721 (2023)
29. Zhang, L., Rao, A., Agrawala, M.: Adding conditional control to text-to-image diffusion models. In: Proceedings of the IEEE/CVF International Conference on Computer Vision, pp. 3836–3847 (2023)

Reg-TTA3D: Better Regression Makes Better Test-Time Adaptive 3D Object Detection

Jiakang Yuan[1], Bo Zhang[2], Kaixiong Gong[3], Xiangyu Yue[3], Botian Shi[2], Yu Qiao[2], and Tao Chen[1(✉)]

[1] School of Information Science and Technology, Fudan University, Shanghai, China
jkyuan22@m.fudan.edu.cn, eetchen@fudan.edu.cn
[2] Shanghai Artificial Intelligence Laboratory, Shanghai, China
[3] The Chinese University of Hong Kong, Central Ave, Hong Kong, China

Abstract. Domain Adaptation (DA) has been widely explored and made significant progress on cross-domain 3D tasks recently. Despite being effective, existing works fail to deal with rapidly changing domains due to the unpredictable test time scenarios and meanwhile fast response time requirement. Thus, we explore a new task named test-time domain adaptive 3D object detection and propose Reg-TTA3D, a pseudo-label-based test-time adaptative 3D object detection method. By investigating the factor that limits the detection accuracy, we find that regression is essential in this task. To make better regression, we first design a noise-consistency pseudo-label generation process to filter pseudo-labels with instability under noise interference and obtain reliable pseudo-labels. Then, confidence-guided regression refinement is introduced, which uses the box regression results of high-confidence boxes to supervise boxes with relatively low confidence, further making the predicted box size gradually approach the distribution of the target domain. Finally, to better update the regression layer and alleviate the class-imbalance issue, a class-balance EMA updating strategy is proposed. Experimental results on multiple cross-domain scenarios including cross-beam, cross-location, and cross-weather demonstrate that Reg-TTA3D can achieve comparable or even better performance compared to unsupervised domain adaptation works by only updating less than **0.1% parameters** within less than **1% time**.

Keywords: Test-time adaptation · 3D object detection · Regression

1 Introduction

LiDAR-based 3D object detection [12,18,20,21,32,36,43] aims to predict the location and category of each object based on LiDAR point clouds and has

Supplementary Information The online version contains supplementary material available at https://doi.org/10.1007/978-3-031-72775-7_12.

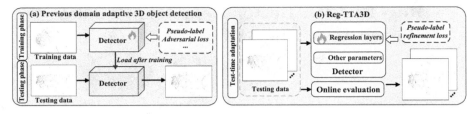

Fig. 1. Comparison between (a) previous domain adaptive 3D object detection and (b) Reg-TTA3D. Previous DA works perform training and testing in a split way. In contrast, Reg-TTA3D tackles domain discrepancies in an online manner and can deal with domain changes rapidly.

attracted increasing attention recently due to its practical applications to autonomous driving. Although significant progress has been made, the stability of detectors under environmental changes, namely cross-domain 3D object detection, remains a challenge. However, the stability is important in practical usage since the surrounding environment dynamically changes. For example, weather and geographical environment changes may be encountered in road scenes.

Inspired by unsupervised domain adaption (UDA) techniques in 2D cross-domain tasks [3,5,8,13,39–41,44,45], some works [6,15,30,31,33,38] begin to deal with 3D domain discrepancies. ST3D [33] is a pioneer in using self-training to gradually transfer the knowledge learned from the source domain to the target domain. ReDB [6] attempts to deal with multi-class problems by designing cross-domain examination and overlapped boxes counting. Despite achieving large performance gains, the existing UDA model still requires a long training time to be adapted to the target domain, lacking the ability to rapidly deal with environmental changes in an online manner as shown in Fig. 1. Actually, in real applications, models are expected to be continuously updated at a low cost and quickly adapt to changing domains.

To make the model quickly adapt to the changing environment in an online manner, test-time adaptation (TTA) [1,14,16,27,28] is introduced to adapt the model during test-time with only source model and unlabeled target data. Table 1 shows differences between TTA and other adaptation methods. In fact, the TTA technique has been explored in 2D classification tasks such as TENT [27] and SHOT [14] recently, but it is still under-explored in 3D vision fields, especially in detection tasks. To verify the effectiveness of previous TTA methods in the domain adaptive 3D object detection tasks, we conduct experiments by directly applying such methods (*e.g.*, TENT [27], SHOT [14], MATE [16]) to our baseline detector (*i.e.*, SECOND-IoU [32]). As shown in Table @reftab:SOTAcomparison, directly applying previous methods results in a large performance drop, even significantly behind the results of source only which directly evaluates with the pre-trained source model. For instance, TENT [27] only achieves 31.34%/19.47% in AP_{BEV}/AP_{3D}, with 15.30%/9.36% declination comparing to source only results.

Accordingly, previous TTA methods are hard to apply to domain adaptive 3D detection tasks due to two major reasons. Firstly, from the perspective of

Table 1. Comparison of different settings to alleviate domain discrepancies.

Setting	Source data	Target data	Online
Fine-tuning	✗	x^t, y^t	✗
Unsupervised domain adaptation	x^s, y^s	x^t	✗
Source-free domain adaptation	✗	x^t	✗
Test-time adaptation	✗	x^t	✓

updating parameter selection, previous methods often only update batch normalization (BN) layers to alleviate the domain shifts in the statistical information of network. However, this is not well suited for 3D detection task that also needs rich localization information. Besides, the optimization objectives of previous methods are designed for classification tasks, which would make the model optimize for extracting features with more category semantic information rather than regression information.

In order to tackle the challenges mentioned above, in this work, we propose Reg-TTA3D to perform test-time adaptation on 3D object detection tasks. To deal with the problem of updated parameter selection, we conduct extensive experiments and observe that high classification accuracy is already achieved by the source model, and the main factor limiting the performance improvement is the inaccurate regression results. Further, by only adapting regression layers, the model can achieve high performance at a low training cost as shown in Table 2 which motivates us to focus on regression adaptation in this work. Therefore, to make better regression, we propose noise-consistency pseudo-label generation (NPG) and confidence-guided regression refinement (CRR). The former module aims to filter boxes with instability under noise interference, *i.e.*, noisy point clouds, while the latter one makes the box regression results align better with the distribution of the target domain box. Finally, we introduce a class-balanced EMA updating strategy (CBU) that can effectively update the regression layers, to improve the performance on multi-class simultaneously.

Our contribution can be summarized as follows:

1. We present a new task, namely test-time domain adaptive 3D object detection, and propose Reg-TTA3D to solve this task from the novel view of focusing on adapting only regression parameters.
2. We propose a noise-consistency pseudo-label generation process, a confidence-guided regression refinement loss, and a class-balanced EMA updating strategy, to obtain better detection results and boost the model's performance in an efficient and effective way.
3. Experiments show that Reg-TTA3D reaches a comparable or even better performance by updating less than 0.1% parameters within less than 1% time compared to UDA works, and significantly outperforms previous TTA methods on multiple cross-domain scenarios.

2 Related Works

2.1 LiDAR-Based 3D Object Detection

With the development of autonomous driving, LiDAR-based 3D object detection has attracted increasing attention. Current prevailing LiDAR-based 3D detectors can be roughly divided into point-based, voxel-based, and point-voxel-based methods according to the point cloud processing procedure. Point-based methods [20,35,43] first sample a subset of point clouds, then extract features and generate proposals from sampled points. Among them, PointRCNN [20] uses a two-stage framework that is composed of a bottom-up proposal generation process and a proposal refinement process. In contrast, voxel-based methods [7,12,32,36,46] first divide disordered points into regular grids which are then fed into the convolutional network. SECOND [32] is a pioneer in using the 3D sparse backbone to extract features. CenterPoint [36] introduces a center-based detector and achieves promising performance. Point-voxel-based [18,19] methods integrate the features extracted from points and voxels to obtain better feature representations. However, existing 3D detectors fail to deal with domain gaps which is essential in practical usage.

2.2 Test-Time Adaptation

Test-time adaptation [1,4,10,22,24,27,28] aims to rapidly adapt the source model to target domain without accessing source domain data during test time, which is more practical in the real world since it can handle dynamic domain shifts. TENT [27] explores test-time adaptation for the first time and updates the batch normalization layers using designed entropy minimization loss. SHOT [14] deals with such a challenging task by maximizing the mutual information. More recently, MATE [16] extends test-time adaptation to 3D point clouds and finds that the model can be adapted effectively by masking and reconstruction.

Although TTA has been extensively explored in 2D scenarios, it is still under-explored in 3D tasks, especially in large-scale point cloud scenarios (*e.g.*, autonomous driving scenarios). Besides, most previous TTA works mainly focus on classification tasks, such as classification [1,16,27] and segmentation [22], and they fail to tackle detection tasks since detection considers classification and box regression at the same time. In contrast, Reg-TTA3D considers TTA for 3D object detection tasks from the perspective of regression, which is more suitable for 3D detection tasks.

2.3 Domain Adaptation for 3D Object Detection

3D detectors often suffer large performance drops when training and testing under different domains, which makes them unstable when the surrounding environment dynamically changes. Recently, researchers have tried to tackle such a problem using domain adaptation techniques [6,17,29,33,34,38]. SN [29] proposes a statistical method to align the size of bounding boxes between source and target domains. ST3D [33] and ST3D++ [34] design a self-training pipeline and mitigate domain gaps using a memory bank and curriculum data augmentation. To deal with the class imbalance issue, ReDB [6] proposes a pseudo-label generation pipeline to obtain reliable, diverse, and class-balanced pseudo-labels. However, the existing works cannot deal with the emergency changes in the environment due to the time-consuming and large amount of calculation.

Table 2. Performance of updating different parameters.

Tuning layers	Waymo → KITTI	nuScenes → KITTI	Parameters
Batch normalization	52.84 / 31.77	31.86 / 18.25	9.08k
Classification layers	52.82 / 36.37	31.51 / 19.76	1.17k
Regression layers	59.38 / 45.54	**33.63 / 21.80**	2.73k
Detection head	59.62 / 46.08	32.51 / 20.39	4.68k
All parameters	**60.78 / 46.71**	33.23 / 19.60	12.12M

3 Method

The overall framework of Reg-TTA3D is shown in Fig. 2. To better illustrate our proposed method, we start by introducing the problem formulation and our baseline in Sect. 3.1. Then, we give an analysis of the updated parameters selection in Sect. 3.2. Finally, we detail each module of Reg-TTA3D and give the overall objective and adaptation strategy in Sect. 3.3 and Sect. 3.4.

3.1 Preliminary

Problem Formulation. Following typical test-time adaptation setting [22,27], given the model M_S pre-trained on the source domain $D_S = \{(x_i^S, y_i^S)\}_{i=1}^{n_S}$ and an unlabeled target set $D_T = \{x_j^T\}_{j=1}^{n_T}$, where the source and target domain follow different distributions, our goal is to adapt the source model to the target domain without accessing source domain data in an online manner. During adaptation, in each timestep, a batch of unlabeled data $[x_1^t, x_2^t, ..., x_N^t]$ is sampled from D_T to update the model, where N is batch size.

Baseline Introduction. Following [6,34], we use SECOND-IoU [32] as our baseline detector which is composed of a feature extractor (*i.e.*, 3D and 2D backbone) to obtain bird's eye view (BEV) features, a region proposal network (RPN) to generate proposals and a detection head to provide the classification and regression results. The detector is trained using the following loss function:

$$\mathcal{L}_{det} = \mathcal{L}_{rpn} + \mathcal{L}_{rcnn}, \tag{1}$$

where $\mathcal{L}_{rpn}$ represents the loss of RPN and $\mathcal{L}_{rcnn}$ denotes box refinement loss.

3.2 Updated Parameters Selection

The aim of Reg-TTA3D is to transfer the source model to the target domain rapidly so that the model can tackle the continuous changes in the surrounding environment. To achieve this goal, the amount of updated parameters we select needs to be relatively small, and the accuracy needs to be largely improved by only updating selected parameters.

Previous TTA works [22,27] mainly focus on adapting batch normalization (BN) layers to mitigate the domain discrepancies, since BN layers contain the statistics of domains. Such an adaptation strategy is effective for domain adaptive classification tasks, as more accurate semantic representations can be obtained, which is especially important for classification. However, only adapting BN layers cannot rapidly transfer the source detector to the target domain since detection tasks contain both classification and regression. Some works [16,26] try to solve this problem by updating the entire model but suffer from huge computational costs. To explore an effective and efficient method, by conducting extensive experiments, we find that the model can adapt to the target domain efficiently by only updating regression layers. As shown in Table 2, we fine-tune

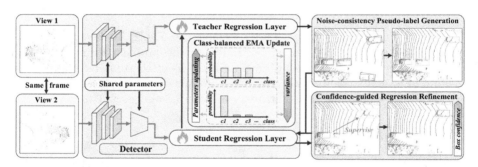

Fig. 2. The overall framework of the proposed Reg-TTA3D, which employs SECOND-IoU [32] as the baseline detector and consists of noisy-consistency pseudo-label generation process (NPG), confidence-guided regression refinement (CRR) and class-balanced EMA updating strategy. The data are first fed into the teacher model to generate pseudo-labels as the supervision to the student model using NPG. Then, the student model is trained using data from another view and supervised by detection loss and refinement loss obtained by CRR. Finally, the class-balanced EMA updating strategy is used to update the teacher model. **Note that the only difference between the teacher model and the student model lies in the regression layers.**

the parameters using 1% KITTI training data for 1 epoch and observe that only fine-tuning the regression layers can reach relatively high performance and even surpass the performance of fine-tuning the entire model. This is because in the autonomous driving scenario, there exists a large difference between categoriessuch as car and pedestrian, the source model can already achieve high classification accuracy and the main factor for limiting the detection accuracy is the inaccurate regression results. Based on the observation, in this work, we focus on studying how to only adapt the parameters of regression layers and make better regression.

3.3 Reg-TTA3D

As mentioned above, the quality of box regression is essential to TTA for 3D object detection. To make better regression, we propose noise-consistency pseudo-label generation to obtain better regression supervision, and confidence-guided regression refinement to further refine the regression results. Besides, to effectively update the regression layers, we design a class-balanced EMA updating strategy to avoid overfitting the model to a single class.

Noise-Consistency Pseudo-label Generation. As mentioned in [34,38], the pseudo-labels directly generated by the source model contain a lot of noise due to domain discrepancies. Previous methods try to alleviate pseudo-label noise by weighting pseudo-labels obtained in different epochs [33,34], or adding the groundtruth instances from the source domain to target domain frames [6]. However, these methods cannot apply to test-time adaptation due to the inaccessibility of source data and online adaptation.

Inspired by test-time augmentation methods which comprehensively consider the model predictions under different data augmentation methods to provide final results. Specifically, the model's prediction of real objects is relatively stable, while being uncertain for the noise regions. To this end, we propose noise-consistency pseudo-label generation to obtain more accurate pseudo-labels, by measuring the stability of model predictions when adding random noise to regions of interest (RoIs). In detail, given proposals generated by the teacher model $B^t = \{(c_x^t, c_y^t, c_z^t, l^t, w^t, h^t, \theta^t)_i\}_{i=1}^n$, its corresponding features $\{f_i^t\}_{i=1}^n$ and confidence scores $\{s_i^t\}_{i=1}^n$ where (c_x^t, c_y^t, c_z^t) and (l^t, w^t, h^t) denote the center and the size of the box respectively, θ^t is the rotation angle and n is the number of boxes. We can obtain boxes with noise $B^{t'} = \{(c_x^{t'}, c_y^{t'}, c_z^{t'}, l^{t'}, w^{t'}, h^{t'}, \theta^{t'})\}_{i=1}^n$, corresponding features $\{f_i^{t'}\}_{i=1}^n$ and scores $\{s_i^{t'}\}_{i=1}^n$, by adding random noise to boxes' size and angle. To measure the stability of RoIs, we calculate the inconsistency rate as follows:

$$R_{inconsistency,i} = 1 - \frac{|s_i^t - s_i^{t'}|}{\text{IoU}(b_i^t, b_i^{t'})}, \tag{2}$$

where IoU($b_i^t, b_i^{t'}$) denotes the intersection of union (IoU) between $b_i^t \in B^t$ and $b_i^{t'} \in B^{t'}$. A high inconsistency rate represents that the model prediction is unstable to the current object which is more likely to be a noise region. Therefore, we filter the boxes with high inconsistency rate by a threshold τ to get boxes and scores $\hat{B}, \{\hat{s}_i\}_{i=1}^n$ and $\hat{B}', \{\hat{s}_i'\}_{i=1}^n$ with stability under noise interference. Then, we ensemble $\hat{B}$ and $\hat{B}'$ by confidence scores as follows:

$$\tilde{B} = \{\tilde{b}_i\}, \quad \tilde{b}_i = \begin{cases} \hat{b}_i, & \hat{s}_i > \hat{s}_i', \hat{b}_i \in \hat{B} \\ \hat{b}_i', & \hat{s}_i < \hat{s}_i', \hat{b}_i' \in \hat{B}' \end{cases}, \tag{3}$$

After obtaining $\tilde{B}$, the boxes are filtered by non-maximum suppression (NMS) with a relatively high confidence threshold, to remove redundant boxes and boxes with low confidence. Finally, the pseudo-labels B^{npg} and $\{l_i^{npg}\}_{i=1}^n$ with high stability and confidence can be acquired where $\{l_i^{npg}\}_{i=1}^n$ is classification results.

Confidence-Guided Regression Refinement. Although lots of boxes containing noisy point cloud can be removed by measuring the stability under noise injection, there still exist the problem of inaccurate box regression results. This is because the distributions of bounding box size are inconsistent between source and target domains as mentioned in [29,37,42], and it is difficult to mitigate such domain discrepancy with only NPG. By analyzing the distribution of predicted bounding boxes, we find that boxes with higher confidence scores align better with the distribution of the target domain boxes as shown in Fig. 3, since precise boxes correspond to high-quality RoI features that are less disturbed by background representations. Based on this observation, we design a confidence-guided regression refinement to obtain more precise regression results of box size.

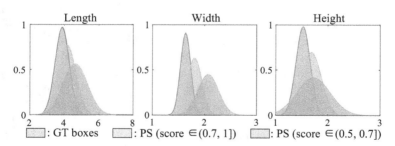

Fig. 3. Distribution of pseudo box size under different thresholds. PS denotes pseudo boxes.

In particular, given predicted boxes generated by the student model $B^s = \{(c_x^s, c_y^s, c_z^s, l^s, w^s, h^s, \theta^s)_j\}_{j=1}^k$ and scores $\{s_j^s\}_{j=1}^k$ of a batch of frames, where k is the total number of predicted boxes. We sort boxes belonging to the same class according to scores in descending order, and sample a proportion of boxes with high scores to calculate the mean value of sample boxes size $(\overline{l}^s, \overline{w}^s, \overline{h}^s)$. Please refer to our supplementary material for the study of the proportion of sampled boxes. Once obtained $(\overline{l}^s, \overline{w}^s, \overline{h}^s)$, we can use it as a supervision signal to make the box regression value closer to the distribution of the target domain as follows:

$$\mathcal{L}_{refine} = \mathbb{E}_{i=1}^n \Sigma_{d \in [d_l, d_w, d_h]} d,$$

$$where \quad d = \begin{cases} 0, & |(z^s - \overline{z}^s)| < p \\ (z^s - \overline{z}^s)^2, & |(z^s - \overline{z}^s)| \geq p \end{cases}, \quad (4)$$

where $z \in [l^s, w^s, h^s]$, $\overline{z} \in [\overline{l}^s, \overline{w}^s, \overline{h}^s]$, n is batch size and p is set 0.1 to allow the size of objects to fluctuate within a certain range. The model can predict more accurate box size when using $\mathcal{L}_{refine}$ for optimization.

Class-Balanced EMA Updating. To make the model more stable, we update the teacher regression layers by exponential moving average (EMA). However, the model tends to overfit to the category with a large number of instances (*i.e.*, car). As a result, the detection performance on categories with few instances will decrease. To make the model better balance the performance improvement for each class, we propose a class-balanced EMA updating strategy to make the model update more when the class is balanced, while updating less when a single category dominates. Specifically, given the pseudo-labels obtained from NPG $B^{npg} = \{b_i^{npg} \in \mathbb{R}^7\}$ and its corresponding class labels $\{c_i^{npg}\}$. We calculate the proportion $\{p_c\}_{c=1}^C$ of each category in pseudo-labels according to $\{c_i^{npg}\}$, where C is the number of total categories and $\Sigma_{c=1}^C p_c = 1$. Then the variance of $\{p_c\}_{c=1}^C$ is calculated and normalized to $[0.99, 0.999]$ to obtain the momentum α. A high value of α means that a severe class-imbalance issue exists in the current batch and the model should update less. Finally the parameter of the teacher regression layer ω_{reg}^t can be updated by the following formula:

$$\omega_{reg}^t \leftarrow \alpha \omega_{reg}^t + (1 - \alpha) \omega_{reg}^s, \quad (5)$$

where ω_{reg}^s is the parameters of student regression layers.

3.4 Overall Objective and Test-Time Adaptation Strategy

Overall Objective. The overall objective to optimize the student regression layers can be formulated as follows:

$$\mathcal{L}_{total} = \mathcal{L}_{det} + \mathcal{L}_{refine}, \quad (6)$$

where $\mathcal{L}_{det}$ and $\mathcal{L}_{refine}$ are defined in Eq. (1) and Eq. (4).

Test-Time Adaptation Strategy. The test-time adaptation strategy of Reg-TTA3D can be summarized in Algorithm 1.

Algorithm 1. Single iteration of Reg-TTA3D strategy

Input: A batch of frames $\{x_i\}_{i=1}^n$ and $\{\dot{x}_i\}_{i=1}^n$, where $\dot{x}_i$ is obtained by performing data augmentation on x_i.
Output: Predictions of current batch B^t and $\{l_i^t\}_{i=1}^n$
1: Get the predictions $B^t = \{b_i^t \in \mathbb{R}^7\}_{i=1}^n$ and $\{l_i^t\}_{i=1}^n$ of $\{x_i\}_{i=1}^n$ using teacher model, B^t and $\{l_i^t\}_{i=1}^n$ are the detection results of current batch.
2: Use NPG to obtain boxes B^{npg} and $\{l_i^{npg}\}_{i=1}^n$ with stability under noise interference as pseudo-labels.
3: Get the predictions of student model B^s and $\{s_i^s\}_{i=1}^n$ of $\{\dot{x}_i\}_{i=1}^n$, supervise the student model by B^{npg} and $\{l_i^{npg}\}_{i=1}^n$ to get $\mathcal{L}_{det}$.
4: Calculate $\mathcal{L}_{refine}$ using CRR.
5: Optimize the student model using Eq. (6).
6: Update the teacher model using CBU and Eq. (5).
7: **Return:** Current predictions B^t and $\{l_i^t\}_{i=1}^n$.

---

4 Experiments

4.1 Experimental Setup

Datasets and Cross-Domain Settings. We conduct experiments on four commonly-used datasets: Waymo [23], KITTI [9], nuScenes [2] and nuScenes-C [11]. To verify the effectiveness of Reg-TTA3D, we consider three cross-domain settings including cross-location (*i.e.*, Waymo → KITTI), cross-beam (*i.e.*, Waymo → nuScenes, nuScenes → KITTI) and cross-weather (*i.e.*, nuScenes → nuScenes-C). Note that in the cross-weather setting, we use the medium-difficulty fog and snow scenarios in the nuScenes-C dataset. Following ReDB [6], in the first two cross-domain settings, we use the KITTI evaluation metric to evaluate for three common classes in autonomous driving scenarios (*i.e.*, car, pedestrian, and cyclist), we report the average precision in both 3D (*i.e.*, AP_{3D}) and BEV (*i.e.*, AP_{BEV}) over 40 recall positions, with IoU threshold 0.7 for car and 0.5 for pedestrian and cyclist. In the cross-weather setting, we use the official nuScenes evaluation metric and report mean average precision (mAP).

Implementation Details. Following [6], we evaluate the proposed Reg-TTA3D on SECOND-IoU [32], and the source model is trained for all categories using random object scaling (ROS) augmentation method. We set hyperparameter τ to 1.5 and p to 0.1 for all cross-domain settings and use Adam optimizer with learning rate of 1×10^{-3}. We utilize commonly-used data augmentation methods (*e.g.*, random world flip, random world rotation, random world scaling) to get $\{\dot{x}_i\}_{i=1}^n$. Our code is built on 3DTrans [25].

Comparison Baselines. To verify the effectiveness of Reg-TTA3D, we mainly compare our methods to unsupervised domain adaptive 3D object detection and previous test-time adaptation methods that we apply to 3D object detection. **UDA methods:** (1) ST3D [33] is a self-training method that continuously updates the pseudo-labels, (2) ST3D++ [34] provides more insight analysis and extends ST3D to get better results, (3) ReDB [6] is a pioneer to perform domain adaptive 3D detection on multiple categories. **TTA methods:** (1) TENT [27] proposes an entropy minimization method which focuses on adapting BN layers, (2) SHOT [14] optimizes feature extraction module using both information maximization and pseudo-labeling. (3) MATE [16] uses masked autoencoder to apply test-time adaptation to 3D classification.

4.2 Main Results

Results on Cross-Location and Cross-Beam Scenarios. We first evaluate Reg-TTA3D on cross-location and cross-beam scenarios and compare the results with previous UDA methods. As shown in Table @reftab:SOTAcomparison, Reg-TTA3D can greatly improve the detection accuracy on Waymo $\rightarrow$ KITTI and nuScenes $\rightarrow$ KITTI settings, largely narrow the performance gap between source only and oracle. For example, in the Waymo $\rightarrow$ KITTI setting, our Reg-TTA3D can achieve 61.78% / 49.96% in AP_{BEV} / AP_{3D}, even surpass UDA methods such as ST3D [33] and ST3D++ [34] and it is comparable to the state-of-the-art UDA method (*i.e.*, 61.14% / 50.10% in AP_{BEV} / AP_{3D} reported in ReDB [6]). In nuScenes $\rightarrow$ KITTI and Waymo $\rightarrow$ nuScenes cross-domain scenarios, the performance of Reg-TTA3D is also better than some UDA methods (*e.g.*, ST3D [33]) which needs source domain data and training for multiple epochs. In contrast, Reg-TTA3D only needs target data and one epoch of adaptation which greatly reduces the time consumption.

Further, we compare Reg-TTA3D with the previous TTA methods in Table @reftab:SOTAcomparison. We find that previous TTA methods suffer performance drops when applied to 3D detection tasks (*e.g.*, TENT [27] reduce the detection accuracy to 31.34% / 19.47% in the Waymo $\rightarrow$ KITTI setting). This is mainly because the detection task is composed of classification and regression and existing methods only consider the classification tasks. As a result, the model will concentrate on extracting effective representations for the classification tasks and ignore the localization information. However, our Reg-TTA3D fully considers the challenges in detection tasks and can improve the performance on multiple categories.

Results on the Cross-Weather Scenario. To more comprehensively evaluate the effectiveness of our proposed method, we conduct experiments in a more realistic scenario (*i.e.*, cross-weather). Since nuScenes-C dataset only has the validation set, we only compare Reg-TTA3D with existing TTA methods. It can be seen in Table @reftab:nusspsnusc that Reg-TTA3D can improve the performance on both snow and fog scenarios (*e.g.*, 45.68% $\rightarrow$ 47.09% mAP on the Normal $\rightarrow$ Snow setting). The results verify that our proposed method can handle dynamic environment changes in road scenes and can be applied in more practical scenarios.

4.3 Insight Analyses

Sensitivity to Detector Architecture. Following [6], to verify the effectiveness of Reg-TTA3D on different types of detectors, we further conduct experiments on nuScenes $\rightarrow$ KITTI cross-domain scenario using PointRCNN [20] as

Table 3. Performance comparisons (AP$_{3D}$) with PointRCNN as baseline detector on nuScenes → KITTI task. ‡ indicates the results reported in the original paper. Red: Best UDA method, **Blue** : Best TTA method.

Method	Setting	Car			Pedestrian			Cyclist			Average		
		Easy	Mod.	Hard	Easy	Mod.	Hard	Easy	Mod.	Hard	Easy	Mod.	Hard
Source Only	-	42.77	32.11	28.75	43.26	37.29	33.16	3.28	4.09	4.18	29.77	24.60	22.03
SN		66.56	50.32	45.92	42.96	37.15	32.45	9.07	7.57	7.42	39.53	31.68	28.60
ST3D		48.85	41.90	38.92	45.67	38.71	33.09	26.50	19.35	18.38	40.34	33.32	30.13
ST3D++	UDA	60.45	49.36	45.88	50.77	42.43	36.64	27.20	17.94	16.52	46.14	36.58	33.01
ReDB		71.45	57.9	53.91	52.32	44.33	37.95	45.13	32.93	31.05	56.30	45.05	40.97
SF-UDA^{3D} ‡	SFDA	68.80	49.80	45.00	-	-	-	-	-	-	-	-	-
TENT		42.55	32.41	27.61	30.71	26.09	23.78	4.24	3.66	3.32	25.83	20.70	18.23
SHOT	TTA	30.09	23.09	20.41	19.24	16.29	14.31	2.43	1.70	1.51	17.25	13.69	12.07
Reg-TTA3D		**65.20**	**50.87**	**46.72**	**45.80**	**42.55**	**34.90**	**40.25**	**27.87**	**25.29**	**50.41**	**40.43**	**35.63**

the baseline detector. As show in Table 3, our Reg-TTA3D outperforms all TTA methods and some UDA methods. Note that compared to the source-free domain adaptation method (*i.e.*, SF-UDA^{3D} [17]) which is inaccessible to the source data and need to adapt for multiple epochs, our Reg-TTA3D adapted on multi-class can achieve a better result on the car category (*e.g.*, 50.87% compared to 49.80% reported by SF-UDA^{3D} in moderate level) in one epoch of adaptation even if SF-UDA^{3D} is trained on the unfair single class. The results further show that our proposed methods can applied to different detector architectures.

Ablation Studies. To further verify the effectiveness of Reg-TTA3D, we conduct extensive experiments to investigate the impact of each component we proposed including NPG, CRR and CBU. As shown in Table 4, the improvement (*i.e.*, 0.92% AP$_{3D}$) of using naive pseudo-labels is relatively small compared to the performance of source only since pseudo-labels contain lots of noise samples and the regression results are not accurate due to domain discrepancies. Reg-TTA can effectively handle such a problem by proposing NPG and CRR modules, with 2.90% and 3.60% improvement respectively compared to source only. Besides, when combining these two modules, the performance can be further improved because NPG focuses on filter noise samples and CRR concentrates on making more accurate regression results. Further, by combining with CBU which makes the model update in a class-balance manner, the performance can be further boosted and achieve 49.96% AP$_{3D}$.

Sensitivity to Hyperparameters. We ablate the impact of hyperparameters τ (*i.e.*, the threshold of inconsistency rate) and p in this subsection. As shown in Table 5, we first set τ to 1.0, 1.5, 2.0 respectively and observe that the fluctuation of the performance is 1.34% / 1.08%. Then, by setting p to different values (*i.e.*, 0.05, 0.1, 0.15), we find that there is only little fluctuation and the best performance can achieved when set to 0.1. The results further show that our method is robust to these hyperparameters.

Table 4. Ablation studies on each component. PS, NPG, CRR and CBU represent using naive pseudo-labels, noise-consistency pseudo-label generation, confidence-guided regression refinement and class-balanced EMA updating, respectively. The experiments are conducted on Waymo → KITTI scenario using SECOND-IoU as the baseline detector.

PS	NPG	CRR	CBU	mean AP$_{BEV}$	mean AP$_{3D}$
-	-	-	-	57.89	42.01
✓	-	-	-	58.73	42.93
✓	✓	-	-	58.89	44.91
✓	-	✓	-	59.84	45.61
✓	-	-	✓	59.73	44.05
✓	✓	✓	-	60.57	47.97
✓	✓	-	✓	60.12	46.62
✓	-	✓	✓	60.59	48.39
✓	✓	✓	✓	**61.78**	**49.96**

Table 5. Ablation studies of hyperparameters τ and p. The experiments are conducted on Waymo → KITTI scenario using SECOND-IoU. Note that when ablate on τ, we set $p = 0.1$ and when ablate on p, we set $\tau = 0.15$.

τ	AP$_{BEV}$ / AP$_{3D}$
1.0	60.92 / 49.12
1.5	**61.78 / 49.96**
2.0	60.44 / 48.88
p	AP$_{BEV}$ / AP$_{3D}$
0.05	60.29 / 48.26
0.10	**61.78 / 49.96**
0.15	61.56 / 49.63

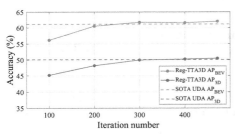

Fig. 4. The relationship between detection accuracy and the iteration number.

Table 6. Ablation studies on costs. The experiments are conducted on W → K scenario using SECOND-IoU. We use Tesla A100 to conduct the experiments.

Method	Time Cost	N_GPUs	Param.	Results
ST3D	4 h	4	12.12M	56.60 / 44.70
ST3D++	3.5 h	4	12.12M	59.42 / 47.97
ReDB	3.2 h	4	12.12M	61.14 / **50.10**
TENT	**4 min**	1	9.08K	31.34 / 19.47
MATE	6 min	1	0.71M	52.94 / 35.92
Reg-TTA3D	4.5 min	1	**2.73K**	**61.78** / 49.96

Analysis on Pseudo-labeling Accuracy. In this section, we measure the accuracy of each iteration during adaptation. Figure 4 reflects the detection accuracy on Waymo → KITTI as iteration increases. It can be seen that the accuracy continuously improves since the model is gradually adapted to the target domain which also verifies the stability of Reg-TTA3D during adaptation.

Analysis on Costs. Further, we report the time cost and total updated parameters in Table 6. Notably, Reg-TTA3D achieves comparable results to UDA methods adapting only less than 0.1% of parameters within less than 1% of time. When compared to TTA methods, our proposed method can achieve much higher performance with less updated parameters and comparable time cost which further verifies that our Reg-TTA3D is an effective and efficient method.

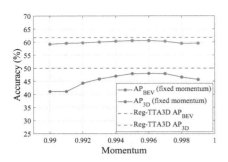 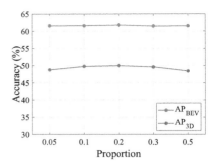

Fig. 5. The impact of momentum and comparisons with CBU.

Fig. 6. Ablation studies on the proportion of selected samples.

Ablation Studies of the Momentum. We ablate the momentum in EMA updating and compare it to our class-balance EMA updating strategy. Specifically, we conduct the experiment for momentum between 0.99 and 0.999 at 0.001 intervals. As shown in Fig. 5, the result demonstrates that CBU can better update the regression layer due to it can balance update the model and improve the performance on multiple classes at the same time.

Ablation Studies on the Proportion of Selected Samples in CRR. In our manuscript, we use 20% boxes with the highest scores to supervise box sizes with relatively low scores. Here, we analyze the impact on the proportion of selected samples. As shown in Fig. 6, when the proportion is relatively high, the average box size will be distributed far from the target domain, and when set to relatively small values, the small number of samples results in the averages being affected by individual instances. Based on the experiments and analysis, we set the proportion to 20%.

5 Conclusion

In this work, for the first time, we explore test-time adaptation in 3D object detection tasks and propose Reg-TTA3D, which handles rapid changes in the surrounding environment from the perspective of updating parameters selection and adaptation algorithm. By adapting using the proposed noise-consistency pseudo-label generation process, confidence-guided regression refinement loss and class-balanced EMA updating strategy and only updating regression layers, Reg-TTA3D can achieve comparable performance to some UDA methods at a low cost.

6 Limitation

Although the proposed Reg-TTA3D can handle dynamic changes in the surrounding environments such as continuously changing weather. It cannot deal with emergencies such as some corner cases which is also important in real-world autonomous driving.

Acknowledgement. This work is supported by National Key Research and Development Program of China (No. 2022ZD0160101), National Natural Science Foundation of China (No. 62071127, and 62101137), Shanghai Natural Science Foundation (No. 23ZR1402900), Shanghai Municipal Science and Technology Major Project (No. 2021SHZDZX0103). This work is also supported by the National Key R&D Program of China (Grant No. 2022ZD0160104), and Shanghai Rising Star Program (Grant No. 23QD1401000). The computations in this research were performed using the CFFF platform of Fudan University.

References

1. Boudiaf, M., Mueller, R., Ben Ayed, I., Bertinetto, L.: Parameter-free online test-time adaptation. In: Proceedings of the IEEE/CVF Conference on Computer Vision and Pattern Recognition, pp. 8344–8353 (2022)
2. Caesar, H., et al.: nuScenes: a multimodal dataset for autonomous driving. In: Proceedings of the IEEE/CVF Conference on Computer Vision and Pattern Recognition, pp. 11621–11631 (2020)
3. Cai, Q., Pan, Y., Ngo, C.W., Tian, X., Duan, L., Yao, T.: Exploring object relation in mean teacher for cross-domain detection. In: Proceedings of the IEEE/CVF Conference on Computer Vision and Pattern Recognition, pp. 11457–11466 (2019)
4. Chen, D., Wang, D., Darrell, T., Ebrahimi, S.: Contrastive test-time adaptation. In: Proceedings of the IEEE/CVF Conference on Computer Vision and Pattern Recognition, pp. 295–305 (2022)
5. Chen, Y., Li, W., Sakaridis, C., Dai, D., Van Gool, L.: Domain adaptive faster R-CNN for object detection in the wild. In: Proceedings of the IEEE Conference on Computer Vision and Pattern Recognition, pp. 3339–3348 (2018)
6. Chen, Z., Luo, Y., Huang, Z., Wang, Z., Baktashmotlagh, M.: Revisiting domain-adaptive 3D object detection by reliable, diverse and class-balanced pseudo-labeling. arXiv preprint arXiv:2307.07944 (2023)
7. Deng, J., Shi, S., Li, P., Zhou, W., Zhang, Y., Li, H.: Voxel R-CNN: towards high performance voxel-based 3D object detection. In: Proceedings of the AAAI Conference on Artificial Intelligence, pp. 1201–1209 (2021)
8. Ganin, Y., Lempitsky, V.: Unsupervised domain adaptation by backpropagation. In: International Conference on Machine Learning, pp. 1180–1189. PMLR (2015)
9. Geiger, A., Lenz, P., Urtasun, R.: Are we ready for autonomous driving? The KITTI vision benchmark suite. In: Proceedings of the IEEE/CVF Conference on Computer Vision and Pattern Recognition, pp. 3354–3361 (2012)
10. Goyal, S., Sun, M., Raghunathan, A., Kolter, J.Z.: Test time adaptation via conjugate pseudo-labels. In: Advances in Neural Information Processing Systems, vol. 35, pp. 6204–6218 (2022)
11. Kong, L., et al.: Robo3D: towards robust and reliable 3D perception against corruptions. In: Proceedings of the IEEE/CVF International Conference on Computer Vision (ICCV), pp. 19994–20006 (2023)
12. Lang, A.H., Vora, S., Caesar, H., Zhou, L., Yang, J., Beijbom, O.: PointPillars: fast encoders for object detection from point clouds. In: Proceedings of the IEEE/CVF Conference on Computer Vision and Pattern Recognition, pp. 12697–12705 (2019)

13. Liang, J., et al.: Pareto domain adaptation. In: Advances in Neural Information Processing Systems, vol. 34, pp. 12917–12929 (2021)
14. Liang, J., Hu, D., Feng, J.: Do we really need to access the source data? Source hypothesis transfer for unsupervised domain adaptation. In: International Conference on Machine Learning, pp. 6028–6039. PMLR (2020)
15. Luo, Z., et al.: Unsupervised domain adaptive 3D detection with multi-level consistency. In: Proceedings of the IEEE/CVF International Conference on Computer Vision, pp. 8866–8875 (2021)
16. Mirza, M.J., et al.: MATE: masked autoencoders are online 3d test-time learners. In: Proceedings of the IEEE/CVF International Conference on Computer Vision, pp. 16709–16718 (2023)
17. Saltori, C., Lathuiliére, S., Sebe, N., Ricci, E., Galasso, F.: SF-UDA 3D: source-free unsupervised domain adaptation for lidar-based 3D object detection. In: 2020 International Conference on 3D Vision (3DV), pp. 771–780. IEEE (2020)
18. Shi, S., et al.: PV-RCNN: point-voxel feature set abstraction for 3D object detection. In: Proceedings of the IEEE/CVF Conference on Computer Vision and Pattern Recognition, pp. 10529–10538 (2020)
19. Shi, S., et al.: PV-RCNN++: point-voxel feature set abstraction with local vector representation for 3D object detection. Int. J. Comput. Vision **131**(2), 531–551 (2023)
20. Shi, S., Wang, X., Li, H.: PointRCNN: 3D object proposal generation and detection from point cloud. In: Proceedings of the IEEE/CVF Conference on Computer Vision and Pattern Recognition, pp. 770–779 (2019)
21. Shi, S., Wang, Z., Shi, J., Wang, X., Li, H.: From points to parts: 3D object detection from point cloud with part-aware and part-aggregation network. IEEE Trans. Pattern Anal. Mach. Intell. **43**(8), 2647–2664 (2020)
22. Shin, I., et al.: MM-TTA: multi-modal test-time adaptation for 3D semantic segmentation. In: Proceedings of the IEEE/CVF Conference on Computer Vision and Pattern Recognition, pp. 16928–16937 (2022)
23. Sun, P., et al.: Scalability in perception for autonomous driving: Waymo open dataset. In: Proceedings of the IEEE/CVF Conference on Computer Vision and Pattern Recognition, pp. 2446–2454 (2020)
24. Sun, Y., Wang, X., Liu, Z., Miller, J., Efros, A., Hardt, M.: Test-time training with self-supervision for generalization under distribution shifts. In: International Conference on Machine Learning, pp. 9229–9248. PMLR (2020)
25. Team, D.D.: 3DTrans: an open-source codebase for exploring transferable autonomous driving perception task (2023). https://github.com/PJLab-ADG/3DTrans
26. Vs, V., Oza, P., Patel, V.M.: Towards online domain adaptive object detection. In: Proceedings of the IEEE/CVF Winter Conference on Applications of Computer Vision, pp. 478–488 (2023)
27. Wang, D., Shelhamer, E., Liu, S., Olshausen, B., Darrell, T.: Tent: fully test-time adaptation by entropy minimization. arXiv preprint arXiv:2006.10726 (2020)
28. Wang, Q., Fink, O., Van Gool, L., Dai, D.: Continual test-time domain adaptation. In: Proceedings of the IEEE/CVF Conference on Computer Vision and Pattern Recognition, pp. 7201–7211 (2022)
29. Wang, Y., et al.: Train in Germany, test in the USA: making 3D object detectors generalize. In: Proceedings of the IEEE/CVF Conference on Computer Vision and Pattern Recognition, pp. 11713–11723 (2020)

30. Wei, Y., Wei, Z., Rao, Y., Li, J., Zhou, J., Lu, J.: LiDAR distillation: bridging the beam-induced domain gap for 3d object detection. arXiv preprint arXiv:2203.14956 (2022)
31. Xu, Q., Zhou, Y., Wang, W., Qi, C.R., Anguelov, D.: SPG: unsupervised domain adaptation for 3D object detection via semantic point generation. In: Proceedings of the IEEE/CVF International Conference on Computer Vision, pp. 15446–15456 (2021)
32. Yan, Y., Mao, Y., Li, B.: SECOND: sparsely embedded convolutional detection. Sensors **18**(10), 3337 (2018)
33. Yang, J., Shi, S., Wang, Z., Li, H., Qi, X.: ST3D: self-training for unsupervised domain adaptation on 3D object detection. In: Proceedings of the IEEE/CVF Conference on Computer Vision and Pattern Recognition, pp. 10368–10378 (2021)
34. Yang, J., Shi, S., Wang, Z., Li, H., Qi, X.: ST3D++: denoised self-training for unsupervised domain adaptation on 3D object detection. IEEE Trans. Pattern Anal. Mach. Intell. **45**(5), 6354–6371 (2022)
35. Yang, Z., Sun, Y., Liu, S., Shen, X., Jia, J.: STD: sparse-to-dense 3D object detector for point cloud. In: Proceedings of the IEEE/CVF International Conference on Computer Vision, pp. 1951–1960 (2019)
36. Yin, T., Zhou, X., Krahenbuhl, P.: Center-based 3D object detection and tracking. In: Proceedings of the IEEE/CVF Conference on Computer Vision and Pattern Recognition, pp. 11784–11793 (2021)
37. Yuan, J., et al.: AD-PT: autonomous driving pre-training with large-scale point cloud dataset. arXiv preprint arXiv:2306.00612 (2023)
38. Yuan, J., et al.: BI3D: bi-domain active learning for cross-domain 3D object detection. arXiv preprint arXiv:2303.05886 (2023)
39. Yue, X., Zhang, Y., Zhao, S., Sangiovanni-Vincentelli, A., Keutzer, K., Gong, B.: Domain randomization and pyramid consistency: simulation-to-real generalization without accessing target domain data. In: Proceedings of the IEEE/CVF International Conference on Computer Vision, pp. 2100–2110 (2019)
40. Yue, X., et al.: Prototypical cross-domain self-supervised learning for few-shot unsupervised domain adaptation. In: Proceedings of the IEEE/CVF Conference on Computer Vision and Pattern Recognition, pp. 13834–13844 (2021)
41. Zhang, B., Chen, T., Wang, B., Li, R.: Joint distribution alignment via adversarial learning for domain adaptive object detection. IEEE Trans. Multimedia (2021)
42. Zhang, B., Yuan, J., Shi, B., Chen, T., Li, Y., Qiao, Y.: Uni3D: a unified baseline for multi-dataset 3D object detection. arXiv preprint arXiv:2303.06880 (2023)
43. Zhang, Y., Hu, Q., Xu, G., Ma, Y., Wan, J., Guo, Y.: Not all points are equal: learning highly efficient point-based detectors for 3D LiDAR point clouds. In: Proceedings of the IEEE/CVF Conference on Computer Vision and Pattern Recognition, pp. 18953–18962 (2022)
44. Zhao, S., et al.: Multi-source domain adaptation for semantic segmentation. In: Advances in Neural Information Processing Systems, vol. 32 (2019)
45. Zheng, Y., Huang, D., Liu, S., Wang, Y.: Cross-domain object detection through coarse-to-fine feature adaptation. In: Proceedings of the IEEE/CVF Conference on Computer Vision and Pattern Recognition, pp. 13766–13775 (2020)
46. Zhou, Y., Tuzel, O.: VoxelNet: end-to-end learning for point cloud based 3D object detection. In: Proceedings of the IEEE Conference on Computer Vision and Pattern Recognition, pp. 4490–4499 (2018)

ShapeLLM: Universal 3D Object Understanding for Embodied Interaction

Zekun Qi[1,2], Runpei Dong[1,2], Shaochen Zhang[1], Haoran Geng[3], Chunrui Han[4], Zheng Ge[4], Li Yi[5,6,7(✉)], and Kaisheng Ma[5(✉)]

[1] Xi'an Jiaotong University, Xi'an, China
[2] IIISCT, Xi'an, China
[3] Peking University, Beijing, China
[4] MEGVII, Beijing, China
[5] IIIS, Tsinghua University, Beijing, China
ericyi0124@gmail.com
[6] Shanghai AI Laboratory, Shanghai, China
[7] Shanghai Qi Zhi Institute, Shanghai, China
https://qizekun.github.io/shapellm/

Abstract. This paper presents SHAPELLM, the first 3D Multimodal Large Language Model (LLM) designed for embodied interaction, exploring a universal 3D object understanding with 3D point clouds and languages. SHAPELLM is built upon an improved 3D encoder by extending SHAPE RECON [101] to RECON++ that benefits from multi-view image distillation for enhanced geometry understanding. By utilizing RECON++ as the 3D point cloud input encoder for LLMs, SHAPELLM is trained on constructed instruction-following data and tested on our newly human-curated benchmark, 3D MM-Vet. RECON++ and SHAPELLM achieve state-of-the-art performance in 3D geometry understanding and language-unified 3D interaction tasks, such as embodied visual grounding.

Keywords: 3D Point Clouds · Large Language Models · Embodied Intelligence · 3D Representation Learning · Zero-shot Learning

1 Introduction

3D shape understanding, serving as a fundamental capability for molding intelligent systems in both digital and physical worlds, has witnessed tremendous progress in graphics, vision, augmented reality, and embodied robotics. However,

---

R. Dong—Project lead.
Work done during Z. Qi and R. Dong's internships at MEGVII & IIISCT.

Supplementary Information The online version contains supplementary material available at https://doi.org/10.1007/978-3-031-72775-7_13.

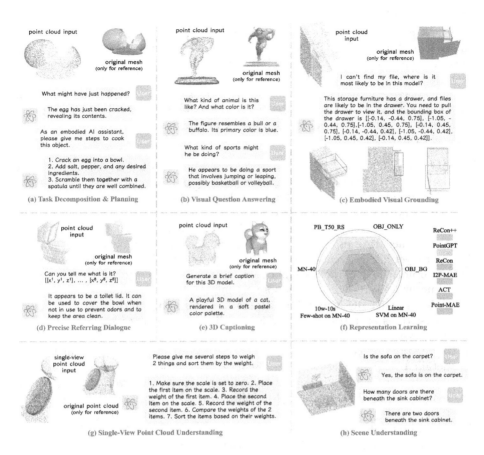

Fig. 1. Demonstrations of ShapeLLM and ReCon++. We present SHAPELLM, the first 3D LLM designed for embodied interaction and spatial intelligence.

to be effectively deployed by real-world agents, several critical criteria must be fulfilled: **(i)** Sufficient 3D *geometry* information needs to be captured for accurate spatial and structure processing [7,10,61,98]. **(ii)** Models should be endowed with a foundational knowledge of the *embodied interaction* fashion with objects - often physically - for functional comprehension [40,49–51,62,97,155,156]. **(iii)** A *universal interface* is required as a bridge between information encoding and decoding, which could help translate high-order instructions for agent reactions like dialogue response and embodied feedback [23,56,157].

Recent advancements in Large Language Models (LLMs) [8,88,106,107,121] have demonstrated unprecedented success of foundational knowledge and unified reasoning capabilities across tasks [5,18,24,27,29,54,57,60,96]. It makes it possible to utilize language as a *universal interface* that enables the comprehensive *commonsense knowledge* embedded in LLMs to enhance understanding of

3D shapes. This is particularly evident in *physically-grounded* tasks, where the wealth of commonsense knowledge simplifies the interpretation of an object's functionality, mobility, and dynamics, *etc.* However, the aforementioned challenges remain when incorporating LLMs for 3D object understanding—especially *embodied interaction* that relies on precise *geometry*—currently under-explored.

The question is: *What makes better 3D representations that bridge language models and interaction-oriented 3D object understanding?* In this work, we introduce SHAPELLM that meets the requirements, which is established based on the following three designing policies:

i. **3D Point Clouds as Inputs.** Some concurrent works [41] recently propose to use point cloud-rendered images [149] as multimodal LLMs' inputs and demonstrate effectiveness. However, these works fail to achieve accurate 3D geometry understanding and often suffer from a well-known visual hallucination issue [69,109,159]. Compared to 2D images, 3D point clouds provide a more accurate representation of the physical environment, encapsulating sparse yet highly precise geometric data [1,28,99]. Moreover, 3D point clouds are crucial in facilitating embodied interactions necessitating accurate 3D structures like 6-DoF object pose estimation [67,123,125,129,134].

ii. **Selective Multi-View Distillation.** Interacting with objects typically necessitates an intricate 3D understanding that involves knowledge at various levels and granularities. For instance, a whole-part *high-level* semantic understanding is needed for interactions like opening a large cabinet, while detailed, *high-resolution* (*i.e.*, *low-level*) semantics are crucial for smaller objects like manipulating a drawer handle [142]. However, existing works mainly distill single-view high-resolution object features from 2D foundation models [105], providing a complementary understanding [28,101,136]. The potential of multi-view images, which offer abundant multi-level features due to view variation and geometry consistency [6,45,47,61,80,114], is often neglected. SHAPELLM extends SHAPE RECON [101] to RECON++ as the 3D encoder by integrating multi-view distillation. To enable the model to selectively distill views that enhance optimization and generalization, inspired by DETR [9], RECON++ is optimized through adaptive selective matching using the Hungarian algorithm [64].

iii. **3D Visual Instruction Tuning.** Instruction tuning has been proven effective in improving LLMs' alignment capability [90,93]. To realize various 3D understanding tasks with a universal language interface, SHAPELLM is trained through instruction-following tuning on constructed language-output data. However, similar to 2D visual instruction tuning [3,73], the data-desert issue [28] is even worse since no object-level VQA data is available, unlike 2D [72]. To validate the efficacy of SHAPELLM, we first construct ~45K instruction-following data using the advanced GPT-4V(ision) [89] on the processed Objaverse dataset [25] and 30K embodied part understanding data from GAPartNet [36] for supervised fine-tuning. Following MM-Vet [144], we further develop a novel evaluation benchmark named 3D MM-Vet. This benchmark is designed to assess the core

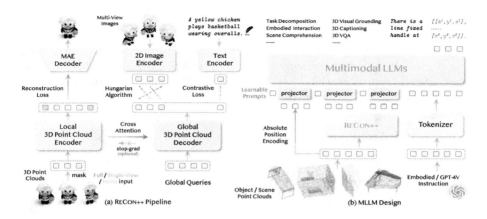

Fig. 2. Overview of our ShapeLLM framework. (a) The introduced RECON++ pipeline incorporates the required 3D encoder. (b) The comprehensive design of the MLLM, featuring an instruction-mode tokenizer and the integration of an aligned multimodal representation, equips the MLLM with the capability to effectively handle 3D vision language tasks.

vision-language capabilities, including embodied interaction in a 3D context, thereby stimulating future research. The 3D MM-Vet benchmark comprises 59 diverse Internet[1] 3D objects and 232 human-written question-answer pairs (Fig. 2).

Through extensive experimentation, we first demonstrate that our improved 3D encoder RECON++ sets a new state-of-the-art representation transferring on both downstream fine-tuned and zero-shot 3D object recognition. Specifically, RECON++ has obtained **95.25%** and **95.0%** fine-tuned accuracy on ScanObjectNN and ModelNet40, surpassing previous best records by **+1.85%** on the most challenging ScanObjectNN. Besides, RECON++ achieved **53.7%** and **65.4%** zero-shot accuracy on Objaverse-LVIS and ScanObjectNN, which is **+0.6%** and **+1.6%** higher than previous best. By utilizing our RECON++ as SHAPELLM's 3D encoder, SHAPELLM successfully unifies various downstream tasks, including *3D captioning, 3D VQA, embodied task planning & decomposition, 3D embodied visual grounding*, and *3D precise referring dialogue* (See Fig. 1). On our newly constructed 3D MM-Vet benchmark, **42.7%** and **49.3%** Total accuracy have been achieved by SHAPELLM-7B and SHAPELLM-13B, surpassing previous best records [133] that also uses 3D point clouds by **+2.1%** and **+5.1%**, respectively. This work initiates a first step towards leveraging LLMs for embodied object interaction, and we hope our SHAPELLM and proposed 3D MM-Vet benchmark could spur more related future research.

---

[1] URL & License.

2 ShapeLLM

In this section, we first introduce the overall architecture of SHAPELLM. Then, we delve into two critical challenges faced in interactive 3D understanding: data desert [28] and representation of 3D point clouds. We present the detailed design of our method to tackle these challenges, respectively.

2.1 Overall Architecture

The main objective of this work is interactive 3D understanding by using the LLM as a universal interface. Drawing inspiration from recent work in visual understanding [73], the proposed SHAPELLM consists a pre-trained 3D encoder and an LLM for effective 3D representation learning and understanding, respectively. Specifically, we adopt LLaMA [121] as our LLM, building upon the success of previous work [22,27,73]. As for the 3D encoder, we propose a novel 3D model named RECON++ based on the recent work SHAPE RECON [101] with multiple improvements as the 3D understanding generally demands more information, such as accurate spatial and multi-view details, etc. To ensure compatibility with the LLM inputs, the representation of a 3D object obtained from RECON++ undergoes a linear projection before being fed into the LLM. To further improve low-level geometry understanding, which benefits tasks like 6-DoF pose estimation, we append the absolute position encoding (APE) obtained by

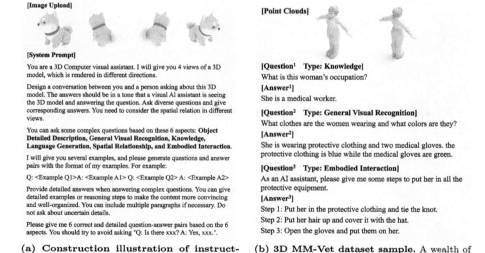

(a) **Construction illustration of instruct-following data using GPT-4V [89].** Four perspective views are input into GPT-4V. In-context prompts focusing on different topics are explicitly incorporated to ensure data diversity.

(b) **3D MM-Vet dataset sample.** A wealth of precise evaluation metrics enable a comprehensive assessment.

Fig. 3. **Qualitative visualization** of the instruction-following and 3D MM-Vet data.

linear projection of 3D coordinates. Besides, we use prefix-tuning with learnable prompts [27,28,59,66] to adaptively modulate the different semantics of APE and RECON++ representations.

2.2 How to Alleviate Interactive 3D Understanding *Data Desert*?

Most published 3D data is typically presented as 3D object-caption pairs, lacking an interactive style. Although a few concurrent works [46,133] have attempted to construct interactive 3D understanding datasets, the questions-and-answers (Q&As) are primarily based on annotated captions, often providing a limited perspective without sufficient details. Additionally, those works have generally been limited to semantic understanding without considering embodied interaction. To address these limitations, our work constructs question-and-answer pairs based on multi-view images of a 3D object using GPT-4V(ision) [89]. For data diversity, we explicitly introduce six aspects as prompts, as illustrated Fig. 3a. In the following, we provide the details about data collection and construction regarding *general semantic understanding* and *embodied object understanding*, respectively.

Data. Objaverse-LVIS [25,83] and GAPartNet [36] are data sources. Objaverse-LVIS covers 1,156 LVIS [42] categories, and we sample Top-10 "likes"[2] 3D objects per category and generate Q&A pairs per sample. After filtering out noisy Q&As, we obtain ~45K instruction-following samples. We use 12 categories from GAPartNet by removing "Remote" to avoid too many tiny boxes, which leads to filtered ~30K Q&A samples constructed from ~8K parts of the ~4K objects states covering ~1.1K different objects.

General Semantic Understanding. This aims to enhance the model's generalization abilities in visual recognition, knowledge integration, spatial understanding, and other aspects. We prompt GPT4-V to generate Q&As in six different aspects based on images captured from four different views, as illustrated in Fig. 3a.

Embodied Object Understanding. A comprehensive understanding of the spatial positions and semantics at the part level is crucial to facilitate effective object grasping and interaction in embodied scenarios. Fortunately, the GAPartNet [36] provides rich part annotations, including semantics and poses, which are instrumental in constructing instruction-tuning data for embodied interactive parts of a subject. Specifically, given a 3D object, questions are formulated based on the semantics of its different parts, and answers are constructed in both the semantics and 3D positions. The positions are represented as 6-DoF 3D bounding boxes in a straightened Python multidimensional list format, denoted as $[[x_1, y_1, z_1], [x_2, y_2, z_2], \ldots, [x_8, y_8, z_8]]$, to meet characteristics of the textual dialogues response in LLMs. The canonical space of the object determines the sequence of coordinates. Using bounding box coordinates leverages the inherent spatial relationship, allowing LLMs to readily learn these patterns and generate

---

[2] "Likes" statistics can be found at Sketchfab.

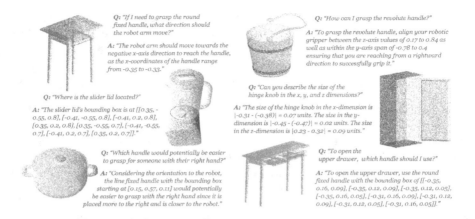

Fig. 4. Qualitative examples of the embodied interaction data.

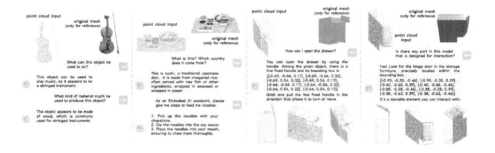

Fig. 5. Selected multimodal dialogue examples. SHAPELLM possesses robust capabilities in knowledge representation, reasoning, and instruction-following dialogue. With its powerful point cloud encoder RECON++, SHAPELLM can even make accurate predictions about minute interactive components, *e.g.*, handle.

accurate output coordinates. This approach can offer specific position information for embodied manipulation, as illustrated in Fig. 4.

2.3 ReCon++: Scaling Up 3D Representation Learning

Interaction with objects such as object grasping [76,123,134] typically requires accurate perception of 3D shape information at multi-level and multi-granularity. This imposes heightened requirements on 3D representations, calling for a higher standard of a holistic understanding of 3D geometry.

However, existing 3D cross-modal representation learning methods [74,137] mainly distill high-resolution object features from single-view 2D foundation models, resulting in a unilateral shape understanding. Besides, they generally employ multi-view images as data augmentation, imposing the learned representation to the average representation of all views. Thus, the accurate 3D shape information is missing. Recently, SHAPE RECON [101] utilizes contrast guided by reconstruction to address the pattern disparities between local masked data modeling and global cross-modal alignment. This results in remarkable performance in various tasks, including transfer learning, zero-shot classification, and part segmentation. However, its potential is hindered by the scarcity of pretraining data [10].

To address the above limitations, this paper proposes RECON++ with multiple improvements. First, multi-view image query tokens collaboratively comprehend the semantic information of 3D objects across different views, encompassing both RGB images and depth maps. Considering the disorderliness of pretraining data in terms of pose, we propose a cross-modal alignment method based on *bipartite matching*, which implicitly learns the pose estimation of 3D objects. Second, we *scale up* the parameters of SHAPE RECON and broaden the scale of the pretraining dataset [15, 25, 83] for robust 3D representations.

Denote N as the number of multi-view images, I_i is the image feature from i-th view, and Q_i represents the global query of i-th view. Following DETR [9], we search for an optimal permutation σ of N elements with the lowest cost:

$$\hat{\sigma} = \arg\min_{\sigma} \sum_{i}^{N} \mathcal{L}_{\text{match}}(I_i, Q_{\sigma(i)}), \tag{1}$$

where $\mathcal{L}_{\text{match}}(I_i, Q_{\sigma(i)})$ is a pair-wise matching cost between i-th view image features I_i and matched query $Q_{\sigma(i)}$ with the permutation σ. In practice, we employ cosine similarity as the matching cost. In this fashion, the query of each view is learned to gather accurate 3D shape information from the 3D point clouds. Concatenating the features from the local 3D point cloud encoder and global 3D point cloud decoder together provides comprehensive information for 3D understanding of multimodal LLMs.

3 3D MM-Vet: Benchmarking 3D Comprehension

A wide range of diverse visual-language capabilities is essential to develop a multimodal large language model tailored for embodied scenarios, particularly addressing task and action planning.

The model's proficiency in processing point clouds enables it to perform general recognition tasks effortlessly, demonstrating a broad understanding of colored point clouds. This capability serves as the groundwork for more intricate tasks. Beyond 3D recognition, the LLM should exhibit competence in addressing tasks in real-world embodied scenarios. This entails unifying the aforementioned abilities to generate decomposed task actions step-by-step in an instruction-following fashion, addressing specific problems.

Hence, to formulate an evaluation system aligned with the aforementioned task description, we establish a multi-level evaluation task system encompassing four-level tasks: **General Recognition, Knowledge and Language Generation, Spatial Awareness**, and **Embodied Interaction**. This framework

Table 1. Fine-tuned 3D recognition on ScanObjectNN and ModelNet40. Overall accuracy (%) with voting [78] is reported. †: Results with a post-pretraining stage [15].

Method	ScanObjectNN			ModelNet40	
	OBJ_BG	OBJ_ONLY	PB_T50_RS	1k P	8k P
Supervised Learning Only					
PointNet [98]	73.3	79.2	68.0	89.2	90.8
PointNet++ [98]	82.3	84.3	77.9	90.7	91.9
DGCNN [126]	82.8	86.2	78.1	92.9	-
PointMLP [85]	-	-	85.4	94.5	-
PointNeXt [104]	-	-	87.7	94.0	-
with Self-Supervised Representation Learning					
Point-BERT [145]	87.43	88.12	83.07	93.2	93.8
Point-MAE [92]	90.02	88.29	85.18	93.8	94.0
Point-M2AE [148]	91.22	88.81	86.43	94.0	-
Point2Vec [146]	91.2	90.4	87.5	94.8	-
ACT [28]	93.29	91.91	88.21	93.7	94.0
TAP [127]	-	-	88.5	94.0	-
VPP [102]	93.11	91.91	89.28	94.1	94.3
I2P-MAE [151]	94.15	91.57	90.11	94.1	-
ULIP-2 [137]	-	-	91.5	-	-
SHAPE RECON [101]	95.35	93.80	91.26	94.5	94.7
PointGPT-B† [15]	95.8	95.2	91.9	94.4	94.6
PointGPT-L† [15]	97.2	96.6	93.4	94.7	94.9
ReCon++-B†	**98.62**	**96.21**	**93.34**	**94.6**	**94.8**
ReCon++-L†	**98.80**	**97.59**	**95.25**	**94.8**	**95.0**

systematically and comprehensively assesses the model's proficiency in information comprehension and language generation when processing interactive objects. The detailed descriptions of the tasks are listed as follows:

i. **General Recognition:** Following MM-Vet [144], we assess the fundamental comprehension abilities of LLMs involving both coarse- and fine-grained aspects. Coarse-grained recognition focuses on basic object attributes such as color, shape, action, *etc.* While fine-grained recognition delves into details like subparts and counting, *etc.*

ii. **Knowledge Capability & Language Generation:** To examine the models' capacity to understand and utilize knowledge, drawing inspiration from MMBench [79], we integrate its reasoning components. This includes knowl-

Table 2. Zero-shot 3D recognition on Objaverse-LVIS [25], ModelNet40 [132] and ScanObjectNN [122]. Ensembled [74]: pretraining with four datasets, Objaverse [25], ShapeNet [10], ABO [20] and 3D-FUTURE [31]. †: Uni3D employs a larger EVA-CLIP-E [117] teacher, while other methods employ OpenCLIP-bigG [58].

Method	Objaverse-LVIS			ModelNet40			ScanObjectNN		
	Top1	Top3	Top5	Top1	Top3	Top5	Top1	Top3	Top5
2D Inference without 3D Training									
PointCLIP [149]	1.9	4.1	5.8	19.3	28.6	34.8	10.5	20.8	30.6
PointCLIPv2 [161]	4.7	9.5	12.9	63.6	77.9	85.0	42.2	63.3	74.5
Trained on ShapeNet									
ReCon [101]	1.1	2.7	3.7	61.2	73.9	78.1	42.3	62.5	75.6
CLIP2Point [53]	2.7	5.8	7.9	49.5	71.3	81.2	25.5	44.6	59.4
ULIP [136]	6.2	13.6	17.9	60.4	79.0	84.4	51.5	71.1	80.2
OpenShape [74]	10.8	20.2	25.0	70.3	86.9	91.3	47.2	72.4	84.7
TAMM [153]	13.7	24.2	29.2	73.1	88.5	91.9	54.8	74.5	83.3
MixCon3D [32]	22.3	37.5	44.3	72.6	87.1	91.3	52.6	69.9	78.7
Trained on Ensembled									
ULIP-2 [137]	26.8	44.8	52.6	75.1	88.1	93.2	51.6	72.5	82.3
OpenShape [74]	46.8	69.1	77.0	84.4	96.5	98.0	52.2	79.7	88.7
TAMM [153]	50.7	73.2	80.6	85.0	96.6	98.1	55.7	80.7	88.9
MixCon3D [32]	52.5	74.5	81.2	**86.8**	**96.9**	**98.3**	58.6	80.3	89.2
Uni3D-B† [158]	51.7	74.1	80.8	86.3	96.5	97.9	**63.8**	**82.7**	90.2
Uni3D-L† [158]	53.1	75.0	81.5	86.3	**96.8**	**98.3**	58.2	81.8	89.4
ReCon++-B	**53.2**	**75.3**	**81.5**	86.5	94.7	95.8	63.6	80.2	**90.6**
ReCon++-L	**53.7**	**75.8**	**82.0**	87.3	95.4	96.1	**65.4**	**84.1**	**89.7**

edge spanning natural and social reasoning, physical properties, sequential prediction, math, *etc.*, evaluating gauges whether multimodal LLMs possess the requisite expertise and capacity to solve intricate tasks. We utilize customized prompts to stimulate models and extract detailed responses to evaluate language generation.

iii. **Spatial Awareness:** In 3D, spatial awareness holds heightened significance compared to 2D due to the provided geometry information. The point clouds contain location information crucial for discerning spatial relationships between different parts. In 2D, achieving the same information intensity level would necessitate multi-view images. Therefore, our evaluation includes questions probing the ability of LLMs to understand spatial relations.

iv. **Embodied Interaction:** The utilization scope of MLLMs extends into the field of embodied interaction, facilitated by the utilization of instruction-following data. Our evaluation system tests their capacity by formally requesting LLMs to provide execution steps toward an instruction. This approach aims to establish connections for handling Embodied Interaction tasks [29,54].

To prevent any overlap with training data, our collection of 3D models is sourced exclusively from Turbosquid [113], a platform not included in the acquisition lists of Objaverse [25] and ShapeNet [10]. We meticulously curated a dataset of 59 3D models, generating 232 Q&As for evaluation purposes. In our pursuit of a precise assessment of single-task capabilities, each question is designed to test only one specific capacity outlined earlier. Every question is paired with a corresponding answer tailored to the particular 3D model, serving as the ground truth. Dataset samples are illustrated in Fig. 3b. More details and analysis can be found in the supplemental material.

4 Experiments

4.1 3D Representation Transferring with ReCon++

Fine-Tuned 3D Object Recognition. In Table 1, we first evaluate the representation transfer learning capabilities of self-supervised RECON++ by fine-tuning on ScanObjectNN [122] and ModelNet [132], which are currently the two most challenging 3D object datasets. ScanObjectNN is a collection of ~15K 3D object point clouds from the real-world scene dataset ScanNet [21], which involves 15 categories. ModelNet is one of the most classical 3D object datasets collected from clean 3D CAD models, which includes ~12K meshed 3D CAD models covering 40 categories. Following PointGPT [15], we adopt the intermediate fine-tuning strategy and use the post-pretraining stage to transfer the general semantics learned through self-supervised pretraining on ShapeNetCore [10]. For a fair comparison, our Base and Large models adopt the same architecture as PointGPT regarding layers, hidden size, and attention heads. Table 1 shows that: (i) RECON++ exhibits representation performance significantly surpassing that of other baselines, achieving state-of-the-art results. (ii) Particularly, RECON++ achieves a remarkable accuracy of 95.25% on the most challenging ScanObjectNN PB_T50_RS benchmark, boosting the Transformer baseline by +16.14%.

Zero-Shot 3D Open-World Recognition. Similar to CLIP [105], our model aligns the feature space of languages and other modalities, which results in a zero-shot open-world recognition capability. In Table 2, we compare the zero-shot 3D open-world object recognition models to evaluate the generalizable recognition capability. Following OpenShape [74], we evaluate on ModelNet [132], ScanObjectNN [122], and Objaverse-LVIS [25]. Objaverse-LVIS is a benchmark involving ~47K clean 3D models of 1,156 LVIS categories [42]. We compare RECON++

Table 3. Zero-shot 3D multimodal comprehension of *core VL capabilities in 3D context* on 3D MM-Vet. Rec: General Visual Recognition, Know: Knowledge, Gen: Language Generation, Spat: Spatial Awareness, Emb: Embodied Interaction.

Method	Input	Rec	Know	Gen	Spat	Emb	Total
LLaVA-13B [73]	1-View 2D Image	40.0	55.3	51.3	43.2	51.1	47.9
DreamLLM-7B [27]	4-View 2D Image	42.2	54.4	50.8	48.9	54.5	50.3
GPT-4V [89]	1-View 2D Image	53.7	59.5	61.1	54.7	59.0	57.4
GPT-4V [89]	4-View 2D Image	65.1	69.1	61.4	52.9	65.5	63.4
PointBind&LLM [41]	3D Point Cloud	16.9	13.0	18.5	32.9	40.4	23.5
PointLLM-7B [133]	3D Point Cloud	40.6	49.5	34.3	29.1	48.7	41.2
PointLLM-13B [133]	3D Point Cloud	46.6	48.3	38.8	45.2	50.9	46.6
ShapeLLM-7B	3D Point Cloud	**45.7**	**42.7**	**43.4**	**39.9**	**64.5**	**47.4**
ShapeLLM-13B	3D Point Cloud	**46.8**	**53.0**	**53.9**	**45.3**	**68.4**	**53.1**

with 2D inference methods, ShapeNet pretrained methods, and "Ensembled" datasets-pretrained methods. It can be concluded from Table 2: i) Compared to 2D inference and ShapeNet-pretrained methods, RECON++ demonstrates significantly superior performance, showing the necessity of *3D point clouds as inputs* and *scaling up*. ii) Compared to state-of-the-art methods trained on "Ensembled" datasets, RECON++ demonstrates superior or on-par performance across all benchmarks. Notably, RECON++-L achieves a remarkable Top-1 accuracy, which is +0.6% and +7.2% higher than Uni3D-L on the most challenging Objaverse-LVIS and ScanObjectNN benchmarks, respectively.

4.2 Multimodal Comprehension with ShapeLLM

Quantitative Analysis. To assess the comprehensive capabilities of SHAPELLM, we first quantitatively compare various baselines and our model on the proposed 3D MM-Vet using GPT-4. Following ModelNet-C [108] and ModelNet40-C [115], we construct 3D MM-Vet-C to benchmark the robustness against 3D corruptions.

i) 3D MM-Vet. Table 3 shows the detailed results of SHAPELLM on different tasks of 3D MM-Vet. It is observed that SHAPELLM significantly outperforms PointLLM [133] across various metrics, particularly in Embodied Tasks. This substantiates our model's versatile capability in addressing real-world tasks.

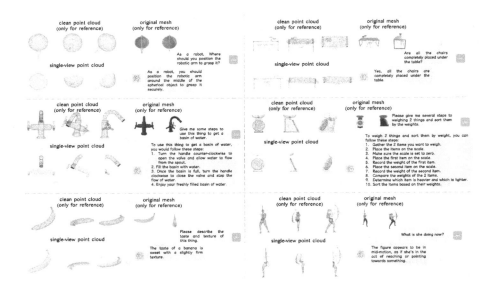

Fig. 6. 3D multimodal dialogue using *single-view* point cloud inputs. All answers are generated by SHAPELLM-13B with single-view occluded inputs. SHAPELLM achieves outstanding robustness against such commonly met occlusion in the real world.

ii) 3D MM-Vet-C. Following the ModelNet-C [108] and ModelNet40-C [115], we construct 3D MM-Vet-C to benchmark the robustness against 3D corruptions. Table 4 compares robustness against "single-view", "jitter", and "rotate" corruptions, which are the most common in real scenarios. The "single-view" issue is the most critical challenge since obtaining the complete point clouds is non-trivial, similar to multi-view images. Therefore, everyday real-world robots only get single-view 3D perceptions with sensors such as RGB-D [43]. The results demonstrate significantly superior robustness of SHAPELLM, indicating stronger potential in real-world applicability.

Table 4. Zero-shot 3D multimodal comprehension of *robustness* on 3D MM-Vet-C. `Clean`: no corruptions. `Single-View`: randomly select a camera viewpoint within the unit sphere and generate a **single viewpoint** within the FoV on polar coordinates. `Jitter`: Gaussian jittering with noise $\epsilon \sim \mathcal{N}(0, \sigma^2)$ and $\sigma = 0.01$. `Rotate`: random SO(3) rotation sampling over X-Y-Z Euler angle $(\alpha, \beta, \gamma) \sim \mathcal{U}(-\theta, \theta)$ and $\theta = \pi/6$.

Method	3D MM-Vet-C Variants			
	Clean	Single-View	Jitter	Rotate
PointBind&LLM [41]	23.5	20.4	19.7	19.5
PointLLM-7B [133]	41.2	33.6	38.8	40.6
PointLLM-13B [133]	46.6	41.3	42.3	44.2
ShapeLLM-7B	**47.4**	**38.3**	**45.8**	**42.7**
ShapeLLM-13B	**53.1**	**43.6**	**47.8**	**49.3**

Table 5. **3D referring expression grounding** on GAPartNet [36]. Accuracy with an IoU threshold of 0.25 is reported. †: Fine-tuned on GAPartNet images. ‡: Inference with 3 in-context demonstrations.

Method	Input	🗄	🔧	📦	⏲	🗃	🖥	Avg
LLaVA-13B [73]	1-View 2D Image	0.0	0.0	0.0	0.0	0.0	0.0	0.0
LLaVA-13B [73]	4-View 2D Image	0.0	0.0	0.0	0.0	0.0	0.0	0.0
LLaVA-13B† [73]	1-View 2D Image	1.8	9.3	3.8	0.0	2.1	11.1	4.4
LLaVA-13B† [73]	4-View 2D Image	2.5	13.7	7.7	0.0	4.3	11.1	6.2
GPT-4V [89]	4-View 2D Image	0.0	0.0	0.0	0.0	0.0	0.0	0.0
GPT-4V‡ [89]	4-View 2D Image	0.1	1.6	0.0	0.0	0.0	0.0	0.3
ShapeLLM-7B	3D Point Cloud	**5.9**	**25.8**	**11.5**	**3.4**	**5.1**	**11.1**	**10.5**
ShapeLLM-13B	3D Point Cloud	**7.6**	**26.7**	**11.5**	**6.7**	**6.8**	**11.1**	**11.7**

Qualitative Analysis Figure 5 illustrates qualitative examples of SHAPELLM in *multimodal dialogue*. SHAPELLM can support general VQA, embodied task and action planning, and 6-DoF pose estimation. Notably, LLMs easily grasp such patterns and consistently produce valid coordinates due to the strict spatial relationship inherent in 6-DoF bounding box coordinates. Figure 6 shows the examples of SHAPELLM-13B's response using *single-view point cloud inputs*, demonstrating surprisingly outstanding robustness in processing such occlusion.

5 Discussions

5.1 Is ShapeLLM Grounded in Physical Worlds?

Table 5 compares SHAPELLM with image-only methods on 3D referring expression grounding (REG) of 6-DoF poses on GAPartNet [36]. The results show that: i) Image-only methods cannot perform zero-shot geometry-necessary 6-DoF pose estimation. ii) Compared to image-only methods with 2D to 6-DoF pose estimation fine-tuning or in-context prompting, SHAPELLM still performs significantly better. It demonstrates the necessity of geometry and the difficulty of the ill-posed 2D to 6-DoF pose estimation problem, as well as the importance of using 3D point clouds as input for spatial intelligence.

5.2 Can ShapeLLM Generalize to Unseen Objects?

Figure 7 shows the part understanding examples of unseen objects. While SHAPELLM's 6-DoF pose estimation and spatial awareness are trained on GAPartNet, which primarily consists of *indoor articulated furniture* objects. It has demonstrated promising generalization potential of spatial understanding on the *open-world objects*, paving ways for scaling up spatial awareness training.

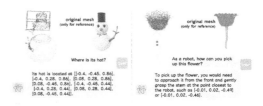

Fig. 7. Part understanding examples of unseen objects beyond GAPartNet.

6 Related Works

Interaction-Oriented 3D Understanding. Interaction with 3D objects typically involves concept-only interaction and physical-grounded interaction [12]. The former works focus on 3D perception and semantic parsing, such as 3D object recognition and scene perception [30,81,98,99,126,128]. By utilizing language for open-ended interaction in 3D, a number of works demonstrate successful 3D scene QA [84,141], grounding [13], and captioning [14]. Recently, some works propose to utilize foundation models like LLMs or CLIP for open-ended 3D object recognition [28,74,103,149,161] and scene segmentation [94,147]. Guo & Zhang *et al.* [41] utilizes ImageBind [38] and LLaMA-Adapter [150] to realize point cloud-based interactive QA. Following LLaVA, PointLLM [133] conducts supervised fine-tuning by constructing a visual instruction-following dataset. Other works focus on scene-level tasks utilizing comprehensive 2D features [52,162] or 3D features distilled from 2D images into LLMs [46,52,162]. The second kind of interaction typically requires physical understanding in 3D, such as part understanding [36,75,82,86], 6-DoF pose estimation [67,77,125,129,142], particularly useful for human-object interaction (HOI) and robotic manipulation [17,34–37,39,68,76,87,100,110,123,134,143] and complex robotic planning [11,26,29,55,71,112]. In this work, we focus on both physical and conceptual interactions with 3D shapes for embodied understanding.

Multimodal Large Language Models. Multimodal comprehension, which allows human interaction with textual and visual elements, has witnessed significant advancements, particularly in extending LLMs like LLaMA [19,120,121]. The early efforts predominantly revolved around integrating LLMs with various downstream systems by employing it as an agent [4,44,70,111,119,124,130,138,139]. Significant success has been demonstrated within this plugin-style framework. Due to the remarkable capabilities of LLMs, aligning the visual semantic space with language through parameter-efficient tuning [2,48,65,140,150,160] and instruction tuning [22,27,73,135] has emerged as the prevailing approach in current research. To further enhance interactive capabilities, some approaches have been developed towards visual-interactive multimodal comprehension by precisely referring to instruction tuning [16,95,152,154]. Another family advances the developments of LLMs endowed with content creation beyond comprehension [27,33,63,91,116,118,131].

7 Conclusions

This paper introduces SHAPELLM, the first 3D MLLM for embodied interaction, excelling in generalizable recognition and interaction comprehension. We present RECON++, a novel 3D point cloud encoder leveraging multi-view distillation and advanced 3D representation learning, forming the basis for SHAPELLM. We perform 3D visual instruction tuning on curated instruction-following data for broad and embodied comprehension. Additionally, we establish 3D MM-Vet, a benchmark to evaluate four levels of capacity in embodied interaction scenarios, from fundamental recognition to control statement generation.

Acknowledgments. The work was supported by the Dushi Program from Tsinghua University, the National Key R&D Program of China (2022YFB2804103), and the National Science and Technology Major Project of China (2023ZD0121300).

References

1. Achlioptas, P., Diamanti, O., Mitliagkas, I., Guibas, L.J.: Learning representations and generative models for 3D point clouds. In: International Conference on Machine Learning (ICML) (2018)
2. Alayrac, J., et al.: Flamingo: a visual language model for few-shot learning. In: Advances in Neural Information Processing Systems (NeurIPS) (2022)
3. Bai, Y., et al.: Sequential modeling enables scalable learning for large vision models. In: IEEE/CVF Conference on Computer Vision and Pattern Recognition (CVPR) (2024)
4. Betker, J., et al.: Improving image generation with better captions (2023)
5. Bommasani, R., et al.: On the opportunities and risks of foundation models. CoRR abs/2108.07258 (2021)
6. Bradski, G., Grossberg, S.: Recognition of 3-D objects from multiple 2-D views by a self-organizing neural architecture. In: Cherkassky, V., Friedman, J.H., Wechsler, H. (eds.) NATO ASI Series, vol. 136, pp. 349–375. Springer, Heidelberg (1994). https://doi.org/10.1007/978-3-642-79119-2_17
7. Bronstein, A.M., Bronstein, M.M., Guibas, L.J., Ovsjanikov, M.: Shape google: geometric words and expressions for invariant shape retrieval. ACM Trans. Graph. **30**(1), 1:1–1:20 (2011)
8. Brown, T.B., et al.: Language models are few-shot learners. In: Advances in Neural Information Processing Systems (NeurIPS) (2020)
9. Carion, N., Massa, F., Synnaeve, G., Usunier, N., Kirillov, A., Zagoruyko, S.: End-to-end object detection with transformers. In: Vedaldi, A., Bischof, H., Brox, T., Frahm, J.-M. (eds.) ECCV 2020. LNCS, vol. 12346, pp. 213–229. Springer, Cham (2020). https://doi.org/10.1007/978-3-030-58452-8_13

10. Chang, A.X., et al.: ShapeNet: an information-rich 3D model repository. CoRR abs/1512.03012 (2015)
11. Chang, M., et al.: GOAT: GO to any thing. In: Robotics: Science and Systems (RSS) (2024)
12. Chen, B., et al.: SpatialVLM: endowing vision-language models with spatial reasoning capabilities. In: IEEE/CVF Conference on Computer Vision and Pattern Recognition (CVPR) (2024)
13. Chen, D.Z., Chang, A.X., Nießner, M.: ScanRefer: 3D object localization in RGB-D scans using natural language. In: Vedaldi, A., Bischof, H., Brox, T., Frahm, J.-M. (eds.) ECCV 2020. LNCS, vol. 12365, pp. 202–221. Springer, Cham (2020). https://doi.org/10.1007/978-3-030-58565-5_13
14. Chen, D.Z., Gholami, A., Nießner, M., Chang, A.X.: Scan2Cap: context-aware dense captioning in RGB-D scans. In: IEEE/CVF Conference on Computer Vision and Pattern Recognition (CVPR) (2021)
15. Chen, G., Wang, M., Yang, Y., Yu, K., Yuan, L., Yue, Y.: PointGPT: autoregressively generative pre-training from point clouds. In: Advances in Neural Information Processing Systems (NeurIPS) (2023)
16. Chen, K., Zhang, Z., Zeng, W., Zhang, R., Zhu, F., Zhao, R.: Shikra: unleashing multimodal LLM's referential dialogue magic. CoRR abs/2306.15195 (2023)
17. Chen, S., Garcia, R., Laptev, I., Schmid, C.: SUGAR: pre-training 3D visual representations for robotics. In: Proceedings of the IEEE/CVF Conference on Computer Vision and Pattern Recognition, pp. 18049–18060 (2024)
18. Chen, X., et al.: PaLI-X: on scaling up a multilingual vision and language model. In: International Conference on Learning Representations (ICLR) (2023)
19. Chiang, W.L., et al.: Vicuna: an open-source chatbot impressing GPT-4 with 90%* ChatGPT quality, March 2023. https://lmsys.org/blog/2023-03-30-vicuna/
20. Collins, J., et al.: ABO: dataset and benchmarks for real-world 3D object understanding. In: IEEE/CVF Conference on Computer Vision and Pattern Recognition (CVPR) (2022)
21. Dai, A., Chang, A.X., Savva, M., Halber, M., Funkhouser, T., Nießner, M.: ScanNet: richly-annotated 3D reconstructions of indoor scenes. In: IEEE/CVF Conference on Computer Vision and Pattern Recognition (CVPR) (2017)
22. Dai, W., et al.: InstructBLIP: towards general-purpose vision-language models with instruction tuning. In: Advances in Neural Information Processing Systems (NeurIPS) (2023)
23. Das, A., et al.: Visual dialog. IEEE Trans. Pattern Anal. Mach. Intell. (TPAMI) **41**(5), 1242–1256 (2019)
24. Davison, J., Feldman, J., Rush, A.M.: Commonsense knowledge mining from pretrained models. In: Proceedings of the 2019 Conference on Empirical Methods in Natural Language Processing and the 9th International Joint Conference on Natural Language Processing, EMNLP-IJCNLP 2019, Hong Kong, China, 3–7 November 2019 (2019)
25. Deitke, M., et al.: Objaverse: a universe of annotated 3d objects. In: IEEE/CVF Conference on Computer Vision and Pattern Recognition (CVPR) (2023)
26. Ding, Y., Zhang, X., Paxton, C., Zhang, S.: Task and motion planning with large language models for object rearrangement. In: IEEE/RSJ International Conference on Intelligent Robots and Systems (IROS) (2023)
27. Dong, R., et al.: DreamLLM: synergistic multimodal comprehension and creation. In: International Conference on Learning Representations (ICLR) (2024)

28. Dong, R., et al.: Autoencoders as cross-modal teachers: can pretrained 2D image transformers help 3D representation learning? In: International Conference on Learning Representations (ICLR) (2023)
29. Driess, D., et al.: PaLM-E: an embodied multimodal language model. In: International Conference on Machine Learning (ICML) (2023)
30. Fan, G., Qi, Z., Shi, W., Ma, K.: Point-GCC: universal self-supervised 3D scene pre-training via geometry-color contrast. CoRR abs/2305.19623 (2023)
31. Fu, H., et al.: 3D-future: 3D furniture shape with texture. Int. J. Comput. Vision **129**, 3313–3337 (2021)
32. Gao, Y., Wang, Z., Zheng, W.S., Xie, C., Zhou, Y.: Sculpting holistic 3D representation in contrastive language-image-3D pre-training. In: IEEE/CVF Conference on Computer Vision and Pattern Recognition (CVPR) (2024)
33. Ge, Y., Ge, Y., Zeng, Z., Wang, X., Shan, Y.: Planting a SEED of vision in large language model. In: International Conference on Learning Representations (ICLR) (2024)
34. Geng, H., Li, Z., Geng, Y., Chen, J., Dong, H., Wang, H.: PartManip: learning cross-category generalizable part manipulation policy from point cloud observations. In: IEEE/CVF Conference on Computer Vision and Pattern Recognition (CVPR) (2023)
35. Geng, H., Wei, S., Deng, C., Shen, B., Wang, H., Guibas, L.: SAGE: bridging semantic and actionable parts for generalizable articulated-object manipulation under language instructions. In: Robotics: Science and Systems (RSS) (2024)
36. Geng, H., et al.: GAPartNet: cross-category domain-generalizable object perception and manipulation via generalizable and actionable parts. In: IEEE/CVF Conference on Computer Vision and Pattern Recognition (CVPR) (2023)
37. Geng, Y., An, B., Geng, H., Chen, Y., Yang, Y., Dong, H.: RLAfford: end-to-end affordance learning for robotic manipulation. In: IEEE International Conference on Robotics and Automation (ICRA) (2023)
38. Girdhar, R., et al.: ImageBind: one embedding space to bind them all. In: Proceedings of the IEEE/CVF Conference on Computer Vision and Pattern Recognition, pp. 15180–15190 (2023)
39. Gong, R., et al.: ARNOLD: a benchmark for language-grounded task learning with continuous states in realistic 3D scenes. In: International Conference on Computer Vision (ICCV) (2023)
40. Grabner, H., Gall, J., Gool, L.V.: What makes a chair a chair? In: IEEE/CVF Conference on Computer Vision and Pattern Recognition (CVPR) (2011)
41. Guo, Z., et al.: Point-bind & point-LLM: aligning point cloud with multi-modality for 3D understanding, generation, and instruction following. CoRR abs/2309.00615 (2023)
42. Gupta, A., Dollar, P., Girshick, R.: LVIS: a dataset for large vocabulary instance segmentation. In: IEEE/CVF Conference on Computer Vision and Pattern Recognition (CVPR) (2019)
43. Gupta, S., Girshick, R., Arbeláez, P., Malik, J.: Learning rich features from RGB-D images for object detection and segmentation. In: Fleet, D., Pajdla, T., Schiele, B., Tuytelaars, T. (eds.) ECCV 2014. LNCS, vol. 8695, pp. 345–360. Springer, Cham (2014). https://doi.org/10.1007/978-3-319-10584-0_23
44. Gupta, T., Kembhavi, A.: Visual programming: compositional visual reasoning without training. In: IEEE/CVF Conference on Computer Vision and Pattern Recognition (CVPR) (2023)

45. Hamdi, A., Giancola, S., Ghanem, B.: MVTN: multi-view transformation network for 3D shape recognition. In: International Conference on Computer Vision (ICCV), pp. 1–11. IEEE (2021)
46. Hong, Y., et al.: 3D-LLM: injecting the 3D world into large language models. In: Advances in Neural Information Processing Systems (NeurIPS) (2023)
47. Hou, J., Xie, S., Graham, B., Dai, A., Nießner, M.: Pri3D: can 3D priors help 2D representation learning? In: International Conference on Computer Vision (ICCV), pp. 5673–5682. IEEE (2021)
48. Hu, E.J., et al.: LoRA: low-rank adaptation of large language models. In: International Conference on Learning Representations (ICLR) (2022)
49. Hu, R., van Kaick, O., Wu, B., Huang, H., Shamir, A., Zhang, H.: Learning how objects function via co-analysis of interactions. ACM Trans. Graph. **35**(4), 47:1–47:13 (2016)
50. Hu, R., Li, W., van Kaick, O., Shamir, A., Zhang, H., Huang, H.: Learning to predict part mobility from a single static snapshot. ACM Trans. Graph. **36**(6), 227:1–227:13 (2017)
51. Hu, R., Zhu, C., van Kaick, O., Liu, L., Shamir, A., Zhang, H.: Interaction context (ICON): towards a geometric functionality descriptor. ACM Trans. Graph. **34**(4), 83:1–83:12 (2015)
52. Huang, J., et al.: An embodied generalist agent in 3D world. In: International Conference on Machine Learning (ICML) (2024)
53. Huang, T., et al.: CLIP2Point: transfer CLIP to point cloud classification with image-depth pre-training. In: International Conference on Computer Vision (ICCV) (2023)
54. Huang, W., Mordatch, I., Pathak, D.: One policy to control them all: shared modular policies for agent-agnostic control. In: International Conference on Machine Learning (ICML) (2020)
55. Huang, W., Wang, C., Zhang, R., Li, Y., Wu, J., Fei-Fei, L.: VoxPoser: composable 3D value maps for robotic manipulation with language models. In: Annual Conference on Robot Learnin (CoRL) (2023)
56. Huang, W., et al.: Inner monologue: embodied reasoning through planning with language models. In: Annual Conference on Robot Learning (CoRL) (2022)
57. Ichter, B., et al.: Do as I can, not as I say: grounding language in robotic affordances. In: Annual Conference on Robot Learnin (CoRL) (2022)
58. Ilharco, G., et al.: OpenCLIP, July 2021
59. Jia, M., et al.: Visual prompt tuning. In: Avidan, S., Brostow, G., Cissé, M., Farinella, G.M., Hassner, T. (eds.) ECCV 2022. LNCS, vol. 13693, pp. 709–727. Springer, Cham (2022). https://doi.org/10.1007/978-3-031-19827-4_41
60. Jiang, Z., Xu, F.F., Araki, J., Neubig, G.: How can we know what language models know. Trans. Assoc. Comput. Linguistics **8**, 423–438 (2020)
61. Kanade, T., Okutomi, M.: A stereo matching algorithm with an adaptive window: theory and experiment. IEEE Trans. Pattern Anal. Mach. Intell. **16**(9), 920–932 (1994)
62. Kim, V.G., Chaudhuri, S., Guibas, L.J., Funkhouser, T.A.: Shape2Pose: human-centric shape analysis. ACM Trans. Graph. **33**(4), 120:1–120:12 (2014)
63. Koh, J.Y., Fried, D., Salakhutdinov, R.: Generating images with multimodal language models. In: Advances in Neural Information Processing Systems (NeurIPS) (2023)
64. Kuhn, H.W.: The Hungarian method for the assignment problem. Naval Res. Logistics Q. **2**(1–2), 83–97 (1955)

65. Li, J., Li, D., Savarese, S., Hoi, S.C.H.: BLIP-2: bootstrapping language-image pre-training with frozen image encoders and large language models. In: International Conference on Machine Learning (ICML) (2023)
66. Li, X.L., Liang, P.: Prefix-tuning: optimizing continuous prompts for generation. In: Proceedings of the 59th Annual Meeting of the Association for Computational Linguistics and the 11th International Joint Conference on Natural Language Processing (Volume 1: Long Papers) (2021)
67. Li, X., Wang, H., Yi, L., Guibas, L.J., Abbott, A.L., Song, S.: Category-level articulated object pose estimation. In: IEEE/CVF Conference on Computer Vision and Pattern Recognition (CVPR) (2020)
68. Li, X., et al.: ManipLLM: embodied multimodal large language model for object-centric robotic manipulation (2023)
69. Li, Y., Du, Y., Zhou, K., Wang, J., Zhao, X., Wen, J.R.: Evaluating object hallucination in large vision-language models. In: Proceedings of the 2023 Conference on Empirical Methods in Natural Language Processing, pp. 292–305. Association for Computational Linguistics, Singapore (2023)
70. Liang, Y., et al.: TaskMatrix.AI: completing tasks by connecting foundation models with millions of APIs. Intell. Comput. **3**, 0063 (2024)
71. Lin, K., Agia, C., Migimatsu, T., Pavone, M., Bohg, J.: Text2Motion: from natural language instructions to feasible plans. Auton. Robot. **47**(8), 1345–1365 (2023)
72. Liu, H., Li, C., Li, Y., Lee, Y.J.: Improved baselines with visual instruction tuning. In: IEEE/CVF Conference on Computer Vision and Pattern Recognition (CVPR) (2024)
73. Liu, H., Li, C., Wu, Q., Lee, Y.J.: Visual instruction tuning. In: Advances in Neural Information Processing Systems (NeurIPS) (2023)
74. Liu, M., et al.: OpenShape: scaling up 3D shape representation towards open-world understanding. In: Advances in Neural Information Processing Systems (NeurIPS) (2023)
75. Liu, X., Wang, B., Wang, H., Yi, L.: Few-shot physically-aware articulated mesh generation via hierarchical deformation. In: International Conference on Computer Vision (ICCV) (2023)
76. Liu, X., Yi, L.: GeneOH diffusion: towards generalizable hand-object interaction denoising via denoising diffusion. In: International Conference on Learning Representations (ICLR) (2024)
77. Liu, X., Zhang, J., Hu, R., Huang, H., Wang, H., Yi, L.: Self-supervised category-level articulated object pose estimation with part-level SE(3) equivariance. In: International Conference on Learning Representations (ICLR) (2023)
78. Liu, Y., Fan, B., Xiang, S., Pan, C.: Relation-shape convolutional neural network for point cloud analysis. In: IEEE/CVF Conference on Computer Vision and Pattern Recognition (CVPR) (2019)
79. Liu, Y., et al.: MMBench: is your multi-modal model an all-around player? CoRR abs/2307.06281 (2023)
80. Liu, Y., et al.: SyncDreamer: generating multiview-consistent images from a single-view image. In: International Conference on Learning Representations (ICLR) (2024)
81. Liu, Y., Chen, J., Zhang, Z., Huang, J., Yi, L.: LeaF: learning frames for 4D point cloud sequence understanding. In: International Conference on Computer Vision (ICCV) (2023)
82. Lu, C., et al.: Beyond holistic object recognition: enriching image understanding with part states. In: IEEE/CVF Conference on Computer Vision and Pattern Recognition (CVPR) (2018)

83. Luo, T., Rockwell, C., Lee, H., Johnson, J.: Scalable 3D captioning with pretrained models. In: Advances in Neural Information Processing Systems (NeurIPS) (2023)
84. Ma, X., et al.: SQA3D: situated question answering in 3D scenes. In: International Conference on Learning Representations (ICLR) (2023)
85. Ma, X., Qin, C., You, H., Ran, H., Fu, Y.: Rethinking network design and local geometry in point cloud: a simple residual MLP framework. In: International Conference on Learning Representations (ICLR). OpenReview.net (2022)
86. Mo, K., et al.: PartNet: a large-scale benchmark for fine-grained and hierarchical part-level 3D object understanding. In: IEEE/CVF Conference on Computer Vision and Pattern Recognition (CVPR) (2019)
87. Mu, Y., et al.: EmbodiedGPT: vision-language pre-training via embodied chain of thought. In: Advances in Neural Information Processing Systems (NeurIPS) (2023)
88. OpenAI: GPT-4 technical report. CoRR abs/2303.08774 (2023). https://openai.com/research/gpt-4
89. OpenAI: GPT-4V(ision) system card (2023). https://openai.com/research/gpt-4v-system-card
90. Ouyang, L., et al.: Training language models to follow instructions with human feedback. In: Advances in Neural Information Processing Systems (NeurIPS) (2022)
91. Pan, X., Dong, L., Huang, S., Peng, Z., Chen, W., Wei, F.: Kosmos-G: generating images in context with multimodal large language models. In: International Conference on Learning Representations (ICLR) (2024)
92. Pang, Y., Wang, W., Tay, F.E.H., Liu, W., Tian, Y., Yuan, L.: Masked autoencoders for point cloud self-supervised learning. In: Avidan, S., Brostow, G., Cissé, M., Farinella, G.M., Hassner, T. (eds.) ECCV 2022. LNCS, vol. 13662, pp. 604–621. Springer, Cham (2022). https://doi.org/10.1007/978-3-031-20086-1_35
93. Peng, B., Li, C., He, P., Galley, M., Gao, J.: Instruction tuning with GPT-4. CoRR abs/2304.03277 (2023)
94. Peng, S., Genova, K., Jiang, C.M., Tagliasacchi, A., Pollefeys, M., Funkhouser, T.A.: OpenScene: 3D scene understanding with open vocabularies. In: IEEE/CVF Conference on Computer Vision and Pattern Recognition (CVPR) (2023)
95. Peng, Z., et al.: KOSMOS-2: grounding multimodal large language models to the world. CoRR abs/2306.14824 (2023)
96. Petroni, F., et al.: Language models as knowledge bases? In: Proceedings of the 2019 Conference on Empirical Methods in Natural Language Processing and the 9th International Joint Conference on Natural Language Processing, EMNLP-IJCNLP 2019, Hong Kong, China, 3–7 November 2019 (2019)
97. Pirk, S., et al.: Understanding and exploiting object interaction landscapes. ACM Trans. Graph. **36**(3), 31:1–31:14 (2017)
98. Qi, C.R., Su, H., Mo, K., Guibas, L.J.: PointNet: deep learning on point sets for 3D classification and segmentation. In: IEEE/CVF Conference on Computer Vision and Pattern Recognition (CVPR), pp. 77–85 (2017)
99. Qi, C.R., Yi, L., Su, H., Guibas, L.J.: PointNet++: deep hierarchical feature learning on point sets in a metric space. In: Advances in Neural Information Processing Systems, pp. 5099–5108 (2017)
100. Qi, H., Kumar, A., Calandra, R., Ma, Y., Malik, J.: In-hand object rotation via rapid motor adaptation. In: Annual Conference on Robot Learning (CoRL) (2023)
101. Qi, Z., et al.: Contrast with reconstruct: contrastive 3D representation learning guided by generative pretraining. In: International Conference on Machine Learning (ICML) (2023)

102. Qi, Z., Yu, M., Dong, R., Ma, K.: VPP: efficient conditional 3D generation via voxel-point progressive representation. In: Advances in Neural Information Processing Systems (NeurIPS) (2023)
103. Qi, Z., et al.: GPT4Point: a unified framework for point-language understanding and generation. In: Proceedings of the IEEE/CVF Conference on Computer Vision and Pattern Recognition, pp. 26417–26427 (2024)
104. Qian, G., et al.: PointNeXt: revisiting PointNet++ with improved training and scaling strategies. In: Advances in Neural Information Processing Systems (NeurIPS) (2022)
105. Radford, A., et al.: Learning transferable visual models from natural language supervision. In: International Conference on Machine Learning (ICML). Proceedings of Machine Learning Research, vol. 139, pp. 8748–8763. PMLR (2021)
106. Radford, A., Narasimhan, K., Salimans, T., Sutskever, I.: Improving language understanding by generative pre-training (2018)
107. Radford, A., Wu, J., Child, R., Luan, D., Amodei, D., Sutskever, I.: Language models are unsupervised multitask learners. OpenAI Blog **1**(8), 9 (2019)
108. Ren, J., Pan, L., Liu, Z.: Benchmarking and analyzing point cloud classification under corruptions. In: International Conference on Machine Learning (ICML) (2022)
109. Rohrbach, A., Hendricks, L.A., Burns, K., Darrell, T., Saenko, K.: Object hallucination in image captioning. In: Proceedings of the 2018 Conference on Empirical Methods in Natural Language Processing, Brussels, Belgium, 31 October–4 November 2018 (2018)
110. Shen, W., Yang, G., Yu, A., Wong, J., Kaelbling, L.P., Isola, P.: Distilled feature fields enable few-shot language-guided manipulation. In: Annual Conference on Robot Learning (CoRL) (2023)
111. Shen, Y., Song, K., Tan, X., Li, D., Lu, W., Zhuang, Y.: HuggingGPT: solving AI tasks with ChatGPT and its friends in HuggingFace. In: Advances in Neural Information Processing Systems (NeurIPS) (2023)
112. Shi, H., Xu, H., Clarke, S., Li, Y., Wu, J.: RoboCook: long-horizon elasto-plastic object manipulation with diverse tools. In: Annual Conference on Robot Learning (CoRL) (2023)
113. Shutterstock: Turbosquid. https://www.turbosquid.com/
114. Su, H., Maji, S., Kalogerakis, E., Learned-Miller, E.G.: Multi-view convolutional neural networks for 3D shape recognition. In: International Conference on Computer Vision (ICCV) (2015)
115. Sun, J., Zhang, Q., Kailkhura, B., Yu, Z., Xiao, C., Mao, Z.M.: ModelNet40-C: a robustness benchmark for 3D point cloud recognition under corruption. In: ICLR 2022 Workshop on Socially Responsible Machine Learning (2022)
116. Sun, Q., et al.: Generative multimodal models are in-context learners. In: IEEE/CVF Conference on Computer Vision and Pattern Recognition (CVPR) (2024)
117. Sun, Q., Fang, Y., Wu, L., Wang, X., Cao, Y.: EVA-CLIP: improved training techniques for CLIP at scale. CoRR abs/2303.15389 (2023)
118. Sun, Q., et al.: Emu: generative pretraining in multimodality. In: International Conference on Learning Representations (ICLR) (2024)
119. Surís, D., Menon, S., Vondrick, C.: ViperGPT: visual inference via Python execution for reasoning. In: International Conference on Computer Vision (ICCV) (2023)
120. Taori, R., et al.: Stanford Alpaca: an instruction-following LLaMA model (2023). https://github.com/tatsu-lab/stanford_alpaca

121. Touvron, H., et al.: LLaMA: open and efficient foundation language models. CoRR abs/2302.13971 (2023)
122. Uy, M.A., Pham, Q.H., Hua, B.S., Nguyen, T., Yeung, S.K.: Revisiting point cloud classification: a new benchmark dataset and classification model on real-world data. In: IEEE/CVF Conference on Computer Vision and Pattern Recognition (CVPR), pp. 1588–1597 (2019)
123. Wan, W., et al.: UniDexGrasp++: improving dexterous grasping policy learning via geometry-aware curriculum and iterative generalist-specialist learning. In: International Conference on Computer Vision (ICCV) (2023)
124. Wang, G., et al.: Voyager: an open-ended embodied agent with large language models. T. Mach. Learn. Res. (TMLR) (2024)
125. Wang, H., Sridhar, S., Huang, J., Valentin, J., Song, S., Guibas, L.J.: Normalized object coordinate space for category-level 6d object pose and size estimation. In: IEEE/CVF Conference on Computer Vision and Pattern Recognition (CVPR) (2019)
126. Wang, Y., Sun, Y., Liu, Z., Sarma, S.E., Bronstein, M.M., Solomon, J.M.: Dynamic graph CNN for learning on point clouds. ACM Trans. Graph. **38**(5), 146:1–146:12 (2019)
127. Wang, Z., Yu, X., Rao, Y., Zhou, J., Lu, J.: Take-a-photo: 3D-to-2D generative pre-training of point cloud models. In: International Conference on Computer Vision (ICCV) (2023)
128. Wen, H., Liu, Y., Huang, J., Duan, B., Yi, L.: Point primitive transformer for long-term 4D point cloud video understanding. In: Avidan, S., Brostow, G., Cissé, M., Farinella, G.M., Hassner, T. (eds.) ECCV 2022. LNCS, vol. 13689, pp. 19–35. Springer, Cham (2022). https://doi.org/10.1007/978-3-031-19818-2_2
129. Weng, Y., et al.: CAPTRA: category-level pose tracking for rigid and articulated objects from point clouds. In: International Conference on Computer Vision (ICCV) (2021)
130. Wu, C., Yin, S., Qi, W., Wang, X., Tang, Z., Duan, N.: Visual ChatGPT: talking, drawing and editing with visual foundation models. CoRR abs/2303.04671 (2023)
131. Wu, S., Fei, H., Qu, L., Ji, W., Chua, T.: Next-GPT: any-to-any multimodal LLM. In: International Conference on Machine Learning (ICML) (2024)
132. Wu, Z., et al.: 3D ShapeNets: a deep representation for volumetric shapes. In: IEEE/CVF Conference on Computer Vision and Pattern Recognition (CVPR), pp. 1912–1920 (2015)
133. Xu, R., Wang, X., Wang, T., Chen, Y., Pang, J., Lin, D.: PointLLM: empowering large language models to understand point clouds. CoRR abs/2308.16911 (2023)
134. Xu, Y., et al.: UniDexGrasp: universal robotic dexterous grasping via learning diverse proposal generation and goal-conditioned policy. In: IEEE/CVF Conference on Computer Vision and Pattern Recognition (CVPR) (2023)
135. Xu, Z., Shen, Y., Huang, L.: MULTIINSTRUCT: improving multi-modal zero-shot learning via instruction tuning. In: Proceedings of the 61st Annual Meeting of the Association for Computational Linguistics (ACL) (Volume 1: Long Papers) (2023)
136. Xue, L., et al.: ULIP: learning unified representation of language, image and point cloud for 3D understanding. In: IEEE/CVF Conference on Computer Vision and Pattern Recognition (CVPR) (2023)
137. Xue, L., et al.: ULIP-2: towards scalable multimodal pre-training for 3D understanding. In: IEEE/CVF Conference on Computer Vision and Pattern Recognition (CVPR) (2024)

138. Yang, R., et al.: GPT4Tools: teaching large language model to use tools via self-instruction. In: Advances in Neural Information Processing Systems (NeurIPS) (2023)
139. Yang, Z., et al.: MM-REACT: prompting ChatGPT for multimodal reasoning and action. CoRR abs/2303.11381 (2023)
140. Ye, Q., et al.: mPLUG-Owl: modularization empowers large language models with multimodality. CoRR abs/2304.14178 (2023)
141. Ye, S., Chen, D., Han, S., Liao, J.: 3D question answering. IEEE Trans. Vis. Comput. Graph. (2022)
142. Yi, L., Huang, H., Liu, D., Kalogerakis, E., Su, H., Guibas, L.J.: Deep part induction from articulated object pairs. ACM Trans. Graph. **37**(6), 209 (2018)
143. You, Y., Shen, B., Deng, C., Geng, H., Wang, H., Guibas, L.J.: Make a donut: language-guided hierarchical EMD-space planning for zero-shot deformable object manipulation. CoRR abs/2311.02787 (2023)
144. Yu, W., et al.: MM-Vet: evaluating large multimodal models for integrated capabilities. In: International Conference on Machine Learning (ICML) (2024)
145. Yu, X., Tang, L., Rao, Y., Huang, T., Zhou, J., Lu, J.: Point-BERT: pre-training 3D point cloud transformers with masked point modeling. In: IEEE/CVF Conference on Computer Vision and Pattern Recognition (CVPR) (2022)
146. Zeid, K.A., Schult, J., Hermans, A., Leibe, B.: Point2Vec for self-supervised representation learning on point clouds. In: Köthe, U., Rother, C. (eds.) DAGM GCPR 2023. LNCS, vol. 14264, pp. 131–146. Springer, Cham (2023). https://doi.org/10.1007/978-3-031-54605-1_9
147. Zhang, J., Dong, R., Ma, K.: CLIP-FO3D: learning free open-world 3D scene representations from 2D dense CLIP. In: International Conference on Computer Vision (ICCV Workshop) (2023)
148. Zhang, R., et al.: Point-m2AE: multi-scale masked autoencoders for hierarchical point cloud pre-training. In: Advances in Neural Information Processing Systems (NeurIPS) (2022)
149. Zhang, R., et al.: PointCLIP: point cloud understanding by CLIP. In: IEEE/CVF Conference on Computer Vision and Pattern Recognition (CVPR) (2022)
150. Zhang, R., et al.: LLaMA-adapter: efficient fine-tuning of language models with zero-init attention. In: International Conference on Learning Representations (ICLR) (2024)
151. Zhang, R., Wang, L., Qiao, Y., Gao, P., Li, H.: Learning 3D representations from 2D pre-trained models via image-to-point masked autoencoders. In: IEEE/CVF Conference on Computer Vision and Pattern Recognition (CVPR) (2023)
152. Zhang, S., et al.: GPT4RoI: instruction tuning large language model on region-of-interest. CoRR abs/2307.03601 (2023)
153. Zhang, Z., Cao, S., Wang, Y.: TAMM: TriAdapter multi-modal learning for 3D shape understanding. In: IEEE/CVF Conference on Computer Vision and Pattern Recognition (CVPR) (2024)
154. Zhao, L., et al.: ChatSpot: bootstrapping multimodal LLMs via precise referring instruction tuning. In: International Joint Conference on Artificial Intelligence (IJCAI) (2024)
155. Zhao, X., Wang, H., Komura, T.: Indexing 3D scenes using the interaction bisector surface. ACM Trans. Graph. **33**(3), 22:1–22:14 (2014)
156. Zheng, J., Zheng, Q., Fang, L., Liu, Y., Yi, L.: CAMS: canonicalized manipulation spaces for category-level functional hand-object manipulation synthesis. In: IEEE/CVF Conference on Computer Vision and Pattern Recognition (CVPR) (2023)

157. Zheng, L., et al.: Judging LLM-as-a-judge with MT-bench and chatbot arena. In: Advances in Neural Information Processing Systems (NeurIPS) (2024)
158. Zhou, J., Wang, J., Ma, B., Liu, Y., Huang, T., Wang, X.: Uni3D: exploring unified 3D representation at scale. In: International Conference on Learning Representations (ICLR) (2024)
159. Zhou, Y., et al.: Analyzing and mitigating object hallucination in large vision-language models. In: International Conference on Learning Representations (ICLR) (2024)
160. Zhu, D., Chen, J., Shen, X., Li, X., Elhoseiny, M.: MiniGPT-4: enhancing vision-language understanding with advanced large language models. In: International Conference on Learning Representations (ICLR) (2024)
161. Zhu, X., et al.: PointCLIP V2: prompting CLIP and GPT for powerful 3D open-world learning. In: International Conference on Computer Vision (ICCV) (2023)
162. Zhu, Z., Ma, X., Chen, Y., Deng, Z., Huang, S., Li, Q.: 3D-VisTA: pre-trained transformer for 3D vision and text alignment. In: International Conference on Computer Vision (ICCV) (2023)

Content-Aware Radiance Fields: Aligning Model Complexity with Scene Intricacy Through Learned Bitwidth Quantization

Weihang Liu[1], Xue Xian Zheng[2], Jingyi Yu[1,3], and Xin Lou[1,3(✉)]

[1] ShanghaiTech University, Shanghai, China
louxin@shanghaitech.edu.cn
[2] King Abdullah University of Science and Technology, Thuwal, Saudi Arabia
[3] Key Laboratory of Intelligent Perception and Human-Machine Collaboration, Shanghai, China

Abstract. The recent popular radiance field models, exemplified by Neural Radiance Fields (NeRF), Instant-NGP and 3D Gaussian Splatting, are designed to represent 3D content by that training models for each individual scene. This unique characteristic of scene representation and per-scene training distinguishes radiance field models from other neural models, because complex scenes necessitate models with higher representational capacity and vice versa. In this paper, we propose content-aware radiance fields, aligning the model complexity with the scene intricacies through Adversarial Content-Aware Quantization (A-CAQ). Specifically, we make the bitwidth of parameters differentiable and trainable, tailored to the unique characteristics of specific scenes and requirements. The proposed framework has been assessed on Instant-NGP, a well-known NeRF variant and evaluated using various datasets. Experimental results demonstrate a notable reduction in computational complexity, while preserving the requisite reconstruction and rendering quality, making it beneficial for practical deployment of radiance fields models. Codes are available at https://github.com/WeihangLiu2024/Content_Aware_NeRF.

Keywords: Radiance fields · Content-aware · Quantization · Model complexity

1 Introduction

Triggered by the phenomenal Neural Radiance Fields (NeRF) [23], the idea of representing 3D scenes using trainable models has been widely adopted in many

---

W. Liu, X. X. Zheng—Equal contribution.

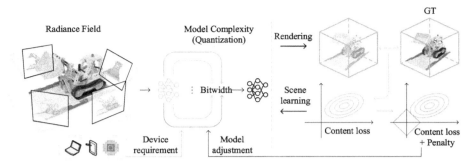

Fig. 1. Overview of content-aware radiance fields. In this work, model complexity is aligned with scene intricacy through learned bitwidth quantization.

reconstruction and rendering applications. Different from traditional explicit representations like meshes and point clouds, radiance field techniques encode a 3D scene with learnable parameters, which are obtained by training models based on sparse samples of the scene. Despite its advantages, the high-quality representation and rendering offered by radiance fields come at the expense of significant computational complexity. This challenge has been extensively studied, leading to numerous research endeavors aimed at exploring more computational efficient radiance field models [12,22,28,33].

While many variants of NeRF have significantly improved efficiency, they typically compress all scenes to a single fixed scale. Very few of them consider dealing with the scene contents differently, i.e., employing a *content-aware* strategy to align model complexity with scene intricacy. This principle, as illustrated in Fig. 1, forms the cornerstone of our work in this paper. It is unique to radiance fields due to their distinctive attributes of scene representation and per-scene training. Intuitively, detailed scenes rich in geometry and texture necessitate more sophisticated radiance field models to capture their nuances. While on the other hand, encoding less complex scenes with simpler models not only suffices for adequate representation but also enhances the efficiency by reducing computational and memory requirements.

Quantization has been proven a fundamental yet effective technique for reducing model complexity. The selection of parameter bitwidth is crucial for balancing computation and performance, bridging the gap between content information loss and computational efficiency. Recent works have investigated this technique on radiance field models by quantizing them to a pre-defined fixed bitwidth [27] as well as learning quantization range [40]. Nevertheless, these methods often rely on extensive human expertise to select appropriate quantization parameters, resorting to trial and error approaches. Moreover, reducing the bitwidth with controllable performance loss is impractical within their frameworks.

In this work, we introduce the concept of content-aware radiance fields, which adaptively quantize models by exploiting the content differences among scenes and the features of each layer. The quantization bitwidth of radiance models,

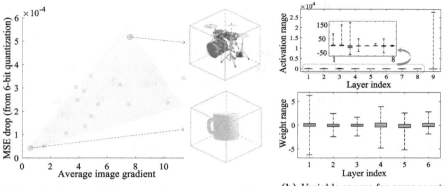

(a) Quantization sensitivity w.r.t. average image gradient.　(b) Variable ranges for components.

Fig. 2. The insights of the proposed LBQ. (a) The correlation between quantization sensitivity and scene complexity measured using average image gradient of the training set. Scenes with complex (simple) geometry and texture suffer more (less) accuracy degradation from quantization. (b) Exhibits the notable distinction of variables' distribution among different components. Those distributed in large (small) range is required to be quantized with high (low) bitwidth.

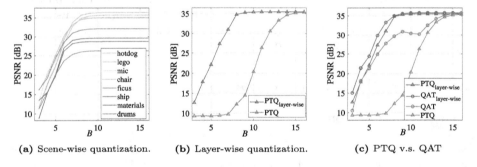

(a) Scene-wise quantization.　(b) Layer-wise quantization.　(c) PTQ v.s. QAT

Fig. 3. The insights of the proposed A-CAQ. (a) Layer-wise quantization results for different scenes with various bitwidths, which reveals content-aware characteristics of quantization effects. (b) Results of layer-wise and non layer-wise quantization for the "lego" scene. The huge accuracy gap verifies the significance of mixed-precision models. (c) QAT alleviates performance degradation as quantization error expands.

which directly correlates with computational complexity, are therefore content-aware, i.e., intricate scenes are accommodated with higher bitwidth models, while simpler scenes utilize lower bitwidth counterparts to reduce computational complexity. The key insights for our method are illustrated in Fig. 2, where average image gradient is used as an estimator for the complexity of the scenes. Results in Fig. 2a indicate that scenes with complex structure and texture suffer more from quantization than those with lower complexity. Besides, to achieve integer-only inference in rendering, all layers are required to be quantized by considering the significant statistical distinction of output features through the entire pipeline (shown in Fig. 2b). Moreover, the results in Fig. 3 demonstrate that scene-wise as well as layer-wise quantization is beneficial and quantization-aware training (QAT) can effectively combat quantization degradation.

Specifically, to establish the connection between quantization sensitivity and bitwidth, we propose Learned Bitwidth Quantization (LBQ) framework illustrated in Fig. 4, facilitating learnable bitwidth based on reconstructed contents. To fully extract the representation capability of quantized models, Adversarial Content-Aware Quantization (A-CAQ) algorithm is further proposed, which searches scene-dependent bitwidth and optimizes radiance fields simultaneously. The main contributions of this work, as illustrated in Fig. 1, are as follows:

- We introduce content-aware radiance fields that aligns the complexity of models with the intricacies of scenes through quantization, showing that the parameter bitwidth bridges the gap between the content information loss and computational efficiency.
- We introduce the LBQ framework which models the integer bitwidth with "soft bitwidth". This approach makes bitwidth differentiable from the content information loss during training, thereby obviating extensive human expertise to select bitwidths through a trial and error approach.
- We propose A-CAQ that integrates LBQ to penalize layer-wise bitwidth, enabling dynamic learning of lower bitwidth with negligible reconstruction and rendering quality loss. Experimental results on various datasets confirm the superiority of the proposed method.

2 Related Work and Motivation

Scene Representations and Radiance Fields. Various follow-up works on NeRF have focused specifically on improving computational efficiency. Rendering based on simplified representations is much more efficient while maintaining quality, which motivates studies of compression for scene primitives. A number of recent works achieve this goal from different perspectives [6,9,11,14,21,24,29]. Specifically, by employing a learnable codebook along with a compact MLP [24,29], Instant-NGP mitigates representation workload of network parameters, resulting in superior results in terms of both rendering quality and speed among NeRF variants.

While content information is easily acquired by explicit primitives, it is challenging for implicit representations. VQ-AD [29] and SHACIRA [14] introduced scalable compression of trainable feature grid which render content with different qualities. Recursive NeRF [37] proposes to dynamically grow the network, considering complexity variations of different patches within one scene. However, none of these works are metric-oriented or consider the complexity of different scenes. The search for the most suitable compression, considering both content and available resources, remains an empirical task.

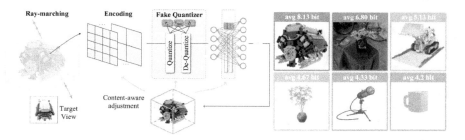

Fig. 4. Overview of the LBQ quantization framework. The fake quantizers are inserted into different components including encoding and MLPs, which are parameterized with variable range scale, upper bound and bitwidth. Quantization error is simulated by quantize and de-quantize procedure and quantization parameters are updated with direction indicated by gradient descent. Examples with different complexity are given on the right.

Model Quantization. Network quantization has been demonstrated as a potent technique for compressing neural models [4,7,13,26,38]. Generally, Post-Training Quantization (PTQ) [8,18] and Quantization-Aware Training (QAT) [41] are two common methods for model quantization. Some recent works endeavor to dynamically allocate bitwidths based on the features of each layer, resulting in mixed-precision quantization [5,15,34]. These mixed-precision methods have succeed in obtaining efficient network for various applications [15,17].

One significant limitation of existing PTQ and QAT is that they quantize pre-trained models using hyper-defined bitwidths. Re-quantizing models to other bitwidths can be challenging, as it suffer from differences in statistical characteristics of weights and activations [19]. One straightforward solution is to find the optimal bitwidth by considering specific application requirements. CADyQ [17] employ this idea to quantize super-resolution (SR) image network based on image gradient and standard deviation of features. Nevertheless, different from SR net based on convolutional neural networks (CNNs), there is no explicit pattern for MLPs features or 3D scene complexity measurement for neural fields. Therefore, we propose a novel framework that enables bitwidth differentiation, leveraging content-related Mean Squared Error (MSE) to adeptly select optimal scene-dependent layer-wise bitwidth.

3 Proposed Method

3.1 Preliminary

Radiance Fields with Learnable Codebook. Instant-NGP, a well-known NeRF variants incorporating a learnable multi-resolution hash table, is selected as our baseline model due to its inclusion of prevalent feature modalities in radiance fields, namely spatial levels-of-detail (LOD) feature grid and MLPs.

Specifically, it approximates the mapping from continuous 5D vectors to opacities σ and view-dependent colors **c**, denoted as

$$F : (\mathbf{x}, \mathbf{d}) \to (\sigma, \mathbf{c}), \tag{1}$$

where **x** and **d** are 3D location coordinate (of sample points) and 2D view direction respectively. Given a multi-resolution feature table with trainable parameters $\boldsymbol{\Theta}$, the location coordinates are encoded as $\mathbf{y_x} = \text{enc}(\mathbf{x}, \boldsymbol{\Theta})$, and the view directions are encoded as $\mathbf{y_d}$ with spherical harmonics [24]. The encoded vector is then fed to the MLP to produce the color and opacity as

$$(\sigma, \mathbf{c}) = m(\mathbf{y}, \boldsymbol{\Phi}), \tag{2}$$

where $\mathbf{y} = (\mathbf{y_x}, \mathbf{y_d})$, $\boldsymbol{\Phi}$ denotes the weights of MLPs. With the colors and opacities of all sample points, the expected color along a ray **r** can be calculated using volume rendering as

$$\mathbf{C}(\mathbf{r}) = \int_{t_n}^{t_f} T(t)\sigma(\mathbf{r}(t))\mathbf{c}(\mathbf{r}(t), \mathbf{d})dt, \text{ where } T(t) = \exp(-\int_{t_n}^{t} \sigma(\mathbf{r}(s))ds), \tag{3}$$

$\mathbf{r}(t) = \mathbf{o} + t\mathbf{d}$ are sampled points along the ray defined by origin **o** and direction **d**.

Quantization. To attain integer-only inference, quantization of both parameters (*i.e.* $\boldsymbol{\Omega} = \{\boldsymbol{\Theta}, \boldsymbol{\Phi}\}$) and activations for each layer is imperative. As shown in Fig. 4, fake quantizers can be inserted to simulate quantization error in training procedure. Given the data to be quantized **v**, along with the step size s and zero-point offset Z, the function of a fake quantizer is to dequantize the quantized data $\hat{\mathbf{V}}$. This operation can be expressed as

$$\hat{\mathbf{v}} = s(\hat{\mathbf{V}} - Z) = s\left[\text{clamp}\left(\left\lfloor\frac{\mathbf{v}}{s}\right\rceil + Z; q_{\min}, q_{\max}\right) - Z\right], \tag{4}$$

where $\lfloor\cdot\rceil$ is the round to nearest operator, and clamp($\cdot$) is defined as

$$\text{clamp}(x; a, c) = \begin{cases} a & x < a \\ x & a \leq x \leq c \\ c & x > c \end{cases}. \tag{5}$$

Based on Eq. (4), quantization training can be conducted to mitigate introduced quantization noise.

3.2 Differentiable Quantization for Radiance Fields

Learned Bitwidth Quantization (LBQ). The quantization training model has been established in Sect. 3.1. In this section, we initially reveal that the prevalent QAT-based quantization technique is inherently limited to fixed-bitwidth scenarios, followed by the introduction of the innovative LBQ scheme used in our content-aware quantization framework.

Existing QAT schemes, such as LSQ [10], LSQ+ [3], aim to directly train the step size and zero-point offset as

$$s = \frac{v_{\max} - v_{\min}}{q_{\max} - q_{\min}} = \frac{r_v}{r_q}, \tag{6}$$

$$Z = \left\lfloor q_{\max} - \frac{v_{\max}}{s} \right\rceil = \left\lfloor q_{\max} - \frac{v_{\max}}{r_v} r_q \right\rceil, \tag{7}$$

where r_v and $r_q = 2^B - 1$ are the range scales of $\mathbf{v}$ and $\hat{\mathbf{V}}$, respectively. To make the "round" operation differentiable, the Straight-Trough-Estimator (STE) [2] $\partial \lfloor x \rceil / \partial x = 1$ is assumed. It indicates that s and Z can only be regarded as leaf nodes with fixed bitwidth, where r_q is considered a constant.

To further make bitwidth scalable, we introduce floating point "soft bitwidth" b, where $B = \lfloor b \rfloor$. Moreover, r_v, $v_{\max}$ and b are selected as trainable quantization parameters, where r_v and $v_{\max}$ serve as replacements of step size and offset, respectively. The partial derivatives of Eq. (4) can then be derived as in Table 1.

Table 1. Derivatives to key parameters.

Variable range	$\partial \hat{v}/\partial r_v$	$\partial \hat{v}/\partial b$	$\partial \hat{v}/\partial v_{\max}$
$v_{\min} \leq v \leq v_{\max}$	$(s \cdot \lfloor v/s \rceil - v)/r_v$	$(v - s \cdot \lfloor v/s \rceil) \cdot 2^B \ln 2 / r_q$	0
$v > v_{\max}$	$1 - v_{\max}/r_v - Z/r_q$	$(v_{\max} - r_v + sZ) \cdot 2^B \ln 2 / r_q$	1
$v < v_{\min}$	$-v_{\max}/r_v - Z/r_q$		

The foregoing equations elucidate the operational mechanism of the proposed differentiable quantization:

– The parameter $v_{\max}$ represents the offsets of the quantization range, calibrated solely upon the occurrence of overflow, as explicated in Table 1.
– Both r_v and b are updated in response to observed quantization errors calculated with the specific term $(v - s \cdot \lfloor v/s \rceil)$ in Table 1.

This proposed method makes bitwidth learnable during the training procedure.

Quantization Schemes for Radiance Field Models. The criterion for selecting quantization parameters is to minimize the error summation generated from rounding and overflow [25]. Considering statistical properties and inference efficiency, we establish three different quantization schemes for different components within the radiance field pipeline.

Neural Weights. [3] indicates that symmetric signed quantization is highly recommended for neural weights which are empirically distributed symmetrically around zero. More importantly, quantizing weights in this manner introduces no additional computational overhead during inference.

ReLU and Exponential Activations. As these functions always produce positive output, we use asymmetric unsigned quantization. Similar settings are found in [10]. Based on this configuration, lower bitwidth can be reached.

Positional Encoding (PE) and Others. To deal with components lacking significant statistical features, we introduce $v_{\max}$ to represent trainable offset. [25] proves that this configuration will not introduce any additional computational overhead during inference. The learnable codebook is quantized in this manner.

The training details about these schemes are provided in Table 2. Calculations of gradients w.r.t. three different sets of trainable quantization parameters can be found in *Supplementary materials*.

Table 2. Quantization schemes for different components in radiance field pipelines.

Module name	$v_{\max}$	r_v	b	$[q_{\min}, q_{\max}]$
Neural weights	N/A	trainable	trainable	$[-2^{B-1}, 2^{B-1} - 1]$
ReLU and exponential	N/A	trainable	trainable	$[0, 2^B - 1]$
PE and others	trainable	trainable	trainable	$[0, 2^B - 1]$

3.3 Adversarial Content-Aware Quantization (A-CAQ)

As we have illustrated in Table 2, the impact of quantization varies depending on the contents of scenes. To leverage this characteristic effectively, it is crucial to extract content-related information from radiance fields. While it is straightforward for explicit representations such as 3D Gaussian Splatting (3DGS) [20], it presents a formidable challenge for neural fields, primarily owing to the implicit nature of the patterns encapsulated within MLPs. Nevertheless, MSE between rendered view and ground truth emerges as a fortuitous revelation, serving as a compelling indicator of the precision achieved in reconstructing the 3D content. Therefore, the MSE loss assumes a pivotal role for ascertaining content-awareness of radiance field modeling.

Utilizing LBQ proposed in Sect. 3.2, scene-dependent layer-wise quantization schemes can therefore be learned from MSE. However, bitwidth cannot adhere to the same objective function that guides other parameters. The discrepancy arises from the accuracy degradation inevitable introduced by reducing bitwidth, which operates in adversarial manner compared to optimizing MSE. To address this problem, we proposed to define bitwidth learning loss as

$$\mathcal{L}^{\text{bit}} = \sqrt{||\mathcal{L}^{\text{NeRF}} - \mathcal{L}^{\text{metric}}||} + \sum_{i \in \mathcal{M}} \epsilon_i B_i, \quad (8)$$

where $\mathcal{M} = 1, 2, ..., M$ is the indexes of layers or components, $\mathcal{L}^{\text{NeRF}}$, defined as

$$\mathcal{L}^{\text{NeRF}} = \sum_{\ell \in \mathcal{R}} ||\hat{C}(\ell) - C(\ell)||_{\text{F}}^2, \tag{9}$$

is the MSE loss function used for training radiance field [23], $\mathcal{R}$ is the set of rays in one batch, and the term $\mathcal{L}^{\text{metric}}$ is a hyper-defined metric representing the minimal accuracy requirement. By employing various loss metrics, rendering quality can be controlled, transferring redundant accuracy to efficiency. Weighted bitwidth penalties are further introduced to remove redundant bitwidth that have minimal impact on the overall loss. In addition, the weights ϵ_i offer greater flexibility, enabling assignment according to specific requirements. For example, higher penalty could be allocated to the bitwidth of the learnable codebook to obtain memory-efficient quantization schemes.

Dynamic bitwidth search can be achieved by solely optimizing Eq. (8). However, this approach results in learning bitwidth in a PTQ manner. Directly searching for bitwidth in the QAT space is impractical, as QAT accuracy can only be obtained through trial and error. As illustrated in Fig. 3c, the performance gap between PTQ and QAT grows when quantizing to lower bitwidth. To mitigate this degradation, we propose A-CAQ, a multi-task learning-based method expressed as

$$\mathcal{L}^{\text{A-CAQ}} = \min_{\mathcal{Q}} \mathcal{L}^{\text{NeRF}} + \min_{\mathbf{b}} \mathcal{L}^{\text{bit}}. \tag{10}$$

where $\mathcal{Q} = \{\mathbf{\Omega}, \mathbf{v}_{\max}, \mathbf{r}_v\}$. This task can be effectively solved by optimizing Eq. (8) and Eq. (9) alternatively. Detailed pseudo codes are found in *Supplementary Materials*.

The interpretation of Eq. 10 primarily pertains to adversarial learning: minimizing Eq. 8 leads to a lower bitwidth solution as well as higher $\mathcal{L}^{\text{NeRF}}$. And a lower bitwidth provides $\mathcal{L}^{\text{NeRF}}$ with more potential to alleviate accuracy degradation, thereby creating more search space to achieve a lower bitwidth solution.

4 Experiments

Experiments are conducted to demonstrate the effectiveness and versatility of our content-aware quantization framework for radiance fields. We first provide the implementation details, including model and training configurations, in Sect. 4.1, followed by quantitative and qualitative results in Sect. 4.2 and Sect. 4.3, respectively. Ablation studies and complexity analysis are presented to validate the effectiveness of the proposed algorithm in Sect. 4.4 and Sect. 4.5, respectively.

4.1 Implementation Details

Models. As mentioned in Sect. 3.1, our proposed quantization framework is evaluated on Instant-NGP [31], a well-known NeRF variant. All activations and parameters are quantized to facilitate integer-only inference. The bitwidth is constrained within the range of [2, 32], as binary quantization consistently yields nonsensical rendering results. Quantization schemes for each component are presented in Table 2.

Training Details. For layer-wise quantization where bitwidth is learned individually for each component, feature quantization rate (FQR) [17], defined as $\text{FQR} = \frac{\sum_{i \in \mathcal{M}} B_i}{M}$, is introduced to measure the quantization performance. To perform training in quantization mode, quantization parameters are initialized using the simple PTQ method. All components are initially quantized to 8-bit except for exponential activation, which is quantized to 32-bit due to its extensive range scale (see Fig. 2b).

As we introduce per-defined $\mathcal{L}^{\text{metric}}$, the compression rate can be manipulated according to specific requirements. Based on the average full precision loss among training set $\mathcal{L}_{\text{fp}}^{\text{NeRF}}$, we conduct experiments with two scenarios:

Minimal Degradation Bitwidth Learning (MDL). For application with high-fidelity requirements, we need to maintain accuracy while minimizing bitwidth. The metric is thus defined as

$$\mathcal{L}^{\text{metric}} = \mathcal{L}_{\text{fp}}^{\text{NeRF}}. \tag{11}$$

Metric-Guided Bitwidth Learning (MGL). As numerous studies strive for ever-higher accuracy, there inevitably arises surplus accuracy for applications with varying precision requirements, such as LOD and rendering on resource-constrained edge devices. The surplus accuracy can be traded off for efficiency by quantizing into lower bitwidth. In this case, the metric is defined as

$$\mathcal{L}^{\text{metric}} > \mathcal{L}_{\text{fp}}^{\text{NeRF}}. \tag{12}$$

Table 3. A-CAQ results for metric-guided bitwidth learning on different datasets including Synthetic-NeRF [23], RTMV [32] and Mip-NeRF360 [1].

Dataset	Full precision		MDL		MGL ($10^{-3.2}$)		MGL (10^{-3})	
	PSNR↑	FQR↓	PSNR↑	FQR↓	PSNR↑	FQR↓	PSNR↑	FQR↓
chair	35.06	32.00	34.57	7.60	29.63	4.80	27.23	4.60
V8	27.68	32.00	27.39	8.00	27.39	7.40	26.16	5.67
bonsai	28.13	32.00	27.49	7.00	27.73	7.07	26.50	5.73

4.2 Quantitative Results

In this section, we evaluate the performance of our A-CAQ algorithm on different datasets, including Synthetic-NeRF [23], RTMV [32], and Mip-NeRF360 [1]. These datasets comprise scenes with varying levels of complexity.

Table 4. Quantitative comparisons. Instant-NGP quantized with PTQ [27], LSQ+ [3,40,43], and A-CAQ are compared. The FQR and PSNR are reported to measure the complexity and accuracy, respectively. The results demonstrate that proposed methods succeed in reducing FQR while minimizing accuracy degradation.

Method		Synthetic-NeRF		Mip-NeRF360		RTMV (V8)	
		FQR$_\downarrow$	PSNR$_\uparrow$	FQR$_\downarrow$	PSNR$_\uparrow$	FQR$_\downarrow$	PSNR$_\uparrow$
	NGP	32.00	32.42	32.00	25.55	32.00	27.68
MDL	NGP-PTQ [27]	9.60	31.98	9.60	25.38	9.60	27.29
	NGP-LSQ+ [3,40,43]	9.60	32.11	9.60	25.48	9.60	27.40
	NGP-A-CAQ (Ours)	7.76	32.00	7.11	25.30	8.00	27.39
MGL ($10^{-3.2}$)	NGP-PTQ	6.60	22.29	7.60	22.62	7.60	22.96
	NGP-LSQ+	6.60	25.06	7.60	23.84	7.60	24.50
	NGP-A-CAQ (Ours)	5.33	27.58	6.86	25.18	7.40	27.39
MGL (10^{-3})	NGP-PTQ	5.60	17.75	6.60	17.54	5.60	10.54
	NGP-LSQ+	5.60	20.26	6.60	22.44	5.60	14.36
	NGP-A-CAQ (Ours)	4.74	25.96	6.58	24.74	5.67	26.16

The experiments begin with various accuracy requirements to demonstrate the capability and flexibility of the metric-guided feature. The experimental results are presented in Table 3. Quantization schemes of different accuracy can be effectively learned with different $\mathcal{L}^{\text{metric}}$. MSE loss is selected as a general metric here. Note that other quality metrics such as PSNR, Structural Similarity Index (SSIM [35]) and Learned Perceptual Image Path Similarity (LPIPS [42]) can also be used according to the specific requirements of different applications.

To validate the effectiveness of our A-CAQ in consistently producing efficient quantization schemes while maintaining requisite quality, we compare it with PTQ and LSQ+ schemes with different bitwidth configurations, which are widely used for radiance field model quantization [27,40,43]. As shown in Table 4, the proposed methods can identify optimal quantization schemes for various scenarios. For high-fidelity applications (labeled as MDL), integer-only rendering is achieved with negligible PSNR loss (<0.5 dB), while the FQR is significantly reduced. In resource-constrained scenarios (labeled as MGL), our method achieves state-of-the-art results for both accuracy and efficiency.

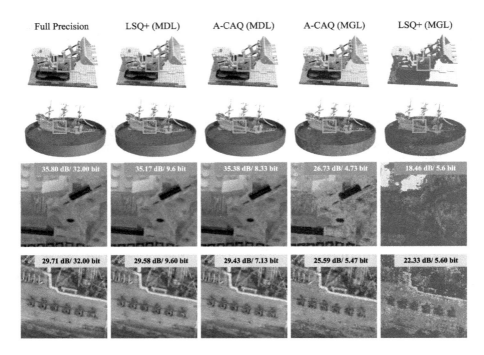

Fig. 5. Qualitative results of the "lego" and "ship" datasets from Synthetic-NeRF. Proposed A-CAQ outputs visual clean results for both MDL and MGL scenarios.

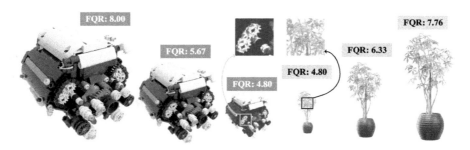

Fig. 6. Quantized LOD. Our proposed method can compress scenes in LOD style. The detail levels can be easily manipulated considering available resources and required accuracy.

4.3 Qualitative Results

We also present qualitative results and comparisons with other widely used methods. As depicted in Fig. 5, our A-CAQ consistently yields visually clean results across various loss metrics while requiring fewer bits. In contrast, existing quantization paradigms introduce significant distortion with comparable bitwidth

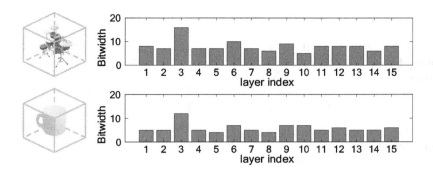

Fig. 7. Visualization of content-aware layer-wise quantization. The total number of components in our radiance field pipeline $M = 15$ including codebook and MLPs.

configurations, as they are unable to discern quantization sensitivities among different scenes and layers.

To demonstrate the effectiveness and flexibility of our method, LOD rendering results are presented in Fig. 6. Our dynamic quantization scheme can simultaneously filter and compress radiance field models, proving advantageous for LOD tasks [29]. Furthermore, quantization results for scenes with varying complexities are illustrated in Fig. 7. Content complexity influences the quantization sensitivity for output features of each layer, which can be distinguished with the proposed A-CAQ algorithm.

4.4 Ablation Study

Scene-Wise and Layer-Wise Quantization. To verify the significance of scene-wise and layer-wise quantization, we compare our proposed quantization framework with fixed-bitwidth QAT schemes, whose bitwidths are determined through a trial and error approach. Results are presented in Table 5. Quantizing all scenes with a fixed bitwidth throughout the entire pipeline results in performance loss and resource wastage (model **(1a)**). To accommodate the varying complexities observed in different scenes, a manual adjustment of scene-specific bitwidth is conducted, which mitigates the degradation in quantization (model **(1b)**). Utilizing A-CAQ, scene-dependent layer-wise quantization bitwidths are effectively learned. Following the MDL approach, the lowest average bitwidth schemes are learned while maintaining accuracy (model **(Ours, 1e)**). As a comparison, we quantize all scenes with the same conservative scheme: quantizing each layer with the highest layer-wise bitwidth learned by A-CAQ among all scenes (model **(Ours, 1d)**). This achieves the highest PSNR among all models while sacrificing extra model complexity.

Adversarial Learning. Our A-CAQ presents a multi-task learning problem. Here, we analyze the effects of two optimization tasks individually, which are reported in Table 6. Quantization without any post-processing results in the worst accuracy drop with the highest computational consumption (model (**2a**)). Post-training can, in turn, alleviate the performance degradation (model (**2b**)). Minimizing $\mathcal{L}^{\text{bit}}$ allows for learning the layer-wise quantization bitwidth, which searches for the lowest bitwidth in the PTQ performance space (model (**2c**)). Our proposed method can achieve much lower bitwidths according to the required accuracy as we search for bitwidth in the QAT performance space.

Table 5. Ablation study on layer-wise and scene-wise quantization.

	layer	scene	FQR↓	PSNR↑
(1a)	✗	✗	14.00	31.56
(1b)	✗	✓	14.27	32.04
(LSQ+, 1c)	✓	✗	9.60	32.11
(Ours, 1d)	✓	✗	8.93	32.30
(Ours, 1e)	✓	✓	7.76	32.00

Table 6. Ablation study on the adversarial losses: $\mathcal{L}^{\text{NeRF}}$ and $\mathcal{L}^{\text{bit}}$.

loss			MDL		MGL	
	$\mathcal{L}^{\text{NeRF}}$	$\mathcal{L}^{\text{bit}}$	FQR↓	PSNR↑	FQR↓	PSNR↑
(2a)	✗	✗	9.60	31.98	6.60	22.29
(2b)	✓	✗	9.60	32.11	6.60	25.06
(2c)	✗	✓	8.07	32.00	6.43	27.95
Ours	✓	✓	7.76	32.00	5.33	27.58

4.5 Complexity Analysis

Our proposed quantization method can determine the optimal content-dependent bitwidth for both high-fidelity (MDL) and resource-constrained (MGL) scenarios. To demonstrate how bitwidth influences resource overhead, we use the number of operations, weighted by bitwidth (BitOps) [36], involved in multiplications as a metric for computational complexity which has been widely employed for mixed-precision network systems [16,30,39]. Additionally, model sizes are listed to reflect memory consumption.

The results presented in Table 7 demonstrate the efficiency of our method. In the MDL scenario, our framework achieves a reduction of ∼90.78% in BitOps compared to the baseline, and ∼11.66% reduction compared to quantization with LSQ+. Regarding memory consumption, our proposed A-CAQ achieves a reduction of ∼77.68% compared to the baseline, and ∼10.74% reduction compared to LSQ+. In the MGL scenario, the reduction expands to ∼94.45% and ∼16.17% in BitOps compared to NGP and NGP-LSQ+ respectively. Moreover, there is a reduction of ∼89.43% and ∼32.34% in storage compared to NGP and NGP-LSQ+ respectively. Notably, our method also achieves a 0.9 dB increase in PSNR compared to NGP-LSQ+.

Table 7. Complexity analysis of A-CAQ for inference. BitOps [36] for rendering one 800 × 800 image and model storage are measured for time and space consumption, respectively.

	MDL			MGL		
	PSNR↑	BitOps↓ [T]	Storage↓ [MB]	PSNR↑	BitOps↓	Storage↓
NGP [24]	32.42	71.01	46.56	-	-	-
NGP-LSQ+ [3,40,43]	32.11	7.41	11.64	25.06	4.70	7.28
NGP-A-CAQ (Ours)	32.00	6.55	10.39	25.96	3.94	4.92

5 Conclusion

In this work, we introduce the concept of content-aware radiance field and explore the relationship between 3D scene complexity and quantization schemes. This motivates us to propose the A-CAQ algorithm, which learns bitwidth using gradients obtained by automatically perceiving the scene. Our method dynamically allocates layer-wise and scene-wise bitwidths. Experimental results demonstrate that the proposed algorithm significantly reduces model complexity for various scenarios through quantization, under different requirements.

Limitations. This study explores the novel concept of content-aware radiance fields, deftly integrating mixed-precision quantization into its framework. Despite the progress made, the realm of content-aware radiance fields beckons with unexplored territories ripe for detailed examination. For instance, by conducting a targeted search of network architecture aligned with reliable indicators that reflect the complexity of 3D scenes, it becomes feasible to construct a more comprehensive content-aware radiance field framework.

Acknowledgements. This work was supported by the Central Guided Local Science and Technology Foundation of China (YDZX20223100001001).

References

1. Barron, J.T., Mildenhall, B., Verbin, D., Srinivasan, P.P., Hedman, P.: Mip-NeRF 360: unbounded anti-aliased neural radiance fields. In: CVPR, pp. 5855–5864 (2022). https://doi.org/10.1109/CVPR52688.2022.00539
2. Bengio, Y., Léonard, N., Courville, A.: Estimating or propagating gradients through stochastic neurons for conditional computation. arXiv preprint arXiv:1308.3432 (2013)
3. Bhalgat, Y., Lee, J., Nagel, M., Blankevoort, T., Kwak, N.: LSQ+: improving low-bit quantization through learnable offsets and better initialization. In: CVPR, pp. 696–697 (2020). https://doi.org/10.1109/CVPRW50498.2020.00356

4. Cai, Z., He, X., Sun, J., Vasconcelos, N.: Deep learning with low precision by half-wave gaussian quantization. In: CVPR, pp. 5918–5926 (2017). https://doi.org/10.1109/CVPR.2017.574
5. Cai, Z., Vasconcelos, N.: Rethinking differentiable search for mixed-precision neural networks. In: CVPR, pp. 2349–2358 (2020). https://doi.org/10.1109/CVPR42600.2020.00242
6. Chen, A., Xu, Z., Geiger, A., Yu, J., Su, H.: TensoRF: Tensorial Radiance Fields. In: Avidan, S., Brostow, G., Cissé, M., Farinella, G.M., Hassner, T. (eds) ECCV 2022. LNCS, vol. 13692, pp. 333–350. Springer, Cham (2022). https://doi.org/10.1007/978-3-031-19824-3_20
7. Choi, Y., El-Khamy, M., Lee, J.: Towards the limit of network quantization. arXiv preprint arXiv:1612.01543 (2016)
8. Choukroun, Y., Kravchik, E., Yang, F., Kisilev, P.: Low-bit quantization of neural networks for efficient inference. In: 2019 IEEE/CVF International Conference on Computer Vision Workshop (ICCVW), pp. 3009–3018 (2019). https://doi.org/10.1109/ICCVW.2019.00363
9. Deng, C.L., Tartaglione, E.: Compressing explicit voxel grid representations: fast NeRFs become also small. In: Proceedings of the IEEE/CVF Winter Conference on Applications of Computer Vision, pp. 1236–1245 (2023). https://doi.org/10.1109/WACV56688.2023.00129
10. Esser, S.K., McKinstry, J.L., Bablani, D., Appuswamy, R., Modha, D.S.: Learned Step Size Quantization. In: ICLR (2020)
11. Fridovich-Keil, S., Yu, A., Tancik, M., Chen, Q., Recht, B., Kanazawa, A.: Plenoxels: radiance fields without neural networks. In: CVPR, pp. 5501–5510 (2022). https://doi.org/10.1109/CVPR52688.2022.00542
12. Garbin, S.J., Kowalski, M., Johnson, M., Shotton, J., Valentin, J.: FastNeRF: high-fidelity neural rendering at 200FPS. In: ICCV, pp. 14346–14355 (2021). https://doi.org/10.1109/ICCV48922.2021.01408
13. Gholami, A., Kim, S., Dong, Z., Yao, Z., Mahoney, M.W., Keutzer, K.: A survey of quantization methods for efficient neural network inference. In: Low-Power Computer Vision, pp. 291–326. Chapman and Hall/CRC (2022)
14. Girish, S., Shrivastava, A., Gupta, K.: SHACIRA: scalable HAsh-grid compression for implicit neural representations. In: Proceedings of the IEEE/CVF International Conference on Computer Vision, pp. 17513–17524 (2023). https://doi.org/10.1109/ICCV51070.2023.01606
15. Guo, L., Fei, W., Dai, W., Li, C., Zou, J., Xiong, H.: Mixed-precision quantization of U-net for medical image segmentation. In: 2022 IEEE International Symposium on Circuits and Systems (ISCAS), pp. 2871–2875 (2022). https://doi.org/10.1109/ISCAS48785.2022.9937283
16. Guo, Z., et al.: Single path one-shot neural architecture search with uniform sampling. In: Vedaldi, A., Bischof, H., Brox, T., Frahm, J.-M. (eds.) ECCV 2020. LNCS, vol. 12361, pp. 544–560. Springer, Cham (2020). https://doi.org/10.1007/978-3-030-58517-4_32
17. Hong, C., Baik, S., Kim, H., Nah, S., Lee, K.M.: CADyQ: content-aware dynamic quantization for image super-resolution. In: Avidan, S., Brostow, G., Cissé, M., Farinella, G.M., Hassner, T. (eds) ECCV 2022. LNCS, vol. 13667, pp. 367–383. Springer, Cham (2022). https://doi.org/10.1007/978-3-031-20071-7_22
18. Hubara, I., Nahshan, Y., Hanani, Y., Banner, R., Soudry, D.: Accurate post training quantization with small calibration sets. In: International Conference on Machine Learning, pp. 4466–4475. PMLR (2021)

19. Jin, Q., Yang, L., Liao, Z.: AdaBits: neural network quantization with adaptive bit-widths. In: CVPR, pp. 2146–2156 (2020). https://doi.org/10.1109/CVPR42600.2020.00222
20. Kerbl, B., Kopanas, G., Leimkühler, T., Drettakis, G.: 3D Gaussian splatting for real-time radiance field rendering. ACM Trans. Graph. **42**(4) (2023). https://doi.org/10.1145/3592433
21. Li, L., Shen, Z., Wang, Z., Shen, L., Bo, L.: Compressing volumetric radiance fields to 1 MB. In: CVPR, pp. 4222–4231 (2023). https://doi.org/10.1109/CVPR52729.2023.00411
22. Luo, H., et al.: Convolutional neural opacity radiance fields. In: 2021 IEEE International Conference on Computational Photography (ICCP), pp. 1–12 (2021). https://doi.org/10.1109/ICCP51581.2021.9466273
23. Mildenhall, B., Srinivasan, P.P., Tancik, M., Barron, J.T., Ramamoorthi, R., Ng, R.: NeRF: representing scenes as neural radiance fields for view synthesis. In: Vedaldi, A., Bischof, H., Brox, T., Frahm, J.-M. (eds.) ECCV 2020. LNCS, vol. 12346, pp. 405–421. Springer, Cham (2020). https://doi.org/10.1007/978-3-030-58452-8_24
24. Müller, T., Evans, A., Schied, C., Keller, A.: Instant neural graphics primitives with a multiresolution hash encoding. ACM Trans. Graph. **41**(4) (2022). https://doi.org/10.1145/3528223.3530127
25. Nagel, M., Fournarakis, M., Amjad, R.A., Bondarenko, Y., Van Baalen, M., Blankevoort, T.: A white paper on neural network quantization. arXiv preprint arXiv:2106.08295 (2021)
26. Polino, A., Pascanu, R., Alistarh, D.: Model compression via distillation and quantization. arXiv preprint arXiv:1802.05668 (2018). https://doi.org/10.48550/arXiv.1802.05668
27. Rao, C., et al.: ICARUS: a specialized architecture for neural radiance fields rendering. ACM Trans. Graph. **41**(6), 1–14 (2022). https://doi.org/10.1145/3550454.3555505
28. Reiser, C., Peng, S., Liao, Y., Geiger, A.: KiloNeRF: speeding up neural radiance fields with thousands of tiny MLPs. In: Proceedings of the IEEE/CVF International Conference on Computer Vision, pp. 14335–14345 (2021). https://doi.org/10.1109/ICCV48922.2021.01407
29. Takikawa, T., et al.: Variable bitrate neural fields. In: ACM SIGGRAPH 2022 Conference Proceedings, pp. 1–9 (2022). https://doi.org/10.1145/3528233.3530727
30. Tang, C., et al.: Mixed-precision neural network quantization via learned layer-wise importance. In: Avidan, S., Brostow, G., Cissé, M., Farinella, G.M., Hassner, T. (eds) ECCV 2022. LNCS, vol. 13671, pp. 259–275. Springer, Cham (2022). https://doi.org/10.1007/978-3-031-20083-0_16
31. Tang, J.: Torch-NGP: a PyTorch implementation of instant-NGP (2022). https://github.com/ashawkey/torch-ngp
32. Tremblay, J., et al.: RTMV: a ray-traced multi-view synthetic dataset for novel view synthesis. arXiv preprint arXiv:2205.07058 (2022)
33. Wadhwani, K., Kojima, T.: SqueezeNeRF: further factorized FastNeRF for memory-efficient inference. In: Proceedings of the IEEE/CVF Conference on Computer Vision and Pattern Recognition, pp. 2717–2725 (2022). https://doi.org/10.1109/CVPRW56347.2022.00307
34. Wang, K., Liu, Z., Lin, Y., Lin, J., Han, S.: HAQ: hardware-aware automated quantization with mixed precision. In: CVPR, pp. 8612–8620 (2019). https://doi.org/10.1109/CVPR.2019.00881

35. Wang, Z., Bovik, A., Sheikh, H., Simoncelli, E.: Image quality assessment: from error visibility to structural similarity. IEEE Trans. Image Process. **13**(4), 600–612 (2004). https://doi.org/10.1109/TIP.2003.819861
36. Wu, B., Wang, Y., Zhang, P., Tian, Y., Vajda, P., Keutzer, K.: Mixed precision quantization of ConvNets via differentiable neural architecture search. arXiv preprint arXiv:1812.00090 (2018)
37. Yang, G.W., Zhou, W.Y., Peng, H.Y., Liang, D., Mu, T.J., Hu, S.M.: Recursive-NeRF: an efficient and dynamically growing NeRF. IEEE Trans. Visual Comput. Graph. (2022). https://doi.org/10.1109/TVCG.2022.3204608
38. Yang, J., et al.: Quantization networks. In: CVPR, June 2019
39. Yang, L., Jin, Q.: FracBits: mixed precision quantization via fractional bit-widths. In: Proceedings of the AAAI Conference on Artificial Intelligence, pp. 10612–10620 (2021). https://doi.org/10.1609/aaai.v35i12.17269
40. Ye, Z., Hu, Q., Zhao, T., Zhou, W., Cheng, J.: MCUNeRF: packing NeRF into an MCU with 1MB memory. In: Proceedings of the 31st ACM International Conference on Multimedia, pp. 9082–9092 (2023). https://doi.org/10.1145/3581783.3612109
41. Youn, J., Song, J., Kim, H.S., Bahk, S.: Bitwidth-adaptive quantization-aware neural network training: a meta-learning approach. In: Avidan, S., Brostow, G., Cissé, M., Farinella, G.M., Hassner, T. (eds.) ECCV 2022. LNCS, vol. 13672, pp. 208–224. Springer, Cham (2022). https://doi.org/10.1007/978-3-031-19775-8_13
42. Zhang, R., Isola, P., Efros, A.A., Shechtman, E., Wang, O.: The unreasonable effectiveness of deep features as a perceptual metric. In: CVPR, pp. 586–595 (2018). https://doi.org/10.1109/CVPR.2018.00068
43. Zhao, T., Chen, J., Leng, C., Cheng, J.: TinyNeRF: towards 100x compression of voxel radiance fields. In: Proceedings of the AAAI Conference on Artificial Intelligence, pp. 3588–3596 (2023). https://doi.org/10.1609/aaai.v37i3.25469

Finding Visual Task Vectors

Alberto Hojel[1], Yutong Bai[1], Trevor Darrell[1], Amir Globerson[2,3], and Amir Bar[1,2(✉)]

[1] UC Berkeley, Berkeley, USA
amirb4r@gmail.com
[2] Tel Aviv University, Tel Aviv, Israel
[3] Google Research, Tel Aviv, Israel

Abstract. Visual Prompting is a technique for teaching models to perform a visual task via in-context examples, without any additional training. In this work, we analyze the activations of MAE-VQGAN, a recent Visual Prompting model [4], and find *Task Vectors*, activations that encode task-specific information. Equipped with this insight, we demonstrate that it is possible to identify the Task Vectors and use them to guide the network towards performing different tasks without having to provide any in-context input-output examples. To find Task Vectors, we compute the mean activations of the attention heads in the model per task and use the REINFORCE [43] algorithm to patch into a subset of them with a new query image. The resulting Task Vectors guide the model towards performing the task better than the original model. (For code and models see www.github.com/alhojel/visual_task_vectors).

1 Introduction

In-context learning (ICL) is an emergent capability of large neural networks, first discovered in GPT-3 [6], which allows models to adapt to novel downstream tasks specified in the user's prompt. In computer vision, Visual ICL (known as Visual Prompting [2–4,51]) is still at its infancy but it is increasingly becoming more popular due to the appeal of using a single model to perform various downstream tasks without specific finetuning or change in the model weights.

In this work, we ask how in-context learning works in computer vision. While this question has yet to be explored, there has been a significant body of research in Natural Language Processing (NLP) trying to explain this phenomenon [1,7,13–15]. Most recently, Hendel et al. [17] suggested that LLMs encode *Task Vectors*, these are vectors that can be patched into the network activation space and replace the ICL examples while resulting in a similar functionality. Concurrently, Todd et al. [38] discovered *Function Vectors*, activations of transformer attention heads that carry task representations. Our work is inspired by these observations and we aim to study ICL in computer vision.

Supplementary Information The online version contains supplementary material available at https://doi.org/10.1007/978-3-031-72775-7_15.

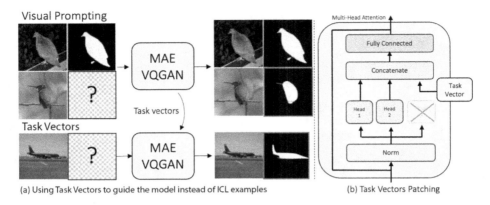

Fig. 1. Visual Prompting models like MAE-VQGAN [4] require input-output example(s) to describe the desired task in their forward pass. We analyze the model activations and find *Task Vectors*, activations that encode task information that can be reused to control the task the model performs (see Fig. 1a). Specifically, we tap into activations of individual attention heads and replace their outputs with Task Vectors to guide the model to the desired task (see Fig. 1b). Surprisingly, the resulting models perform better than the original model while removing the need for input-output examples. This confirms that *Task Vectors* exist in the network activation space and they can guide the model to perform the desired task.

Do Task Vectors exist in computer vision models as well? To build intuition, we start by exploring the activation space of MAE-VQGAN [4]. Intuitively, we look for intermediate activations that are invariant to changes within a task, but have high variance across different tasks. We use this simple and fast-to-compute metric to rank activations according to their "taskness". Using this score, we find that the activation space of certain attention heads can perfectly cluster the data by tasks, hinting at the existence of *Visual Task Vectors*.

Based on this insight, we hypothesize that Visual ICL models create Task Vectors too, and aim to find them. However, finding Visual Task Vectors by relying on previous approaches is challenging. For example, both [17,38] restricted their search space to the output activations of the last token in the prompt sequence (colored in red in the following example: ``Banana:B, Apple:''). This approach is natural for text as it is processed sequentially by autoregressive models like LLaMA [39], but with images (see Fig. 1a, "Visual Prompting" input image), it is not obvious what token activations hold task information, and architectures like MAE-VQGAN [4] do not process image tokens sequentially. This alone increases the search space significantly because multiple tokens might hold Task Vectors. Additionally, while Hendel et al. [17] and Todd et al. [38] differ in the activation space in which they perform interventions, both methods evaluate single interventions, ignoring emergent second-order effects.

To identify Task Vectors, we first compute the mean attention head outputs for each token position across a set of task examples (Fig. 1a). We then replace attention head outputs at selected positions with these pre-calculated means, using a new query as input. Each position is modeled as a random variable, and we use

REINFORCE [43] to optimize patching positions, guiding the model towards the desired task. This approach yields competitive performance across various tasks, validating the existence of Task Vectors. Our method effectively explores higher-order effects and complex interactions between attention heads in different layers, improving upon previous approaches that examine activations in isolation.

Our contributions are as follows. We show evidence for the existence of *Visual Task Vectors* and propose a practical way to identify them. Moreover, we find that patching the resulting Task Vectors guides the model towards the desired task with better performance compared to the original in-context prompts while reducing the number of FLOPS needed for a forward pass by 22.5%.

2 Related Work

2.1 Visual Prompting

Visual Prompting [2–4,18,46,51] is a class of approaches to adapt computer vision models to downstream tasks, inspired by the success of prompting in NLP [6]. Approaches like [2,18] seek to improve task-specific performance by adding trainable prompt vectors to the model. Other Visual Prompting approaches allow a model to handle various vision tasks [3,4,46,51] by introducing visual examples or text at the time of inference. Such prompting is related to the way in-context learning [22,25,42,45] operates in language models [6,33,41]. In fact, trainable prompts and in-context learning can be viewed as two complementary approaches for "describing" a task to a model [21]. Our goal here is to better understand the underlying mechanism of Visual ICL, and we analyze the MAE-VQGAN model presented in [4].

2.2 Explainability

Causal Interventions [5,28,31,32] and Activation Patching [48] are valuable tools for understanding complex neural networks' internal mechanisms, enhancing model interpretability [29,37,50]. These methods enable systematic examination of how models encode information and represent high-level concepts [24,44,49]. By manipulating internal states or inputs and observing output changes, they reveal causal structures and effects driving model predictions [12]. In this work, we employ Activation Patching [48] to improve Visual Prompting models' guidance for various computer vision tasks through targeted interventions.

2.3 Task Vectors

In [11,17,38], Task Vectors or Function Vectors are sets of latent activations derived from particular positions inside a transformer [40] model which serve as internal representations of a task implicitly described by an ICL prompt with input-output demonstrations. These latent activations at particular positions can consequently be used in a new forward pass in the absence of the ICL prompt (or with a corrupted prompt) while still managing to guide the model to perform the desired task. The investigation of Task Vectors aligns with broader efforts in the field to make neural networks more adaptable and tailored to specific tasks [23,26] as well as boosting the performance [19,30,47] by gaining a deeper understanding of how different attention heads within a model contribute to its overall function. Our work is the first to explore Task Vectors in computer vision.

3 Methods

Our goal is to understand in-context learning for computer vision and how existing models can be adapted to different downstream tasks at inference time. We focus on the MAE-VQGAN model [4], a variant of MAE [16] with a Vision Transformer [9] encoder-decoder architecture. Given an input-output example (x_s, y_s) and a new query x_q, to succeed in an ICL task, the model F must implicitly apply the transformation from x_s to y_s over x_q to produce y_q:

$$y_q = F(x_s, y_s, x_q) \tag{1}$$

Based on observations from NLP [17,38], we hypothesize that computer vision models also encode latent Task Vectors in their activation space during the forward pass. This requires the model to implicitly map the ICL example into a set of latent Task Vectors $z = \{z_i\}_{i=1}^{k}$ which we derive from the attention head outputs for different tokens across the model's layers (the internal representations). The original function F can then be decomposed into extracting the Task Vectors by a function G and applying F on the query while fixing the computed task activations $z \in \mathbb{R}^d$, where d is the hidden dimension of the model.

$$z = G(x_s, y_s) \qquad y_q = F(x_q | z) \tag{2}$$

We now describe how Task Vectors can be derived from the internal representations of a model.

3.1 Computing Mean Activations

Let $i = (l, m, k)$ denote the position in the model, where l, m, and k are the attention block, head, and token indices respectively. Define $H_i : (x_s, y_s, x_q) \to \mathbb{R}^d$ as the function that outputs the activation at position i for a given demonstration and query triplet. Let $(x_s, y_s, x_q) \sim \mathbf{D_{task_j}}$ be a triplet of input-output example and a query from task j. We compute the intermediate activation $h_{task_j}^i$ via H_i and denote its mean activation as $\mu_{i,j}$:

$$h_{task_j}^i = H_i(x_s, y_s, x_q) \qquad \mu_{i,j} = \mathbb{E}[h_{task_j}^i] \tag{3}$$

3.2 Scoring Activations

Intuitively, every task vector is an activation that changes across different tasks but remains relatively invariant to changes within a specific task. We define the data distribution $(x'_s, y'_s, x'_q) \sim \mathbf{D_{all\_tasks}}$ as the union of all task-specific distributions, and the intermediate activations $h_{all}^i = H_i(x_s', y_s', x_q')$. The scoring function $\rho_{token}(i)$ is defined as the ratio of inter-task to intra-task variance as $\rho_{token}(i)$ where Var($\cdot$) denotes the variance, $h[e]$ is the e-th element of vector h, and n is the number of tasks.

$$\rho_{token}(i) = \frac{\sum_{e=1}^{d} \text{Var}(h_{all}^i[e])}{\frac{1}{n}\sum_{j=1}^{n}\sum_{e=1}^{d} \text{Var}(h_{task_j}^i[e])} \tag{4}$$

Computing these is efficient, requiring only forward passes across batches of data. This scoring function identifies "taskness" in activations (see Fig. 2, and Table 1), but may also highlight other task-variant activations (e.g., color histograms varying across tasks but consistent within tasks like binary segmentation maps versus colored images). We use these scores mainly to build intuition, analyze task clusters in the activation space (Sect. 5), and develop a simple baseline ("Greedy Random Search", see Suppl. 11.1). However, we employ a more robust general approach to identifying high-order Visual Task Vectors, as described in the following section.

3.3 Finding Visual Task Vectors via REINFORCE

How can Task Vectors be identified? An exhaustive search over all subsets of activations is intractable. For example, MAE-VQGAN [4] which we utilize here has 32 attention blocks, each with 16 heads, therefore, if we consider these to be the potential set of activations then we need to search over 2^{512} options, evaluating each option over a held-out validation set.

For every task j we can apply the following procedure to find the Task Vectors. Recall that $\{\mu_{i,j}\}$ denotes the mean activations and $F(\cdot)$ is a pretrained visual prompting model. Denote $\alpha_{i,j} \sim Bernoulli(\sigma(\theta_{ij}))$ as random variables that signify whether the mean task activation $\mu_{i,j}$ is a task vector of task j placed in activation position i and θ_{ij} is a learned weight followed by the sigmoid function to ensure it is in $[0, 1]$.

Denote z_j as the set of Task Vectors for task j: $z_j = \{\mu_{i,j}|\alpha_{i,j} = 1\}$. Given pairs of input-output task demonstrations $x_q, y_q \sim \mathbf{D_{task_j}}$,[1] we want to find the set of Task Vectors that minimizes the loss function:

$$L(\theta) = E_{z_j \sim p_\theta} L(y_q, F(x_q|z_j)) \tag{5}$$

where p_θ denotes the sampling distribution of z_j and L denotes the loss function that suits the particular task j. Note that differently than in Eq. 1, if we find a good set of Task Vectors z_j, then we no longer need to condition F over additional input-output examples.

To find a set of Task Vectors, we need to estimate the parameters θ. Since the sampling distribution p_θ depends on θ, it is natural to use the REINFORCE algorithm [43]. The key idea in REINFORCE is the observation that:

$$\nabla L(\theta) = E_{z_j \sim p_\theta} L(y_q, F(x_q|z_j)) \nabla \log p_\theta(z_j)$$

---

[1] We overload the definition of $\mathbf{D_{task_j}}$ to avoid notation clutter.

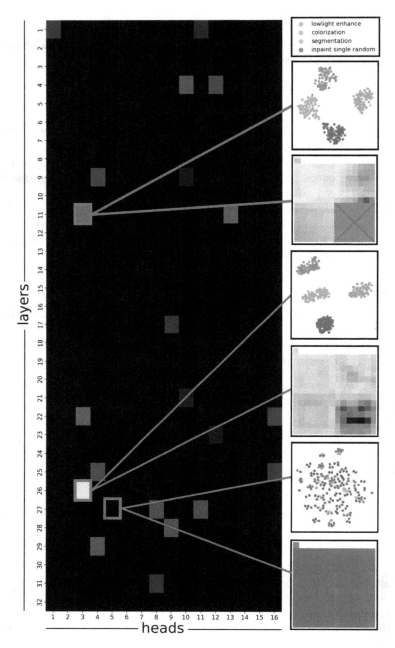

Fig. 2. Activation Scoring Analysis. Individual scores ($\rho_{token}(i)$) aggregated per Attention Head (left) for the encoder and decoder. Individual token scores of specific heads, along with the t-SNE [27] clustering of head activations of different tasks (right).

Thus, we can approximate $\nabla L(\theta)$ by sampling z_j and averaging the above equation. After iteratively optimizing with gradient descent, we select the final task vector positions by sampling a set of z_j.

Instead of learning Task Vectors placements α_{ij} for each individual task, we can revise the procedure above to learn a single placement by defining random variables $\alpha_i \sim Bernoulli(\sigma(\theta_i))$ that accommodate all tasks. In this setting, training requires drawing examples across tasks with their respective task-wise mean activations and task specific loss function. We refer to this setting as *multi-task* patching. Furthermore, the search can be applied at different granularities, such as patching token groups from the same quadrant, all tokens in an attention head, or entire layers. We discuss these design choices in the next section.

4 Experiments

Our experiments explore if activation patching of Task Vectors can make a model perform the desired visual task as well as or better than its original one-shot performance. We describe the implementation of MAE-VQGAN, prompting schemes, baselines for comparison, visual tasks, and conducted experiments.

4.1 Implementation Details

MAE-VQGAN [4]. An MAE [16] with a ViT-L [9] backbone. The decoder predicts a distribution over a VQGAN [10] codebook to output images with better visual quality. We used the pretrained checkpoint from [4] where it was trained over the Computer Vision Figures [4] dataset and ImageNet [35].

Finding Task Vectors. Similar to one-shot (Sect. 4.4), we use a 2×2 image grid with the prompt. For task vectors, we embed only the query in the bottom left quadrant. The model reconstructs only the output part (bottom right). We patchify the query image at 112×112 resolution, apply bottom left quadrant positional encodings, feed patches into the encoder, and process them with the decoder alongside bottom right quadrant mask tokens to obtain the result.

We intervene on attention head outputs by replacing them with mean activations at positions identified using REINFORCE [43]. Initially, we set $\theta_{ij} = -1$. To reduce the search space, we group patching positions into three categories: CLS (1 token), bottom left quadrant (49 query image patch tokens), and bottom right quadrant (49 MASK tokens). In each iteration, we sample 32 times from the Bernoulli distribution for each of 10 images, patching positions where the sampled value is 1. This results in 320 executions of task vector conditioned MAE-VQGAN per iteration. We optimize Bernoulli parameters as outlined in Sect. 3.3 using Adam [20] with a learning rate of 0.1. The algorithm runs for 600 steps, selecting the best checkpoint every 50 steps based on evaluation on a held-out test set.

4.2 Activation Scoring Analysis

Here our goal is to evaluate the Activation Scoring step (outlined in Sect. 3.2), specifically whether high scoring activations indeed correspond to Task Vectors.

Collecting Activations. To compute activation scores, we first run the model's forward pass in a one-shot setting across different tasks (Sect. 3.1). We use 100 prompts and queries from Pascal 5i [36] training set, ensuring reasonable one-shot performance. We save the activations for each task j and position $i = (l, m, k)$, where l, m, and k are the attention block, head, and token indices respectively. We then compute the mean activation $\mu_{i,j}$ and score $\rho_{token}(i)$.

Evaluation via Clustering. Next, we wish to analyze if $\rho_{token}(i)$ indeed captures "taskness". Intuitively, we expect layers that capture task information to succeed in clustering activations by task. To assess this, we analyze the clustering performance of vectors with high-ranking activations versus those marked with low scores. We measure the clustering performance using common clustering metrics like the Silhouette Score [34] and the Davies-Bouldin Score [8]. Finally, we also perform a qualitative analysis by visualizing the representations on a t-SNE [27] plot, coloring each data point by it's task label.

4.3 Downstream Tasks

We evaluate the performance on standard image-to-image tasks like Foreground Segmentation, Low Light Enhancement, In-painting, and Colorization.

Dataset. We utilize Pascal-5i [36], consisting of 4 image splits (346–725 images each) with segmentation masks. For evaluation, We sample 1000 prompt-query pairs per split from the validation set. For Activation Scoring and methods to find Task Vectors, we use only training set examples.

Foreground Segmentation. We use Pascal-5i [36] segmentation masks and report mean IOU (mIOU) across four splits.

Low Light Enhancement. We multiply Pascal-5i image color channels by 0.5 for input, using the original as output. We report Mean Squared Error (MSE).

Inpainting. We mask a random 25% square region (1/8 area) of each image for input. We report MSE using the original image as output.

Colorization. To obtain input-output pairs, we convert an image to grayscale and denote it as the input, and have the output be the original image. For evaluation, we report the MSE metric.

4.4 Baselines

One-Shot Prompting. We follow the basic one-shot setup in [4]. Specifically, we construct a grid-like image structure with an input-output demonstration, a query, and a masked output region which are embedded into a 2 × 2 grid. We feed this grid image to the model to obtain the output prediction which we use for evaluation purposes.

Causal Mediation Analysis. We compare our methodology with the Causal Mediation Analysis methodology as presented in [38] as a baseline. We select the top 25% of activations with the highest causal score across 10 images.

Greedy Random Search. We compare our methodology to an iterative greedy random search algorithm (GRS) used to select Task Vectors based on the activation scoring metric proposed in Sect. 3.2. This serves as a baseline and is outlined in the Supplementary Materials (Sect. 11.1).

4.5 Ablations

In this section, we describe the set of experiments conducted to validate our implementation choices of our REINFORCE [43] method.

Task Vectors Location in Encoder vs. Decoder. We hypothesize that task implementation spans both encoder and decoder. To test this, we apply interventions to the encoder only, decoder only, and the whole network. We report mIoU on four Segmentation task splits to assess the necessity of interventions in both parts for effective task implementation.

Patching Granularity. We explore intervention granularities by grouping token positions to reduce dimensionality and optimization search space. Positions $i = (l, m, k)$ are grouped into spatial quadrants or attention heads. To balance precision and search space, we use three granularity levels: individual tokens, quadrants, and attention heads, and report performance across four tasks.

4.6 Task Vector Patching

We investigate whether patching Task Vectors into a model's forward pass, without ICL demonstrations, can achieve performance comparable to one-shot prompting. Our experiments include:

Task-Specific. For each of the four tasks we execute our task-specific method, and the baselines of Causal Mediation Analysis (CMA) and Greedy Random Search (GRS) with 10 images and report the results on corresponding tasks.

Multi-task. We apply our method in a multi-task scenario, selecting two images per task and evaluating alongside CMA and GRS baselines. We jointly assess loss across all tasks, normalizing per task to ensure equal weighting and prevent single-task dominance. We include an identity-copy task to maintain a consistent batch size of 10 for fair comparisons.

We also provide the following baselines to benchmark the performance of the three algorithms, and to validate the necessity of top k score ranking the layers and performing the iterative search for the GRS method:

MAE-VQGAN. We compare to the original model's one-shot performance.

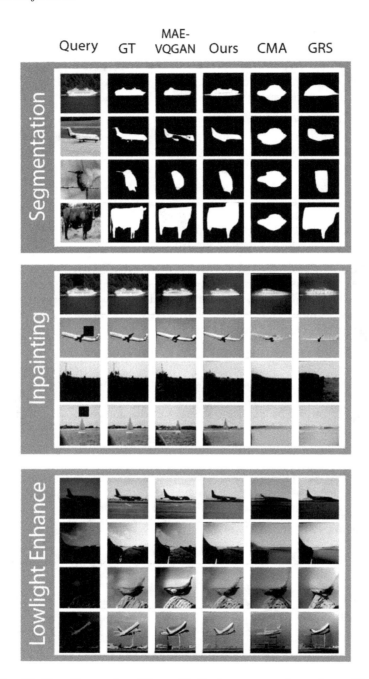

Fig. 3. Qualitative Examples. We qualitatively compare the task-specific variants of our methodology's results with the original model and the CMA and GRS baselines. Our patching methodology performs better than the original MAE-VQGAN model.

Table 1. Task Clustering Quality. Clustering Scores of Different Attention Heads, ranked by our Activation Score (see Sect. 3). This indicates that higher Activation Scores indeed correlate with better clustering by tasks.

	(Layer, Head)	Our Score ↑	Silhouette Score ↑	Davies-Bouldin Score ↓
High 1	(26, 3)	**2.1663**	**0.3583**	**1.2744**
High 2	(11, 3)	1.0827	0.2692	1.5567
Random 1	(4, 3)	0.2329	0.0708	4.1062
Random 2	(18, 16)	0.1259	0.0369	4.3256
Low 1	(2, 16)	0.0221	-0.0518	21.8265
Low 2	(2, 12)	0.0264	-0.0334	13.5982

Random Quadrants. We patch into randomly sampled positions across the whole network (same total amount of patches as task-specific and multi-task).

GRS Across Random K Layers. We execute the search algorithm on a random set of k layers, instead of score-ranked layers, to validate the necessity of our proposed scoring method.

Patching into Top Quadrants. We patch into the top quadrants based on their scoring (same total amount of patches as task-specific and multi-task). That is, we directly patch into areas with high scores naively without the iterative refining steps of the GRS, to validate its utility.

We report results across four splits for all four tasks, comparing our variants to the original MAE-VQGAN model and random baselines. Supplementary Materials (Sect. 4) include initial explorations of REINFORCE-powered Language Task Vectors with Llama7B, vector addition to create composable tasks, and others.

5 Results

5.1 Activation Scoring Analysis

We compute and display $\rho_{token}(i)$ per head on a heatmap, with layers on the y-axis and head indices on the x-axis (Fig. 2). This highlights heads that potentially hold Task Vectors. We select top heads—(26, 3) and (11, 3)—and a lower-ranked head (27, 5). For each, we visualize activation clustering and individual Activation Score per token.

Clustering Visualization. Highly-ranked heads (e.g., (26,3)) show clear task-based clustering, while low-ranked heads like (27,5) show many small clusters with different tasks, likely based on input semantics. The activations are projected onto 2D using t-SNE [27] with colors indicating tasks.

Score-per-Token Heatmap. We display $\rho_{token}(i)$ values for each token, reflecting spatial positioning in a 2 × 2 grid. For heads of interest, these values are

Table 2. Quantitative Analysis. Results comparison across different tasks and splits, indicating the effectiveness of our task-specific model.

Model	Segmentation ↑ (Mean ± STD)	Lowlight Enhance ↓ (Mean ± STD)	Colorization ↓ (Mean ± STD)	In-painting ↓ (Mean ± STD)
Original MAE-VQGAN	0.338 ± 0.033	0.685 ± 0.032	0.618 ± 0.027	0.550 ± 0.042
Random Quadrants	0.170 ± 0.061	3.000 ± 0.967	3.025 ± 1.190	2.350 ± 0.955
Random K Layers	0.090 ± 0.007	1.825 ± 0.043	0.568 ± 0.022	0.875 ± 0.100
Top Quadrants	0.150 ± 0.023	4.875 ± 0.228	4.250 ± 0.269	3.900 ± 0.141
CMA (Task-specific)	0.230 ± 0.012	0.825 ± 0.043	0.895 ± 0.063	1.750 ± 0.112
CMA (Multi-task)	0.150 ± 0.017	1.400 ± 0.122	1.130 ± 0.075	1.225 ± 0.083
GRS (Task-specific)	0.320 ± 0.021	0.600 ± 0.025	0.555 ± 0.025	0.580 ± 0.047
GRS (Multi-task)	0.323 ± 0.019	0.515 ± 0.032	0.568 ± 0.029	0.605 ± 0.036
Ours (Multi-task)	0.325 ± 0.026	0.492 ± 0.025	0.502 ± 0.036	0.558 ± 0.022
Ours (Task-specific)	**0.353** ± 0.028	**0.458** ± 0.032	**0.453** ± 0.036	**0.480** ± 0.022

shown on a heatmap. The CLS token is in the top left, followed by values for quadrants (x_s, y_s, x_q, y_q). The encoder lacks y_q tokens (bottom right), marked as X. Token values within a head vary widely but show consistency within quadrants, supporting quadrant-based token patching.

Quantitative Clustering Analysis. To validate our scoring-based clustering quality observations, we report Silhouette and Davies-Bouldin scores for the two highest-ranked heads, two randomly sampled heads, and the two lowest-ranked heads (Table 1). Heads scored highly by our method show high-quality clustering scores, while low-scored heads show poor clustering. Randomly selected heads received intermediate scores, further supporting our methodology.

5.2 Task Vector Patching

We present the performance of our task-specific and multi-task methods compared to the original model and baselines. Task Vector interventions improve visual ICL task performance over the original model. Table 2 shows mean and variance across 4 splits (full results in Table 5, Supplementary Materials). Our task-specific models outperform MAE-VQGAN in all tasks, with GRS models surpassing it in some. Our multi-task models excel in all tasks except Segmentation, where they match MAE-VQGAN. GRS multi-task matches the original model, while CMA improves upon random baselines but falls short of our method and GRS.

We validate the importance of ranking layers by $\rho_{layer}(l)$ (Suppl. 11.1) and selecting top-k for greedy random search performance. Random K layers generally underperform, except in colorization. Simply patching top quadrants without GRS also yields poor results. Qualitative results for Segmentation, Lowlight Enhancement, and In-painting are presented. We compare task-specific models of our method with original MAE-VQGAN, CMA, and GRS baselines (Fig. 3), and visualize task-specific and multi-task variants against CMA and GRS (Fig. 4).

Table 3. Optimal Patching Granularity. Patching into Tokens (T), Quadrants (Q), or Heads (H)

Model	Segmentation ↑ (Mean ± STD)	Lowlight Enhance ↓ (Mean ± STD)	Colorization ↓ (Mean ± STD)	In-painting ↓ (Mean ± STD)
Ours T (Task-specific)	**0.350 ± 0.025**	0.495 ± 0.036	**0.453 ± 0.036**	0.485 ± 0.018
Ours Q (Task-specific)	0.338 ± 0.023	**0.458 ± 0.032**	0.465 ± 0.036	**0.480 ± 0.022**
Ours H (Task-specific)	0.245 ± 0.015	0.942 ± 0.070	0.857 ± 0.069	0.885 ± 0.067
Ours T (Multi-task)	0.318 ± 0.023	0.510 ± 0.028	0.510 ± 0.039	0.565 ± 0.025
Ours Q (Multi-task)	**0.325 ± 0.026**	**0.492 ± 0.025**	**0.502 ± 0.036**	**0.558 ± 0.022**
Ours H (Multi-task)	0.253 ± 0.013	1.105 ± 0.064	0.998 ± 0.069	0.980 ± 0.082

Table 4. Isolating Task Locations. Patching into Encoder only, Decoder only, and both.

Model	Segmentation ↑			
	Split 0	Split 1	Split 2	Split 3
Encoder (Task-specific)	0.09	0.14	0.14	0.13
Decoder (Task-specific)	0.32	0.34	0.29	0.29
Both (Task-specific)	**0.35**	**0.35**	**0.31**	**0.29**

5.3 Ablations

Task Vectors Location in Encoder vs. Decoder. We compare interventions isolated to the encoder, decoder, and throughout the whole network. Results show that in-context task learning utilizes both components, with the decoder playing a more crucial role. Intervening in both components is essential for task implementation, supporting our hypothesis of distributed computation with cascading effects throughout the network (see Table 4).

Patching Granularity. Inspired by quadrant patterns in per-token scoring, we explore optimal token grouping to reduce search space dimensionality. Quadrant grouping improves performance for Segmentation and Colorization, while token-level granularity is better for Lowlight Enhancement and In-painting (see Table 3 for mean and variance across 4 splits, and Table 6 in Supplementary Materials for individual split evaluations).

6 Limitations

While we focused on Task Vectors, other important vector-types might exist, for example, vectors capturing image structure and ordering. We evaluated performance using MSE (except mIoU for Segmentation) after decoding VQGAN tokens. Future work could explore direct evaluation in VQGAN token space using cross entropy loss for potentially more accurate results.

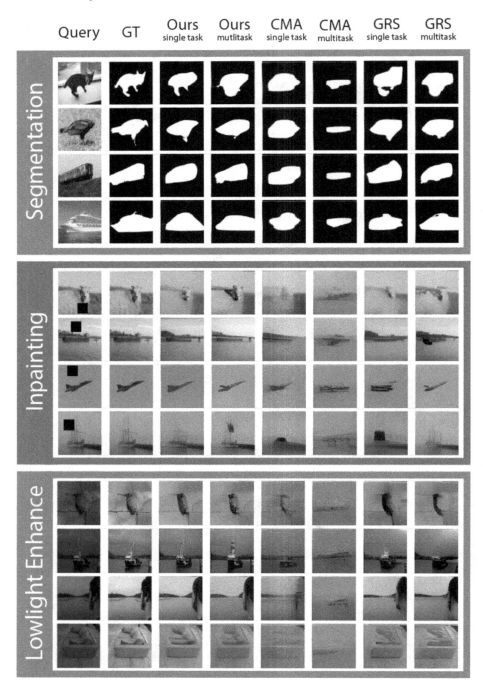

Fig. 4. Qualitative Examples. We qualitatively compare the task-specific and multi-task variants of our methodology with the CMA and GRS baselines. Our patching methodology performs better than the original MAE-VQGAN model.

7 Conclusion

In this work we explore the internal mechanisms of visual in-context learning and devise an algorithm to identify Task Vectors, activations present in transformers that can replace the in-context examples to guide the model into performing a specific task. We confirm our approach by adapting MAE-VQGAN to perform tasks without including the ICL demonstration in the prompt by patching the Task Vectors identified. We find that different than in NLP, in computer vision Task Vectors are distributed throughout the network's encoder and decoder.

Acknowledgments. This project has received funding from the European Research Council (ERC) under the European Unions Horizon 2020 research and innovation programme (grant ERC HOLI 819080). Prof. Darrell's group was supported in part by DoD including DARPA's LwLL and/or SemaFor programs, as well as BAIR's industrial alliance programs. This work was completed in partial fulfillment for the Ph.D degree of the last author.

References

1. Akyürek, E., Schuurmans, D., Andreas, J., Ma, T., Zhou, D.: What learning algorithm is in-context learning? Investigations with linear models. arXiv preprint arXiv:2211.15661 (2022)
2. Bahng, H., Jahanian, A., Sankaranarayanan, S., Isola, P.: Exploring visual prompts for adapting large-scale models. arXiv preprint arXiv:2203.17274 (2022)
3. Bai, Y., et al.: Sequential modeling enables scalable learning for large vision models. arXiv preprint arXiv:2312.00785 (2023)
4. Bar, A., Gandelsman, Y., Darrell, T., Globerson, A., Efros, A.: Visual prompting via image inpainting. In: Advances in Neural Information Processing Systems, vol. 35, pp. 25005–25017 (2022)
5. Bau, D., et al.: Gan dissection: visualizing and understanding generative adversarial networks. arXiv preprint arXiv:1811.10597 (2018)
6. Brown, T., et al.: Language models are few-shot learners. In: Advances in Neural Information Processing Systems, vol. 33, pp. 1877–1901 (2020)
7. Dai, D., Sun, Y., Dong, L., Hao, Y., Sui, Z., Wei, F.: Why can GPT learn in-context? Language models secretly perform gradient descent as meta optimizers. arXiv preprint arXiv:2212.10559 (2022)
8. Davies, D., Bouldin, D.: A cluster separation measure. IEEE Trans. Pattern Anal. Mach. Intell. **PAMI-1**, 224–227 (1979). https://doi.org/10.1109/TPAMI.1979.4766909
9. Dosovitskiy, A., et al.: An image is worth 16x16 words: transformers for image recognition at scale (2021)
10. Esser, P., Rombach, R., Ommer, B.: Taming transformers for high-resolution image synthesis. In: Proceedings of the IEEE/CVF Conference on Computer Vision and Pattern Recognition (CVPR), pp. 12873–12883, June 2021
11. Ferry, Q.R., Ching, J., Kawai, T.: Emergence and function of abstract representations in self-supervised transformers. arXiv preprint arXiv:2312.05361 (2023)
12. Gandelsman, Y., Efros, A.A., Steinhardt, J.: Interpreting CLIP's image representation via text-based decomposition. arXiv preprint arXiv:2310.05916 (2023)

13. Garg, S., Tsipras, D., Liang, P.S., Valiant, G.: What can transformers learn in-context? A case study of simple function classes. In: Advances in Neural Information Processing Systems, vol. 35, pp. 30583–30598 (2022)
14. Hahn, M., Goyal, N.: A theory of emergent in-context learning as implicit structure induction. arXiv preprint arXiv:2303.07971 (2023)
15. Han, C., Wang, Z., Zhao, H., Ji, H.: In-context learning of large language models explained as kernel regression. arXiv preprint arXiv:2305.12766 (2023)
16. He, K., Chen, X., Xie, S., Li, Y., Dollár, P., Girshick, R.B.: Masked autoencoders are scalable vision learners. CoRR abs/2111.06377 (2021). https://arxiv.org/abs/2111.06377
17. Hendel, R., Geva, M., Globerson, A.: In-context learning creates task vectors. arXiv preprint arXiv:2310.15916 (2023)
18. Jia, M., et al.: Visual prompt tuning. In: Avidan, S., Brostow, G., Cissé, M., Farinella, G.M., Hassner, T. (eds.) ECCV 2022. LNCS, vol. 13693, pp. 709–727. Springer, Cham (2022). https://doi.org/10.1007/978-3-031-19827-4_41
19. Jin, Z., et al.: Cutting off the head ends the conflict: a mechanism for interpreting and mitigating knowledge conflicts in language models. arXiv preprint arXiv:2402.18154 (2024)
20. Kingma, D.P., Ba, J.: Adam: a method for stochastic optimization (2017)
21. Li, X.L., Liang, P.: Prefix-tuning: optimizing continuous prompts for generation. arXiv preprint arXiv:2101.00190 (2021)
22. Liu, J., Shen, D., Zhang, Y., Dolan, B., Carin, L., Chen, W.: What makes good in-context examples for GPT-3? arXiv preprint arXiv:2101.06804 (2021)
23. Liu, S., Xing, L., Zou, J.: In-context vectors: making in context learning more effective and controllable through latent space steering. arXiv preprint arXiv:2311.06668 (2023)
24. Lu, S., Schuff, H., Gurevych, I.: How are prompts different in terms of sensitivity? arXiv preprint arXiv:2311.07230 (2023)
25. Lu, Y., Bartolo, M., Moore, A., Riedel, S., Stenetorp, P.: Fantastically ordered prompts and where to find them: overcoming few-shot prompt order sensitivity. arXiv preprint arXiv:2104.08786 (2021)
26. Luo, H., Specia, L.: From understanding to utilization: a survey on explainability for large language models. arXiv preprint arXiv:2401.12874 (2024)
27. van der Maaten, L., Hinton, G.: Visualizing data using t-SNE. J. Mach. Learn. Res. **9**(86), 2579–2605 (2008). http://jmlr.org/papers/v9/vandermaaten08a.html
28. Meng, K., Bau, D., Andonian, A., Belinkov, Y.: Locating and editing factual associations in GPT. In: Advances in Neural Information Processing Systems, vol. 35, pp. 17359–17372 (2022)
29. Moraffah, R., Karami, M., Guo, R., Raglin, A., Liu, H.: Causal interpretability for machine learning-problems, methods and evaluation. ACM SIGKDD Explor. Newsl. **22**(1), 18–33 (2020)
30. Palit, V., Pandey, R., Arora, A., Liang, P.P.: Towards vision-language mechanistic interpretability: a causal tracing tool for blip. In: Proceedings of the IEEE/CVF International Conference on Computer Vision, pp. 2856–2861 (2023)
31. Park, K., Choe, Y.J., Veitch, V.: The linear representation hypothesis and the geometry of large language models. arXiv preprint arXiv:2311.03658 (2023)
32. Pearl, J.: Direct and indirect effects. In: Probabilistic and Causal Inference: The Works of Judea Pearl, pp. 373–392 (2022)
33. Radford, A., Wu, J., Child, R., Luan, D., Amodei, D., Sutskever, I., et al.: Language models are unsupervised multitask learners. OpenAI Blog **1**(8), 9 (2019)

34. Rousseeuw, P.J.: Silhouettes: a graphical aid to the interpretation and validation of cluster analysis. J. Comput. Appl. Math. **20**, 53–65 (1987). https://doi.org/10.1016/0377-0427(87)90125-7. https://www.sciencedirect.com/science/article/pii/0377042787901257
35. Russakovsky, O., et al.: ImageNet large scale visual recognition challenge (2015)
36. Shaban, A., Bansal, S., Liu, Z., Essa, I., Boots, B.: One-shot learning for semantic segmentation. arXiv preprint arXiv:1709.03410 (2017)
37. Singh, C., Inala, J.P., Galley, M., Caruana, R., Gao, J.: Rethinking interpretability in the era of large language models. arXiv preprint arXiv:2402.01761 (2024)
38. Todd, E., Li, M.L., Sharma, A.S., Mueller, A., Wallace, B.C., Bau, D.: Function vectors in large language models. arXiv preprint arXiv:2310.15213 (2023)
39. Touvron, H., et al.: LLaMA: open and efficient foundation language models. arXiv preprint arXiv:2302.13971 (2023)
40. Vaswani, A., et al.: Attention is all you need. CoRR abs/1706.03762 (2017). http://arxiv.org/abs/1706.03762
41. Wang, B., Komatsuzaki, A.: GPT-J-6B: a 6 billion parameter autoregressive language model (2021)
42. Wei, J., et al.: Chain-of-thought prompting elicits reasoning in large language models. In: Advances in Neural Information Processing Systems, vol. 35, pp. 24824–24837 (2022)
43. Williams, R.J.: Simple statistical gradient-following algorithms for connectionist reinforcement learning. Mach. Learn. **8**, 229–256 (1992)
44. Wu, X., Varshney, L.R.: Transformer-based causal language models perform clustering. arXiv preprint arXiv:2402.12151 (2024)
45. Xie, S.M., Raghunathan, A., Liang, P., Ma, T.: An explanation of in-context learning as implicit Bayesian inference. arXiv preprint arXiv:2111.02080 (2021)
46. Xu, J., et al.: IMProv: inpainting-based multimodal prompting for computer vision tasks. arXiv preprint arXiv:2312.01771 (2023)
47. Xu, S., Dong, W., Guo, Z., Wu, X., Xiong, D.: Exploring multilingual human value concepts in large language models: is value alignment consistent, transferable and controllable across languages? arXiv preprint arXiv:2402.18120 (2024)
48. Zhang, F., Nanda, N.: Towards best practices of activation patching in language models: metrics and methods. arXiv preprint arXiv:2309.16042 (2023)
49. Zhang, K., Lv, A., Chen, Y., Ha, H., Xu, T., Yan, R.: Batch-ICL: effective, efficient, and order-agnostic in-context learning. arXiv preprint arXiv:2401.06469 (2024)
50. Zhang, Y., Tiňo, P., Leonardis, A., Tang, K.: A survey on neural network interpretability. IEEE Trans. Emerging Top. Comput. Intell. **5**(5), 726–742 (2021)
51. Zhang, Y., Zhou, K., Liu, Z.: What makes good examples for visual in-context learning? In: Advances in Neural Information Processing Systems, vol. 36 (2024)

Connecting Consistency Distillation to Score Distillation for Text-to-3D Generation

Zongrui Li[1,2], Minghui Hu[2], Qian Zheng[3,4(✉)], and Xudong Jiang[2]

[1] Rapid-Rich Object Search (ROSE) Lab, Nanyang Technological University, Singapore, Singapore
{ZONGRUI001,e200008}@e.ntu.edu.sg
[2] School of Electrical and Electronic Engineering, Nanyang Technological University, Singapore, Singapore
EXDJiang@ntu.edu.sg
[3] College of Computer Science and Technology, Zhejiang University, Hangzhou, China
[4] The State Key Lab of Brain-Machine Intelligence, Zhejiang University, Hangzhou, China
qianzheng@zju.edu.cn

Abstract. Although recent advancements in text-to-3D generation have significantly improved generation quality, issues like limited level of detail and low fidelity still persist, which requires further improvement. To understand the essence of those issues, we thoroughly analyze current score distillation methods by connecting theories of consistency distillation to score distillation. Based on the insights acquired through analysis, we propose an optimization framework, Guided Consistency Sampling (GCS), integrated with 3D Gaussian Splatting (3DGS) to alleviate those issues. Additionally, we have observed the persistent oversaturation in the rendered views of generated 3D assets. From experiments, we find that it is caused by unwanted accumulated brightness in 3DGS during optimization. To mitigate this issue, we introduce a Brightness-Equalized Generation (BEG) scheme in 3DGS rendering. Experimental results demonstrate that our approach generates 3D assets with more details and higher fidelity than state-of-the-art methods. The codes are released at https://github.com/LMozart/ECCV2024-GCS-BEG.

Keywords: Text-to-3D Generation · Score Distillation Sampling · Consistency Model

---

Z. Li and M. Hu—Equal contribution.

Supplementary Information The online version contains supplementary material available at https://doi.org/10.1007/978-3-031-72775-7_16.

1 Introduction

Text-to-3D generation [4,11,17,27,29,43,44] has gained substantial attention due to its great potential and indispensable role in many applications, such as gaming, filmmaking, and architecture. While an end-to-end text-to-3D generative model [11,27] is often difficult to train and lacks versatility due to the limited 3D assets in the training set, distillation 3D assets from a well-trained 2D generative model (*e.g.*, diffusion model (DM) [9,30]) has become a popular option due to its data-free feature. A typical work, DreamFusion [29], transfers knowledge from the pre-trained diffusion model to a learnable 3D representation through Score Distillation Sampling (SDS), generating text-aligned 3D assets.

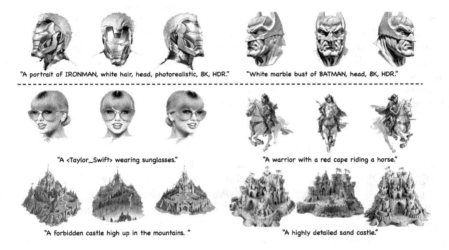

Fig. 1. Text-to-3D generation results of the proposed **Guided Consistency Sampling (GCS)** and **Brightness-equalized Generation**. Each 3D asset is distilled from a pre-trained 2D diffusion model and demonstrated with three different views. The results below the dotted line are generated by the fine-tuned diffusion models.

Nevertheless, as highlighted in many subsequent works [17,39,43–45], SDS suffers from a low level of detail in generation results. Such an issue stems from the poor generalization ability of the distillation method [43] and the inherent randomness in the sampling process [17,44]. Although work like [43] is proposed to enhance the generalization ability, it necessitates fine-tuning the diffusion model during training, dramatically extending the training time. By contrast, more recent works [17,44] aim to mitigate the randomness in the sampling process by introducing Probability Flow Ordinary Differential Equations (PF-ODEs) [38], and speed up the distillation process through 3D Gaussian Splatting (3DGS) [13], resulting in cost-efficient, high-quality 3D generation.

Despite the effectiveness of score distillation methods, issues like limited detail and low fidelity remain in the generated 3D asset. To reveal the essence

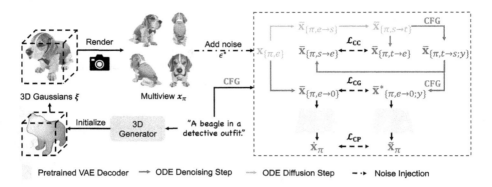

Fig. 2. An overview of the proposed GCS. We first initialize the 3D representation via the pre-trained 3D generator. For each training epoch, we randomly render a batch of views $\mathbf{x}_\pi$ and diffuse them to $\mathbf{x}_{\{\pi,e\}}$ with a fixed noise ϵ^*. We then apply the ODE diffusion process to gradually add noise to the $\mathbf{x}_{\{\pi,e\}}$ and transfer it to $\bar{\mathbf{x}}_{\{\pi,e\to s\}}$ and $\bar{\mathbf{x}}_{\{\pi,s\to t\}}$. In the denoising path, we conduct conditional and unconditional denoising steps, as shown in the figure. Eventually, we calculate the $\mathcal{L}_{\text{GCS}}$ (Eq. 17) to update the parameters of 3D representation (3DGS). Note that we add '*' on $\bar{\mathbf{x}}_{\{\pi,e\to 0;y\}}$ to indicate that it is obtained from different sampling trajectories.

of those issues, we conduct an in-depth analysis to connect consistency distillation [14,37,51], a method that distills information from a pre-trained DM through PF-ODEs, to score distillation. Such a novel view helps us identify potential causes of those issues in the score distillation process: 1) the inherent distillation errors of PF-ODEs, 2) ineffective conditional guidance accounting for classifier-free guidance (CFG) effects [10], and 3) lack of constraints in pixel domain to avoid out-of-distribution problem, requiring further improvements.

Except for identified issues in score distillation, there is another noteworthy concern in the 3D rendering process, *i.e*, the over-saturated views for generated 3D asset [17,43,44]. While the cause of such an issue is correlated with types of 3D representation, this paper particularly focuses on one (*i.e*, 3DGS) due to its remarkable reduction in optimization time. Through experiments, we observe that the brightness in generated 2D views accumulates to the next epoch during distillation, eventually leading to over-saturation.

To achieve better performance and overcome the aforementioned issues, this paper proposes **Guided Consistency Sampling** (GCS) with **Brightness-equalized Generation** (BEG) for 3DGS-based text-to-3D generation (Fig. 2). The proposed GCS includes three components to solve the identified issues in score distillation: a compact consistency loss reduces distillation errors, a conditional guidance loss provides more effective conditional guidance, and a constraint on pixel domain enhances the fidelity of the 3D assets. Additionally, the proposed BEG helps regularize the accumulation of brightness in 3DGS, which significantly alleviates over-saturation issues. Examples of generated 3D assets can be found in Fig. 1. In summary, our contributions are as follows:

1. We identify three problems in the PF-ODEs-based score distillation method by connecting consistency distillation to score distillation.
2. We propose **Guided Consistency Sampling**, an improved score distillation method, to enhance the details and fidelity of the generated 3D asset.
3. We find an accumulated brightness issue in the 3DGS-based rendering process that causes over-saturation and propose **Brightness-equalized Generation** to alleviate it in the 3DGS-based rendering process.

2 Related Work

Text-to-3D Generation by Score Distillation. SDS, introduced in Dream-Fusion [29], and concurrent Score Jacobian Chain (SJC) proposed in [42], have been regarded as milestone works in distilling information from a 2D pre-trained diffusion models to differentiable 3D representations, which inspiring many subsequent Text-to-3D generation methods [4,5,7,12,18,19,24,33,35,39–41,43,44,47–49,53] to improve the quality of generated 3D assets. Generally, they either endow the DM with the pose-aware generative ability [19,24,35,41,48,49] to alleviate the Janus problem [1] for a more consistent 3D asset, or redesign the distillation process and optimization objectives [4,5,12,18,43,44,47,53] to reduce the distillation error from the 2D pre-trained diffusion model for a higher level of details. While the performance of the former methods may be restricted by the inherent limitations of SDS [17], the latter methods dedicated to overcoming these limitations have recently gained increasing attention. For instance, ProlificDreamer [43] proposes sampling from the parameter distribution instead of seeking the optimal solution for high-fidelity generation. CSD [10] introduces insights into the effects of SDS's components and proposes improved strategies. More recent works like [17,44] introduce PF-ODEs into the SDS for a stable, deterministic distillation process. After thoroughly investigating recent works, we recognize a connection between consistency distillation and score distillation, which motivates us to analyze and redesign the optimization objectives in view of consistency distillation. Although concurrent work [44] also highlighted the effectiveness of text-to-3D generation using consistency distillation, our analysis delves deeper into such connection from a broader perspective. We also compare with the findings of [44] in Table 1. Another concurrent study [23] established a relationship between SDS and bridge matching [6,15], leading to an improvement similar to that seen in our method. Guided by acquired insights, we propose GCS that integrates techniques in CM to improve generation quality and gain state-of-the-art performance.

Differentiable 3D Representations, such as NeRF [17,25,29,43], 3DGS [7, 13,39], or learnable intrinsic that integrated with differentiable geometry [4, 34] and textures [2,3,8,16], can be optimized by minimizing the loss between rendered and ground truth images, establishing a connection between 3D and 2D representations. Text-to-3D by distillation relies heavily on differentiable 3D representations. While NeRF is well-known to be slow to optimize, we use 3DGS considering its efficiency and comparable performance with NeRF.

3 Preliminary

Diffusion Models (DM) [9,31] apply a forward process to perturb data samples $\mathbf{x}_0$ drawn from a data distribution $p_{\text{data}}(\mathbf{x})$ by the Gaussian noise $\mathcal{N}(\mathbf{0}, \sigma_t^2 \mathbf{I})$ with time-dependent variance σ_t, where $t \in [0, T]$. As such, the distribution of noisy samples $\mathbf{x}_t$ at time-step t is:

$$p_t(\mathbf{x}_t \mid \mathbf{x}_0) = \mathcal{N}(\mathbf{x}_t; \mathbf{x}_0, \sigma_t^2 \mathbf{I}), \tag{1}$$

which can be re-parameterized to $\mathbf{x}_t = \alpha_t \mathbf{x}_0 + \sigma_t \epsilon_t$, where $\epsilon_t \sim \mathcal{N}(\mathbf{0}, \mathbf{I})$. A reverse process can denoise the noisy samples from $\mathbf{x}_t$ to $\mathbf{x}_0$, expressed as variational inference of Markov processes [9]. For all diffusion processes, there exists a corresponding deterministic process with trajectory sharing the same marginal distribution, known as PF-ODE [38]:

$$\text{PF-ODE: } d\mathbf{x}_t = \left[\mathbf{f}(\mathbf{x}_t, t) - \frac{1}{2} g(t)^2 \nabla_{\mathbf{x}_t} \log p_t(\mathbf{x}_t) \right] dt, \tag{2}$$

where $\mathbf{f}(\mathbf{x}_t, t)$ is the drift coefficients, dt is the infinitesimal negative timestep, $\nabla_{\mathbf{x}_t} \log p_t(\mathbf{x}_t)$ is the score function of $p_t(\mathbf{x}_t)$, estimated by a neural network $\epsilon_\theta(\mathbf{x}_t, t)$. An ODE solver (*e.g.*, DDIM [36], DPM-Solver [20], etc.) can be used to derive the solution of ODE, identical as a sampling process. The sampling trajectory of ODE is deterministic, and the randomness comes from the initial value, which is more stable than a random sampling process described by the inverse Stochastic Differentiable Equation (SDE) [38].

Consistency Model (CM) [37] is proposed to facilitate a single-step or low number of function evaluations (NFEs) [28] generation by distilling knowledge from pre-trained DM models. It defines a one-step generator $\boldsymbol{f}_\theta(.;.)$ with trainable parameters θ that directly predicts the denoised image $\mathbf{x}_0$ given t and $\mathbf{x}_t$, constrained by some boundary conditions [37]. $\boldsymbol{f}_\theta(.;.)$ is trained by minimizing consistency distillation loss [37] defined as:

$$\mathcal{L}_{\text{CD}}(\boldsymbol{\theta}, \boldsymbol{\theta}^-) = \mathbb{E}\left[\omega(t) \left\| \boldsymbol{f}_\theta(\mathbf{x}_{t_{n+1}}; t_{n+1}) - \boldsymbol{f}_{\boldsymbol{\theta}^-}(\bar{\mathbf{x}}_{\{t_{n+1} \to t_n\}}; t_n) \right\|_2^2 \right], \tag{3}$$

where, $0 = t_1 < t_2 \cdots < t_N = T$, $\bar{\mathbf{x}}_{\{t_{n+1} \to t_n\}}$ is calculated given ODE solver $\Phi(.)$ as $\bar{\mathbf{x}}_{\{t_{n+1} \to t_n\}} = \Phi(\mathbf{x}_{t_{n+1}}; t_{n+1}, t_n)$, $\boldsymbol{\theta}^-$ is updated during training process through an exponential moving average (EMA) strategy [38]. The ultimate goal of CM is to maintain the self-consistency condition along the trajectory $\{\mathbf{x}_t\}_{t \in [0,T]}$, satisfying,

$$\boldsymbol{f}(\mathbf{x}_t; t) = \boldsymbol{f}(\mathbf{x}_{t'}; t') \quad \forall t, t' \in [0, T]. \tag{4}$$

Subsequent works of CM [10,14,51] integrate Classifier-free guidance (CFG) [10] or jumps along PF-ODE trajectory into the CM for high-fidelity samples given low or high NFEs, respectively. Inspired by those works, we seek to improve the performance of SDS further in view of CM given their connection.

4 Methodology

In Sect. 4.1, we analyze current advances [17,29,43,44,47] in a unified framework obtained by connecting consistency distillation [14,37,51] to score distillation. We revisited and identified three problems in the current PF-ODEs-based score distillation from the perspective of consistency distillation. We further proposed our solution, i.e., Guided Consistency Sampling (GCS), and explained in detail in Sect. 4.2. While GCS is integrated with 3DGS, we observe the accumulated brightness causing over-saturation in 3DGS, as explained in Sect. 4.3. We propose Brightness Equalized Generation (BEG) to alleviate this issue.

Table 1. Specific design space employed by proposed GCS and selected 2D diffusion-based score distillation text-to-3D generation method.

	DreamFusion [29]	ProlificDreamer [43]	Consistent3D [44]
3D Representation	NeRF/GS	NeRF	NeRF/GS
CFG Weight	–	7.5+	50+
Objective	SDS (Eq. 7)	VSD (Eq. 9)	CDS (Eq. 15)
Objective #1: Generator Loss			
term A	$\mathbf{x}_\pi$	$F_\phi\left(\mathbf{x}_{\{\pi,t\}}; t, y\right)$	$F_\theta\left(\mathbf{x}_{\{\pi,t\}}; t, y\right)$
term B	$F_\theta\left(\mathbf{x}_{\{\pi,t\}}; t, y\right)$	$F_\theta\left(\mathbf{x}_{\{\pi,t\}}; t, y\right)$	$F_\theta\left(\bar{\mathbf{x}}_{\{\pi,t \to s; y\}}; s, y\right)$
	LucidDreamer [17]	DreamFusion w/ CFG [29]	Ours
3D Representation	NeRF/GS	NeRF/GS	GS w/BEG
CFG Weight	7.5+	100+	7.5+
Objective	ISD (Eq. 13)	SDS w/ CFG (Eq. 10)	GCS (Eqs. 20, 22, 23)
Objective #1: Generator Loss			
term A	–	$\mathbf{x}_\pi$	$G_\theta\left(\hat{\mathbf{x}}_{\{\pi,t\}}; t, e, \emptyset\right)$
term B	–	$F_\theta\left(\mathbf{x}_{\{\pi,t\}}; t, y\right)$	$F_\theta\left(\bar{\mathbf{x}}_{\{\pi,t \to s; y\}}; s, e, \emptyset\right)$
Objective #2: Classifier Loss			
term A	$F_\theta\left(\hat{\mathbf{x}}_{\{\pi,s\}}; s, \emptyset\right)$	$F_\theta\left(\mathbf{x}_{\{\pi,t\}}; t, \emptyset\right)$	$F_\theta\left(\hat{\mathbf{x}}_{\{\pi,e\}}; e, \emptyset\right)$
term B	$F_\theta\left(\hat{\mathbf{x}}_{\{\pi,t\}}; t, y\right)$	$F_\theta\left(\mathbf{x}_{\{\pi,t\}}; t, y\right)$	$F_\theta\left(G_\theta\left(\hat{\mathbf{x}}_{\{\pi,t\}}; t, e, y\right); e, y\right)$
Pixel Constraint		+SDS on pixel domain [53]	$+\mathcal{L}_{\mathrm{CP}}$ (Eq. 23)

4.1 Connecting Consistency Distillation to Score Distillation

Through theoretical analysis, we find that consistency and score distillation share a similar optimization objective. While CM enforces a *self-consistency* of a PF-ODE sampling trajectory (Eq. 3), SDS and its variants (score distillation) facilitate a *cross-consistency*[1] between the diffusion trajectory of the rendered view and the denoising trajectory (usually with text condition) of a noisy sample. This cross-consistency is maintained by minimizing the gap between a sample on the diffusion and denoising trajectory. The score distillation method diversifies in choosing those samples to form the optimization objectives (see Table 1

---

[1] We define *cross-consistency* as an alignment between the diffusion trajectory of the rendered view and the text-conditioned sampling trajectory.

for a summary). In the following part, we unify the optimization objective of the score distillation method by sample ($\mathbf{x}$) instead of score (ϵ) to connect consistency distillation to score distillation based on their similarity.

In vanilla SDS, given a camera poses π, a 3D representation ξ can be projected to a specific 2D view, noted as $\mathbf{x}_\pi = g(\pi, \xi)$. The optimization objective of SDS on 3D parameters ξ is given as:

$$\min_{\xi} \mathcal{L}_{\text{SDS}}(\xi) = \mathbb{E}_{t,\pi} \left[\omega(t) \left\| \epsilon_\theta \left(\mathbf{x}_t; t, y \right) - \epsilon \right\|_2^2 \right], \tag{5}$$

where $\epsilon_\theta(\mathbf{x}_t; t, y)$ is the predicted noise given $\mathbf{x}_t$, time-step t, and y; $\omega(t)$ is a t related weight function. As suggested in [32,52,53], SDS loss is equivalent to:

$$\mathcal{L}_{\text{SDS}}(\xi) = \mathbb{E}_{t,\pi} \left[\omega(t) \left\| \epsilon_\theta \left(\mathbf{x}_{\{\pi,t\}}; t, y \right) - \epsilon \right\|_2^2 \right], \tag{6}$$

$$= \mathbb{E}_{t,\pi} \left[c(t) \left\| \mathbf{x}_{\{\pi,t\}} - F_\theta(\mathbf{x}_{\{\pi,t\}}; t, y) \right\|_2^2 \right], \tag{7}$$

$$=: \mathcal{L}_{\text{Distill}}(\xi), \tag{8}$$

where $\mathbf{x}_{\{\pi,t\}} = \alpha_t \mathbf{x}_\pi + \sigma_t \epsilon$, $F_\theta(\mathbf{x}_{\{\pi,t\}}; t, y) = \frac{\mathbf{x}_{\{\pi,t\}} - \sigma_t \epsilon_\theta(\mathbf{x}_{\{\pi,t\}}; t, y)}{\alpha_t}$, $c(t)$ is another t related weight function. In CM [9,22], $\boldsymbol{f}_\theta(.; .)$ can be parameterize as $F_\theta(.; .)$, which makes Eq. 3 similar to Eq. 7 in formula. As stated in [36,44], we interpret minimizing $\mathcal{L}_{\text{SDS}}$ as facilitating a cross-consistency of the stochastic and deterministic trajectory. The variance of SDS injects different conditions on samples. For instance, replacing $\mathbf{x}_\pi$ by samples from another fine-tuning diffusion model, noted as $F_\phi(\mathbf{x}_{\{\pi,t\}}; t, y)$ in Eq. 7, we have *Variational Score Distillation* (VSD) [43] as:

$$\mathcal{L}_{\text{VSD}}(\xi) = \mathbb{E}_{t,\pi} \left[c(t) \left\| F_\phi(\mathbf{x}_{\{\pi,t\}}; t, y) - F_\theta(\mathbf{x}_{\{\pi,t\}}; t, y) \right\|_2^2 \right], \tag{9}$$

which distills information by enforcing cross-consistency of the trajectory parameterized by ϕ and θ. In practice, CFG with a higher guidance weight w (*e.g.*, $w = 100$) is imperative in SDS. Specifically, the expression of SDS with CFG in the form of distillation loss [10,47] is:

$$\mathcal{L}_{\text{Distill}}^{\text{CFG}}(\xi) := \mathbb{E}_{t,\pi}[c(t) \| \underbrace{[\mathbf{x}_\pi - F_\theta(\mathbf{x}_{\{\pi,t\}}; t, y)]}_{\text{generator loss}}$$

$$+ w \underbrace{[F_\theta(\mathbf{x}_{\{\pi,t\}}; t, \emptyset) - F_\theta(\mathbf{x}_{\{\pi,t\}}; t, y)]}_{\text{classifier loss}} \|_2^2]. \tag{10}$$

Notably, we extend the definition in [47], noting the generator loss as a guidance to make $\mathbf{x}_\pi$ more close to the prior distribution and the classifier loss as an update direction for $\mathbf{x}_\pi$ to align with the text-condition. As studied in [47], the driving force of SDS is the classifier loss. By omitting the generator loss, they derive *Classifier Score Distillation* (CSD), given:

$$\mathcal{L}_{\text{CSD}}(\xi) := \mathbb{E}_{t,\pi}[c(t) \| [F_\theta(\mathbf{x}_{\{\pi,t\}}; t, \emptyset) - F_\theta(\mathbf{x}_{\{\pi,t\}}; t, y)] \|_2^2]. \tag{11}$$

Additionally, as shown in [17], $\mathbf{x}_{\{\pi,t\}}$ could be replaced by the deterministic status derived from DDIM to improve the distillation quality:

$$\hat{\mathbf{x}}_{\{\pi,t\}} = \sum_{k=0}^{s} \alpha_{(k+\delta_k)} \left(\frac{1}{\alpha_k} \mathbf{x}_k - \frac{\sigma_{(k+\delta_k)}}{\alpha_{(k+\delta_k)}} \epsilon_\theta(\mathbf{x}_k; k, \emptyset) \right) + \sigma_{(k+\delta_k)} \epsilon_\theta(\mathbf{x}_k; k, \emptyset), \quad (12)$$

where $t > s$ with $t = s + \delta_s$. Similar to derive Eq. 8, based on Eq. 12, we can express *Interval Score Distillation* (ISD) [17] as follows:

$$\mathcal{L}_{\text{ISD}}(\xi) = \mathbb{E}_{t,s,\pi} \left[c(t) \left\| F_\theta(\hat{\mathbf{x}}_{\{\pi,s\}}; s, \emptyset) - F_\theta(\hat{\mathbf{x}}_{\{\pi,t\}}; t, y) \right\|_2^2 \right], \quad (13)$$

which can be regarded as a classifier loss. With CFG, an additional generator loss is combined with ISD,

$$\mathcal{L}_{\text{ISD}}^{\text{CFG}}(\xi) = \mathbb{E}_{t,s,\pi}[c(t) \| [F_\theta(\hat{\mathbf{x}}_{\{\pi,s\}}; s, \emptyset) - F_\theta(\hat{\mathbf{x}}_{\{\pi,t\}}; t, \emptyset)] \\ + w[F_\theta(\hat{\mathbf{x}}_{\{\pi,t\}}; t, \emptyset) - F_\theta(\hat{\mathbf{x}}_{\{\pi,t\}}; t, y)], \|_2^2], \quad (14)$$

we find the generator loss in Eq. 14 is highly correlated to a *Consistency Distillation Sampling* (CDS) studied in [44]:

$$\mathcal{L}_{\text{CDS}}(\xi) = \mathbb{E}_{t,s,\pi} \left[c(t) \| F_\theta(\mathbf{x}_{\{\pi,t\}}; t, y) - F_\theta(\bar{\mathbf{x}}_{\{\pi,t \to s; y\}}; s, y) \|_2^2 \right], \quad (15)$$

where, $\bar{\mathbf{x}}_{\{\pi,t \to s; y\}} = F_\theta(\mathbf{x}_{\{\pi,t\}}; t, s, y) = \Phi(\mathbf{x}_{\{\pi,t\}}; t, s, y)$. Particularly, CDS is a special case as it mainly guides the generated views to match a particular origin of PF-ODE trajectory from the prior distribution through self-consistency. In such case, an upper bound for the distillation error can be deduced [44]:

$$\|\mathbf{x}_\pi - \mathbf{x}_0\|_2 = \mathcal{O}\left((\Delta t)^p\right) T. \quad (16)$$

where $\mathbf{x}_0 \sim p_{\text{data}}(\mathbf{x})$ is an real image, $\Delta t = \max\{|\delta_k|\}$, $k \in [0,...,s]$ and $t = s + \delta_s$. With CFG, CDS can further facilitate a cross-consistency of text and null condition trajectories. However, it performs poorly with low CFG weight [44], which needs further improvement. Inspired by the similarity of CM and the optimization objectives of the score distillation method, we are motivated to reformulate the optimization objective in SDS by extending the theories in CM to improve text-to-3D generation quality. We sum up our solution in three aspects: 1) we reduce the error bound suggested Eq. 16 to improve the distillation quality; 2) we provide more reliable guidance during distillation accounting for CFG effects; 3) we implement constraints on pixel domain to enhance the fidelity.

4.2 Guided Consistency Sampling

In this section, we introduce the Guided Consistency Sampling (GCS), which includes three parts of objectives: a Compact Consistency (CC) loss, a Conditional Guidance (CG) score, and a Constraint on Pixel domain (CP):

$$\mathcal{L}_{\text{GCS}}(\xi) = \mathcal{L}_{\text{CC}}(\xi) + \mathcal{L}_{\text{CG}}(\xi) + \mathcal{L}_{\text{CP}}(\xi). \quad (17)$$

Compact Consistency (CC) Loss aims to improve further the self-consistency of PF-ODE denoising trajectory, which eventually reduces the distillation error bound in Eq. 16 for a more aligned distribution of rendered views. Inspired by [14,51], we define a solution function:

$$G_\theta(\mathbf{x}_t; t, s, y) := \mathbf{x}_t + \int_t^s \frac{\mathbf{x}_u - \mathbb{E}[\mathbf{x}|\mathbf{x}_u]}{u} du. \quad (18)$$

which $G_\theta(\mathbf{x}_t; t, s, y)$ solves the PF-ODE from initial time t to a final time s according to exponential integrator [20,21,50]. Owing to the fact that G is intractable as it can only be obtained by calculating s partial derivative at time t, we follow the scheme in [51] and adhere to the first-order definition[2] of DPM-Solver [20], re-parameterised G as:

$$G_\theta(\mathbf{x}_t; t, s, y) = \frac{\sigma_s}{\sigma_t}\mathbf{x}_t - \alpha_s(e^{-h} - 1)\epsilon_\theta(\mathbf{x}_t, t, y), \quad (19)$$

where $h = \lambda_s - \lambda_t$ with the log-SNR λ defined as $\lambda = \log(\alpha/\sigma)$ and ϵ_θ is the prediction from the network. According to the previous definition, we mathematically define Compact Consistency Loss, a critical part of GCS, to improve the details of the 3D asset:

$$\mathcal{L}_{\text{CC}}(\xi) = \mathbb{E}_{t,s,e,\pi}\left[\|G_\theta(\hat{\mathbf{x}}_{\{\pi,t\}}; t, e, \emptyset) - G_\theta(\bar{\mathbf{x}}_{\{\pi,t\to s;y\}}; s, e, \emptyset)\|_2^2\right], \quad (20)$$

where $t > s > e$, and $\hat{\mathbf{x}}_{\{\pi,t\}}$ is obtained by DDIM inversion from $\mathbf{x}_{\{\pi,e\}}$, calculated as $\mathbf{x}_{\{\pi,e\}} = \alpha_e \mathbf{x}_\pi + \sigma_e \epsilon^*$ to $\bar{\mathbf{x}}_{\{\pi,e\to s\}}$ and eventually to $\bar{\mathbf{x}}_{\{\pi,s\to t\}}$. A null condition is applied to all DDIM inversion steps. Notably, ϵ^* is the random noise that will only be sampled once and kept fixed in practice[3]. During the training process, the parameters in the pre-trained model are frozen. In this condition, we substantiated Lemma 1, which extends the premises established in [37,44].

Lemma 1 ([14,37,44,51]). *Let $\Delta t = \max\{|\delta_k|\}$, $k \in [0,...,n_s]$, where n_s is the index of δ at time step s, and $F_\theta(\cdot,\cdot)$ is the origin prediction function grounded on the empirical PF-ODE. Assume F_θ satisfies the Lipschitz condition, if there is a $\mathbf{x}_\pi$ satisfying $\mathcal{L}_{\text{CC}}(\xi) = 0$, given an image $\mathbf{x}_0 \sim p_{data}(\mathbf{x})$, for any $t, s, e \in [0,...,T]$ with $t > s > e$, we have:*

$$\sup_{t,e,\mathbf{x}_\pi} \|\hat{\mathbf{x}}_{\{\pi,e\}}, \hat{\mathbf{x}}_{\{0,e\}}\|_2 = \mathcal{O}\left((\Delta t)^p\right)(T - e), \quad (21)$$

$\hat{\mathbf{x}}_{\{0,e\}}$ *is the distribution of $\mathbf{x}_0$ diffused to time e, p is the order of ODE solver.*

Proof. The proof is provided in the Appendix for completeness.

---

[2] For the effect of higher orders, please refer to the Appendix for more details.
[3] We use ϵ^* to ensure a valid diffusion process at low time step and reduce the computational cost, compared to [17].

Fig. 3. Generated views by using $\mathcal{L}_{CC}$ with different CFG strategies at a low CFG weight ($w = 7.5$). While $\mathcal{L}_{CC}$ (left) implements CFG in only one denoising step, $\mathcal{L}_{CC}^*$ (right) applies CFG in every ODE denoising step.

Lemma 1 validates that CC achieves a lower upper limit on error margins than CDS (Eq. 16) for a better distillation from the pre-trained model. Considering the importance of CFG in generating high-quality contents, another issue worth noting is whether it is practical to integrate CFG (or Perp-Neg [1]) into a generator loss $\mathcal{L}_{CC}$. In CDS, it applies CFG in every ODE denoising step, which leads to a large accumulated error magnified by CFG [22,26]. Instead, we only implement CFG in one step [22] (see Fig. 2), leading to lower accumulated errors. This strategy significantly improves the generation quality with low CFG weight (see Fig. 3). However, we still observe artifacts in the generated 3D model, likely due to a lack of effective classifier loss that provides text-conditional guidance.

Conditional Guidance (CG) Loss is proposed to accommodate guidance while maintaining a lower accumulated error, which we interpret as conditional guidance. Inspired by [10,17,26], we facilitate an alignment between the unconditional trajectory ($\mathcal{T}_{\{e \to 0, \emptyset\}}$ from a noisy sample at e to 0) and the text-conditional trajectory ($\mathcal{T}_{\{t \to e \to 0, y\}}$ derived from a noisy sample $\hat{\mathbf{x}}_{\{\pi, t\}}$ then sampling down to e and 0), separately. Mathematically, the CG score can be expressed as:

$$\mathcal{L}_{CG}(\xi) = \mathbb{E}_{t,e,\pi} \left[\| F_\theta(\hat{\mathbf{x}}_{\{\pi,e\}}; e, \emptyset) - F_\theta(G_\theta(\hat{\mathbf{x}}_{\{\pi,t\}}; t, e, y); e, y) \|_2^2 \right], \quad (22)$$

where $F_\theta(\mathbf{x}; t, y)$ predicts the $\mathbf{x}_0$ from time step t given the conditional information y as shown in Eq. 8, and $G_\theta(\mathbf{x}; t, e, y)$ access the midpoint e of the trajectory from initial timestep t as explained in Eq. 19. In practice, we can integrate CFG into $\mathcal{L}_{CG}$ for better performance. We believe the proposed $\mathcal{L}_{CG}$ provides more reliable guidance with less accumulated error, as it avoids long navigation along the trajectory. Specifically, the gap between $\mathcal{T}_{\{e \to 0, \emptyset\}}$ and $\mathcal{T}_{\{t \to e \to 0, y\}}$ is a more precise guidance since it directly affects $\hat{\mathbf{x}}_{\{\pi,e\}}$, which is a more similar sample to $\mathbf{x}_\pi$ [26] due to the DDIM inversion.

It is also interesting to see that $\mathcal{L}_{ISD}$ is equivalent to $\mathcal{L}_{CG}$ when set $e = 0$ in computing $F_\theta(G_\theta(\hat{\mathbf{x}}_{\{\pi,t\}}; t, e, y); e, y)$. We intuitively explain the effect of a midpoint e to reduce the error of one-step denoising from a noisy sample conditioned at large t. Eventually, we find $\mathcal{L}_{CC}$ and $\mathcal{L}_{CG}$ work together improving the quality of generated 3D asset.

Constraint on Pixel (CP) Domain aims to achieve a closer resemblance in the pixel domain for $\mathcal{L}_{CG}$. While a satisfactory $\mathbf{x}_\pi$ is obtained in the latent domain, whether such an $\mathbf{x}_\pi$ is equally satisfactory in the pixel domain appears to have been overlooked. In fact, the optimized $\mathbf{x}_\pi$ may not adhere to the prior distribution stipulations of VAE, leading to out-of-distribution artifacts. Consequently, this leads to unrealistic color in the decoded images (*e.g.*, a paint-like color in the generated results of LucidDreamer [17]). Inspired by similar methods proposed in [53], we further calculate $\mathcal{L}_{\text{CG}}$ in the pixel domain for enhanced supervision. In this context, given an image decoder $\mathcal{D}$ along with $\hat{\mathbf{x}}_{\{\pi,e\}}$, the definition of CP can be articulated as follows:

$$\mathcal{L}_{\text{CP}}(\xi) = \mathbb{E}_{t,e,\pi}\left[\|\mathcal{D}(\dot{\mathbf{x}}_\pi) - \mathcal{D}(\tilde{\mathbf{x}}_\pi)\|_2^2\right], \tag{23}$$

$$\dot{\mathbf{x}}_\pi = F_\theta(G_\theta(\hat{\mathbf{x}}_{\{\pi,t\}}; t, e, y); e, y), \tag{24}$$

$$\tilde{\mathbf{x}}_\pi = F_\theta(\hat{\mathbf{x}}_{\{\pi,e\}}; e, \emptyset). \tag{25}$$

Our empirical observations indicate that optimizing CP yields notable enhancements in color fidelity, albeit with increased computational cost and oversaturation. Consequently, we suggest an additional strategy outlined in Sect. 4.3, aimed at alleviating over-saturation. Extensive experiments lead us to conclude that the CP term is constructive in bolstering the overall quality of 3D assets.

4.3 Brightness-Equalized Generation

A common problem in previous works is that the generated 3D assets often suffer from over-saturation, especially in those that use Gaussian Splatting as the 3D representation to be trained. We experimentally find that the high brightness region (highlight) generated in the current training epoch is carried over to the next epoch[4]. That is, the brightness of the highlight points accumulates during training, eventually leading to over-saturation. To alleviate this issue, we propose resetting the brightness of the Gaussian[5] adaptively according to the exposure status of the generated view. Specifically, in each training epoch, we calculate the $m^{\text{th}}(85)$ percentile of each image $\mathbf{x}_\pi^i$ in an image batch $\mathbf{X}_\pi = \{\mathbf{x}_\pi^i, i \in [1, ..., B]\}$, noted as $P_m = \{p_m^i, i \in [1, ..., B]\}$, where B is the batch size. We then find the maximum value in P_m and reset the brightness of Gaussian to $T_\text{B} = 0.8$ of the current brightness if $\max(P_m) > T_\text{GS}$, $T_\text{GS} = 0.9$. Such a simple method will significantly alleviate the over-saturation issues and facilitate brightness equalization.

5 Experiments

5.1 Implementation Details

We implement the GCS with PyTorch and train it on an A800 GPU with the Adam optimizer. Some hyperparameters (the learning rates, camera positions,

---

[4] A visualization of the brightness accumulation can be found in the Appendix.
[5] We use the average of RGB channels as the brightness.

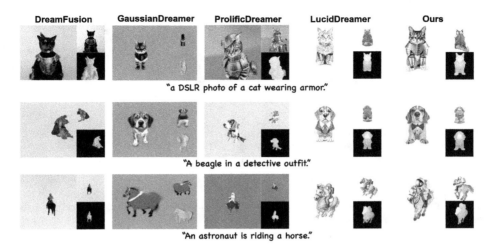

Fig. 4. Qualitative comparison among the proposed and other methods on text-to-3D generation results. From left to right, results generated by DreamFusion [29], GaussianDreamer [46], ProlificDreamer [43], LucidDreamer [17], and the proposed method, with a CFG weight 100, 100, 7, 7, 7, respectively. For each sub-figure, left: main view, right-top: back view, right-bottom: normal/depth map of the 3D asset.

rendering resolution, *etc..*) are similar to that in [17]. We use 5000 epochs as [43] for total training iterations, which takes about one hour per scene.

Time Schedule. In the experiment, t is sampled from a uniform distribution $\mathcal{U}(20, 500 + \delta_{\text{warm}})$, where δ_{warm} is linearly decreasing from 480 to 0 in the first 1500 epochs. $s = t - \delta$, where $\delta = 100$ for most cases. e is sampled from a uniform distribution as $\mathcal{U}(s - \delta, s - \frac{\delta}{10})$.

Initialization. We apply Shap-E [11] and Point-E [27] to initialize the 3D Gaussians, which are widely used in 3DGS-based text-to-3D generation [17,44].

5.2 Text-to-3D Generation

Qualitative Comparison. We have showcased diverse views of the 3D assets generated by our versatile approach, indicating that we can not only generate highly detailed figures but also create reasonable outputs to align with the complex prompt (results shown in the first row). To comprehensively assess the efficacy of our proposed Guided Conditional Sampling (GCS), we conducted a comparative analysis against the current state-of-the-art in GS-based [17,46] and baseline NeRF-based [29,43] methods in text-to-3D generation, as shown in Fig. 4. To ensure a fair comparison, all listed methods implement Stable Diffusion 2.1 as the base model. Other hyperparameters were configured following the default settings of the respective methods. The results indicate that previous methods often struggled with insufficient details, prominent artifacts (*e.g.*, Janus problem [1]), or over-saturation. In contrast, our approach exhibits notable

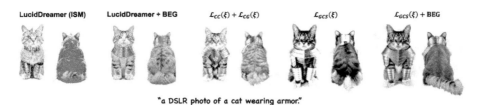

Fig. 5. Ablation study of proposed components. LucidDreamer [17] serves as the baseline, we demonstrate the results under settings: (a) LucidDreamer [17] + BEG, (b) $\mathcal{L}_{\mathrm{CC}}(\xi) + \mathcal{L}_{\mathrm{CG}}(\xi)$, (c) $\mathcal{L}_{\mathrm{GCS}}(\xi)$, and (d) full mode ($\mathcal{L}_{\mathrm{GCS}}(\xi)$+ BEG) from left to right.

improvements in detail (*e.g.*, the armor of the cat), fidelity, and exposure. Specifically, we can generate highly detailed textures, achieve photorealistic effects (such as lights on the cat's armor), and maintain more consistent colors (such as the cat's fur color) even with a low CFG weight, illustrating the effectiveness of our proposed GCS and the brightness-equalization techniques employed during training. Additional results can be found in the Appendix.

Quantitative Comparison. We present the average CLIP Score, FID Score, and user preference of five comparative methods [17,29,39,43,46] and the proposed GCS on 120 generated views across 30 prompts selected from the Dream-Fusion gallery[6] in Table 2. For CLIP Score computation, we follow the steps presented in [29]. Results show that our performance is comparable to that of VSD [43] in the evaluation of prompt alignment, with much fewer instances of Janus problems [1], as evidenced in the qualitative comparison shown in Fig. 4. The proposed approach exhibits improved 3D consistency and fidelity for the 3D asset, resulting in a higher user preference in prompt alignment.

For FID, we follow similar steps in [43] to calculate FID between 120 views of 3D object[7] and 50k Stable Diffusion 2.1 generated images. We implement Perp-Neg when necessary for different objects. The results show that the generated views of the proposed method are more similar to the prior distribution compared to other methods, indicating that the proposed GCS reduces distillation error.

User Study. We further conduct a user study among 30 volunteers for a more comprehensive evaluation. Specifically, we asked the volunteers to evaluate the generation quality regarding *brightness* (Q1), *prompt alignment* (Q2), and *fidelity* (Q3). Those volunteers are also required to indicate their preference based on rendered 360° videos of 5 objects randomly sampled from 30 examples within six methods. We shuffled the presented order of the generated results to avoid any form of leakage. The results of user preference in percentage form shown in Table 2 indicate that our method performs best in all aspects. An example question can be found in the Appendix.

---

[6] https://dreamfusion3d.github.io/gallery.html.
[7] Views are generated with fixed elevation (0°, 3.6k views generated in total) and varying azimuth (uniformly covering 360°).

Table 2. Quantitative comparison regarding CLIP Score, FID Score, and user preference on three given questions (**Q1**, **Q2**, and **Q3**). The **bold** (<u>underline</u>) number indicates the best (second-best) results.

	GaussianDreamer	DreamGaussian	DreamFusion	LucidDreamer	ProlificDreamer	Ours
CLIP Score↑	30.29	28.65	30.68	31.30	**33.17**	<u>32.37</u>
FID Score↓	133.48	215.89	137.22	<u>109.57</u>	119.53	**103.40**
User Preference: Q1↑[a]	9.47	1.05	6.32	<u>24.21</u>	12.63	**46.32**
User Preference: Q2↑[b]	10.53	1.05	6.32	<u>30.53</u>	7.37	**44.21**
User Preference: Q3↑[c]	9.47	2.11	6.32	<u>26.32</u>	11.58	**44.21**

[a]**Q1**: From the perspective of **color brightness**, which of the following methods produces the most balanced brightness (e.g., no over-saturation, moderate saturation)?
[b]**Q2**: From the perspective of **prompt alignment**, which of the following methods align most with the prompt?
[c]**Q3**: From the perspective of **fidelity**, which of the following methods produces the most realistic objects?

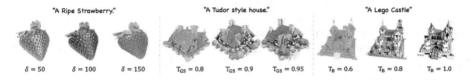

Fig. 6. Ablation study of hyperparameters' effects (left: δ, middle: T_{GS}, right: T_{B})

5.3 Ablation Study

We conducted an ablation study to assess the effects of parameter setup (including δ, T_{GS}, and T_{B}), the proposed Guided Consistency Sampling (GCS) loss, and the brightness-equalized generation method (BEG).

Effect of Time Interval δ is evaluated by an ablation study on δ shown in Fig. 6. We find that large δ causes over-smoothness (e.g., low level of detail on the strawberry's surface), while small δ causes color distortion (e.g., unrealistic color in strawberry). 100 is set to generate visual-pleasant results.

Effect of Different Loss and BEG[8]. As illustrated in Fig. 5, we compared the full model's ($\mathcal{L}_{\text{GCS}}(\xi)$ + BEG) generated results with those from Lucid-Dreamer [17], and two alternative models: $\mathcal{L}_{\text{CC}}(\xi) + \mathcal{L}_{\text{CG}}(\xi)$ and $\mathcal{L}_{\text{GCS}}(\xi)$ without BEG method, with a CFG weight set to 7.5. Compared to LucidDreamer [17], the full model successfully generated more intricate details (e.g., armor's patterns and higher armor coverage). While there was a slight alleviation in color distortion (inconsistency of fur color between the back and front of the cat), the issue may persist. Consequently, we introduced $\mathcal{L}_{\text{CP}}$ to mitigate color distortion further (comparing $\mathcal{L}_{\text{CC}}(\xi) + \mathcal{L}_{\text{CG}}(\xi)$ and $\mathcal{L}_{\text{GCS}}(\xi)$).

---

[8] We conduct additional ablation studies regarding CC and CG to analyze the individual effect on generated 3D assets in the Appendix.

However, we observed a significant increase in overall brightness during optimization. The implementation of the BEG method notably alleviates this issue, as evidenced in the qualitative comparison between $\mathcal{L}_{\text{GCS}}(\xi)$ and the full model in Fig. 5. We find that BEG will only affect the overall brightness, but not the fidelity, as shown in a qualitative comparison between LucidDreamer [17] and LucidDreamer [17] + BEG. The effect of BEG is controlled by T_B and T_GS. Specifically, large (small) T_B causes over-saturation (color distortion) results (see the comparison in Fig. 6). $T_\text{B} = 0.8$ works for general cases. Small T_GS will reset the brightness too frequently to cause artifacts, $T_\text{GS} = 0.9$ is a suitable setup.

6 Conclusion

In this work, we connect consistency distillation to score distillation. From this connection, we propose Guided Consistency Sampling (GCS), an optimization framework that includes three parts: Compact Consistency (CC) loss for improved generator loss, Conditional Guidance (CG) score to enhance conditional guidance, and Constraints in the Pixel domain (CP) for improved fidelity. In addition, we innovate the Brightness-equalized Generation (BEG) to tackle the over-saturation issue in 3DGS training. The proposed approach improves the details, quality, and lighting effects of the generated 3D assets.

Acknowledgement. This work is supported by Rapid-Rich Object Search (ROSE) Lab, Nanyang Technological University, Singapore, and the State Key Lab of Brain-Machine Intelligence, Zhejiang University, Hangzhou, China.

References

1. Armandpour, M., Zheng, H., Sadeghian, A., Sadeghian, A., Zhou, M.: Re-imagine the negative prompt algorithm: transform 2d diffusion into 3D, alleviate Janus problem and beyond. arXiv preprint arXiv:2304.04968 (2023)
2. Bensadoun, R., et al.: Meta 3D TextureGen: fast and consistent texture generation for 3d objects. arXiv preprint arXiv:2407.02430 (2024)
3. Burley, B., Studios, Walt Disney Animation: Physically-based shading at Disney. In: ACM SIGGRAPH (2012)
4. Chen, R., Chen, Y., Jiao, N., Jia, K.: Fantasia3D: disentangling geometry and appearance for high-quality text-to-3D content creation. In: International Conference on Computer Vision (2023)
5. Chen, Y., et al.: It3D: improved text-to-3D generation with explicit view synthesis. In: AAAI (2024)
6. Chen, Y., Georgiou, T.T., Pavon, M.: On the relation between optimal transport and Schrödinger bridges: a stochastic control viewpoint. J. Optim. Theory Appl. **169**, 671–691 (2016)
7. Chen, Z., Wang, F., Wang, Y., Liu, H.: Text-to-3D using gaussian splatting. In: IEEE Conference on Computer Vision and Pattern Recognition (2024)
8. Deng, K., et al.: FlashTex: fast relightable mesh texturing with lightcontrolnet. arXiv preprint arXiv:2402.13251 (2024)

9. Ho, J., Jain, A., Abbeel, P.: Denoising diffusion probabilistic models. In: Advances in Neural Information Processing Systems (2020)
10. Ho, J., Salimans, T.: Classifier-free diffusion guidance. arXiv preprint arXiv:2207.12598 (2022)
11. Jun, H., Nichol, A.: Shap-E: generating conditional 3D implicit functions. arXiv preprint arXiv:2305.02463 (2023)
12. Katzir, O., Patashnik, O., Cohen-Or, D., Lischinski, D.: Noise-free score distillation. arXiv preprint arXiv:2310.17590 (2023)
13. Kerbl, B., Kopanas, G., Leimkühler, T., Drettakis, G.: 3D gaussian splatting for real-time radiance field rendering. ACM Trans. Graph. **42**, 1–14 (2023)
14. Kim, D., et al.: Consistency trajectory models: learning probability flow ode trajectory of diffusion. In: The International Conference on Learning Representations (2024)
15. Léonard, C.: A survey of the Schrödinger problem and some of its connections with optimal transport. arXiv preprint arXiv:1308.0215 (2013)
16. Li, Z., Zheng, Q., Shi, B., Pan, G., Jiang, X.: DANI-Net: uncalibrated photometric stereo by differentiable shadow handling, anisotropic reflectance modeling, and neural inverse rendering. In: IEEE Computer Society Conference on Computer Vision and Pattern Recognition (2023)
17. Liang, Y., Yang, X., Lin, J., Li, H., Xu, X., Chen, Y.: LucidDreamer: towards high-fidelity text-to-3D generation via interval score matching. arXiv preprint arXiv:2311.11284 (2023)
18. Lin, C.H., et al.: Magic3D: high-resolution text-to-3D content creation. In: IEEE Conference on Computer Vision and Pattern Recognition (2023)
19. Liu, R., Wu, R., Van Hoorick, B., Tokmakov, P., Zakharov, S., Vondrick, C.: Zero-1-to-3: Zero-shot one image to 3D object. In: The International Conference on Computer Vision (2023)
20. Lu, C., Zhou, Y., Bao, F., Chen, J., Li, C., Zhu, J.: DPM-Solver: a fast ode solver for diffusion probabilistic model sampling in around 10 steps. In: Advances in Neural Information Processing Systems (2022)
21. Lu, C., Zhou, Y., Bao, F., Chen, J., Li, C., Zhu, J.: DPM-Solver++: fast solver for guided sampling of diffusion probabilistic models. arXiv preprint arXiv:2211.01095 (2022)
22. Luo, S., Tan, Y., Huang, L., Li, J., Zhao, H.: Latent consistency models: synthesizing high-resolution images with few-step inference. arXiv preprint arXiv:2310.04378 (2023)
23. McAllister, D., et al.: Rethinking score distillation as a bridge between image distributions. arXiv preprint arXiv:2406.09417 (2024)
24. Metzer, G., Richardson, E., Patashnik, O., Giryes, R., Cohen-Or, D.: Latent-NeRF for shape-guided generation of 3D shapes and textures. In: Proceedings of the IEEE/CVF Conference on Computer Vision and Pattern Recognition, pp. 12663–12673 (2023)
25. Mildenhall, B., Srinivasan, P.P., Tancik, M., Barron, J.T., Ramamoorthi, R., Ng, R.: NeRF: representing scenes as neural radiance fields for view synthesis. In: Vedaldi, A., Bischof, H., Brox, T., Frahm, J.-M. (eds.) ECCV 2020. LNCS, vol. 12346, pp. 405–421. Springer, Cham (2020). https://doi.org/10.1007/978-3-030-58452-8_24
26. Mokady, R., Hertz, A., Aberman, K., Pritch, Y., Cohen-Or, D.: Null-text inversion for editing real images using guided diffusion models. 2023 IEEE. In: CVF Conference on Computer Vision and Pattern Recognition (CVPR), pp. 6038–6047 (2022)

27. Nichol, A., Jun, H., Dhariwal, P., Mishkin, P., Chen, M.: Point-E: a system for generating 3D point clouds from complex prompts. arXiv preprint arXiv:2212.08751 (2022)
28. Pham, H.A.: Reduction of function evaluation in differential evolution using nearest neighbor comparison. Vietnam J. Comput. Sci. **2**, 121–131 (2015)
29. Poole, B., Jain, A., Barron, J.T., Mildenhall, B.: DreamFusion: text-to-3D using 2D diffusion. In: The International Conference on Learning Representations (2022)
30. Rombach, R., Blattmann, A., Lorenz, D., Esser, P., Ommer, B.: High-resolution image synthesis with latent diffusion models. In: IEEE Conference on Computer Vision and Pattern Recognition (2022)
31. Saharia, C., et al.: Photorealistic text-to-image diffusion models with deep language understanding. In: Advances in Neural Information Processing Systems (2022)
32. Sauer, A., Lorenz, D., Blattmann, A., Rombach, R.: Adversarial diffusion distillation. arXiv preprint arXiv:2311.17042 (2023)
33. Seo, J., et al.: Let 2D diffusion model know 3D-consistency for robust text-to-3D generation. arXiv preprint arXiv:2303.07937 (2023)
34. Shen, T., Gao, J., Yin, K., Liu, M.Y., Fidler, S.: Deep marching tetrahedra: a hybrid representation for high-resolution 3D shape synthesis. In: Advances in Neural Information Processing Systems (2021)
35. Shi, Y., Wang, P., Ye, J., Long, M., Li, K., Yang, X.: MVDream: multi-view diffusion for 3D generation. arXiv preprint arXiv:2308.16512 (2023)
36. Song, J., Meng, C., Ermon, S.: Denoising diffusion implicit models. arXiv preprint arXiv:2010.02502 (2020)
37. Song, Y., Dhariwal, P., Chen, M., Sutskever, I.: Consistency models. arXiv preprint arXiv:2303.01469 (2023)
38. Song, Y., Sohl-Dickstein, J., Kingma, D.P., Kumar, A., Ermon, S., Poole, B.: Score-based generative modeling through stochastic differential equations. arXiv preprint arXiv:2011.13456 (2020)
39. Tang, J., Ren, J., Zhou, H., Liu, Z., Zeng, G.: DreamGaussian: generative gaussian splatting for efficient 3D content creation. arXiv preprint arXiv:2309.16653 (2023)
40. Tang, J., et al.: Make-it-3D: high-fidelity 3D creation from a single image with diffusion prior. In: IEEE Conference on Computer Vision and Pattern Recognition (2023)
41. Tsalicoglou, C., Manhardt, F., Tonioni, A., Niemeyer, M., Tombari, F.: TextMesh: generation of realistic 3D meshes from text prompts. In: International Conference on 3D Vision (3DV) (2024)
42. Wang, H., Du, X., Li, J., Yeh, R.A., Shakhnarovich, G.: Score Jacobian chaining: lifting pretrained 2D diffusion models for 3D generation. In: IEEE Conference on Computer Vision and Pattern Recognition (2023)
43. Wang, Z., et al.: ProlificEDreamer: high-fidelity and diverse text-to-3D generation with variational score distillation. In: Advances in Neural Information Processing Systems (2024)
44. Wu, Z., Zhou, P., Yi, X., Yuan, X., Zhang, H.: Consistent3D: towards consistent high-fidelity text-to-3D generation with deterministic sampling prior. arXiv preprint arXiv:2401.09050 (2024)
45. Yang, X., et al.: Learn to optimize denoising scores for 3D generation: a unified and improved diffusion prior on NeRF and 3D gaussian splatting. arXiv preprint arXiv:2312.04820 (2023)
46. Yi, T., et al.: GaussianDreamer: fast generation from text to 3D gaussian splatting with point cloud priors. In: IEEE Conference on Computer Vision and Pattern Recognition (2024)

47. Yu, X., Guo, Y.C., Li, Y., Liang, D., Zhang, S.H., Qi, X.: Text-to-3D with classifier score distillation. arXiv preprint arXiv:2310.19415 (2023)
48. Zhang, H., et al.: AvatarVerse: high-quality & stable 3D avatar creation from text and pose. In: AAAI (2024)
49. Zhang, J., et al.: AvatarStudio: high-fidelity and animatable 3D avatar creation from text. arXiv preprint arXiv:2311.17917 (2023)
50. Zhang, Q., Chen, Y.: Fast sampling of diffusion models with exponential integrator. arXiv preprint arXiv:2204.13902 (2022)
51. Zheng, J., et al.: Trajectory consistency distillation. arXiv preprint arXiv:2402.19159 (2024)
52. Zhou, Z., Tulsiani, S.: SparseFusion: distilling view-conditioned diffusion for 3D reconstruction. In: Proceedings of the IEEE/CVF Conference on Computer Vision and Pattern Recognition, pp. 12588–12597 (2023)
53. Zhu, J., Zhuang, P.: HiFA: high-fidelity text-to-3D with advanced diffusion guidance. arXiv preprint arXiv:2305.18766 (2023)

Event Camera Data Dense Pre-training

Yan Yang[2], Liyuan Pan[1(✉)], and Liu Liu[3]

[1] School of CSAT, Beijing Institute of Technology, Beijing, China
Liyuan.Pan@bit.edu.cn
[2] BDSI, Australian National University, Canberra, Australia
Yan.Yang@anu.edu.au
[3] KooMap Department, Huawei, Beijing, China
liuliu33@huawei.com

Abstract. This paper introduces a self-supervised learning framework designed for pre-training neural networks tailored to dense prediction tasks using event camera data. Our approach utilizes solely event data for training.

Transferring achievements from dense RGB pre-training directly to event camera data yields subpar performance. This is attributed to the spatial sparsity inherent in an event image (converted from event data), where many pixels do not contain information. To mitigate this sparsity issue, we encode an event image into event patch features, automatically mine contextual similarity relationships among patches, group the patch features into distinctive contexts, and enforce context-to-context similarities to learn discriminative event features.

For training our framework, we curate a synthetic event camera dataset featuring diverse scene and motion patterns. Transfer learning performance on downstream dense prediction tasks illustrates the superiority of our method over state-of-the-art approaches.

Keywords: Event Camera Data · Self-supervised Learning · Dense Prediction

1 Introduction

An event camera asynchronously records pixel-wise brightness changes of a scene [18]. In contrast to conventional RGB cameras that capture all pixel intensities at a fixed frame rate, event cameras offer a high dynamic range and microsecond temporal resolution, and is robust to lighting changes and motion blur, showing promising applications in diverse vision tasks [4,23,43,53].

This paper addresses the task of pre-training neural networks with event camera data for dense prediction tasks, including segmentation, depth estimation, and optical flow estimation. Our self-supervised method is pre-trained solely with event camera data. One can simply transfer our pre-trained model for dense prediction tasks. Please refer to Fig. 1 for the performance comparisons.

The direct way to pre-training is supervised training, using dense annotations for event data. However, due to the scarcity of dense annotations [4,20,53], training large-scale networks becomes challenging [12, 15].

An alternative to supervised pre-training is self-supervised learning for event camera data [47,51], which has been proposed very recently. These approaches necessitate paired RGB images and event data, enforcing image-level embedding similarities between RGB images and event data. This form of RGB-guided pre-training directs networks to focus on the overall structure of events, neglecting intricate pixel-level features that are crucial for dense prediction tasks.

Next to pre-training is transferring the achievements of dense RGB pre-training [28,42] to event camera data. One may first convert event camera data to an event image [47], split the image into patches, and then learn fine-grained patch features by enforcing patch-to-patch similarities in a self-supervised learning framework. While feasible, this baseline approach is constrained as event images are sparse, containing patches with little to no infor-

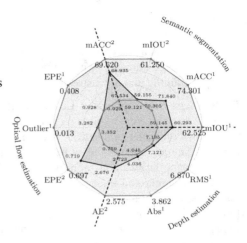

Fig. 1. Comparison of our scores with respect to the second-best and third-best scores for semantic segmentation [1,4,20,39], optical flow estimation [20,21,53], and depth estimation [53]. Superscripts besides evaluation metrics are used to differentiate benchmark datasets for a specific task. (Color figure online)

mation, often from the meaningless background. The sparsity diminishes the discriminativeness of an event patch, introduces background noise/bias to the patch feature learning, and makes training unstable.

Inspired by the above discriminative self-supervised approaches that learn features at the image and patch level, we show that fine-grained event features can be learned by enforcing context-level similarities among patches. Our motivation is described below.

Given an event image, humans can recognize objects (*e.g.*, buildings and trees) by considering multiple similar pixels. In essence, a group of event image pixels contains sufficient information to make them discriminative. Inspired by this insight, we propose to automatically mine the contextual similarity relationship among patches, group patch features into discriminative contexts, and enforce context-to-context similarities. This context-level similarity, requiring no manual annotation, not only promotes stable training but also empowers the model to achieve highly accurate dense predictions.

Our contributions are summarized as follows:

1. A self-supervised framework for pre-training a backbone network for event camera dense prediction tasks. The pre-trained model can be transferred to diverse downstream dense prediction tasks;
2. Introduction of a context-level similarity loss to address the sparsity issue of event data for learning discriminative event features;
3. Construction of a pre-training dataset based on the TartanAir dataset [41], covering diverse scenes and motion patterns to facilitate network training;
4. State-of-the-art performance on standard event benchmark datasets for dense prediction tasks.

2 Related Works

We survey recent advancements in self-supervised learning frameworks applied to RGB and event image domains. We then provide an overview of event datasets used for network pre-training and downstream task fine-tuning.

RGB Image Self-supervised Learning. Research in self-supervised learning generally falls into three categories: i) contrastive learning. Images are augmented into multiple views for instance discrimination. By defining a matching pair (*e.g.*, views from the same image), the similarity between them is maximized [9,22]. Some works also enforce dissimilarity among non-matching pairs [7,8,10,25,45]; ii) masked image modeling. With unmasked image patches, the networks are trained to reconstruct masked ones. The reconstruction targets can be represented as intensity values of patch pixels [24,46], discrete indices assigned by an image tokenizer [3,16,35], or patch embeddings obtained from pre-trained vision foundation models [17,36]; iii) self-distillation. This category can be considered as an extension of contrastive learning from instances to groups [5,6], and is usually combined with MIM [33,52]. The similarity between matching image pairs is optimized by minimizing a cross-entropy loss, while MIM is optionally performed. For adapting self-supervised learning frameworks to dense prediction tasks, objectives at the patch/region level are proposed to maximize the similarity between matching patches [2,28,42,48]. However, the spatial sparsity interferes with the patch-level objective and turns the network pre-training unstable, as most event image patches, containing little to no events, provide meaningless supervision signals.

Event Image Self-supervised Learning. Explorations of self-supervised learning on event data remain in an early stage. Existing works [47,51] primarily leverage a pre-trained CLIP network [36] and paired RGB images for training, guiding the event network to have similar outputs with the RGB network (i.e., the image encoder of CLIP) in feature space. Because an event image is more similar to its paired RGB image at a high-level than at a low-level [49], these approaches concentrate on capturing the overall structures of the event image. This explains their substantial performance improvements in object recognition tasks for event data while lagging in various dense prediction tasks. In this paper, we do not require paired RGB images and pre-trained RGB networks, and focus on pre-training a versatile network by utilizing solely event data for diverse dense prediction tasks on event datasets.

Event Datasets. Event cameras are bio-inspired sensors that pixel-wisely record spatial location, time, and polarity of brightness changes in a scene as an event sequence. One of the largest-scale event datasets covering diverse scenes is the N-ImageNet dataset [27]. It is built by moving an event camera to observe RGB images (from the ImageNet-1K dataset [14]) rendered by a monitor, and inherits scene diversity from the ImageNet-1K dataset. Existing event image self-supervised learning frameworks favor leveraging the N-ImageNet dataset for pre-training, enabling transfer learning for tasks such as object recognition [11,27,34,38], depth estimation [53], semantic segmentation [4,20], and optical flow estimations [20,53]. This paper focuses on pre-training a network for the three dense prediction tasks. Moreover, considering the limited motion patterns in the N-ImageNet dataset [27], which are square, vertical, and horizontal, we curate a synthetic event dataset containing diverse motion patterns and scenes for pre-training.

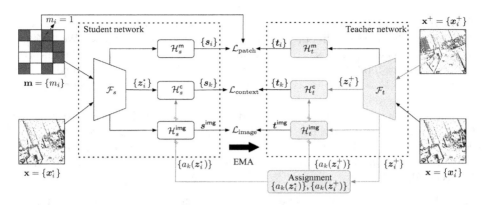

Fig. 2. Overall architecture. During pre-training, our approach takes an event image $\mathbf{x}^+$ and its affine-transformed counterpart $\mathbf{x}^*$ as inputs, producing a pre-trained backbone network $\mathcal{F}_s$. A teacher network (colored by red boxes) and a student network are employed in the self-supervised training stage. Event images $\mathbf{x}^+$ and $\mathbf{x}^*$ are tiled into N patches, denoted as $\mathbf{x}^+ = \{\mathbf{x}_i^+\}$ and $\mathbf{x}^* = \{\mathbf{x}_i^*\}, i = 1, ..., N$. We randomly mask some patches of $\mathbf{x}^*$ given to the student, but leave $\mathbf{x}^*$ intact for the teacher. Patch-wise binary masks are represented by $\mathbf{m} = \{m_i\}$. Three similarity constraints are imposed based on output patch-wise features from the student and teacher backbones, respectively. They are: i) patch-level similarity. Patch-wise features of masked $\mathbf{x}^*$ and $\mathbf{x}^*$ are separately projected by heads $\mathcal{H}_s^m$ in the student network and $\mathcal{H}_t^m$ in the teacher network, obtaining embeddings $\{s_i\}$ and $\{t_i\}$. To reconstruct masked patch embeddings, we employ a cross-entropy loss $\mathcal{L}_{\text{patch}}$; ii) context-level similarity. Features $\{z_i^+\}$ from the teacher network are assigned to K contexts, obtaining assignments $\{a_k(z_i^+)\}$. $a_k(z_i^+)$ denotes the membership of the feature z_i^+ to k-th context. The assignments of student features $\{z_i^*\}$ are computed by directly transferring $a_k(z_i^+)$ with an affine transformation. With the assignments $\{a_k(z_i^+)\}$ and $\{a_k(z_i^*)\}$, we collect and pool all features assigned to each context using heads $\mathcal{H}_s^c$ and $\mathcal{H}_t^c$, generating context embeddings $\{s_k\}$ and $\{t_k\}$. A cross-entropy loss $\mathcal{L}_{\text{context}}$ is used to learn masked context embeddings. The forward passes from $\mathbf{x}^+$ are colored in blue, and the blocked lines mean crosslines; iii) image-level similarity. $\{z_i^*\}$ and $\{z_i^+\}$ are initially pooled separately and subsequently projected by the heads $\mathcal{H}_s^{\text{img}}$ and $\mathcal{H}_t^{\text{img}}$ into global image embeddings s^{img} and t^{img}. A cross-entropy loss $\mathcal{L}_{\text{image}}$ is used to encourage image-level similarity. (Color figure online)

3 Method

We present our self-supervised method in this section. Our network is trained end-to-end, and the overall architecture is shown in Fig. 2.

Overall Architecture. We aim to learn discriminative features from event data for dense prediction tasks, such as optical flow estimation. Sharing similarities with the learning process of DINOv2 [33], we convert raw events to an image [54], and construct two event images $\mathbf{x}^+$ and its augmentation $\mathbf{x}^*$. The two images are then fed into teacher and student networks to learn features, followed by enforcing similarities between the features of $\mathbf{x}^+$ and $\mathbf{x}^*$. We enforce three types of feature similarities: i) patch-level similarity; ii) context-level similarity; iii) image-level similarity. Details of our components are provided below.

Event Image Augmentations. We perform a 2D affine transformation on $\mathbf{x}^+$, followed by GaussianBlur and ColorJitter [47], to create a distorted event image $\mathbf{x}^*$. We tile each image into N patches, i.e., $\mathbf{x}^+ = \{\mathbf{x}_i^+\}$ and $\mathbf{x}^* = \{\mathbf{x}_i^*\}, i = 1, ..., N$. The linearity of the affine transformation establishes pixel correspondences between $\mathbf{x}^+$ and $\mathbf{x}^*$. For each pixel in $\mathbf{x}^*$, we can find its corresponding pixel in $\mathbf{x}^+$, enabling context-level feature learning.

Image patches $\{\mathbf{x}_i^+\}$ and $\{\mathbf{x}_i^*\}$ are fed to the teacher and student networks for feature extraction. In the training stage, the student network is optimized by gradient descent. To avoid model collapse, the teacher network is kept as a momentum of the student network, and its parameters are updated with an exponential moving average (EMA) [25].

Patch-Level Similarity. We randomly mask some patches of $\mathbf{x}^*$ given to the student, but leave $\mathbf{x}^*$ intact for the teacher. The goal is to reconstruct masked patch embeddings, utilizing a cross-entropy loss between the patch features of both networks on each masked patch. This objective, introduced by [33], is briefly summarized below.

A patch-level binary mask $\mathbf{m} = \{m_i\}, i = 1, ..., N$ is randomly sampled. For $\mathbf{x}_i^*$, it is masked and replaced by a [MASK] token if $m_i = 1$. The unmasked patches and [MASK] tokens are fed to the student network $\mathcal{F}_s$ to extract features, and a feature projection head $\mathcal{H}_s^m$ is employed to obtain patch embeddings $\{\mathbf{s}_i\} = \mathcal{H}_s^m(\mathcal{F}_s(\mathbf{x}^*, \mathbf{m}))$.

Without masking, patches $\{\mathbf{x}_i^*\}$ are fed to the teacher network $\mathcal{F}_t$ to extract features, followed by a feature projection head $\mathcal{H}_t$ to extract patch embeddings $\{\mathbf{t}_i\} = \mathcal{H}_t(\mathcal{F}_t(\mathbf{x}^*))$. The patch-level similarity objective is

$$\mathcal{L}_{\text{patch}} = \frac{1}{\|\mathbf{m}\|} \sum_{\substack{i=1 \\ m_i=1}}^{N} \text{CE}(\mathbf{t}_i, \mathbf{s}_i) \ , \quad (1) \quad \text{CE}(\mathbf{t}, \mathbf{s}) = -\langle \mathcal{P}(\mathbf{t}), \log \mathcal{P}(\mathbf{s}) \rangle \ , \quad (2)$$

where $\|\cdot\|$ is the L1 norm that computes the number of masked patches. $\text{CE}(\cdot, \cdot)$ is the cross-entropy loss. $\mathcal{P}(\cdot)$ is the Softmax function that normalizes the patch embedding to a distribution. $\langle \cdot, \cdot \rangle$ is the dot product.

Context-Level Similarity. Reconstructing each masked patch embedding independently is prone to generating noisy embeddings. This is due to the sparsity of an event image. An event patch contains little information, and many patches are from a meaningless background (see Fig. 6). To overcome the limitations of independently reconstructing masked patch embeddings, we propose to mine contextual relationships among patch embeddings on the fly, and learn embeddings with context conditioning. We provide an overview in Fig. 3.

Specifically, we perform K-means clustering on patch features $\{\mathbf{z}_i^+\} = \mathcal{F}_t(\mathbf{x}^+)$ of the teacher network, generating K cluster centers (i.e., contexts) and assignments $a_k(\mathbf{z}_i^+)$. $a_k(\mathbf{z}_i^+)$ denotes the membership of the feature $\mathbf{z}_i^+$ to k-th context, i.e., it is 1 if $\mathbf{z}_i^+$ is closest to k-th context and 0 otherwise.

For each context, features assigned to it are aggregated by an attention pooling network [36], generating a context embedding $\mathbf{s}_k$. Collecting all context embeddings, we have embeddings $\{\mathbf{s}_k\}, k = 1, ..., K$, describing features $\mathcal{F}_t(\mathbf{x}^+)$ of the teacher.

For patch features $\{z_i^*\} = \mathcal{F}_s(\mathbf{x}^*, \mathbf{m})$ of the student network, we use the same cluster centers. Due to the linearity of affine transformation, we can easily obtain the correspondence between patches $\{x_i^*\}$ and $\{x_i^+\}$, and directly transfer the assignments $\{a_k(z_i^+)\}$ to get $\{a_k(z_i^*)\}$. Given assignments $a_k(z_i^*)$, we follow the same pipeline to aggregate features $\{z_i^*\}$ into context embeddings $\{t_k\}, k = 1, ..., K$.

By using adaptively mined contexts, such as roads and buildings, as proxies, we overcome the sparsity limitation of enforcing event patch-level similarity. In essence, we aim to enforce the similarity between a group of patches belonging to the same context. The context-level similarity loss $\mathcal{L}_{\text{context}}$ is defined below

$$\mathcal{L}_{\text{context}} = \frac{1}{K} \sum_{k=1}^{K} \text{CE}(t_k, s_k) . \quad (3)$$

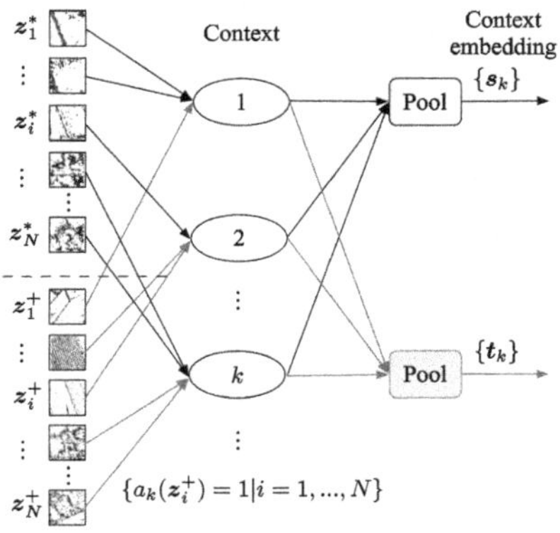

Fig. 3. Context assignment and aggregation. Given patch features $\{z_i^*\}$ and $\{z_i^+\}$, we perform K-means clustering to mine K contexts, and obtain the patch-to-context assignments $\{a_k(z_i^*)\}$ and $\{a_k(z_i^+)\}$, respectively. For the k-th context, $\{z_i^+\}$ assigned to it $\{a_k(z_i^+) = 1 | i = 1, ..., N\}$ are pooled into a context embeddings t_k. Similarly, $\{z_i^*\}$ are pooled into context embeddings $\{s_k\}$. The red box and blue lines denote components of our teacher network and forward passes of $\{z_i^+\}$, respectively. (Color figure online)

Image-Level Similarity. We aim to reconstruct masked image embedding of $\mathbf{x}^*$, by adding a cross-entropy loss between the image features of student and teacher networks on $\mathbf{x}^*$ and $\mathbf{x}^+$.

Patch features $\{z_i^*\}$ and $\{z_i^+\}$ from the student and teacher network $\mathcal{F}_s$ and $\mathcal{F}_t$ are pooled and fed to feature projection heads $\mathcal{H}_s^{\text{img}}$ and $\mathcal{H}_t^{\text{img}}$, generating image-level feature embeddings s^{img} and t^{img}, respectively. The image-level similarity objective is

$$\mathcal{L}_{\text{image}} = \text{CE}\left(t^{\text{img}}, s^{\text{img}}\right) . \quad (4)$$

Pre-training Objective. Our network is trained end-to-end. By using λ_1 and λ_2 hyperparameters for balancing losses, we optimize the following objective,

$$\mathcal{L}_{\text{total}} = \mathcal{L}_{\text{patch}} + \lambda_1 \mathcal{L}_{\text{context}} + \lambda_2 \mathcal{L}_{\text{image}} . \quad (5)$$

4 Experiments

Pre-training Dataset. To pre-train our network, we synthesize an E-TartanAir event camera dataset from the TartanAir dataset [41]. The TartanAir dataset is collected in photo-realistic simulation environments, featuring various light conditions, weather, and moving objects. It has 1037 sequences with RGB frames of 480 × 640 resolution. Different with N-ImageNet dataset [27] that has limited motion patterns, our E-TartanAir contains diverse motion patterns and scenes.

Table 1. Comparison of semantic segmentation accuracies on the DDD17 [1,4] and DSEC datasets [20,39]. Mean interaction over union (mIoU (%)) and mean class accuracy (mACC (%)) are used as evaluation metrics. '#Param', 'Pre. Dataset', and 'Pre. Epo.' respectively denote the number of backbone parameters, pre-training dataset, and pre-training epoch.

Method	Backbone	#Param	Pre. Dataset	Pre. Epo.	DDD17		DSEC	
					mIOU↑	mACC↑	mIOU↑	mACC↑
The best performance in the literature.								
ESS [39]	-	-	-	-	61.370	70.874	53.295	62.942
Self-supervised ResNets.								
SimCLR [7]	ResNet50	23M	ImageNet-1K	100	57.218	69.154	59.062	66.807
MoCo-v2 [8]	ResNet50	23M	ImageNet-1K	200	58.284	65.563	59.090	66.900
DenseCL [42]	ResNet50	23M	ImageNet-1K	200	57.969	71.840	59.121	68.935
ECDP [47]	ResNet50	23M	N-ImageNet	300	59.145	70.176	59.155	67.534
Ours	ResNet50	23M	E-TartanAir	300	**62.912**	**74.015**	60.641	69.502
Self-supervised Transformers.								
MoCo-v3 [10]	ViT-S/16	21M	ImageNet-1K	300	53.654	68.122	49.211	57.133
BeiT [3]	ViT-B/16	86M	ImageNet-1K	800	52.391	61.950	51.899	59.660
IBoT [52]	ViT-S/16	21M	ImageNet-1K	800	53.652	61.607	50.822	59.377
MAE [24]	ViT-B/16	86M	ImageNet-1K	800	53.758	64.783	51.958	59.839
SelfPatch [48]	ViT-S/16	21M	ImageNet-1K	300	54.287	62.821	51.475	59.164
DINOv2 [33]	ViT-S/16	21M	LVD-142M	-	53.846	64.500	52.165	59.795
CIM [29]	ViT-B/16	86M	ImageNet-1K	300	54.013	63.926	51.582	59.628
ECDP [47]	ViT-S/16	21M	N-ImageNet	300	54.663	66.077	52.517	60.553
ESViT [28]	Swin-T/7	28M	ImageNet-1K	300	60.293	70.305	56.517	63.798
Ours	ViT-S/16	21M	E-TartanAir	300	55.729	64.771	56.378	66.000
Ours	Swin-T/7	28M	E-TartanAir	300	62.525	74.301	**61.250**	**69.620**

Implementation Details. We adopt ResNet50 [26], ViT-S/16 [15], and Swin-T/7 architectures as our backbones. The architectures of our projection heads follow [33,36]. Our model is pre-trained for 300 epochs with batch size 1024. We set λ_1 and λ_2 to 0.1 and 0.9, respectively. The number of clusters is set to 8. Our code is available at here.

Baselines. Our method is compared against two groups of methods: i) transfer learning of self-supervised pre-training. The initial weights of state-of-the-art methods are obtained in a self-supervised manner using the ImageNet-1K [14], N-ImageNet [27], or LVD-142M dataset [33]; ii) previous best. We compare with state-of-the-art methods specific to each downstream task, namely, semantic segmentation, flow estimation, and depth estimation. In tables, the symbols '↓' and '↑' indicate that a higher or lower value of a metric is preferable, respectively. The symbol '-' denotes unavailability.

4.1 Semantic Segmentation

Settings. Following the setup of [47], we evaluate on the DDD17 [1,4] and DSEC dataset [20,39] for semantic segmentation. The two datasets contain selected intervals of multiple event sequences, covering 6 and 11 semantic classes, respectively.

Table 2. Comparison of optical flow estimation accuracies on the MVSEC dataset [53]. Endpoint error (EPE) and outlier ratios (%) [47] are used as evaluation metrics. Pixels with EPE above 3 and 5% of the ground truth optical flow magnitudes are deemed as outliers [32].

Method	Backbone	indoor_flying1 EPE↓	indoor_flying1 Outlier↓	indoor_flying2 EPE↓	indoor_flying2 Outlier↓	indoor_flying3 EPE↓	indoor_flying3 Outlier↓
The best performance in the literature							
DCEIFlow [40]	–	0.748	0.597	1.388	8.015	1.132	5.294
Self-supervised ResNets.							
SimCLR [7]	ResNet50	0.646	0.488	1.445	9.331	1.188	5.507
MoCo-v2 [8]	ResNet50	0.612	0.459	1.359	8.683	1.130	5.201
ECDP [47]	ResNet50	0.604	0.354	1.352	8.572	1.122	5.263
DenseCL	ResNet50	0.634	0.529	1.349	7.596	1.130	5.176
Ours	ResNet50	0.413	0.055	0.489	0.041	0.462	0.007
Self-supervised Transformers.							
MoCo-v3 [10]	ViT-S/16	0.648	0.744	1.361	8.660	1.119	5.594
BeiT [3]	ViT-B/16	0.613	0.438	1.159	5.622	1.013	4.654
iBoT [52]	ViT-S/16	0.630	0.562	1.259	6.752	1.038	4.912
MAE [24]	ViT-B/16	0.613	0.167	1.293	6.952	1.109	4.635
SelfPatch [48]	ViT-S/16	0.623	0.317	1.337	7.894	1.097	5.286
DINOv2 [33]	ViT-S/16	0.602	0.325	1.196	6.185	0.990	4.333
CIM [29]	ViT-B/16	0.625	0.491	1.332	8.926	1.040	4.869
ECDP [47]	ViT-S/16	0.614	0.046	1.261	6.689	1.001	3.111
ESViT [28]	Swin-T/7	0.812	1.224	1.338	8.316	1.078	5.185
Ours	ViT-S/16	0.508	0.112	0.691	0.290	0.610	0.075
Ours	Swin-T/7	**0.362**	**0.035**	**0.445**	**0.002**	**0.417**	**0.001**

Table 3. Comparisons of optical flow estimation accuracies on the DSEC dataset [20]. Note that IDNet, ranking first previously, maintains anonymity at the time of submission. According to the DSEC leaderboard, we present results for 1/2/3-pixel error (1/2/3-PE), end-point error (EPE), and angular error (AE). All data is sourced from the online benchmark at the time of submission.

Methods	1PE↓	2PE↓	3PE↓	EPE↓	AE↓
E-RAFT [21]	12.742	4.740	2.684	0.788	2.851
MultiCM [37]	76.570	48.480	30.855	3.472	13.983
E-Flowformer [30]	11.225	4.102	2.446	0.759	2.676
TMA [31]	10.863	3.972	2.301	0.743	2.684
OF_EV_SNN [13]	53.671	20.238	10.308	1.707	6.338
IDNet [44]	10.069	3.497	2.036	0.719	2.723
Ours (ResNet50)	9.013	3.290	1.983	0.701	2.611
Ours (ViT-S/16)	9.288	3.339	2.005	0.714	2.615
Ours (Swin-T/7)	**8.887**	**3.199**	**1.958**	**0.697**	**2.575**

Table 4. Comparison of depth estimation accuracies on the MVSEC dataset [53]. Averaged scores across all sequences with a cutoff threshold at 30 m are reported. Threshold accuracy ($\delta 1$, $\delta 2$, and $\delta 3$), absolute error (Abs), root mean squared error (RMS), and root mean squared logarithmic error (RMSlog) are used as evaluation metrics. The inputs of HMNet[1] are events, and HMNet[2] additionally takes RGB frames as inputs.

Method	Backbone	$\delta 1\uparrow$	$\delta 2\uparrow$	$\delta 3\uparrow$	Abs$\downarrow$	RMS$\downarrow$	RMSlog$\downarrow$
The best performance in the literature.							
HMNet[1] [23]	–	0.626	0.818	0.912	2.882	4.772	0.361
HMNet[2] [23]	–	0.628	0.803	0.905	2.908	4.858	0.359
Self-supervised ResNets.							
SimCLR [7]	ResNet50	0.633	0.822	0.918	2.886	4.612	0.351
MoCo-v2 [8]	ResNet50	0.647	0.827	0.919	2.817	4.556	0.346
ECDP [47]	ResNet50	0.651	0.829	0.921	2.798	4.530	0.343
DenseCL [42]	ResNet50	0.649	0.826	0.920	2.813	4.541	0.344
Ours	ResNet50	0.649	**0.837**	**0.931**	2.713	4.302	**0.330**
Self-supervised Transformers.							
MoCo-v3 [10]	ViT-S/16	0.630	0.814	0.909	3.043	4.817	0.362
BeiT [3]	ViT-B/16	0.622	0.805	0.903	3.147	4.965	0.372
iBoT [52]	ViT-S/16	0.623	0.816	0.912	2.998	4.736	0.360
MAE [24]	ViT-B/16	0.612	0.802	0.900	3.214	5.075	0.377
SelfPatch [48]	ViT-S/16	0.605	0.801	0.900	3.435	5.067	0.380
DINOv2 [33]	ViT-S/16	0.612	0.805	0.903	3.181	5.030	0.375
CIM [29]	ViT-B/16	0.625	0.808	0.904	3.108	4.906	0.370
ECDP [47]	ViT-S/16	0.614	0.802	0.899	3.228	5.104	0.378
ESViT [28]	Swin-T/7	0.644	0.829	0.923	2.796	4.482	0.342
Ours	ViT-S/16	0.649	0.827	0.920	2.815	4.476	0.343
Ours	Swin-T/7	**0.658**	**0.837**	0.928	**2.658**	**4.257**	**0.330**

Results. Table 1 gives the comparisons on the DDD17 and DSEC datasets. Consistently, our method surpasses the state-of-the-art methods within the backbone groups of ResNet50, ViT-S/16, and Swin-T/7, and achieves a better performance than the methods pre-trained with a larger ViT-B/16 backbone. For example, our method with a Swin-T/7 backbone achieves mIoU/mACC scores at 62.525%/74.301% and 61.250%/69.620% on the DDD17 and DSEC datasets, respectively, outperforming all other methods. Even though DINOv2 [33] is trained on the huge LVD-142M dataset, our method significantly outperforms it.

4.2 Flow Estimation

Settings. We compare our method with state-of-the-art methods on the MVSEC dataset [53]. End-point error (EPE) and outlier ratios (%) are used as evaluation metrics [40,

Table 5. Comparison of depth estimation accuracies on the MVSEC dataset [53]. Averaged scores across all sequences are reported. Threshold accuracy ($\delta 1$, $\delta 2$, and $\delta 3$), absolute error (Abs), root mean squared error (RMS), and root mean squared logarithmic error (RMSlog) are used as evaluation metrics. The inputs of HMNet[1] are events, and HMNet[2] additionally takes RGB frames as inputs.

Method	Backbone	$\delta 1\uparrow$	$\delta 2\uparrow$	$\delta 3\uparrow$	Abs$\downarrow$	RMS$\downarrow$	RMSlog$\downarrow$
The best performance in the literature.							
HMNet[1] [23]	–	0.588	0.784	0.889	4.171	7.534	0.397
HMNet[2] [23]	–	0.582	0.754	0.860	4.614	8.602	0.430
Self-supervised ResNets.							
SimCLR [7]	ResNet50	0.594	0.789	0.897	4.176	7.343	0.386
MoCo-v2 [8]	ResNet50	0.609	0.797	0.901	4.045	7.135	0.377
ECDP [47]	ResNet50	0.611	0.797	0.901	4.061	7.197	0.377
DenseCL [42]	ResNet50	0.610	0.798	0.903	4.036	7.121	0.375
Ours	ResNet50	0.612	**0.809**	**0.915**	3.889	**6.805**	**0.359**
Self-supervised Transformers.							
MoCo-v3 [10]	ViT-S/16	0.590	0.782	0.891	4.313	7.466	0.394
BeiT [3]	ViT-B/16	0.584	0.775	0.886	4.398	7.562	0.402
iBoT [52]	ViT-S/16	0.583	0.782	0.892	4.309	7.521	0.394
MAE [24]	ViT-B/16	0.575	0.772	0.884	4.449	7.601	0.405
SelfPatch [48]	ViT-S/16	0.567	0.768	0.882	4.515	7.735	0.410
DINOv2 [33]	ViT-S/16	0.575	0.774	0.885	4.449	7.653	0.406
CIM [29]	ViT-B/16	0.585	0.777	0.888	4.356	7.495	0.398
ECDP [47]	ViT-S/16	0.576	0.772	0.883	4.491	7.680	0.406
ESViT [28]	Swin-T/7	0.604	0.796	0.903	4.083	7.219	0.377
Ours	ViT-S/16	0.610	0.800	0.906	3.987	6.957	0.369
Ours	Swin-T/7	**0.618**	0.806	0.912	**3.862**	6.870	0.360

47]. In accordance with [47], the evaluations are performed on the 'indoor_flying1', 'indoor_flying2', and 'indoor_flying3' sequences.

Additionally, our method is evaluated on the DSEC-Flow benchmark[1] [20,21], securing the first-place position at the time of submission.

Results. Table 2 presents the comparisons on MVSEC dataset. Among the three different backbone groups, we have the most accurate flow estimation by pre-training with a Swin-T/7 backbone, and the EPE and outlier ratios on the three sequences are 0.362/0.035%, 0.445/0.002%, and 0.417/0.001%, respectively, which are significantly better than all other methods.

---

[1] https://dsec.ifi.uzh.ch/uzh/dsec-flow-optical-flow-benchmark/.

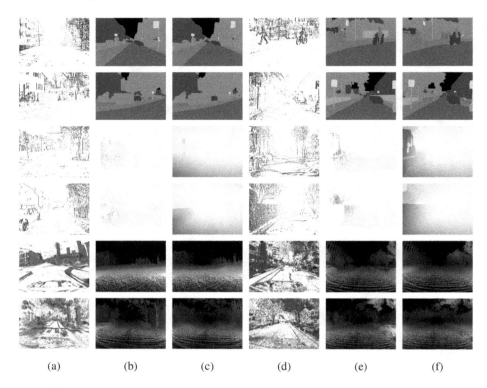

Fig. 4. Qualitative comparison examples of dense predictions, namely, semantic segmentation (1st-2nd rows), optical flow estimation (3rd-4th rows), and depth estimation (5th-6th rows). (a) and (d): event images. Red and blue pixels depict positive and negative events, respectively. (b) and (e): ground-truth labels. (c) and (f): our model predictions. The brightness of depth maps in the 5th row of (b) and (c) is enhanced for visualization. (Color figure online)

Results for the DSEC-Flow benchmark are given in Table 3. Compared with the method IDNet [44], previously holding the top position, our method achieves superior optical flow estimation accuracy. For example, our method with a ResNet50, ViT-S/16, and Swin-T/7 backbone respectively improves the state-of-the-art EPE/AE scores from 0.719/2.723 to 0.701/2.611, 0.714/2.615, and 0.697/2.575.

4.3 Depth Estimation

Settings. We evaluate the performance of our methods for depth estimation on the MVSEC dataset [53]. Following [19], the evaluations are performed on the 'outdoor_day1', 'outdoor_night1', 'outdoor_night2', and 'outdoor_night3' sequences.

Results. The comparisons of our methods and state-of-the-art methods with and without a cutoff threshold at 30 m are given in Table 4 and Table 5, respectively. Though the previous best method HMNet [23] performs supervised pre-training using ground-truth depth before fine-tuning on the MVSEC dataset, all our methods outperform it. For

Table 6. (a)–(c) Comparison of state-of-the-art methods pre-trained on the N-imageNet datasets with backbones of ResNet50, ViT-S/16, Swin-T/7. (d)–(f) Comparison of state-of-the-art methods pre-trained on the E-TartanAir datasets with backbones of ResNet50, ViT-S/16, Swin-T/7.

(a) All methods are pre-trained using the ResNet50 backbone, on the N-ImageNet dataset.		(b) All methods are pre-trained using the ViT-S/16 backbone, on the N-ImageNet dataset.		(c) All methods are pre-trained using the Swin-T/7 backbone, on the N-ImageNet dataset.	
Method	mIOU↑ mACC↑	Method	mIOU↑ mACC↑	Method	mIOU↑ mACC↑
SelfPatch	57.881 64.916	SelfPatch	50.442 58.452	SelfPatch	52.997 59.928
ESViT	57.796 64.928	ESViT	51.011 58.902	ESViT	53.051 60.094
ECDP	59.155 67.534	ECDP	52.517 60.553	ECDP	55.842 63.548
Ours	**60.243 69.195**	Ours	**54.897 62.527**	Ours	**56.654 65.250**
(d) All methods are pre-trained using the ResNet50 backbone, on the E-TartanAir dataset.		(e) All methods are pre-trained using the ViT-S/16 backbone, on the E-TartanAir dataset.		(f) All methods are pre-trained using the Swin-T/7 backbone, on the E-TartanAir dataset.	
Method	mIOU↑ mACC↑	Method	mIOU↑ mACC↑	Method	mIOU↑ mACC↑
SelfPatch	58.365 65.180	SelfPatch	52.347 59.947	SelfPatch	57.243 66.070
ESViT	59.058 65.879	ESViT	51.945 60.470	ESViT	58.593 65.746
ECDP	59.572 68.317	ECDP	53.229 61.712	ECDP	56.568 64.234
Ours	**60.641 69.502**	Ours	**55.729 64.771**	Ours	**61.250 69.620**

example, in Table 5, the averaged root mean squared error of HMNet is 7.534, while the errors of our methods with ResNet50, ViT-S/16, and Swin-T/7 backbones are 6.805, 6.957, and 6.870, respectively.

Sample prediction results of our method on the semantic segmentation, optical flow estimation, and depth estimation tasks are provided in Fig. 4.

4.4 Discussions

We perform ablations on the DSEC semantic segmentation dataset [20,39] to study our model components. We set the pre-training backbone and dataset to the Swin-T/7 and E-TartanAir dataset, except where otherwise indicated.

Pre-training Datasets and Backbones. We investigate the impact of different pre-training datasets and backbones. We pre-train state-of-the-art methods with backbones of ResNet50, ViT-S/16, and Swin-T/7 respectively on the N-ImageNet and E-TartanAir datasets. The results are given in Table 6.

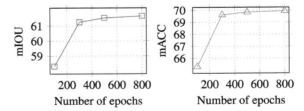

Fig. 5. Comparison of the number of pre-training epochs.

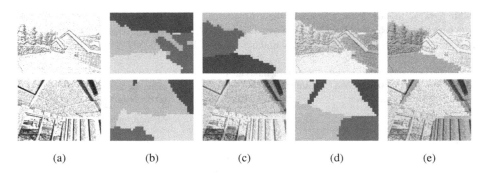

Fig. 6. Sample results of patches belonging to different contexts on the E-TartanAir dataset. (a): input event images. (b): mined context labels (without enforcing the context-level similarity). (c): mined context labels (enforcing the context-level similarity). (d) and (e): blends of the event image with context labels from (b) and (c) for visualization purposes, respectively.

We have the following observations: i) Our proposed E-TartanAir dataset outperforms the N-ImageNet dataset. The performance of each method gets improved, when changing the pretraining dataset from N-ImageNet to E-TartanAir; ii) Our method consistently achieves the best performance under each setting, and the best performance of our method is achieved with the Swin-T/7 backbone and E-TartanAir dataset. Furthermore, our method gains performance improvements by at least 0.812% mIOU and 1.185% mACC scores over the state-of-the-art methods, with the same pre-training backbone and dataset. Note that ESViT exhibits numerical instability during pre-training, and the restart strategy from [50] is used for pre-training.

Pre-training Epochs. We explore the impact of pre-training epochs, ranging from 100 to 800, and the results are presented in Fig. 5. Limited performance improvements in mIOU and mACC scores are observed after 300 epochs, prompting us to set the pre-training epoch number to 300.

Context-Level Similarity. To check the effectiveness of our context-level similarity loss, we pre-train several networks without using it, varying our backbone network and pre-training dataset. Results in Table 7 reveal that a network pre-trained with $\mathcal{L}_{\text{context}}$ consistently outperforms its counterpart pre-trained without using $\mathcal{L}_{\text{context}}$. For example, for networks pre-trained on the E-tartanAir dataset with the Swin-T/7 backbone, without using $\mathcal{L}_{\text{context}}$ in pre-training, the mIOU/mACC scores are 55.556%/63.486%, which are lower than our best scores of 61.250%/69.620%. This justifies the effectiveness of the proposed context-level similarity loss.

Table 7. Comparison of the performance of networks trained with and without using the proposed context-level similarity loss $\mathcal{L}_{\text{context}}$. Using $\mathcal{L}_{\text{context}}$ consistently improves accuracies. 'Pre. Dataset' and '#Param' respectively denote the pre-training dataset and the number of backbone parameters.

Pre. Dataset	Backbone	#Param	mIOU↑	mACC↑
$\mathcal{L}_{\text{patch}} + \mathcal{L}_{\text{image}}$				
N-ImageNet	ResNet50	23M	58.308	65.597
N-ImageNet	ViT-S/16	21M	53.706	61.328
N-ImageNet	Swin-T/7	28M	54.905	63.271
E-TartanAir	ResNet50	23M	58.687	66.171
E-TartanAir	ViT-S/16	21M	54.193	61.711
E-TartanAir	Swin-T/7	28M	55.556	63.486
$\mathcal{L}_{\text{patch}} + \mathcal{L}_{\text{context}} + \mathcal{L}_{\text{image}}$				
N-ImageNet	ResNet50	23M	60.243	69.195
N-ImageNet	ViT-S/16	21M	54.897	62.527
N-ImageNet	Swin-T/7	28M	56.654	65.250
E-TartanAir	ResNet50	23M	60.641	69.502
E-TartanAir	ViT-S/16	21M	55.729	64.771
E-TartanAir	Swin-T/7	28M	**61.250**	**69.620**

Sample results of patches belonging to different contexts are given in Fig. 6. For example, in the 1st row, our method successfully mines contexts (tree, building, ground, and sky) in an event image, and groups patches with the same semantics.

Number of Contexts. Our model utilizes K context embeddings by aggregating patch features. To study the impact of contexts, we train our model with a different number of contexts. Results in Fig. 7 indicate that the best performance is achieved with 8 contexts. Increasing the number of contexts results in inferior performance. Due to the sparsity of event camera data, for large context numbers, many contexts aggregate features from event patches with little to no events. This results in noisy context embeddings, interferes with the training process, and hinders the network from learning discriminative event features.

Generalization Ability of Context-Level Similarity. To further demonstrate the effectiveness of the proposed $\mathcal{L}_{\text{context}}$, we add $\mathcal{L}_{\text{context}}$ to the objective function of the state-of-the-art event data pre-training method, ECDP [47]. The mIOU/mAcc scores of ECDP are increased by a large margin, having

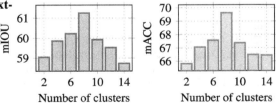

Fig. 7. Comparison of the number of contexts.

improvements from 52.517%/60.553% to 53.826%/61.008%. The improvements validate the generalization ability of $\mathcal{L}_{\text{context}}$.

Limitation. Although our self-supervised pre-trained network has achieved state-of-the-art performance across various dense prediction tasks, it necessitates task-specific fine-tuning to refine pre-trained network weights. However, we believe that our self-supervised learning exploration helps to learn task-agnostic pre-trained representations.

5 Conclusion and Broader Impact

We present a neural network trained for dense prediction tasks using an event camera. Our self-supervised learning method enforces three levels of similarity constraints: patch-level, context-level, and image-level. Our key insight is enforcing context similarity from event patch embeddings to pre-train our model. The proposed context-level similarity effectively addresses the sparsity problem of event data, resulting in state-of-the-art performance on semantic segmentation, optical flow, and depth estimation benchmarks. We believe that our dense pre-training techniques deserve a position in highly accurate event-based dense predictions.

Broader Impact. By aligning event data with paired RGB frames, our pre-training framework is promising to be extended to an event-vision-language foundation model. We hope it inspires future work.

Acknowledgements. Liyuan Pan's work was supported in part by the Beijing Institute of Technology Research Fund Program for Young Scholars, BIT Special-Zone, and National Natural Science Foundation of China 62302045.

References

1. Alonso, I., Murillo, A.C.: EV-SegNet: semantic segmentation for event-based cameras. In: IEEE Conference on Computer Vision and Pattern Recognition Workshops, CVPR Workshops 2019, Long Beach, CA, USA, 16–20 June 2019, pp. 1624–1633. Computer Vision Foundation / IEEE (2019). https://doi.org/10.1109/CVPRW.2019.00205. http://openaccess.thecvf.com/content_CVPRW_2019/html/EventVision/Alonso_EV-SegNet_Semantic_Segmentation_for_Event-Based_Cameras_CVPRW_2019_paper.html
2. Bai, Y., Chen, X., Kirillov, A., Yuille, A.L., Berg, A.C.: Point-level region contrast for object detection pre-training. In: IEEE/CVF Conference on Computer Vision and Pattern Recognition, CVPR 2022, New Orleans, LA, USA, 18–24 June 2022, pp. 16040–16049. IEEE (2022). https://doi.org/10.1109/CVPR52688.2022.01559
3. Bao, H., Dong, L., Piao, S., Wei, F.: Beit: BERT pre-training of image transformers. In: The Tenth International Conference on Learning Representations, ICLR 2022, Virtual Event, 25–29 April 2022. OpenReview.net (2022). https://openreview.net/forum?id=p-BhZSz59o4
4. Binas, J., Neil, D., Liu, S., Delbrück, T.: DDD17: end-to-end DAVIS driving dataset. CoRR **abs/1711.01458** (2017). http://arxiv.org/abs/1711.01458

5. Caron, M., Misra, I., Mairal, J., Goyal, P., Bojanowski, P., Joulin, A.: Unsupervised learning of visual features by contrasting cluster assignments. In: Larochelle, H., Ranzato, M., Hadsell, R., Balcan, M., Lin, H. (eds.) Advances in Neural Information Processing Systems 33: Annual Conference on Neural Information Processing Systems 2020, NeurIPS 2020, 6–12 December 2020, virtual (2020). https://proceedings.neurips.cc/paper/2020/hash/70feb62b69f16e0238f741fab228fec2-Abstract.html
6. Caron, M., et al.: Emerging properties in self-supervised vision transformers. In: 2021 IEEE/CVF International Conference on Computer Vision, ICCV 2021, Montreal, QC, Canada, 10–17 October 2021, pp. 9630–9640. IEEE (2021). https://doi.org/10.1109/ICCV48922.2021.00951
7. Chen, T., Kornblith, S., Norouzi, M., Hinton, G.E.: A simple framework for contrastive learning of visual representations. In: Proceedings of the 37th International Conference on Machine Learning, ICML 2020, 13–18 July 2020, Virtual Event. Proceedings of Machine Learning Research, vol. 119, pp. 1597–1607. PMLR (2020). http://proceedings.mlr.press/v119/chen20j.html
8. Chen, X., Fan, H., Girshick, R., He, K.: Improved baselines with momentum contrastive learning. arXiv preprint arXiv:2003.04297 (2020)
9. Chen, X., He, K.: Exploring simple Siamese representation learning. In: IEEE Conference on Computer Vision and Pattern Recognition, CVPR 2021, virtual, 19–25 June 2021, pp. 15750–15758. Computer Vision Foundation / IEEE (2021). https://doi.org/10.1109/CVPR46437.2021.01549. https://openaccess.thecvf.com/content/CVPR2021/html/Chen_Exploring_Simple_Siamese_Representation_Learning_CVPR_2021_paper.html
10. Chen*, X., Xie*, S., He, K.: An empirical study of training self-supervised vision transformers. arXiv preprint arXiv:2104.02057 (2021)
11. Cheng, W., Luo, H., Yang, W., Yu, L., Li, W.: Structure-aware network for lane marker extraction with dynamic vision sensor. CoRR **abs/2008.06204** (2020). https://arxiv.org/abs/2008.06204
12. Cherti, M., et al.: Reproducible scaling laws for contrastive language-image learning. In: IEEE/CVF Conference on Computer Vision and Pattern Recognition, CVPR 2023, Vancouver, BC, Canada, 17–24 June 2023, pp. 2818–2829. IEEE (2023). https://doi.org/10.1109/CVPR52729.2023.00276
13. Cuadrado, J., Rancon, U., Cottereau, B., Barranco, F., Masquelier, T.: Optical flow estimation from event-based cameras and spiking neural networks. Front. Neurosci. **17** (2023). https://doi.org/10.3389/fnins.2023.1160034
14. Deng, J., Dong, W., Socher, R., Li, L., Li, K., Fei-Fei, L.: ImageNet: a large-scale hierarchical image database. In: 2009 IEEE Computer Society Conference on Computer Vision and Pattern Recognition (CVPR 2009)x, Miami, Florida, USA 20-25, pp. 248–255. IEEE Computer Society (2009). https://doi.org/10.1109/CVPR.2009.5206848
15. Dosovitskiy, A., et al.: An image is worth 16x16 words: transformers for image recognition at scale. In: 9th International Conference on Learning Representations, ICLR 2021, Virtual Event, Austria, 3–7 May 2021. OpenReview.net (2021). https://openreview.net/forum?id=YicbFdNTTy
16. Esser, P., Rombach, R., Ommer, B.: Taming transformers for high-resolution image synthesis. In: IEEE Conference on Computer Vision and Pattern Recognition, CVPR 2021, virtual, 19–25 June 2021, pp. 12873–12883. Computer Vision Foundation / IEEE (2021). https://doi.org/10.1109/CVPR46437.2021.01268. https://openaccess.thecvf.com/content/CVPR2021/html/Esser_Taming_Transformers_for_High-Resolution_Image_Synthesis_CVPR_2021_paper.html

17. Fang, Y., et al.: EVA: exploring the limits of masked visual representation learning at scale. In: IEEE/CVF Conference on Computer Vision and Pattern Recognition, CVPR 2023, Vancouver, BC, Canada, 17–24 June 2023, pp. 19358–19369. IEEE (2023). https://doi.org/10.1109/CVPR52729.2023.01855
18. Gallego, G., et al.: Event-based vision: a survey. IEEE Trans. Pattern Anal. Mach. Intell. **44**(1), 154–180 (2022). https://doi.org/10.1109/TPAMI.2020.3008413
19. Gehrig, D., Rüegg, M., Gehrig, M., Hidalgo-Carrió, J., Scaramuzza, D.: Combining events and frames using recurrent asynchronous multimodal networks for monocular depth prediction. IEEE Robot. Autom. Lett. **6**(2), 2822–2829 (2021). https://doi.org/10.1109/LRA.2021.3060707
20. Gehrig, M., Aarents, W., Gehrig, D., Scaramuzza, D.: DSEC: a stereo event camera dataset for driving scenarios. IEEE Robotics Autom. Lett. **6**(3), 4947–4954 (2021). https://doi.org/10.1109/LRA.2021.3068942
21. Gehrig, M., Millhäusler, M., Gehrig, D., Scaramuzza, D.: E-RAFT: dense optical flow from event cameras. In: International Conference on 3D Vision, 3DV 2021, London, United Kingdom, 1–3 December 2021, pp. 197–206. IEEE (2021). https://doi.org/10.1109/3DV53792.2021.00030
22. Grill, J., et al.: Bootstrap your own latent - a new approach to self-supervised learning. In: Larochelle, H., Ranzato, M., Hadsell, R., Balcan, M., Lin, H. (eds.) Advances in Neural Information Processing Systems 33: Annual Conference on Neural Information Processing Systems 2020, NeurIPS 2020, 6–12 December 2020, virtual (2020). https://proceedings.neurips.cc/paper/2020/hash/f3ada80d5c4ee70142b17b8192b2958e-Abstract.html
23. Hamaguchi, R., Furukawa, Y., Onishi, M., Sakurada, K.: Hierarchical neural memory network for low latency event processing. In: IEEE/CVF Conference on Computer Vision and Pattern Recognition, CVPR 2023, Vancouver, BC, Canada, 17–24 June 2023, pp. 22867–22876. IEEE (2023). https://doi.org/10.1109/CVPR52729.2023.02190
24. He, K., Chen, X., Xie, S., Li, Y., Dollár, P., Girshick, R.B.: Masked autoencoders are scalable vision learners. In: IEEE/CVF Conference on Computer Vision and Pattern Recognition, CVPR 2022, New Orleans, LA, USA, 18–24 June 2022, pp. 15979–15988. IEEE (2022). https://doi.org/10.1109/CVPR52688.2022.01553
25. He, K., Fan, H., Wu, Y., Xie, S., Girshick, R.B.: Momentum contrast for unsupervised visual representation learning. In: 2020 IEEE/CVF Conference on Computer Vision and Pattern Recognition, CVPR 2020, Seattle, WA, USA, 13–19 June 2020, pp. 9726–9735. Computer Vision Foundation / IEEE (2020). https://doi.org/10.1109/CVPR42600.2020.00975
26. He, K., Zhang, X., Ren, S., Sun, J.: Deep residual learning for image recognition. CoRR **abs/1512.03385** (2015). http://arxiv.org/abs/1512.03385
27. Kim, J., Bae, J., Park, G., Zhang, D., Kim, Y.M.: N-ImageNet: towards robust, fine-grained object recognition with event cameras. In: Proceedings of the IEEE/CVF International Conference on Computer Vision (ICCV), pp. 2146–2156, October 2021
28. Li, C., et al.: Efficient self-supervised vision transformers for representation learning. In: The Tenth International Conference on Learning Representations, ICLR 2022, Virtual Event, 25–29 April 2022. OpenReview.net (2022). https://openreview.net/forum?id=fVu3o-YUGQK
29. Li, W., Xie, J., Loy, C.C.: Correlational image modeling for self-supervised visual pre-training. In: IEEE/CVF Conference on Computer Vision and Pattern Recognition, CVPR 2023, Vancouver, BC, Canada, 17–24 June 2023, pp. 15105–15115. IEEE (2023). https://doi.org/10.1109/CVPR52729.2023.01450
30. Li, Y., et al.: BlinkFlow: a dataset to push the limits of event-based optical flow estimation. CoRR **abs/2303.07716** (2023). https://doi.org/10.48550/arXiv.2303.07716
31. Liu, H., et al.: TMA: temporal motion aggregation for event-based optical flow. In: ICCV (2023)

32. Menze, M., Heipke, C., Geiger, A.: Joint 3d estimation of vehicles and scene flow. ISPRS Ann. Photogram. Remote Sens. Spat. Inf. Sci. **II-3/W5**, 427–434 (2015). https://doi.org/10.5194/isprsannals-II-3-W5-427-2015
33. Oquab, M., et al.: DINOv2: learning robust visual features without supervision. CoRR **abs/2304.07193** (2023). https://doi.org/10.48550/arXiv.2304.07193
34. Orchard, G., Jayawant, A., Cohen, G., Thakor, N.V.: Converting static image datasets to spiking neuromorphic datasets using saccades. CoRR **abs/1507.07629** (2015). http://arxiv.org/abs/1507.07629
35. Peng, Z., Dong, L., Bao, H., Ye, Q., Wei, F.: BEiT v2: masked image modeling with vector-quantized visual tokenizers. CoRR **abs/2208.06366** (2022). https://doi.org/10.48550/arXiv.2208.06366
36. Radford, A., et al.: Learning transferable visual models from natural language supervision. In: Meila, M., Zhang, T. (eds.) Proceedings of the 38th International Conference on Machine Learning, ICML 2021, 18–24 July 2021, Virtual Event. Proceedings of Machine Learning Research, vol. 139, pp. 8748–8763. PMLR (2021). http://proceedings.mlr.press/v139/radford21a.html
37. Shiba, S., Aoki, Y., Gallego, G.: Secrets of event-based optical flow. In: Avidan, S., Brostow, G.J., Cissé, M., Farinella, G.M., Hassner, T. (eds.) ECCV 2022, Part XVIII. LNCS, vol. 13678, pp. 628–645. Springer, Cham (2022). https://doi.org/10.1007/978-3-031-19797-0_36
38. Sironi, A., Brambilla, M., Bourdis, N., Lagorce, X., Benosman, R.: HATS: histograms of averaged time surfaces for robust event-based object classification. In: 2018 IEEE Conference on Computer Vision and Pattern Recognition, CVPR 2018, Salt Lake City, UT, USA, 18–22 June 2018, pp. 1731–1740. Computer Vision Foundation / IEEE Computer Society (2018). https://doi.org/10.1109/CVPR.2018.00186. http://openaccess.thecvf.com/content_cvpr_2018/html/Sironi_HATS_Histograms_of_CVPR_2018_paper.html
39. Sun, Z., Messikommer, N., Gehrig, D., Scaramuzza, D.: ESS: learning event-based semantic segmentation from still images. In: Avidan, S., Brostow, G.J., Cissé, M., Farinella, G.M., Hassner, T. (eds.) ECCV 2022, Part XXXIV. LNCS, vol. 13694, pp. 341–357. Springer, Cham (2022). https://doi.org/10.1007/978-3-031-19830-4_20
40. Wan, Z., Dai, Y., Mao, Y.: Learning dense and continuous optical flow from an event camera. IEEE Trans. Image Process. **31**, 7237–7251 (2022). https://doi.org/10.1109/TIP.2022.3220938
41. Wang, W., et al.: TartanAir: a dataset to push the limits of visual SLAM. In: IEEE/RSJ International Conference on Intelligent Robots and Systems, IROS 2020, Las Vegas, NV, USA, 24 October 2020–24 January 2021, pp. 4909–4916. IEEE (2020). https://doi.org/10.1109/IROS45743.2020.9341801
42. Wang, X., Zhang, R., Shen, C., Kong, T., Li, L.: Dense contrastive learning for self-supervised visual pre-training. In: IEEE Conference on Computer Vision and Pattern Recognition, CVPR 2021, virtual, 19–25 June 2021, pp. 3024–3033. Computer Vision Foundation / IEEE (2021). https://doi.org/10.1109/CVPR46437.2021.00304. https://openaccess.thecvf.com/content/CVPR2021/html/Wang_Dense_Contrastive_Learning_for_Self-Supervised_Visual_Pre-Training_CVPR_2021_paper.html
43. Weikersdorfer, D., Adrian, D.B., Cremers, D., Conradt, J.: Event-based 3d SLAM with a depth-augmented dynamic vision sensor. In: 2014 IEEE International Conference on Robotics and Automation, ICRA 2014, Hong Kong, China, May 31–June 7 2014, pp. 359–364. IEEE (2014). https://doi.org/10.1109/ICRA.2014.6906882
44. Wu, Y., Paredes-Vallés, F., de Croon, G.C.H.E.: Lightweight event-based optical flow estimation via iterative deblurring. In: Proceedings of IEEE International Conference on Robotics and Automation (ICRA 2024), May 2024, to Appear

45. Wu, Z., Xiong, Y., Yu, S.X., Lin, D.: Unsupervised feature learning via non-parametric instance discrimination. In: 2018 IEEE Conference on Computer Vision and Pattern Recognition, CVPR 2018, Salt Lake City, UT, USA, 18–22 June 2018, pp. 3733–3742. Computer Vision Foundation / IEEE Computer Society (2018). https://doi.org/10.1109/CVPR.2018.00393. http://openaccess.thecvf.com/content_cvpr_2018/html/Wu_Unsupervised_Feature_Learning_CVPR_2018_paper.html
46. Xie, Z., et al.: SimMIM: a simple framework for masked image modeling. In: IEEE/CVF Conference on Computer Vision and Pattern Recognition, CVPR 2022, New Orleans, LA, USA, 18–24 June 2022, pp. 9643–9653. IEEE (2022). https://doi.org/10.1109/CVPR52688.2022.00943
47. Yang, Y., Pan, L., Liu, L.: Event camera data pre-training. In: Proceedings of the IEEE/CVF International Conference on Computer Vision (ICCV), pp. 10699–10709, October 2023
48. Yun, S., Lee, H., Kim, J., Shin, J.: Patch-level representation learning for self-supervised vision transformers. In: IEEE/CVF Conference on Computer Vision and Pattern Recognition, CVPR 2022, New Orleans, LA, USA, 18–24 June 2022, pp. 8344–8353. IEEE (2022). https://doi.org/10.1109/CVPR52688.2022.00817
49. Zhang, D., Ding, Q., Duan, P., Zhou, C., Shi, B.: Data association between event streams and intensity frames under diverse baselines. In: Avidan, S., Brostow, G.J., Cissé, M., Farinella, G.M., Hassner, T. (eds.) ECCV 2022, Part VII. LNCS, vol. 13667, pp. 72–90. Springer, Cham (2022). https://doi.org/10.1007/978-3-031-20071-7_5
50. Zhang, S., et al.: OPT: open pre-trained transformer language models. CoRR **abs/2205.01068** (2022). https://doi.org/10.48550/ARXIV.2205.01068. https://doi.org/10.48550/arXiv.2205.01068
51. Zhou, J., Zheng, X., Lyu, Y., Wang, L.: E-CLIP: towards label-efficient event-based open-world understanding by CLIP. CoRR **abs/2308.03135** (2023). https://doi.org/10.48550/arXiv.2308.03135
52. Zhou, J., et al.: Image BERT pre-training with online tokenizer. In: The Tenth International Conference on Learning Representations, ICLR 2022, Virtual Event, 25–29 April 2022. OpenReview.net (2022). https://openreview.net/forum?id=ydopy-e6Dg
53. Zhu, A.Z., Thakur, D., Özaslan, T., Pfrommer, B., Kumar, V., Daniilidis, K.: The multi vehicle stereo event camera dataset: an event camera dataset for 3D perception. CoRR **abs/1801.10202** (2018). http://arxiv.org/abs/1801.10202
54. Zhu, A.Z., Yuan, L., Chaney, K., Daniilidis, K.: Unsupervised event-based learning of optical flow, depth, and egomotion. CoRR **abs/1812.08156** (2018). http://arxiv.org/abs/1812.08156

Distractors-Immune Representation Learning with Cross-Modal Contrastive Regularization for Change Captioning

Yunbin Tu[1], Liang Li[2(✉)], Li Su[1(✉)], Chenggang Yan[3], and Qingming Huang[1]

[1] University of Chinese Academy of Sciences, Beijing, China
tuyunbin22@mails.ucas.ac.cn, {suli,qmhuang}@ucas.ac.cn
[2] Key Laboratory of AI Safety of CAS, Institute of Computing Technology, Chinese Academy of Sciences (CAS), Beijing, China
liang.li@ict.ac.cn
[3] Lishui Institute of HDU, Hangzhou Dianzi University (HDU), Hangzhou, China

Abstract. Change captioning aims to succinctly describe the semantic change between a pair of similar images, while being immune to distractors (illumination and viewpoint changes). Under these distractors, unchanged objects often appear pseudo changes about location and scale, and certain objects might overlap others, resulting in perturbational and discrimination-degraded features between two images. However, most existing methods directly capture the difference between them, which risk obtaining error-prone difference features. In this paper, we propose a distractors-immune representation learning network that correlates the corresponding channels of two image representations and decorrelates different ones in a self-supervised manner, thus attaining a pair of stable image representations under distractors. Then, the model can better interact them to capture the reliable difference features for caption generation. To yield words based on the most related difference features, we further design a cross-modal contrastive regularization, which regularizes the cross-modal alignment by maximizing the contrastive alignment between the attended difference features and generated words. Extensive experiments show that our method outperforms the state-of-the-art methods on four public datasets. The code is available at https://github.com/tuyunbin/DIRL.

Keywords: Change Captioning · Distractors-Immune Representation Learning · Cross-modal Contrastive Regularization

1 Introduction

Recently, researchers have witnessed the great success toward vision-language understanding and generation [21,30,43,44]. As an emerging task, change captioning [27,36,46] is to describe what has semantically changed between two

---

Supplementary Information The online version contains supplementary material available at https://doi.org/10.1007/978-3-031-72775-7_18.

© The Author(s), under exclusive license to Springer Nature Switzerland AG 2025
A. Leonardis et al. (Eds.): ECCV 2024, LNCS 15101, pp. 311–328, 2025.
https://doi.org/10.1007/978-3-031-72775-7_18

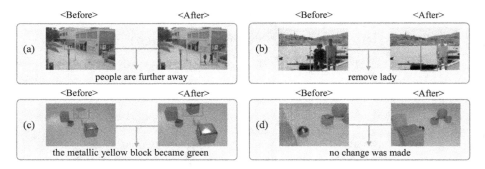

Fig. 1. The examples of change captioning under different scenarios. The first and second cases show that object moving and dropping, respectively. The third one shows the color change under distractors (viewpoint and illumination changes), where the real change is overwhelmed by pseudo changes. The last one shows with only distractors. Changed objects are shown in red boxes. (Color figure online)

similar images in natural language (Fig. 1(a)–(d)). This task has a wide range of practical applications, such as generating reports for surveillance area changes [10] and pathological changes between medical images [17], as well as providing visually-impaired users with the explanations of image editing effects [29].

This task also poses a formidable challenge: models should be powerful enough to not only understand the contents of two images, but also describe the semantic change between them, while being immune to distractors (viewpoint and illumination changes). In a dynamic environment, two images of the same scene are usually obtained in the presence of distractors. In this situation, unchanged objects between two images often appear obviously pseudo changes about the scale and location (Fig. 1(c) (d)), where the features of same objects within the image pair might be perturbational. Especially, as a viewpoint drastically changes, certain objects might partially overlap the others, *e.g.*, the brown ball is partially occluded by the red block in the "after" image of Fig. 1(d). Accordingly, the feature discrimination might be weakened under this viewpoint. In short, there are perturbational and discrimination-degraded features between two image representations under distractors.

Most previous methods directly subtract between such two image representations [9,24,28,38] or compute their feature similarity [25,26,45,47], which risk capturing error-prone difference features in between. The latest works SCORER [35] and SMART [34] overcome the distractors by implementing contrastive learning between similar/dissimilar image pairs. By maximizing the alignment of similar ones, both methods make the features of unchanged objects non-perturbational under distractors. However, both methods disregard the feature discrimination-degraded problem, leaving the influence of distractors upon a pair of image representations remained.

In this paper, we propose a **D**istractors-**I**mmune **R**epresentation **L**earning (DIRL) network to attain a pair of stable image representations that are non-perturbational and discriminative under distractors, for robust change captioning. Concretely, given two raw image representations, DIRL first computes a channel correlation matrix between them. Next, DIRL implements cross-channel decorrelation to optimize this matrix to be close to the identity matrix, *i.e.*, making the corresponding channels of two image representations have similar semantics, while enforcing different channels to be independent. In this self-supervised fashion, DIRL not only alleviates feature perturbation between two images, but also enhances feature discrimination for each image. As such, the model can sufficiently interact the two stable image representations to infer their reliable difference features, which are then translated into a linguistic sentence via a transformer decoder.

During sentence generation, as the changed object typically occurs in a local region with weak feature, traditional attention mechanisms used in mainstream methods [14,25,31] may lead to less satisfactory cross-modal alignment. To facilitate correct alignment, we further design a **C**ross-modal **C**ontrastive **R**egularization (CCR), which is built upon the cross-attention module of transformer decoder. When the cross-attention module yields the attended difference features, CCR further regularizes them by maximizing the contrastive alignment between them and generated words, which helps the decoder generate the change caption based on the most related difference features.

Our Key Contributions are: **(1)** We propose the novel DIRL network to learn a pair of stable image representations under distractors by correlating their corresponding channels and decorrelating different channels. Based on the two distractors-immune representations, the model is able to learn the robust difference features between them for caption generation. **(2)** We design the CCR to regularize the cross-modal alignment by maximizing the contrastive alignment between the features of attended difference and generated words. This helps the decoder generate words based on the most related difference features. **(3)** Our method performs favourably against state-of-the-art methods on four public datasets with different change scenarios.

2 Related Work

Change captioning is a new task in the vision-and-language area [4,6,12,16, 32,39]. The pioneer work [13] releases the dataset about changes in surveillance scenarios with potential illumination changes. Afterwards, Tan *et al.* [29] collect a dataset about image editing scenes. Both works [14,24] develop two datasets for simulating distractors (viewpoint and illumination changes), where the unchanged objects' scale and location would illustrate obviously pseudo changes.

To learn the difference features under distractors, prior works [9,20,24] directly subtract between two image representations, which are hard to generalize to unaligned image pairs. SGCC [18] introduces 3D information of object depths to overcome viewpoint changes. Recent works [14,26,31,47] first measure the feature similarity to summarize the shared features and then remove them

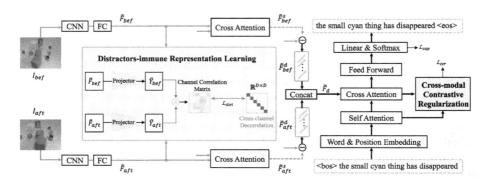

Fig. 2. The framework of our method, where the core blocks are **distractors-immune representation learning** network and **cross-modal contrastive regularization**. FC and Concat are short for the fully-connect layer and concatenation operation.

to obtain difference features. On the other hand, the works [7,25,45] correlate the similar features of two images to implicitly deduce the difference features. However, the above methods do not well address the influence of distractors upon image representations. The latest works SCORER [35] and SMART [34] try contrastive learning between similar/dissimilar image pairs, which model the alignment between similar ones to handle distractors. There are two differences between both works and our DIRL. First, SCORER and SMART capture relations between all paired/unpaired images in a batch, whereas DIRL computes the correlation between the feature channels of each image pair. Second, SCORER and SMART only make the features between two images non-perturbational under distractors, while DIRL additionally considers enhancing feature discrimination for each image to better recognize semantic changes.

During caption generation, since the changed object typically appears in a local region with weak feature, it is difficult to learn reliable alignment via current attention modules in prevalent methods. To address this issue, SMART [34] employs part-of-speech to guide the decoder to dynamically use visual information. These works [9,14,35] propose a cycle consistence module to enforce correct cross-modal alignment. The other works [3,7,45] design a pre-training and fine-tuning strategy to strengthen the alignment. In this paper, we regularize cross-modal alignment by proposing CCR, which maximizes the contrastive alignment between the generated words and attended difference features. We find that two works in single-image captioning [6,44] tried the similar idea. Both works model the coarse-grained cross-modal contrastive regularization between the input image and generated caption. Instead, our CCR reuses the attended difference features; establishes the fine-grained contrastive regularization between them and generated words during decoding. In this way, the model can use the most relevant difference features to predict next words.

3 Methodology

As shown in Fig. 2, the overall framework of our method includes following parts. First, a pre-trained CNN model extracts the raw representations of an image pair, which are then transformed into a low-dimensional space. Next, the two image representations are fed into the proposed DIRL to make them non-perturbational and discriminative under distractors. Subsequently, a cross-attention module interacts the two stable representations to mine the shared features, which are removed to learn the difference features for decoding into a sentence. During sentence generation, the designed CCR regularizes the cross-modal alignment, so as to help the decoder generate the sentence based on the most related difference features.

3.1 Image Pair Feature Extraction

Given a pair of images "before" I_{bef} and "after" I_{aft}, we first use a pre-trained CNN to extract their feature representations. They are denoted as F_{bef} and F_{aft}, where $F_o \in \mathbb{R}^{C \times H \times W}$. C, H, W indicate the number of channels, height, and width. Then, we project them into a low-dimensional space of $\mathbb{R}^D$ by a 2D-convolution:

$$\tilde{F}_o = \mathrm{conv}_2(F_o) + \mathrm{pos}(F_o), \tag{1}$$

where $o \in (bef, aft)$. pos is the learnable position encoding function.

3.2 Distractors-Immune Representation Learning

After obtaining the two position-embedded image representations, we propose a DIRL network to make them non-perturbational and discriminative under distractors in a self-supervised manner. Concretely, we first project $\tilde{F}_{bef}$ and $\tilde{F}_{aft}$ into a common embedding space with shared parameters:

$$\tilde{Y}_o = \mathrm{MLP}\left(\tilde{F}_o\right), o \in (bef, aft), \tag{2}$$

where MLP is a two-layer multi-layer perceptron with the ReLU activation function in between. Then, we compute a channel correlation matrix between $\tilde{Y}_{bef}$ and $\tilde{Y}_{aft}$ along the batch dimension:

$$\mathcal{C}_{ij} = \frac{\sum_b \tilde{y}_{b,i}^{bef} \tilde{y}_{b,j}^{aft}}{\sqrt{\sum_b \left(\tilde{y}_{b,i}^{bef}\right)^2} \sqrt{\sum_b \left(\tilde{y}_{b,j}^{aft}\right)^2}}, \tag{3}$$

where b indexes batch samples and i, j index the channel dimension of features. $\mathcal{C} \in \mathbb{R}^{D \times D}$ is a channel correlation matrix. Inspired by the recent self-supervised learning methods [42,48], we perform cross-channel decorrelation to enforce this matrix to be close to the identity matrix, which is implemented by using the ℓ_2-norm minimization:

$$\mathcal{L}_{dirl} = \sum_i (1 - \mathcal{C}_{ii})^2 + \alpha \sum_i \sum_{j \neq i} \mathcal{C}_{ij}^2. \tag{4}$$

The first term equates the diagonal of $\mathcal{C}$ to one, *i.e.*, the corresponding channels of two image representations will be correlated and thus have similar semantics under distractors. The second term equates the off diagonal of $\mathcal{C}$ to zero, *i.e.*, the different channels will be decorrelated. This enhances the discrimination of each image representation. α is a trade-off parameter to balance the importance between two terms, which is discussed in the supplementary material.

3.3 Difference Representation Learning

The difference features are modeled based on the two distractors-immune image representations. As shown in Fig. 1, most objects are identical between two images. Hence, it is natural to extract their shared features to infer the difference features. For a powerful model, it should capture all potential changes w.r.t. both images. To this end, we first learn the difference features for each image and then combine them to construct the difference features between two images.

Specifically, we first compute the shared features on each image by the multi-head cross-attention (MHCA) mechanism [40]:

$$\begin{aligned}\tilde{F}_{bef}^s &= \text{MHCA}\left(\tilde{F}_{bef}, \tilde{F}_{aft}, \tilde{F}_{aft}\right), \\ \tilde{F}_{aft}^s &= \text{MHCA}\left(\tilde{F}_{aft}, \tilde{F}_{bef}, \tilde{F}_{bef}\right).\end{aligned} \tag{5}$$

Subsequently, we subtract each from the corresponding image representation to compute the difference features on each image, respectively:

$$\begin{aligned}\tilde{F}_{bef}^d &= \tilde{F}_{bef} - \tilde{F}_{bef}^s, \\ \tilde{F}_{aft}^d &= \tilde{F}_{aft} - \tilde{F}_{aft}^s.\end{aligned} \tag{6}$$

Both $\tilde{F}_{bef}^d$ and $\tilde{F}_{aft}^d$ are then concatenated as the omni-representation of difference features between two images, which is implemented by a fully-connected layer with the ReLU activation function:

$$\tilde{F}_d = \text{ReLU}\left(\left[\tilde{F}_{bef}^d; \tilde{F}_{aft}^d\right] W_c + b_c\right), \tag{7}$$

where [;] is a concatenation operation.

3.4 Caption Generation

After learning $\tilde{F}_d \in \mathbb{R}^{HW \times D}$, we use a standard transformer decoder [40] to translate it into a sentence. First, we obtain the embedding features of m words (ground-truth words during training, predicted words during inference). Then, we use the multi-head self-attention (MHSA) layer to model relationships among these word embedding features $E[W] = \{E[w_1], ..., E[w_m]\}$:

$$\hat{E}[W] = \text{MHSA}\ (E[W], E[W], E[W]). \tag{8}$$

Subsequently, we use the relation-embedded word features $\hat{E}[W]$ to attend to the related difference features from $\tilde{F}_d$ based on the MHCA layer:

$$\hat{V} = \text{MHCA}\ (E[\hat{W}], \tilde{F}_d, \tilde{F}_d). \tag{9}$$

Next, the $\hat{V} \in \mathbb{R}^{HW \times D}$ is passed to a feed-forward network (FFN) to obtain the enhanced difference features:

$$\hat{V}' = \text{LayerNorm}(\hat{V} + \text{FFN}(\hat{V})). \tag{10}$$

Finally, the probability distributions of target words are calculated via a single hidden layer:

$$W = \text{Softmax}\left(\hat{V}' W_h + b_h\right), \tag{11}$$

where $W_h \in \mathbb{R}^{D \times \nu}$ and $b_h \in \mathbb{R}^\nu$ are the learnable parameters. ν is the dimension of vocabulary size.

3.5 Cross-Modal Contrastive Regularization

It is often the case that the changed object appears in a local area with weak feature, so it is difficult for the cross-attention module to directly align the captured difference features with word features. To this end, we devise the CCR to regularize the cross-modal alignment, which reuses the attented difference features $\hat{V}$ computed in Eq. (9) and maximizes the contrastive alignment between them and generated words $\hat{E}[W]$. Specifically, we first compute the global representation for $\hat{E}[W]$ and $\hat{V}$ via mean-pooling operation, respectively:

$$\begin{aligned}\tilde{E}[W] &= \frac{1}{m}\hat{E}[W], \\ \tilde{V} &= \frac{1}{HW}\hat{V},\end{aligned} \tag{12}$$

where m and HW are the length of $\hat{E}[W]$ and $\hat{V}$. Then, given a training batch, we sample B pairs of global representations of generated words and attended difference features. For k-th global word representation $\tilde{E}[W]_k$, k-th global difference representation $\tilde{V}_k$ is its positive, while the others will be the negatives in this batch. The similarity between two global representations is computed by dot-product:

$$\text{sim}\left(\tilde{E}[W], \tilde{V}\right) = \tilde{E}[W] \cdot \tilde{V}^\top, \tag{13}$$

Next, we leverage the InfoNCE loss [22] to pull semantically close pairs of $\tilde{E}[W]_k$ and $\tilde{V}_k$ together and push away non-related pairs:

$$\mathcal{L}_{t2v} = -\frac{1}{B}\sum_k^B \log \frac{e^{(\text{sim}(\tilde{E}[W]_k, \tilde{V}_k)/\tau)}}{\sum_r^B e^{(\text{sim}(\tilde{E}[W]_k, \tilde{V}_r)/\tau)}},$$

$$\mathcal{L}_{v2t} = -\frac{1}{B}\sum_k^B \log \frac{e^{(\text{sim}(\tilde{V}_k, \tilde{E}[W]_k)/\tau)}}{\sum_r^B e^{(\text{sim}(\tilde{V}_k, \tilde{E}[W]_r)/\tau)}}, \qquad (14)$$

$$\mathcal{L}_{ccr} = \frac{1}{2}(\mathcal{L}_{t2v} + \mathcal{L}_{v2t}),$$

where τ is the temperature hyper-parameter. This loss formulates a self-supervisory signal to regularize the cross-modal alignment during caption generation, so as to improve the quality of generated captions.

3.6 Joint Training

We train the overall architecture end-to-end by maximizing the likelihood of the observed word sequence. Given the ground-truth words $(w_1^*, \ldots, w_m^*)$, we minimize the negative log-likelihood loss:

$$\mathcal{L}_{cap}(\theta) = -\sum_{t=1}^m \log p_\theta\left(w_t^* \mid w_{<t}^*\right), \qquad (15)$$

where $p_\theta\left(w_t^* \mid w_{<t}^*\right)$ is computed by Eq. (11), and θ are all the learnable parameters. In addition, our method is self-supervised by the losses of DIRL and CCR. Thus, the total loss is defined as:

$$\mathcal{L} = \mathcal{L}_{cap} + \lambda_d \mathcal{L}_{dirl} + \lambda_c \mathcal{L}_{ccr}, \qquad (16)$$

where λ_d and λ_c are the trade-off parameters that are discussed in Sect. 4.6.

4 Experiments

4.1 Datasets and Evaluation Metrics

Spot-the-Diff [13] has 13,192 image pairs from surveillance cameras, where each image pair includes an underlying illumination change. According to the official split, we split it into training, validation, and testing with a ratio of 8:1:1.

CLEVR-Change [24] has 79,606 image pairs and 493,735 captions. It contains five change types, *i.e.*, "Color", "Texture", "Add", "Drop", and "Move", as well as moderate distractors. We use the official split of 67,660 for training, 3,976 for validation, and 7,970 for testing, respectively.

CLEVR-DC [14] has 48,000 image pairs with the same change types as CLEVR-Change, but this dataset includes drastic distractors. We use the official split with 85% for training, 5% for validation, and 10% for testing.

Image Editing Request [29] is comprised of 3,939 image pairs with 5,695 editing instructions. We use the official split with 3,061 image pairs for training, 383 for validation, and 495 for testing, respectively.

Evaluation Metrics: We follow the state-of-the-art methods to use the following five metrics for evaluating the quality of generated captions: BLEU-4 [23], METEOR [2], ROUGE-L [19], CIDEr [41] and SPICE [1]. We compute all the results by the Microsoft COCO evaluation server [5].

4.2 Implementation Details

For fair-comparison, we follow current SOTA methods to utilize a pre-trained ResNet-101 [8] to extract the features of a pair of images, with the dimension of $1024 \times 14 \times 14$. We project them into a lower dimension of 512. We set the hidden size of the model and word embedding size as 512 and 300. Temperature τ in Eq. (14) is set to 0.5. We train the model to converge with 10K iterations in total. Adam optimizer [15] is employed to minimize the negative log-likelihood loss of Eq. (16). During inference, we use the greedy decoding strategy to generate captions. More details are shown in the supplementary material.

Table 1. Comparison with SOTA methods on the Spot-the-Diff dataset.

Model	BLEU-4	METEOR	ROUGE-L	CIDEr	SPICE
M-VAM [26] (ECCV 2020)	10.1	12.4	31.3	38.1	14.0
DUDA+TIRG [9] (CVPR 2021)	8.1	12.5	29.9	34.5	–
R^{3}Net+SSP [37] (EMNLP 2021)	–	13.1	<u>32.6</u>	36.6	18.8
VACC [14] (ICCV 2021)	9.7	12.6	32.1	41.5	–
MCCFormers-D [25] (ICCV 2021)	10.0	12.4	–	**43.1**	18.3
IFDC [11] (TMM 2022)	8.7	11.7	30.2	37.0	–
I3N-TD [47] (TMM 2023)	**10.3**	13.0	31.5	<u>42.7</u>	18.6
VARD-Trans [31] (TIP 2023)	–	12.5	29.3	30.3	17.3
SCORER+CBR [35] (ICCV 2023)	<u>10.2</u>	13.2	–	38.9	18.4
SMART [34] (TPAMI 2024)	–	<u>13.5</u>	31.6	39.4	<u>19.0</u>
DIRL+CCR (Ours)	**10.3**	**13.8**	**32.8**	40.9	**19.9**

Table 2. Comparison with SOTA methods on the CLEVR-Change dataset.

Model	BLEU-4	METEOR	ROUGE-L	CIDEr	SPICE
DUDA [24] (ICCV 2019)	42.9	29.7	–	94.6	19.9
DUDA+TIRG [9] (CVPR 2021)	49.9	34.3	65.4	101.3	27.9
MCCFormers-D [25] (CVPR 2021)	53.3	37.1	70.8	119.1	30.4
R^3Net+SSP [37] (EMNLP 2021)	52.7	36.2	69.8	116.6	30.3
IFDC [11] (TMM 2022)	47.2	29.3	63.7	105.4	–
I3N [47] (TMM 2023)	53.1	37.0	70.8	117.0	32.1
NCT [33] (TMM 2023)	53.1	36.5	70.7	118.4	30.9
VARD-Trans [31] (TIP 2023)	53.6	36.7	71.0	119.1	30.5
SCORER+CBR [35] (ICCV 2023)	<u>54.4</u>	<u>37.6</u>	71.7	<u>122.4</u>	31.6
SMART [34] (TPAMI 2024)	54.3	37.4	<u>71.8</u>	**123.6**	**32.0**
DIRL+CCR (Ours)	**54.6**	**38.1**	**71.9**	**123.6**	<u>31.8</u>

4.3 Performance Comparison

Results on the Spot-the-Diff Dataset. This dataset contains image pairs from surveillance cameras with underlying illumination changes. We compare DIRL+CCR with the state-of-the-art methods: M-VAM [26], DUDA+TIRG [9], R^3Net+SSP [37], VACC [14], MCCFormers-D [25], IFDC [11], I3N-TD [47], VARD-Trans [31], SCORER+CBR [35], and SMART [34].

The results are shown in Table 1. Our DIRL+CCR obtains the best results on most metrics. Compared with the latest works SCORER+CBR and SMART, DIRL+CCR surpasses them on all metrics, especially with increases of 8.2% and 4.7% on SPICE. As the image pairs on this dataset are from surveillance cameras, most objects have no good postures, which makes the model difficult to locate and describe changed objects. In this situation, our DIRL+CCR also achieves an encouraging performance, which shows its robustness.

Results on the CLEVR-Change Dataset. This dataset contains basic geometric objects and moderate distractors. We evaluate DIRL+CCR under the setting of semantic change & distractors and detailed change categories. The comparison state-of-the-art methods are: DUDA [24], DUDA+TIRG [9], R^3Net+SSP [37], MCCFormers-D [25], IFDC [11], I3N [47], VARD-Trans [31], SCORER+CBR [35], and SMART [34].

The results are shown in Table 2 and 3. In the both tables, our DIRL+CCR gains the superior results under the setting of semantic change & distractors, and obtains best results on most change categories. The recent work NCT [33] is a match-based method under the architecture of transformer, which directly correlates two image representations to extract the shared features and then removes them to compute the difference features. We find that DIRL+CCR outperforms it on all metrics. This comparison validates the effectiveness of per-

Table 3. Evaluation on CLEVR-Change with varied change categories by METEOR.

Model	METEOR				
	Color	Texture	Add	Drop	Move
DUDA [24] (ICCV 2019)	32.8	27.3	33.4	31.4	23.5
DUDA+TIRG [9] (CVPR 2021)	36.1	30.4	37.8	36.7	27.0
R^3Net+SSP [37] (EMNLP 2021)	38.9	35.5	38.0	37.5	30.9
IFDC [11] (TMM 2022)	33.1	27.9	36.2	31.4	31.2
I3N [47] (TMM 2023)	39.9	36.7	<u>39.9</u>	**38.1**	30.6
NCT [33] (TMM 2023)	39.1	36.3	39.0	37.2	30.5
SMART [34] (TPAMI 2024)	<u>40.2</u>	<u>37.8</u>	39.3	**38.1**	<u>31.5</u>
DIRL+CCR (Ours)	**40.7**	**38.2**	**40.0**	<u>37.9</u>	**33.5**

Table 4. Comparison with SOTA methods on the CLEVR-DC dataset.

Model	BLEU-4	METEOR	ROUGE-L	CIDEr	SPICE
DUDA [24] (ICCV 2019)	40.3	27.1	–	56.7	16.1
M-VAM [26] (ECCV 2020)	40.9	27.1	–	60.1	15.8
VACC [14] (ICCV 2021)	45.0	29.3	–	71.7	**17.6**
NCT [33] (TMM 2023)	47.5	<u>32.5</u>	65.1	76.9	15.6
VARD-Trans [31] (TIP 2023)	48.3	32.4	–	77.6	15.4
SCORER+CBR [35] (ICCV 2023)	<u>49.4</u>	**33.4**	<u>66.1</u>	<u>83.7</u>	16.2
DIRL+CCR (Ours)	**51.4**	32.3	**66.3**	**84.1**	<u>16.8</u>

forming distractors-immune representation learning before change localization, and cross-modal alignment regularizing during caption generation.

Results on the CLEVR-DC Dataset. The experiment is conducted on CLEVR-DC with extreme distractors. We compare with the following state-of-the-art methods: DUDA [24], M-VAM [26], VACC [14], NCT [33], VARD-Trans [31], and SCORER+CBR [35]. The results are shown in Table 4.

We note that the overall performance of our DIRL+CCR is better. Specifically, DIRL+CCR obtains the best results on BLEU-4, ROUGE-L, and CIDEr metrics, while achieving comparable performance against the state-of-the-art methods on the other metrics. Especially, DIRL+CCR obtains relative improvement of 8.2%, 6.4% and 4.0% on BLEU-4 against the recent works NCT, VARD-Trans and SCORER+CBR, respectively. The results validate the robustness of our method under the disturbance of drastic distractors.

Results on the Image Editing Request Dataset. To further validate the generalization, we conduct the experiment on the scenario of image editing. The comparison methods are DUDA [24], Dynamic rel-att [29], BDLSCR [45], NCT [33], VARD-Trans [31], SCORER+CBR [35], and SMART [34]. Besides,

Table 5. Comparison with the SOTA methods on Image Editing Request. "*" indicates that the model is trained by pre-training and fine-tuning strategy.

Model	BLEU-4	METEOR	ROUGE-L	CIDEr
VIXEN-QFormer* [3] (AAAI 2024)	7.9	14.4	33.5	35.4
VIXEN-CLIP* [3] (AAAI 2024)	8.6	**15.4**	**42.5**	**38.1**
DUDA [24] (ICCV 2019)	6.5	12.4	37.3	22.8
Dyn rel-att [29] (ACL 2019)	6.7	12.8	37.5	26.4
BDLSCR [45] (IJIS 2022)	6.9	14.6	38.5	27.7
NCT [33] (TMM 2023)	8.1	15.0	38.8	34.2
VARD-Trans [31] (TIP 2023)	10.0	14.8	39.0	35.7
SCORER+CBR [35] (ICCV 2023)	10.0	15.0	39.6	33.4
SMART [34] (TPAMI 2024)	10.5	**15.2**	39.1	**37.8**
DIRL+CCR (Ours)	**10.9**	15.0	**41.0**	34.1

we compare our method with the latest work VIXEN [3]. VIXEN is first pretrained on a large-scale image editing dataset (which has not been released yet) and fine-tuned on the Image Editing Request dataset.

The results are shown in Table 5. Compared with the end-to-end training methods, our DIRL+CCR achieves the best results on two metrics and is on par with the SOTA methods on other two metrics. Compared with VIXEN, our DIRL+CCR also achieves a competitive performance. The comparison results validate the effectiveness and generalization of our method.

Performance Analysis. In brief, compared to the state-of-the-art methods, our method achieves superior results in different scenarios. This benefits from that 1) DIRL can make the paired image representations non-perturbational and discriminative under distractors. This helps learn robust difference features for caption generation. 2) CCR regularizes the cross-modal alignment by maximizing the contrastive alignment between the features of generated words and attended difference, thus helping improve the quality of yielded captions.

Table 6. Ablation study about DIRL and CCR on the CLEVR-DC dataset.

Ablation	DIRL	CCR	BLEU-4	ROUGE-L	CIDEr	SPICE
Transformer	×	×	48.9	65.6	79.6	15.7
Transformer	✓	×	50.5	65.8	81.8	16.2
Transformer	×	✓	49.3	65.5	82.7	16.4
Transformer	✓	✓	**51.4**	**66.3**	**84.1**	**16.8**

Table 7. Ablation study about DIRL on the CLEVR-DC dataset.

Ablation	non-perturbation	discrimination	BLEU-4	ROUGE-L	CIDEr	SPICE
Transformer	✗	✗	48.9	65.6	79.6	15.7
DIRL	✓	✗	**50.7**	65.2	79.3	**16.2**
DIRL	✗	✓	50.1	65.1	79.5	**16.2**
DIRL	✓	✓	50.5	**65.8**	**81.8**	**16.2**

4.4 Ablation Study

Ablation Study of Each Module. To figure out the contribution of the proposed DIRL and CCR, we conduct the ablation study on the CLEVR-DC dataset (with extreme distractors). The baseline model is a vanilla Transformer. It directly interacts two image representations to extract share features, which are then removed from image representations to attain the difference features for caption generation. The results shown in Table 6.

When we augment the vanilla Transformer with DIRL, it yields a performance boost. This validates DIRL can learn the two image representations that are non-perturbational and discriminative under the extreme distractors. As such, the model can better mine the shared features to learn reliable difference features. Besides, the performance of vanilla Transformer is enhanced by equipping it with CCR, which verifies that CCR can regularize the cross-modal alignment. By using both DIRL and CCR, the vanilla Transformer achieves best performance, in particular with the improvements of 5.7% and 7.0% on CIDEr and SPICE metrics. This shows that each module not only plays its unique role, but also supplements each other. In Fig. 3, we visualize the change localization and captioning results of DIRL+CCR and vanilla Transformer under distractors, where the unchanged objects appear obviously pseudo changes about scale and location. The visualization indicates that with the help of DIRL and CCR, the model can pinpoint and describe the actually changed object.

Ablation Study of DIRL. DIRL aims to learn two stable image representations from the perspectives of non-perturbation and discrimination, where the former is achieved by the first term of Eq. (4); the latter is achieved by the second term. To validate our claim, we conduct the ablative study for non-perturbation and discrimination, respectively. The results are shown in Table 7. The performance of DIRL with either characteristic does not obtain much gains against the vanilla Transformer. Our conjecture is that the influence of distractors is not well solved. By jointly modeling the non-perturbation and discrimination, the performance of DIRL is significantly improved, which shows that each characteristic is key to handle distractors. That is, only if the model learns both, can it make the paired image representations stable under distractors. As such, the model can capture the reliable difference features for caption generation.

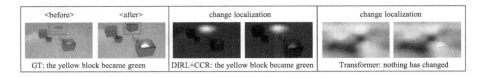

Fig. 3. Visualization of change localization and captioning results of DIRL+CCR and Transformer. GT is short for ground-truth and changed objects are shown in red boxes. (Color figure online)

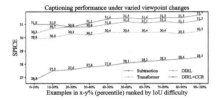

Fig. 4. Visualization of captioning performance under varied viewpoint changes.

Fig. 5. The effects of two trade-off parameters of λ_d and λ_c on CLEVR-DC.

4.5 Measuring Captioning Performance Under Distractors

To verify whether our method is immune to distractors, we measure the model's captioning performance under varied viewpoint changes. The varied viewpoints are computed by the IoU of the bounding boxes of the objects of two images, where the more the camera changes its position, the less the bounding boxes overlap. That is, lower IoU means higher difficulty. The results are shown in Fig. 4, where the compared models are Subtraction (direct subtraction between two image representations), vanilla Transformer, DIRL and DIRL+CCR. We find that the performance of Subtraction is the lowest, showing that direct subtraction generalizes poorly to distractors. DIRL apparently surpasses Transformer, validating it does make the representations of image pair immune to distractors. Further, the performance of DIRL further boosts by augmenting it with CCR, which demonstrates that CCR plays a role in enhancing cross-modal alignment and thus improves the captioning quality under distractors.

4.6 Study on the Trade-Off Parameters λ_d and λ_c

We discuss the effects of two trade-off parameters λ_d and λ_c in Eq. (16). Figure 5 shows the results of DIRL and DIRL+CCR with different λ_d and λ_c on the CLEVR-DC dataset. We first discuss λ_d. As its value increases or decreases, the performance of DIRL changes. Based on the results, we set λ_d as 0.03. Then, we fix λ_d to discuss λ_c and set it as 0.05. Similarly, we further discuss λ_d and λ_c on the other three datasets, and set them as 0.5 and 0.3 on the CLEVR-Change dataset; 0.5 and 0.004 on the Spot-the-Diff dataset; 0.001 and 0.05 on the Image Editing Request dataset.

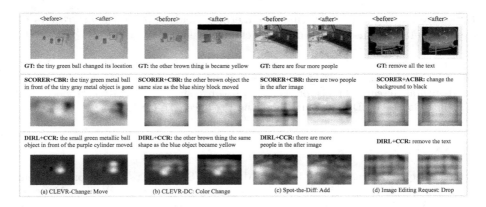

Fig. 6. Qualitative analysis on four datasets. We visualize the ground-truth captions (GT), generated captions and change localization results of SCORER+CBR [35] and our DIRL+CCR. The ground-truth changes are shown in red boxes. (Color figure online)

4.7 Qualitative Analysis

We conduct qualitative analysis about change localization and caption generation on the four datasets, as shown in Fig. 6(a)–(d). The compared method is SCORER+CBR [35], which first obtains view-invariant representations and then interacts both to compute the difference features, as well as improving captioning quality by cross-model backward reasoning. In these examples, the change localization maps of SCORER+CBR are relatively scattered and cannot focus on genuine changes. Hence, it fails to generate the correct sentences to describe these changes. By contrast, our DIRL+CCR successfully recognizes and captions the changes. This superiority benefits from making two image representations non-perturbational and discriminative under distractors. As such, our method can better pinpoint the actually changed object between them. During caption generation, CCR further regularizes cross-modal alignment, so as to make the target words generated based on the most related difference features. More qualitative examples are shown in the supplementary material.

5 Conclusion

In this paper, we propose the DIRL to learn a pair of non-perturbational and discriminative image representations under distractors, by correlating their corresponding channels and decorrelating different channels. As such, the model can sufficiently mine their shared features to learn the robust difference features for caption generation. Further, we design the CCR to regularize the cross-modal alignment by maximizing the contrastive alignment between the attended difference features and generated words, which helps improve captioning performance.

Extensive experiments show that our method yields state-of-the-art results on the four public datasets with different change scenes.

Acknowledgements. This work was supported in part by National Natural Science Foundation of China: 62322211, 61931008, 62236008, 62336008, U21B2038, 62225207, Fundamental Research Funds for the Central Universities (E2ET1104), "Pionee" and "Leading Goose" R&D Program of Zhejiang Province (2024C01023).

References

1. Anderson, P., Fernando, B., Johnson, M., Gould, S.: SPICE: semantic propositional image caption evaluation. In: Leibe, B., Matas, J., Sebe, N., Welling, M. (eds.) ECCV 2016. LNCS, vol. 9909, pp. 382–398. Springer, Cham (2016). https://doi.org/10.1007/978-3-319-46454-1_24
2. Banerjee, S., Lavie, A.: METEOR: an automatic metric for MT evaluation with improved correlation with human judgments. In: Proceedings of the ACL Workshop on Intrinsic and Extrinsic Evaluation Measures for Machine Translation and/or Summarization, pp. 65–72 (2005)
3. Black, A., Shi, J., Fai, Y., Bui, T., Collomosse, J.: VIXEN: visual text comparison network for image difference captioning. In: AAAI (2024)
4. Chen, J., Li, L., Su, L., Zha, Z.j., Huang, Q.: Prompt-enhanced multiple instance learning for weakly supervised video anomaly detection. In: CVPR, pp. 18319–18329 (2024)
5. Chen, X., et al.: Microsoft COCO captions: data collection and evaluation server. arXiv preprint arXiv:1504.00325 (2015)
6. Cho, J., Yoon, S., Kale, A., Dernoncourt, F., Bui, T., Bansal, M.: Fine-grained image captioning with clip reward. In: Findings of NAACL, pp. 517–527 (2022)
7. Guo, Z., Wang, T.J., Laaksonen, J.: CLIP4IDC: CLIP for image difference captioning. In: AACL, pp. 33–42 (2022)
8. He, K., Zhang, X., Ren, S., Sun, J.: Deep residual learning for image recognition. In: CVPR, pp. 770–778 (2016)
9. Hosseinzadeh, M., Wang, Y.: Image change captioning by learning from an auxiliary task. In: CVPR, pp. 2725–2734 (2021)
10. Hoxha, G., Chouaf, S., Melgani, F., Smara, Y.: Change captioning: a new paradigm for multitemporal remote sensing image analysis. IEEE Trans. Geosci. Remote Sens. **60**, 1–14 (2022)
11. Huang, Q., et al.: Image difference captioning with instance-level fine-grained feature representation. IEEE Trans. Multimedia **24**, 2004–2017 (2022)
12. Islam, M.M., Ho, N., Yang, X., Nagarajan, T., Torresani, L., Bertasius, G.: Video recap: recursive captioning of hour-long videos. In: CVPR (2024)
13. Jhamtani, H., Berg-Kirkpatrick, T.: Learning to describe differences between pairs of similar images. In: EMNLP, pp. 4024–4034 (2018)
14. Kim, H., Kim, J., Lee, H., Park, H., Kim, G.: Agnostic change captioning with cycle consistency. In: ICCV, pp. 2095–2104 (2021)
15. Kingma, D.P., Ba, J.: Adam: a method for stochastic optimization. arXiv preprint arXiv:1412.6980 (2014)
16. Li, L., Gao, X., Deng, J., Tu, Y., Zha, Z.J., Huang, Q.: Long short-term relation transformer with global gating for video captioning. IEEE Trans. Image Process. **31**, 2726–2738 (2022)

17. Li, M., Lin, B., Chen, Z., Lin, H., Liang, X., Chang, X.: Dynamic graph enhanced contrastive learning for chest x-ray report generation. In: CVPR, pp. 3334–3343 (2023)
18. Liao, Z., Huang, Q., Liang, Y., Fu, M., Cai, Y., Li, Q.: Scene graph with 3d information for change captioning. In: ACM MM, pp. 5074–5082 (2021)
19. Lin, C.Y.: Rouge: a package for automatic evaluation of summaries. In: Text Summarization Branches Out, pp. 74–81 (2004)
20. Liu, C., Zhao, R., Chen, H., Zou, Z., Shi, Z.: Remote sensing image change captioning with dual-branch transformers: a new method and a large scale dataset. IEEE Trans. Geosci. Remote Sens. **60**, 1–20 (2022)
21. Liu, X., et al.: Entity-enhanced adaptive reconstruction network for weakly supervised referring expression grounding. IEEE Trans. Pattern Anal. Mach. Intell. **45**(3), 3003–3018 (2023)
22. Oord, A.v.d., Li, Y., Vinyals, O.: Representation learning with contrastive predictive coding. arXiv preprint arXiv:1807.03748 (2018)
23. Papineni, K., Roukos, S., Ward, T., Zhu, W.J.: BLEU: a method for automatic evaluation of machine translation. In: ACL, pp. 311–318 (2002)
24. Park, D.H., Darrell, T., Rohrbach, A.: Robust change captioning. In: ICCV, pp. 4624–4633 (2019)
25. Qiu, Y., et al.: Describing and localizing multiple changes with transformers. In: ICCV, pp. 1971–1980 (2021)
26. Shi, X., Yang, X., Gu, J., Joty, S., Cai, J.: Finding it at another side: a viewpoint-adapted matching encoder for change captioning. In: Vedaldi, A., Bischof, H., Brox, T., Frahm, J.-M. (eds.) ECCV 2020. LNCS, vol. 12359, pp. 574–590. Springer, Cham (2020). https://doi.org/10.1007/978-3-030-58568-6_34
27. Sun, Y., Qiu, Y., Khan, M., Matsuzawa, F., Iwata, K.: The STVchrono dataset: towards continuous change recognition in time. In: CVPR, pp. 14111–14120 (2024)
28. Sun, Y., et al.: Bidirectional difference locating and semantic consistency reasoning for change captioning. Int. J. Intell. Syst. **37**(5), 2969–2987 (2022)
29. Tan, H., Dernoncourt, F., Lin, Z., Bui, T., Bansal, M.: Expressing visual relationships via language. In: ACL, pp. 1873–1883 (2019)
30. Tang, W., Li, L., Liu, X., Jin, L., Tang, J., Li, Z.: Context disentangling and prototype inheriting for robust visual grounding. IEEE Trans. Pattern Anal. Mach. Intell. **46**(5), 3213–3229 (2024)
31. Tu, Y., Li, L., Su, L., Du, J., Lu, K., Huang, Q.: Viewpoint-adaptive representation disentanglement network for change captioning. IEEE Trans. Image Process. **32**, 2620–2635 (2023)
32. Tu, Y., et al.: I^{2}transformer: intra-and inter-relation embedding transformer for TV show captioning. IEEE Trans. Image Process. **31**, 3565–3577 (2022)
33. Tu, Y., Li, L., Su, L., Lu, K., Huang, Q.: Neighborhood contrastive transformer for change captioning. IEEE Trans. Multimedia **25**, 9518–9529 (2023)
34. Tu, Y., Li, L., Su, L., Zha, Z.J., Huang, Q.: Smart: syntax-calibrated multi-aspect relation transformer for change captioning. IEEE Trans. Pattern Anal. Mach. Intell. **46**(7), 4926–4943 (2024)
35. Tu, Y., Li, L., Su, L., Zha, Z.J., Yan, C., Huang, Q.: Self-supervised cross-view representation reconstruction for change captioning. In: ICCV, pp. 2805–2815 (2023)
36. Tu, Y., Li, L., Su, L., Zha, Z.J., Yan, C., Huang, Q.: Context-aware difference distilling for multi-change captioning. In: ACL (2024)
37. Tu, Y., Li, L., Yan, C., Gao, S., Yu, Z.: R^{3}Net:relation-embedded representation reconstruction network for change captioning. In: EMNLP, pp. 9319–9329 (2021)

38. Tu, Y., et al.: Semantic relation-aware difference representation learning for change captioning. In: Findings of ACL, pp. 63–73 (2021)
39. Tu, Y., Zhou, C., Guo, J., Li, H., Gao, S., Yu, Z.: Relation-aware attention for video captioning via graph learning. Pattern Recogn. **136**, 109204 (2023)
40. Vaswani, A., et al.: Attention is all you need. In: NeurIPS, pp. 5998–6008 (2017)
41. Vedantam, R., Lawrence Zitnick, C., Parikh, D.: CIDEr: consensus-based image description evaluation. In: CVPR, pp. 4566–4575 (2015)
42. Wang, Q., Zhang, Y., Zheng, Y., Pan, P., Hua, X.S.: Disentangled representation learning for text-video retrieval. arXiv preprint arXiv:2203.07111 (2022)
43. Xiao, J., Li, L., Lv, H., Wang, S., Huang, Q.: R&B: region and boundary aware zero-shot grounded text-to-image generation. ICLR (2024)
44. Yang, C.F., Tsai, Y.H.H., Fan, W.C., Salakhutdinov, R.R., Morency, L.P., Wang, F.: Paraphrasing is all you need for novel object captioning. In: NeurIPS, vol. 35, pp. 6492–6504 (2022)
45. Yao, L., Wang, W., Jin, Q.: Image difference captioning with pre-training and contrastive learning. In: AAAI (2022)
46. Yue, S., Tu, Y., Li, L., Gao, S., Yu, Z.: Multi-grained representation aggregating transformer with gating cycle for change captioning. ACM Trans. Multimedia Comput. Commun. Appl. (2024)
47. Yue, S., Tu, Y., Li, L., Yang, Y., Gao, S., Yu, Z.: I3N: intra- and inter-representation interaction network for change captioning. IEEE Trans. Multimedia **25**, 8828–8841 (2023)
48. Zbontar, J., Jing, L., Misra, I., LeCun, Y., Deny, S.: Barlow twins: self-supervised learning via redundancy reduction. In: ICML, pp. 12310–12320 (2021)

Rethinking Image-to-Video Adaptation: An Object-Centric Perspective

Rui Qian[1], Shuangrui Ding[1], and Dahua Lin[1,2,3]

[1] The Chinese University of Hong Kong, Hong Kong, China
[2] Shanghai Artificial Intelligence Laboratory, Shanghai, China
[3] Centre for Perceptual and Interactive Intelligence (CPII), Hong Kong, China
{qr021,ds023,dhlin}@ie.cuhk.edu.hk

Abstract. Image-to-video adaptation seeks to efficiently adapt image models for use in the video domain. Instead of finetuning the entire image backbone, many image-to-video adaptation paradigms use lightweight adapters for temporal modeling on top of the spatial module. However, these attempts are subject to limitations in efficiency and interpretability. In this paper, we propose a novel and efficient image-to-video adaptation strategy from the object-centric perspective. Inspired by human perception, which identifies objects as key components for video understanding, we integrate a proxy task of object discovery into image-to-video transfer learning. Specifically, we adopt slot attention with learnable queries to distill each frame into a compact set of object tokens. These object-centric tokens are then processed through object-time interaction layers to model object state changes across time. Integrated with two novel object-level losses, we demonstrate the feasibility of performing efficient temporal reasoning solely on the compressed object-centric representations for video downstream tasks. Our method achieves state-of-the-art performance with fewer tunable parameters, only 5% of fully finetuned models and 50% of efficient tuning methods, on action recognition benchmarks. In addition, our model performs favorably in zero-shot video object segmentation without further retraining or object annotations, proving the effectiveness of object-centric video understanding.

Keywords: Image-to-video Adaptation · Object-centric Learning

1 Introduction

Recently, large-scale image pre-training has revolutionized the computer vision community, with the large foundation models dominating various image tasks [34,56,69,78,87,94]. The remarkable success of these models can be attributed to the enormous amounts of image data [28,56] or image-text

Supplementary Information The online version contains supplementary material available at https://doi.org/10.1007/978-3-031-72775-7_19.

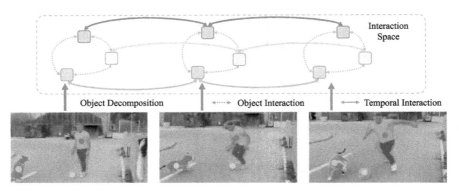

Fig. 1. Illustration of our proposed object-centric video understanding pipeline for image-to-video adaptation. Specifically, we first parse each frame into several object components (represented in different colors) to form an interaction space. Then we respectively establish inter-object interactions within each frame (represented by dotted arrows), and temporal state changes of individual objects (depicted by solid arrows).

pairs [34,69,78]. However, when we extend this discussion to the video domain, it becomes non-trivial. Acquiring large video datasets and managing high computational costs make training video foundation models from scratch a significant challenge. To circumvent these difficulties, the community has developed an innovative concept dubbed image-to-video adaptation. This approach aims to leverage the power of pre-trained image models and apply them to video data, enhancing both performance and efficiency. The image-to-video adaptation establishes a connection between effective methodologies in image models and the realm of video understanding, thereby opening new avenues for research and exploration.

Motivated by parameter-efficient transfer learning in natural language processing [31,32,41,89], a series of recent studies [36,46,47,55,57,86] have adopted an efficient tuning approach for image-to-video adaptation. These approaches maintain the image pre-trained model in a frozen state and only finetune a minimal set of additional parameters. For example, [47,55,57,86] insert learnable spatio-temporal modules into Transformer encoder blocks. However, this formulation encounters two vital limitations. Firstly, tuning the in-between parameters in the shallow layers still requires a considerable amount of GPU memory during gradient backpropagation. Secondly, the introduction of basic temporal modules, like attention layers, makes temporal reasoning on highly redundant features and lacks the essential inductive bias to effectively benefit the image-to-video adaptation, thereby presenting challenges in interpreting the training's success.

To tackle these challenges, we propose to transfer object knowledge from frozen image pre-trained models to videos for more efficient image-to-video adaptation. In essence, our approach entails two main steps: decoupling each frame into distinct object components, and subsequently identifying object state changes across time to facilitate comprehensive video comprehension as illustrated in Fig. 1. Specifically, we first pass the video frames through the image pre-trained model to obtain frame-wise features. Then, inspired by recent works on object discovery [51,82,83], we employ slot attention [51] with learnable queries [85] to parse each frame into a compact set of object tokens, serving as compressed representations for each frame. Thereafter, we feed these object tokens into object-time interaction layers to model the object state changes, which vividly depict the temporal dynamics in the video. Finally, we apply a linear head to the latent features representing these object state changes to produce the video-level prediction for action recognition. Unlike the existing works on object-centric representations, our method does not rely on object annotations [30,80,92], extra detectors [88] or additional modalities [91]. Instead, we develop two simple object-level losses to assist in distilling effective object-centric representations from image foundation models and establishing meaningful object state changes.

In this way, our method effectively mitigates the limitations of the previous work. Firstly, our method exhibits a notable enhancement in memory efficiency compared to conventional approaches. By exclusively finetuning the additional parameters on top of the image pre-trained model, we successfully reduce the memory required for gradient backpropagation. Additionally, compressing the video frames into object tokens significantly diminishes the computation requirements for temporal modeling, making our method more accessible and scalable. Secondly, by utilizing object-centric reasoning in the form of learnable object tokens and establishing the temporal state changes of individual objects, we inject strong inductive bias into the learning process that goes beyond basic temporal modules like attention layers. We demonstrate that our model can localize the specific objects in the video as a side-product, providing an insightful interpretation of the image-to-video adaptation.

To summarize, our contributions are as follows. (1) We propose an efficient object-centric method to adapt image pre-trained models to the video domain. By parsing video frames into object tokens and establishing object state changes across time, we enhance temporal perception with reduced computation redundancy. (2) We achieve state-of-the-art performance on action recognition, presenting higher efficiency than existing efficient tuning counterparts. (3) Our method effectively discovers different objects in videos without object-level annotations and achieves robust zero-shot video object segmentation.

2 Related Work

Video Action Recognition is a fundamental problem in computer vision. Compared to image tasks, the extra temporal relationships between different frames are crucial for video understanding. To this end, prevalent methods build on existing image models [20,29,49], and make architectural modifications to enable temporal modeling [2,10,15,16,19,21,45,50,62,64,65,74,75]. These works either fully finetune the whole model [2,5,10,21] or directly train on large-scale video data from scratch [66,73], requiring huge computation cost. More recently, some works propose to recognize actions from the perspective of state changes [1,54,59,71,79]. This novel idea regards actions as transformations that make changes to the objects and environments. However, most of these works directly model the state changes on the representations of the entire scene, where different semantics and objects are highly entangled. It is non-trivial to accurately perceive the state changes of different objects. In this work, we explicitly decouple different objects from the video scene, and respectively establish the state changes on individual objects to assist action recognition.

Object-Centric Video Representation Learning aims at establishing robust representations for different objects to assist video understanding. To achieve this goal, a series of works borrow the idea from object discovery and employ iterative slot attention [51] to decompose different objects [3,6,12,17, 18,37,39,63,85]. However, they are specifically trained for segmentation and can only deal with low-level segmentation tasks and lack the ability for high-level understanding. Another line of works uses the object-related features as a complement to the original video representations to enhance object awareness [24,30,53,92,93]. These methods achieve promising results on multiple action benchmarks [25,26,53], but still have two major limitations. First, they depend on bounding box annotations [30,80,92], additional detectors or modalities [88,91] to identify different objects, restricting the scalability to unseen data. Second, the identified objects provide plentiful cues for video analysis, yet these methods still fuse the disentangled object representations with the original video features, imposing limitations on efficiency. In contrast, our process is free of object annotations or handcrafted priors but only adopts simple contrastive loss to distill general object knowledge from image foundation models and decouple object components in video frames. Then we directly model temporal state changes on compact object representations, enabling more intuitive and efficient video understanding.

Efficient Tuning becomes a tending direction with the development of large foundation models [7,14,69,94]. It is proposed in natural language processing to reduce the trainable parameters and improve efficiency when transferring pre-trained models to downstream tasks [31,32,41,89]. Common strategies include inserting task-specific adapters into Transformer encoders [31,60,61], attaching prompt tokens to the input [42,48,67,70] and learning low-rank approximations [32]. Recently, the efficient tuning is increasingly explored in computer vision [4,11,22,35,44,72,95,96]. By freezing the image pre-trained model, [11,72] tune adapters with learnable parameters for efficient adaptation. [4,22,35,44] develop prompt tuning to adapt the frozen image model to various image tasks. Later, [36,46,47,55,57,58,68,77,86] take one step further to adapt image models to videos with temporal dynamics. In order to equip the model with temporal perception ability, [47,57,86] insert spatio-temporal modules into Transformer encoder blocks, but the tunable parameters in shallow Transformer layers require huge GPU memory in gradient backpropagation. [58,68] require two pathways to process spatial and temporal information. [36,46] attach spatio-temporal atten-

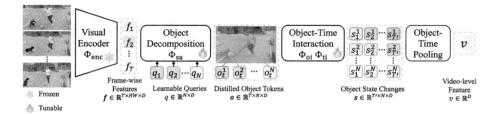

Fig. 2. An overview of our object-centric image-to-video adaptation framework. We use a frozen image pre-trained model to extract frame-wise features and pass them through a lightweight temporal fusion block. Then we employ slot attention with learnable queries to decompose each frame into a compact set of object tokens. Thereafter, we develop object-time interaction layers to establish inter-object interactions and build temporal state changes of individual objects, which are then pooled into the video-level feature for action recognition.

tion or motion decoders to the frame-wise features, but there exists high redundancy in attention computation [5,73]. Contrarily, our method retains the backbone architecture without modifications, distills object knowledge from image foundation models, and captures temporal state changes of discovered objects for video downstream tasks.

3 Method

The overall architecture of our object-centric image-to-video adaptation is shown in Fig. 2. We use a frozen pre-trained Vision Transformer [20] as the backbone to extract frame-wise features from the video clip. Then we apply slot attention with learnable query tokens to identify different object components and pass the object tokens through object-time interaction layers to model the temporal state changes of individual objects. The obtained object state change vectors encode rich semantics and essential temporal dynamics in the given video, and are subsequently pooled into the final video-level feature for downstream tasks.

3.1 Object Decomposition

To start with, given a video clip $x = \{x_1, x_2, \cdots, x_T\}$ consisting of T frames, we use a frozen image pre-trained Vision Transformer Φ_{enc} as the backbone to extract frame-wise features:

$$p_t, f_t = \Phi_{\text{enc}}(x_t), \quad p_t \in \mathbb{R}^D, \quad f_t \in \mathbb{R}^{HW \times D}, \tag{1}$$

where p_t and f_t denote the output cls token and feature map from the last Transformer block. H, W, D respectively denote height, width, and channel dimension. These image pre-trained features encode rich but highly-entangled semantics. And the next step is to parse these frame features into different object components which are crucial for video perception and understanding [24,30,80,92].

Inspired by recent works on unsupervised object discovery and segmentation [51,82,83], we develop a variant of slot attention on top of the feature map f_t for object decomposition in each frame. To be specific, instead of random sampling from a prior distribution, we use N learnable query tokens $q \in \mathbb{R}^{N \times D}$ to initialize the slot vectors $S \in \mathbb{R}^{N \times D}$. The intuition of this design is to encourage each query token to capture specific semantics or concepts [33,83] and produce temporally coherent object decompositions, which provide essential inductive bias for temporal reasoning in later stages. Then at each slot attention iteration, following [51], we use three learnable linear transformations to project the slot vectors S and frame features f_t into query, key and value, i.e., $Q \in \mathbb{R}^{N \times D}$, $K \in \mathbb{R}^{HW \times D}$ and $V \in \mathbb{R}^{HW \times D}$. After that, we compute the attention matrix $\tilde{A} \in \mathbb{R}^{N \times HW}$ with softmax normalization along the slot dimension as follows:

$$\tilde{A}_{i,j} := \frac{e^{A_{i,j}}}{\sum_{l=1}^{N} e^{A_{l,j}}}, \quad \text{where} \quad A := \frac{1}{\sqrt{D}} Q K^T. \quad (2)$$

We use the weighted mean to aggregate the value and pass them through Gated Recurrent Unit (GRU) to update the slot vectors:

$$S := \text{GRU}(\text{inputs} = MV, \text{states} = S), \quad \text{where} \quad M_{i,j} := \frac{\tilde{A}_{i,j}}{\sum_{l=1}^{HW} \tilde{A}_{i,l}}. \quad (3)$$

We iterate the routing process three times, and take the final slot attention weights as potential object segmentation masks. The slot vectors from the final iteration are treated as the object tokens that distill object knowledge from the pre-trained image features. Note that we perform slot attention Φ_{sa} on each frame in parallel, and denote the object tokens of each time stamp as:

$$o = \{o_1 \oplus o_2 \oplus ... \oplus o_T\} \in \mathbb{R}^{T \times N \times D}, \quad (4)$$

where $o_t = \Phi_{\text{sa}}(q, \tilde{f}_t) \in \mathbb{R}^{N \times D}$, and $\oplus$ denotes concatenation operation along time dimension. Since this attention mechanism introduces competition among slots, the obtained object tokens o_t are expected to take over distinct semantic parts and separate each frame into N different components.

Surprisingly, we find that the distilled tokens and corresponding attention matrices perform remarkably well in *zero-shot* video object segmentation tasks. They are even comparable to some pixel-level supervised counterparts, as demonstrated in Sect. 4.3. This result verifies that our method effectively extracts robust object-centric features for temporal reasoning in the later stage. Note that since videos often contain partial object semantics, there tend to be some slots that carry background information and interact with the foreground object tokens for comprehensive video analysis.

3.2 Object-Time Interaction

After obtaining the object tokens that reveal potential object semantics, we use them as compressed representations of video frames and further model the temporal dynamics from an object-centric perspective [30,93]. Our intuition is that in continuous frames, the object tokens, $o_t^n \in \mathbb{R}^D, t = 1, 2, ..., T$, with the same index n tend to represent the same objects across time as illustrated in Fig. 1. Hence, it is feasible to temporally track each object and conclude the object state changes based on the learned object tokens, thereby assisting the process of video understanding.

To achieve this goal, we first apply self-attention to the object tokens within the same frame to introduce inter-object interactions at each time stamp:

$$\tilde{o}_t = \Phi_{\text{oi}}(o_t) \in \mathbb{R}^{N \times D}, \tag{5}$$

where Φ_{oi} is the object interaction module instantiated by a standard Transformer encoder layer consisting of linear projections, self-attention calculation and a feed forward network. The resulting vector $\tilde{o}_t^n$ not only encodes the attribute of object n but also captures its relationships with contextual objects, serving as a representation for the latent state of object n at time stamp t.

Considering that the temporal state changes are a concrete manifestation of the effects of actions, they serve as an important cue for analyzing the events occurring in a video and assisting video understanding [1,59,71,79]. Hence, based on the vectors representing latent object states, we take one step further to look into multiple time stamps and explicitly establish the state changes of each object. Without loss of generality, we take the object with index n for illustration. Given the object states at different time stamps, we define a temporal interval δ to sample the initial and final states, i.e., $\tilde{o}_t^n$ as the initial state and $\tilde{o}_{t+\delta}^n$ as the final state. We concatenate them along channel dimension and pass them through the temporal interaction layer Φ_{ti} to model the object state change between time t and $t + \delta$:

$$s_t^n = \Phi_{\text{ti}}\left([\tilde{o}_t^n \oplus \tilde{o}_{t+\delta}^n]\right) \in \mathbb{R}^D, \quad t = \{1, 2, ..., T - \delta\}, \tag{6}$$

where $\oplus$ denotes channel-wise concatenation. We instantiate Φ_{ti} as a multi-layer perceptron of channel dimension $2D$-D-D. In this way, for each object, we obtain $T' = T - \delta$ state change vectors. The state change matrix covering all objects and time stamps can be denoted as $s \in \mathbb{R}^{T' \times N \times D}$.

Since the state change is a continuous process within a video sequence, it is feasible to conclude the overall state change of each object throughout the entire video by simple temporal average pooling on the established matrix s. Thereafter, we further apply average pooling on object dimension to aggregate the state change information of all objects as the final video-level representation, $v \in \mathbb{R}^D$, for downstream action recognition.

Discussion. Unlike the existing works that use object features as complements to original video features [24,30,53,92,93] or model state changes on the highly entangled scene representations [1,59,71,79], we perform temporal reasoning solely on distilled object tokens. Our object-centric state change modeling reduces redundancy, mitigates ambiguity and results in more fine-grained cues for video understanding. And comparing with the prevalent image-to-video adaptation methods, they introduce dense temporal fusion operations, e.g., temporal convolutions [46,47,57] and attention [46,47,58,68,86], to mix multi-frame features for action recognition. In contrast, we leverage the object state changes to perceive temporal dynamics, which closely resemble human logic and provide an effective inductive bias to the model. To be specific, it does not require dense fusion or traversal of all frames. Instead, we respectively model state changes on separated objects across sparsely sampled frames, enabling more efficient scene parsing and temporal perception as verified in Sect. 4.4.

3.3 Training

In training, we freeze the parameters in the image pre-trained model Φ_{enc}, and tune parameters that are responsible for object decomposition and temporal perception, i.e., Φ_{temp}, Φ_{oi} and Φ_{ti}. Besides the standard cross-entropy loss for action classification, we introduce two object-level losses to further facilitate object decomposition and temporal state change modeling.

Object Distillation Loss. Considering that the image pre-trained foundation models tend to encode rich semantics in the given scene [9,56,69], the output cls token p_t is expected to encompass the majority of the semantic components in frame t. Hence, we use a contrastive grounding loss [23,27,90] to align the learned object tokens o_t and the cls token p_t, encouraging o_t to cover all objects in frame t. Particularly, we first define the correspondence score between the o_t and p_t as:

$$c(o_t, p_t) = \sum_{n=1}^{N} \frac{e^{\langle o_t^n, p_t \rangle / \tau}}{\sum_{l=1}^{N} e^{\langle o_t^l, p_t \rangle / \tau}} \langle o_t^n, p_t \rangle, \tag{7}$$

where $\langle o_t^n, p_t \rangle = \frac{o_t^n p_t}{\|o_t^n\| \|p_t\|}$ denotes cosine similarity calculation, τ is the temperature hyper-parameter. This correspondence score adaptively aggregates each object component and avoids penalizing unrelated object semantics. Thereafter, we randomly sample cls token vectors p' from other videos within the mini-batch to formulate a negative sample pool $\mathcal{N}$ and apply the contrastive loss to maximize the agreement between the aligned object and cls token pairs:

$$\mathcal{L}_{obj} = -\sum_{t=1}^{T} \log \frac{e^{c(o_t, p_t)/\tau}}{e^{c(o_t, p_t)/\tau} + \sum_{p' \sim \mathcal{N}} e^{c(o_t, p')/\tau}}. \tag{8}$$

This self-supervised alignment objective encourages the object tokens to cover diverse semantics and distill object knowledge from the image pre-trained model.

Temporal Reasoning Loss. In our architecture, we analyze temporal dynamics from the perspective of object state changes. Therefore, it is crucial to formulate a learning objective that guides the temporal interaction module Φ_{ti} to accurately capture the state changes between different time stamps. To achieve this goal, our intuition is that the same objects undergoing the same type of actions should possess consistent state changes. Conversely, different actions will result in distinct state changes for the same object, and different objects will exhibit diverse state changes under the same action. Based on this, we develop a margin loss to guide object state change modeling. In detail, given s_t^n as query, we define the videos within the mini-batch that have the same action labels with the current video to form positive pool $\mathcal{P}$, and those of different action labels as negative pool $\mathcal{N}$. We sample the state change vectors of the same object and the same action, i.e., $\{\hat{s}_t^n | \hat{s} \in \mathcal{P}\}$, as positive pairs, which should be aligned with s_t^n. As for the negatives, we respectively sample the state change vectors of the same object but different actions, i.e., $\{s_t'^n | s' \in \mathcal{N}\}$, and those from the same video but of different objects, i.e., $\{s_t^m, m \neq n\}$, as *hard negatives* for discrimination. The margin loss with a margin hyper-parameter λ is formulated as:

$$\mathcal{L}_{temp} = \sum_{t=1}^{T'} \sum_{n=1}^{N} \{\sum_{\hat{s} \in \mathcal{P}} \|s_t^n - \hat{s}_t^n\|_2 \\ + \sum_{s' \in \mathcal{N}} \max(\lambda - \|s_t^n - s_t'^n\|_2, 0) + \sum_{\substack{m=1 \\ m \neq n}}^{N} \max(\lambda - \|s_t^n - s_t^m\|_2, 0)\}, \quad (9)$$

where $\|\cdot\|_2$ denotes $\mathcal{L}_2$ distance. This term enforces the model to discriminate diverse object states and capture unique dynamic effects of each action category.

Overall Objectives. Besides the above object-level losses, for action recognition tasks, we apply a linear classification head on the video-level representation v with a supervised classification loss $\mathcal{L}_{cls}$ in the form of cross-entropy. In this way, we formulate the overall loss function as:

$$\mathcal{L} = \mathcal{L}_{obj} + \mathcal{L}_{temp} + \mathcal{L}_{cls}. \quad (10)$$

4 Experiments

4.1 Implementation Details

We train and evaluate on three popular action recognition datasets: Kinetics-400 (K-400) [38], Something-Something-v2 (SSv2) [25], and Epic-Kitchens-100 (EK100) [13]. Additionally, we assess zero-shot video object segmentation on challenging multiple object segmentation dataset, DAVIS-2017-UVOS [8]. We use the frozen CLIP [69] pre-trained Vision Transformer [20] as Φ_{enc} for framewise feature extraction. In training, we crop each frame into 224×224, set the number of query tokens $N = 8$, and the time interval for object state change $\delta = T/4$ in default. We use AdamW optimizer [52] with an initial learning rate 5×10^{-4} and mini-batch size 512 to update the tunable parameters.

4.2 Video Action Recognition

In this section, we present the comparisons between our method and recent state-of-the-art on K-400, SSv2 and EK100 in terms of the number of tunable parameters, inference GFLOPs and action recognition accuracy.

Table 1. Results on Kinetics-400. We report the pre-train model initialization, inference GFLOPs, the number of tunable parameters (M), Top-1 and Top-5 accuracy. The 'Views' indicates Frame Number × Temporal Clips × Spatial Crops.

Method	Pre-train	Views	GFLOPs	Tunable	Top-1	Top-5
Full finetuning						
TimeSformer-L [5]	IN-21K	64 × 1 × 3	7140	121	80.7	94.7
VideoSwin-L [50]	IN-21K	32 × 4 × 3	7248	197	83.1	95.9
MTV-L [84]	JFT	32 × 4 × 3	18050	876	84.3	96.3
PromptCLIP A7 [36]	CLIP	16 × 5 × 1	–	–	76.8	93.5
ActionCLIP [77]	CLIP	32 × 10 × 3	16890	142	83.8	97.1
X-CLIP-L/14 [55]	CLIP	8 × 4 × 3	7890	420	87.1	97.6
Efficient tuning						
EVL ViT-L/14 [46]	CLIP	32 × 3 × 1	8088	59	87.3	–
ST-Adapter ViT-B/16 [57]	CLIP	32 × 3 × 1	1821	7.2	82.7	96.2
STAN-self-B/16 [47]	CLIP	16 × 3 × 1	1187	–	84.9	96.8
AIM ViT-B/16 [86]	CLIP	32 × 3 × 1	2428	11	84.7	96.7
DualPath ViT-B/16 [58]	CLIP	32 × 3 × 1	710	10	85.4	97.1
DiST ViT-B/16 [68]	CLIP	32 × 3 × 1	1950	–	85.0	97.0
DiST ViT-L/14 [68]	CLIP	32 × 3 × 1	8490	–	88.0	97.9
Ours ViT-B/16	CLIP	8 × 3 × 1	281	6.5	85.8	97.3
Ours ViT-B/16	CLIP	16 × 3 × 1	562	6.5	86.1	97.5
Ours ViT-B/16	CLIP	32 × 3 × 1	1123	6.5	86.2	97.5
Ours ViT-L/14	CLIP	32 × 3 × 1	5068	21	**88.5**	**98.0**

K-400. We first show the results on K-400 in Table 1 and conclude three observations. (1) Compared to fully finetuned models, our method achieves superior performance with significantly fewer tunable parameters. For example, our method of ViT-L/14 with only 21M tunable parameters outperforms all the fully finetuned approaches that require over 100M tunable parameters. (2) Our method with only a visual encoder achieves better results than the multimodal methods, ActionCLIP [77] and X-CLIP [55], which employ an extra text branch to guide the finetuning process. (3) Among the efficient tuning methods [46,47,57,58,68,86], EVL [46] attaches a motion decoder with multiple temporal convolution and attention layers for temporal modeling. AIM [86], ST-Adapter [57] and STAN [47] insert temporal blocks into intermediate CLIP encoder layers. DualPath [58] and DiST [68] introduce an additional temporal pathway to capture temporal dynamics. Our formulation models temporal relationships on decomposed objects in the form of state changes, which reduces computation redundancy but attains superior performance. For example, on ViT-B/16, our method with only an 8-frame input yields superior results to

Table 2. Results on Something-Something-v2. K-400†/K-600† indicates the model is pre-trained on both IN-21K and K-400/K-600.

Method	Pre-train	Views	GFLOPs	Tunable	Top-1	Top-5
Full finetuning						
TimeSformer-L [5]	IN-21K	64 × 1 × 1	7140	121	62.4	–
MTV-B [84]	IN-21K	32 × 4 × 3	4790	310	67.6	90.4
ViViT-L/16×2 [2]	K-400†	16 × 4 × 3	11892	311	65.4	89.8
VideoSwin-B [50]	K-400†	32 × 1 × 1	963	89	69.6	92.7
MViTv2-L (312↑) [43]	K-400†	32 × 1 × 3	8484	213	73.3	94.1
Efficient tuning						
EVL ViT-L/14 [46]	CLIP	32 × 1 × 3	9641	175	66.7	–
ST-Adapter ViT-B/16 [57]	CLIP	32 × 3 × 1	955	7.2	69.5	92.6
STAN-self-B/16 [47]	CLIP	16 × 3 × 1	1376	–	69.5	92.7
AIM ViT-B/16 [86]	CLIP	32 × 1 × 3	2496	14	69.1	92.2
DualPath ViT-B/16 [58]	CLIP	32 × 1 × 3	716	13	70.3	92.9
DiST ViT-B/16 [68]	CLIP	32 × 3 × 1	1950	–	70.9	92.1
DiST ViT-L/14 [68]	CLIP	32 × 3 × 1	8490	–	73.1	93.2
Ours ViT-B/16	CLIP	8 × 1 × 3	281	6.5	71.2	93.1
Ours ViT-B/16	CLIP	16 × 1 × 3	562	6.5	71.6	93.3
Ours ViT-B/16	CLIP	32 × 1 × 3	1123	6.5	71.8	93.4
Ours ViT-L/14	CLIP	32 × 1 × 3	5068	21	**73.6**	**94.3**

other approaches that employ a 32-frame input. On ViT-L/14, we reach the state-of-the-art results with 40% fewer GFLOPs (5068 vs 8490) than DiST [68].

SSv2. When extending to SSv2 which requires stronger temporal modeling, the observations are consistent. As depicted in Table 2, our object-centric formulation achieves state-of-the-art results and presents higher efficiency than both full finetuning and efficient tuning methods. In particular, our method with ViT-L/14 surpasses the advanced MViTv2-L [43] and DiST [68] with significantly reduced GFLOPs (5068 vs 8484, 8490). Moreover, on ViT-B/16, our method with only 8 frames as input outperforms others using 32 frames. It is non-trivial to achieve this on such a motion-heavy dataset, which in return reveals the significance of our temporal modeling in the form of object state changes. On the one hand, it enables accurate temporal perception with sparser sampling. On the other hand, our temporal modeling on individual objects eliminates the need for dense temporal fusion, substantially reducing computational redundancy.

EK100. On challenging egocentric EK100 that requires long-term object interactions, our method showcases substantial advantages over existing efficient tuning counterparts [46,57,68] in Table 3. The underlying reasons are two-fold. First, our method involves explicit object decomposition, which helps to capture the primary objects in egocentric videos and contributes to the significant improvements in nouns. Second, our temporal modeling on individual objects reduces the interference of camera motions and enforces the model to focus on the dynamics of each object, thus facilitating verb perception.

4.3 Zero-Shot Video Object Segmentation

To verify whether our method understands videos in an object-centric way and interpret why the model works, we evaluate unsupervised video object segmentation in addition to action recognition. Particularly, we use the frozen ViT-B/16 backbone and adopt slot attention weights $\{\tilde{A}_i \in \mathbb{R}^{HW}\}_{i=1}^{N}$ in Eq. 2 as object masks candidate without bells and whistles. We report $\mathcal{J}\&\mathcal{F}$ on challenging multiple object discovery benchmark, DAVIS-2017-UVOS [8]. Note that we directly apply the model trained on K-400 or SSv2 to video object segmentation in a *zero-shot* manner.

Table 3. Results on Epic-Kitchens-100 Verb, Noun and Action. All compared methods take an input of 8 × 1 × 3.

Method	Verb	Noun	Action
EVL ViT-B/16 [46]	62.7	51.0	37.7
ST-Adapter ViT-B/16 [57]	67.7	55.0	–
DiST ViT-B/16 [68]	69.5	58.1	45.8
DiST ViT-L/14 [68]	70.7	61.6	48.9
Ours ViT-B/16	71.1	60.9	49.0
Ours ViT-L/14	**73.2**	**64.5**	**51.7**

Table 4. Results on multiple object segmentation on DAVIS-2017-UVOS.

Model	$\mathcal{J}\&\mathcal{F}$	$\mathcal{J}$	$\mathcal{F}$
OCLR [82]	39.6	38.2	41.1
SMTC [63]	40.5	36.4	44.6
BA [17]	43.9	39.2	48.6
RVOS [76]	41.2	36.8	45.7
Ours SSv2	41.0	37.6	44.4
Ours K-400	44.7	42.3	47.1
Ours K-400 CRF	**49.3**	**46.5**	**52.1**

Quantitative Results. We present the comparison with recent state-of-the-arts without object annotations [63,81,82] as well as a fully supervised baseline [76] (denoted in grey) in Table 4. Among the compared works without mask annotations, OCLR [82] takes optical flow as input to distinguish motion patterns of diverse objects. SMTC [63] designs two-stage slot attention to produce temporally consistent object segmentations. BA [17] also utilizes image foundation models like DION [9] to provide spatio-temporal correspondence and use clustering to predict the object segmentation masks. On the one hand, since K-400 covers more diverse scenes and objects than SSv2, it leads to better generalization ability in this zero-shot segmentation scenario. On the other hand, our formulation trained on K-400 achieves superior results to these counterparts, even surpassing the simple supervised baseline RVOS [76]. This demonstrates the feasibility of transferring object knowledge from image foundation models to the video domain, and our object-centric temporal reasoning further facilitates temporally coherent object decomposition. However, the gap to the state-of-the-art supervised method is still large because the mask annotations lead to much more accurate segmentation boundaries. A straightforward way is to apply Conditional Random Field (CRF) [40] for refinement, which leads to 4.6 points improvement on $\mathcal{J}\&\mathcal{F}$ score. It is worth further exploration in the future.

Qualitative Results. We also visualize some examples of the generated segmentation masks in Fig. 3, where each column presents the frames within the same video and the same color denotes the masks generated by the same slot query. Generally, our model discriminates different object semantics in videos. Each distilled object token can capture specific semantics or concepts that generalize across diverse time stamps and scenes. For example, the token marked in red focuses on humans, the blue one emphasizes quadruped animals, and the yellow one identifies two-wheeled vehicles. This temporally coherent object decomposition provides an effective inductive bias for subsequent temporal reasoning, leading to more efficient temporal modeling.

4.4 Ablation Study

In this section, we use the CLIP ViT-B/16 as Φ_{enc} with an input of 16 frames for all ablative experiments unless otherwise specified. More details and extra ablation studies are presented in Supplementary Materials.

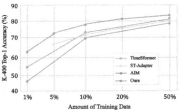

Fig. 3. Visualization of object decomposition in the form of segmentation. Each column presents three frames in a video. The same color denotes the objects identified by the same object token. (Color figure online)

Fig. 4. Comparison on training data efficiency. Our method demonstrates advantages, especially with a small amount of training data.

Training Losses. We explore the effectiveness of our newly introduced object distillation loss $\mathcal{L}_{obj}$ and temporal reasoning loss $\mathcal{L}_{temp}$ in Table 5. We observe that $\mathcal{L}_{obj}$ leads to dramatic improvements on all three datasets, K-400, SSv2, and DAVIS-2017-UVOS. This is because $\mathcal{L}_{obj}$ offers rich semantic knowledge reference from the image foundation model to facilitate object identification. And better object decomposition further enhances high-level action recognition. As for $\mathcal{L}_{temp}$, it guides the model to capture the state changes on each object. Without this guidance, relying solely on the action classification loss is insufficient to direct the temporal module to perceive representative temporal dynamics, thus resulting in a drastic performance drop on the motion-heavy dataset SSv2. And notably, this temporal loss also leads to improvements on DAVIS, which verifies that the temporal perception in return bolsters object decomposition.

Table 5. Ablation studies on training loss items $\mathcal{L}_{obj}$ and $\mathcal{L}_{temp}$. Note that $\mathcal{L}_{cls}$ is always applied in training.

$\mathcal{L}_{obj}$	$\mathcal{L}_{temp}$	K-400	SSv2	DAVIS
✗	✗	70.4	44.8	27.3
✓	✗	80.3	55.2	39.5
✗	✓	73.5	52.1	30.5
✓	✓	86.1	71.6	44.7

Table 6. Ablation studies on the number of frames T and sampling interval δ.

T	δ	GFLOPs	K-400	SSv2
4	1	4.2	85.4	70.7
8	2	8.4	85.8	71.2
8	4	8.2	85.8	71.1
16	2	18.0	86.1	71.6
16	4	17.8	86.1	71.6

Temporal Sampling Strategy. We also study the effect of different temporal sampling strategies in training. Specifically, we vary the number of frames per clip T and the time interval for object state change modeling δ. In Table 6, we display the Top-1 accuracy on K-400 and SSv2, as well as the computation GFLOPs excluding ViT backbone for more intuitive comparison. An exciting phenomenon is that even on SSv2 that requires strong temporal modeling, we achieve promising results with very sparse sampling, i.e., only 4 frames as input with significantly reduced computation. Our method is generally robust to the temporal sampling hyper-parameters, demonstrating the effectiveness of our temporal reasoning on compressed object tokens.

Training Data Efficiency. Figure 4 presents K-400 action recognition performance with various amounts of training data. For fair comparison with fully finetuned TimeSformer [5] as well as the efficient tuning methods, ST-Adapter [57] and AIM [86], we use CLIP ViT-B/16 with an input of 8 frames. Our method exhibits superior data efficiency to existing works. And the advantage becomes more significant with less training data, demonstrating the robust capacity of our object-centric formulation to transfer image pre-trained knowledge to videos.

5 Conclusion

In this work, we propose an object-centric formulation to adapt the image pretrained models to the video domain. Specifically, we employ a frozen image pre-trained model to extract frame-wise features and use slot attention with learnable queries to parse each frame into a compact set of object tokens. Then we apply object-time interaction to these object tokens to explicitly establish object state changes across time span. By using the distilled object tokens as the compressed representations of video frames, we have reduced computation costs and improved temporal reasoning capabilities especially under sparse sampling scenarios. Our method demonstrates state-of-the-art performance across action recognition benchmarks while realizing robust zero-shot video object segmentation, signifying a promising direction for future research in video analysis.

Limitations and Social Negative Impact. Our work demonstrates that the object-centric formulation serves as a reliable compression of video frames, which reduces redundancy and enables efficient temporal perception. But in some applications like a video QA system, it would be better to distill useful information according to specific questions. Incorporating questions as a condition to slot attention might address this problem. And there is a potential negative impact in privacy concerns. This work involves processing large amounts of visual data, which could include sensitive information and raise privacy concerns.

Acknowledgements. This work is partially supported by the National Key R&D Program of China (2022ZD0160201), Shanghai Artificial Intelligence Laboratory, the Centre for Perceptual and Interactive Intelligence (CPII) Ltd under the Innovation and Technology Commission (ITC)'s InnoHK.

References

1. Alayrac, J.B., Miech, A., Laptev, I., Sivic, J., et al.: Multi-task learning of object states and state-modifying actions from web videos. IEEE Trans. Pattern Anal. Mach. Intell. **46**, 5114–5130 (2024)
2. Arnab, A., Dehghani, M., Heigold, G., Sun, C., Lučić, M., Schmid, C.: VIVIT: a video vision transformer. In: Proceedings of the IEEE/CVF International Conference on Computer Vision, pp. 6836–6846 (2021)
3. Aydemir, G., Xie, W., Guney, F.: Self-supervised object-centric learning for videos. In: Thirty-Seventh Conference on Neural Information Processing Systems (2023). https://openreview.net/forum?id=919tWtJPXe
4. Bahng, H., Jahanian, A., Sankaranarayanan, S., Isola, P.: Visual prompting: modifying pixel space to adapt pre-trained models. arXiv preprint arXiv:2203.17274 (2022)
5. Bertasius, G., Wang, H., Torresani, L.: Is space-time attention all you need for video understanding? In: ICML, vol. 2, p. 4 (2021)
6. Besbinar, B., Frossard, P.: Self-supervision by prediction for object discovery in videos. In: 2021 IEEE International Conference on Image Processing (ICIP), pp. 1509–1513. IEEE (2021)
7. Brown, T., et al.: Language models are few-shot learners. Adv. Neural. Inf. Process. Syst. **33**, 1877–1901 (2020)
8. Caelles, S., Pont-Tuset, J., Perazzi, F., Montes, A., Maninis, K.K., Van Gool, L.: The 2019 DAVIS challenge on VOS: unsupervised multi-object segmentation. arXiv preprint arXiv:1905.00737 (2019)
9. Caron, M., et al.: Emerging properties in self-supervised vision transformers. In: Proceedings of the IEEE/CVF International Conference on Computer Vision, pp. 9650–9660 (2021)

10. Carreira, J., Zisserman, A.: Quo Vadis, action recognition? A new model and the kinetics dataset. In: Proceedings of the IEEE Conference on Computer Vision and Pattern Recognition, pp. 6299–6308 (2017)
11. Chen, S., et al.: AdaptFormer: adapting vision transformers for scalable visual recognition. Adv. Neural. Inf. Process. Syst. **35**, 16664–16678 (2022)
12. Crawford, E., Pineau, J.: Exploiting spatial invariance for scalable unsupervised object tracking. In: Proceedings of the AAAI Conference on Artificial Intelligence, vol. 34, pp. 3684–3692 (2020)
13. Damen, D., et al.: Rescaling egocentric vision. arXiv preprint arXiv:2006.13256 (2020)
14. Devlin, J., Chang, M.W., Lee, K., Toutanova, K.: BERT: pre-training of deep bidirectional transformers for language understanding. arXiv preprint arXiv:1810.04805 (2018)
15. Ding, S., et al.: Motion-aware contrastive video representation learning via foreground-background merging. In: Proceedings of the IEEE/CVF Conference on Computer Vision and Pattern Recognition, pp. 9716–9726 (2022)
16. Ding, S., Qian, R., Xiong, H.: Dual contrastive learning for spatio-temporal representation. In: Proceedings of the 30th ACM International Conference on Multimedia, pp. 5649–5658 (2022)
17. Ding, S., Qian, R., Xu, H., Lin, D., Xiong, H.: Betrayed by attention: a simple yet effective approach for self-supervised video object segmentation. arXiv preprint arXiv:2311.17893 (2023)
18. Ding, S., et al.: Motion-inductive self-supervised object discovery in videos. arXiv preprint arXiv:2210.00221 (2022)
19. Ding, S., Zhao, P., Zhang, X., Qian, R., Xiong, H., Tian, Q.: Prune spatio-temporal tokens by semantic-aware temporal accumulation. In: Proceedings of the IEEE/CVF International Conference on Computer Vision, pp. 16945–16956 (2023)
20. Dosovitskiy, A., et al.: An image is worth 16x16 words: transformers for image recognition at scale. arXiv preprint arXiv:2010.11929 (2020)
21. Feichtenhofer, C., Fan, H., Malik, J., He, K.: SlowFast networks for video recognition. In: Proceedings of the IEEE/CVF International Conference on Computer Vision, pp. 6202–6211 (2019)
22. Feng, C., et al.: PromptDet: towards open-vocabulary detection using uncurated images. In: Avidan, S., Brostow, G., Cissé, M., Farinella, G.M., Hassner, T. (eds.) ECCV 2022, Part IX. LNCS, vol. 13669, pp. 701–717. Springer, Cham (2022). https://doi.org/10.1007/978-3-031-20077-9_41
23. Ghiasi, G., Gu, X., Cui, Y., Lin, T.Y.: Scaling open-vocabulary image segmentation with image-level labels. In: Avidan, S., Brostow, G., Cissé, M., Farinella, G.M., Hassner, T. (eds.) ECCV 2022, Part XXXVI. LNCS, vol. 13696, pp. 540–557. Springer, Cham (2022). https://doi.org/10.1007/978-3-031-20059-5_31
24. Girdhar, R., Carreira, J., Doersch, C., Zisserman, A.: Video action transformer network. In: Proceedings of the IEEE/CVF Conference on Computer Vision and Pattern Recognition, pp. 244–253 (2019)
25. Goyal, R., et al.: The "something something" video database for learning and evaluating visual common sense. In: Proceedings of the IEEE International Conference on Computer Vision, pp. 5842–5850 (2017)
26. Gu, C., et al.: AVA: a video dataset of spatio-temporally localized atomic visual actions. In: Proceedings of the IEEE Conference on Computer Vision and Pattern Recognition, pp. 6047–6056 (2018)

27. Gupta, T., Vahdat, A., Chechik, G., Yang, X., Kautz, J., Hoiem, D.: Contrastive learning for weakly supervised phrase grounding. In: Vedaldi, A., Bischof, H., Brox, T., Frahm, J.-M. (eds.) ECCV 2020, Part III. LNCS, vol. 12348, pp. 752–768. Springer, Cham (2020). https://doi.org/10.1007/978-3-030-58580-8_44
28. He, K., Chen, X., Xie, S., Li, Y., Dollár, P., Girshick, R.: Masked autoencoders are scalable vision learners. In: Proceedings of the IEEE/CVF Conference on Computer Vision and Pattern Recognition, pp. 16000–16009 (2022)
29. He, K., Zhang, X., Ren, S., Sun, J.: Deep residual learning for image recognition. In: Proceedings of the IEEE Conference on Computer Vision and Pattern Recognition, pp. 770–778 (2016)
30. Herzig, R., et al.: Object-region video transformers. In: Proceedings of the IEEE/CVF Conference on Computer Vision and Pattern Recognition, pp. 3148–3159 (2022)
31. Houlsby, N., et al.: Parameter-efficient transfer learning for NLP. In: International Conference on Machine Learning, pp. 2790–2799. PMLR (2019)
32. Hu, E.J., et al.: LoRA: low-rank adaptation of large language models. arXiv preprint arXiv:2106.09685 (2021)
33. Jia, B., Liu, Y., Huang, S.: Unsupervised object-centric learning with bi-level optimized query slot attention. arXiv preprint arXiv:2210.08990 (2022)
34. Jia, C., et al.: Scaling up visual and vision-language representation learning with noisy text supervision. In: International Conference on Machine Learning, pp. 4904–4916. PMLR (2021)
35. Jia, M., et al.: Visual prompt tuning. In: Avidan, S., Brostow, G., Cissé, M., Farinella, G.M., Hassner, T. (eds.) ECCV 2022, Part XXXIII. LNCS, vol. 13693, pp. 709–727. Springer, Cham (2022). https://doi.org/10.1007/978-3-031-19827-4_41
36. Ju, C., Han, T., Zheng, K., Zhang, Y., Xie, W.: Prompting visual-language models for efficient video understanding. In: Avidan, S., Brostow, G., Cissé, M., Farinella, G.M., Hassner, T. (eds.) ECCV 2022, Part XXXV. LNCS, vol. 13695, pp. 105–124. Springer, Cham (2022). https://doi.org/10.1007/978-3-031-19833-5_7
37. Kabra, R., et al.: SIMONe: view-invariant, temporally-abstracted object representations via unsupervised video decomposition. Adv. Neural. Inf. Process. Syst. **34**, 20146–20159 (2021)
38. Kay, W., et al.: The kinetics human action video dataset. arXiv preprint arXiv:1705.06950 (2017)
39. Kipf, T., et al.: Conditional object-centric learning from video. In: International Conference on Learning Representations (2022). https://openreview.net/forum?id=aD7uesX1GF_
40. Lafferty, J., McCallum, A., Pereira, F.C.: Conditional random fields: probabilistic models for segmenting and labeling sequence data (2001)
41. Lester, B., Al-Rfou, R., Constant, N.: The power of scale for parameter-efficient prompt tuning. arXiv preprint arXiv:2104.08691 (2021)
42. Li, X.L., Liang, P.: Prefix-tuning: optimizing continuous prompts for generation. arXiv preprint arXiv:2101.00190 (2021)
43. Li, Y., et al.: MViTv2: improved multiscale vision transformers for classification and detection. In: Proceedings of the IEEE/CVF Conference on Computer Vision and Pattern Recognition, pp. 4804–4814 (2022)
44. Liang, F., et al.: Open-vocabulary semantic segmentation with mask-adapted clip. arXiv preprint arXiv:2210.04150 (2022)
45. Lin, Z., et al.: Space: unsupervised object-oriented scene representation via spatial attention and decomposition. arXiv preprint arXiv:2001.02407 (2020)

46. Lin, Z., et al.: Frozen clip models are efficient video learners. In: Avidan, S., Brostow, G., Cissé, M., Farinella, G.M., Hassner, T. (eds.) ECCV 2022. LNCS, vol. 13695, pp. 388–404. Springer, Cham (2022). https://doi.org/10.1007/978-3-031-19833-5_23
47. Liu, R., Huang, J., Li, G., Feng, J., Wu, X., Li, T.H.: Revisiting temporal modeling for clip-based image-to-video knowledge transferring. In: Proceedings of the IEEE/CVF Conference on Computer Vision and Pattern Recognition (CVPR), pp. 6555–6564, June 2023
48. Liu, X., et al.: P-tuning V2: prompt tuning can be comparable to fine-tuning universally across scales and tasks. arXiv preprint arXiv:2110.07602 (2021)
49. Liu, Z., et al.: Swin transformer: hierarchical vision transformer using shifted windows. In: Proceedings of the IEEE/CVF International Conference on Computer Vision, pp. 10012–10022 (2021)
50. Liu, Z., et al.: Video swin transformer. In: Proceedings of the IEEE/CVF Conference on Computer Vision and Pattern Recognition, pp. 3202–3211 (2022)
51. Locatello, F., et al.: Object-centric learning with slot attention. Adv. Neural. Inf. Process. Syst. **33**, 11525–11538 (2020)
52. Loshchilov, I., Hutter, F.: Decoupled weight decay regularization. In: International Conference on Learning Representations (2018)
53. Materzynska, J., Xiao, T., Herzig, R., Xu, H., Wang, X., Darrell, T.: Somethingelse: compositional action recognition with spatial-temporal interaction networks. In: Proceedings of the IEEE/CVF Conference on Computer Vision and Pattern Recognition, pp. 1049–1059 (2020)
54. Nagarajan, T., Grauman, K.: Attributes as operators: factorizing unseen attribute-object compositions. In: Ferrari, V., Hebert, M., Sminchisescu, C., Weiss, Y. (eds.) ECCV 2018. LNCS, vol. 11205, pp. 172–190. Springer, Cham (2018). https://doi.org/10.1007/978-3-030-01246-5_11
55. Ni, B., et al.: Expanding language-image pretrained models for general video recognition. In: Avidan, S., Brostow, G., Cissé, M., Farinella, G.M., Hassner, T. (eds.) ECCV 2022, Part IV. LNCS, vol. 13664, pp. 1–18. Springer, Cham (2022). https://doi.org/10.1007/978-3-031-19772-7_1
56. Oquab, M., et al.: DINOv2: learning robust visual features without supervision. arXiv preprint arXiv:2304.07193 (2023)
57. Pan, J., Lin, Z., Zhu, X., Shao, J., Li, H.: ST-Adapter: parameter-efficient image-to-video transfer learning for action recognition. arXiv preprint arXiv:2206.13559 (2022)
58. Park, J., Lee, J., Sohn, K.: Dual-path adaptation from image to video transformers. In: Proceedings of the IEEE/CVF Conference on Computer Vision and Pattern Recognition (CVPR), pp. 2203–2213, June 2023
59. Peh, E., Parmar, P., Fernando, B.: Learning to visually connect actions and their effects. arXiv preprint arXiv:2401.10805 (2024)
60. Pfeiffer, J., Kamath, A., Rücklé, A., Cho, K., Gurevych, I.: AdapterFusion: non-destructive task composition for transfer learning. In: Proceedings of the 16th Conference of the European Chapter of the Association for Computational Linguistics: Main Volume, pp. 487–503 (2021)
61. Pfeiffer, J., et al.: AdapterHub: a framework for adapting transformers. arXiv preprint arXiv:2007.07779 (2020)
62. Qian, R., Ding, S., Liu, X., Lin, D.: Static and dynamic concepts for self-supervised video representation learning. In: Avidan, S., Brostow, G., Cissé, M., Farinella, G.M., Hassner, T. (eds.) ECCV 2022. LNCS, vol. 13686, pp. 145–164. Springer, Cham (2022). https://doi.org/10.1007/978-3-031-19809-0_9

63. Qian, R., Ding, S., Liu, X., Lin, D.: Semantics meets temporal correspondence: self-supervised object-centric learning in videos. In: Proceedings of the IEEE/CVF International Conference on Computer Vision, pp. 16675–16687 (2023)
64. Qian, R., et al.: Streaming long video understanding with large language models. arXiv preprint arXiv:2405.16009 (2024)
65. Qian, R., et al.: Enhancing self-supervised video representation learning via multi-level feature optimization. In: Proceedings of the IEEE/CVF International Conference on Computer Vision, pp. 7990–8001 (2021)
66. Qian, R., et al.: Spatiotemporal contrastive video representation learning. In: Proceedings of the IEEE/CVF Conference on Computer Vision and Pattern Recognition, pp. 6964–6974 (2021)
67. Qin, G., Eisner, J.: Learning how to ask: querying LMS with mixtures of soft prompts. arXiv preprint arXiv:2104.06599 (2021)
68. Qing, Z., et al.: Disentangling spatial and temporal learning for efficient image-to-video transfer learning. In: Proceedings of the IEEE/CVF International Conference on Computer Vision (ICCV), pp. 13934–13944, October 2023
69. Radford, A., et al.: Learning transferable visual models from natural language supervision. In: International Conference on Machine Learning, pp. 8748–8763. PMLR (2021)
70. Shin, T., Razeghi, Y., Logan IV, R.L., Wallace, E., Singh, S.: AutoPrompt: eliciting knowledge from language models with automatically generated prompts. arXiv preprint arXiv:2010.15980 (2020)
71. Souček, T., Alayrac, J.B., Miech, A., Laptev, I., Sivic, J.: Look for the change: learning object states and state-modifying actions from untrimmed web videos. In: Proceedings of the IEEE/CVF Conference on Computer Vision and Pattern Recognition, pp. 13956–13966 (2022)
72. Sung, Y.L., Cho, J., Bansal, M.: VL-adapter: parameter-efficient transfer learning for vision-and-language tasks. In: Proceedings of the IEEE/CVF Conference on Computer Vision and Pattern Recognition, pp. 5227–5237 (2022)
73. Tong, Z., Song, Y., Wang, J., Wang, L.: VideoMAE: masked autoencoders are data-efficient learners for self-supervised video pre-training. Adv. Neural. Inf. Process. Syst. **35**, 10078–10093 (2022)
74. Tran, D., Bourdev, L., Fergus, R., Torresani, L., Paluri, M.: Learning spatiotemporal features with 3d convolutional networks. In: Proceedings of the IEEE International Conference on Computer Vision, pp. 4489–4497 (2015)
75. Tran, D., Wang, H., Torresani, L., Ray, J., LeCun, Y., Paluri, M.: A closer look at spatiotemporal convolutions for action recognition. In: Proceedings of the IEEE Conference on Computer Vision and Pattern Recognition, pp. 6450–6459 (2018)
76. Ventura, C., Bellver, M., Girbau, A., Salvador, A., Marques, F., Giro-i Nieto, X.: RVOS: end-to-end recurrent network for video object segmentation. In: Proceedings of the IEEE/CVF Conference on Computer Vision and Pattern Recognition, pp. 5277–5286 (2019)
77. Wang, M., Xing, J., Liu, Y.: ActionCLIP: a new paradigm for video action recognition. arXiv preprint arXiv:2109.08472 (2021)
78. Wang, W., et al.: Image as a foreign language: Beit pretraining for vision and vision-language tasks. In: Proceedings of the IEEE/CVF Conference on Computer Vision and Pattern Recognition, pp. 19175–19186 (2023)
79. Wang, X., Farhadi, A., Gupta, A.: Actions~transformations. In: Proceedings of the IEEE Conference on Computer Vision and Pattern Recognition, pp. 2658–2667 (2016)

80. Wang, X., Gupta, A.: Videos as space-time region graphs. In: Ferrari, V., Hebert, M., Sminchisescu, C., Weiss, Y. (eds.) ECCV 2018. LNCS, vol. 11209, pp. 413–431. Springer, Cham (2018). https://doi.org/10.1007/978-3-030-01228-1_25
81. Wang, X., Misra, I., Zeng, Z., Girdhar, R., Darrell, T.: VideoCutLER: surprisingly simple unsupervised video instance segmentation. arXiv preprint arXiv:2308.14710 (2023)
82. Xie, J., Xie, W., Zisserman, A.: Segmenting moving objects via an object-centric layered representation. In: Advances in Neural Information Processing Systems (2022)
83. Xu, J., et al.: GroupViT: semantic segmentation emerges from text supervision. In: Proceedings of the IEEE/CVF Conference on Computer Vision and Pattern Recognition, pp. 18134–18144 (2022)
84. Yan, S., et al.: Multiview transformers for video recognition. In: Proceedings of the IEEE/CVF Conference on Computer Vision and Pattern Recognition, pp. 3333–3343 (2022)
85. Yang, C., Lamdouar, H., Lu, E., Zisserman, A., Xie, W.: Self-supervised video object segmentation by motion grouping. In: Proceedings of the IEEE/CVF International Conference on Computer Vision (ICCV), pp. 7177–7188, October 2021
86. Yang, T., Zhu, Y., Xie, Y., Zhang, A., Chen, C., Li, M.: AIM: adapting image models for efficient video action recognition. In: The Eleventh International Conference on Learning Representations (2023). https://openreview.net/forum?id=CIoSZ_HKHS7
87. Yuan, L., et al.: Florence: a new foundation model for computer vision. arXiv preprint arXiv:2111.11432 (2021)
88. Zadaianchuk, A., Kleindessner, M., Zhu, Y., Locatello, F., Brox, T.: Unsupervised semantic segmentation with self-supervised object-centric representations. arXiv preprint arXiv:2207.05027 (2022)
89. Zaken, E.B., Ravfogel, S., Goldberg, Y.: BitFit: simple parameter-efficient fine-tuning for transformer-based masked language-models. arXiv preprint arXiv:2106.10199 (2021)
90. Zareian, A., Rosa, K.D., Hu, D.H., Chang, S.F.: Open-vocabulary object detection using captions. In: Proceedings of the IEEE/CVF Conference on Computer Vision and Pattern Recognition, pp. 14393–14402 (2021)
91. Zhang, C., et al.: Object-centric video representation for long-term action anticipation. In: Proceedings of the IEEE/CVF Winter Conference on Applications of Computer Vision, pp. 6751–6761 (2024)
92. Zhang, C., Gupta, A., Zisserman, A.: Is an object-centric video representation beneficial for transfer? In: Proceedings of the Asian Conference on Computer Vision, pp. 1976–1994 (2022)
93. Zhang, Y., Tokmakov, P., Hebert, M., Schmid, C.: A structured model for action detection. In: Proceedings of the IEEE/CVF Conference on Computer Vision and Pattern Recognition, pp. 9975–9984 (2019)
94. Zhou, J., et al.: iBOT: image BERT pre-training with online tokenizer. arXiv preprint arXiv:2111.07832 (2021)
95. Zhou, K., Yang, J., Loy, C.C., Liu, Z.: Conditional prompt learning for vision-language models. In: Proceedings of the IEEE/CVF Conference on Computer Vision and Pattern Recognition, pp. 16816–16825 (2022)
96. Zhou, K., Yang, J., Loy, C.C., Liu, Z.: Learning to prompt for vision-language models. Int. J. Comput. Vision **130**(9), 2337–2348 (2022)

Layer-Wise Relevance Propagation with Conservation Property for ResNet

Seitaro Otsuki[1]([✉]) , Tsumugi Iida[1], Félix Doublet[1], Tsubasa Hirakawa[2] ,
Takayoshi Yamashita[2] , Hironobu Fujiyoshi[2] , and Komei Sugiura[1]

[1] Keio University, Minato, Japan
{otsu8sei14,tiida,felixdoublet,komei.sugiura}@keio.jp
[2] Chubu University, Kasugai, Japan
hirakawa@mprg.cs.chubu.ac.jp, {takayoshi,fujiyoshi}@isc.chubu.ac.jp

Abstract. The transparent formulation of explanation methods is essential for elucidating the predictions of neural networks, which are typically black-box models. Layer-wise Relevance Propagation (LRP) is a well-established method that transparently traces the flow of a model's prediction backward through its architecture by backpropagating relevance scores. However, the conventional LRP does not fully consider the existence of skip connections, and thus its application to the widely used ResNet architecture has not been thoroughly explored. In this study, we extend LRP to ResNet models by introducing Relevance Splitting at points where the output from a skip connection converges with that from a residual block. Our formulation guarantees the conservation property throughout the process, thereby preserving the integrity of the generated explanations. To evaluate the effectiveness of our approach, we conduct experiments on ImageNet and the Caltech-UCSD Birds-200-2011 dataset. Our method achieves superior performance to that of baseline methods on standard evaluation metrics such as the Insertion-Deletion score while maintaining its conservation property. We will release our code for further research at https://5ei74r0.github.io/lrp-for-resnet.page/

Keywords: Explainable AI · Interpretability · Conservation Property

1 Introduction

The widespread adoption of neural networks underscores the critical importance of explainability of these models [30,35]. Indeed, the European Parliament's AI Act, promulgated in December 2023, proclaims that AI systems must be safe and transparent [18]. This strengthens the need to develop the methods to generate appropriate and meaningful explanations of neural network models. Current methods often lack transparency, making the interpretation of the results a nontrivial task [19]. Additionally, the black-box nature of neural network models

Supplementary Information The online version contains supplementary material available at https://doi.org/10.1007/978-3-031-72775-7_20.

© The Author(s), under exclusive license to Springer Nature Switzerland AG 2025
A. Leonardis et al. (Eds.): ECCV 2024, LNCS 15101, pp. 349–364, 2025.
https://doi.org/10.1007/978-3-031-72775-7_20

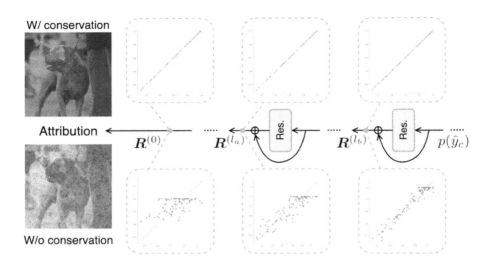

Fig. 1. We propose LRP for ResNet. By formulating Relevance Splitting at a point where the output from a skip connection converges with that from a residual block, we extend LRP—originally designed for propagating relevance between two consecutive layers—to the ResNet architecture while guaranteeing its conservation property, thereby preserving the integrity of the explanation process.

sometimes masks the underlying logic of their inference processes. This opacity presents significant challenges in verifying the validity of the models' predictions. To address this issue, Explainable AI has been proposed as a means of making the models more transparent and thus promoting the application of AI in critical domains [31].

The generation of visual explanations within neural networks poses a significant challenge, necessitating the precise extraction of critical areas. For example, the current Layer-wise Relevance Propagation (LRP) [3] implementation encounters notable challenges when applied to ResNet [10] architectures. ResNet's residual connections create multiple non linear relevance pathways, which cannot be handled by the typical relevance attribution process of LRP. Consequently, the explanations generated by LRP for ResNet models often give unreliable results as shown in Sect. 5.2. Therefore, while LRP offers a valuable framework for relevance attribution in neural networks, its limitations, particularly in handling architectures like ResNet, pose a challenge. This underscores the pressing need for advanced methodologies in this domain.

In explaining inference processes of black-box models, it is crucial that the explanation method is transparently formulated. Although several explanation methods applicable to ResNet, such as Grad-CAM [33], RISE [25], and LIME [30], offer valuable insights, they often lack transparency. For instance, Grad-CAM relies on the output of the last convolutional layer for its explanation, which may obscure the contributions of earlier model structures. Although Grad-CAM effectively highlights relevant areas in the input image for the model's inference, it falls short in elucidating how the earlier layers and their interconnections influence these findings. This limitation prevents a comprehensive understanding of the model's internal inference mechanisms.

By contrast, LRP [3] is a transparent explanation method that traces the flow of a model's prediction back through its architecture by backpropagating relevance scores. A key aspect of LRP is its conservation property, which ensures that the total relevance score received by an unit is equal to the total it redistributes. However, the application of LRP to CNN models has been limited. Although it has been applied to traditional models such as VGG [36], LRP's application to the widely used [6,15,27–29,44,45] ResNet architecture has not been sufficiently discussed.

In this study, we extend LRP to models with residual connections, such as ResNet, by formulating a novel relevance propagation rule that maintains the conservation property. This adaptation allows high-quality, transparent explanations to be generated for ResNet models, thereby contributing to a deeper understanding and improved interpretability of complex neural networks. A comprehensive summary of our method is shown in Fig. 1.

Our method is potentially applicable to other models featuring residual connections. Previous studies [1,2] have applied LRP to models such as LSTM [11] and transformers [41]. However, these adaptations have not addressed relevance propagation in the presence of residual connections. Specifically, the application of LRP to transformers has ignored the existence of residual connections in its relevance propagation [1]. Our key contributions are as follows:

- We extend LRP to models with residual connections by introducing Relevance Splitting at points where the output from a skip connection converges with that from a residual block.
- The conservation property is guaranteed throughout the proposed process, thereby preserving the integrity of the relevance propagation mechanism.
- To mitigate the issue of overconcentration of generated attributions within irrelevant regions, we introduce Heat Quantization.
- Our method demonstrate superior performance to that of baseline methods on standard evaluation metrics, including the Insertion-Deletion score.
- We investigate improved designs for relevance propagation within residual blocks and skip connections through an ablation study and demonstrate the significance of propagating relevance through skip connections and employing Ratio-Based Relevance Splitting.

2 Related Work

There has been widespread research on generating visual explanations for neural network models [3–5,9,14,22,33,43]. General post-hoc explanation methods are primarily categorized into three types based on the method whereby they generate explanations: perturbation, backpropagation, and approximation.

Perturbation Methods. Perturbation methods such as RISE [25], extremal perturbations [8], and SHAP [17] generate explanations by deliberately modifying the input images. These methods are effective, but often entail multiple interactions with black-box models, resulting in time-consuming processes that pose a significant drawback for practical applications.

Backpropagation Methods. Backpropagation methods [3,33–35,37,39,40] leverage the backpropagation algorithm to produce gradient or gradient-related explanations. Integrated Gradient [40] utilizes the gradient theorem to allocate attributions to input features. Grad-CAM [33] calculates explanations by summing CNN activations across channels, with weights derived from the average of the corresponding gradients. Rule-based backpropagation methods such as DeepLift [35] and Layer-wise Relevance Propagation (LRP) [3], which propagate scores using distinct methodologies. LRP is a well-established method that guarantees the conservation property; however, its proper application for ResNet is currently limited. Although the original LRP is applicable to simple skip connections with identity weights, we cannot directly apply the original LRP to skip connections with upscaling or downscaling mappings.

In this study, we integrates LRP into the ResNet architecture. This integration marks a significant advance over the traditional application of LRP, which lacks support for models with residual connections.

Approximation Methods. Approximation methods [24,30] employ an external entity to elucidate inference processes of black-box models. These methods create understandable approximations or explanations of how the black-box model works, usually focusing on specific predictions or decisions. However, their separation from the black-box models can pose challenges in accurately capturing the models' intricate working.

Incorporating Modules for Explanation. Another body of research [9,12,13,21] has investigated the direct incorporation of modules designed for generating explanations into the model architecture. For example, the Attention Branch Network [9] enhances image recognition performance by simultaneously generating explanations. However, the integration of such explanation-centric modules adds a layer of complexity, potentially reducing the model's overall transparency.

Datasets. In the pursuit of generating visual explanations for image classification tasks, standard datasets such as ImageNet [7], CIFAR-10 and CIFAR-100 [16] are commonly employed. In addition to these general datasets, domain-specific datasets such as Caltech-UCSD Birds-200-2011 [42] and the Indian Diabetic Retinopathy Image Dataset (IDRiD) [26] are often used. CUB features annotated bird images from 200 species, along with 15 part attributes per species. IDRiD contains fundus images with annotations of varying severity levels of fundus hemorrhage as evaluated by medical professionals.

3 Background: Layer-Wise Relevance Propagation

In the following sections, we first briefly review Layer-wise Relevance Propagation (LRP) [3] before discussing how it should be extended to ResNet and introducing several relevance propagation rules.

LRP is an explanation method that propagates a model's prediction back through the network via relevance scores. Its key feature is the propagation

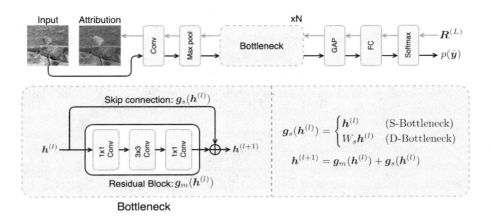

Fig. 2. LRP for ResNet. **Top:** LRP propagates the relevance score backward to generate an attribution map corresponding to the input image. We focus on the relevance score propagation through the Bottleneck module, which incorporates a residual connection. **Bottom:** Architecture of the Bottleneck module. The D-Bottleneck employs a linear projection, in its skip connection for dimension matching. ReLU activation functions and batch normalization layers are omitted for simplicity.

property, ensuring that the sum of relevance scores received by an unit equals the sum redistributed by the same unit.

3.1 LRP Between Two Consecutive Layers

Consider $\boldsymbol{h}^{(l)} \in \mathbb{R}^D$ and $\boldsymbol{h}^{(l+1)} \in \mathbb{R}^E$ as the intermediate features in the l-th and $(l+1)$-th consecutive layers, respectively. Let $\boldsymbol{f_\theta}$ be the function parametrized by $\boldsymbol{\theta}$ which maps $\boldsymbol{h}^{(l)}$ to $\boldsymbol{h}^{(l+1)}$, and define $\boldsymbol{R}^{(l)}$ and z_{ij} as the relevance scores of $\boldsymbol{h}^{(l)}$ and the contribution from $h_i^{(l)}$ to $h_j^{(l+1)}$, respectively.

The relevance scores propagating through $\boldsymbol{f_\theta}$ from all $h_j^{(l+1)}$ to $h_i^{(l)}$ can then be expressed as:

$$R_i^{(l)} = \sum_{j=1}^{E} \frac{z_{ij}}{\sum_{k=1}^{D} z_{kj}} R_j^{(l+1)}, \tag{1}$$

where $R_j^{(l+1)}$ and $R_i^{(l)}$ denote the relevance scores of the j-th element of $\boldsymbol{h}^{(l+1)}$ and the i-th element of $\boldsymbol{h}^{(l)}$, respectively. Here, the quantity z_{ij} can be formulated in various ways. The denominator $\sum_{k=1}^{D} z_{kj}$ ensures the conservation of the total relevance score between two consecutive layers.

In this study, we employ the z^+-Rule [20] for the relevance propagation where we can view $\boldsymbol{f_\theta}$ as a linear projection. The relevance propagation through linear projection $\boldsymbol{f}$ in the z^+-Rule can be written as:

$$R_i^{(l)} = \sum_{j=1}^{E} \frac{w_{ji}^+ h_i^{(l)}}{\sum_{k=1}^{D} w_{jk}^+ h_k^{(l)}} R_j^{(l+1)}, \tag{2}$$

where $\boldsymbol{f}(\boldsymbol{h}^{(l)}) = W\boldsymbol{h}^{(l)}$, $w_{ji}^+ = \max(0, w_{ji})$, and w_{ji} denotes the (j,i)-th element of $W \in \mathbb{R}^{E \times D}$. In this context, W^+ is defined as the matrix consisting of the non-negative elements of W, that is, $W_{ij}^+ = \max(0, W_{ij})$ for all i and j. This allows us to reformulate the equation as:

$$R_i^{(l)} = \sum_{j=1}^{E} \frac{\frac{\partial f_j^+}{\partial h_i^{(l)}}(\boldsymbol{h}^{(l)}) h_i^{(l)}}{f_j^+(\boldsymbol{h}^{(l)})} R_j^{(l+1)}, \tag{3}$$

where $\boldsymbol{f}^+(\boldsymbol{h}^{(l)}) = W^+ \boldsymbol{h}^{(l)}$.

We apply this rule to convolution, global average pooling (GAP), max pooling, and fully-connected layers, as these operations can be formulated as linear projections with specific weight matrices. For ReLU activation functions and batch normalization (BN) layers, the relevance scores are passed through without any modifications.

4 Method

By introducing a novel relevance propagation rule for models with residual connections, we extend Layer-wise Relevance Propagation (LRP) [3] to be applicable in such models, specifically developing an LRP for ResNet [10] models. This extension defines the calculation method of LRP for models with residual connections. Consequently, our approach is widely applicable to models possessing residual blocks. Figure 2 provides an overview of the ResNet architecture and the associated application of LRP.

4.1 General Formulation

In this study, we focus on the task of visualizing important regions in an image as a visual explanation of the model's prediction. In this task, the visual explanation should focus on the pixels that contributed to the model's decision. Figure 3 shows a typical sample of the task. The left and right panels show the input image and the visual explanation, respectively. The input is an image $\boldsymbol{x} \in \mathbb{R}^{c^{(0)} \times h^{(0)} \times w^{(0)}}$, where $c^{(0)}, h^{(0)}$, and $w^{(0)}$ denote the number of channels, height and width of the input image, respectively. The output $p(\hat{\boldsymbol{y}}) \in [0,1]^C$ denotes the predicted probability for each class, where C is the number of classes. Additionally, the importance of each pixel is obtained as an attribution $\boldsymbol{\alpha} \in \mathbb{R}^{h^{(0)} \times w^{(0)}}$ which is used as a visual explanation. To quantitatively evaluate our method, we use the Insertion, Deletion, and Insertion-Deletion scores [25]. In this study, we assume that the model is based on a ResNet architecture.

"Water ouzel" Relevant attribution

Fig. 3. Left: Typical sample of an input image from ImageNet [7]. **Right:** the corresponding attribution as a visual explanation.

4.2 Architecture of ResNet50

Before discussing the extension of LRP to ResNet models, we briefly review the architecture of ResNet, with a focus on ResNet50. ResNet50 consists of a convolution layer, a BN layer, a max pooling layer, 16 Bottleneck modules, a GAP layer, and a fully-connected layer followed by a softmax function. Each Bottleneck module consists of a skip connection and a residual block. The residual block is structured with three convolution layers: a 1×1 convolution for dimension reduction, a 3×3 convolution for processing, and another 1×1 convolution to restore or increase the dimension. These layers are sequentially followed by BN and ReLU activation functions.

We treat two types of Bottleneck modules: the Simple Bottleneck (S-Bottleneck) module and the Downsampling Bottleneck (D-Bottleneck) module. The S-Bottleneck module uses an identity mapping in its skip connection. In contrast, the D-Bottleneck module employs a linear projection in its skip connection for dimension matching. In this study, we mainly focus on how relevance should be propagated through these Bottleneck modules.

4.3 LRP for Bottleneck Modules

In this section, we discuss and formulate several potential relevance propagation rules for Bottleneck modules. As seen in Eq. (1), LRP originally defines a propagation rule between two consecutive layers. However, this rule does not account for the existence of a skip connection that bridges nonconsecutive layers. This raises a fundamental question: how should we propagate the relevance scores at the point where the output of the skip connection converges with that of the residual block?

As a preliminary step, we first divide the relevance score $\boldsymbol{R}^{(l+1)}$ into two parts: $\boldsymbol{R}_s$ for propagation through the skip connection and $\boldsymbol{R}_m$ for the mainstream of the residual block. In this context, s in $\boldsymbol{R}_s$ signifies the skip connection, and m in $\boldsymbol{R}_m$ denotes the mainstream of the residual block. To adhere to the conservation property, we impose the constraint: $\boldsymbol{R}^{(l)} = \boldsymbol{R}^{(l+1)} = \boldsymbol{R}_s + \boldsymbol{R}_m$.

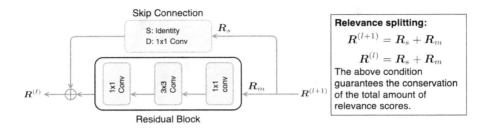

Fig. 4. Architecture of the Bottleneck block in ResNet and our Relevance Splitting approach. We introduce Relevance Splitting to consider the existence of skip connections in the relevance propagation of LRP.

We formulate two approaches for dividing $\boldsymbol{R}_s$ and $\boldsymbol{R}_m$: Symmetric Splitting and Ratio-Based Splitting, as illustrated in Fig. 4. Symmetric Splitting divides $\boldsymbol{R}^{(l+1)}$ equally as follows:

$$(R_s)_i = (R_m)_i = \frac{R_i^{(l+1)}}{2}. \quad (4)$$

This method is straightforward, but does not account for the varying contributions of individual elements in the intermediate feature $\boldsymbol{h}^{(l+1)}$, which could lead to a less-nuanced propagation of relevance.

Ratio-Based Splitting is a more thorough approach. Let $\boldsymbol{h}_s$ and $\boldsymbol{h}_m$ represent the outputs of the skip connection and the residual block, respectively. We divide $\boldsymbol{R}^{(l+1)}$ to satisfy the following conditions:

$$(R_s)_i = \frac{R_i^{(l+1)} \cdot |(h_s)_i|}{|(h_m)_i| + |(h_s)_i|}, \quad (R_m)_i = \frac{R_i^{(l+1)} \cdot |(h_m)_i|}{|(h_m)_i| + |(h_s)_i|}. \quad (5)$$

This approach accounts for the ratio of the absolute values of $\boldsymbol{h}_s$ and $\boldsymbol{h}_m$. When the skip connection is an identity mapping, $\boldsymbol{h}_s = \boldsymbol{h}^{(l)}$, and $\boldsymbol{h}_m$ is equivalent to $\boldsymbol{h}^{(l+1)} - \boldsymbol{h}^{(l)}$, representing the change in features attributable to the model's parameters. In such cases, the greater the absolute value of an element in $\boldsymbol{h}_m$, the more significant the model's contribution to that element. This approach naturally satisfies the conservation property:

$$(R_s)_i + (R_m)_i = \frac{R_i^{(l+1)} \cdot |(h_s)_i|}{|(h_m)_i| + |(h_s)_i|} + \frac{R_i^{(l+1)} \cdot |(h_m)_i|}{|(h_m)_i| + |(h_s)_i|} \quad (6)$$

$$= \frac{R_i^{(l+1)} \cdot (|(h_m)_i| + |(h_s)_i|)}{|(h_m)_i| + |(h_s)_i|} = R_i^{(l+1)}. \quad (7)$$

4.4 LRP for Two Types of Skip Connections

As mentioned in Sect. 4.2, there are two types of Bottleneck modules, each with a distinct type of skip connection. Specifically, the S-Bottleneck module features an identity mapping in its skip connection, whereas the D-Bottleneck module uses a linear projection.

This distinction leads to another question: should the two types of skip connections, i.e., identity mapping and linear projection, be treated equally? It appears logical to propagate relevance scores through linear projections in skip connections, as they transform the input through multiplication with their parameter matrices. These operations in the D-Bottleneck modules, although primarily intended to downsample the input to match the output dimension, can selectively emphasize important input elements through a linear projection. In contrast, a skip connection with an identity mapping does not perform any transformation. This suggests that there might be room to apply different relevance propagation approaches for the two types of skip connections. One approach entails setting $\boldsymbol{R}_s = 0$ for skip connections with identity mappings, thereby propagating all relevance through the residual block. The other approach involves applying the Relevance Splitting method discussed in Sect. 4.3, even to skip connections with identity mappings.

Based on the results of preliminary experiments, our proposed explanation method employs Ratio-Based Splitting and applies it to all skip connections, including those with identity mappings. To evaluate the effectiveness of this strategy, we conduct ablation studies comparing these conditions (see Sect. 5.4).

By sequentially applying the described propagation rules and backpropagating the relevance scores, we can compute the relevance score $\boldsymbol{R}^{(0)}$ for the input $\boldsymbol{x}$. Similar to existing LRP methods, $\boldsymbol{R}^{(0)}$ has the same shape as $\boldsymbol{x}$, and its channel-wise sum, denoted as $\boldsymbol{\alpha}_R$, can be directly used as an attribution map. However, $\boldsymbol{\alpha}_R$ tends to excessively concentrate on irrelevant regions. To mitigate this tendency, we quantize the values in $\boldsymbol{\alpha}_R$ to obtain a more even distribution of attribution, resulting in the final attribution map $\boldsymbol{\alpha}$. We refer to this operation as Heat Quantization, formulated as follows:

$$\alpha_{i,j} = (\alpha_R)_{\min} + \left\lfloor \frac{(\alpha_R)_{i,j} - (\alpha_R)_{\min}}{((\alpha_R)_{\max} - (\alpha_R)_{\min})/Q} \right\rfloor Q, \qquad (8)$$

where Q denotes the number of quantization bins. We set $Q = 8$.

5 Experiments and Results

5.1 Experimental Setup

We used the Caltech-UCSD Birds-200-2011 (CUB) dataset [42] and the validation set of ImageNet [7] (ILSVRC) 2012 to evaluate our method. These datasets were chosen because they are standard datasets for visual explanation generation tasks. The CUB dataset contains 11,788 images from 200 classes of bird species. The validation set of ImageNet consists of 50,000 images from 1,000 classes.

For the experiments on the CUB dataset and ImageNet, we employed ResNet-50 [10] that was trained on the CUB dataset and pretrained on ImageNet. Detailed information about the training setup, data preprocessing, and hardware specifications can be found in the supplementary material.

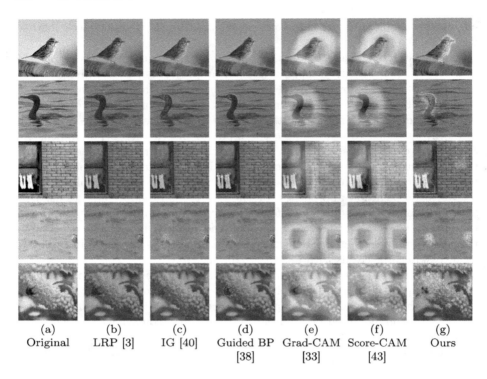

Fig. 5. Qualitative Results: Attribution produced by each explanation method for the prediction of ResNet50 with respect to the ground-truth classes (top to bottom): "Brandt Cormorant," "Savannah Sparrow," "Sock," "Bustard," and "Bee." IG and Guided BP denote Integrated Gradients and Guided BackPropagation, respectively.

5.2 Qualitative Analysis

Figure 5 presents the qualitative results. Column (a) displays the original images. Columns (b)–(f) depict the attribution maps generated by the baseline methods, overlaid on the original images. Column (g) represents the results generated by the proposed method. Columns (b), (c), and (d) show the explanations generated by Layer-wise Relevance Propagation (LRP) [3], Integrated Gradients [40], and Guided BackPropagation [38], respectively. Attribution maps generated by these explanation methods were often noisy or did not sufficiently highlight relevant regions, as demonstrated by the qualitative results. Columns (e) and (f) show the explanations generated by Grad-CAM [33] and Score-CAM [43], respectively. Both of their results have attention regions that encompass the whole of the relevant objects, but also focus on the background surrounding them. In contrast, the attribution maps generated by the proposed method specifically target the relevant objects with detailed focus and demonstrate minimal attention to background regions, thus yielding more appropriate explanations.

Table 1. Quantitative results on ImageNet and the CUB dataset. IG and Guided BP denote Integrated Gradients and Guided BackPropagation, respectively. Ins. and Del. denote Insertion and Deletion score, respectively. The best results are marked in bold.

Dataset Metric [%]	CUB [42]			ImageNet [7]		
Method	Ins. (↑)	Del. (↓)	ID score (↑)	Ins. (↑)	Del. (↓)	ID score (↑)
LRP [3]	5.8 ± 0.2	4.7 ± 0.1	1.1 ± 0.0	9.5	8.3	1.1
IG [40]	2.0 ± 0.1	1.5 ± 0.1	0.6 ± 0.0	5.2	6.2	−1.1
Guided BP [38]	4.2 ± 0.2	1.4 ± 0.1	2.8 ± 0.2	11.5	5.7	5.7
Grad-CAM [33]	50.8 ± 1.5	5.5 ± 0.4	45.3 ± 1.1	49.7	12.6	37.1
Score-CAM [43]	51.1 ± 1.7	5.4 ± 0.4	45.7 ± 1.4	48.8	13.3	35.5
Ours	**59.5 ± 1.0**	**1.4 ± 0.0**	**58.2 ± 1.0**	**56.3**	**1.8**	**54.5**

5.3 Quantitative Comparison Against Baselines

Table 1 presents the quantitative results of a comparison between several baseline methods and the proposed method. For the experiments on the CUB dataset, we conducted five experiments using each method and computed the mean and standard deviation as the final results. For the experiments on the ImageNet, we conducted a single experimental run with a pretrained model. The following methods were selected as baselines: LRP, Integrated Gradients, Guided BackPropagation, Grad-CAM, and Score-CAM. We selected these methods (except LRP) because they are standard methods that have been successfully applied to models with skip connections.

To quantitatively evaluate our method, we employed the Insertion, Deletion, and Insertion-Deletion (ID) scores [23,25,32]. These are standard evaluation metrics for explanation generation tasks, and we consider the ID score as the primary evaluation metric. The Insertion and Deletion scores were calculated as the area under the Insertion and Deletion curves, respectively. The ID score is defined as the difference between the Insertion and Deletion scores. Please find details of these metrics in the supplementary material. For empirical evaluation, we randomly selected 1,000 samples from the target dataset, ensuring equal representation from each class.

As listed in Table 1, our method achieved an ID score of 0.582 in the experiments on the CUB dataset. The corresponding ID scores of LRP, Integrated Gradients, Guided BackPropagation, Grad-CAM, and Score-CAM are 0.011, 0.006, 0.028, 0.453 and 0.457, respectively. The proposed method outperformed the best baseline method, Score-CAM, by 0.125 in terms of the ID score and achieved the best performance in terms of both the Insertion and Deletion scores. Furthermore, as listed in Table 1, in the ImageNet experiments, our method outperformed all the baselines on the ID score. Specifically, it exceeded the highest-scoring baseline method, Grad-CAM, by 0.174 on the ID score, and

Table 2. Comparison of propagation rules for the Bottleneck modules. "Include Identical" denotes the condition in which the relevance score for skip connections with identity mapping is not set to 0. Insertion, Deletion and ID scores were calculated on ImageNet. The highest scores are marked in bold.

Method	Include identical	Relevance splitting	Insertion (↑)	Deletion (↓)	ID Score (↑)
(i)		Symmetric	0.543	0.033	0.510
(ii)	✓	Symmetric	0.553	0.036	0.517
(iii)		Ratio-Based	0.543	0.033	0.510
(iv)	✓	Ratio-Based	**0.563**	**0.018**	**0.545**

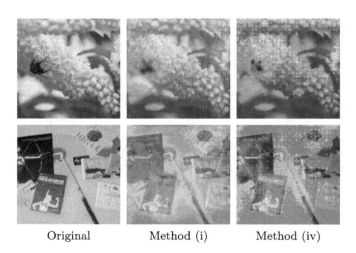

Original   Method (i)   Method (iv)

Fig. 6. Qualitative results from the ablation study of propagation rules for the Bottleneck modules. Methods (i) and (iv) refer to the methods described in Table 2. Method (iv) exhibits a more concentrated attribution towards relevant objects than Method (i).

again achieved the best performances in terms of both the Insertion and Deletion scores.

5.4 Ablation Study of Relevance Propagation Rules for Bottleneck Modules

As discussed in Sects. 4.3 and 4.4, we formulated several types of relevance propagation rules for Bottleneck modules. This section describes the results of an ablation study focusing on the key design elements within these rules. Table 2 presents the quantitative results of this ablation study.

The ID score under Method (iv) is 0.545, which surpasses the score under Method (iii) by 0.035. This indicates the effectiveness of allocating relevance to the skip connections with identity mappings. Furthermore, the ID score under Method (iv) exceeds that of Method (ii) by 0.028. This suggests that adopt-

Table 3. Quantitative results of the ablation study of Heat Quantization (HQ). These experimental runs were conducted on ImageNet. The best results are highlighted in bold.

Method	Insertion (↑)	Deletion (↓)	ID Score (↑)
(i) w/o HQ	0.442	0.066	0.376
(ii) w/ HQ	**0.543**	**0.033**	**0.510**

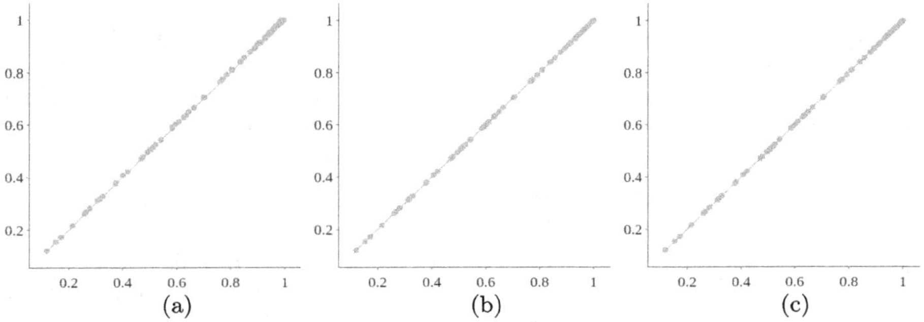

Fig. 7. Visualization of the sum of relevance scores backpropagated to three critical checkpoints within ResNet50: (a) input of the entire network, (b) input of the first Bottleneck block, and (c) input of the last Bottleneck block. In each panel, the horizontal axis and the vertical axis represent the sum of relevance score at the corresponding checkpoint and model's output predictions, respectively.

ing Ratio-Based Splitting, which considers the element-wise proportion of the output of a skip connection and a residual connection, results in higher-quality attribution maps than those generated by Symmetric Splitting.

The performance drop from Method (iv) to Method (iii) is greater than that from Method (iv) to Method (ii). This indicates that allocating relevance to skip connections with identity mappings makes a greater contribution to the enhancement in the quality of the attribution maps.

When examining Method (iv), there is a notable reduction in the Deletion score compared with Methods (i)–(iii). This implies that Ratio-Based Splitting while also propagating relevance to skip connections with identity mappings prevents the attribution from dispersing to non-relevant objects such as the background. Indeed, as shown in Fig. 6, when these two key design elements are used simultaneously, there is a visible reduction in the dispersion of attribution to non-relevant backgrounds compared with when neither element is used.

Furthermore, we conducted an ablation study to evaluate the Heat Quantization technique. Table 3 presents the quantitative results of it. The ID score under Method (ii) is 0.510, surpassing that of Method (i) by 0.134. This indicates that Heat Quantization successfully improves the quality of the generated attribution maps.

5.5 Empirical Analysis of Conservation Property

For an explanation method to be considered valid, the sum of the contributions from all inputs must equal the model's output, thereby ensuring that the entirety of the prediction's attribution is accounted for. The conservation property establishes this condition and is therefore very important in providing a rigorous foundation for the interpretability of complex models [1,3,35,40]. In this section, we show that our implementation strictly guarantees the conservation property.

We conducted an empirical analysis using 100 samples extracted from ImageNet to rigorously assess the conservation property of our method. For each sample, we visualized the sum of the relevance scores backpropagated to three checkpoints: (a) the input, (b) the input of the first Bottleneck block, and (c) the input of the last Bottleneck block. Figure 7 plots these sums against the model's output predictions. The plotted points are precisely aligned with $p(\hat{y}_c) = \sum_i R_i^{(l)}$, where $p(\hat{y}_c)$ represents the model's predicted probability for class c and $\sum_i R_i^{(l)}$ represents the accumulated relevance scores at each layer l. This precise alignment across all sampled points at all checkpoints demonstrates that our explanation method rigorously adheres to the conservation property. The uniformity of these results demonstrates our method's integrity of the relevance propagation.

6 Conclusion

We focused on the task of visualizing important regions in an image as a visual explanation of the model's decisions. We extended Layer-wise Relevance Propagation (LRP) [3] to models with residual connections such as ResNet [10] by introducing Relevance Splitting at points where the output from a skip connection converges with that from a residual block. Moreover, the conservation property is guaranteed throughout the proposed process, thereby preserving the integrity of the relevance propagation mechanism. Additionally, we introduced the Heat Quantization to mitigate the issue of overconcentration of generated attributions within irrelevant regions. Our method demonstrated superior performance to that of baseline methods on standard evaluation metrics, including the ID score. Furthermore, we investigated improved designs for relevance propagation within residual blocks and skip connections through an ablation study on relevance propagation within the Bottleneck module and demonstrated the significance of propagating relevance through skip connections and employing Ratio-Based Relevance Splitting.

Limitations and Future Directions. Although our experiments were conducted on ResNet, our method has the potential to be applied to various models with residual connections. Consequently, further experiments could be undertaken on other models, such as transformers [41]. Additionally, our method could potentially be adapted for use in different modalities, such as natural language processing, by applying it to models specific to these modalities. We leave these further extensions for future study.

Acknowledgements. This work was partially supported by JSPS KAKENHI Grant Number 23H03478, JST CREST, and NEDO.

References

1. Ali, A., Schnake, T., Eberle, O., et al.: XAI for transformers: better explanations through conservative propagation. In: ICML, pp. 435–451 (2022)
2. Arras, L., Montavon, G., Müller, R., et al.: Explaining recurrent neural network predictions in sentiment analysis. In: WASSA, pp. 159–168 (2017)
3. Bach, S., et al.: On pixel-wise explanations for non-linear classifier decisions by layer-wise relevance propagation. PLoS ONE **10**(7), 1–46 (2015)
4. Binder, A., et al.: Layer-wise relevance propagation for neural networks with local renormalization layers. In: ICANN, pp. 63–71 (2016)
5. Chefer, H., Gur, S., Wolf, L.: Transformer interpretability beyond attention visualization. In: CVPR, pp. 782–791 (2021)
6. Chen, S., Sun, P., Song, Y., Luo, P.: DiffusionDet: diffusion model for object detection. In: ICCV, pp. 19773–19786 (2023)
7. Deng, J., Dong, W., Socher, R., Li, L.J., Li, K., Fei-Fei, L.: ImageNet: a large-scale hierarchical image database. In: CVPR, pp. 248–255 (2009)
8. Fong, R.C., Vedaldi, A.: Interpretable explanations of black boxes by meaningful perturbation. In: ICCV, pp. 3429–3437 (2017)
9. Fukui, H., Hirakawa, T., et al.: Attention branch network: learning of attention mechanism for visual explanation. In: CVPR, pp. 10705–10714 (2019)
10. He, K., Zhang, X., Ren, S., Sun, J.: Deep residual learning for image recognition. In: CVPR, pp. 770–778 (2016)
11. Hochreiter, S., Schmidhuber, J.: Long short-term memory. Neural Comput. **9**(8), 1735–1780 (1997)
12. Iida, T., Komatsu, T., Kaneda, K., et al.: Visual explanation generation based on lambda attention branch networks. In: ACCV, pp. 3536–3551 (2022)
13. Itaya, H., et al.: Visual explanation using attention mechanism in actor-critic-based deep reinforcement learning. In: IJCNN, pp. 1–10 (2021)
14. Jacovi, A., Schuff, H., Adel, H., Vu, N.T., et al.: Neighboring words affect human interpretation of saliency explanations. In: ACL, pp. 11816–11833 (2023)
15. Kamath, A., Singh, M., LeCun, Y., Synnaeve, G., Misra, I., Carion, N.: MDETR - modulated detection for end-to-end multi-modal understanding. In: ICCV, pp. 1780–1790 (2021)
16. Krizhevsky, A., Nair, V., Hinton, G.: Learning multiple layers of features from tiny images. University of Toronto, Technical report (2009)
17. Lundberg, S., Lee, I.: A unified approach to interpreting model predictions. In: NeurIPS, pp. 4765–4774 (2017)
18. Madiaga: Artificial Intelligence Act (2023). https://www.europarl.europa.eu/RegData/etudes/BRIE/2021/698792/EPRS_BRI(2021)698792_EN.pdf
19. Molnar, C., Casalicchio, G., et al.: Interpretable machine learning – a brief history, state-of-the-art and challenges. In: ECML PKDD 2020 Workshops, pp. 417–431 (2020)
20. Montavon, G., Lapuschkin, S., et al.: Explaining nonlinear classification decisions with deep Taylor decomposition. Pattern Recogn. **65**, 211–222 (2017)
21. Ogura, T., et al.: Alleviating the burden of labeling: sentence generation by attention branch encoder-decoder network. RA-L **5**(4), 5945–5952 (2020)

22. Pan, B., Panda, R., Jiang, Y., et al.: IA-RED^2: interpretability-aware redundancy reduction for vision transformers. In: NeurIPS, pp. 24898–24911 (2021)
23. Pan, D., Li, X., Zhu, D.: Explaining deep neural network models with adversarial gradient integration. In: IJCAI (2021)
24. Parekh, J., Mozharovskyi, P., d'Alché-Buc, F.: A framework to learn with interpretation. In: NeurIPS, pp. 24273–24285 (2021)
25. Petsiuk, V., Das, A., Saenko, K.: RISE: randomized input sampling for explanation of black-box models. In: BMVC, pp. 151–164 (2018)
26. Porwal, P., Pachade, S., Kokare, M., et al.: IDRiD: diabetic retinopathy – segmentation and grading challenge. Med. Image Anal. **59**(101561) (2020)
27. Radford, A., et al.: Learning transferable visual models from natural language supervision. In: ICML, pp. 8748–8763 (2021)
28. Reed, S., et al.: A generalist agent. In: TMLR 2022 (2022)
29. Ren, S., He, K., et al.: Faster R-CNN: towards real-time object detection with region proposal networks. IEEE Trans. PAMI **39**(6), 1137–1149 (2017)
30. Ribeiro, M., Singh, S., et al.: "Why Should I Trust You?": explaining the predictions of any classifier. In: KDD, pp. 1135–1144 (2016)
31. Saeed, W., Omlin, C.: Explainable AI (XAI): a systematic meta-survey of current challenges and future opportunities. Knowl.-Based Syst. **263**, 110273 (2023)
32. Samek, W., Binder, A., Montavon, G., Lapuschkin, S., Müller, K.R.: Evaluating the visualization of what a deep neural network has learned. IEEE Trans. Neural Netw. Learn. Syst. **28**(11), 2660–2673 (2017)
33. Selvaraju, R., et al.: Grad-CAM: visual explanations from deep networks via gradient-based localization. In: ICCV, pp. 618–626 (2017)
34. Shrikumar, A., Greenside, P., Shcherbina, A., Kundaje, A.: Not just a black box: learning important features through propagating activation differences. arXiv preprint arXiv:1605.01713 (2016)
35. Shrikumar, A., et al.: Learning important features through propagating activation differences. In: ICML, vol. 70, pp. 3145–3153 (2017)
36. Simonyan, K., Zisserman, A.: Very deep convolutional networks for large-scale image recognition. In: ICLR, pp. 1–14 (2015)
37. Simonyan, K., Vedaldi, A., et al.: Deep inside convolutional networks: visualising image classification models and saliency maps. In: ICLR, pp. 1–8 (2014)
38. Springenberg, J., Dosovitskiy, A., Brox, T., Riedmiller, M.: Striving for simplicity: the all convolutional net. In: ICLR (Workshop Track) (2015)
39. Srinivas, S., Fleuret, F.: Full-gradient representation for neural network visualization. In: NeurIPS, vol. 32 (2019)
40. Sundararajan, M., Taly, A., Yan, Q.: Axiomatic attribution for deep networks. In: ICML, pp. 3319–3328 (2017)
41. Vaswani, A., et al.: Attention is all you need. In: NeurIPS, pp. 5998–6008 (2017)
42. Wah, C., Branson, S., et al.: The Caltech-UCSD birds-200-2011 dataset. Technical report. CNS-TR-2011-001, California Institute of Technology (2011)
43. Wang, H., Wang, Z., et al.: Score-CAM: score-weighted visual explanations for convolutional neural networks. In: CVPR, pp. 24–25 (2020)
44. Wang, W., et al.: VisionLLM: large language model is also an open-ended decoder for vision-centric tasks. In: NeurIPS, pp. 61501–61513 (2023)
45. Zhou, X., Girdhar, R., Joulin, A., et al.: Detecting twenty-thousand classes using image-level supervision. In: Avidan, S., Brostow, G., Cissé, M., Farinella, G.M., Hassner, T. (eds.) ECCV 2022. LNCS, vol. 13669, pp. 350–368. Springer, Cham (2022). https://doi.org/10.1007/978-3-031-20077-9_21

DECap: Towards Generalized Explicit Caption Editing via Diffusion Mechanism

Zhen Wang[1], Xinyun Jiang[2], Jun Xiao[1], Tao Chen[3], and Long Chen[4](✉)

[1] Zhejiang University, Hangzhou, China
{zju_wangzhen,junx}@zju.edu.cn
[2] MIT, Cambridge, USA
sunsh16e@mit.edu
[3] Tongdun Technology, Hangzhou, China
[4] HKUST, Kowloon, Hong Kong
longchen@ust.hk

Abstract. Explicit Caption Editing (ECE)—refining reference image captions through a sequence of explicit edit operations (*e.g.*, KEEP, DETELE)—has raised significant attention due to its explainable and human-like nature. After training with carefully designed reference and ground-truth caption pairs, state-of-the-art ECE models exhibit limited generalization ability beyond the original training data distribution, *i.e.*, they are tailored to refine content details only in in-domain samples but fail to correct errors in out-of-domain samples. To this end, we propose a new Diffusion-based Explicit Caption editing method: **DECap**. Specifically, we reformulate the ECE task as a denoising process under the diffusion mechanism, and introduce innovative edit-based noising and denoising processes. Thanks to this design, the noising process can help to eliminate the need for meticulous paired data selection by directly introducing word-level noises for training, learning diverse distribution over input reference caption. The denoising process involves the explicit predictions of edit operations and corresponding content words, refining reference captions through iterative step-wise editing. To further efficiently implement our diffusion process and improve the inference speed, DECap discards the prevalent multi-stage design and directly generates edit operations and content words simultaneously. Extensive ablations have demonstrated the strong generalization ability of DECap in various scenarios. More interestingly, it even shows great potential in improving the quality and controllability of caption generation.

Keywords: Explicit Caption Editing · Image Captioning · Diffusion Model · Out-of-Distribution (OOD) Evaluation

---

Work was done when Zhen Wang visited HKUST.

Supplementary Information The online version contains supplementary material available at https://doi.org/10.1007/978-3-031-72775-7_21.

1 Introduction

Explicit Caption Editing (ECE), emerging as a novel task within the broader domain of caption generation, has raised increasing attention from the multimodal learning community [28]. As shown in Fig. 1(a)(c), given an image and a reference caption (Ref-Cap), ECE aims to explicitly predict a sequence of edit operations, which can translate the Ref-Cap to ground-truth caption (GT-Cap). Compared to conventional image captioning methods which generate captions from scratch [2,6,20,27,30], ECE aims to enhance the quality of existing captions in a more explainable, efficient, and human-like manner.

Currently, existing ECE methods primarily rely on two prevalent benchmarks for model training and evaluation, *i.e.*, COCO-EE and Flickr30K-EE [28]. Specifically, both datasets are carefully constructed to emphasize the refinement of content details while preserving the original caption structure. As shown in Fig. 1, each ECE instance consists of an image along with a Ref-Cap (*e.g.*, two birds standing on a bench near the water) and a corresponding GT-Cap (man sitting on a bench overlooking the ocean). For this in-domain sample, state-of-the-art ECE models can effectively improve the quality of the Ref-Cap. By "in-domain", we mean that the test set of existing ECE benchmarks has

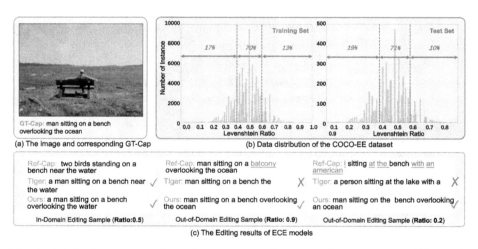

Fig. 1. (a) An image example and its corresponding ground-truth caption (GT-Cap). (b) Data distribution of the COCO-EE dataset [28]. The distribution of the training set and test set are very similar, where most of the editing instances have ratios ranging from 0.4 to 0.6. (c) Editing results of state-of-the-art ECE model TIger and our DECap. The *in-domain* Ref-Cap sample is from the COCO-EE test set, and *out-of-domain* Ref-Cap samples are constructed by replacing the GT-Cap with other words, *e.g.*, predicted by BERT (or sentences generated by pretrained captioning models).

a similar distribution with its training set (Fig. 1(b))[1]. However, we found that existing ECE models have limited generalization ability when faced with out-of-domain samples. Take the model TIger [28] as an example, given a highly similar Ref-Cap with a single wrong word (man sitting on a balcony overlooking the ocean), although it corrected the wrong word, it also removes other accurate words. Meanwhile, when faced with more irrelevant Ref-Caps (e.g., i sitting at the bench with an american), TIger even fails to correct all errors or introduce sufficient accurate details. Obviously, this limited generalization ability will limit their utilization in real-world scenarios, as we hope our ECE models can help to edit or refine different sentences.

To address this limitation, we propose a novel diffusion-based ECE model, denoted as **DECap**, which reformulates the ECE task as a series of denoising process steps. Specifically, we design an edit-based noising process that constructs editing samples by introducing word-level noises (i.e., random words) directly into the GT-Caps to obtain Ref-Caps. This noising process is parameterized by the distributions over both edit operations (e.g., KEEP, DELETE, INSERT, and REPLACE) and caption lengths, which can not only avoid the meticulous selection of Ref-GT caption pairs but also help ECE models to learn a more adaptable distribution over Ref-Caps, capturing a broader spectrum of editing scenarios. Then, we train model DECap to refine Ref-Caps through an edit-based denoising process, which contains the iterative predictions of edit operations and content words. Meanwhile, DECap discards the prevalent multi-stage architecture designs and directly generates edit operations and content words simultaneously, which can significantly accelerate the inference speed with simple Transformer encoder architectures. Extensive ablations have demonstrated that DECap can not only achieve outstanding editing performance on the challenging ECE benchmarks but can also further enhance the quality of model-generated captions. Meanwhile, it even achieves competitive caption generation performance with existing diffusion-based image captioning models. Furthermore, DECap even shows potential for word-level controllable captioning, which is beyond the ability of existing controllable captioning models [5,8,9,29]. In summary, *DECap realizes a strong generalization ability across various in-domain and out-of-domain editing scenarios, and showcases great potential in improving the quality and controllability of caption generation, keeping the strong explainable ability.* **With such abilities, our DECap can serve as an innovative and uniform framework that can achieve both caption editing and generation.**

In summary, we make several contributions in this paper: 1) To the best of our knowledge, we are the first work to point out the poor generalization issues of existing ECE models, and propose a series of caption editing scenarios for generalization ability evaluation. 2) DECap is the first diffusion-based ECE model, which pioneers the use of the discrete diffusion mechanism for ECE. 3) DECap shows strong generalization ability across various editing scenarios,

---

[1] In this paper, we use the Levenshtein ratio (ratio) to quantify the similarity between two captions by considering their lengths and edit distances. The range of ratio is from 0 to 1, where a higher value indicates higher similarity..

achieving outstanding performance. 4) DECap has a much faster inference speed than existing ECE methods.

2 Related Work

Explicit Caption Editing (ECE). Given the image, ECE aims to refine existing Ref-Caps through a sequence of edit operations, which was first proposed by Wang et al. [28]. Specifically, by realizing refinement under the explicit traceable editing path composed of different edit operations, this task encourages models to enhance the caption quality in a more explainable and efficient manner. However, existing ECE benchmarks are carefully designed, targeting the refinement of specific content details, which leads to limited model generalization ability across diverse real-world editing scenarios beyond the training data distribution. Meanwhile, existing editing models [19,22,28] tend to perform editing with multiple sub-modules sequentially. For example, conducting the insertion operation by first predicting the ADD operation, then applying another module to predict the specific word that needs to be added. In this paper, we construct Ref-Caps by directly noising the GT-Caps at word-level through a novel edit-based noising process, allowing the model to capture various editing scenarios during training. We further optimize model architecture to predict both edit operations and content words parallelly, which can significantly accelerate the editing speed.

Diffusion-Based Captioning Models. Taking inspiration from the remarkable achievements of diffusion models in image generation [3,23,31], several pioneering works have applied the diffusion mechanism for caption generation. Existing diffusion-based captioning works can be categorized into two types: 1) **Continuous Diffusion**: They aim to convert discrete words into continuous vectors (e.g., word embeddings [11] and binary bits [7,18]) and apply the diffusion process with Gaussian noises. 2) **Discrete Diffusion**: They aim to extend the diffusion process to discrete state spaces by directly noising and denoising sentences at the token level, such as gradually replacing tokens in the caption with a specific [MASK] token and treating the denoising process as a multi-step mask prediction task starting from an all [MASK] sequence [32]. As the first diffusion-based ECE model, in contrast to iterative mask replacement, which only trains the ability to predict texts for [MASK] tokens, our edit-based noising and denoising process can help our model to learn a more flexible way of editing (e.g., insertion, deletion, and replacement) by different edit operations. Meanwhile, our model shows its great potential in directly editing random word sequences, which achieves competitive performance to diffusion-based captioning models.

3 Diffusion-Based Explicit Caption Editing

In this section, we first give a brief introduction of the task ECE and the preliminaries about discrete diffusion mechanism in Sect. 3.1. Then, we show the edit-based noising and denoising process in Sect. 3.2. We introduce our model architecture in Sect. 3.3. Lastly, we demonstrate the details of training objectives and inference process in Sect. 3.4.

3.1 Task Formulation and Preliminaries

Explicit Caption Editing. Given an image I and a reference caption (Ref-Cap) $\boldsymbol{x}^r = \{w_r^1, ..., w_r^n\}$ with n words, ECE aims to predict a sequence of m edit operations $E = \{e^1, ..., e^m\}$ to translate the Ref-Cap close to the ground-truth caption (GT-Cap) $\boldsymbol{x}_0 = \{w_0^1, ..., w_0^k\}$ with k words.

Edit Operations. Normally, different ECE models may utilize different edit operations. While early models mainly focus on the reservation (*e.g.*, KEEP) and deletion (*e.g.*, DELETE) of existing contents, and the insertion (*e.g.*, ADD, INSERT) of new contents, subsequent works [19,22] have demonstrated that incorporating replacement can improve editing performance more efficiently. Acknowledging this established insight and without loss of generality, in this paper, we utilize the four Levenshtein edit operations[2] for both the noising and denoising process, including: 1) KEEP, the keep operation preserves the current word unchanged; 2) DELETE, the deletion operation removes the current word; 3) INSERT, the insertion operation adds a new word after the current word; 4) REPLACE, the replacement operation overwrites the current word with a new word.

Discrete Diffusion Mechanism. For diffusion models in the discrete state spaces for text generation, each word of sentence $\boldsymbol{x}_t$ is a discrete random variable with K categories, where K is the word vocabulary size. Denoting $\boldsymbol{x}_t$ as a stack of one-hot vectors, the noising process is written as:

$$q(\boldsymbol{x}_t|\boldsymbol{x}_{t-1}) = Cat(\boldsymbol{x}_t; p = \boldsymbol{x}_{t-1}\boldsymbol{Q}_t), \tag{1}$$

where $Cat(\cdot)$ is a categorical distribution and $\boldsymbol{Q}_t$ is a transition matrix applied to each word in the sentence independently: $[\boldsymbol{Q}_t]_{i,j} = q(w_t = j|w_{t-1} = i)$. Existing discrete diffusion text generation works [3,12,32] mainly follow the noising strategy of BERT [10], where each word stays unchanged or has some probability transitions to the [MASK] token or other random words from the vocabulary. Meanwhile, they incorporate an absorbing state for their diffusion model as the [MASK] token with transition probability β_t:

$$[\boldsymbol{Q}_t]_{i,j} = \begin{cases} 1 & \text{if } i = j = \text{[MASK]}, \\ \beta_t & \text{if } j = \text{[MASK]}, i \neq \text{[MASK]}, \\ 1 - \beta_t & \text{if } i = j \neq \text{[MASK]}. \end{cases} \tag{2}$$

After a sufficient number of noising steps, this Markov process converges to a stationary distribution $q(\boldsymbol{x}_T)$ where all words are replaced by the [MASK] token. Discrete diffusion works then train their models to predict target words for [MASK] tokens as the denoise process $p_\theta(\boldsymbol{x}_{t-1}|\boldsymbol{x}_t, t)$, and generate the sentence by performing a series of denoising steps from an all [MASK] token sequence:

$$P_\theta(\boldsymbol{x}_0) = \prod_{t=1}^T p_\theta(\boldsymbol{x}_{t-1}|\boldsymbol{x}_t, t). \tag{3}$$

---

[2] We are open to investigating other operations (*e.g.*, REORDER) in the future.

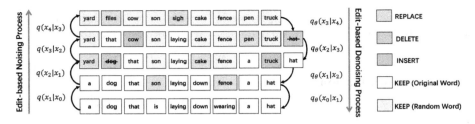

Fig. 2. Edit-based noising process for DECap. Blue represents the REPLACE operation, red represents the DELETE operation, purple represents the INSERT operation, white and grey represent the KEEP operation for original word and random word respectively. (Color figure online)

3.2 Discrete Diffusion for ECE

Taking inspiration from the discrete process where a noised sentence is iteratively refined into the target, we reformulate ECE training with a discrete diffusion mechanism and parameterize the noising and denoising process by way of sampled discrete edit operations applied over the caption words. *This noising and denoising process can clearly mitigate the need for paired Ref-GT caption pairs, as we only need to conduct the diffusion process on original GT-Caps.*

Edit-Based Noising Process. Different from directly transiting one word to another, the edit-based noising process gradually adds word-level noises to the caption x_{t-1} based on different edit operations. For any time step $t \in (0, T]$, the edit-based noising process is defined as

$$q(x_t|x_{t-1}) = p(x_t|x_{t-1}, E_t^N) \cdot Cat(E_t^N; p = x_{t-1}Q_t), \qquad (4)$$

where $Cat(\cdot)$ is a categorical distribution and Q_t here is a transition matrix assigning edit operation for each word in the caption x_{t-1} independently: $[Q_t]_{i,j} = q(e_t = j|w_{t-1} = i)$. Subsequently, $E_t^N = \{e_t^1, e_t^2, ..., e_t^l\}$ is a sequence of noising edit operations that has same length with caption $x_{t-1} = \{w_{t-1}^1, w_{t-1}^2, ..., w_{t-1}^l\}^3$, where each edit operation e_t^i is operated on the corresponding word w_{t-1}^i to get x_t. Specifically, Q_t is parameterized by the distribution over both edit types and GT-Cap length k with an absorbing state as the random word (RW).

$$[Q_t]_{i,j} = \begin{cases} 1 & \text{if } j = \text{KEEP}, \quad i = \text{RW}, \\ \alpha_t^k & \text{if } j = \text{REPLACE}, i \neq \text{RW}, \\ \beta_t^k & \text{if } j = \text{DELETE}, \quad i \neq \text{RW}, \\ \gamma_t^k & \text{if } j = \text{INSERT}, \quad i \neq \text{RW}, \\ 1 - \alpha_t^k - \beta_t^k - \gamma_t^k, & \text{if } j = \text{KEEP}, \quad i \neq \text{RW}. \end{cases} \qquad (5)$$

---

[3] Generally, captions' length may vary in different steps. For simplicity, we slightly use l to denote the length of all other x_t captions in this paper.

Subsequently, as the example shown in Fig. 2, being operated with e_t, each word w_{t-1} has a probability of α_t^k to be replaced by another random word, has a probability of β_t^k to be removed from the caption, and has a probability of γ_t^k to be added with a random word after it, leaving the probability of $\delta_t^k = 1 - \alpha_t^k - \beta_t^k - \gamma_t^k$ to be unchanged. Accordingly, the distribution over the GT-Cap length k can ensure a smooth increase of noised words for each noising step from $\boldsymbol{x}_0$ to $\boldsymbol{x}_T$.

The distribution over edit types ensures the balance between different noising operations and the learning of different denoising abilities: 1) To learn the ability to INSERT new words, we remove words by DELETE operation. 2) To learn the ability to DELETE incorrect words, we add random words by INSERT operation. 3) To learn the ability to directly REPLACE incorrect words, we change current words into random words by REPLACE operation. 4) To learn the ability to KEEP correct content, we leave the correct words unchanged by KEEP operation. Meanwhile, if the word has already been noised into the random word, it will not be re-noised again. Through sufficient noising steps T, the caption will be noised into a random word sequence.

Edit-Based Denoising Process. The edit-based denoising process aims to iteratively edit $\boldsymbol{x}_T$ to $\boldsymbol{x}_0$ by predicting appropriate edit operations. Specifically, given the image I and the caption $\boldsymbol{x}_t = \{w_t^1, w_t^2, ..., w_t^l\}$, we model this edit-based denoising process with the explicit prediction of both edit operations and content words transforming $\boldsymbol{x}_t$ to $\boldsymbol{x}_{t-1}$:

$$p_\theta(\boldsymbol{x}_{t-1}|\boldsymbol{x}_t, t, I) = p(\boldsymbol{x}_{t-1}|\boldsymbol{x}_t, E_t^D, C_t) \cdot p(E_t^D, C_t|\boldsymbol{x}_t, t, I),$$

where p_θ parameterized the model to predict a sequence of denoising edit operations $E_t^D = \{e_t^1, e_t^2, ..., e_t^l\}$, together with a sequence of content words $C_t = \{c_t^1, c_t^2, ..., c_t^l\}$ which all have the same length with $\boldsymbol{x}_t$. As the example shown in Fig. 3, the denoising step transforms the caption $\boldsymbol{x}_t$ to $\boldsymbol{x}_{t-1}$ based on the edit operations and predicted words, i.e., for each word w_t^i, we keep the original word if it is predicted operation e_t^i is KEEP, remove the word if it is predicted operation e_t^i is DELETE, copy the original word and add a new word c_t^i after it if predicted operation e_t^i is INSERT, and replace it with a new word c_t^i if predicted operation e_t^i is REPLACE. We then feed the output of this step into the model and perform the next denoising step. Following this, we can generate the caption by performing a series of denoising steps from an all random word sequence:

$$P_\theta(\boldsymbol{x}_0) = \prod_{t=1}^T p_\theta(\boldsymbol{x}_{t-1}|\boldsymbol{x}_t, t, I). \tag{6}$$

3.3 Transformer-Based Model Architecture

The DECap is built based on the standard Transformer [25] architecture, which has strong representation encoding abilities. To facilitate the denoising process, we further construct DECap with a parallelized system for the efficient generation of both edit operations and content words.

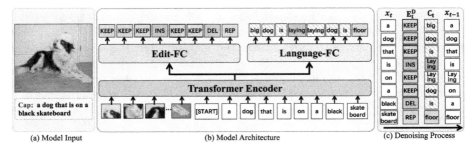

Fig. 3. The edit-based denoising step and architecture of DECap. DECap will predict a sequence of edit operations and content words to transform the caption. Specifically, contents words are used only when the predicted corresponding edit operation is INSERT or REPLACE, while the rest of the predicted words are abandoned, *i.e.*, the shaded words.

Feature Extraction. Given an image I and caption x_t, we construct the input for the model as a sequence of visual tokens and word tokens. Specifically, we encode the image I into visual tokens through pre-trained visual backbones such as CLIP. The word tokens are represented by the sum of word embedding, position encoding, and segment encoding. Meanwhile, following previous works [3,13,15], we encode the time step t as a sinusoidal embedding the same way as the position encoding, adding it to the word tokens.

Model Architecture. As shown in Fig. 3, given the visual-word token sequence with a connecting token, *e.g.*, [START], we first utilize the Transformer encoder blocks with self-attention and co-attention layers to learn the multi-modal representations of each token. We then use two simple yet effective FC layers to predict the edit operation and content word for each word token. Specifically, by feeding the hidden states of word tokens as input, 1) the Edit-FC generates the edit operation sequence E_t^D by making a four-category classification for each word, *i.e.*, $e_t \in \{\text{REPLACE}, \text{DELETE}, \text{INSERT}, \text{KEEP}\}$. 2) In parallel, the Language-FC maps each hidden state to a distribution over the vocabulary to predict specific words to generate the content word sequence C_t. Following the denoising step in Sect. 3.2, we then transform the caption x_t to x_{t-1} based on the edit operations and content words for next step.

3.4 Training Objectives and Inference

Training. Following previous discrete diffusion works [3,32], we train the model to directly predict the original ground-truth caption x_0 for caption x_t:

$$\mathcal{L} = \mathcal{L}_{Edit} + \mathcal{L}_{Languge} = -(\log p_\theta(E_t^G|\boldsymbol{x}_t, t, I) + \log p_\theta(C_t^G|\boldsymbol{x}_t, t, I)), \quad (7)$$

where E_t^G and C_t^G are ground truth edit operations and content words constructed based on the $\boldsymbol{x}_0$ and $\boldsymbol{x}_t$. $\mathcal{L}_{Edit}$ and $\mathcal{L}_{Languge}$ are cross-entropy loss over the distribution of predicted edit operations and content words, and $\mathcal{L}_{Languge}$ is only trained to predict content words for the input words assigned with INSERT and REPLACE operations.

Inference. Given image and caption x_t ($t \in (t, T]$), the model predicts x_{t-1}, x_{t-2} iteratively for t denoising steps, and produces the final result of x_0.

4 Experiments

4.1 Experimental Setup

Datasets. We evaluated DECap on both caption editing (*i.e.*, **COCO-EE** [28], **Flickr30K-EE**) and caption generation benchmarks (*i.e.*, **COCO** [17][4]).

Evaluation Metrics. We utilized all prevalent accuracy-based metrics including BLEU-N [21], METEOR [4], ROUGE-L [16], CIDEr-D [26], and SPICE [1]. Meanwhile, we also computed CLIP-Score and the inference time (See Footnote 4).

Table 1. The *in-domain* evaluation on the COCO-EE test set. All ECE models were trained on the COCO-EE training set. "Ref-Caps" denotes the initial quality of given reference captions. "TIger-N" denotes the TIger trained with noised unpaired data.

Model	Unpaired Data	Step	Quality Evaluation							Inference
			B-1	B-2	B-3	B-4	R	C	S	CLIP-Score	Time(ms)
Ref-Caps	—	—	50.0	37.1	27.7	19.5	48.2	129.9	18.9	0.6997	—
TIger [28]	✗	4	50.3	38.5	29.4	22.3	53.1	176.7	31.4	0.7269	614.23
TIger-N [28]	✓	4	51.8	38.6	28.9	20.7	49.6	145.0	21.8	0.7097	611.72
DECap	✓	4	55.5	41.7	31.5	23.3	52.7	173.7	29.8	0.7439	277.30
DECap	✓	5	56.0	**42.0**	**31.6**	**23.5**	53.0	176.2	31.4	0.7498	335.45
DECap	✓	6	**56.1**	41.9	31.4	23.4	**53.1**	**177.0**	**32.2**	**0.7522**	409.99

4.2 Generalization Ability in ECE

In this subsection, we evaluated the generalization ability of our model with both in-domain and out-of-domain evaluation on the COCO-EE. Specifically, we try to answer four research questions: 1) **Q1:** Does DECap perform well on the existing in-domain benchmark? 2) **Q2:** Does DECap perform well on reference captions with different noisy levels (*i.e.*, Levenshtein ratios (See Footnote 1))? **Q3:** Does DECap can further boost the performance of "good" reference captions from other captioning models? 3) **Q4:** Does DECap perform well on pure random reference captions? It is worth noting that DECap only used the unpaired data (*i.e.*, image and GT-Cap without Ref-Cap) while the state-of-the-art TIger [28]

---
[4] Due to the limited space, more details are left in the Appendix.

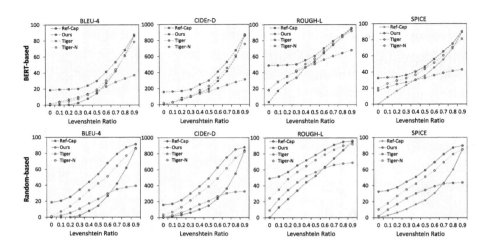

Fig. 4. Performance on two kinds of out-of-domain GT-based reference captions constructed from COCO-EE test set. All models were trained on the COCO-EE training set. "Ref-Caps" denotes the initial quality of given reference captions, and "TIger-N" denotes the TIger trained with unpaired data.

was trained with the complete editing instance. For a more fair comparison, we also trained TIger with the synthesized noised unpaired data (denoted as TIger-N).

In-Domain Evaluation: COCO-EE (Q1)

Settings. Since the COCO-EE dataset was carefully designed to emphasize the refinement of content details, its training set and test set have similar distribution (*i.e.*, Ratio (See Footnote 1) around 0.5 for most instances), thus we directly compared the performance of each model on the COCO-EE test set as the in-domain evaluation. We evaluated edited captions against their single GT-Cap.

Results. The in-domain evaluation results are reported in Table 1. From the table, we can observe: 1) For the quality evaluation, TIger achieves its best performance using four editing steps, while our DECap achieves competitive results with the same step (*e.g.*, better BLEU scores but slightly lower CIDEr-D score). With more editing steps, DECap can further improve the quality of captions, outperforming TIger on all metrics. *It is worth noting that TIger was even trained on the in-domain Ref-GT caption pairs.* 2) TIger-N achieves limited quality improvement on the in-domain samples. 3) For the efficiency evaluation, DECap achieves significantly faster inference speed than TIger even with more editing steps. This is because DeCap predicts edit operations and content words simultaneously but TIger needs to conduct editing by three sequential modules.

OOD: GT-Based Reference Caption (Q2)

Setting. The GT-based reference captions were constructed based on the GT-Caps in the COCO-EE test set. We systematically replaced words in GT-Caps

Table 2. Performance of ECE models editing model-generated captions on COCO test set. All ECE models were trained on the COCO-EE training set. "TIger-N" denotes the TIger trained with unpaired data.

Model	Quality Evaluation							
	B-1	B-2	B-3	B-4	R	C	S	CLIP-Score
Up-Down [2]	74.9	58.6	45.2	35.1	55.9	109.9	20.0	0.7359
+ TIger [28] ↓	70.3	54.9	41.2	30.9	54.3	95.7	17.1	0.7285
+ TIger-N [28] ↓	74.8	58.5	45.1	34.8	55.8	109.5	19.9	0.7356
+ **DECap** ↑	**75.1**	**58.9**	**45.5**	**35.2**	**56.3**	**112.3**	**20.2**	**0.7432**
Transformer [24]	75.2	58.9	45.6	35.5	56.0	112.8	20.6	0.7452
+ TIger [28] ↓	70.0	55.0	41.5	31.1	54.4	97.1	17.3	0.7313
+ TIger-N [28] ↓	75.1	58.8	45.4	35.2	55.9	112.4	20.4	0.7443
+ **DECap** ↑	**75.5**	**59.2**	**45.8**	**35.6**	**56.2**	**114.1**	**20.7**	**0.7472**
BLIP [14]	79.7	64.9	51.4	40.4	60.6	136.7	24.3	0.7734
+ TIger [28] ↓	73.5	58.7	44.7	33.7	56.3	104.0	18.2	0.7377
+ TIger-N [28] ↓	79.5	64.5	51.0	39.9	60.3	133.0	23.7	0.7690
+ **DECap** ↑	**79.9**	**65.1**	**51.7**	**40.6**	**60.7**	**138.1**	**24.4**	**0.7738**

with other words, resulting in the creation of various out-of-domain Ref-Caps. They varied in terms of their Levenshtein ratios, ranging from 0.9 (*i.e.*, with only a few incorrect words) to 0.0 (*i.e.*, where all words were wrong). Specifically, we constructed two kinds of GT-based reference captions: 1) **BERT-based**. We first replaced the GT words with the special [MASK] tokens and then utilized the pretrained BERT [10] model to predict other words different from GT words. 2) **Random-based**. We directly replaced GT words with other random words. We evaluated edited captions against their single GT-Cap.

Results. As shown in Fig. 4, For models trained with unpaired data, our model successfully improves the quality of all kinds of the GT-based Ref captions (*i.e.*, BERT- and Random-based) and surpasses TIger-N. In contrast, TIger struggles when editing Ref captions with either "minor" or "severe" errors, and even degrading the captions' quality (*e.g.*, Ref-Caps with ratio larger than 0.6 (See Footnote 4)) by inadvertently removing accurate words or failing to introduce accurate details.

OOD: Model-Generated Reference Caption (Q3)

Setting. We explored the models' generalization ability to further improve quality of captions generated by captioning models. We utilized captions generated by effective captioning models [2,14,24] on the COCO test set as reference captions. We evaluated edited captions against their corresponding five GT-Caps.

Results. From Table 2, we can observe: 1) Our model successfully improves the quality of captions generated by captioning models. 2) Both TIger and TIger-N fail to do so and even degrading the caption's quality. 3) Notably, while existing captioning models fall short of achieving comparable performance with the powerful vision-language pretrained models (*e.g.*, BLIP [14]), our DECap trained solely on COCO-EE, demonstrates its unique editing ability to further enhance the quality of captions generated by BLIP.

OOD: Pure Random Reference Caption (Q4)

Setting. To further evaluate the models' generalization ability without utilizing any GT captions, we constructed pure random reference captions based on the COCO test set. Specifically, each editing instance consists of a single image and a Ref-Cap with ten random words. We evaluated the edited captions against their corresponding five GT-Caps. All results are reported in Table 3.

Results. From Table 3, we can observe: 1) Given the image, all the models achieve their best performance with ten editing steps, DECap successfully edits the sentence with all random words into a coherent caption. In contrast, both TIger and TIger-N face challenges in doing so. 2) For efficiency metrics, DECap achieves significantly faster inference speed compared to TIger and TIger-N.

Table 3. Performance of ECE models on pure random (ten random words) reference captions constructed based on the COCO test set. All models were trained on the COCO-EE training set. "TIger-N" denotes TIger trained with noised unpaired data.

Model	Unpaired Data	Step	Quality Evaluation								Inference Time(ms)
			B-1	B-2	B-3	B-4	R	C	S	CLIP-Score	
TIger [28]	✗	10	14.7	4.6	1.9	0.9	13.5	3.0	1.2	0.5158	1413.16
TIger-N [28]	✔	10	7.2	5.9	4.5	3.5	29.1	23.5	4.8	0.6116	1417.09
DECap	✔	10	**74.7**	**57.4**	**42.1**	**30.0**	**55.3**	**102.5**	**19.6**	**0.7501**	684.32

4.3 Conventional Caption Generation Ability

Surprised by the results of editing model-generated captions and pure random reference captions, we further investigated DECap's capacity for directly generating captions.

Settings. We compared DECap with SOTA diffusion-based captioning methods on the COCO dataset, especially the discrete diffusion-based captioning model DDCap [32]. We trained DECap on the COCO training set with a vocabulary size of 23,531 together with different diffusion steps (10 and 15). During testing, we constructed input instances consisting of a single image from the COCO test set and a Ref-Cap with ten random words. The edited captions were then evaluated against the corresponding five GT-Caps.

Results. As shown in Table 4: 1) Within ten editing steps, DECap achieves superior performance on key metrics (*e.g.*, CIDEr-D and SPICE) compared with other diffusion-based captioning works which even need more diffusion steps.

While it falls slightly behind SCD-Net on BLEU-N, it's important to note that CIDEr and SPICE metrics are specifically designed for captioning evaluation and are better aligned with human judgments (than BLEU-N). 2) DECap achieves a significantly faster inference speed compared with another discrete diffusion-based model DDCap (675.80 vs. 3282.58 ms). 3) Additionally, DECap can further boost the performance with more editing steps (*e.g.*, 121.2 on CIDEr-D with 15 steps) and keep the inference speed at a reasonable level (*i.e.*, 933.10 ms). These results suggest the remarkable potential of DECap in improving the quality of caption generation in a more explainable and efficient edit-based manner.

Table 4. Comparison between our DECap and state-of-the-art diffusion-based captioning models on the COCO test set. All models were trained on the COCO training set. The **best** and second best results are denoted with corresponding formats.

Model	Step	B-1	B-2	B-3	B-4	M	R	C	S
Continuous Diffusion									
Bit Diffusion [7]	20	—	—	—	34.7	—	58.0	115.0	—
SCD-Net [18]	50	**79.0**	**63.4**	**49.1**	**37.3**	28.1	58.0	118.0	21.6
Discrete Diffusion									
DDCap [32]	20	—	—	—	35.0	28.2	57.4	117.8	21.7
DECap	10	78.0	61.4	46.4	34.5	28.6	58.0	119.0	21.9
DECap	15	78.5	62.2	47.4	35.3	**29.0**	**58.4**	**121.2**	**22.7**

4.4 Potential Ability: Controllable Captioning

Building on the remarkable generalization ability exhibited by DECap in both caption editing and generation, we further conducted a preliminary exploration of its potential for controllability. Compared to existing CIC methods, which offer only coarse control over contents and structures, we can achieve precise and explicit control over caption generation through predefined control words.

Settings. We constructed input instances consisting of a single image from the COCO test set and a sentence with ten random words. We replaced several random words with specific control words (*e.g.*, objects and attributes) at predefined positions based on the visual information of images.

Results. As shown in Fig. 5, DECap is capable of editing sentences based on input control words, *i.e.*, all generated captions follow the order of the given control words with guaranteed fluency. Meanwhile, DECap shows its reasoning ability to generate relevant semantic content based on the control words: 1) Given the attributes (*e.g.*, color), DECap can generate specific contents with these attributes (*e.g.*, "red" → "helmet", "green" → "grass" and "brown" → "sheep"). 2) Given objects, DECap can generate further descriptions or related

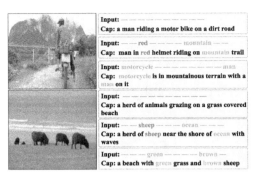

Fig. 5. Controllability of DECap. The grey lines represent random words from the vocabulary, and other colored words represent the manually placed control words. (Color figure online)

Table 5. Performance of DECap on the COCO test set with different numbers of input random words (RW).

RW	B-4	M	R	C	S
8	32.4	26.3	56.5	109.7	20.1
9	**35.5**	27.8	58.0	118.1	21.5
10	34.5	28.6	**58.0**	**119.0**	21.9
11	32.9	28.9	57.1	115.7	22.4
12	31.2	**29.0**	56.1	109.3	**22.7**

Table 6. Performance of DECap on the COCO test set when with different distributions of noising edit types.

Distribution	B-4	M	R	C	S
$\alpha = \beta = \gamma$	34.0	**28.6**	57.8	117.4	21.8
$\alpha > \beta = \gamma$	**34.5**	**28.6**	**58.0**	**119.0**	**21.9**
$\beta = \gamma = 0$	33.8	25.8	57.8	116.6	21.7

objects (e.g., "mountain" → "trail" and "ocean" → "wave"). These results indicate the potential of DECap to enhance controllability and diversity, achieving a more direct and word-level control beyond existing CIC methods.

4.5 Ablation Study

Number of Random Words. In this section, we run a set of ablations about the influence of different numbers of random words on caption generation. We utilized the DECap trained with diffusion step $T = 10$ from Sect. 4.3, and constructed input instances consisting of an image from the COCO test set and a Ref-Cap with n random words, where $n \in \{8, 9, 10, 11, 12\}$. As shown in Table 5, DECap's performance consistently improves as the number of random words increases from 8 to 10 and starts to decline beyond 10 (See Footnote 4). DECap achieves its highest CIDEr-D score when editing sentences with 10 random words as the average length of GT captions in COCO is around 10. We thus selected 10 words as a balanced choice for caption generation.

Distribution of Edit Types. As discussed in Sect. 3.2, the distribution over edit types plays a crucial role in balancing different noising operations and training diverse denoising abilities. Therefore, we examined the impact of varying distribution settings for the edit types within the edit-based noising process. Specifically, the probabilities for the noising edit operations REPLACE, DELETE, and INSERT are denoted as α, β, and γ, respectively. We perform ablations by imposing global control over these probabilities (See Footnote 4). We trained the DECap on the COCO training set with different distributions of edit types with the same diffusion step $T = 10$. During testing, we constructed input instances consisting of a single image from the COCO test set and a Ref-Cap with ten random words. From Table 6, we can observe: 1) DECap performs better when

Editing Step 1
In: a large heard of zebra standing in the grass
E/C: KEP DEL KEP KEP INS(are) KEP KEP KEP REP(dirt)
Out: a heard of zebra are standing in the dirt

Editing Step 2
In: a heard of zebra are standing in the dirt
E/C: KEP KEP KEP REP(zebras) KEP KEP KEP KEP INS(sand)
Out: a heard of zebras are standing in the dirt sand

Fig. 6. Editing process of DECap. "E" and "C" denote the edit operation and corresponding content word, respectively.

emphasizing the denoising ability of the replacement operation compared to an even distribution of edit types. 2) Training DECap exclusively for the replacement operation, neglecting deletion and insertion abilities, leads to a noticeable decline in caption quality. 3) $\alpha > \beta = \gamma$ could be a sensible choice for caption generation. Importantly, our method allows for flexible adaptation, enabling us to set different edit type distributions tailored to specific tasks or requirements.

Visualization. Figure 6 shows a two-step editing example (See Footnote 4).

5 Conclusion

In this paper, we pointed out the challenge of limited generalization ability in existing ECE models. And we proposed a novel diffusion-based ECE model, DECap, which reformulates ECE with a discrete diffusion mechanism, incorporating an innovative edit-based noising and denoising process. Extensive results have demonstrated DECap's strong generalization ability and potential as a uniform framework for both caption editing and generation. Moving forward, we are going to: 1) extend DECap into other modalities beyond images, e.g., video; 2) explore advanced techniques for finer controllability of DECap's editing process.

Acknowledgment. This work was supported by the National Natural Science Foundation of China (62337001) and the Fundamental Research Funds for the Central Universities (226202400058). Long Chen was supported by HKUST Special Support for Young Faculty (F0927) and HKUST Sports Science and Technology Research Grant (SSTRG24EG04).

References

1. Anderson, P., Fernando, B., Johnson, M., Gould, S.: SPICE: semantic propositional image caption evaluation. In: Leibe, B., Matas, J., Sebe, N., Welling, M. (eds.) ECCV 2016. LNCS, vol. 9909, pp. 382–398. Springer, Cham (2016). https://doi.org/10.1007/978-3-319-46454-1_24
2. Anderson, P., et al.: Bottom-up and top-down attention for image captioning and visual question answering. In: CVPR, pp. 6077–6086 (2018)

3. Austin, J., Johnson, D.D., Ho, J., Tarlow, D., Van Den Berg, R.: Structured denoising diffusion models in discrete state-spaces. In: NeurIPS, vol. 34, pp. 17981–17993 (2021)
4. Banerjee, S., Lavie, A.: METEOR: an automatic metric for MT evaluation with improved correlation with human judgments. In: ACL Workshop, pp. 65–72 (2005)
5. Chen, L., Jiang, Z., Xiao, J., Liu, W.: Human-like controllable image captioning with verb-specific semantic roles. In: CVPR, pp. 16846–16856 (2021)
6. Chen, L., et al.: SCA-CNN: spatial and channel-wise attention in convolutional networks for image captioning. In: CVPR, pp. 5659–5667 (2017)
7. Chen, T., Zhang, R., Hinton, G.: Analog bits: generating discrete data using diffusion models with self-conditioning. arXiv (2022)
8. Cornia, M., Baraldi, L., Cucchiara, R.: Show, control and tell: a framework for generating controllable and grounded captions. In: CVPR, pp. 8307–8316 (2019)
9. Deng, C., Ding, N., Tan, M., Wu, Q.: Length-controllable image captioning. In: Vedaldi, A., Bischof, H., Brox, T., Frahm, J.-M. (eds.) ECCV 2020. LNCS, vol. 12358, pp. 712–729. Springer, Cham (2020). https://doi.org/10.1007/978-3-030-58601-0_42
10. Devlin, J., Chang, M.W., Lee, K., Toutanova, K.: BERT: pre-training of deep bidirectional transformers for language understanding. arXiv (2018)
11. He, Y., Cai, Z., Gan, X., Chang, B.: DiffCap: exploring continuous diffusion on image captioning. arXiv (2023)
12. He, Z., Sun, T., Wang, K., Huang, X., Qiu, X.: DiffusionBERT: improving generative masked language models with diffusion models. arXiv (2022)
13. Ho, J., Jain, A., Abbeel, P.: Denoising diffusion probabilistic models. In: NeurIPS, vol. 33, pp. 6840–6851 (2020)
14. Li, J., Li, D., Xiong, C., Hoi, S.: BLIP: bootstrapping language-image pre-training for unified vision-language understanding and generation. arXiv (2022)
15. Li, X., Thickstun, J., Gulrajani, I., Liang, P.S., Hashimoto, T.B.: Diffusion-LM improves controllable text generation. In: NeurIPS, vol. 35, pp. 4328–4343 (2022)
16. Lin, C.Y.: ROUGE: a package for automatic evaluation of summaries. In: ACL Workshop, pp. 74–81 (2004)
17. Lin, T.-Y., et al.: Microsoft COCO: common objects in context. In: Fleet, D., Pajdla, T., Schiele, B., Tuytelaars, T. (eds.) ECCV 2014. LNCS, vol. 8693, pp. 740–755. Springer, Cham (2014). https://doi.org/10.1007/978-3-319-10602-1_48
18. Luo, J., et al.: Semantic-conditional diffusion networks for image captioning. In: CVPR, pp. 23359–23368 (2023)
19. Mallinson, J., Severyn, A., Malmi, E., Garrido, G.: FELIX: flexible text editing through tagging and insertion. arXiv (2020)
20. Mao, Y., et al.: Rethinking the reference-based distinctive image captioning. In: ACM MM, pp. 4374–4384 (2022)
21. Papineni, K., Roukos, S., Ward, T., Zhu, W.J.: BLEU: a method for automatic evaluation of machine translation. In: ACL, pp. 311–318 (2002)
22. Reid, M., Neubig, G.: Learning to model editing processes. arXiv (2022)
23. Rombach, R., Blattmann, A., Lorenz, D., Esser, P., Ommer, B.: High-resolution image synthesis with latent diffusion models. In: CVPR, pp. 10684–10695 (2022)
24. Sharma, P., Ding, N., Goodman, S., Soricut, R.: Conceptual captions: a cleaned, hypernymed, image alt-text dataset for automatic image captioning. In: ACL, pp. 2556–2565 (2018)
25. Vaswani, A., et al.: Attention is all you need. In: NeurIPS, vol. 30 (2017)
26. Vedantam, R., Lawrence Zitnick, C., Parikh, D.: CIDEr: consensus-based image description evaluation. In: CVPR, pp. 4566–4575 (2015)

27. Vinyals, O., Toshev, A., Bengio, S., Erhan, D.: Show and tell: a neural image caption generator. In: CVPR, pp. 3156–3164 (2015)
28. Wang, Z., et al.: Explicit image caption editing. In: Avidan, S., Brostow, G., Cissé, M., Farinella, G.M., Hassner, T. (eds.) ECCV 2022. LNCS, vol. 13696, pp. 113–129. Springer, Cham (2022). https://doi.org/10.1007/978-3-031-20059-5_7
29. Wang, Z., Xiao, J., Zhuang, Y., Gao, F., Shao, J., Chen, L.: Learning combinatorial prompts for universal controllable image captioning. IJCV, 1–22 (2024)
30. Xu, K., et al.: Show, attend and tell: neural image caption generation with visual attention. In: ICML, pp. 2048–2057 (2015)
31. Xu, Y., Wang, Z., Xiao, J., Liu, W., Chen, L.: FreeTuner: any subject in any style with training-free diffusion. arXiv preprint arXiv:2405.14201 (2024)
32. Zhu, Z., et al.: Exploring discrete diffusion models for image captioning. arXiv (2022)

EgoLifter: Open-World 3D Segmentation for Egocentric Perception

Qiao Gu[1,2(✉)], Zhaoyang Lv[2], Duncan Frost[2], Simon Green[2], Julian Straub[2], and Chris Sweeney[2]

[1] University of Toronto, Toronto, ON M5S 1A1, Canada
q.gu@mail.utoronto.ca
[2] Meta Reality Labs, Redmond, WA 98052, USA
{zhaoyang,frost,simongreen,jstraub,sweeneychris}@meta.com

Abstract. In this paper we present *EgoLifter*, a novel system that can automatically segment scenes captured from egocentric sensors into a complete decomposition of individual 3D objects. The system is specifically designed for egocentric data where scenes contain hundreds of objects captured from natural (non-scanning) motion. *EgoLifter* adopts 3D Gaussians as the underlying representation of 3D scenes and objects and uses segmentation masks from the Segment Anything Model (SAM) as weak supervision to learn flexible and promptable definitions of object instances free of any specific object taxonomy. To handle the challenge of dynamic objects in ego-centric videos, we design a transient prediction module that learns to filter out dynamic objects in the 3D reconstruction. The result is a fully automatic pipeline that is able to reconstruct 3D object instances as collections of 3D Gaussians that collectively compose the entire scene. We created a new benchmark on the Aria Digital Twin dataset that quantitatively demonstrates its state-of-the-art performance in open-world 3D segmentation from natural egocentric input. We run *EgoLifter* on various egocentric activity datasets which shows the promise of the method for 3D egocentric perception at scale. Please visit project page at https://egolifter.github.io/.

Keywords: Egocentric Perception · Open-world Segmentation · 3D Reconstruction

1 Introduction

The rise of personal wearable devices has led to the increased importance of egocentric machine perception algorithms capable of understanding the physical

---

Work done during internship at Reality Labs, Meta.

EgoLifter: Open-World 3D Segmentation for Egocentric Perception

Fig. 1. *EgoLifter* solves 3D reconstruction and open-world segmentation simultaneously from egocentric videos. *EgoLifter* augments 3D Gaussian Splatting [18] with instance features and lifts open-world 2D segmentation by contrastive learning, where 3D Gaussians belonging to the same objects are learned to have similar features. In this way, *EgoLifter* solves the multi-view mask association problem and establishes a consistent 3D representation that can be decomposed into object instances. *EgoLifter* enables multiple downstream applications including detection, segmentation, 3D object extraction and scene editing. See project webpage for animated visualizations.

3D world around the user. Egocentric videos directly reflect the way humans see the world and contain important information about the physical surroundings and how the human user interacts with them. The specific characteristics of egocentric motion, however, present challenges for 3D computer vision and machine perception algorithms. Unlike datasets captured with deliberate "scanning" motions, egocentric videos are not guaranteed to provide complete coverage of the scene. This makes reconstruction challenging due to limited or missing multi-view observations. The specific content found in egocentric videos also presents challenges to conventional reconstruction and perception algorithms. An average adult interacts with hundreds of different objects many thousands of times per day [5]. Egocentric videos capturing this frequent human-object interaction thus contain a huge amount of dynamic motion with challenging occlusions. A system capable of providing useful scene understanding from egocentric data must therefore be able to recognize hundreds of different objects while being robust to sparse and rapid dynamics.

To tackle the above challenges, we propose *EgoLifter*, a novel egocentric 3D perception algorithm that simultaneously solves reconstruction and open-world 3D instance segmentation from egocentric videos. We represent the geometry of the scene using 3D Gaussians [18] that are trained to minimize photometric reconstruction of the input images. To learn a flexible decomposition of objects that make up the scene we leverage SAM [22] for its strong understanding of objects in 2D and lift these object priors into 3D using contrastive learning. Specifically, 3D Gaussians are augmented with additional N-channel feature embeddings that are rasterized into feature images. These features are then learned to encode the object segmentation information by contrastive lifting [2]. This technique allows us to learn a flexible embedding with useful object priors that can be used for several downstream tasks.

To handle the difficulties brought by the dynamic objects in egocentric videos, we design *EgoLifter* to focus on reconstructing the static part of the 3D scene. *EgoLifter* learns a transient prediction network to filter out the dynamic objects from the reconstruction process. This network does not need extra supervision and is optimized together with 3D Gaussian Splatting using solely the photometric reconstruction losses. We show that the transient prediction module not only helps with photorealistic 3D reconstruction but also results in cleaner lifted features and better segmentation performance.

EgoLifter is able to reconstruct a 3D scene while decomposing it into 3D object instances without the need for any human annotation. The method is evaluated on several egocentric video datasets. The experiments demonstrate strong 3D reconstruction and open-world segmentation results. We also showcase several qualitative applications including 3D object extraction and scene editing. The contributions of this paper can be summarized as follows:

- We demonstrate *EgoLifter*, the first system that can enable open-world 3D understanding from natural dynamic egocentric videos.
- By lifting output from recent image foundation models to 3D Gaussian Splatting, *EgoLifter* achieve strong open-world 3D instance segmentation performance without the need for expensive data annotation or extra training.
- We propose a transient prediction network, which filters out transient objects from the 3D reconstruction results. By doing so, we achieve improved performance on both reconstruction and segmentation of static objects.
- We set up the first benchmark of dynamic egocentric video data and quantitatively demonstrate the leading performance of *EgoLifter*. On several large-scale egocentric video datasets, *EgoLifter* showcases the ability to decompose a 3D scene into a set of 3D object instances, which opens up promising directions for egocentric video understanding in AR/VR applications.

2 Related Work

2.1 3D Gaussian Models

3D Gaussian Splatting (3DGS) [18] has emerged as a powerful algorithm for novel view synthesis by 3D volumetric neural rendering. It has shown promising performance in many applications, like 3D content generation [4,51,62], SLAM [17,30,57] and autonomous driving [58,65]. Recent work extend 3DGS to dynamic scene reconstruction [7,27,56,59,60]. The pioneering work from Luiten et al. [27] first learns a static 3DGS using the multi-view observations at the initial timestep and then updates it by the observations at the following timesteps. Later work [56,60] reconstructs dynamic scenes by deforming a canonical 3DGS using a time-conditioned deformation network. Another line of work [7,59] extends 3D Gaussians to 4D, with an additional variance dimension in time. While they show promising results in dynamic 3D reconstruction, they typically require training videos from multiple static cameras. However, in egocentric perception, there are only one or few cameras with a narrow baseline. As we show in the experiments, dynamic 3DGS struggles to track dynamic objects and results in floaters that harm instance segmentation feature learning.

2.2 Open-World 3D Segmentation

Recent research on open-world 3D segmentation [9,13,15,16,19,23,24,31,38,44,45,52,54] has focused on lifting outputs from 2D open-world models - large, powerful models that are trained on Internet-scale datasets and can generalize to a wide range of concepts [22,35,40,42]. These approaches transfer the ability of powerful 2D models to 3D, require no training on 3D models, and alleviate the need for large-scale 3D datasets that are expensive to collect. Early work [16,19,38] lifts dense 2D feature maps to 3D representations by multi-view feature fusion, where each position in 3D is associated with a feature vector. This allows queries in fine granularity over 3D space, but it also incurs high memory usage. Other work [12,26,49] builds object-decomposed 3D maps using 2D open-world detection or segmentation models [22,25], where each 3D object is reconstructed separately and has a single feature vector. This approach provides structured 3D scene understanding in the form of object maps or scene graphs but the scene decomposition is predefined and the granularity does not vary according to the query at inference time. Recently, another work [2] lifts 2D instance segmentation to 3D by contrastive learning. It augments NeRF [33] with an extra feature map output and optimizes it such that pixels belonging to the same 2D segmentation mask are pulled closer and otherwise pushed apart. In this way, multi-view association of 2D segmentation is solved in an implicit manner and the resulting feature map allows instance segmentation by either user queries or clustering algorithms.

Concurrent Work. We briefly review several recent and unpublished preprints that further explore topics in this direction using techniques similar to ours. Concurrently, OmniSeg3D [63] and GARField [20] follow the idea of [2], and focus on learning 3D hierarchical segmentation. They both take advantage of the multi-scale outputs from SAM [22] and incorporate the scales into the lifted features. GaussianGrouping [61] also approaches the open-world 3D segmentation problem but they rely on a 2D video object tracker for multi-view association instead of directly using 2D segmentation via contrastive learning. Similar to our improvement on 3DGS, FMGS [66], LangSplat [39] and Feature3DGS [64] also augment 3DGS with feature rendering. They learn to embed the dense features from foundation models [36,40] into 3DGS such that the 3D scenes can be segmented by language queries. While the concurrent work collectively also achieves 3D reconstruction with the open-world segmentation ability, *EgoLifter* is the first to explicitly handle the dynamic objects that are commonly present in the real-world and especially in egocentric videos. We demonstrate this is a challenge in real-world scenarios and show improvements on it brought by *EgoLifter*.

2.3 3D Reconstruction from Egocentric Videos

NeuralDiff [55] first approached the problem of training an egocentric radiance field reconstruction by decomposing NeRF into three branches, which capture ego actor, dynamic objects, and static background respectively as inductive biases. EPIC-Fields [53] propose an augmented benchmark using 3D reconstruction by augmenting the EPIC-Kitchen [6] dataset using neural reconstruction. They also provide comprehensive reconstruction evaluations of several baseline methods [10,29,55]. Recently, two datasets for egocentric perception, Aria Digital Twin (ADT) dataset [37] and Aria Everyday Activities (AEA) Dataset [28], have been released. Collected by Project Aria devices [8], both datasets feature egocentric video sequences with human actions and contain multimodal data streams and high-quality 3D information. ADT also provides extensive ground truth annotations using a motion capture system. Preliminary studies on egocentric 3D reconstruction have been conducted on these new datasets [28,48] and demonstrate the challenges posed by dynamic motion. In contrast, this paper tackles the challenges in egocentric 3D reconstruction and proposes to filter out transient objects in the videos. Compared to all existing work, we are the first work that holistically tackles the challenges in reconstruction and open-world scene understanding, and set up the quantitative benchmark to systematically evaluate performance in egocentric videos.

3 Method

3.1 3D Gaussian Splatting with Feature Rendering

3D Gaussian Splatting (3DGS) [18] has shown state-of-the-art results in 3D reconstruction and novel view synthesis. However, the original design only reconstructs the color radiance in RGB space and is not able to capture the rich semantic information in a 3D scene. In *EgoLifter*, we augment 3DGS to also

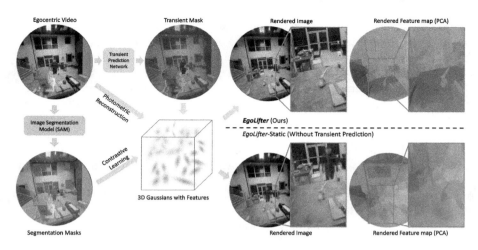

Fig. 2. Naive 3D reconstruction from egocentric videos creates a lot of "floaters" in the reconstruction and leads to blurry rendered images and erroneous instance features (bottom right). *EgoLifter* tackles this problem using a transient prediction network, which predicts a probability mask of transient objects in the image and guides the reconstruction process. In this way, *EgoLifter* gets a much cleaner reconstruction of the static background in both RGB and feature space (top right), which in turn leads to better object decomposition of 3D scenes.

render a feature map of arbitrary dimension in a differentiable manner, which enables us to encode high-dimensional features in the learned 3D scenes and lift segmentation from 2D to 3D. These additional feature channels are used to learn object instance semantics in addition to photometric reconstruction.

Formally, 3DGS represents a 3D scene by a set of N colored 3D Gaussians $\mathcal{S} = \{\Theta_i | i = 1, \cdots, N\}$, with location and shape represented by a center position $\mathbf{p}_i \in \mathbb{R}^3$, an anisotropic 3D covariance $\mathbf{s}_i \in \mathbb{R}^3$ and a rotation quaternion $\mathbf{q}_i \in \mathbb{R}^4$. The radiance of each 3D Gaussian is described by an opacity parameter $\alpha_i \in \mathbb{R}$ and a color vector $\mathbf{c}_i$, parameterized by spherical harmonics (SH) coefficients. In *EgoLifter*, we additionally associate each 3D Gaussian with an extra feature vector $\mathbf{f} \in \mathbb{R}^d$, and thus the optimizable parameter set for i-th Gaussian is $\Theta_i = \{\mathbf{p}_i, \mathbf{s}_i, \mathbf{q}_i, \alpha_i, \mathbf{c}_i, \mathbf{f}_i\}$.

To train 3DGS for 3D reconstruction, a set of M observations $\{\mathbf{I}_j, \theta_j | j = 1, \cdots, M\}$ is used, where $\mathbf{I}_j$ is an RGB image and θ_j is the corresponding camera parameters. During the differentiable rendering process, all 3D Gaussians are splatted onto the 2D image plane according to θ_j and α-blended to get a rendered image $\hat{\mathbf{I}}_j$. Then the photometric loss is computed between the rendered image $\hat{\mathbf{I}}_j$ and the corresponding ground truth RGB image $\mathbf{I}_j$ as

$$\mathcal{L}_{\text{RGB}}(\mathbf{I}_j, \hat{\mathbf{I}}_j) = \mathcal{L}_{\text{MSE}}(\mathbf{I}_j, \hat{\mathbf{I}}_j) = \sum_{u \in \Omega} \|\mathbf{I}_j[u] - f(\hat{\mathbf{I}}_j[u])\|_2^2, \quad (1)$$

where $\mathcal{L}_{\text{MSE}}$ is the mean-squared-error (MSE) loss, Ω is set of all coordinates on the image and $\mathbf{I}_j[u]$ denotes the pixel value of $\mathbf{I}_j$ at coordinate u. $f(\cdot)$ is an image formation model that applies special properties of the camera (e.g. vignetting, radius of valid pixels) on the rendered image. By optimizing $\mathcal{L}_{RGB}$, the location, shape, and color parameters of 3D Gaussians are updated to reconstruct the geometry and appearance of the 3D scene. A density control mechanism is also used to split or prune 3D Gaussians during the training process [18].

In *EgoLifter*, we also implement the differentiable feature rendering pipeline similar to that of the RGB images, which renders to a 2D feature map $\hat{\mathbf{F}} \in \mathbb{R}^{H \times W \times d}$ according to the camera information. During the training process, the feature vectors are supervised by segmentation output obtained from 2D images and jointly optimized with the location and color parameters of each Gaussian. We also include gradients of feature learning for the density control process in learning 3DGS. More details may be found in the supplementary material.

3.2 Learning Instance Features by Contrastive Loss

Egocentric videos capture a huge number of different objects in everyday activities, and some of them may not exist in any 3D datasets for training. Therefore, egocentric 3D perception requires an ability to generalize to unseen categories (open-world) which we propose to achieve by lifting the output from 2D instance segmentation models. The key insight is that 2D instance masks from images of different views can be associated to form a consistent 3D object instance and that this can be done together with the 3D reconstruction process. Recent work has approached this problem using linear assignment [46], video object tracking [61], and incremental matching [12, 26, 49].

To achieve open-world 3D segmentation, we use $\mathbf{f}$ as instance features to capture the lifted segmentation and their similarity to indicate whether a set of Gaussians should be considered as the same object instance. Inspired by Contrastive Lift [2], we adopt supervised contrastive learning, which pulls the rendered features belonging to the same mask closer and pushes those of different masks further apart. Formally, given a training image $\mathbf{I}_j$, we use a 2D segmentation model to extract a set of instance masks $\mathcal{M}_j = \{\mathbf{M}_j^k | k = 1, \cdots, m_i\}$ from $\mathbf{I}_j$. The feature map $\hat{\mathbf{F}}_j$ at the corresponding camera pose θ_j is then rendered, and the contrastive loss is computed over a set of pixel coordinates $\mathcal{U}$, for which we use a uniformly sampled set of pixels $\mathcal{U} \subset \Omega$ due to GPU memory constraint. The contrastive loss is formulated as

$$\mathcal{L}_{\text{contr}}(\hat{\mathbf{F}}_j, \mathcal{M}_j) = -\frac{1}{|\mathcal{U}|} \sum_{u \in \mathcal{U}} \log \frac{\sum_{u' \in \mathcal{U}^+} \exp(\text{sim}(\hat{\mathbf{F}}_j[u], \hat{\mathbf{F}}_j[u']; \gamma))}{\sum_{u' \in \mathcal{U}} \exp(\text{sim}(\hat{\mathbf{F}}_j[u], \hat{\mathbf{F}}_j[u']; \gamma))}, \quad (2)$$

where $\mathcal{U}^+$ is the set of pixels that belong to the same instance mask as u and $\hat{\mathbf{F}}_j[u]$ denotes the feature vector of the $\hat{\mathbf{F}}_j$ at coordinate u. We use a Gaussian RBF kernel as the similarity function, i.e. $\text{sim}(f_1, f_2; \gamma) = \exp(-\gamma \|f_1 - f_2\|_2^2)$.

In the contrastive loss, pixels on the same instance mask are considered as positive pairs and will have similar features during training. Note that since the

2D segmentation model does not output consistent object instance IDs across different views, the contrastive loss is computed individually on each image. This weak supervision allows the model to maintain a flexible definition of object instances without hard assignments and is key to learning multi-view consistent instance features for 3D Gaussians that enables flexible open-world 3D segmentation.

3.3 Transient Prediction for Egocentric 3D Reconstruction

Egocentric videos contain a lot of dynamic objects that cause many inconsistencies among 3D views. As we show in Fig. 2, the original 3DGS algorithm on the egocentric videos results in many floaters and harms the results of both reconstruction and feature learning. In *EgoLifter*, we propose to filter out transient phenomena in the egocentric 3D reconstruction, by predicting a transient probability mask from the input image, which is used to guide the 3DGS reconstruction process.

Specifically, we employ a transient prediction network $G(\mathbf{I}_j)$, which takes in the training image $\mathbf{I}_j$ and outputs a probability mask $\mathbf{P}_j \in \mathbb{R}^{H \times W}$ whose value indicates the probability of each pixel being on a transient object. Then $\mathbf{P}_j$ is used to weigh the reconstruction loss during training, such that when a pixel is considered transient, it is filtered out in reconstruction. Therefore the reconstruction loss from Eq. (1) is adapted to

$$\mathcal{L}_{\text{RGB-w}}(\mathbf{I}_j, \hat{\mathbf{I}}_j, \mathbf{P}_j) = \sum_{u \in \Omega}(1 - \mathbf{P}_j[u])\|\mathbf{I}_j[u] - \hat{\mathbf{I}}_j[u]\|_2^2, \tag{3}$$

where the pixels with lower transient probability will contribute more to the reconstruction loss. As most of the objects in egocentric videos remain static, we also apply an L-1 regularization loss on the predicted $\mathbf{P}_j$ as $\mathcal{L}_{\text{reg}}(\mathbf{P}_j) = \sum_{p \in \mathbf{P}_j} |p|$. This regularization also helps avoid the trivial solution where $\mathbf{P}_j$ equals one and all pixels are considered transient. The transient mask $\mathbf{P}_j$ is also used to guide contrastive learning for lifting instance segmentation, where the pixel set $\mathcal{U}$ is only sampled on pixels with the probability of being transient less than a threshold δ. As shown in Fig. 2 and Fig. 3, this transient filtering also helps learn cleaner instance features and thus better segmentation results.

In summary, the overall training loss on image $\mathbf{I}_j$ is a weighted sum as

$$\mathcal{L} = \lambda_1 \mathcal{L}_{\text{RGB-w}}(\mathbf{I}_j, \hat{\mathbf{I}}_j, \mathbf{P}_j) + \lambda_2 \mathcal{L}_{\text{contr}}(\hat{\mathbf{F}}_j, \mathcal{M}_j) + \lambda_3 \mathcal{L}_{\text{reg}}(\mathbf{P}_j), \tag{4}$$

with λ_1, λ_2 and λ_3 as hyperparameters.

3.4 Open-World Segmentation

After training, instance features $\mathbf{f}$ capture the similarities among 3D Gaussians, and can be used for open-world segmentation in two ways, query-based and clustering-based. In query-based open-world segmentation, one or few clicks on the object of interest are provided and a query feature vector is computed as the averaged features rendered at these pixels. Then a set of 2D pixels or a set of 3D Gaussians can be obtained by thresholding their Euclidean distances from the query feature, from which a 2D segmentation mask or a 3D bounding box can be estimated. In clustering-based segmentation, an HDBSCAN clustering algorithm [32] is performed to assign 3D Gaussians into different groups, which gives a full decomposition of the 3D scene into a set of individual objects. In our experiments, query-based segmentation is used for quantitative evaluation, and clustering-based mainly for qualitative results.

4 Experiments

Implementation. We use a U-Net [43] with the pretrained MobileNet-v3 [14] backbone as the transient prediction network G. The input to G is first resized to 224×224 and then we resize its output back to the original resolution using bilinear interpolation. We use feature dimension $d = 16$, threshold $\delta = 0.5$, temperature $\gamma = 0.01$, and loss weights $\lambda_1 = 1$, $\lambda_2 = 0.1$ and $\lambda_3 = 0.01$. The 3DGS is trained using the Adam optimizer [21] with the same setting and the same density control schedule as in [18]. The transient prediction network is optimized by another Adam optimizer with an initial learning rate of 1×10^{-5}. *EgoLifter* is agnostic to the specific 2D instance segmentation method, and we use the Segment Anything Model (SAM) [22] for its remarkable instance segmentation performance.

Datasets. We evaluate *EgoLifter* on the following egocentric datasets:

- **Aria Digital Twin (ADT)** [37] provides 3D ground truth for objects paired with egocentric videos, which we used to evaluate *EgoLifter* quantitatively. ADT dataset contains 200 egocentric video sequences of daily activities, captured using Aria glasses. ADT also uses a high-quality simulator and motion capture devices for extensive ground truth annotations, including 3D object bounding boxes and 2D segmentation masks for all frames. ADT does not contain an off-the-shelf setting for scene reconstruction or open-world 3D segmentation. We create the evaluation benchmark using the GT 2D masks and 3D bounding boxes by reprocessing the 3D annotations. Note that only the RGB images are used during training, and for contrastive learning, we used the masks obtained by SAM [22].
- **Aria Everyday Activities (AEA) Dataset** [28] provides 143 egocentric videos of various daily activities performed by multiple wearers in five different indoor locations. Different from ADT, AEA contains more natural video activity recordings but does not offer 3D annotations. For each location, multiple sequences of different activities are captured at different times but aligned in the same 3D coordinate space. Different frames or recordings may observe the same local space at different time with various dynamic actions, which represent significant challenges in reconstruction. We group all

Table 1. Quantitative evaluation of 2D instance segmentation (measured in mIoU) and novel view synthesis (measured in PNSR) on the ADT dataset. The evaluations are conducted on the frames in the novel subset of each scene.

Evaluation Object set	mIoU (In-view)			mIoU (Cross-view)			PSNR		
	Static	Dynamic	All	Static	Dynamic	All	Static	Dynamic	All
SAM [22]	54.51	32.77	50.69	–	–	–	–	–	–
Gaussian Grouping [61]	35.68	30.76	34.81	23.79	11.33	21.58	21.29	14.99	19.97
EgoLifter-Static	55.67	**39.61**	52.86	51.29	18.67	45.49	21.37	15.32	20.16
EgoLifter-Deform	54.23	38.62	51.49	51.10	18.02	45.22	21.16	**15.39**	19.93
EgoLifter (Ours)	**58.15**	37.74	**54.57**	**55.27**	**19.14**	**48.84**	**22.14**	14.37	**20.28**

daily videos in each location and run *EgoLifter* for each spatial environment. The longest aggregated video in one location (Location 2) contains 2.3 h of video recording and a total of 170K RGB frames. The dataset demonstrates our method can not only tackle diverse dynamic activities, but also produce scene understanding at large scale in space and time.

- **Ego-Exo4D [11] dataset** is a large and diverse dataset containing over one thousand hours of videos captured simultaneously by egocentric and exocentric cameras. Ego-Exo4D videos capture humans performing a wide range of activities. We qualitatively evaluate *EgoLifter* on the egocentric videos of Ego-Exo4D.

We use the same process for all Project Aria videos. Since Aria glasses use fisheye cameras, we undistort the captured images first before training. We use the image formation function $f(\cdot)$ in Eq. (1) to capture the vignetting effect and the radius of valid pixels, according to the specifications of the camera on Aria glasses. We use high-frequency 6DoF trajectories to acquire RGB camera poses and the semi-dense point clouds provided by each dataset through the Project Aria Machine Perception Services (MPS).

Baselines. We compare *EgoLifter* to the following baselines.

- **SAM [22]** masks serve as input to *EgoLifter*. The comparison on segmentation between *EgoLifter* and SAM shows the benefits of multi-view fusion of 2D masks. As we will discuss in Sect. 4.1, SAM only allows prompts from the same image (in-view query), while *EgoLifter* enables segmentation prompts from different views (cross-view query) and 3D segmentation.
- **Gaussian Grouping [61]** also lifts the 2D segmentation masks into 3D Gaussians. Instead of the contrastive loss, Gaussian Grouping uses a video object tracker to associate the masks from different views and employs a linear layer for identity classification. Gaussian Grouping does not handle the dynamic objects in 3D scenes.

Ablations. We further provide two variants of *EgoLifter* in particular to study the impact of reconstruction backbone.

- ***EgoLifter-Static*** disabled the transient prediction network. A vanilla static 3DGS [18] is learned to reconstruct the scene. We use the same method to lift and segment 3D features.
- ***EgoLifter-Deform*** uses a dynamic variant of 3DGS [60] instead of the transient prediction network to handle the dynamics in the scene. Similar to [60], *EgoLifter*-Deform learns a canonical 3DGS and a network to predict the location and shape of each canonical 3D Gaussian at different timestamps.

We also compare *EgoLifter* with NeRF-based methods [50,63] on the ADT dataset and evaluate *EgoLifter* on non-egocentric datasets [1,47]. Please see the supplementary material for the results.

4.1 Benchmark Setup on ADT

We use the ADT dataset [37] for the quantitative evaluation. We use 16 video sequences from ADT, and the frames of each sequence are split into **seen** and **novel** subsets. The seen subset are used for training and validation, while the novel subset contains a chunk of consecutive frames separate from the seen subset and is only used for testing. The evaluation on the novel subset reflects the performance on novel views. The objects in each video sequence are also tagged **dynamic** and **static** according to whether they move. Each model is trained and evaluated on one video sequence separately. We evaluate the performance of the query-based open-world 2D instance segmentation and 3D instance detection tasks, as described in Sect. 3.4. For the Gaussian Grouping baseline [61], we use their learned identity encoding for extracting query features and computing similarity maps. Please refer to Appendix for more details of the evaluation settings, the exact sequence IDs and splits we used.

Open-World 2D Instance Segmentation. We adopt two settings in terms of query sampling for 2D evaluation, namely **in-view** and **cross-view**. In both settings, a similarity map is computed between the query feature and the rendered feature image. The instance segmentation mask is then obtained by cutting off the similarity map using a threshold that maximizes the IoU with respect to the GT mask, which resembles the process of a human user adjusting the threshold to get the desired object. In the **in-view** setting, the query feature is sampled from one pixel on the rendered feature map in the same camera view. For a fair comparison, SAM [22] in this setup takes in the rendered images from the trained 3DGS and the same query pixel as the segmentation prompt. For each prompt, SAM predicts multiple instance masks at different scales, from which we also use the GT mask to pick one that maximizes the IoU for evaluation. The **cross-view** setting follows the prompt propagation evaluation used in the literature [3,34,41,63]. We randomly sample 5 pixels from the training images (in the seen subset) on the object, and their average feature is used as the query for segmentation on the novel subset. To summarize, the in-view setting evaluates how well features group into objects after being rendered into a feature map, and the cross-view setting evaluates how well the learned feature represents each object instance in 3D and how they generalize to novel views.

Open-World 3D Instance Detection. For 3D evaluation, we use the same query feature obtained in the above cross-view setting. The similarity map is computed between the query feature and the learned 3D Gaussians, from which a subset of 3D Gaussians are obtained by thresholding, and a 3D bounding box is estimated based on their coordinates. The 3D detection performance is evaluated by the IoU between the estimated and the GT 3D bounding boxes. We also select the threshold that maximizes the IoU with the GT bounding box. We only evaluate the 3D static objects in each scene.

Novel View Synthesis. We evaluate the synthesized frames in the novel subset using PSNR metric. We use "All" to indicate full-frame view synthesis. We also separately evaluate the pixels on dynamic and static regions using the provided the 2D ground truth dynamic motion mask.

4.2 Quantitative Results on ADT

The quantitative results are reported in Table 1 and 2. As shown in Table 1, *EgoLifter* consistently outperforms all other baselines and variants in the reconstruction and segmentation of static object instances in novel views. Since transient objects are deliberately filtered out during training, *EgoLifter* has slightly worse performance on dynamic objects, However, the improvements on static objects outweigh the drops on transient ones in egocentric videos and *EgoLifter* still achieves the best overall results in all settings. Similarly, this trend also holds in 3D, and *EgoLifter* has the best 3D detection performance as shown in Table 2.

Table 2. 3D instance detection performance for the static objects in the ADT dataset.

Method	mIoU
Gaussian Grouping [61]	7.48
EgoLifter-Static	21.10
EgoLifter-Deform	20.58
EgoLifter (**Ours**)	**23.11**

4.3 Qualitative Results on Diverse Egocentric Datasets

In Fig. 3, we visualize the qualitative results on several egocentric datasets [11, 28, 37]. Please refer to project webpage for the videos rendered by *EgoLifter* with comparison to baselines. As shown in Fig. 3, without transient prediction, *Ego-Lifter*-Static creates 3D Gaussians to overfit the dynamic observations in some training views. However, since dynamic objects are not geometrically consistent and may be at a different location in other views, these 3D Gaussians become floaters that explain dynamics in a ghostly way, harming both the rendering and segmentation quality. In contrast, *EgoLifter* correctly identifies the dynamic objects in each image using transient prediction and filters them out in

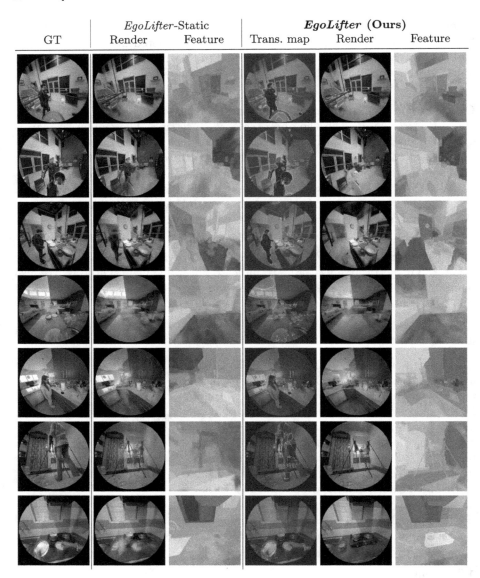

Fig. 3. RGB images and feature maps (colored by PCA) rendered by the *EgoLifter* Static baseline and *EgoLifter*. The predicted transient maps (Trans. map) from *EgoLifter* are also visualized, with red color indicating a high probability of being transient. Note that the baseline puts ghostly floaters on the region of transient objects, but *EgoLifter* filters them out and gives a cleaner reconstruction of both RGB images and feature maps. Rows 1–3 are from ADT, rows 4–5 from AEA, and rows 6–7 from Ego-Exo4D.(Color figure online)

the reconstruction. The resulting cleaner reconstruction leads to better results in novel view synthesis and segmentation, as we have already seen quantitatively in Sect. 4.2. We also compare the qualitative results with Gaussian Grouping [61] in Fig. 4, from which we can see that Gaussian Grouping not only struggles with floaters associated with transient objects but also has a less clean feature map even on the static region. We hypothesize this is because our contrastive loss helps learn more cohesive identity features than the classification loss used in [61]. This also explains why *EgoLifter*-Static significantly outperforms Gaussian Grouping in segmentation metrics as shown in Table 1 and 2.

4.4 3D Object Extraction and Scene Editing

Based on the features learned by *EgoLifter*, we can decompose a 3D scene into individual 3D objects, by querying or clustering over the feature space. Each extracted 3D object is represented by a set of 3D Gaussians which can be photorealistically rendered. In Fig. 5, we show the visualization of 3D objects extracted from a scene in the ADT dataset. This can further enable scene editing applications by adding, removing, or transforming these objects over the 3D space. In Fig. 1, we demonstrate one example of background recovery by removing all segmented 3D objects from the table.

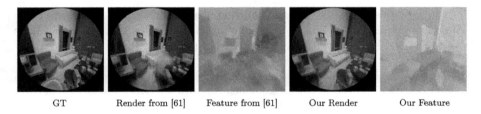

GT   Render from [61]   Feature from [61]   Our Render   Our Feature

Fig. 4. Rendered images and feature maps (visualised in PCA colors) by Gaussian Grouping [61] and *EgoLifter* (Ours).

Rendered Image and Features from *EgoLifter*   Extracted 3D Objects

Fig. 5. Individual 3D object can be extracted by querying or clustering over the 3D features from *EgoLifter*. Note object reconstructions are not perfect since each object might be partial observable in the egocentric videos rather than scanned intentionally.

5 Conclusion and Limitation

We present *EgoLifter*, a novel algorithm that simultaneously solves the 3D reconstruction and open-world segmentation problem for in-the-wild egocentric perception. By lifting the 2D segmentation into 3D Gaussian Splatting, *EgoLifter* achieves strong open-world 2D/3D segmentation performance with no 3D data annotation. To handle the rapid and sparse dynamics in egocentric videos, we employ a transient prediction network to filter out transient objects and get more accurate 3D reconstruction. *EgoLifter* is evaluated on several challenging egocentric datasets and outperforms other existing baselines. The representations obtained by *EgoLifter* can also be used for several downstream tasks like 3D object asset extraction and scene editing, showing great potential for personal wearable devices and AR/VR applications.

Limitations: We observe the transient prediction module may mix the regions that are hard to reconstruct with transient objects. As shown in rows (4) and (5) of Fig. 3, the transient prediction module predicts a high probability for pixels on the windows, which have over-exposed pixels that are hard to be reconstructed from LDR images. In this case, *EgoLifter* learns to filter them out to improve reconstruction on that region. Besides, the performance of *EgoLifter* may also be dependent on the underlying 2D segmentation model. *EgoLifter* is not able to segment an object if the 2D model consistently fails on it.

Potential Negative Impact: 3D object digitization for egocentric videos in the wild may pose a risk to privacy considerations. Ownership of digital object rights of physical objects is also a challenging and complex topic that will have to be addressed as AR/VR becomes more ubiquitous.

Acknowledgments. The authors thank the Project Aria team for providing open-source support and Nickolas Charron for helping with the ADT dataset evaluation. The authors also thank Fangzhou Hong, Kevin QH Lin, Hyo Jin Kim, Lingni Ma, Lambert Mathias, Pierre Moulon, Richard Newcombe, Tianwei Shen, Nan Yang, and Wang Zhao for discussions and useful feedback over the course of this project.

References

1. Barron, J.T., Mildenhall, B., Verbin, D., Srinivasan, P.P., Hedman, P.: Mip-NeRF 360: unbounded anti-aliased neural radiance fields. In: Proceedings of the IEEE/CVF Conference on Computer Vision and Pattern Recognition, pp. 5470–5479 (2022)
2. Bhalgat, Y., Laina, I., Henriques, J.F., Zisserman, A., Vedaldi, A.: Contrastive lift: 3d object instance segmentation by slow-fast contrastive fusion. In: Advances in Neural Information Processing Systems (2023)
3. Cen, J., et al.: Segment anything in 3D with NerFS. Adv. Neural. Inf. Process. Syst. **36**, 25971–25990 (2023)
4. Chen, Z., Wang, F., Wang, Y., Liu, H.: Text-to-3d using gaussian splatting. In: Proceedings of the IEEE/CVF Conference on Computer Vision and Pattern Recognition, pp. 21401–21412 (2024)

5. Crabtree, A., Tolmie, P.: A day in the life of things in the home. In: Proceedings of the 19th ACM Conference on Computer-Supported Cooperative Work & Social Computing, pp. 1738–1750 (2016)
6. Damen, D., et al.: Scaling egocentric vision: the epic-kitchens dataset. In: Ferrari, V., Hebert, M., Sminchisescu, C., Weiss, Y. (eds.) ECCV 2018. LNCS, vol. 11208, pp. 753–771. Springer, Cham (2018). https://doi.org/10.1007/978-3-030-01225-0_44
7. Duan, Y., Wei, F., Dai, Q., He, Y., Chen, W., Chen, B.: 4d-rotor gaussian splatting: towards efficient novel view synthesis for dynamic scenes. In: ACM SIGGRAPH 2024 Conference Papers, pp. 1–11 (2024)
8. Engel, J., et al.: Project aria: a new tool for egocentric multi-modal AI research. arXiv preprint arXiv:2308.13561 (2023)
9. Engelmann, F., Manhardt, F., Niemeyer, M., Tateno, K., Pollefeys, M., Tombari, F.: OpenNeRF: open set 3D neural scene segmentation with pixel-wise features and rendered novel views. In: International Conference on Learning Representations (2024)
10. Gao, H., Li, R., Tulsiani, S., Russell, B., Kanazawa, A.: Monocular dynamic view synthesis: a reality check. Adv. Neural. Inf. Process. Syst. **35**, 33768–33780 (2022)
11. Grauman, K., et al.: Ego-Exo4D: understanding skilled human activity from first- and third-person perspectives. In: Proceedings of the IEEE/CVF Conference on Computer Vision and Pattern Recognition, pp. 19383–19400 (2024)
12. Gu, Q., et al.: ConceptGraphs: open-vocabulary 3d scene graphs for perception and planning. In: IEEE International Conference on Robotics and Automation (2023)
13. Gu, Q., Okorn, B., Held, D.: OSSID: online self-supervised instance detection by (and for) pose estimation. IEEE Robot. Autom. Lett. **7**(2), 3022–3029 (2022)
14. Howard, A., et al.: Searching for mobileNetV3. In: Proceedings of the IEEE/CVF International Conference on Computer Vision, pp. 1314–1324 (2019)
15. Huang, C., Mees, O., Zeng, A., Burgard, W.: Audio visual language maps for robot navigation. In: Proceedings of the International Symposium on Experimental Robotics (ISER), Chiang Mai, Thailand (2023)
16. Jatavallabhula, K.M., et al.: ConceptFusion: open-set multimodal 3D mapping. In: Robotics: Science and Systems (2023)
17. Keetha, N., et al.: SplaTAM: splat track & map 3D Gaussians for dense RGB-D SLAM. In: Proceedings of the IEEE/CVF Conference on Computer Vision and Pattern Recognition, pp. 21357–21366 (2024)
18. Kerbl, B., Kopanas, G., Leimkühler, T., Drettakis, G.: 3D Gaussian splatting for real-time radiance field rendering. ACM Trans. Graph. (ToG) **42**(4), 1–14 (2023)
19. Kerr, J., Kim, C.M., Goldberg, K., Kanazawa, A., Tancik, M.: LERF: language embedded radiance fields. In: International Conference on Computer Vision (ICCV) (2023)
20. Kim, C.M., Wu, M., Kerr, J., Goldberg, K., Tancik, M., Kanazawa, A.: Garfield: group anything with radiance fields. In: Proceedings of the IEEE/CVF Conference on Computer Vision and Pattern Recognition, pp. 21530–21539 (2024)
21. Kingma, D.P., Ba, J.: Adam: a method for stochastic optimization. In: International Conference on Learning Representations (2015)
22. Kirillov, A., et al.: Segment anything. In: Proceedings of the IEEE/CVF International Conference on Computer Vision, pp. 4015–4026 (2023)
23. Kobayashi, S., Matsumoto, E., Sitzmann, V.: Decomposing nerf for editing via feature field distillation. Adv. Neural. Inf. Process. Syst. **35**, 23311–23330 (2022)
24. Liu, K., et al.: 3D open-vocabulary segmentation with foundation models. arXiv preprint arXiv:2305.14093 (2023)

25. Liu, S., et al.: Grounding DINO: marrying DINO with grounded pre-training for open-set object detection. arXiv preprint arXiv:2303.05499 (2023)
26. Lu, S., Chang, H., Jing, E.P., Boularias, A., Bekris, K.: OVIR-3D: open-vocabulary 3D instance retrieval without training on 3D data. In: Conference on Robot Learning, pp. 1610–1620. PMLR (2023)
27. Luiten, J., Kopanas, G., Leibe, B., Ramanan, D.: Dynamic 3D Gaussians: tracking by persistent dynamic view synthesis. In: 2024 International Conference on 3D Vision (3DV), pp. 800–809. IEEE (2024)
28. Lv, Z., et al.: Aria everyday activities dataset. arXiv preprint arXiv:2402.13349 (2024)
29. Martin-Brualla, R., Radwan, N., Sajjadi, M.S., Barron, J.T., Dosovitskiy, A., Duckworth, D.: NeRF in the wild: neural radiance fields for unconstrained photo collections. In: Proceedings of the IEEE/CVF Conference on Computer Vision and Pattern Recognition, pp. 7210–7219 (2021)
30. Matsuki, H., Murai, R., Kelly, P.H., Davison, A.J.: Gaussian splatting slam. In: Proceedings of the IEEE/CVF Conference on Computer Vision and Pattern Recognition, pp. 18039–18048 (2024)
31. Mazur, K., Sucar, E., Davison, A.J.: Feature-realistic neural fusion for real-time, open set scene understanding. In: 2023 IEEE International Conference on Robotics and Automation (ICRA), pp. 8201–8207. IEEE (2023)
32. McInnes, L., Healy, J., Astels, S.: HDBSCAN: hierarchical density based clustering. J. Open Source Softw. **2**(11), 205 (2017)
33. Mildenhall, B., Srinivasan, P.P., Tancik, M., Barron, J.T., Ramamoorthi, R., Ng, R.: NeRF: representing scenes as neural radiance fields for view synthesis. Commun. ACM **65**(1), 99–106 (2021)
34. Mirzaei, A., et al.: SPIn-NeRF: multiview segmentation and perceptual inpainting with neural radiance fields. In: Proceedings of the IEEE/CVF Conference on Computer Vision and Pattern Recognition, pp. 20669–20679 (2023)
35. OpenAI: GPT-4 technical report. arXiv preprint arXiv:2303.08774 (2023)
36. Oquab, M., et al.: DINOv2: learning robust visual features without supervision. arXiv preprint arXiv:2304.07193 (2023)
37. Pan, X., et al.: Aria digital twin: a new benchmark dataset for egocentric 3D machine perception. In: Proceedings of the IEEE/CVF International Conference on Computer Vision, pp. 20133–20143 (2023)
38. Peng, S., et al.: OpenScene: 3D scene understanding with open vocabularies. In: Proceedings of the IEEE/CVF Conference on Computer Vision and Pattern Recognition, pp. 815–824 (2023)
39. Qin, M., Li, W., Zhou, J., Wang, H., Pfister, H.: LangSplat: 3D language gaussian splatting. In: Proceedings of the IEEE/CVF Conference on Computer Vision and Pattern Recognition, pp. 20051–20060 (2024)
40. Radford, A., et al.: Learning transferable visual models from natural language supervision. In: International Conference on Machine Learning, pp. 8748–8763. PMLR (2021)
41. Ren, Z., Agarwala, A., Russell, B., Schwing, A.G., Wang, O.: Neural volumetric object selection. In: Proceedings of the IEEE/CVF Conference on Computer Vision and Pattern Recognition, pp. 6133–6142 (2022)
42. Rombach, R., Blattmann, A., Lorenz, D., Esser, P., Ommer, B.: High-resolution image synthesis with latent diffusion models. In: Proceedings of the IEEE/CVF Conference on Computer Vision and Pattern Recognition, pp. 10684–10695 (2022)

43. Ronneberger, O., Fischer, P., Brox, T.: U-Net: convolutional networks for biomedical image segmentation. In: Navab, N., Hornegger, J., Wells, W.M., Frangi, A.F. (eds.) MICCAI 2015, Part III. LNCS, vol. 9351, pp. 234–241. Springer, Cham (2015). https://doi.org/10.1007/978-3-319-24574-4_28
44. Shafiullah, N.M.M., Paxton, C., Pinto, L., Chintala, S., Szlam, A.: Clip-fields: weakly supervised semantic fields for robotic memory. In: Bekris, K.E., Hauser, K., Herbert, S.L., Yu, J. (eds.) Robotics: Science and Systems (2023)
45. Shen, W., Yang, G., Yu, A., Wong, J., Kaelbling, L.P., Isola, P.: Distilled feature fields enable few-shot language-guided manipulation. In: 7th Annual Conference on Robot Learning (2023)
46. Siddiqui, Y., et al.: Panoptic lifting for 3D scene understanding with neural fields. In: Proceedings of the IEEE/CVF Conference on Computer Vision and Pattern Recognition, pp. 9043–9052 (2023)
47. Straub, J., et al.: The replica dataset: a digital replica of indoor spaces. arXiv preprint arXiv:1906.05797 (2019)
48. Sun, J., Qiu, J., Zheng, C., Tucker, J., Yu, J., Schwager, M.: Aria-NeRF: multimodal egocentric view synthesis. arXiv preprint arXiv:2311.06455 (2023)
49. Takmaz, A., Fedele, E., Sumner, R.W., Pollefeys, M., Tombari, F., Engelmann, F.: OpenMask3D: open-vocabulary 3D instance segmentation. In: Advances in Neural Information Processing Systems (NeurIPS) (2023)
50. Tancik, M., et al.: Nerfstudio: a modular framework for neural radiance field development. In: ACM SIGGRAPH 2023 Conference Proceedings, pp. 1–12 (2023)
51. Tang, J., Ren, J., Zhou, H., Liu, Z., Zeng, G.: DreamGaussian: generative gaussian splatting for efficient 3D content creation. arXiv preprint arXiv:2309.16653 (2023)
52. Tsagkas, N., Mac Aodha, O., Lu, C.X.: Vl-fields: towards language-grounded neural implicit spatial representations. arXiv preprint arXiv:2305.12427 (2023)
53. Tscherlnezki, V., et al.: Epic fields: marrying 3D geometry and video understanding. Adv. Neural Inf. Process. Syst. **36** (2024)
54. Tschernezki, V., Laina, I., Larlus, D., Vedaldi, A.: Neural feature fusion fields: 3D distillation of self-supervised 2D image representations. In: International Conference on 3D Vision (3DV). IEEE (2022)
55. Tschernezki, V., Larlus, D., Vedaldi, A.: NeuralDiff: segmenting 3D objects that move in egocentric videos. In: 2021 International Conference on 3D Vision (3DV), pp. 910–919. IEEE (2021)
56. Wu, G., et al.: 4D Gaussian splatting for real-time dynamic scene rendering. In: Proceedings of the IEEE/CVF Conference on Computer Vision and Pattern Recognition, pp. 20310–20320 (2024)
57. Yan, C., et al.: GS-SLAM: dense visual slam with 3D Gaussian splatting. In: Proceedings of the IEEE/CVF Conference on Computer Vision and Pattern Recognition, pp. 19595–19604 (2024)
58. Yan, Y., et al.: Street Gaussians for modeling dynamic urban scenes. arXiv preprint arXiv:2401.01339 (2024)
59. Yang, Z., Yang, H., Pan, Z., Zhu, X., Zhang, L.: Real-time photorealistic dynamic scene representation and rendering with 4D Gaussian splatting. arXiv preprint arXiv:2310.10642 (2023)
60. Yang, Z., Gao, X., Zhou, W., Jiao, S., Zhang, Y., Jin, X.: Deformable 3D Gaussians for high-fidelity monocular dynamic scene reconstruction. In: Proceedings of the IEEE/CVF Conference on Computer Vision and Pattern Recognition, pp. 20331–20341 (2024)
61. Ye, M., Danelljan, M., Yu, F., Ke, L.: Gaussian grouping: segment and edit anything in 3D scenes. arXiv preprint arXiv:2312.00732 (2023)

62. Yi, T., et al.: GaussianDreamer: fast generation from text to 3D Gaussians by bridging 2D and 3D diffusion models. In: Proceedings of the IEEE/CVF Conference on Computer Vision and Pattern Recognition (2024)
63. Ying, H., et al.: OmniSeg3D: omniversal 3D segmentation via hierarchical contrastive learning. In: Proceedings of the IEEE/CVF Conference on Computer Vision and Pattern Recognition, pp. 20612–20622 (2024)
64. Zhou, S., et al.: Feature 3DGS: supercharging 3D Gaussian splatting to enable distilled feature fields. In: Proceedings of the IEEE/CVF Conference on Computer Vision and Pattern Recognition, pp. 21676–21685 (2024)
65. Zhou, X., Lin, Z., Shan, X., Wang, Y., Sun, D., Yang, M.H.: DrivingGaussian: composite Gaussian splatting for surrounding dynamic autonomous driving scenes. In: Proceedings of the IEEE/CVF Conference on Computer Vision and Pattern Recognition, pp. 21634–21643 (2024)
66. Zuo, X., Samangouei, P., Zhou, Y., Di, Y., Li, M.: FMGS: foundation model embedded 3D gaussian splatting for holistic 3D scene understanding. arXiv preprint arXiv:2401.01970 (2024)

MEVG: Multi-event Video Generation with Text-to-Video Models

Gyeongrok Oh[1](✉), Jaehwan Jeong[1], Sieun Kim[1], Wonmin Byeon[2], Jinkyu Kim[1], Sungwoong Kim[1], and Sangpil Kim[1]

[1] Korea University, Seoul, South Korea
dhrudfhr98@korea.ac.kr
[2] NVIDIA, Santa Clara, USA

Abstract. We introduce a novel diffusion-based video generation method, generating a video showing multiple events given multiple individual sentences from the user. Our method does not require a large-scale video dataset since our method uses a pre-trained diffusion-based text-to-video generative model without a fine-tuning process. Specifically, we propose a last frame-aware diffusion process to preserve visual coherence between consecutive videos where each video consists of different events by initializing the latent and simultaneously adjusting noise in the latent to enhance the motion dynamic in a generated video. Furthermore, we find that the iterative update of latent vectors by referring to all the preceding frames maintains the global appearance across the frames in a video clip. To handle dynamic text input for video generation, we utilize a novel prompt generator that transfers course text messages from the user into the multiple optimal prompts for the text-to-video diffusion model. Extensive experiments and user studies show that our proposed method is superior to other video-generative models in terms of temporal coherency of content and semantics. Video examples are available on our project page: https://kuai-lab.github.io/eccv2024mevg.

Keywords: Multi-event video generation · Training-free · Diffusion model

1 Introduction

Deep generative models in the computer vision community gain a significant spotlight due to their unprecedented performance. Especially, text-to-image generation model (T2I) [7,13,23,31–33] has successfully produced high-quality images with complex text descriptions. However, the complexity of spatial-temporal relations for modeling motion dynamics, light conditions, and scene transitions for video generation with deep generative models requires huge computational resources and large-scale text-video paired datasets. Despite these challenges, recent methods [2,15,20–22,24,34] for text-to-video generation achieve data and cost-efficient training by leveraging the pre-trained text-to-image generative models [7,32]. Although spatial-temporal modeling aided by prior knowledge from text-image pairs helps to generate high-quality frames and capture semantically complex descriptions, it falls short in addressing real-world video comprehensively.

In essence, videos in the wild consist of consecutive events with dynamic movements, backgrounds, objects, and viewpoint changes over time. However, existing approaches mainly generate a video with a single prompt that disregards semantic transitions from event to event and restrictively expresses the entire story when the story consists of multiple events. Multi-event-based video generation requires three significant criteria: 1) smooth transition between each video clip, 2) semantic alignment between the prompt from the user and the generated video, and 3) ensuring diversity of the content and motion in the video.

Recently, some studies [10,38] have made progress in embracing the multiple descriptions that contain chronological sequence events. Despite this breakthrough, substantial training efforts are necessary using extensive text-video datasets because they present the additional networks to make multi-prompt video generation. The concurrent work [41] leverages the pre-trained text-to-video generation model (T2V). However, the overlapped denoised process on the consecutive distinct prompts induces visual degradation and significant inconsistency between the background and objects. Moreover, traditional long-term T2V methods [2,10,21,34,50] hierarchically generate a video by bridging the gap between each keyframe, followed by generating keyframes given a single description. This hierarchical video generation process makes it challenging to incorporate the multiple-time variant prompts since it comprehensively generates the global video content in the beginning (Fig. 1).

Our method, multi-event video generation method (MEVG), is delicately designed to generate a video clip consisting of multiple events without any video data nor fine-tuning process. MEVG successively enforces the temporal coherence between independently generated video clips given multiple event descriptions. By building upon the publicly released diffusion-based video generation model, we effectively utilize the pre-trained single-prompt T2V generative model[1] to generate complex scenario videos. Specifically, to preserve the visual coherence between time variable prompts in video generation while producing realistic and diverse motion, we introduce two novel techniques: a last frame-aware latent vector initialization method and a structure-guided sampling strategy for the diffusion-based generative model. First, the last frame-aware latent vector initialization stage includes (i) *dynamic noise*, which diversifies motion across frames, and (ii) *last frame-aware inversion*, which guides to generate consistent contents between prompts. *Structure-guided sampling* further improves the visual consistency by progressively updating the latent code during the sampling process. In addition, to incorporate sequentially structured prompts, we

---

[1] We used [15]. From our experiments, our proposed method can be applied to any kind of diffusion-based T2V pre-trained model.

Fig. 1. An example of multi-event video generation. MEVG produces impressive output that corresponds to the given prompts and consists of chronologically continuous events.

leverage the Large Language Model (LLM) as a prompt generator. Since a single story has sequentially incorporated events within one sentence, LLM separates a complex story into multiple prompts, each having only one event.

From our extensive experiments and user studies, we demonstrate that our proposed methods generate realistic videos that include three representative types of change: object motion, background, and complex content changes. Moreover, we examine the effectiveness and legitimacy of each proposed method by conducting ablation studies. To summarize, our main contributions are as follows:

– Our proposed diffusion-based video generation method generates a video consisting of multiple events without requiring any training or additional video data.
– We present a last-frame aware initialization method and dynamic noise adjustment strategy for the latent vector that enhances temporal and semantic consistency between individual videos where each video shows a different event.
– We present a novel prompt generator that transforms course text inputs into optimal text instructions for a text-to-video generative model, ensuring coherence of semantic transitions in the generated video.
– We show that our proposed video generation method outperforms the other zero-shot video generation methods in reflecting multiple events while maintaining visually coherent content for video generation.

2 Related Work

Text-to-Video Generation. Text-to-video (T2V) generation has shown remarkable progress. Three primary methodologies are utilized in the field of computer vision. A Generative Adversarial Network (GAN) [3,35,37,40,42,49] is a well-known algorithm to generate diverse video from a noise vector utilizing a generator and discriminator. Another approach is auto-regressive transformers [10,21,43–45,48] that leverage discrete representation to depict the motion dynamics. Recently, diffusion-based methods [1,2,5,9,11,15,17,19,34,39,51] have shown significant progress in learning data distribution while iteratively removing noise from the initial gaussian noise.

Long-term video generation has recently been a popular topic in the computer vision community. Auto-regressive approaches [10,21,26] leveraging transformer architecture show plausible results in long-term video generation. However, they require massive training costs and datasets. Furthermore, although TATS [10] and Phenaki [38] can generate videos driven by a sequence of prompts, accumulated errors over time cause drastic changes in video content and visual quality degradation due to the auto-regressive property. Several works [2,11,34,39,50] based on the diffusion model leverage temporal interpolation networks and masked strategies for generating smoother videos. VidRD [12] directly utilizes the previous initial latent code to expand the video.

Note that most previous works focused on video generation from a single prompt or an event. However, in this work, we tackle the multi-event video generation task, which consists of consecutive events in a long-term video. Animate-A-Story [14] utilizes abundant real-world video corresponding to each story for natural motion. Moreover, SEINE [6] focuses on the transition from short to long video generation models by utilizing millions of datasets. Gen-L-Video [41], another approach, uses overlapped frames between two successive prompts. Although this strategy makes the outcomes more realistic, undesirable contents occur due to the overlapping denoising process.

Zero-shot approach. FateZero [29] and INFUSION [25] edit video by leveraging the pre-trained image diffusion models while ensuring temporal consistency. These methods utilize attention maps and spatial features to preserve the structure and temporal coherence over the frame. In the generation field, Text2Video-Zero [24] synthesizes the video to keep the global structure across the sequence of frames without video data while encoding the motion dynamics to provide diverse movement. To encode the movement, latent codes enclose the motion dynamics with direction parameters. Free-bloom [22] and DirecT2V [20] are distinct approaches employing the text-to-image models while sharing a similar conceptual framework. Both utilize the Large Language Model (LLM) in order to maintain the semantic information for each generated frame. DirecT2V leverages self-attention to preserve the appearance of the video; in addition, Free-bloom leverages joint distribution to sample the initial code for consistent frames.

Inspired by these approaches, we leverage a pre-trained text-to-video (T2V) generation model to extend a short, monotonous video into an exciting video

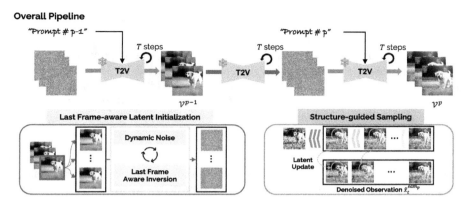

Fig. 2. MEVG synthesizes the consecutive video clips corresponding to distinct prompts. The overall pipeline comprises two major components: last frame-aware latent initialization and structure-guided sampling. First, in the last frame-aware latent initialization, the pre-trained text-to-video generation model adopts the repeated frame as an input to invert into the initial latent code with two novel techniques: *dynamic noise* and *last frame-aware inversion*. Second, *structure-guided sampling* enforces continuity within a video clip by updating the latent code.

containing variable events. To pursue the naturalness of results, challenges exist in maintaining visual coherence and guaranteeing diversity. Therefore, we adjust the latent code near the preceding video and grant dynamic changes through gradual perturbation.

3 Method

We propose a novel pipeline to generate a temporally and semantically coherent video conditioned on multiple event-based prompts. Specifically, our goal is to generate multiple video clips without disturbing the natural flow and recurrent video pattern across the semantic transitions in the given prompts. In this section, we first provide the introductory diffusion model that is the basis for our research and an overview of our proposed pipeline (see Sect. 3.1 and Sect. 3.2). Next, we present the technical details of our two main components: (i) *last frame-aware latent initialization*, (ii) *structure-guided sampling*, in Sect. 3.3 and Sect. 3.4. Finally, we introduce the prompt generator, harnessing the powerful ability of Large Language Model (LLM) to handle a complex story containing multiple meaningful events (Sect. 3.5).

3.1 Preliminaries

DDPM. The diffusion probabilistic model [18] has two components: a forward diffusion process and a backward diffusion process. In the forward process, data distribution transforms into noise distribution by adding noise iteratively. For every timestep t, noise $\epsilon \sim N(0, I)$ diffract the original data x utilizing the variance schedule β_t as follows:

$$x_t = \sqrt{\bar{\alpha}_t} x + \sqrt{1 - \bar{\alpha}_t} \epsilon, \tag{1}$$

where $\alpha_t = 1 - \beta_t$ and $\bar{\alpha}_t = \prod_{i=0}^{t} \alpha_i$. In training time, the diffusion model predicts noise during every step to reconstruct the original data distribution. This reverse process $q(x_{t-1}|x_t)$ is parameterized as follows:

$$p_\theta(x_{t-1}|x_t) := \mathcal{N}(x_{t-1}; \mu_\theta(x_t, t), \Sigma_\theta(x_t, t)). \tag{2}$$

DDIM. DDIM [36] is a variant of DDPM that leverages the non-Markovian manner instead of the Markov chain. DDIM sampling strategy makes the diffusion process to be deterministic, which can be written as follows:

$$x_{t-1} = \underbrace{\sqrt{\bar{\alpha}_{t-1}} \left(\frac{x_t - \sqrt{1 - \bar{\alpha}_t} \epsilon_\theta(x_t, t)}{\sqrt{\bar{\alpha}_t}} \right)}_{\hat{x}_t} + \sqrt{1 - \bar{\alpha}_{t-1} - \sigma_t^2} \underbrace{\cdot \epsilon_\theta(x_t, t)}_{\epsilon_t} - \sigma_t n_t, \tag{3}$$

where $\hat{x}_t$ indicates the denoised observation of x_0 at each diffusion step t, ϵ_t denotes the predicted noise at each time step t and σ controls whether the model is stochastic or deterministic. In this paper, we use the modified DDIM Inversion to strengthen the naturalness of the video, maintaining overall visual consistency despite the semantic changes along the temporal axis in the given prompts.

3.2 MEVG Pipeline

We outline our proposed MEVG that utilizes the former video clip to generate subsequent video clips considering the given prompts (see Fig. 2). Our method is built upon the latent video diffusion model [15], which leverages the low-dimensional latent space $x \in \mathbb{R}^{F \times c \times h \times w}$, where c, h, and w denote latent space dimension and F indicates the total number of frames. The video output $\mathcal{V} \in \mathbb{R}^{F \times 3 \times H \times W}$ are obtained by passing the latent code x through the decoder $\mathcal{D}$, where $H \times W$ is the resolution of the frame.

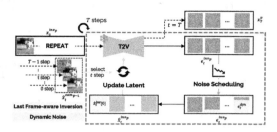

Fig. 3. Last Frame-aware Latent Initialization Initial latent code is crucial for maintaining global geometric structure. We apply two techniques performing different roles: (i) *dynamic noise* tailors flexibility differentially across each frame, and (ii) *last frame-aware inversion* restricts the model to minimize the divergence of the entire frames from the content of the preceding video clip.

To get the final result V = $\{\mathcal{V}^p\}_{p=0}^{\mathcal{P}-1}$, where $\mathcal{P}$ denotes the number of given prompts, we first sample the video conditioned on the first prompt. An initial video clip is created by Gaussian distribution $x_T \sim \mathcal{N}(0, I)$ as well as a static image as conditional guidance to capture the essential visual content corresponding to user intention. After generating the initial video clip, we extend a preceding video in accordance with the semantic context of the subsequent prompt, driven by two major elements: *last frame-aware latent initialization* and *structure-guided sampling*. Last frame-aware latent initialization is the initialization process of the noise latent that helps to preserve the spatial information while generating more diverse contents. Structure-guided sampling then enforces motion consistency between frames during the backward diffusion process.

3.3 Last Frame-Aware Latent Initialization

The inversion technique, which reconstructs the initial latent code from the visual input (e.g., image or video), is used in real-world applications [8,27,47], where accurate spatial layout or visual content reconstruction is important. Video generation should have visual coherence across the entire sequence of frames while adapting the movement of objects and the background transition. To achieve this goal, we aim to find the optimal latent code that helps preserve global coherence between the generated videos for the previous and next prompts and maintains an ability to adapt to the changes. However, the existing approaches [12,41] present repetitive video patterns (e.g., similar camera movement and object position) and awkward scene transition caused by overlapped content in a single frame.

To solve these challenges, we reuse the generated video (essentially the last frame) from the previous prompt to generate the frames for the new prompt. Basically, the last frame of the previously generated video is copied over the entire sequence as an initial conditioning input. We then propose dynamic noise to enforce the diversity of the generated video. This process preserves the overall visual contents, such as an object and background across the video, and also improves the generation diversity.

Dynamic Noise. To generate diverse motion of the object and smooth transition of the background given prompts, we add the video noise prior similar to [11]. Essentially, the noise vector $\epsilon_t^{dyn} \sim \mathcal{N}(0, \frac{1}{1+\kappa^2}I)$ is added to the predicted noise $\epsilon_t^{inv_p}$ during inversion stage for next prompt p. κ regulates the dynamics and variability of the frames within a single video segment; $\kappa \to 0$ increases video variations.

Algorithm 1: Last Frame-aware Latent Initialization

Input: latent code of the the previous prompt x_0^{p-1}, denoised observation of the last frame from the previous prompt $\{\hat{x}_t^{sam_{p-1}}[-1]\}_{t=0}^{T-1}$, noise scheduling function $\mathcal{F}(\cdot)$, and pre-trained T2V model **T2V**$(\cdot)$

Result: Initial latent code x_T^p of next prompt p
// $T =$ Number of diffusion steps
// $N =$ Number of frames
// $p =$ Index of next prompt

1   $x_0^{inv_p} \leftarrow \text{REPEAT}(x_0^{p-1}[-1])$
2   **for** t *in* $0, ..., T-1$ **do**
3     $\epsilon_t^{inv_p} \leftarrow \textbf{T2V}(x_t^{inv_p}, t)$
     // Dynamic Noise
4   **for** n *in* $0, ..., N-1$ **do**
5       $\kappa_n \leftarrow \mathcal{F}(n)$
6       $\epsilon_t^{dyn} \sim \mathcal{N}(0, \frac{1}{1+\kappa_n^2}I)$
7       $\epsilon_t^{inv_p}[n] \leftarrow \frac{\kappa_n}{\sqrt{1+\kappa_n^2}}\epsilon_t^{inv_p}[n] + \epsilon_t^{dyn}$
8   **end**
9     $\hat{x}_t^{inv_p} \leftarrow (x_t^{inv_p} - \sqrt{1-\bar{\alpha}_t}\epsilon_t^{inv_p})/\sqrt{\bar{\alpha}_t}$
     // Last Frame-aware Inversion
10    $\mathcal{L}_{\text{LFAI}} = \|\hat{x}_t^{sam_{p-1}}[-1] - \hat{x}_t^{inv_p}[0]\|_2^2$
11    $\hat{x}_t^{inv_p} \leftarrow \hat{x}_t^{inv_p} - \delta_{\text{LFAI}}\nabla_{\hat{x}_t}\mathcal{L}_{\text{LFAI}}$
12    $x_{t+1}^{inv_p} \leftarrow \sqrt{\bar{\alpha}_{t+1}}\hat{x}_t^{inv_p} + \sqrt{1-\bar{\alpha}_{t+1}}\epsilon_t^{inv_p}$
13 **end**
14 $x_T^p = x_T^{inv_p}$

Since the beginning of the new video should be similar to the preceding video clip, and then more changes occur toward the end of the video, we design a noise scheduling function $\mathcal{F} = \exp(-x)$ that monotonically decreases the κ. κ_n corresponding to the frame index n is determined by:

$$\kappa_n = \mathcal{F}(n), \quad 0 \leq n < N, \qquad (4)$$

where N is the total number of frames within one video clip. Finally, the predicted noise $\epsilon_t^{inv_p}$ is obtained as follows:

$$\epsilon_t^{inv_p}[n] = \frac{\kappa_n}{\sqrt{1+\kappa_n^2}}\epsilon_t^{inv_p}[n] + \epsilon_t^{dyn}, \quad 0 \leq n < N, \qquad (5)$$

where $[\cdot]$ denotes the index of frame.

Last Frame-aware Inversion. While *dynamic noise* helps to generate diverse video contents, they cause a temporal inconsistency problem between individual video clips conditioned on consecutive prompts. *Last frame-aware inversion* coerces to maintain a visual correlation between different video clips guided by the denoised observation $\hat{x}_t$. Since the denoised observation $\hat{x}_t$ is the predicted noise-free latent at diffusion step t (see Eq. 3), it contains a sketchy spatial layout and video context. We regularize the initial frame of the current video clip $\hat{x}_t^{inv_p}[0]$ using the denoised observation of the last frame $\hat{x}_t^{sam_{p-1}}[-1]$ from the previous clip. This process ensures the visual consistency between two video clips. We minimize the objective $\mathcal{L}_{\text{LFAI}}$ using L2 loss as follows:

$$\mathcal{L}_{\text{LFAI}} = ||\hat{x}_t^{sam_{p-1}}[-1] - \hat{x}_t^{inv_p}[0]||_2^2. \tag{6}$$

It basically aligns the denoised observations between the sampling process for the previous prompt and the inversion process for the next prompt at each diffusion step t. After all, we update the $\hat{x}_t^{inv_p}$, the denoised observation during the inversion procedure, along the direction that minimizes the $\mathcal{L}_{\text{LFAI}}$ and δ_{LFAI} controls the guidance strength.

Consequently, through this procedure, we maintain the flexibility allowed by the *dynamic noise* and regularize the overall visual content by employing the denoised observation $\hat{x}_t$. We present the procedure of *last frame-aware latent initialization* in Algorithm 1 to facilitate understanding, and the overall procedure are illustrated in Fig. 3.

3.4 Structure-Guided Sampling

The video clip is generated for the next prompt using the initial latent x_T^p produced at the previous step. Although the video clip for the new prompt should preserve the appearance of the previous video clip by using the last frame-aware initial latent, undesirable variation in scene texture and object placement often occurs due to the stochastic nature of the sampling process. To improve the visual consistency within a video clip, we progressively update the predicted original $\hat{x}_t^{sam_p}$ of the current video clip in the sampling process. Specifically, we formulate the objective as follows:

$$\mathcal{L}_{\text{SGS}} = ||\hat{x}_t^{sam_p}[1:n] - \hat{x}_t^{sam_p}[:n-1]||_2^2, \tag{7}$$

where $n \in \{1, ..., N\}$. Note that for the first frame ($n = 0$), we compute $\mathcal{L}_{SGS}$ using the denoised observation of the last frame from the previous prompt $\{\hat{x}_t^{sam_{p-1}}[-1]\}_{t=0}^{T-1}$. Finally, we update $\hat{x}_t^{sam_p}$ as follows:

$$\hat{x}_t^{sam_p} \leftarrow \hat{x}_t^{sam_p} - \delta_{\text{SGS}} \nabla_{\hat{x}_t} \mathcal{L}_{\text{SGS}}, \tag{8}$$

where δ_{SGS} is responsible for guidance scale. Equation 7 and Eq. 8 are iteratively conducted frame-by-frame at each diffusion step. Guidance on the denoised observation leads to a similar global geometric structure between frames within a single video clip.

3.5 Prompt Generator

In real-world scenarios, multiple sequential events can be described in one sentence or paragraph. For instance, *"The dog runs across the wide field, then comes to a halt, yawns softly, and lies down."*. However, existing long-video generation models [10,15,21] are designed to generate only a single event. Therefore, the generated video does not reflect the entire text when the prompt contains multiple events. To address this issue, the Large Language Model (LLM) has been utilized to generate an appropriate input for the pre-trained T2V models. We introduce a prompt generator to segment the comprehensive description into the prescribed textual format. We put the exemplar and guidelines in the supplementary materials.

4 Experiments

4.1 Implementation Details

We directly leverage the released pre-trained text-to-video generation model [15] to generate a multi-text conditioned video. In our experiments, we generate multi-text conditioned video, and each video clip consists of 16 frames with

Fig. 4. Generation results on given prompts by our method and baseline models. T2V-Zero and DirecT2V build upon the T2I pre-trained model. In contrast, VidRD and Gen-L-Video leverage the same foundation model utilized in our experiments.

Table 1. Compared with baseline methods in terms of two primary categories: automatic metric and human evaluation. Note that we use **bold** to highlight the best scores, and underline indicates the second-best scores.

Method	Automatic Metric		Human Evaluation			
	CLIP-Text ↑	CLIP-Image ↑	Temporal ↑	Semantic ↑	Realism ↑	Preference ↑
T2V-Zero [24]	**0.322**	0.808	<u>3.61</u>	<u>3.59</u>	3.45	<u>3.47</u>
DirecT2V [20]	0.301	0.898	2.96	3.04	3.01	3.30
Gen-L-Video [41]	0.308	<u>0.953</u>	3.35	3.38	3.37	3.05
VidRD [12]	0.287	0.951	3.40	3.43	<u>3.56</u>	3.14
Ours	<u>0.309</u>	**0.957**	**3.82**	**3.71**	**3.68**	**3.68**

256×256 resolution. Moreover, we employ ChatGPT [28], which is a Large Language Model, to separate the complex scenarios into individual prompts that comprise the sequence of events. We set each guidance weight δ_{LFAI} and δ_{SGS} to 1000 and 7 in our experiments. All experiments are performed on a single NVIDIA GeForce RTX 3090.

4.2 Qualitative Results

We provide qualitative comparisons along with other recent multi-prompts video generation methods [12,41], including zero-shot video generation methods [20,24] which leveraging frame-level descriptions. In the supplementary material and the project page, we provide additional videos for frame-level and video-level comparison to show more qualitative examples. As shown in Fig. 4, our proposed method achieves better video quality, especially in two points: naturalness and temporal coherence. First, compared with T2V-Zero [24] and DirecT2V [20], we observe that only leveraging the image-based model fails to generate reasonable video flow in terms of naturalness; thus, utilizing the video-based approaches is necessary. Second, generated videos by our method show a strong visual relation between each video segment without the recurrent video pattern. To be more specific, visual examples of VidRD [12] exhibit recurrent video patterns between two distinct video clips; e.g., the movement pattern of a man within each video segment mirrors the previous one. Additionally, although Gen-L-Video [41] generates more diverse movement, they can not preserve the structure coherence of objects, and the background is not stable across the entire frame. On the other hand, our method not only smoothly bridges the gap between two individual video clips but also maintains the overall temporal coherence of the content.

4.3 Quantitative Results

Automatic Metrics. We report the CLIP-Text score [16,30] that represents the alignment between given prompts and outputs, and CLIP-Image score [4,9,29,46] that shows the similarity between two consecutive frames. We measure the metrics over the 30 scenarios, each consisting of multiple prompts. For a fair evaluation, we randomly sampled 20 videos per scenario. As shown in Table 1, our method generally outperforms the other state-of-the-art methods. Among the baseline, the generated video of Text2Video-Zero (T2V-Zero) aligns well with the semantics of given prompts as this approach yields a single frame corresponding to a prompt of the same video segments. However, they fail to generate temporally coherent video. In contrast, DirecT2V [20], which utilizes the frame-specific descriptions sharing high-level stories, shows a higher CLIP-Image score, whereas we observe a decrease in performance for the CLIP-Text score. Comparison with text-to-video-based approaches exhibits a relatively low difference in all metrics due to the common foundation model [15]. The CLIP-Text score of VidRD [12] is notably lower than the baselines. The visual content is substantially maintained without a significant performance drop in CLIP-Image score since the VidRD directly utilizes the latent code of the previous video clip with minimal deviations from the sampling step. Gen-L-Video [41] performs well in capturing the meaning of the prompts. However, the global content variations during the sampling process caused by overlapping prompts lead to a decrease in similarity between consecutive frames.

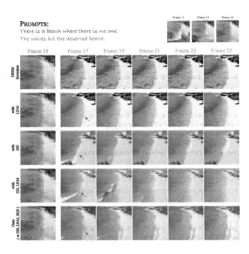

Fig. 5. Generated video clips *with* and *without* proposed our modules. The red arrow and yellow box highlight the visual changes between distinct video clips.

Human Evaluation. We recruited 100 participants through Amazon Mechanical Turk (AMT) to evaluate five models: T2V-Zero [24], DirecT2V [20], Gen-L-Video [41], VidRD [12], and our method. We employ a Likert scale ranging from 1 (low quality) to 5 (high quality). Participants score each method considering temporal consistency, semantic alignment, realism, and preference over 30 videos generated by different scenarios. As clearly indicated in Table 1, generated videos from our method significantly outperform other state-of-the-art approaches in all four criteria, regardless of individual frame quality. In particular, based on human evaluation results, we observe that preserving the identity of the object and background is crucial for human preference. When compared with text-to-video-based methods, temporal inconsistency between each video clip caused by the semantic transition of given prompts results in lower human evaluation scores in spite of the same foundation model as ours.

4.4 Ablation Studies

Effectiveness of Proposed Methods. We qualitatively show the effectiveness of *last frame-aware inversion* (LFAI), *dynamic noise* (DN), and *structure-guided sampling* (SGS), as shown in Fig. 5. We utilize the basic inversion strategy [36] as a base model. Although basic DDIM inversion somewhat preserves the overall structure of visual content between each video clip, the detailed texture and background show severe changes. DN model relaxes this problem but shows the disconnection between two video clips since the DN module gives flexibility to the model by using the i.i.d noise in the inversion procedure. Combining two modules, LFAI and DN, generates natural-looking videos since LFAI module preserves the structure of the content in the previous frame. However, we find that the stochastic characteristics of the sampling process introduce slight fluctuations in the videos, and iterative update of latent code (SGS) during the sampling process is beneficial in enhancing the realism of video content, as shown in the fifth row at Fig. 5. Moreover, as reflected in Tab. 2, we provide an additional human evaluation to validate the impact of our proposed module by human judges.

Analysis on Dynamic Noise. Figure 6 indicates that κ controls the flexibility of the frame sequence. We modify the noise scheduling function $\mathcal{F}$ into the static value to validate the effectiveness of our method. When κ applies to the entire frames as a smaller value, we figure out that the latter frames can not preserve the geometric structure. However, frozen

Table 2. We report user study results on ablation studies using four different criteria: Temporal, Semantics, Realism, and Preference. Note that we use **bold** to highlight the best scores, and underline indicates the second-best scores.

Method			Human Evaluation			
LFAI	DN	SGS	Temporal ↑	Semantics ↑	Realism ↑	Preference ↑
-	-	-	3.42	3.40	3.45	2.89
✓	-	-	3.48	3.40	3.50	<u>2.93</u>
-	✓	-	3.53	3.46	<u>3.63</u>	2.51
✓	✓	-	<u>3.61</u>	**3.58**	3.58	2.78
✓	✓	✓	**3.70**	<u>3.47</u>	**3.69**	**3.27**

video is observed when *kappa* is set to a high value. As a result, we achieve the smooth transition and flexibility between consecutive video clips by adopting the scheduling function $\mathcal{F}$, which decrease steadily (Table 2).

Analysis on Last Frame-aware Inversion. We introduce the *last frame-aware inversion* to prevent the visual inconsistency in terms of object spatial location and scene texture driven by the Dynamic Noise. In the LFAI process, we guide the first frame to adjust the initial latent code that is correlated to the geometric structure of the previous video clip. Here, we explore the influence of the number of frames that offer guidance by the last frame of the previous video clip. As shown in Fig. 7, we observe that only the first frame is sufficient to maintain the visual structure at the beginning of the frames. On the contrary, the increase in the number of affected frames makes the stationary movement in

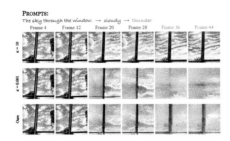
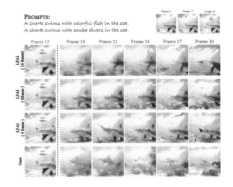

Fig. 6. Ablation study for validating the noise schedule. Note that the first and second-row leverage constant value over the entire frames.

Fig. 7. Effectiveness of adjusting the number of influenced frames.

objects; e.g., the shark only moves on the right side. Conversely, the restriction of the affected frame as only a single one gives increased flexibility to the subsequent frames. This flexibility enhances their ability to effectively convey the meaning of the subsequent prompts and generate a diverse range of movement.

4.5 Applications

Video Generation with Large Language Model (LLM). In real-world scenarios, more intricate descriptions are generally used, which have the time-variant events in a single narrative. *Prompt generator* (see Sect. 3.5) to separate into the individual prompts for handling the consecutive events. As shown in Fig. 8 (left), the visual examples indicate the entire frame ensures temporal consistency while reflecting the overall storyline.

Image and Multi-event-based Video Generation. Our proposed MEVG is capable of generating video with a given image and multi-text, multi-text-image-to-video generation (MTI2V). For generating video, we first encode the seeding image using the encoder into the latent vector and duplicate it as the number of frames. Then, we follow MEVG pipeline. Figure 8 (right) demonstrates that the generated video successfully preserves the visual appearance and structure of the object in the reference image and shows temporal coherence along the given prompts.

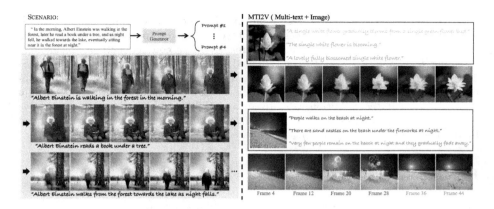

Fig. 8. (Left) Example that leverages the Large Language Model (LLM). Given the complex scenario, our prompt generator split into each individual prompt using the predefined instructions. **(Right)** Example of our results conditioned on multiple prompts and given image.

5 Conclusion

We introduced a novel method that generates multi-text-based videos by taking temporally consecutive descriptions. Specifically, we propose two techniques, *last frame-aware latent initialization* and *structure-guided sampling*, to preserve the visual and temporal consistency in the generated video. Our proposed method can generate much more natural and temporally coherent videos than the other state-of-the-art methods with qualitative and quantitative results. Our pipeline also handles a single story containing time-variant events by utilizing the Large Language Model (LLM). In addition, our proposed method can generate videos conditioned on both the multi-prompts and a reference image and can be used in various applications.

Limitation and Future Work. Although our proposed method yields promising outcomes in preserving visual consistency and generation diversity over the distinct prompts, there exists potential for future works as listed: 1) Our method requires a certain text format as our model inherits the characteristics of a pretrained single-prompt video generator and 2) The absence of benchmark datasets for multi-text video generation makes it hard to conduct a quantitative evaluation. Video generation with diverse input conditions and curating multi-prompts video datasets are promising future directions.

Acknowledgements. This research was supported by Culture, Sports and Tourism R&D Program through the Korea Creative Content Agency grant funded by the Ministry of Culture, Sports and Tourism in 2024((International Collaborative Research and Global Talent Development for the Development of Copyright Management and Protection Technologies for Generative AI, RS-2024-00345025, 25%),(Research on neural watermark technology for copyright protection of generative AI 3D content, RS-2024-00348469, 25%),(Development of sketch-based semantic 3D modeling technology for creating user-centric Metaverse content spaces for indoor spaces, RS-2023-00227409, 10%))), Institute of Information & communications Technology Planning & Evaluation (IITP) grant funded by the Korea government(MSIT)(RS-2019-II190079, 10%), Artificial intelligence industrial convergence cluster development project funded by the Ministry of Science and ICT(MSIT, Korea)& Gwangju Metropolitan City(contribution rate: 15%), and IITP under the Leading Generative AI Human Resources Development(IITP-2024-RS-2024-00397085, 15%) grant funded by the Korea government(MSIT).

References

1. Bar-Tal, O., et al.: Lumiere: a space-time diffusion model for video generation. arXiv preprint arXiv:2401.12945 (2024)
2. Blattmann, A., et al.: Align your Latents: high-resolution video synthesis with latent diffusion models. In: Proceedings of the IEEE/CVF Conference on Computer Vision and Pattern Recognition, pp. 22563–22575 (2023)
3. Brooks, T., et al.: Generating long videos of dynamic scenes. Adv. Neural. Inf. Process. Syst. **35**, 31769–31781 (2022)
4. Ceylan, D., Huang, C.H.P., Mitra, N.J.: Pix2video: video editing using image diffusion. In: Proceedings of the IEEE/CVF International Conference on Computer Vision (ICCV), pp. 23206–23217 (2023)
5. Chen, H., et al.: Videocrafter2: overcoming data limitations for high-quality video diffusion models. In: Proceedings of the IEEE/CVF Conference on Computer Vision and Pattern Recognition, pp. 7310–7320 (2024)
6. Chen, X., et al.: Seine: short-to-long video diffusion model for generative transition and prediction. In: The Twelfth International Conference on Learning Representations (2023)
7. Ding, M., Zheng, W., Hong, W., Tang, J.: CogView2: faster and better text-to-image generation via hierarchical transformers. Adv. Neural. Inf. Process. Syst. **35**, 16890–16902 (2022)
8. Dong, W., Xue, S., Duan, X., Han, S.: Prompt tuning inversion for text-driven image editing using diffusion models. arXiv preprint arXiv:2305.04441 (2023)
9. Esser, P., Chiu, J., Atighehchian, P., Granskog, J., Germanidis, A.: Structure and content-guided video synthesis with diffusion models. In: Proceedings of the IEEE/CVF International Conference on Computer Vision, pp. 7346–7356 (2023)
10. Ge, S., et al.: Long video generation with time-agnostic VQGAN and time-sensitive transformer. In: European Conference on Computer Vision, pp. 102–118. Springer (2022). https://doi.org/10.1007/978-3-031-19790-1_7
11. Ge, S., et al.: Preserve your own correlation: a noise prior for video diffusion models. In: Proceedings of the IEEE/CVF International Conference on Computer Vision, pp. 22930–22941 (2023)
12. Gu, J., et al.: Reuse and Diffuse: iterative denoising for text-to-video generation. arXiv preprint arXiv:2309.03549 (2023)
13. Gu, S., et al.: Vector quantized diffusion model for text-to-image synthesis. In: Proceedings of the IEEE/CVF Conference on Computer Vision and Pattern Recognition, pp. 10696–10706 (2022)
14. He, Y., et al.: Animate-a-story: storytelling with retrieval-augmented video generation. arXiv preprint arXiv:2307.06940 (2023)

15. He, Y., Yang, T., Zhang, Y., Shan, Y., Chen, Q.: Latent video diffusion models for high-fidelity video generation with arbitrary lengths. arXiv preprint arXiv:2211.13221 (2022)
16. Hessel, J., Holtzman, A., Forbes, M., Bras, R.L., Choi, Y.: CLIPScore: a reference-free evaluation metric for image captioning. arXiv preprint arXiv:2104.08718 (2021)
17. Ho, J., et al.: Imagen video: high definition video generation with diffusion models. arXiv preprint arXiv:2210.02303 (2022)
18. Ho, J., Jain, A., Abbeel, P.: Denoising diffusion probabilistic models. Adv. Neural. Inf. Process. Syst. **33**, 6840–6851 (2020)
19. Ho, J., Salimans, T., Gritsenko, A., Chan, W., Norouzi, M., Fleet, D.J.: Video diffusion models (2022)
20. Hong, S., Seo, J., Hong, S., Shin, H., Kim, S.: Large language models are frame-level directors for zero-shot text-to-video generation. arXiv preprint arXiv:2305.14330 (2023)
21. Hong, W., Ding, M., Zheng, W., Liu, X., Tang, J.: CogVideo: large-scale pretraining for text-to-video generation via transformers. arXiv preprint arXiv:2205.15868 (2022)
22. Huang, H., Feng, Y., SibeiYang, C.L.J.: Free-Bloom: zero-shot text-to-video generator with LLM director and LDM animator. arXiv preprint arXiv:2309.14494 **3** (2023)
23. Kang, M., et al.: Scaling up GANs for text-to-image synthesis. In: Proceedings of the IEEE/CVF Conference on Computer Vision and Pattern Recognition, pp. 10124–10134 (2023)
24. Khachatryan, L., et al.: Text2Video-Zero: text-to-image diffusion models are zero-shot video generators. arXiv preprint arXiv:2303.13439 (2023)
25. Khandelwal, A.: InFusion: inject and attention fusion for multi concept zero-shot text-based video editing. In: Proceedings of the IEEE/CVF International Conference on Computer Vision, pp. 3017–3026 (2023)
26. Liang, J., et al.: NUWA-infinity: autoregressive over autoregressive generation for infinite visual synthesis. Adv. Neural. Inf. Process. Syst. **35**, 15420–15432 (2022)
27. Nguyen, T., Li, Y., Ojha, U., Lee, Y.J.: Visual instruction inversion: image editing via visual prompting. arXiv preprint arXiv:2307.14331 (2023)
28. Ouyang, L., et al.: Training language models to follow instructions with human feedback. Adv. Neural. Inf. Process. Syst. **35**, 27730–27744 (2022)
29. Qi, C., et al.: FateZero: fusing attentions for zero-shot text-based video editing. arXiv:2303.09535 (2023)
30. Radford, A., et al.: Learning transferable visual models from natural language supervision. In: International Conference on Machine Learning, pp. 8748–8763. PMLR (2021)
31. Ramesh, A., Dhariwal, P., Nichol, A., Chu, C., Chen, M.: Hierarchical text-conditional image generation with clip latents. arXiv preprint arXiv:2204.06125 (2022)
32. Rombach, R., Blattmann, A., Lorenz, D., Esser, P., Ommer, B.: High-resolution image synthesis with latent diffusion models. In: Proceedings of the IEEE/CVF Conference on Computer Vision and Pattern Recognition, pp. 10684–10695 (2022)
33. Saharia, C., et al.: Photorealistic text-to-image diffusion models with deep language understanding. Adv. Neural. Inf. Process. Syst. **35**, 36479–36494 (2022)
34. Singer, U., et al.: Make-a-Video: text-to-video generation without text-video data. arXiv preprint arXiv:2209.14792 (2022)

35. Skorokhodov, I., Tulyakov, S., Elhoseiny, M.: StyleGAN-v: a continuous video generator with the price, image quality and perks of styleGAN2. In: Proceedings of the IEEE/CVF Conference on Computer Vision and Pattern Recognition, pp. 3626–3636 (2022)
36. Song, J., Meng, C., Ermon, S.: Denoising diffusion implicit models. arXiv preprint arXiv:2010.02502 (2020)
37. Tian, Y., et al.: A good image generator is what you need for high-resolution video synthesis. arXiv preprint arXiv:2104.15069 (2021)
38. Villegas, R., et al.: Phenaki: variable length video generation from open domain textual description. arXiv preprint arXiv:2210.02399 (2022)
39. Voleti, V., Jolicoeur-Martineau, A., Pal, C.: MCVD-masked conditional video diffusion for prediction, generation, and interpolation. Adv. Neural. Inf. Process. Syst. **35**, 23371–23385 (2022)
40. Vondrick, C., Pirsiavash, H., Torralba, A.: Generating videos with scene dynamics. In: Advances in Neural Information Processing Systems, vol. 29 (2016)
41. Wang, F.Y., Chen, W., Song, G., Ye, H.J., Liu, Y., Li, H.: Gen-L-Video: multi-text to long video generation via temporal co-denoising. arXiv preprint arXiv:2305.18264 (2023)
42. Wang, Y., Bilinski, P., Bremond, F., Dantcheva, A.: G3AN: disentangling appearance and motion for video generation. In: Proceedings of the IEEE/CVF Conference on Computer Vision and Pattern Recognition, pp. 5264–5273 (2020)
43. Weissenborn, D., Täckström, O., Uszkoreit, J.: Scaling autoregressive video models. arXiv preprint arXiv:1906.02634 (2019)
44. Wu, C., et al.: GODIVA: generating open-domain videos from natural descriptions. arXiv preprint arXiv:2104.14806 (2021)
45. Wu, C., et al.: NÜWA: visual synthesis pre-training for neural visual world creation. In: Avidan, S., Brostow, G., Cissé, M., Farinella, G.M., Hassner, T. (eds.) Computer Vision – ECCV 2022: 17th European Conference, Tel Aviv, Israel, October 23–27, 2022, Proceedings, Part XVI, pp. 720–736. Springer Nature Switzerland, Cham (2022). https://doi.org/10.1007/978-3-031-19787-1_41
46. Wu, J.Z., et al.: Tune-a-Video: one-shot tuning of image diffusion models for text-to-video generation. In: Proceedings of the IEEE/CVF International Conference on Computer Vision, pp. 7623–7633 (2023)
47. Xing, X., et al.: Inversion-by-Inversion: exemplar-based sketch-to-photo synthesis via stochastic differential equations without training. arXiv preprint arXiv:2308.07665 (2023)
48. Yan, W., Zhang, Y., Abbeel, P., Srinivas, A.: VideoGPT: video generation using VQ-VAE and transformers. arXiv preprint arXiv:2104.10157 (2021)
49. Yu, S., et al.: Generating videos with dynamics-aware implicit generative adversarial networks. arXiv preprint arXiv:2202.10571 (2022)
50. Zhang, D.J., et al.: Show-1: marrying pixel and latent diffusion models for text-to-video generation. arXiv preprint arXiv:2309.15818 (2023)
51. Zhou, D., Wang, W., Yan, H., Lv, W., Zhu, Y., Feng, J.: MagicVideo: efficient video generation with latent diffusion models. arXiv preprint arXiv:2211.11018 (2022)

Open-Vocabulary SAM: Segment and Recognize Twenty-Thousand Classes Interactively

Haobo Yuan[1]($\boxtimes$), Xiangtai Li[1], Chong Zhou[1], Yining Li[2], Kai Chen[2], and Chen Change Loy[1]

[1] S-Lab, Nanyang Technological University, Singapore, Singapore
yuanhaobo@whu.edu.cn, ccloy@ntu.edu.sg
[2] Shanghai AI Laboratory, Shanghai, China
https://www.mmlab-ntu.com/project/ovsam

Abstract. The CLIP and Segment Anything Model (SAM) are remarkable vision foundation models (VFMs). SAM excels in segmentation tasks across diverse domains, whereas CLIP is renowned for its zero-shot recognition capabilities. This paper presents an in-depth exploration of integrating these two models into a unified framework. Specifically, we introduce the Open-Vocabulary SAM, a SAM-inspired model designed for simultaneous interactive segmentation and recognition, leveraging two unique knowledge transfer modules: SAM2CLIP and CLIP2SAM. The former adapts SAM's knowledge into the CLIP via distillation and learnable transformer adapters, while the latter transfers CLIP knowledge into SAM, enhancing its recognition capabilities. Extensive experiments on various datasets and detectors show the effectiveness of Open-Vocabulary SAM in both segmentation and recognition tasks, significantly outperforming the naïve baselines of simply combining SAM and CLIP. Furthermore, aided with image classification data training, our method can segment and recognize approximately 22,000 classes.

Keywords: Scene Understanding · Promptable Segmentation

1 Introduction

The Segment Anything Model (SAM) [30] and CLIP [53] have made significant strides in various vision tasks, showcasing remarkable generalization capabilities in *segmentation* and *recognition*, respectively. SAM, in particular, has been trained with a massive dataset of mask labels, making it highly adaptable to a wide range of downstream tasks through interactive prompts. On the other hand, CLIP's training with billions of text-image pairs has given it an unprecedented ability in zero-shot visual recognition. This has led to numerous studies [18, 64, 72, 78] exploring the extension of CLIP to open vocabulary tasks, such as detection and segmentation.

Supplementary Information The online version contains supplementary material available at https://doi.org/10.1007/978-3-031-72775-7_24.

© The Author(s), under exclusive license to Springer Nature Switzerland AG 2025
A. Leonardis et al. (Eds.): ECCV 2024, LNCS 15101, pp. 419–437, 2025.
https://doi.org/10.1007/978-3-031-72775-7_24

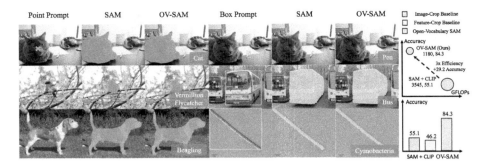

Fig. 1. Open-Vocabulary SAM not only can segment anything with prompts just like SAM but also has the capability of recognition in the real world, like CLIP. With drastically lower computational cost, Open-Vocabulary SAM has a higher recognition performance than directly combining SAM and CLIP with image or feature cropping (measured on the COCO open vocabulary benchmark).

While SAM and CLIP offer considerable advantages, they also have inherent limitations in their original designs. SAM, for instance, lacks the capability to recognize the segments it identifies. Efforts to overcome this by integrating a classification head have been made [32, 83], but these solutions are constrained to specific datasets or closed-set settings. On the other hand, CLIP, which is trained using image-level contrastive losses, faces challenges in adapting its representations for dense prediction tasks. To address this, several studies [18, 57, 66, 67, 76] have investigated ways to align CLIP's representation for dense predictions. However, these approaches tend to be dataset-specific and not universally applicable. For example, some research has focused on open vocabulary segmentation on the ADE-20k [84] dataset, using the COCO [43] dataset for pre-training. Merging SAM and CLIP in a naïve manner, as illustrated in Fig. 2 (a) and (b), proves to be inefficient. This approach incurs substantial computational expenses and yields subpar results, including recognition of small-scale objects, as evidenced by our experimental results.

In this study, we address these challenges with a unified encoder-decoder framework that integrates a CLIP encoder and a SAM decoder, as depicted in Fig. 2 (c). To bridge these two distinct components effectively, we introduce two novel modules, *SAM2CLIP* and *CLIP2SAM*, facilitating dual knowledge transfer. First, we distill knowledge from the SAM encoder to the CLIP encoder using *SAM2CLIP*. This distillation process is uniquely executed not directly on the CLIP encoder, which is kept frozen to maintain its existing knowledge, but rather on a lightweight transformer-like adapter using a pixel-wise distillation loss. The adapter takes multi-scale features as input, with the goal of aligning CLIP features with SAM representation. On the decoding side, the *CLIP2SAM* module transfers knowledge from the frozen CLIP encoder to the SAM decoder. In particular, we design a feature pyramid adapter with a RoIAlign operator to be jointly trained with the SAM decoder. Both modules are lightweight and naturally combine the strengths of CLIP and SAM.

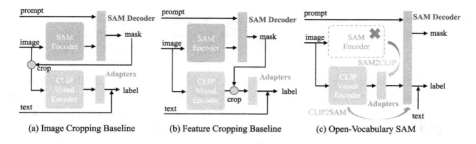

Fig. 2. Comparison of two simple SAM-CLIP combination baselines (a) and (b), and our proposed single encoder architecture (c). The adapters for (a) and (b) are optional and can be replaced with various designs (please refer to Sect. 4.1 for details). Note that, in our method, the SAM encoder will be discarded during inference.

Following the spirit of SAM, we enhance our model's recognition capabilities by harnessing the power of established semantic datasets, including COCO [43], LVIS [19], and ImageNet-22k [11]. This strategy elevates our model to the versatility of SAM, endowing it with enhanced capability to segment and recognize any objects, as shown in Fig. 1. As our approach is an adaptation of SAM, it is flexible enough to be integrated with various detectors, making it suitable for both closed-set and open-set environments.

We conduct extensive experiments across a range of datasets and scenarios, encompassing closed-set and open-vocabulary interactive segmentation. Notably, when compared to basic combined baselines, our approach demonstrates superior performance, achieving over 2% improvement in IoU and 3% in mAP with various detectors on the COCO dataset. In particular, in the case of recognition on LVIS, our approach achieves over 20% improvements over previous adapters. Furthermore, by expanding our approach with a more diverse array of datasets, we have developed a versatile, interactive tool suitable for practical applications. For detailed results, we direct the reader to Sect. 4 and the appendix.

2 Related Work

Vision Language Models (VLMs). Vision-language pre-training has given rise to models with aligned image and text representations [25,26,28,41,53]. Recent studies on contrastive vision-language pre-training [26,53,58,80] have significantly improved the generalization ability of recognition models. Meanwhile, several works [28,33–35] aim to design better optimization goals for downstream multi-modal tasks, including caption and visual question answering. Among these works, CLIP models [53] that are pre-trained on billion-scale image-text pairs have shown impressive zero-shot classification performance on a wide range of datasets. Our goal is to enable SAM to perform recognition tasks with the help of pre-trained VLMs.

Open Vocabulary Dense Prediction. This direction aims to recognize region visual concepts of arbitrary categories described by texts, which includes object detection [18,63,64,69,79], semantic segmentation [37,38,40,71,86,87], and panoptic segmentation [70,75,76]. This necessitates the alignment between region and text representations with the help of VLMs [26,53,58]. For open-vocabulary detection, a series of works [18,57,66,78] distill knowledge from the CLIP models to recognize novel objects. In contrast to distillation-based methods, several works [31,68] directly build object detectors upon frozen CLIP CNNs. For open-vocabulary segmentation, the typical works [12,70,72,76] first generate class-agnostic mask proposals and then classify the proposals with CLIP. Recently, several works [70,76] build the mask generator upon the frozen diffusion model [54] and CLIP model [53]. Meanwhile, several studies [27,47,51,52] focus on class-agnostic segmentation and detection to enrich generalization ability in various domains. However, most approaches are trained and tested on specific datasets. Our approach is based on SAM, which provides a general, interactive tool to support different open vocabulary detectors.

Prompting in Computer Vision. Prompting, originating from in-context learning in natural language processing (NLP) as seen in works like Brown et al. [4] and Rubin et al. [55], leverages a large language model to infer unseen tasks through context-specific input-output pairs. Recent studies [1,3,16,45,61,62,88] have explored in-context learning for visual tasks. Common techniques involve mask image modeling [2,21,77] for cross-task visual prompting, as employed by approaches like Painter [61] and Bar et al. [3]. SAM [30] demonstrates in-context learning through interactive segmentation, using diverse visual prompts like points, boxes, and masks, although it is limited to class-agnostic mask prediction. Meanwhile, other studies [8,17,23,39,42] have concentrated on efficient parameter tuning of visual foundation models, typically focusing on a single model. Our work uniquely bridges two models, CLIP and SAM, exploring their combined potential for enhanced general segmentation and recognition capabilities. In particular, we adopt visual prompts (box, point) as the model's inputs.

Segmentation Anything Model. SAM [30] presents a new data engine and portable model for segmentation. Subsequent research has employed SAM as an interactive segmentation tool for various vision tasks, including grounding [44], tracking [10], distillation [81,85], medical analysis [5,65], and generation [82]. While some studies use SAM and CLIP for segmentation [6,20,36,49,59], none have yet integrated VLMs and SAM into a unified model capable of both segmentation and recognition of novel classes. Our work makes the first attempt to merge the capabilities of VLMs with SAM for enhanced task versatility.

3 Methodology

We first review the SAM, CLIP, and combined baselines in Sect. 3.1. Then, we detail our Open Vocabulary SAM in Sect. 3.2. Last, we present our model's training details and application in Sect. 3.3.

3.1 Preliminaries and Baselines

SAM. SAM is a prompt-driven segmentor. It contains an image encoder, a prompt encoder, and a light-weight mask decoder. Here, we use box prompts as an example. We denote an input image as $X \in \mathbb{R}^{H \times W \times 3}$ and input visual prompts as $P \in \mathbb{R}^{N \times 4}$, where $H \times W$ are the spatial size, N is the number of box prompts. The image encoder is a modified vision transformer (ViT). It encodes an image into dense feature $F_{SAM} \in \mathbb{R}^{\frac{H}{16} \times \frac{W}{16} \times d}$. The prompt encoder encodes P into sparse prompts Q_{sp}. Meanwhile, mask tokens Q_{mask} and an IoU token Q_{IoU} are initialized for the mask decoder.

The mask decoder takes the image feature F, sparse prompts Q_{sp}, mask tokens Q_{mask}, and the IoU token Q_{IoU} as input. All the inputs will be concatenated and encoded with a lightweight two-way transformer. Consequently, each mask token is transformed into a dynamic linear classifier, capable of calculating the foreground mask probability for every sparse prompt. Simultaneously, the IoU token is tasked with predicting the confidence score for each mask. Considering the multi-granular nature of SAM's data annotations, encompassing both instance and part level, Q_{mask} naturally encodes multi-granularity. Our study concentrates exclusively on the object level, which aligns more closely with prevalent real-world applications and datasets such as COCO [43] and LVIS [19].

CLIP. Given an input image X and a corresponding caption C, the CLIP framework processes these modalities to produce respective embeddings: the image embedding E_I, derived from its image encoder, and the text embedding $\mathbf{t}$, obtained from its text encoder. In the context of open-vocabulary object detection and segmentation, CLIP's capability to generalize beyond fixed class labels is leveraged to replace traditional classifiers. For instance, in open-vocabulary detection scenarios, the text embedding $\mathbf{t_c}$ for the c-th object category is generated by inputting the category name into the CLIP text encoder. This process can employ a single template prompt, such as "a photo of {category}," or multiple prompt templates. Subsequently, for a given region embedding r, that is produced by the RoI-Align [22], the classification score for the c-th category is

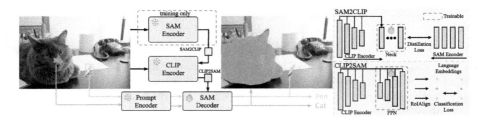

Fig. 3. Illustration of Open-Vocabulary SAM. For training, the SAM encoder is as a teacher network, while SAM2CLIP plays the role of a student network and aligns the knowledge of SAM into CLIP. The CLIP2SAM transfers the CLIP knowledge to the SAM decoder and performs joint segmentation and classification for close-set and open vocabulary settings.

computed as: $p_c = \frac{\exp(\tau \cdot <\mathbf{r},\mathbf{t_c}>)}{\sum_{i=0}^{C} \exp(\tau \cdot <\mathbf{r},\mathbf{t_i}>)}$, where $<\cdot,\cdot>$ denotes the cosine similarity, and τ is a learnable or fixed temperature to re-scale the value.

Combined Baselines. We introduce two different baselines for combining CLIP and SAM, as depicted in Fig. 2 (a) and (b). The first approach, termed the 'cropped image baseline', employs the SAM mask decoder's output to segment and resize the original input image. This processed image then serves as the input for the CLIP image encoder, and, in conjunction with the CLIP text embedding, the mask is classified using the similarities between visual and text embeddings. The second approach referred to as the 'cropped CLIP image feature baseline', employs the same initial CLIP feature extraction step. However, in this method, masks predicted by the SAM decoder are used to crop the CLIP image features. Subsequent pooling of these masked features yields the final label, akin to baseline (a).

While both baselines enable zero-shot inference of images, they exhibit a noticeable knowledge gap on specific datasets. To address this, we draw inspiration from recent advancements in visual prompting or adapters [8,88]. Specifically, we propose incorporating additional learnable tokens as an adapter to fine-tune the model for enhanced performance on downstream datasets. These zero-shot inference capabilities and the fine-tuned models constitute our primary comparison baselines under various experimental conditions, detailed in Sect. 4.1.

3.2 Open Vocabulary SAM

While both baseline models can be enhanced through visual prompting or adapters, as we will discuss in Sect. 4, they face several challenges in real-world applications. First, the requirement for two independent backbones in the combined model increases computational costs (*Prob*.1). Second, SAM and CLIP are trained with distinct objectives - SAM through supervised learning and CLIP via contrastive learning - and there is limited research on knowledge transfer between such diverse architectures (*Prob*.2). Third, despite adapter integration, significant performance gaps remain in recognizing small objects (*Prob*.3). Fourth, there is a lack of exploration into integrating open-vocabulary capabilities for SAM and CLIP, particularly in the context of feature fusion and data scaling (*Prob*.4). Our work aims to solve these problems in a unified yet effective framework.

Unified Architecture. We design a unified architecture for both segmentation and recognition to address *Prob*.1. Specifically, we adopt the frozen CLIP visual encoder as our feature extractor. Then, both SAM's mask decoder and prompt encoder are appended behind the CLIP encoder. The meta-architecture of open-vocabulary SAM is shown in Fig. 2 (c), with the more detailed version shown in Fig. 3. This unified architecture is made possible via the **SAM2CLIP**, which transfers knowledge of SAM to CLIP with distillation, and **CLIP2SAM**, which employs CLIP knowledge and combines the SAM mask decoder for recognition. We have chosen convolution-based visual backbones for the frozen CLIP backbone, aligning with previous studies that have highlighted their superiority in capturing spatial structures [31,73]. The efficacy of different CLIP backbones is further explored in Sect. 4.2.

SAM2CLIP. To resolve *Prob*.2, we design the SAM2CLIP module that bridges the gap in feature representations learned by SAM and CLIP, using adaptation and distillation methods. Through comprehensive experiments, we discovered that employing distillation loss $L_{distill}$ along with transformer-based adapters [14], yields effective results. Specifically, the distillation process involves a simple pixel-wise approach, where SAM-Huge serves as the teacher, and the frozen CLIP equipped with an adapter assumes the student's role. We then implement a per-pixel mean squared error (MSE) loss to align the SAM feature F_{sam} with the CLIP feature E_I, as detailed below:

$$L_{distill} = \text{MSE}(F_{sam}, E_I). \tag{1}$$

We design a multi-scale adapter $A_{sam2clip}$ to align the features from CLIP and SAM. In particular, we take pyramid CLIP features $E_I^i, i = 1, 2, 3$ as the inputs. Such pyramid features contain both high-resolution and semantic information, which is proven crucial for semantic segmentation [29]. The MSE loss is revised as follows:

$$L_{distill} = \text{MSE}(F_{sam}, A_{sam2clip}(\text{Fusion}(E_I^i))), \tag{2}$$

where $A_{sam2clip}$ comprises several transformer layers, and Fusion is achieved by bilinear upsampling and addition.

With SAM2CLIP, we can even achieve comparable segmentation results with the SAM-Huge with much lower computational costs. As detailed in Sect. 4.2, we observe that employing convolution-based methods specifically designed for backbone adaptation [8,15] results in sub-optimal outcomes. The reason for this might be in the inherent architecture of the SAM encoder, which is purely based on ViT. A symmetrical structure is crucial for effective knowledge transfer. With the implementation of SAM2CLIP, we can achieve segmentation results comparable to those of SAM-Huge across various detectors, while significantly reducing computational costs.

CLIP2SAM. This module aims to leverage CLIP's knowledge to enhance the recognition capabilities of the SAM decoder. A straightforward approach involves appending a label token Q_{label} to the existing mask token Q_{mask} and IoU token Q_{IoU}. Using Q_{label}, we introduce a specialized adapter to facilitate the transfer of knowledge from the frozen CLIP to the SAM decoder. Subsequently, the enhanced Q_{label}, combined with the output of the prompt encoder and adapted CLIP features, is fed into a two-way transformer. Following the cross-attention process, the improved Q_{label} undergoes further refinement through a multilayer perceptron (MLP), ensuring better alignment with CLIP's text embedding. The final labels are derived by calculating the distance between the refined label token and the CLIP text embedding.

This design, however, falls short of recognizing small objects (*Prob*.3) since the adaptation only involves the single-scale feature, which is mainly focused on segmentation. We present a simple yet effective solution to handle this issue, introducing a lightweight feature pyramid network (FPN) for CLIP2SAM adaption. As shown in Fig. 3, the pyramid network extracts multi-scale CLIP features as the inputs. Then, we apply the RoI-Align [22] operation to extract region features. Like the R-CNN framework [22], we apply one convolution layer and a MLP to learn the feature embedding without introducing cross-attention in the mask decoder. In particular, for point prompts, we first obtain the corresponding masks via the SAM decoder and obtain the box via the corresponding masks. For box prompts, we can directly send it to the FPN for region feature extraction. Given that our method incorporates only a few convolution layers, it does not significantly increase computational costs compared to the original SAM.

Open Vocabulary. To tackle *Prob*.4, the open-vocabulary challenge, we leverage the knowledge embedded in the frozen CLIP backbone, which aids in recognizing novel and unseen objects during inference. In line with previous studies [31,67], we fuse the learned class scores with those from the frozen CLIP via a geometric mean to leverage information from both the CLIP and CLIP2SAM. Additionally, we investigate various strategies to expand the vocabulary size, such as joint training with multiple datasets, as detailed in Sect. 4.2. Our experimental results show that the model scales effectively with large datasets.

While our approach can address the open vocabulary challenge, it is important to distinguish the setting of Open-Vocabulary SAM from that of previous open vocabulary segmentation methods [6,13,24,74]. Unlike previous techniques, which typically depend on a separate segmentor to produce mask proposals, our method only uses visual prompts, such as boxes and points, to generate masks and labels. Furthermore, our method specifically targets foreground objects, aligning with the ImageNet [11] datasets and the purpose of box prompts. To show the effectiveness of our method, we compare the performance of our method with previous open vocabulary segmentation methods in Sect. 4.1.

3.3 Training and Application

Training and Loss Function. We first use the SA-1B (1%) dataset [30] for training the SAM2CLIP module to transfer SAM's knowledge into open-vocabulary SAM, with the loss $L_{distill}$ (Eq. (2)). Then, we joint train the CLIP2SAM and mask decoder using segmentation mask and label annotations from COCO or LVIS. The final loss function is given as $L = \lambda_{cls}L_{t\_cls} + \lambda_{ce}L_{t\_ce} + \lambda_{dice}L_{t\_dice}$. Here, L_{t_ce} is the Cross-Entropy (CE) loss for mask classification, and L_{t_ce} and L_{t_dice} are mask Cross Entropy (CE) loss and Dice loss [48] for segmentation, respectively. In addition, we adopt joint training with the ImageNet dataset for classifying over 22,000 categories.

Inference and Demo Tools. Our model performs inference like SAM, with points and boxes as visual prompts. Specifically, we test boxes and points as visual prompts for the encoder in Sect. 4. In the appendix, we show a demo of our model, which can segment and recognize with prompts.

4 Experiments

Datasets and Metrics. We mainly use COCO [43] and LVIS [19] datasets for the experiments. Moreover, we also use part of SA-1B [30] data (1%) for SAM2CLIP knowledge transfer. For COCO, we report the results of both close-set and open-vocabulary settings for the instance segmentation task. In particular, following Zareian et al. [79], we split 48 base classes with annotations and 17 target classes without annotations. We use the base class annotations for training. For LVIS datasets, we adopt the open-vocabulary setting and report the results of AP_{rare} for novel classes. For evaluation metrics, we report the accuracy of recognition for reference to evaluate the recognition capability. Meanwhile, each prompt's intersection-over-union (IoU) with its ground truth mask is also adopted to evaluate the segmentation ability of our method. As mentioned in Sect. 3.2, different from previous open vocabulary segmentation tasks, boxes or points serve as visual prompts in our method.

Baselines. As shown in Fig. 2 (a) and (b), based on different adapter designs, we append these adapters to the different locations of the combined models. For example, when using CoOp [88], we append the learnable tokens by combining them with CLIP features. For several convolution-based adapters [8], we add the extra convolution layers along with SAM or CLIP backbone for fair comparison. By default, we adopt SAM-huge and CLIP R50 × 16.

Table 1. Comparison of combined baselines and Open-Vocabulary SAM using visual prompts. "*" indicates using mask center point as prompts, while others indicate using ground truth boxes prompts. IoU_b and IoU_n refer to the average IoU for each mask of base classes and novel classes, respectively.

Method	COCO			LVIS			FLOPs	#Param
	IoU_b	IoU_n	Acc	IoU_b	IoU_n	Acc		
Image-Crop baseline	78.1	81.4	46.2	78.3	81.6	9.6	3,748G	808M
Feature-Crop baseline	78.1	81.4	55.1	78.3	81.6	26.5	3,545G	808M
Image-Crop baseline + CoOp [88]	79.6	82.1	62.0	80.1	82.0	32.1	3,748G	808M
Feature-Crop baseline + CoOp [88]	79.6	82.1	70.9	80.1	82.0	48.2	3,545G	808M
Open-Vocabulary SAM	81.5	84.0	84.3	80.4	83.1	66.6	1,180G	304M
Image-Crop baseline*	60.7	66.7	24.5	53.0	62.3	6.2	3,748G	808M
Feature-Crop baseline*	60.7	66.7	32.1	53.0	62.3	11.0	3,545G	808M
Image-Crop baseline + CoOp [88]*	64.7	66.7	28.2	58.9	64.2	8.3	3,748G	808M
Feature-Crop baseline + CoOp [88]*	64.7	66.7	35.1	58.9	64.2	13.2	3,545G	808M
Open-Vocabulary SAM*	68.4	65.2	76.7	63.6	67.9	60.4	1,180G	304M

Implementation Details. We implement our models in PyTorch [50] with MMDetection [7]. We use 8 A100 GPUs for distributed training. Each mini-batch has two images per GPU. The optimizer is AdamW [46] with a weight

Table 2. Comparison of combined baselines and Open-Vocabulary SAM on prompts generated by the open vocabulary detector. For the LVIS dataset, only 'normal' and 'frequent' classes are in the training set. The labels are generated by each baseline or our method. We adopt Detic [89] as the OV-Detector to provide box prompts.

Method	COCO		LVIS				FLOPs	#Params
	AP_{base}	AP_{novel}	AP	AP_{rare}	AP_{norm}	AP_{freq}	AP		
Image-Crop baseline + CoOp [88]	26.2	31.2	27.3	19.8	18.3	16.3	17.2	3,748G	808M
Feature-Crop baseline + CoOp [88]	28.0	33.8	29.5	24.2	21.4	18.6	20.8	3,545G	808M
Open-Vocabulary SAM	31.1	36.0	32.4	24.0	21.3	22.9	22.4	1,180G	304M

decay of 0.0001. We adopt full image size for a random crop in the pre-training and training process following Cheng et al. [9]. All the class names are transferred into CLIP text embedding, following previous works [18]. We train each model for 12 epochs for fair comparison. Due to the limitation of computation costs, we do not adopt joint SAM data and COCO data training. We first perform training the SAM2CLIP on SA-1B (1%), and then we finetune the model on COCO or LVIS data. Please refer to the supplementary material for more details.

Benchmark Setup. Open Vocabulary SAM is different from both SAM [30] and open vocabulary segmentation methods [13,24,74]. Compared with Open-Vocabulary SAM, SAM [30] cannot recognize segmented objects while open vocabulary segmentation methods [13,24,74] usually rely on a separate segmentor (e.g., Mask-RCNN [22]) instead of prompts like points or boxes. For a complete evaluation of Open-Vocabulary SAM, we set the benchmark as three parts, including 1). Comparison with combined baselines to verify the effectiveness; 2). Comparison with open-vocabulary segmentation methods to evaluate the recognition; 3). Comparision with SAM models to evaluate segmentation.

Table 3. Comparison of Open-Vocabulary SAM with previous methods on open vocabulary instance segmentation. The results presented in the table all use the Mask-RCNN [22] with ResNet-50 backbone. Different from previous works, Open-Vocabulary SAM uses the bounding box as the prompt to generate the mask. We report mask AP50 following base-novel setting as [24].

Method	Venue	Constrained		Generalized		
		Base	Novel	Base	Novel	All
XPM [24]	CVPR'22	42.4	24.0	41.5	21.6	36.3
MaskCLIP [13]	ICML'23	42.8	23.2	42.6	21.7	37.2
MasQCLIP [74]	ICCV'23	40.9	30.1	40.7	28.4	37.5
Open-Vocabulary SAM	(Ours)	41.7	37.5	39.3	39.8	39.4

Table 4. Comparison of mask quality with various detectors on COCO dataset. We report the mask mean AP for comparison. The masks are generated by each method, while the labels are from the corresponding detectors.

Method	Detectors	mAP	AP50	AP75	APS	APM	APL	#Params	FLOPs
SAM-Huge	Faster-RCNN (R50)	35.6	54.9	38.4	17.2	39.1	51.4	641M	3,001G
SAM-Huge (finetuned)	Faster-RCNN (R50)	35.8	55.0	38.4	16.5	38.6	53.0	641M	3,001G
Open-Vocabulary SAM	Faster-RCNN (R50)	35.8	55.6	38.3	16.0	38.9	53.1	304M	1,180G
SAM-Huge	Detic (swin-base)	36.4	57.1	39.4	21.4	40.8	54.6	641M	3,001G
SAM-Huge (finetuned)	Detic (swin-base)	36.8	57.4	39.8	20.8	40.6	55.1	641M	3,001G
Open-Vocabulary SAM	Detic (swin-base)	36.7	57.2	39.7	20.7	40.8	54.9	304M	1,180G
SAM-Huge	ViTDet (Huge)	46.3	72.0	49.8	25.2	45.5	59.6	641M	3,001G
SAM-Huge (finetuned)	ViTDet (Huge)	46.5	72.3	50.3	25.2	45.8	60.1	641M	3,001G
Open-Vocabulary SAM	ViTDet (Huge)	48.8	73.8	52.9	24.8	46.3	64.2	304M	1,180G

Table 5. Scaling up with large-scale datasets.

Datasets	Accuracy	#vocaulary	#images
LVIS	83.1	1,203	99K
V3Det	78.7	13,204	183K
I-21k	44.5	19,167	13M
V3Det + LVIS	82.7	13,844	282K
V3Det + LVIS + I-21k	83.3	25,898	13M
V3Det + LVIS + I-21k + Object365	83.0	25,970	15M

Table 6. The effectiveness of each component. We use Detic [89] as the detector. The labels are generated by the corresponding model. S2C and C2S denote SAM2CLIP and CLIP2SAM respectively. The baseline refers to the image-crop variant.

Setting	AP	AP_{base}	AP_{novel}	FLOPs (G)	#Params (M)
Baseline + CoOp [88]	29.5	28.0	33.8	3,545	808
Our + S2C	28.7	27.3	33.3	1,127	291
Our + S2C + C2S	34.4	33.1	38.0	1,180	304

4.1 Main Results

Comparison with Combined Baselines Using Ground Truth. To avoid the influence of other modules, we first demonstrate the recognition ability of our model in Table 1. Compared to the simple combined approaches, adding adapters with joint co-training leads to better results. However, the recognition ability is still limited on both COCO and LVIS. Our Open-Vocabulary SAM achieves the best results on both boxes and points as visual prompts. We observe more significant gains on LVIS datasets. We argue that LVIS contains more small objects, which is more challenging than COCO. Our method can solve

Prob.2 and lead to over 20% accuracy improvement. Although the segmentation quality is pretty good (about 80 IoU on COCO and LVIS with box prompt), our method still achieves 2% IoU improvements. This indicates the effectiveness of our joint co-training on mask prediction and classification. Compared with boxes as prompts, using points as prompts is more challenging since the location clues of points are much weaker than boxes. However, our approach is still better than combined baselines or them with adapters.

Comparison with Combined Baselines on OV-Detector. In Table 2, we adopt a more challenging setting by using the box prediction from the existing open-vocabulary detector to simulate the interactive segmentation process with deviation. We choose the representative Detic [89] as the open-vocabulary detector. Again, our method also achieves the best performance on both COCO and LVIS datasets. In particular, on COCO, compared with previous works [88], our method achieves 3.0 mask mAP improvements with much lower costs.

Comparison with Open-Vocabulary Segmentation. In Table 3, we compare our method with previous open-vocabulary instance segmentation methods. Open-Vocabulary SAM uses the boxes generated by Mask- RCNN [22] and previous works use the masks of Mask-RCNN. The results show that our Open-Vocabulary SAM performs strongly, especially in the novel classes.

Comparison with SAM on Various Detectors. In Table 4, we test the mask prediction quality of our model and original SAM on different detectors. Our method performs better than the original SAM and performs comparably with fine-tuned SAM. It is worth noting that our Open-Vocabulary SAM has much lower computational costs and parameters than SAM.

Comparison on Interactive Segmentation. In Table 8 (right), we compare interactive segmentation performance. Notably, our approach excels beyond interactive segmentation and can recognize classes in an open-vocabulary setting.

Visualization Comparison. In Fig. 4, we compare our approach with the feature-crop baseline. Our model shows better performance in classifying small and rare object classification and handling occlusion scenarios.

Model as a Zero Shot Annotation Tool. In addition to COCO and LVIS standard datasets training, following the spirit of SAM, we also scale up our model by training it with more data (Table 5). In particular, we adopt more detection data (V3Det [60], Object365 [56]) and classification data (ImageNet22k [11]). Owing to significant costs, we have not conducted comparisons with other baselines for this setting. Rather, we have adapted our method into an interactive annotation tool capable of segmenting and recognizing over 22,000 classes.

Table 7. Ablation studies on COCO open-vocabulary dataset. We use boxes of the ground truth as a prompt to generate masks and labels. (left): Ablation on SAM2CLIP design. (right): Ablation on CLIP2SAM design.

Setting	IoU	FLOPs (G)	#Params (M)
SAM	78.7	3,001	641
Conv-L + MultiScale Neck	78.3	1,313	321
Conv-L + SingleScale Neck	73.6	1,280	307
R50x16 + MultiScale Neck	78.1	1,180	304
R50 + MultiScale Neck	77.3	728	165

Setting	IoU	Acc	AP_{base}	AP_{novel}
SAM2CLIP (baseline)	78.1	54.2	27.6	33.2
+ Cls Token	81.3	79.3	29.8	34.5
+ Cls Token & CLIP MLP fusion	80.9	78.9	29.0	33.9
+ light FPN (Ours)	81.5	84.3	31.1	36.0

Table 8. (left): Ablation study on different CLIP backbone. We test results on the COCO open-vocabulary dataset. We use boxes of the ground truth as a prompt to generate masks and labels. (right): Comparison with other SAM models.

Backbone	IoU	Acc	#FLOPs(G)	#Params (M)
RN50	77.3	50.8	728	165
RN50x16	78.1	55.1	1,180	304
RN50x64	78.1	54.1	2,098	568
ConvNeXt-L	78.3	59.1	1,313	321
ViT-L-14	38.6	14.3	2,294	441

Method	1-IoU (COCO)	cls.	open.
SAM [30] (H)	78.2	-	-
SEEM [90] (T)	73.7	✓	-
Semantic-SAM [32] (T)	76.1	✓	-
OV-SAM (ours)	81.7	✓	✓

4.2 Ablation Studies and Analysis

Effectiveness of SAM2CLIP and CLIP2SAM. We first verify the effectiveness of our proposed two modules in Table 6. We adopt image-crop variant

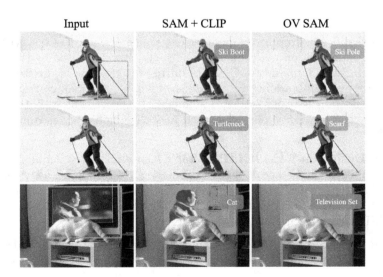

Fig. 4. Visualization Comparison. We compare the mask and classification results of the image-crop baseline (SAM + CLIP) and Open-Vocabulary SAM (OV SAM). The predicted labels are presented on the mask.

Fig. 5. Qualitative results of Open-Vocabulary SAM. In the visualization, boxes refer to box prompts (up), and the red stars refer to the point prompts (middle). The masks can be merged into a segmentation map (bottom). We show the mask and labels generated by the proposed Open-Vocabulary SAM. Our method can segment and recognize open vocabulary objects in diverse scenes identified by prompts. (Color figure online)

of the baseline for comparison. In particular, by sharing a single backbone, we observe a significant drop in the number of parameters and FLOPs, with a little drop in segmentation performance. The slight drop is caused by the domain gap between SAM data and COCO data during the training of the SAM2CLIP module. However, after adding our CLIP2SAM module and joint co-training with mask classification and prediction, a significant improvement in both segmentation and classification is observed, with just a negligible increase in compute cost.

Detailed Design on SAM2CLIP. In Table 7, we explore the detailed design of SAM2CLIP in the first stage of open-vocabulary SAM training. The results show that distillation benefits most when multi-scale features are adopted, suggesting that both high-resolution features and high-level semantics are important to align CLIP's feature with the SAM's feature (Fig. 5).

Detailed Design on CLIP2SAM. In Table 7, we present extensive design for the CLIP2SAM module. We compare two designs: a simple classification token with cross attention (Cls Token) and a combination of this token with mask pooled CLIP feature (CLS Token & CLIP MLP fusion). These designs work better than the combined baseline shown in the first row. Nonetheless, due to resolution constraints, these variants cannot handle small objects well, as shown in Fig. 4. In contrast, our design improves the performance considerably.

Ablation on Different CLIP Backbones. In Table 8, we explore the effect of frozen CLIP visual backbone. We do not add the CLIP2SAM. Motivated by recent works [31,67,73,76], CNN-based CLIPs encapsulate more structural information, which is good for our goal since we have location-sensitive visual prompts. Thus, adopt CNN-based CLIPs, aligned with previous works. As shown in the table, we find ConvNext large achieves the best performance, but we choose RN50 × 16 since it has comparable performance and better efficiency.

5 Conclusion

We present Open Vocabulary SAM, a SAM-inspired method for interactive segmentation and recognition. Unlike previous open-vocabulary detection and segmentation methods, our method explores interactive open-vocabulary segmentation for the first time. Given the user's inputs, such as boxes or points, the proposed approach can interactively segment and label each visual prompt. Compared with the combined baselines and various visual adapters, our proposed CLIP2SAM and SAM2CLIP are both efficient and effective in various settings. Our open vocabulary segmentation is compatible with different detectors, including open-vocabulary detectors and close-set detectors. With more data, our model plays a similar role as SAM, offering an effective annotation tool for both segmentation and instance labeling. In particular, our method can perform large vocabulary segmentation and recognition over 22K classes. We hope our Open-Vocabulary SAM can provide a solid baseline for combining the strengths of different forms of vision foundation models and inspire further research.

Acknowledgements. This study is supported under the RIE2020 Industry Alignment Fund-Industry Collaboration Projects (IAF-ICP) Funding Initiative, as well as cash and in-kind contributions from the industry partner(s). The project is also supported by Singapore MOE AcRF Tier 1 (RG16/21) and the National Key R&D Program of China (No. 2022ZD0161600).

References

1. Balažević, I., Steiner, D., Parthasarathy, N., Arandjelović, R., Hénaff, O.J.: Towards in-context scene understanding. In: NeurIPS (2023)
2. Bao, H., Dong, L., Piao, S., Wei, F.: BEiT: BERT pre-training of image transformers. In: ICLR (2022)
3. Bar, A., Gandelsman, Y., Darrell, T., Globerson, A., Efros, A.: Visual prompting via image inpainting. In: NeurIPS (2022)
4. Brown, T., et al.: Language models are few-shot learners. In: NeurIPS (2020)
5. Chen, C., et al.: MA-SAM: modality-agnostic SAM adaptation for 3D medical image segmentation. arXiv preprint arXiv:2309.08842 (2023)

6. Chen, J., Yang, Z., Zhang, L.: Semantic segment anything (2023)
7. Chen, K., et al.: MMDetection: openMMLab detection toolbox and benchmark. arXiv preprint arXiv:1906.07155 (2019)
8. Chen, Z., et al.: Vision transformer adapter for dense predictions. In: ICLR (2023)
9. Cheng, B., Misra, I., Schwing, A.G., Kirillov, A., Girdhar, R.: Masked-attention mask transformer for universal image segmentation. In: CVPR (2022)
10. Cheng, Y., et al.: Segment and track anything. arXiv preprint arXiv:2305.06558 (2023)
11. Deng, J., Dong, W., Socher, R., Li, L.J., Li, K., Fei-Fei, L.: ImageNet: a large-scale hierarchical image database. In: CVPR (2009)
12. Ding, J., Xue, N., Xia, G.S., Dai, D.: Decoupling zero-shot semantic segmentation. In: CVPR (2022)
13. Ding, Z., Wang, J., Tu, Z.: Open-vocabulary universal image segmentation with MaskCLIP. In: ICML (2023)
14. Dosovitskiy, A., et al.: An image is worth 16×16 words: transformers for image recognition at scale. In: ICLR (2021)
15. Fang, Y., Yang, S., Wang, S., Ge, Y., Shan, Y., Wang, X.: Unleashing vanilla vision transformer with masked image modeling for object detection. In: ICCV (2023)
16. Fang, Z., Li, X., Li, X., Buhmann, J.M., Loy, C.C., Liu, M.: Explore in-context learning for 3D point cloud understanding. In: NeurIPS (2023)
17. Gao, P., et al.: CLIP-Adapter: better vision-language models with feature adapters. Int. J. Comput. Vis. **132**(2), 581–595 (2024). https://doi.org/10.1007/s11263-023-01891-x
18. Gu, X., Lin, T.Y., Kuo, W., Cui, Y.: Open-vocabulary object detection via vision and language knowledge distillation. In: ICLR (2021)
19. Gupta, A., Dollar, P., Girshick, R.: LVIS: a dataset for large vocabulary instance segmentation. In: CVPR (2019)
20. Han, X., et al.: Boosting segment anything model towards open-vocabulary learning. arXiv preprint arXiv:2312.03628 (2023)
21. He, K., Chen, X., Xie, S., Li, Y., Dollár, P., Girshick, R.: Masked autoencoders are scalable vision learners. In: CVPR (2022)
22. He, K., Gkioxari, G., Dollár, P., Girshick, R.: mask R-CNN. In: ICCV (2017)
23. Hu, E.J., et al.: LoRA: low-rank adaptation of large language models. In: ICLR (2022)
24. Huynh, D., Kuen, J., Lin, Z., Gu, J., Elhamifar, E.: Open-vocabulary instance segmentation via robust cross-modal pseudo-labeling. In: CVPR (2022)
25. Jayaraman, D., Grauman, K.: Zero-shot recognition with unreliable attributes. In: NeurIPS (2014)
26. Jia, C., et al.: Scaling up visual and vision-language representation learning with noisy text supervision. In: ICML (2021)
27. Kim, D., Lin, T.Y., Angelova, A., Kweon, I.S., Kuo, W.: Learning open-world object proposals without learning to classify. RA-L (2022)
28. Kim, W., Son, B., Kim, I.: ViLT: vision-and-language transformer without convolution or region supervision. In: ICML (2021)
29. Kirillov, A., He, K., Girshick, R., Rother, C., Dollár, P.: Panoptic segmentation. In: CVPR (2019)
30. Kirillov, A., et al.: Segment anything. In: ICCV (2023)
31. Kuo, W., Cui, Y., Gu, X., Piergiovanni, A.J., Angelova, A.: F-VLM: open-vocabulary object detection upon frozen vision and language models. In: ICLR (2023)

32. Li, F., et al.: Semantic-SAM: segment and recognize anything at any granularity. arXiv preprint arXiv:2307.04767 (2023)
33. Li, J., Li, D., Savarese, S., Hoi, S.: BLIP-2: bootstrapping language-image pre-training with frozen image encoders and large language models. In: ICML (2023)
34. Li, J., Li, D., Xiong, C., Hoi, S.: Blip: bootstrapping language-image pre-training for unified vision-language understanding and generation. In: ICML (2022)
35. Li, J., Selvaraju, R.R., Gotmare, A., Joty, S.R., Xiong, C., Hoi, S.C.: Align before fuse: vision and language representation learning with momentum distillation. In: NeurIPS (2021)
36. Li, S., Cao, J., Ye, P., Ding, Y., Tu, C., Chen, T.: ClipSAM: CLIP and SAM collaboration for zero-shot anomaly segmentation. arXiv preprint arXiv:2401.12665 (2024)
37. Li, X., et al.: Transformer-based visual segmentation: a survey. arXiv pre-print (2023)
38. Li, X., et al.: Semantic flow for fast and accurate scene parsing. In: ECCV (2020)
39. Li, X., et al.: OMG-Seg: is one model good enough for all segmentation? In: CVPR (2024)
40. Li, X., et al.: SFNet: faster and accurate semantic segmentation via semantic flow. Int. J. Comput. Vis. **132**(2), 466–489 (2024). https://doi.org/10.1007/s11263-023-01875-x
41. Li, Y., Fan, H., Hu, R., Feichtenhofer, C., He, K.: Scaling language-image pre-training via masking. In: CVPR (2023)
42. Lian, D., Zhou, D., Feng, J., Wang, X.: Scaling & shifting your features: A new baseline for efficient model tuning. In: NeurIPS (2022)
43. Lin, T.Y., et al.: Microsoft COCO: common objects in context. In: ECCV (2014)
44. Liu, S., et al.: Grounding DINO: marrying DINO with grounded pre-training for open-set object detection. arXiv preprint arXiv:2303.05499 (2023)
45. Liu, Y., Qi, L., Tsai, Y.J., Li, X., Chan, K.C.K., Yang, M.H.: Effective adapter for face recognition in the wild. arXiv preprint (2023)
46. Loshchilov, I., Hutter, F.: Decoupled weight decay regularization. In: ICLR (2019)
47. Maaz, M., Rasheed, H., Khan, S., Khan, F.S., Anwer, R.M., Yang, M.H.: Class-agnostic object detection with multi-modal transformer. In: ECCV (2022)
48. Milletari, F., Navab, N., Ahmadi, S.: V-Net: fully convolutional neural networks for volumetric medical image segmentation. In: 3DV (2016)
49. Pan, T., Tang, L., Wang, X., Shan, S.: Tokenize anything via prompting. arXiv preprint arXiv:2312.09128 (2023)
50. Paszke, A., et al.: PyTorch: an imperative style, high-performance deep learning library. In: NeurIPS (2019)
51. Qi, L., et al.: High-quality entity segmentation. In: ICCV (2023)
52. Qi, L., et al.: Open world entity segmentation. IEEE TPAMI (2022)
53. Radford, A., et al.: Learning transferable visual models from natural language supervision. In: ICML (2021)
54. Rombach, R., Blattmann, A., Lorenz, D., Esser, P., Ommer, B.: High-resolution image synthesis with latent diffusion models. In: CVPR (2022)
55. Rubin, O., Herzig, J., Berant, J.: Learning to retrieve prompts for in-context learning. arXiv:2112.08633 (2021)
56. Shao, S., et al.: Objects365: a large-scale, high-quality dataset for object detection. In: ICCV (2019)
57. Shi, C., Yang, S.: EdaDet: open-vocabulary object detection using early dense alignment. In: ICCV (2023)

58. Sun, Q., Fang, Y., Wu, L., Wang, X., Cao, Y.: EVA-CLIP: improved training techniques for CLIP at scale. arXiv preprint arXiv:2303.15389 (2023)
59. Wang, H., et al.: SAM-CLIP: merging vision foundation models towards semantic and spatial understanding. arXiv preprint arXiv:2310.15308 (2023)
60. Wang, J., et al.: V3Det: vast vocabulary visual detection dataset. In: ICCV (2023)
61. Wang, X., Wang, W., Cao, Y., Shen, C., Huang, T.: Images speak in images: a generalist painter for in-context visual learning. In: CVPR (2023)
62. Wang, X., Zhang, X., Cao, Y., Wang, W., Shen, C., Huang, T.: SegGPT: segmenting everything in context. In: ICCV (2023)
63. Wu, J., et al.: Betrayed by captions: joint caption grounding and generation for open vocabulary instance segmentation. In: ICCV (2023)
64. Wu, J., et al.: Towards open vocabulary learning: a survey. IEEE TPAMI (2024)
65. Wu, J., et al.: Medical SAM Adapter: adapting segment anything model for medical image segmentation. arXiv preprint arXiv:2304.12620 (2023)
66. Wu, S., Zhang, W., Jin, S., Liu, W., Loy, C.C.: Aligning bag of regions for open-vocabulary object detection. In: CVPR (2023)
67. Wu, S., et al.: CLIPSelf: vision transformer distills itself for open-vocabulary dense prediction. In: ICLR (2024)
68. Wu, X., Zhu, F., Zhao, R., Li, H.: CORA: adapting CLIP for open-vocabulary detection with region prompting and anchor pre-matching. In: CVPR (2023)
69. Xie, J., Li, W., Li, X., Liu, Z., Ong, Y.S., Loy, C.C.: MosaicFusion: diffusion models as data augmenters for large vocabulary instance segmentation. arXiv preprint arXiv:2309.13042 (2023)
70. Xu, J., Liu, S., Vahdat, A., Byeon, W., Wang, X., De Mello, S.: Open-vocabulary panoptic segmentation with text-to-image diffusion models. In: CVPR (2023)
71. Xu, M., Zhang, Z., Wei, F., Hu, H., Bai, X.: Side adapter network for open-vocabulary semantic segmentation. In: CVPR (2023)
72. Xu, M., et al.: A simple baseline for open-vocabulary semantic segmentation with pre-trained vision-language model. In: ECCV (2022)
73. Xu, S., et al.: DST-DET: simple dynamic self-training for open-vocabulary object detection. arXiv pre-print (2023)
74. Xu, X., Xiong, T., Ding, Z., Tu, Z.: MasQCLIP for open-vocabulary universal image segmentation. In: ICCV (2023)
75. Yang, J., et al.: Panoptic video scene graph generation. In: CVPR (2023)
76. Yu, Q., He, J., Deng, X., Shen, X., Chen, L.C.: Convolutions die hard: open-vocabulary segmentation with single frozen convolutional CLIP. In: NeurIPS (2023)
77. Yu, X., Tang, L., Rao, Y., Huang, T., Zhou, J., Lu, J.: Point-BERT: pre-training 3D point cloud transformers with masked point modeling. In: CVPR (2022)
78. Zang, Y., Li, W., Zhou, K., Huang, C., Loy, C.C.: Open-vocabulary DETR with conditional matching. In: ECCV (2022)
79. Zareian, A., Rosa, K.D., Hu, D.H., Chang, S.F.: Open-vocabulary object detection using captions. In: CVPR (2021)
80. Zhai, X., et al.: LiT: zero-shot transfer with locked-image text tuning. In: CVPR (2022)
81. Zhang, C., et al.: Faster segment anything: towards lightweight SAM for mobile applications. arXiv preprint arXiv:2306.14289 (2023)
82. Zhang, R., et al.: Personalize segment anything model with one shot. arXiv preprint arXiv:2305.03048 (2023)
83. Zhao, X., et al.: Fast segment anything. arXiv preprint arXiv:2306.12156 (2023)

84. Zhou, B., et al.: Semantic understanding of scenes through the ADE20K dataset. Int. J. Comput. Vis. **127**(3), 302–321 (2018). https://doi.org/10.1007/s11263-018-1140-0
85. Zhou, C., Li, X., Loy, C.C., Dai, B.: EdgeSAM: prompt-in-the-loop distillation for on-device deployment of SAM. arXiv preprint arXiv:2312.06660 (2023)
86. Zhou, C., Loy, C.C., Dai, B.: Extract free dense labels from CLIP. In: ECCV (2022)
87. Zhou, H., et al.: Rethinking evaluation metrics of open-vocabulary segmentaion. arXiv preprint arXiv:2311.03352 (2023)
88. Zhou, K., Yang, J., Loy, C.C., Liu, Z.: Learning to prompt for vision-language models. IJCV (2022). https://doi.org/10.1007/s11263-022-01653-1
89. Zhou, X., Girdhar, R., Joulin, A., Krähenbühl, P., Misra, I.: Detecting twenty-thousand classes using image-level supervision. In: ECCV (2022)
90. Zou, X., et al.: Segment everything everywhere all at once. In: NeurIPS (2023)

Data-to-Model Distillation: Data-Efficient Learning Framework

Ahmad Sajedi[1]($\boxtimes$)[🆔], Samir Khaki[1][🆔], Lucy Z. Liu[2], Ehsan Amjadian[2][🆔], Yuri A. Lawryshyn[1], and Konstantinos N. Plataniotis[1][🆔]

[1] University of Toronto, Toronto, Canada
{ahmad.sajedi,samir.khaki}@mail.utoronto.ca
[2] Royal Bank of Canada (RBC), Toronto, Canada
https://datadistillation.github.io/D2M/

Abstract. Dataset distillation aims to distill the knowledge of a large-scale real dataset into small yet informative synthetic data such that a model trained on it performs as well as a model trained on the full dataset. Despite recent progress, existing dataset distillation methods often struggle with computational efficiency, scalability to complex high-resolution datasets, and generalizability to deep architectures. These approaches typically require retraining when the distillation ratio changes, as knowledge is embedded in raw pixels. In this paper, we propose a novel framework called Data-to-Model Distillation (D2M) to distill the real dataset's knowledge into the learnable parameters of a pre-trained generative model by aligning rich representations extracted from real and generated images. The learned generative model can then produce informative training images for different distillation ratios and deep architectures. Extensive experiments on 15 datasets of varying resolutions show D2M's superior performance, re-distillation efficiency, and cross-architecture generalizability. Our method effectively scales up to high-resolution 128×128 ImageNet-1K. Furthermore, we verify D2M's practical benefits for downstream applications in neural architecture search.

Keywords: Data Distillation · Efficient Training · Image Classification

1 Introduction

Deep learning has been successful in various domains, including computer vision and natural language processing [1,4,25,32,62,63,71,78]; however, it often relies on deep neural networks (DNNs) and large-scale datasets. These requirements

---

A. Sajedi and S. Khaki—Equal contribution.

Supplementary Information The online version contains supplementary material available at https://doi.org/10.1007/978-3-031-72775-7_25.

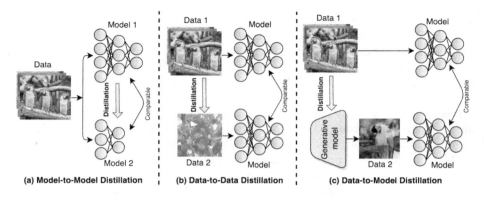

Fig. 1. Different distillation frameworks for efficient learning.

necessitate significant investments in training time, data storage, and electricity consumption, making training impractical for those with limited computational resources. Techniques such as pruning [33,34,44,54], quantization [79,92], and model distillation [27,43] show promise in reducing these computational expenses while preserving performance. *Model distillation* [18,26,27,57,59,64,65,85], in particular, transfers the informative knowledge from a large teacher model to a smaller student one to facilitate model compression (Fig. 1(a)). Recently, *dataset distillation* [5,6,61,77,86,89,90,93] has emerged as a promising data-efficient learning approach that distills knowledge from a large training dataset into a small set of synthetic images. These images enable models to attain test performance comparable to those trained on the original dataset (Fig. 1(b)). Dataset distillation is widely applied in computer vision for applications like neural architecture search [28,70], continual learning [10,21,61,81,87], federated learning [47,49,80], and privacy-preserving [7,14,51].

Traditional data-centric methods, known as *coreset selection*, reduce training costs by selecting a subset of the original dataset based on specific metrics [2,58,67,72,92]. However, these approaches have a deficiency in representation capability and coverage of chosen samples, leading to suboptimal results for classification tasks [58,72,92]. State-of-the-art dataset distillation algorithms overcome these limitations by synthesizing a small yet informative dataset that aligns characteristics between synthetic and real images. Drawing on seminal work [77], researchers have developed various methods, including gradient matching [17,42,87,90], distribution matching [35,61,75,89,91], trajectory matching [5,12,16,22], and kernel-inducing points [50,56,93]. These works initialize a small number of learnable images and update their raw pixel values using different matching strategies during distillation.

Existing dataset distillation studies have shown remarkable performance but suffer from three major drawbacks. First, they require complete retraining of distillation algorithms when the distillation ratio (or the number of images per class) changes, which leads to a computationally demanding re-distillation process. Second, these methods generally struggle with high-resolution, large-scale datasets (e.g.128×128 ImageNet-1K [13]) and tend to distill visually noisy images [5]. Lastly, the distilled dataset often performs poorly on architectures like ResNet [25], DenseNet [30] and ViT [15]. These issues partially stem from parameterizing the synthetic dataset in pixel space. Optimizing raw pixels can capture high-frequency, detailed information that is not essential for downstream tasks and is susceptible to over-fitting to the training architecture [6]. Moreover, employing pixel-level optimization on high-resolution, large-scale data incurs significant computational and memory costs, rendering it unscalable for such datasets.

In this paper, we propose a novel framework, Data-to-Model distillation (D2M), that transfers information from the original dataset into a generative model rather than relying on raw pixel data (Fig. 1(c)). D2M effectively addresses the aforementioned issues by parameterizing the synthetic dataset within the parameter space of a generative model, such as Generative Adversarial Networks (GANs) [3,20]. Following the approach of [61,74,91,93], our method employs a model pool to extract rich representations that are essential for learning. We design embedding matching and prediction matching modules to minimize differences in channel attention maps (at various layers) and output predictions between real and generated images, respectively. As depicted in Fig. 2, these modules facilitate the distillation of different types of knowledge into the generative model using suitable matching losses, all geared towards improving classification performance. Following the distillation stage, our generative model can create training samples for classification tasks with *any* number of images per class from random noises. Unlike other dataset distillation methods, D2M circumvents the need for retraining the distillation when changing the distillation ratio, which offers clear advantages in re-distillation efficiency. As D2M stores information in models rather than pixels, it stands as one of the pioneer distillation methods scalable to high-resolution (256×256) and large-scale datasets. This makes it a promising tool for efficient synthetic data generation, requiring significantly less memory storage. The contributions of our study are:

[**C1**]: We propose a novel framework to distill the knowledge of large-scale datasets into a *parameter space of a generative model* that can produce informative images for classification tasks. The method distills various representations from the real data to provide diverse supervision.

[**C2**]: We conduct extensive experiments on 15 datasets, differing in resolution, label complexity, and application domains. We achieve state-of-the-art results across all settings and enable the distillation of the ImageNet-1K at a resolution of 128×128 with a fixed storage complexity across different settings.

[**C3**]: We show that data-to-model distillation outperforms prior dataset distillation algorithms in re-distillation efficiency and cross-architecture generality. The high-quality images generated by D2M also notably improve downstream applications in neural architecture search.

2 Related Work

Dataset Distillation. Wang et al.[77] introduced dataset distillation to synthesize small-scale data from a large real dataset using meta-learning, minimizing training loss differences between the synthetic and original data. Meta-learning involves bi-level optimization and incurs significant computational costs. To mitigate this, researchers employed kernel methods to facilitate the inner optimization loop, as seen in KIP [55,56], RFAD [50], and FRePo [93]. Other studies adopted surrogate objectives to tackle unrolled optimization problems in meta-learning. DC [90], DSA [87], and IDC [36] align gradients between synthetic and real datasets for distillation. Meanwhile, DM [89], CAFE [75], DAM [61], and ATOM [35] use distribution-matching to alleviate bias caused by original samples with large gradients. Further, MTT [5], FTD [16], and TESLA [12] match model parameter trajectories to improve performance. However, these methods distill information into pixels, leading to a linear growth in computational costs with class numbers and resolutions. This limits scalability with larger datasets and poses challenges for cross-architecture generalization and re-distillation efficiency [5,48,61,75,90]. In contrast, D2M distills the knowledge into the generative model to address the limitations posed by pixel-wise dataset distillation.

Generative Models. This work is closely related to generative models, specifically generative adversarial networks (GANs) [20,40,76,94]. Goodfellow et al. [20] introduced vanilla GANs to generate realistic images that deceive human observers. A variety of GAN models, such as CycleGAN [94], InfoGAN [9], and BigGAN [3], have since emerged for tasks like image manipulation [41,76,94], super-resolution [23,40], and object detection [45,46]. However, these GAN-generated images often prioritize visual realism over informativeness, which may not be ideal for data-efficient classification tasks [6,74,75,88]. Some methods have been proposed to address this by generating samples that can be used to train deep neural networks more efficiently. GLaD [6] distills numerous images into a few intermediate feature vectors in the GAN's latent space to help cross-architecture generalization. Wang et al. [74] introduced a two-stage algorithm for optimizing generative models to create training samples for 10-class classification tasks. Nevertheless, these methods face challenges such as high training costs [74], the necessity of retraining for different distillation ratios [6], and scalability constraints for complex, high-resolution datasets [74]. In contrast, our approach distills discriminative knowledge from the dataset into the *parameter space* of generative models, reducing re-distillation costs and achieving superior performance on standard benchmarks for both low- and high-resolution datasets.

3 Methodology

In this section, we introduce a novel framework called D2M, which distills knowledge from a large-scale training dataset $\mathcal{T} = \{(\boldsymbol{x}_i, y_i)\}_{i=1}^{|\mathcal{T}|}$, with $|\mathcal{T}|$ image-label pairs, into a generative model G. In the distillation phase, we extract comprehensive knowledge from the real dataset $\mathcal{T}$, including embedded representations

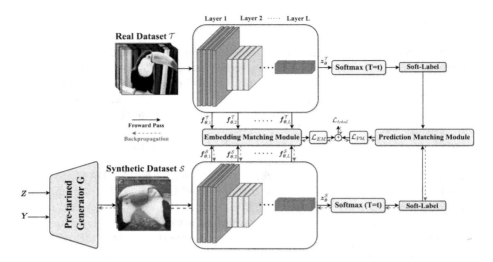

Fig. 2. An overview of the proposed D2M framework. D2M distills the knowledge of large-scale datasets into the parameter space of a pre-trained generator through embedding matching and prediction matching modules. The learned generator can then produce small yet informative training images for the downstream classification tasks. Z and Y represent the random noises and labels, respectively.

and logit prediction, and then distill them into the parameters of a pre-trained generative model. This procedure refines the generative model to synthesize a small yet informative dataset $\mathcal{S} = \{(s_j, y_j)\}_{j=1}^{|\mathcal{S}|}$ that has comparable training power to the real dataset. Following distillation, the learned generative model produces training images from random noises, enabling flexibility in the distillation ratios. These samples are subsequently deployed in classification tasks to evaluate D2M's performance. The overall D2M pipeline is depicted in Fig. 2.

3.1 Data-to-Model Distillation

The proposed D2M differs from data distillation works that directly distill knowledge into raw pixels. In D2M, we distill the information into the learnable parameters of a generative model, offering a new path for efficient learning.

A randomly initialized generative model yields noisy outputs, making it challenging to establish meaningful correspondences with real images during the distillation process. Therefore, we employ a pre-trained Generative Adversarial Network (GAN) with a generator G and a discriminator D, trained using the following standard loss function:

$$\min_G \max_D \mathbb{E}_{\boldsymbol{x} \sim P_\mathcal{T}} \left[\log D(\boldsymbol{x}|y) \right] + \mathbb{E}_{z \sim P_Z} \left[\log(1 - D(G(\boldsymbol{z}|y))) \right], \quad (1)$$

where $P_\mathcal{T}$ and P_Z denote the distributions of real training images and latent vectors, respectively. As highlighted by [88], conventional GANs typically generate images that deviate from the original data distribution and are less informative than real samples for training deep networks. To mitigate this limitation, we introduce a distillation stage to refine the generative model. This refinement enables the model to synthesize images that are more discriminative and better suited to downstream classification tasks.

Distillation Stage. We randomly select B pairs consisting of random noises Z and labels Y in each batch. These pairs are then fed into the pre-trained generative model to produce synthetic images $\mathcal{S}$. Meanwhile, we also randomly pick B real images from the original dataset $\mathcal{T}$ with corresponding labels Y. During each iteration, both real and synthetic batches undergo a forward pass using a randomly chosen neural network $\phi_\theta(\cdot)$ with different initializations from a model pool. This network is responsible for extracting features and predicting classification logits for both real and synthetic images. The network generates feature maps and predicted logits for each dataset, denoted as $\phi_\theta(\mathcal{T}) = [\boldsymbol{f}_{\theta,1}^{\mathcal{T}}, \cdots, \boldsymbol{f}_{\theta,L}^{\mathcal{T}}, \boldsymbol{z}_\theta^{\mathcal{T}}]$ for the real dataset and $\phi_\theta(\mathcal{S}) = [\boldsymbol{f}_{\theta,1}^{\mathcal{S}}, \cdots, \boldsymbol{f}_{\theta,L}^{\mathcal{S}}, \boldsymbol{z}_\theta^{\mathcal{S}}]$ for the synthetic one. The feature map $\boldsymbol{f}_{\theta,l}^{\mathcal{T}}$ and logit $\boldsymbol{z}_\theta^{\mathcal{T}}$ are multi-dimensional arrays obtained from the real dataset in the lth and pre-softmax layers of the network, respectively. Similarly, feature $\boldsymbol{f}_{\theta,l}^{\mathcal{S}}$ and logits $\boldsymbol{z}_\theta^{\mathcal{S}}$ are derived from the synthetic set. We then adopt embedding matching and prediction matching modules to capture the representations of the real training set across various layers.

The embedding matching module aligns the channel-wise attention maps between the real and synthetic sets across different feature extraction layers using the loss function $\mathcal{L}_{\text{EM}}$, formulated as:

$$\mathop{\mathbb{E}}_{\theta \sim P_\theta} \left[\sum_{l=1}^{L} \left\| \mathbb{E}_{\mathcal{T}}\left[\tilde{\boldsymbol{f}}_{\theta,l}^{\mathcal{T}}\right] - \mathbb{E}_{\mathcal{S}}\left[\tilde{\boldsymbol{f}}_{\theta,l}^{\mathcal{S}}\right] \right\|^2 \right], \qquad (2)$$

where P_θ is the distribution of neural network parameters and $\tilde{\boldsymbol{f}}_{\theta,l}^{\mathcal{T}}$ is defined as:

$$\tilde{\boldsymbol{f}}_{\theta,l}^{\mathcal{T}} := \begin{cases} \boldsymbol{a}_{\theta,l}^{\mathcal{T}} & \text{if } l = 1, \cdots, L-1 \\ \boldsymbol{f}_{\theta,L}^{\mathcal{T}} & \text{if } l = L, \end{cases} \qquad (3)$$

in which $\boldsymbol{a}_{\theta,l}^{\mathcal{T}}$ represents the vectorized channel attention maps in the lth layer for the real dataset. In the same way, the $\tilde{\boldsymbol{f}}_{\theta,l}^{\mathcal{S}}$ can be defined for the synthetic set. In particular, we characterize each training image through the use of channel-wise attention maps created by different layers. As noted in [8,35], these attention maps highlight the most discriminative regions of the input image, revealing the network's focus across various layers (early and intermediate) for obtaining information at low- and mid-level representations. The last layer of feature extraction in the neural network contains the highest-level abstract information [52]. This embedded has been shown to effectively capture semantic information from the input data [60,89]. Thus, we leverage the mean square error loss to match the vectorized versions of the final feature maps with real and synthetic data. Our embedding matching uses the most *discriminative* regions of feature maps using the concept of attention, differing significantly from pure feature matching approaches like CAFE [75]. Notably, in cases where ground-truth data distributions are unavailable, we empirically estimate the expectation term in Eq. (2).

Although $\mathcal{L}_{\text{EM}}$ effectively approximates large-scale real data distribution, its matching loss primarily minimizes the mean feature distance within each batch without explicitly constraining the diversity of synthetic images. Therefore, we use a complementary loss as a regularizer to provide more specific supervision. This regularization promotes similarity in the output probability predictions between the two datasets, directly influencing the results of the downstream classification task. Drawing inspiration from knowledge distillation works [27,31,84], we employ logit-based matching to minimize the differences in the *softened* output predictions between real and generated synthetic images. The soft-label predictions introduce additional information to the generative model that can be helpful for classification tasks [83]. The prediction matching loss $\mathcal{L}_{\text{PM}}$ can then be written as follows:

$$\mathbb{E}_{\theta \sim P_\theta} \left[\sum_{i=1}^{B} KL\left(\sigma\left(\frac{z_\theta^{x_i}}{T}\right), \sigma\left(\frac{z_\theta^{s_i}}{T}\right)\right) \right], \quad (4)$$

where KL stands for Kullback-Leibler divergence, $\sigma(\cdot)$ denotes the softmax function, and $z_\theta^{x_i}$ and $z_\theta^{s_i}$ are the prediction logits for real and synthetic images sharing the same label y_i, respectively. The temperature hyperparameter T is used to generate soft-label predictions while regulating the entropy of the output distribution [27,84]. Additional details on the impact of T are discussed in Sect. 4.3. Finally, the training loss for the D2M framework is formulated as the augmented Lagrangian of the two mentioned losses. We learn the *parameters of a generative model* by solving the following optimization problem using SGD:

$$G^* = \arg\min_G \left(\mathcal{L}_{\text{EM}} + \lambda \mathcal{L}_{\text{PM}} \right), \quad (5)$$

where λ serves as a Lagrangian multiplier to balance the gradients of $\mathcal{L}_{\text{EM}}$ and $\mathcal{L}_{\text{PM}}$. The impact of λ is examined in Sect. 4.3 and a learning algorithm is summarized in Alg. 1.

Comparing with Other Generative Prior Methods. Recent generative-based dataset distillation methods typically distill large-scale datasets into the latent space of a specific GAN model [6,88]. For example, GLaD [6] utilizes various intermediate feature spaces of StyleGAN-XL [66] to characterize synthetic datasets. IT-GAN [6] creates synthetic datasets in a BigGAN's latent space [3] by initially inverting the full training set and then fine-tuning the latent representations based on the distillation objective. Both GLaD and IT-GAN necessitate complete retraining when distillation ratios change due to their information containers, which require updates when changing the IPC. In contrast, D2M proposes a novel approach by parameterizing the synthetic dataset within the learnable *parameter space* of any generative model using our carefully designed framework. Unlike GLaD and IT-GAN, whose storage complexity grows linearly with increasing IPCs, our framework maintains a constant storage complexity across

Algorithm 1. Data-to-Model Distillation (D2M)

Input: Real training dataset $\mathcal{T} = \{(\boldsymbol{x}_i, y_i)\}_{i=1}^{|\mathcal{T}|}$
Required: Pre-trained generative model G, Latent vector set Z, Deep neural network $\phi_{\boldsymbol{\theta}}$ from a model pool with parameter distribution $P_{\boldsymbol{\theta}}$, Learning rate η_G, Task balance parameter λ, Number of training iterations I.

1: **for** $i = 1, 2, \cdots, I$ **do**
2:   Sample $\boldsymbol{\theta}$ from $P_{\boldsymbol{\theta}}$
3:   Sample B pairs from the real dataset and the generated images of $G(Z)$
4:   Compute $\mathcal{L}_{\text{EM}}$ and $\mathcal{L}_{\text{PM}}$ using Eqs. 2 and 4
5:   Calculate $\mathcal{L} = \mathcal{L}_{\text{EM}} + \lambda \mathcal{L}_{\text{PM}}$
6:   Update the generator using $G \leftarrow G - \eta_G \nabla_G \mathcal{L}$
7: **end for**

Output: Synthetic dataset $\mathcal{S} = G(Z) = \{(\boldsymbol{s}_i, y_i)\}_{i=1}^{|\mathcal{S}|}$

various distillation ratios. To demonstrate the efficacy of our framework, we empirically compare D2M with GLaD and IT-GAN, consistently outperforming them in all cases, as detailed in Sect. 4.2 and supplementary materials.

4 Experiments

4.1 Experimental Setup

Datasets. In line with prior studies, we evaluate the proposed framework across diverse image classification datasets with varying resolutions and label complexity. Our method is applied to CIFAR10/100 [37] for low-resolution data. The TinyImageNet [39] and ImageNet-1K [13] datasets are resized to 64×64 for medium-resolution ones. For high-resolution data (*i.e.* 128×128), we employ ImageNet-1K and its subsets-ImageNette [29], ImageWoof [29], ImageSquawk [5], ImageFruit [5], and ImageMeow [5]-each emphasizing specific categories such as assorted objects, dog breeds, birds, fruits, and cat breeds. We also use the recently introduced 10-class ImageNet-1K subsets (ImageNet-[A, B, C, D, E]) [6], which were determined through the evaluation performance of a pre-trained ResNet-50 on ImageNet-1K. Finally, we extend D2M to the medical imaging domain, focusing on high-resolution 256×256 images in DermaMNIST [82]. This dataset features dermatoscopic images of typical pigmented skin lesions. More details on the datasets can be found in the supplementary materials.

Network Architectures. Following recent dataset distillation works [61,74,93], we leverage a model pool to mitigate overfitting and offer diverse views for matching tasks. The model pool includes Depth-n ConvNets [19] and ResNet-18/32 [25] with different initialization random seeds [61,93]. To ensure diversity, we randomly select one model from the pool at each iteration. In a Depth-n ConvNet, n blocks precede a fully connected layer, with each block containing a 3×3 convolutional layer (128 filters), instance normalization [73], ReLU activation, and 2×2 average pooling (stride of 2). The number of blocks depends on dataset resolution: $n = 3, 4, 5$, and 6 for 32×32, 64×64, 128×128, and 256×256, respectively. We initialize networks using normal initialization [24] and use BigGAN [3] as a default generative model.

Evaluation Protocol. Our method uniquely allows the dynamic generation of images for training. To ensure fairness with conventional studies like [61,87,89,90], we first measure the end-to-end latency for training each model on the given dataset at a particular IPC. When evaluating our method, we generate images on-the-fly during downstream model training, ensuring that the total time for both generating and training matches the latency constraints set by previous works. This approach explores a tradeoff between model training and image generation on the fly during evaluation. Details on the generation and training time costs are available with our source code[1]. Apart from dynamic generation, we use the same ConvNet networks and optimizers as previous works for valid comparison. We also quantify the computational costs during distillation in GPU hours and the number of learnable parameters. For fairness and reproducibility, we employ publicly available datasets of baselines and train models using our experimental settings. For datasets without specific experiments, we implemented them using the released author codes. Additional implementation details are provided in the supplementary materials.

Table 1. The performance (%) comparison to priors on CIFAR and TinyImageNet. ConvNet is used for evaluation. For more details, refer to the supplementary materials.

Dataset	CIFAR-10			CIFAR-100			Tiny ImageNet		
IPC	1	10	50	1	10	50	1	10	50
Ratio %	0.02	0.2	1	0.2	2	10	0.2	2	10
Random	$14.4_{\pm 2.0}$	$26.0_{\pm 1.2}$	$43.4_{\pm 1.0}$	$4.2_{\pm 0.3}$	$14.6_{\pm 0.5}$	$30.0_{\pm 0.4}$	$1.4_{\pm 0.1}$	$5.0_{\pm 0.2}$	$15.0_{\pm 0.4}$
Herding	$21.5_{\pm 1.2}$	$31.6_{\pm 0.7}$	$40.4_{\pm 0.6}$	$8.4_{\pm 0.3}$	$17.3_{\pm 0.3}$	$33.7_{\pm 0.5}$	$2.8_{\pm 0.2}$	$6.3_{\pm 0.2}$	$16.7_{\pm 0.3}$
Forgetting	$13.5_{\pm 1.2}$	$23.3_{\pm 1.0}$	$23.3_{\pm 1.1}$	$4.5_{\pm 0.2}$	$15.1_{\pm 0.3}$	$30.5_{\pm 0.3}$	$1.6_{\pm 0.1}$	$5.1_{\pm 0.2}$	$15.0_{\pm 0.3}$
DC [90]	$28.3_{\pm 0.5}$	$44.9_{\pm 0.5}$	$53.9_{\pm 0.5}$	$12.8_{\pm 0.3}$	$25.2_{\pm 0.3}$	$30.6_{\pm 0.6}$	$5.3_{\pm 0.1}$	$12.9_{\pm 0.1}$	$12.7_{\pm 0.4}$
DSA [87]	$28.8_{\pm 0.7}$	$52.1_{\pm 0.5}$	$60.6_{\pm 0.5}$	$13.9_{\pm 0.3}$	$32.3_{\pm 0.3}$	$42.8_{\pm 0.4}$	$6.6_{\pm 0.2}$	$16.3_{\pm 0.2}$	$25.3_{\pm 0.2}$
DM [89]	$26.0_{\pm 0.8}$	$48.9_{\pm 0.6}$	$63.0_{\pm 0.4}$	$11.4_{\pm 0.3}$	$29.7_{\pm 0.3}$	$43.6_{\pm 0.4}$	$3.9_{\pm 0.2}$	$12.9_{\pm 0.4}$	$24.1_{\pm 0.3}$
CAFE [75]	$30.3_{\pm 1.1}$	$46.3_{\pm 0.6}$	$55.5_{\pm 0.6}$	$12.9_{\pm 0.3}$	$27.8_{\pm 0.3}$	$37.9_{\pm 0.3}$	-	-	-
CAFE+DSA [75]	$31.6_{\pm 0.8}$	$50.9_{\pm 0.5}$	$62.3_{\pm 0.4}$	$14.0_{\pm 0.3}$	$31.5_{\pm 0.2}$	$42.9_{\pm 0.2}$	-	-	-
DAM [61]	$32.0_{\pm 1.2}$	$54.2_{\pm 0.8}$	$67.0_{\pm 0.4}$	$14.5_{\pm 0.5}$	$34.8_{\pm 0.5}$	$49.4_{\pm 0.3}$	$8.3_{\pm 0.4}$	$18.7_{\pm 0.3}$	$28.7_{\pm 0.3}$
ATOM [35]	$34.8_{\pm 1.0}$	$57.9_{\pm 0.7}$	$68.8_{\pm 0.5}$	$18.1_{\pm 0.4}$	$35.7_{\pm 0.4}$	$50.2_{\pm 0.3}$	$9.1_{\pm 0.2}$	$19.5_{\pm 0.4}$	$29.1_{\pm 0.3}$
IDM [91]	$45.6_{\pm 0.7}$	$58.6_{\pm 0.1}$	$67.5_{\pm 0.1}$	$20.1_{\pm 0.3}$	$45.1_{\pm 0.1}$	$50.0_{\pm 0.2}$	$10.1_{\pm 0.2}$	$21.9_{\pm 0.2}$	$27.7_{\pm 0.2}$
KIP [56]	$49.9_{\pm 0.2}$	$62.7_{\pm 0.3}$	$68.6_{\pm 0.2}$	$15.7_{\pm 0.2}$	$28.3_{\pm 0.1}$	-	-	-	-
FRePo [93]	$46.8_{\pm 0.7}$	$65.5_{\pm 0.4}$	$71.7_{\pm 0.2}$	$28.7_{\pm 0.1}$	$42.5_{\pm 0.2}$	$44.3_{\pm 0.2}$	$15.4_{\pm 0.3}$	$25.4_{\pm 0.2}$	-
MTT [5]	$46.3_{\pm 0.8}$	$65.3_{\pm 0.7}$	$71.6_{\pm 0.2}$	$24.3_{\pm 0.3}$	$40.1_{\pm 0.4}$	$47.7_{\pm 0.2}$	$8.8_{\pm 0.3}$	$23.2_{\pm 0.2}$	$28.0_{\pm 0.3}$
TESLA [12]	$48.5_{\pm 0.8}$	$66.4_{\pm 0.8}$	$72.6_{\pm 0.7}$	$24.8_{\pm 0.4}$	$41.7_{\pm 0.3}$	$47.9_{\pm 0.3}$	-	-	-
D2M (Ours)	$\mathbf{50.2_{\pm 0.3}}$	$\mathbf{67.8_{\pm 0.5}}$	$\mathbf{74.4_{\pm 0.3}}$	$\mathbf{29.8_{\pm 0.4}}$	$\mathbf{46.6_{\pm 0.3}}$	$\mathbf{51.2_{\pm 0.2}}$	$\mathbf{16.7_{\pm 0.2}}$	$\mathbf{26.1_{\pm 0.4}}$	$\mathbf{30.1_{\pm 0.2}}$
Full Dataset		$84.8_{\pm 0.1}$			$56.2_{\pm 0.3}$			$37.6_{\pm 0.4}$	

---

[1] https://github.com/DataDistillation/D2M.

Implementation Details. We deploy the pre-trained BigGAN with the default hyperparameters and learning strategy outlined in [3]. In the distillation, we refine BigGAN using the SGD optimizer with a batch size of $B = 128$ over $K = 60$ epochs while setting the task balance λ and the temperature T to 100 and 4, respectively. We also apply the same set of differentiable augmentations [87] in all experiments, during distillation and evaluation. Experiments are conducted on two RTX A-6000 GPUs with 50 GB of memory. Additional hyperparameter details can be found in the supplementary materials.

4.2 Comparison to State-of-the-Art Methods

CIFAR and Tiny ImageNet. We compare the proposed D2M with 3 coreset selection and 12 dataset distillation approaches on the CIFAR-10/100 [37] and Tiny ImageNet [39] datasets, as detailed in Table 1. D2M consistently outperforms all baselines across different settings for these low- and medium-resolution datasets. Specifically, it achieves up to 88% of CIFAR-10's upper bound performance with only 1% of its training data and up to 70% of the upper bound with 2% of Tiny ImageNet's training data. These results highlight the D2M framework's robustness for data-efficient learning and its ability to distill informative knowledge for improved classification task performance.

Table 2. The performance comparison (%) to state-of-the-art methods on ImageNet-1K [13] and its high-resolution subsets [5].

Dataset	IPC	Random	DAM [61]	GLaD [6]	TESLA [12]	FRePo [93]	**D2M (Ours)**	Full Dataset
ImageNet-1K (64×64)	1	$0.5_{\pm 0.1}$	$2.0_{\pm 0.1}$	-	$7.7_{\pm 0.2}$	$7.5_{\pm 0.3}$	$\mathbf{8.3_{\pm 0.2}}$	$33.8_{\pm 0.3}$
	2	$0.9_{\pm 0.1}$	$2.2_{\pm 0.1}$	-	$10.5_{\pm 0.3}$	$9.7_{\pm 0.2}$	$\mathbf{11.1_{\pm 0.2}}$	
	10	$3.1_{\pm 0.2}$	$6.3_{\pm 0.0}$	-	$17.8_{\pm 1.3}$	-	$\mathbf{18.2_{\pm 0.9}}$	
	50	$7.6_{\pm 1.2}$	$15.5_{\pm 0.2}$	-	$27.9_{\pm 1.2}$	-	$\mathbf{28.3_{\pm 1.1}}$	
ImageNette	1	$23.5_{\pm 4.8}$	$34.7_{\pm 0.9}$	$38.7_{\pm 1.6}$	-	$48.1_{\pm 0.7}$	$\mathbf{49.9_{\pm 0.2}}$	$87.4_{\pm 1.0}$
	10	$47.7_{\pm 2.4}$	$59.4_{\pm 0.4}$	-	-	$66.5_{\pm 0.8}$	$\mathbf{67.7_{\pm 0.3}}$	
ImageWoof	1	$14.2_{\pm 0.9}$	$24.2_{\pm 0.5}$	$23.4_{\pm 1.1}$	-	$29.7_{\pm 0.6}$	$\mathbf{32.3_{\pm 0.4}}$	$67.0_{\pm 1.3}$
	10	$27.0_{\pm 1.9}$	$34.4_{\pm 0.4}$	-	-	$42.2_{\pm 0.9}$	$\mathbf{44.9_{\pm 0.3}}$	
ImageSquawk	1	$21.8_{\pm 0.5}$	$36.4_{\pm 0.8}$	$35.8_{\pm 1.4}$	-	-	$\mathbf{40.3_{\pm 0.5}}$	$87.5_{\pm 0.3}$
	10	$40.2_{\pm 0.4}$	$55.4_{\pm 0.9}$	-	-	-	$\mathbf{59.8_{\pm 0.2}}$	

ImageNet-1K and Its Subsets. ImageNet-1K [13] and its subsets are more challenging than the CIFAR-10/100 and Tiny ImageNet datasets due to their complex label spaces and higher resolutions. Consequently, memory and time constraints prevented previous data distillation works from scaling up to ImageNet-1K at high resolution [5,16,61,87,88,93]. However, D2M overcomes this by distilling information from the original dataset into the generative model's parameters rather than relying on complex raw pixels. This enables D2M, as one of the pioneering works to generate synthetic 128×128 ImageNet-1K images, outperforming Random by a significant margin up to 19.1% (Fig. 3). To understand how well our method performs on high-resolution images, we also extended the evaluation to three high-resolution ImageNet-1K subsets: ImageNette (assorted objects), ImageWoof (dog breeds), and ImageSquawk (birds) datasets, all with a resolution of 128×128. As shown in Table 2, we outperform prior works with the same model architecture across all settings. Notably, D2M improves the performance of the top competitor, FRePo, on the ImageWoof by more than 2.7%. This improvement stems from the fact that D2M effectively distills essential information from the original dataset into the generative model through channel-wise attention matching and softened output prediction alignment. Our ablation studies in Sect. 4.3 further confirm that the performance gain is directly related to the method's ability to generate informative images.

To further evaluate our method's effectiveness, we compare D2M to GLaD [6] using ImageNet-A to ImageNet-E, as well as ImageFruit and ImageMeow (cat breeds). It should be noted that GLaD is a plug-and-play method and can be added to existing dataset distillation studies. Table 3 presents an analysis of performance between D2M and GLaD when applied to previous techniques, namely DC [90], DM [89], and MTT [5]. In each case, we distilled a synthetic dataset with one image per class and then evaluated it on the Depth-5 ConvNet. The results show D2M's superiority across all experimental settings.

Table 3. The performance (%) comparison to prior works on 128×128 ImageNet-Subset (ImageFruit, ImageMeow, and ImageNet-[A, B, C, D, E]) with IPC1.

Methods	ImageFruit	ImageMeow	ImageNet-A	ImageNet-B	ImageNet-C	ImageNet-D	ImageNet-E
DC [90]	$21.0_{\pm 0.9}$	$22.0_{\pm 0.6}$	$43.2_{\pm 0.6}$	$47.2_{\pm 0.7}$	$41.3_{\pm 0.7}$	$34.3_{\pm 1.5}$	$34.9_{\pm 1.5}$
DM [89]	$20.4_{\pm 1.9}$	$20.1_{\pm 1.2}$	$39.4_{\pm 1.8}$	$40.9_{\pm 1.7}$	$39.0_{\pm 1.3}$	$30.8_{\pm 0.9}$	$27.0_{\pm 0.8}$
MTT [5]	$22.4_{\pm 1.1}$	$26.6_{\pm 0.4}$	$51.7_{\pm 0.2}$	$53.3_{\pm 1.0}$	$48.0_{\pm 0.7}$	$43.0_{\pm 0.6}$	$39.5_{\pm 0.9}$
DC+GLaD [6]	$20.7_{\pm 1.1}$	$22.6_{\pm 0.8}$	$44.1_{\pm 2.4}$	$49.2_{\pm 1.1}$	$42.0_{\pm 0.6}$	$35.6_{\pm 0.9}$	$35.8_{\pm 0.9}$
DM+GLaD [6]	$21.8_{\pm 1.8}$	$22.3_{\pm 1.6}$	$41.0_{\pm 1.5}$	$42.9_{\pm 1.9}$	$39.4_{\pm 0.7}$	$33.2_{\pm 1.4}$	$30.3_{\pm 1.3}$
MTT+GLaD [6]	$23.1_{\pm 0.4}$	$26.0_{\pm 1.1}$	$50.7_{\pm 0.4}$	$51.9_{\pm 1.3}$	$44.9_{\pm 0.4}$	$39.9_{\pm 1.7}$	$37.6_{\pm 0.7}$
D2M (Ours)	$\mathbf{24.4}_{\pm 0.5}$	$\mathbf{28.4}_{\pm 0.3}$	$\mathbf{55.2}_{\pm 0.2}$	$\mathbf{56.8}_{\pm 0.6}$	$\mathbf{50.4}_{\pm 0.4}$	$\mathbf{45.2}_{\pm 0.6}$	$\mathbf{41.6}_{\pm 0.8}$

Table 4. GPU hours for re-distillation process from IPC1 to IPC10 and IPC50 on CIFAR-10. Total denotes the GPU hours spent on the distillation for all three IPCs.

Method	Distillation Time (GPU hours)					
	DC [90]	DSA [87]	DM [89]	MTT [5]	DAM [61]	**D2M**
IPC1	0.2	0.2	1.2	1.4	1.2	4.9
IPC1→ IPC10	1.8	2.0	1.5	7.8	1.5	**0**
IPC1→ IPC50	8.3	8.9	4.7	38.2	4.8	**0**
Total	10.3	11.1	7.4	47.4	7.5	**4.9**

Higher Resolution Datasets (256×256). We extend the scaling of D2M to higher resolutions, specifically datasets with a resolution of 256×256. To that end, we conduct experiments on two datasets: ImageSquawk and DermaMNIST. The results, shown in Table 6, demonstrate superior performance across both datasets, surpassing Random by a significant margin.

Distillation Cost Analysis. In efficient learning, it is critical to consider computational costs during the distillation. We compare the re-distillation cost and the number of learnable parameters for D2M against priors in Table 4 and Fig. 3, respectively. When IPCs change, previous methods require dataset re-distillation, incurring substantial computational expenses and increasing GPU hours (Table 4). In contrast, D2M conducts optimization on generative models' parameters rather than raw pixels, requiring only a *single* distillation for varying IPCs. This precludes the need for repeated distillation, saving up to 42.5 GPU hours. Additionally, while the number of learnable parameters in pixel-wise data distillation increases with dataset size and resolution, D2M keeps the cost unchanged across all IPCs, saving up to 35x for IPC50, as shown in Fig. 3.

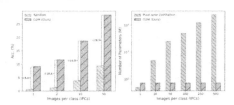

Fig. 3. Performance comparison and count of parameters on 128×128 ImageNet-1K.

Table 5. Cross-architecture performance on CIFAR-10 with 50 images per class.

Method	**Unseen** Evaluation Architecture					Average (%)
	AlexNet [38]	VGG-11 [68]	ResNet-50 [25]	DenseNet-121 [30]	ViT [15]	
DSA [87]	53.7$_{\pm 0.6}$	51.4$_{\pm 1.0}$	24.4$_{\pm 0.5}$	24.6$_{\pm 1.0}$	43.3$_{\pm 0.4}$	39.5$_{\pm 0.7}$
CAFE [75]	34.0$_{\pm 0.6}$	40.5$_{\pm 0.8}$	46.7$_{\pm 1.2}$	49.7$_{\pm 1.3}$	22.7$_{\pm 0.7}$	38.7$_{\pm 0.9}$
MTT [5]	48.2$_{\pm 1.0}$	55.4$_{\pm 0.8}$	61.3$_{\pm 0.4}$	62.3$_{\pm 1.1}$	47.7$_{\pm 0.6}$	55.0$_{\pm 0.8}$
DM [89]	60.1$_{\pm 0.5}$	57.4$_{\pm 0.8}$	51.5$_{\pm 0.4}$	58.9$_{\pm 0.7}$	45.2$_{\pm 0.4}$	54.6$_{\pm 0.6}$
FTD [16]	53.8$_{\pm 0.9}$	58.4$_{\pm 1.6}$	59.9$_{\pm 1.3}$	58.3$_{\pm 1.5}$	49.3$_{\pm 0.7}$	55.9$_{\pm 1.2}$
DAM [61]	63.9$_{\pm 0.9}$	64.8$_{\pm 0.5}$	59.6$_{\pm 0.8}$	62.2$_{\pm 0.4}$	48.2$_{\pm 0.8}$	59.7$_{\pm 0.7}$
D2M (Ours)	**65.4**$_{\pm 0.7}$	**67.3**$_{\pm 0.5}$	**64.8**$_{\pm 0.6}$	**66.1**$_{\pm 0.5}$	**54.7**$_{\pm 0.6}$	**63.6**$_{\pm 0.6}$

4.3 Ablation Studies and Analysis

Cross-Architecture Generalization. Building on prior studies [5,61,75,89,90], we evaluate the ability of D2M-generated images for understanding classification tasks without overfitting to a specific architecture. We use distillation methods to create synthetic data with a default network and evaluate their performance on unseen architectures. Following the settings of [16,61], we experiment

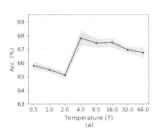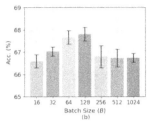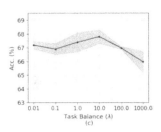

Fig. 4. The effect of (a) temperature, (b) batch size, and (c) task balance on the D2M's performance for CIFAR10 with IPC10.

on the CIFAR-10 dataset with IPC 50 and test generalization with architectures like AlexNet [38], VGG-11 [68], ResNet-50 [25], DenseNet-121 [30], and ViT [15]. The findings in Table 5 show that synthetic images from D2M not only generalize well across various models but also outperform the best competitor on average up to a margin of 3.9%. D2M illustrates the robustness of vision transformers and complex architectures like ResNet-50 and DenseNet-121 due to its powerful generative capability and distilled knowledge. This suggests D2M can identify crucial learning features beyond mere knowledge matching.

Exploring the Effect of Temperature T in D2M. In our framework, soft label predictions enrich the information in the generated synthetic dataset. Inspired by [27,59], we use a temperature parameter T to control the entropy of the output distribution while generating soft label predictions. This ablation explores the impact of temperature on overall performance. As shown in Fig. 4(a), increasing the T value improves D2M's performance up to a certain point ($T = 4$). A lower temperature emphasizes only maximal logits, while a higher value flattens the distribution, focusing on different logits uniformly. Tuning the softmax temperature allows soft labels to reveal extra knowledge like class relationships for training, enabling information flow across classes (*dark knowledge* [27]).

Evaluation of Task Balance λ in D2M. In the D2M framework, we train the generative model by minimizing a linear combination of embedded-based and prediction-based matching losses, as expressed in Eq. (5). The parameter λ serves as the regularizing coefficient, determining the balance between the losses. In Fig. 4(c), we study different values of λ (ranging from 0.01 to 1000) on the CIFAR10 dataset with IPC10 to assess its sensitivity. Our default value of $\lambda = 10$ yields the best results. Values that are too high, such as 1000, reduce the effectiveness of embedding matching, resulting in a loss of informative knowledge. Conversely, too small values compromise regularization and impede the distillation of discriminative label knowledge, which is beneficial for classification tasks.

Exploring the Effect of Loss Components in D2M. We conduct an experiment to evaluate the effect of loss components, $\mathcal{L}_{EM}$ and $\mathcal{L}_{PM}$, on D2M's performance. Conventional generative models strive to produce real-looking images but may lack the informative value of real images for training due to inherent

Table 6. The performance comparison for high-resolution (256×256) datasets.

Dataset	IPC	Random	D2M
ImageSquawk	1	$22.3_{\pm 0.5}$	$\mathbf{48.2_{\pm 0.4}}$
	10	$41.3_{\pm 0.6}$	$\mathbf{63.4_{\pm 0.3}}$
DermaMNIST	1	$21.3_{\pm 1.3}$	$\mathbf{46.4_{\pm 0.4}}$
	10	$38.4_{\pm 1.1}$	$\mathbf{67.2_{\pm 0.3}}$

Table 7. The effect of different generators on D2M's performance for the CIFAR10 dataset.

IPC	1	10	50
CGAN	$49.6_{\pm 0.7}$	$67.4_{\pm 0.3}$	$72.2_{\pm 0.8}$
BigGAN	$50.2_{\pm 0.2}$	$67.8_{\pm 0.5}$	$74.4_{\pm 0.3}$
StyleGAN-XL	$\mathbf{50.7_{\pm 0.5}}$	$\mathbf{68.2_{\pm 0.4}}$	$\mathbf{75.2_{\pm 0.4}}$
CVAE	$49.8_{\pm 0.3}$	$67.6_{\pm 0.2}$	$73.1_{\pm 0.4}$

information loss [61,75,89]. We refine these generators during distillation with different loss designs and observe improvements of 10.3% and 7.9% for $\mathcal{L}_{EM}$ and $\mathcal{L}_{PM}$, respectively, on CIFAR10 with IPC10. This enhancement is attributed to the transfer of informative and discriminative knowledge for classification tasks. Moreover, combining $\mathcal{L}_{EM}$ and $\mathcal{L}_{PM}$ led to a remarkable 12.7% improvement.

Evaluation of Different Generative Models in D2M. We investigate the effect of different pre-trained generators on D2M by deploying BigGAN [3], CGAN [53], StyleGAN-XL [66], and a CVAE model [11,69] on CIFAR-10 with IPC1, 10, and 50. Table 7 demonstrates that all models effectively synthesize images for classification, with StyleGAN-XL outperforming the rest. However, BigGAN was chosen as the default due to its balance of performance and computational efficiency. Details on the computational costs are given in the supplementary.

4.4 Visulization and Application

Synthetic Images. We visualize synthetic images for various datasets and resolutions in Fig. 5. These images are identifiable, although they may contain artificial patterns that improve their informativeness, like discriminative features of the animals. Note that our generative model prioritizes informativeness over realism. More visualizations are available in the supplementary materials.

Neural Architecture Search. The synthetic images can act as a proxy set to accelerate model evaluation in the Neural Architecture Search (NAS). Following [61,90], we conduct NAS on CIFAR-10 with a search space of 720 ConvNets,

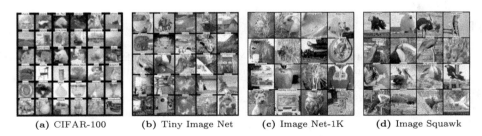

(a) CIFAR-100 (b) Tiny Image Net (c) Image Net-1K (d) Image Squawk

Fig. 5. Example generated images from 32×32 CIFAR-100, 64×64 Tiny ImageNet, 128×128 ImageNet-1K, and 128×128 ImageSquawk.

varying in network depth, width, activation, normalization, and pooling layers. More implementation details can be found in the supplementary materials. We train all architectures on the Random, DSA, CAFE, DM, DAM, and D2M synthetic sets with IPC50 for 200 epochs and on the full dataset for 100 epochs to establish a baseline. We rank architectures based on validation performance and report the testing accuracy of the best-selected model when trained on the test set. Spearman's rank correlation also evaluates the reliability of synthetic images for architecture search, comparing testing performances between models trained on proxy and real training sets. As illustrated in Table 8, D2M-generated images achieve the highest performance (88.8%) and a strong rank correlation (0.80) compared to priors.

Table 8. Neural architecture search on the CIFAR-10 dataset.

	Random	DSA [87]	DM [89]	DAM [61]	CAFE [75]	**D2M (Ours)**	Whole Dataset
Performance (%)	88.9	87.2	87.2	88.8	83.6	**88.9**	89.2
Correlation	0.70	0.66	0.71	0.72	0.59	**0.80**	1.00
Time cost (min)	206.4	206.4	206.6	206.4	206.4	206.5	5168.9
Number of Images	**500**	**500**	**500**	**500**	**500**	**500**	5×10^4

5 Conclusion and Limitations

In this work, we introduce an efficient framework (D2M), which distills information from large-scale training datasets into the parameters of a pre-trained generative model instead of synthetic images. Our generative model employs carefully designed matching modules for distillation to enhance re-distillation efficiency, scalability to high-resolution datasets, and cross-architecture generality. We conduct extensive experiments across various datasets to demonstrate the efficiency and effectiveness of D2M. However, D2M needs more GPU memory in IPC1 compared to data distillation methods (Fig. 3). In the future, we plan to apply a more efficient generative model to small IPCs and explore D2M's potential for other downstream tasks like object detection and segmentation.

References

1. Amer, H., Salamah, A.H., Sajedi, A., Yang, E.H.: High performance convolution using sparsity and patterns for inference in deep convolutional neural networks. arXiv preprint arXiv:2104.08314 (2021)
2. Belouadah, E., Popescu, A.: ScaIL: classifier weights scaling for class incremental learning. In: Proceedings of the IEEE/CVF Winter Conference on Applications of Computer Vision, pp. 1266–1275 (2020)
3. Brock, A., Donahue, J., Simonyan, K.: Large scale GAN training for high fidelity natural image synthesis. In: International Conference on Learning Representations (2018)

4. Brown, T., et al.: Language models are few-shot learners. Adv. Neural. Inf. Process. Syst. **33**, 1877–1901 (2020)
5. Cazenavette, G., Wang, T., Torralba, A., Efros, A.A., Zhu, J.Y.: Dataset distillation by matching training trajectories. In: Proceedings of the IEEE/CVF Conference on Computer Vision and Pattern Recognition, pp. 4750–4759 (2022)
6. Cazenavette, G., Wang, T., Torralba, A., Efros, A.A., Zhu, J.Y.: Generalizing dataset distillation via deep generative prior. In: Proceedings of the IEEE/CVF Conference on Computer Vision and Pattern Recognition, pp. 3739–3748 (2023)
7. Chen, D., Kerkouche, R., Fritz, M.: Private set generation with discriminative information. Adv. Neural. Inf. Process. Syst. **35**, 14678–14690 (2022)
8. Chen, L., et al.: SCA-CNN: spatial and channel-wise attention in convolutional networks for image captioning. In: Proceedings of the IEEE Conference on Computer Vision and Pattern Recognition, pp. 5659–5667 (2017)
9. Chen, X., Duan, Y., Houthooft, R., Schulman, J., Sutskever, I., Abbeel, P.: InfoGAN: interpretable representation learning by information maximizing generative adversarial nets. In: Advances in Neural Information Processing Systems, vol. 29 (2016)
10. Chen, X., Yang, Y., Wang, Z., Mirzasoleiman, B.: Data distillation can be like vodka: distilling more times for better quality. arXiv preprint arXiv:2310.06982 (2023)
11. Child, R.: Very deep VAEs generalize autoregressive models and can outperform them on images. In: International Conference on Learning Representations (2021)
12. Cui, J., Wang, R., Si, S., Hsieh, C.J.: Scaling up dataset distillation to ImageNet-1k with constant memory. In: International Conference on Machine Learning, pp. 6565–6590. PMLR (2023)
13. Deng, J., Dong, W., Socher, R., Li, L.J., Li, K., Fei-Fei, L.: ImageNet: a large-scale hierarchical image database. In: 2009 IEEE Conference on Computer Vision and Pattern Recognition, pp. 248–255. IEEE (2009)
14. Dong, T., Zhao, B., Lyu, L.: Privacy for Free: how does dataset condensation help privacy? In: International Conference on Machine Learning, pp. 5378–5396. PMLR (2022)
15. Dosovitskiy, A., et al.: An image is worth 16×16 words: transformers for image recognition at scale. arXiv preprint arXiv:2010.11929 (2020)
16. Du, J., Jiang, Y., Tan, V.Y., Zhou, J.T., Li, H.: Minimizing the accumulated trajectory error to improve dataset distillation. In: Proceedings of the IEEE/CVF Conference on Computer Vision and Pattern Recognition, pp. 3749–3758 (2023)
17. Du, J., Shi, Q., Zhou, J.T.: Sequential subset matching for dataset distillation. In: Advances in Neural Information Processing Systems, vol. 36 (2024)
18. Furlanello, T., Lipton, Z., Tschannen, M., Itti, L., Anandkumar, A.: Born again neural networks. In: International Conference on Machine Learning, pp. 1607–1616. PMLR (2018)
19. Gidaris, S., Komodakis, N.: Dynamic few-shot visual learning without forgetting. In: Proceedings of the IEEE Conference on Computer Vision and Pattern Recognition, pp. 4367–4375 (2018)
20. Goodfellow, I., et al.: Generative adversarial nets. In: Advances in Neural Information Processing Systems, vol. 27 (2014)
21. Gu, J., Wang, K., Jiang, W., You, Y.: Summarizing stream data for memory-restricted online continual learning. In: Proceedings of the AAAI Conference on Artificial Intelligence (AAAI) (2024)

22. Guo, Z., Wang, K., Cazenavette, G., Li, H., Zhang, K., You, Y.: Towards lossless dataset distillation via difficulty-aligned trajectory matching. In: The Twelfth International Conference on Learning Representations (2024). https://openreview.net/forum?id=rTBL8OhdhH
23. He, J., Shi, W., Chen, K., Fu, L., Dong, C.: GCFSR: a generative and controllable face super resolution method without facial and GAN priors. In: Proceedings of the IEEE/CVF Conference on Computer Vision and Pattern Recognition, pp. 1889–1898 (2022)
24. He, K., Zhang, X., Ren, S., Sun, J.: Delving deep into rectifiers: surpassing human-level performance on ImageNet classification. In: Proceedings of the IEEE International Conference on Computer Vision, pp. 1026–1034 (2015)
25. He, K., Zhang, X., Ren, S., Sun, J.: Deep residual learning for image recognition. In: Proceedings of the IEEE Conference on Computer Vision and Pattern Recognition, pp. 770–778 (2016)
26. He, Y., Xiao, L., Zhou, T.J.: You Only Condense Once: two rules for pruning condensed datasets. In: Proceedings of the Advances in Neural Information Processing Systems (NeurIPS) (2023)
27. Hinton, G., Vinyals, O., Dean, J.: Distilling the knowledge in a neural network. arXiv preprint arXiv:1503.02531 (2015)
28. Ho, J., Ermon, S.: Generative adversarial imitation learning. In: Advances in Neural Information Processing Systems, vol. 29 (2016)
29. Howard, J.: Imagenette: a smaller subset of 10 easily classified classes from ImageNet, and a little more French (2019)
30. Huang, G., Liu, Z., Van Der Maaten, L., Weinberger, K.Q.: Densely connected convolutional networks. In: Proceedings of the IEEE Conference on Computer Vision and Pattern Recognition, pp. 4700–4708 (2017)
31. Jin, Y., Wang, J., Lin, D.: Multi-level logit distillation. In: Proceedings of the IEEE/CVF Conference on Computer Vision and Pattern Recognition, pp. 24276–24285 (2023)
32. Kenton, J.D.M.W.C., Toutanova, L.K.: BERT: pre-training of deep bidirectional transformers for language understanding. In: Proceedings of NAACL-HLT, pp. 4171–4186 (2019)
33. Khaki, S., Luo, W.: CFDP: common frequency domain pruning. In: Proceedings of the IEEE/CVF Conference on Computer Vision and Pattern Recognition, pp. 4714–4723 (2023)
34. Khaki, S., Plataniotis, K.N.: The need for speed: pruning transformers with one recipe. arXiv preprint arXiv:2403.17921 (2024)
35. Khaki, S., Sajedi, A., Wang, K., Liu, L.Z., Lawryshyn, Y.A., Plataniotis, K.N.: ATOM: attention mixer for efficient dataset distillation. In: Proceedings of the IEEE/CVF Conference on Computer Vision and Pattern Recognition, pp. 7692–7702 (2024)
36. Kim, J.H., et al.: Dataset condensation via efficient synthetic-data parameterization. In: International Conference on Machine Learning, pp. 11102–11118. PMLR (2022)
37. Krizhevsky, A., Hinton, G., et al.: Learning multiple layers of features from tiny images (2009)
38. Krizhevsky, A., Sutskever, I., Hinton, G.E.: ImageNet classification with deep convolutional neural networks. Commun. ACM **60**(6), 84–90 (2017)
39. Le, Y., Yang, X.: Tiny ImageNet visual recognition challenge. CS 231N **7**(7), 3 (2015)

40. Ledig, C., et al.: Photo-realistic single image super-resolution using a generative adversarial network. In: Proceedings of the IEEE Conference on Computer Vision and Pattern Recognition, pp. 4681–4690 (2017)
41. Lee, C.H., Liu, Z., Wu, L., Luo, P.: MaskGAN: towards diverse and interactive facial image manipulation. In: Proceedings of the IEEE/CVF Conference on Computer Vision and Pattern Recognition, pp. 5549–5558 (2020)
42. Lee, S., Chun, S., Jung, S., Yun, S., Yoon, S.: Dataset condensation with contrastive signals. In: International Conference on Machine Learning, pp. 12352–12364. PMLR (2022)
43. Li, C., et al.: Knowledge condensation distillation. In: European Conference on Computer Vision, pp. 19–35. Springer (2022). https://doi.org/10.1007/978-3-031-20083-0_2
44. Li, H., Kadav, A., Durdanovic, I., Samet, H., Graf, H.P.: Pruning filters for efficient convnets. In: International Conference on Learning Representations (2017)
45. Li, J., Liang, X., Wei, Y., Xu, T., Feng, J., Yan, S.: Perceptual generative adversarial networks for small object detection. In: Proceedings of the IEEE Conference on Computer Vision and Pattern Recognition, pp. 1222–1230 (2017)
46. Liu, L., Muelly, M., Deng, J., Pfister, T., Li, L.J.: Generative modeling for small-data object detection. In: Proceedings of the IEEE/CVF International Conference on Computer Vision, pp. 6073–6081 (2019)
47. Liu, P., Yu, X., Zhou, J.T.: Meta knowledge condensation for federated learning. In: The Eleventh International Conference on Learning Representations (2022)
48. Liu, S., Wang, K., Yang, X., Ye, J., Wang, X.: Dataset distillation via factorization. Adv. Neural. Inf. Process. Syst. **35**, 1100–1113 (2022)
49. Liu, S., Ye, J., Yu, R., Wang, X.: Slimmable dataset condensation. In: Proceedings of the IEEE/CVF Conference on Computer Vision and Pattern Recognition, pp. 3759–3768 (2023)
50. Loo, N., Hasani, R., Amini, A., Rus, D.: Efficient dataset distillation using random feature approximation. Adv. Neural. Inf. Process. Syst. **35**, 13877–13891 (2022)
51. Loo, N., Hasani, R., Lechner, M., Amini, A., Rus, D.: Understanding reconstruction attacks with the neural tangent kernel and dataset distillation (2023)
52. Ma, C., Huang, J.B., Yang, X., Yang, M.H.: Hierarchical convolutional features for visual tracking. In: Proceedings of the IEEE International Conference on Computer Vision, pp. 3074–3082 (2015)
53. Mirza, M., Osindero, S.: Conditional generative adversarial nets. arXiv preprint arXiv:1411.1784 (2014)
54. Molchanov, P., Tyree, S., Karras, T., Aila, T., Kautz, J.: Pruning convolutional neural networks for resource efficient inference. In: International Conference on Learning Representations (2019)
55. Nguyen, T., Chen, Z., Lee, J.: Dataset meta-learning from kernel-ridge regression. In: International Conference on Learning Representations (2021). https://openreview.net/forum?id=l-PrrQrK0QR
56. Nguyen, T., Novak, R., Xiao, L., Lee, J.: Dataset distillation with infinitely wide convolutional networks. Adv. Neural. Inf. Process. Syst. **34**, 5186–5198 (2021)
57. Park, W., Kim, D., Lu, Y., Cho, M.: Relational knowledge distillation. In: Proceedings of the IEEE/CVF Conference on Computer Vision and Pattern Recognition, pp. 3967–3976 (2019)
58. Rebuffi, S.A., Kolesnikov, A., Sperl, G., Lampert, C.H.: iCaRL: incremental classifier and representation learning. In: Proceedings of the IEEE Conference on Computer Vision and Pattern Recognition, pp. 2001–2010 (2017)

59. Romero, A., Ballas, N., Kahou, S.E., Chassang, A., Gatta, C., Bengio, Y.: FitNets: hints for thin deep nets. arXiv preprint arXiv:1412.6550 (2014)
60. Saito, K., Watanabe, K., Ushiku, Y., Harada, T.: Maximum classifier discrepancy for unsupervised domain adaptation. In: Proceedings of the IEEE Conference on Computer Vision and Pattern Recognition, pp. 3723–3732 (2018)
61. Sajedi, A., Khaki, S., Amjadian, E., Liu, L.Z., Lawryshyn, Y.A., Plataniotis, K.N.: DataDAM: efficient dataset distillation with attention matching. In: Proceedings of the IEEE/CVF International Conference on Computer Vision, pp. 17097–17107 (2023)
62. Sajedi, A., Khaki, S., Lawryshyn, Y.A., Plataniotis, K.N.: ProbMCL: simple probabilistic contrastive learning for multi-label visual classification. In: ICASSP 2024-2024 IEEE International Conference on Acoustics, Speech and Signal Processing (ICASSP), pp. 5115–5119. IEEE (2024)
63. Sajedi, A., Khaki, S., Plataniotis, K.N., Hosseini, M.S.: End-to-end supervised multilabel contrastive learning. arXiv preprint arXiv:2307.03967 (2023)
64. Sajedi, A., Lawryshyn, Y.A., Plataniotis, K.N.: Subclass knowledge distillation with known subclass labels. In: 2022 IEEE 14th Image, Video, and Multidimensional Signal Processing Workshop (IVMSP), pp. 1–5. IEEE (2022)
65. Sajedi, A., Plataniotis, K.N.: On the efficiency of subclass knowledge distillation in classification tasks. arXiv preprint arXiv:2109.05587 (2021)
66. Sauer, A., Schwarz, K., Geiger, A.: StyleGAN-XL: Scaling StyleGAN to large diverse datasets. In: ACM SIGGRAPH 2022 Conference Proceedings, pp. 1–10 (2022)
67. Sener, O., Savarese, S.: Active learning for convolutional neural networks: a core-set approach. In: International Conference on Learning Representations (2018)
68. Simonyan, K., Zisserman, A.: Very deep convolutional networks for large-scale image recognition. arXiv preprint arXiv:1409.1556 (2014)
69. Sohn, K., Lee, H., Yan, X.: Learning structured output representation using deep conditional generative models. In: Advances in Neural Information Processing Systems, vol. 28 (2015)
70. Such, F.P., Rawal, A., Lehman, J., Stanley, K., Clune, J.: Generative teaching networks: accelerating neural architecture search by learning to generate synthetic training data. In: International Conference on Machine Learning, pp. 9206–9216. PMLR (2020)
71. Tan, M., Le, Q.: EfficientNet: rethinking model scaling for convolutional neural networks. In: International Conference on Machine Learning, pp. 6105–6114. PMLR (2019)
72. Toneva, M., Sordoni, A., des Combes, R.T., Trischler, A., Bengio, Y., Gordon, G.J.: An empirical study of example forgetting during deep neural network learning. In: International Conference on Learning Representations (2019)
73. Ulyanov, D., Vedaldi, A., Lempitsky, V.: Instance normalization: the missing ingredient for fast stylization. arXiv preprint arXiv:1607.08022 (2016)
74. Wang, K., Gu, J., Zhou, D., Zhu, Z., Jiang, W., You, Y.: DiM: distilling dataset into generative model. arXiv preprint arXiv:2303.04707 (2023)
75. Wang, K., et al.: Cafe: learning to condense dataset by aligning features. In: Proceedings of the IEEE/CVF Conference on Computer Vision and Pattern Recognition, pp. 12196–12205 (2022)
76. Wang, T.C., Liu, M.Y., Zhu, J.Y., Tao, A., Kautz, J., Catanzaro, B.: High-resolution image synthesis and semantic manipulation with conditional GANs. In: Proceedings of the IEEE Conference on Computer Vision and Pattern Recognition, pp. 8798–8807 (2018)

77. Wang, T., Zhu, J.Y., Torralba, A., Efros, A.A.: Dataset distillation. arXiv preprint arXiv:1811.10959 (2018)
78. Wang, Y.R., Khaki, S., Zheng, W., Hosseini, M.S., Plataniotis, K.N.: CONetV2: efficient auto-channel size optimization for CNNs. In: 2021 20th IEEE International Conference on Machine Learning and Applications (ICMLA), pp. 998–1003 (2021). https://doi.org/10.1109/ICMLA52953.2021.00164
79. Wu, J., Leng, C., Wang, Y., Hu, Q., Cheng, J.: Quantized convolutional neural networks for mobile devices. In: Proceedings of the IEEE Conference on Computer Vision and Pattern Recognition, pp. 4820–4828 (2016)
80. Xiong, Y., Wang, R., Cheng, M., Yu, F., Hsieh, C.J.: FedDM: iterative distribution matching for communication-efficient federated learning. In: Proceedings of the IEEE/CVF Conference on Computer Vision and Pattern Recognition, pp. 16323–16332 (2023)
81. Yang, E., Shen, L., Wang, Z., Liu, T., Guo, G.: An efficient dataset condensation plugin and its application to continual learning. In: Advances in Neural Information Processing Systems, vol. 36 (2024)
82. Yang, J., et al.: MedMNIST v2-a large-scale lightweight benchmark for 2D and 3D biomedical image classification. Sci. Data **10**(1), 41 (2023)
83. Yin, Z., Xing, E., Shen, Z.: Squeeze, recover and relabel: Dataset condensation at imagenet scale from a new perspective. In: Proceedings of the Advances in Neural Information Processing Systems (NeurIPS) (2023)
84. Yuan, L., Tay, F.E., Li, G., Wang, T., Feng, J.: Revisiting knowledge distillation via label smoothing regularization. In: Proceedings of the IEEE/CVF Conference on Computer Vision and Pattern Recognition, pp. 3903–3911 (2020)
85. Zagoruyko, S., Komodakis, N.: Paying more attention to attention: improving the performance of convolutional neural networks via attention transfer. In: International Conference on Learning Representations (2016)
86. Zhang, L., et al.: Accelerating dataset distillation via model augmentation. In: Proceedings of the IEEE/CVF Conference on Computer Vision and Pattern Recognition, pp. 11950–11959 (2023)
87. Zhao, B., Bilen, H.: Dataset condensation with differentiable Siamese augmentation. In: International Conference on Machine Learning, pp. 12674–12685. PMLR (2021)
88. Zhao, B., Bilen, H.: Synthesizing informative training samples with GAN. In: NeurIPS 2022 Workshop on Synthetic Data for Empowering ML Research (2022)
89. Zhao, B., Bilen, H.: Dataset condensation with distribution matching. In: Proceedings of the IEEE/CVF Winter Conference on Applications of Computer Vision, pp. 6514–6523 (2023)
90. Zhao, B., Mopuri, K.R., Bilen, H.: Dataset condensation with gradient matching. In: Ninth International Conference on Learning Representations 2021 (2021)
91. Zhao, G., Li, G., Qin, Y., Yu, Y.: Improved distribution matching for dataset condensation. In: Proceedings of the IEEE/CVF Conference on Computer Vision and Pattern Recognition, pp. 7856–7865 (2023)
92. Zhou, D., et al.: Dataset quantization. In: Proceedings of the IEEE/CVF International Conference on Computer Vision, pp. 17205–17216 (2023)
93. Zhou, Y., Nezhadarya, E., Ba, J.: Dataset distillation using neural feature regression. In: Advances in Neural Information Processing Systems (2022)
94. Zhu, J.Y., Park, T., Isola, P., Efros, A.A.: Unpaired image-to-image translation using cycle-consistent adversarial networks. In: Proceedings of the IEEE International Conference on Computer Vision, pp. 2223–2232 (2017)

DiffuX2CT: Diffusion Learning to Reconstruct CT Images from Biplanar X-Rays

Xuhui Liu[1], Zhi Qiao[2], Runkun Liu[2], Hong Li[1], Juan Zhang[1(✉)], Xiantong Zhen[2(✉)], Zhen Qian[2], and Baochang Zhang[1,3]

[1] Beihang University, Beijing, China
{xhliu,zhang_juan}@buaa.edu.cn
[2] Central Research Institute, United Imaging Healthcare, Beijing, China
zhenxt@gmail.com
[3] Zhongguancun Laboratory, Beijing, China

Abstract. Computed tomography (CT) is widely utilized in clinical settings because it delivers detailed 3D images of the human body. However, performing CT scans is not always feasible due to radiation exposure and limitations in certain surgical environments. As an alternative, reconstructing CT images from ultra-sparse X-rays offers a valuable solution and has gained significant interest in scientific research and medical applications. However, it presents great challenges as it is inherently an ill-posed problem, often compromised by artifacts resulting from overlapping structures in X-ray images. In this paper, we propose *DiffuX2CT*, which models CT reconstruction from orthogonal biplanar X-rays as a conditional diffusion process. *DiffuX2CT* is established with a 3D global coherence denoising model with a new, implicit conditioning mechanism. We realize the conditioning mechanism by a newly designed tri-plane decoupling generator and an implicit neural decoder. By doing so, *DiffuX2CT* achieves structure-controllable reconstruction, which enables 3D structural information to be recovered from 2D X-rays, therefore producing faithful textures in CT images. As an extra contribution, we collect a real-world lumbar CT dataset, called LumbarV, as a new benchmark to verify the clinical significance and performance of CT reconstruction from X-rays. Extensive experiments on this dataset and three more publicly available datasets demonstrate the effectiveness of our proposal.

Keywords: CT Reconstruction from Biplanar X-rays · Implicit Conditioning Mechanism · Tri-plane Representation

1 Introduction

Computed tomography (CT) is one of the widely used imaging modalities in clinical routines. CT images provide detailed and comprehensive views of target

Supplementary Information The online version contains supplementary material available at https://doi.org/10.1007/978-3-031-72775-7_26.

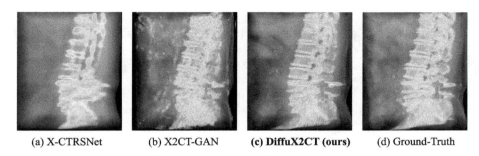

(a) X-CTRSNet   (b) X2CT-GAN   **(c) DiffuX2CT (ours)**   (d) Ground-Truth

Fig. 1. Comparison of different reconstruction methods. The reconstructed CT images are rendered for a more intuitive illustration. Our *DiffuX2CT* significantly outperforms competitors including both generative and regression methods. *DiffuX2CT* generates high-quality and faithful CT images. Notably, it recovers precise bone structures and correctly positioned implants, which are close to ground truth.

areas of interest, thereby playing a crucial role in clinical practice for diagnosis [76]. To acquire high-quality CT images, a full rotational scan with numerous X-ray projections is performed around the body. Standard reconstruction algorithms, such as filtered back projection and iterative reconstruction, facilitate the creation of these volumes [23]. This process allows physicians to accurately visualize the location and shape of structures like bones and implants [64]. However, the extensive data acquisition in CT scans makes it infeasible in most surgical scenarios. In addition, it is also less preferred to perform CT scans due to the higher radiation exposure to patients compared to a single X-ray. Fortunately, reconstruction of CT images from ultra-sparse X-rays offers an attractive alternative [62], which only requires cost-effective X-ray machines and reduces radiation exposure to patients. While X-rays excel in bone contrast and enable rapid assessments [32], they only project objects onto a 2D plane, lacking a complete 3D view of internal structures. Specifically, the limited internal body information in X-rays leads to significant ambiguity and artifacts, as numerous possible CT images could be obtained from the same set of 2D X-ray projections. In addition, the overlapping of objects in X-rays further complicates the reconstruction of CT images.

Due to the significance in both research and clinical practice, great efforts have been made in 3D CT reconstruction from X-rays [19,22,28,31,62,63,83, 85,91]. Early attempts treat it as a regression task, which usually utilizes deep neural networks [57] to directly establish a mapping from X-rays to CT images. However, these methods tend to generate blurry CT images, failing to recover realistic body structures (Fig. 1(a)). This is largely due to regression losses that simply calculate the averaged results of possible predictions. Deep generative models, e.g., generative adversarial networks (GANs) [20], have also been explored, which perform better in learning the distribution of body anatomy from a large training set, thereby delivering better sample quality [28,85]. However, these GAN-based methods fall short of recovering sufficient 3D structure from 2D X-rays, yielding severe artifacts and blurry images (Fig. 1(b)). Recently, diffusion models [25,67] have achieved tremendous success in various synthesis tasks [16,36,44,53,56,68,72,90]. They have also been tried on settings of sparse-view and limited-angle CT reconstruction problems [12,35,41]. Unlike the challenging setting addressed in our work, they still rely on CT scanners, not offering a solution for the extremely ill-posed problem of using biplanar X-rays in CT reconstruction. Moreover, they are developed under 2D network architectures, lacking the ability to establish 3D global coherence in CT images.

This paper presents a new 3D conditional diffusion model, termed *DiffuX2CT*, for CT reconstruction from biplanar orthogonal X-rays. This is different from current diffusion-based CT reconstruction methods, which still rely on CT scans. At a high level, *DiffuX2CT* formulates the reconstruction task as a conditional denoising diffusion process, where biplanar orthogonal X-rays are incorporated as a condition in the diffusion process. *DiffuX2CT* is implemented with a 3D encoder-decoder network to model the 3D prior distribution of body anatomy, ensuring high sample quality and global coherence throughout the 3D CT image. To reduce computation costs, *DiffuX2CT* introduces the shifted-window attention mechanism [45] as a substitute for the self-attention [4,73] operations. Moreover, we propose a new, implicit conditioning mechanism (ICM) to incorporate 3D structure information into CT image generation in the diffusion process. The ICM is realized by a newly designed tri-plane decoupling generator and an implicit neural decoder. The tri-plane decoupling generator individually reconstructs the three 2D planes of the tri-plane representations [9], while preserving their spatial correlations, enabling the effective recovery of 3D structure from 2D biplanar X-rays. Subsequently, the implicit neural decoder maps the tri-plane features and the assigned coordinates to 3D implicit representations for conditioning the synthesis process. In addition, we introduce a geometry projection loss in training *DiffuX2CT* to facilitate structural consistency in the 3D space.

To verify the clinical significance of CT reconstruction using biplanar X-rays, we collect a new real-world lumbar vertebra dataset, called LumbarV, with 268 3D CT images from different patients. Each contains implants inserted in the lumbar vertebrae. The *DiffuX2CT* is capable of reconstructing high-quality CT images, which could assist surgeons in locating the implants and measuring the shape of the vertebrae as shown in Fig. 1(c).

The major contributions of our work are summarized in three folds as follows:

- We propose a new, 3D conditional diffusion model - *DiffuX2CT*, which is the first model that leverages diffusion models with a new tri-plane conditioning mechanism to achieve faithful and high-quality CT image reconstruction from biplanar X-rays.
- We propose the implicit conditioning mechanism (ICM), which is realized by a newly designed tri-plane decoupling generator and an implicit neural decoder. Our implicit conditioning mechanism enables 3D structure information to be recovered for reconstructing CT images.

– We contribute a new real-world lumbar vertebra dataset, by collecting CT images from real clinical practice, which covers a wide range of diseased cases. The dataset will serve as a new benchmark to evaluate the effectiveness and performance of CT reconstruction from biplanar X-rays.

2 Related Work

CT Reconstruction from X-rays. Classical CT reconstruction methods, such as filtered back projection and iterative reconstruction, require a large number of X-ray projections taken from all viewing angles around the body to generate a high-quality volume [23] but encounter severe distortions under ill-posed settings. Subsequent model-based iterative reconstruction (MBIR) methods [27,49,60,74,75] are dedicated to using less number of X-rays and maintaining high sample quality, but they still can not address the unmet demand for real-time imaging while substantially reducing radiation. The success of deep learning methods makes it possible to reconstruct CT images using ultra-sparse X-rays. Regression-based methods [19,22,31,62,63,83,91] directly learn a mapping from X-rays to CT through specialized neural networks with a mean squared error (MSE) loss. [22,62,83] explore the feasibility of using a single X-ray projection to predict a 3D volume representation. X-CTRSNet [19] and XTransCT [31] achieve higher accuracy and show that using biplanar X-ray inputs can be advantageous for CT reconstruction. However, these regression-based methods perform poorly in generating high-quality CT images with fine details. Recently, GAN-based methods [28,85] have been proposed to improve the sample quality of generated CT images. For instance, X2CT-GAN [85] increases the data dimension of 2D X-rays to 3D CT and introduces adversarial loss for realistic CT reconstruction. Despite the ability to generate relatively fine details, these GAN-based methods still suffer from severe artifacts and struggle to capture complex data distributions, yielding unnatural textures.

3D Generative Models. Generative Adversarial Networks (GANs) [20], Variational Auto-Encoders (VAEs) [33], and DMs have been widely used to model the explicit 3D data distribution, such as voxel grids [7,82], point clouds [8,47,84], and meshes [39,46]. The recent advancement of Neural Radiance Fields (NeRF) [5,48] further facilitates the emergence of numerous works on building 3D structures through learning implicit representations [10,11,15,37,52,61,66,89,92]. In particular, EG3D [9] first proposes the concept of hybrid explicit-implicit tri-plane representation for efficient 3D-aware generation from 2D data. Subsequently, Rodin [78] and PVDM [86] generate tri-plane representations for synthesizing 3D digital avatars and videos, respectively. In comparison, our objective is to extract the tri-plane representations from biplanar orthogonal X-rays for capturing the 3D structural information.

Diffusion Probabilistic Models. Deep diffusion models (DMs) are first introduced by Sohl-Dickstein et al. [67] as a novel generative model that generates

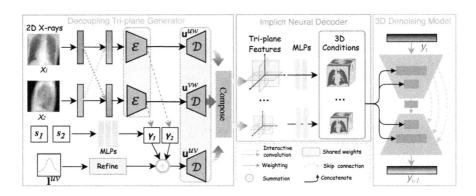

Fig. 2. Overview of our *DiffuX2CT*. Taking 2D biplanar orthogonal X-rays as inputs, the proposed implicit conditioning mechanism is implemented with a tri-plane decoupling generator and an implicit neural decoder to recover 3D structure information. By incorporating the 3D conditions, the 3D conditional denoising model iteratively generates CT images while preserving consistent structures provided in X-rays.

samples by gradually denoising images corrupted by Gaussian noise. Recent DM advances have demonstrated their superior performance in image synthesis, including DDPM [25], DDIM [68], ADM [16], LSGM [72], LDM [56], and DiT [53]. DMs also have achieved state-of-the-art performance in other synthesis tasks, such as text-to-image generation in GLIDE [50], DALLE-2 [55], and Imagen [58], speech synthesis [34,42], and super-resolution in SR3 [59] and IDM [18]. Moreover, DMs have been applied to text-to-3D synthesis in DreamFusion [54] and its follows [38,40,70,77,79,87,93], video synthesis [6,24,26,81,86], text-to-motion generation [71], and other 3D object synthesis in RenderDiffusion [2], diffusion SDF [37], and 3D point cloud generation [47]. Recently, several works [12,13,35,41,69,88] explore their potential in solving inverse problems in medical scenarios, such as sparse-view computed tomography (SV-CT), limited-angle computed tomography (LACT), and compressed-sensing MRI, *etc*. We would like to highlight that our DiffuX2CT is fundamentally different from these diffusion-based CT reconstruction methods in that DiffuX2CT can extract faithful 3D conditions from 2D X-rays, while they still rely on the CT scanners to obtain the 3D conditions. Moreover, their 2D denoising model's architectural design may not effectively handle the global coherence across the CT images.

3 Methodology

This section presents the *DiffuX2CT* approach, a 3D conditional diffusion model with an implicit conditioning mechanism (ICM), as shown in Fig. 2. *DiffuX2CT* is trained to recover faithful 3D structures from biplanar orthogonal X-rays (the front view and the side view) to generate high-quality 3D CT images.

3.1 Problem Statement

Given a CT image $\mathbf{y}$ and paired biplanar X-rays of a front view $\mathbf{x}_1$ and a side view $\mathbf{x}_2$, *DiffuX2CT* aims to learn a parametric approximation to the data distribution $p(\mathbf{y}|\mathbf{x}_1,\mathbf{x}_2)$ through a Markov chain of length T. Following [59], we define the forward Markovian diffusion process q by adding Gaussian noise as:

$$q\left(\mathbf{y}_t|\mathbf{y}_{t-1}\right) = \mathcal{N}\left(\mathbf{y}_t|\sqrt{1-\beta_t}\mathbf{y}_{t-1},\beta_t\mathbf{I}\right), \tag{1}$$

where $\beta_t \in (0,1)$ are the variances of the Gaussian noise in T iterations.

In the training process, *DiffuX2CT* is trained to infer the conditional distributions $p_\theta\left(\mathbf{y}_{t-1}|\mathbf{y}_t,\mathbf{x}_1,\mathbf{x}_2\right)$ to denoise the latent features sequentially. Formally, the inference process can be conducted as a reverse Markovian process from Gaussian noise $\mathbf{y}_T \sim \mathcal{N}(\mathbf{0},\mathbf{I})$ to a target volume $\mathbf{y}_0$ as:

$$p_\theta\left(\mathbf{y}_{t-1}|\mathbf{y}_t,\mathbf{x}_1,\mathbf{x}_2\right) = \mathcal{N}\left(\mathbf{y}_{t-1}|\mu_\theta\left(\mathbf{x}_1,\mathbf{x}_2,\mathbf{y}_t,t\right),\sigma_t^2\mathbf{I}\right). \tag{2}$$

As shown in Fig. 2, a 3D U-Net architecture is adopted as the denoising model that encodes the noisy image $\mathbf{y}_t$ into multi-resolution feature maps. Meanwhile, ICM is designed to recover multi-resolution 3D structural features from 2D X-rays $\mathbf{x}_1,\mathbf{x}_2$ for conditioning the generation process of the denoising model. Overall, *DiffuX2CT* offers an elegant, end-to-end framework that unifies the iterative denoising process with 3D conditional learning from 2D inputs, without involving extra prior models.

3.2 Implicit Conditioning Mechanism

Extracting 3D structural information from 2D X-rays is crucial for reconstructing faithful CT images that have consistent details with the inputs. An intuitive idea is to first extract 2D features of X-rays, and then expand them to 3D conditions by duplicating along the channel dimension, and directly fuse them by pixel-wise addition, as done in X2CT-GAN [85]. This way, 3D structure information would not be well preserved. In contrast, we develop an implicit conditioning mechanism with a tri-plane decoupling generator and an implicit neural function.

Tri-Plane Decoupling Generator. To implement our conditioning mechanism, we introduce a new formulation of tri-plane features for X-rays, which represents the 3D feature using three axis-aligned orthogonal planes, denoted by $\{\mathbf{u}^{uv} \in \mathbb{R}^{H \times W \times C}, \mathbf{u}^{uw} \in \mathbb{R}^{H \times D \times C}, \mathbf{u}^{vw} \in \mathbb{R}^{W \times D \times C}\}$, where H,W,D denote the spatial resolution and C is the number of channels. Given that the input front view $\mathbf{x}_1$ and side view $\mathbf{x}_2$ can be directly utilized for extracting the uw and vw feature planes $\mathbf{x}^{uw}$ and $\mathbf{x}^{vw}$, we propose to generate the tri-plane features in a decoupling manner. Furthermore, to capture the spatial correlation between $\mathbf{x}^{uw}$ and $\mathbf{x}^{vw}$, we design an interleaved convolution to mutually enhance $\mathbf{x}_1$ and $\mathbf{x}_2$. To model the third feature plane $\mathbf{x}^{uv}$, we introduce two view modulators for refining and combining $\mathbf{x}^{uw}$ and $\mathbf{x}^{vw}$. The combination weighted by view modulators inherently guarantees the spatial correlation between $\mathbf{x}^{uv}$ and the other two feature planes. However, reconstructing the absent $\mathbf{x}^{uv}$ solely through these two feature planes would not be sufficient. Hence we include a learnable embedding with a Gaussian perturbation to collect statistical information on the uv plane. Lastly, a lightweight decoder is employed to decode $\mathbf{x}^{uv}$, $\mathbf{x}^{uw}$, and $\mathbf{x}^{vw}$, consequently producing $\mathbf{u}^{uv}, \mathbf{u}^{uw}$, and $\mathbf{u}^{vw}$.

Interleaved Convolution. Since both $\mathbf{x}_1$ and $\mathbf{x}_2$ contain the information of the w axis, we propose to capture their spatial correlated features by incorporating each other's information before inputting them into the implicit encoder. Following a stem layer instantiated to extract initial features, we conduct the axis-wise pooling on $\mathbf{x}_1$ and $\mathbf{x}_2$ to obtain the vectors $\mathbf{x}_1^w, \mathbf{x}_2^w \in \mathbb{R}^{1 \times D \times C}$ along the w axis. They are subsequently expanded to the original 2D dimension by replicating the vectors along row dimension, deriving the auxiliary features $\mathbf{x}_1^{(\cdot)w} \in \mathbb{R}^{W \times D \times C}$, $\mathbf{x}_2^{(\cdot)w} \in \mathbb{R}^{H \times D \times C}$. Then the interleaved convolution and the plane feature encoder are successively applied to process the enhanced features $\hat{\mathbf{x}}_1 = \text{Concat}(\mathbf{x}_1, \mathbf{x}_2^{(\cdot)w})$ and $\hat{\mathbf{x}}_2 = \text{Concat}(\mathbf{x}_2, \mathbf{x}_1^{(\cdot)w})$ for extracting $\mathbf{x}^{uw}$ and $\mathbf{x}^{vw}$.

View Modulators and Learnable Embedding. To obtain the absent uv feature plane, we initialize two view embeddings with $s_1 = (1, 0, 1)$ for $\mathbf{x}_1$ and $s_2 = (0, 1, 1)$ for $\mathbf{x}_2$, and map them to two view modulators γ_1, γ_2, with the adaptive multi-layer perceptrons (MLPs). Subsequently, γ_1 and γ_2 are normalized by the L2 norm, and then used to modulate $\mathbf{x}^{uw}$ and $\mathbf{x}^{vw}$ channel-wisely and fuse them as the raw uv feature plane $\bar{\mathbf{x}}^{uv}$. Formally, the modulation process can be summarized as follows:

$$\bar{\gamma}_1 = \frac{|\gamma_1|}{\sqrt{\gamma_1^{(2} + \gamma_2^2 + \delta}}, \bar{\gamma}_2 = \frac{|\gamma_2|}{\sqrt{\gamma_1^2 + \gamma_2^2 + \delta}}, \quad (3)$$

$$\bar{\mathbf{x}}^{uv} = \bar{\gamma}_1 \cdot \mathbf{x}^{uw} + \bar{\gamma}_2 \cdot \mathbf{x}^{vw}, \quad (4)$$

where $\delta = 1e-8$ is introduced to avoid zero denominators. Furthermore, considering the information loss in uv plane, we adopt a sine distribution encoding **sine** and inject a Gaussian noise disturbance $\mathbf{z}$ to construct a learnable embedding $\mathbf{l}^{uv}$, which can supply additional information for generating the final uv feature plane $\mathbf{u}^{uv}$. As a result, $\mathbf{u}^{uv}$ can be obtained by:

$$\mathbf{x}^{uv} = \bar{\mathbf{x}}^{uv} + \text{CONV}(\mathbf{sine} + \mathbf{z}), \quad (5)$$

where CONV is a convolution layer with activation to refine the initial learnable embedding. Finally, we employ three simple decoders with shared weights, concurrently decoding $\mathbf{x}^{uv}$, $\mathbf{x}^{uw}$, and $\mathbf{x}^{vw}$ to generate the multi-resolution tri-plane features $\mathbf{u}^{(i)} = \{\mathbf{u}^{uv,(i)}, \mathbf{u}^{uw,(i)}, \mathbf{u}^{vw,(i)}\}$, where $i \in \{1, \cdots, N\}$ denotes the index with different outputs, and N is the number of stage depths in the decoders.

Implicit Neural Decoder. Based on the expressive tri-plane features $\mathbf{u}^{(i)}$ with rich explicit 3D information, we aim at recovering the 3D structural information used as the condition for the CT generation. To accomplish this, we leverage the appealing property of implicit neural function by encoding the tri-plane features as a query function based on the coordinates to acquire the 3D structural representation. Specifically, we assign $\mathbf{u}^{(i)}$ with the assumed 3D spatial coordinates $\mathbf{c}^{(i)} \in \mathbb{R}^{(D \times H \times W) \times 3}$ as a reference, where $D \times H \times W$ denotes the number of 3D points. Given the coordinates of each 3D point, we sample three corresponding feature vectors with bilinear interpolation by projecting them onto each plane and aggregating the retrieved vectors via summation. Afterward, the final 3D structural representation $\mathbf{h}^{(i)} \in \mathbb{R}^{D \times H \times W \times C}$ can be derived using a two-layer lightweight MLP decoder by:

$$\mathbf{h}^{(i)} = Reshape\{\mathrm{MLP}[\mathbf{u}^{(i)}(\mathbf{c}^{(i)})]\}, \tag{6}$$

where $\mathbf{u}^{(i)}(\mathbf{c}^{(i)})$ represents the aggregated features of the three vectors projected onto each plane. Note that we do not input the spatial coordinates into the lightweight MLP decoder. As a result, the multi-resolution 3D representations $\mathbf{h}^{(i)}$ can serve as faithful 3D conditions for guiding the CT generation.

3.3 Model Architecture

Given the feature maps $\mathbf{f}_{\mathrm{down}}^{(i)}$ and $\mathbf{f}_{\mathrm{up}}^{(i)}$ from the encoder and decoder of the 3D U-Net denoising model, where $i \in \{1, \cdots, N\}$ and N is the number of stage depths in the 3D U-Net, unlike the cross-attention mechanism that introduces additional computational cost, we directly concatenate them with the 3D conditions $\mathbf{h}^{(i)}$, respectively, so that CT features are synchronously generated according to the structural information in the conditions. While the 3D denoising model possesses more weight parameters compared to the 2D U-Net, it offers superior global coherence throughout the 3D volume. Moreover, the 2D U-Net synthesizes each slice in CT one by one for reconstructing a CT image, leading to a much slower inference time than our 3D model. To reduce the computation cost, we decrease the number of U-Ne layers by setting N to 3 and employ the shifted-window attention layer [45] rather than self-attention in the last stage of the 3D U-Net.

3.4 Training Objective

Noise Prediction Loss. By introducing $\mathbf{x}_1, \mathbf{x}_2$ as the conditions and their corresponding view embeddings s_1, s_2 in the reverse diffusion process, we optimize the denoising model as a noise estimator ϵ_θ to reconstruct the target CT image $\hat{\mathbf{y}}_0$ from a pure noise. The noise prediction loss is defined as:

$$\mathcal{L}_{simple} = E_{(\mathbf{x}_1, \mathbf{x}_2, \mathbf{y})} E_{\epsilon, t} \left[\left\| \epsilon - \epsilon_\theta \left(\mathbf{y}_t, t, \mathbf{x}_1, \mathbf{x}_2, s_1, s_2 \right) \right\|_1^1 \right], \tag{7}$$

where $\epsilon \sim \mathcal{N}(0, 1), t \sim [1, T]$, and $(\mathbf{x}_1, \mathbf{x}_2, \mathbf{y})$ are paired X-rays and CT.

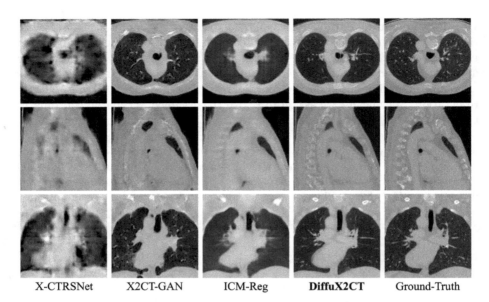

| X-CTRSNet | X2CT-GAN | ICM-Reg | **DiffuX2CT** | Ground-Truth |

Fig. 3. Qulitative comparisons on the CTSpine1K dataset. From top to bottom, we show the axial, sagittal, and coronal views. Best viewed by zoom in.

Geometry Projection Loss. Furthermore, we draw inspiration from [85] and introduce a geometry projection loss $\mathcal{L}_{proj}$. Specifically, given the predicted noise $\hat{\epsilon}$ and the timestep t, we can derive the $\hat{\mathbf{y}}_0$ by:

$$\hat{\mathbf{y}}_0 = (\mathbf{y}_t - (1-\alpha_t) \cdot \hat{\epsilon})/\sqrt{\alpha_t}, \tag{8}$$

where $\alpha_t = \prod_{i=1}^{t}(1-\beta_i)$. To streamline the process and focus on general geometry consistency, we employ three orthogonal projections to calculate the loss:

$$\mathcal{L}_{proj} = \frac{1}{3}(||\mathcal{P}_A(\mathbf{y}_0) - \mathcal{P}_A(\hat{\mathbf{y}}_0)||_1^1 + ||\mathcal{P}_C(\mathbf{y}_0) - \mathcal{P}_C(\hat{\mathbf{y}}_0)||_1^1 + ||\mathcal{P}_S(\mathbf{y}_0) - \mathcal{P}_S(\hat{\mathbf{y}}_0)||_1^1). \tag{9}$$

Putting everything together, we obtain the training objective as follows:

$$\mathcal{L}_{total} = \mathcal{L}_{simple} + \lambda \mathcal{L}_{proj}, \tag{10}$$

where we set $\lambda = 1/T$ in our experiments.

4 Experiments

In addition to the extensive experiments described in this section, we also provide more implementation details and results in the supplementary materials.

4.1 Datasets

We conduct extensive experiments on four datasets, including a newly collected lumbar dataset and three publicly available datasets.

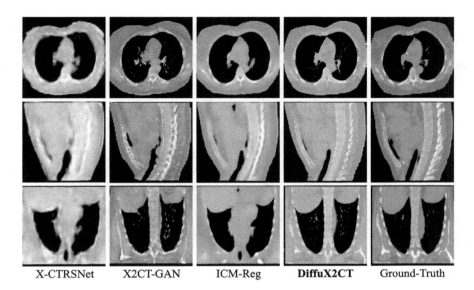

| X-CTRSNet | X2CT-GAN | ICM-Reg | **DiffuX2CT** | Ground-Truth |

Fig. 4. Qulitative comparisons on the LIDC-IDRI dataset. From top to bottom, we show the axial, sagittal, and coronal views. Best viewed by zoom in.

Newly Collected Lumbar Dataset. In our experiments, we collect a new lumbar vertebra dataset, called LumbarV, which comprises 268 3D CT images of different patients. Each patient has implants inserted in the vertebrae. We randomly select 231 CT images for training and leave the remaining 37 CT images for test. To ensure consistency, we initially resampled all CT images to an isotropic voxel resolution of $2 \times 2 \times 2$ mm^3 and then center-cropped them to a fixed size of $128 \times 128 \times 128$. Then, we adopt CT value clipping to limit the CT value range to [-1024, 1500], and min-max normalization to scale data to the [-1, 1] range. To construct the paired data of CT and biplanar X-ray data for training and testing, we follow previous works [62,85] and synthesize the corresponding biplanar X-rays from imagetric CT images using digitally reconstructed radiographs (DRR) [17].

Public Datasets. To make a comprehensive evaluation, we also perform experiments on three publicly available datasets: (1) **LIDC-IDRI dataset** [3] contains 1,018 chest CT images. We follow the work of X2CT-GAN [85] to split 917 CT images for training and 101 for test; (2) **CTSpine1K** [14] is collected from four open sources, including COLONOG [30], HNSCC-3DCT-RT [51], MSD Liver [65], and COVID-19 [21]. We randomly select 912 images for training and the remaining 102 for testing. (3) **CTPelvic1K** [43] is constructed for pelvic bone segmentation with annotation labels for the lumbar spine, sacrum, left hip, and right hip. We delete the duplicate data with CTSpine1K and randomly split 328 of the remaining data as the training set and 50 as the test set.

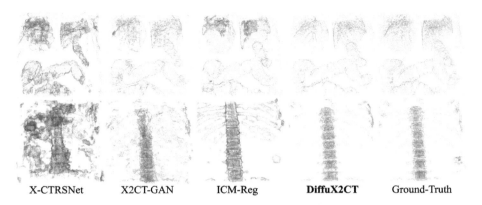

| X-CTRSNet | X2CT-GAN | ICM-Reg | **DiffuX2CT** | Ground-Truth |

Fig. 5. Rendering comparisons on the CTSpine1K datasets. The first and second row show the visualizations of soft tissues and bones, respectively.

4.2 Implementation Details

Compared Methods. We compare our *DiffuX2CT* quantitatively and qualitatively with a regression-based method X-CTRSNet [19] and a GAN-based method X2CT-GAN [85]. For X2CT-GAN, we use the official code provided by the authors to perform experiments. For X-CTRSNet, we reproduce the model code according to the paper, due to the absence of the official code. Particularly, since the proposed ICM formulates an effective way to recover 3D information from the 2D X-rays, we can easily develop a regression model, termed ICM-*Reg*, by combining ICM with a simple 3d decoder. Detailed model architecture is shown in the supplementary.

Training Details. We train our *DiffuX2CT* in an end-to-end manner. Following the vanilla DDPM [25], we use the Adam optimizer with a fixed learning rate of $2e-4$. The number of training iterations is set to $200,000$ for the LumbarV and CTPelvic1K datasets and $800,000$ for the LIDC-IDRI and CTSpine1K datasets. For training ICM-*Reg*, we use the same optimizer to train 200 epochs. We utilize a dropout rate of 0.2 and two 80GB NVIDIA RTX A800 GPUs for all experiments.

Evaluation Metrics. The evaluation metrics include two distortion-based metrics PSNR, SSIM, and two perceptual metrics FID and LPIPS. We calculate PSNR and SSIM in two ways: (1) directly computing the values of 3D data, denoted as PSNR3D and SSIM3D; and (2) averaging the values of all 2D slices along the three axes. It is worth noting that we use the image encoder of MedCLIP [80] trained on the medical dataset, i.e., MIMIC-CXR [29], which is appropriate for medical data to compute FID.

4.3 Comparisons with Prior Arts

We compare with two representative families of reconstruction models, i.e., generative and regression models. The performance of CT reconstruction methods is measured both qualitatively and quantitatively on the four datasets.

Qualitative Results. Fig. 3, Fig. 4, and Fig. 5 show the qualitative comparisons with prior arts on the LIDC-IDRI and CTSpine1K datasets. The results of the regression model X-CTRSNet exhibit inferior sample quality. Although the generative model, X2CT-GAN, can generate sharp textures, it often creates severe artifacts (e.g., incorrect heart shape and lack of aorta and pulmonary artery in Fig. 4) and distortions (i.e. unnatural bone structures in Fig. 5). Our regression version, ICM-*Reg*, is effective in recovering accurate structures but suffers from over-smoothing textures and lacks high-frequency details. In comparison, our *DiffuX2CT* is capable of reconstructing clear CT textures and consistent structure with the ground-truth, far superior to the counterpart methods.

Quantitative Results. Table 1 shows the quantitative comparisons with prior arts on the four datasets. *DiffuX2CT* exhibits the overall best results in terms of distortion-based metrics. This highlights the effectiveness of our implicit conditioning mechanism in extracting faithful structural information from 2D X-rays. Particularly, *DiffuX2CT* performs much better than other methods in terms of SSIM3D, showcasing its strong capacity to reconstruct CT images that are globally similar to the ground-truth in 3D space. Moreover, *DiffuX2CT* surpasses others by a large margin in all perceptual metrics, demonstrating its superiority in reconstructing high-quality CT images. Although the regression-based models, e.g., X-CTRSNet, can deliver promising PSNR since they are typically optimized by directly minimizing the L1 or L2 losses, they often encounter subpar sample quality and deliver lower perception metric scores.

4.4 Further Analysis by 3D Segmentation

To intuitively demonstrate the effectiveness and practical significance of our *DiffuX2CT*, we evaluate the performance in 3D segmentation of the reconstructed CT images on the CTPelvic1K dataset. Specifically, we utilize a well-trained VNet [1] for 3D pelvic segmentation to segment the sacrum, coccyx, left and right pelvis in the CT images. The visualizations of inputs with segmentation masks are shown in Fig. 6. We can clearly observe that *DiffuX2CT* significantly surpasses other methods, capably reconstructing bone structures that align with real CT. Moreover, we use the Dice Similarity Coefficient (DSC) as the evaluation metric for segmentation. Quantitative results of the three objectives are given in Table 2. *DiffuX2CT* outperforms the previous works by a large margin in all objectives, demonstrating its effectiveness in reconstructing CT images with commendably accurate shapes and positions from biplanar X-rays.

Table 1. Quantitative comparisons on the LumbarV and three public datasets.

Dataset	Method	PSNR↑	PSNR3D↑	SSIM↑	SSIM3D↑	FID ↓	LPIPS ↓
LumbarV	X-CTRSNet [19]	31.67	27.98	0.7738	0.7502	75.28	0.2957
	X2CT-GAN [85]	34.27	27.19	0.7456	0.6518	35.04	0.1554
	DiffuX2CT	**34.32**	**28.01**	**0.7984**	**0.8544**	**13.23**	**0.1198**
CTPelvic1K	X-CTRSNet [19]	23.89	23.27	0.6512	0.6103	91.64	0.5065
	X2CT-GAN [85]	23.33	22.64	0.5922	0.5571	28.88	0.2524
	DiffuX2CT	**23.91**	**23.27**	**0.6584**	**0.7650**	**7.78**	**0.1696**
CTSpine1K	X-CTRSNet [19]	21.34	20.84	0.5355	0.5159	255.98	0.3981
	X2CT-GAN [85]	20.60	20.16	0.4841	0.5052	50.63	0.2221
	DiffuX2CT	**21.53**	**21.12**	**0.5924**	**0.7099**	**8.90**	**0.1673**
LIDC-IDRI	X-CTRSNet [19]	23.95	**22.35**	0.6223	0.6228	143.73	0.2784
	X2CT-GAN [85]	26.02	21.17	0.6086	0.6199	16.65	0.1354
	DiffuX2CT	**26.35**	21.15	**0.6872**	**0.7423**	**4.52**	**0.1217**

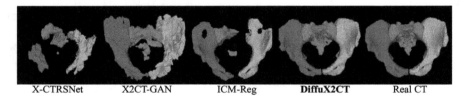

X-CTRSNet   X2CT-GAN   ICM-Reg   **DiffuX2CT**   Real CT

Fig. 6. Comparisons by rendering the 3D segmentation masks of the reconstructed CT images. The illustrated images are selected from the CTPelvic1K dataset. Our *DiffuX2CT* performs the best among compared methods.

Table 2. Quantitative results in terms of DSC for 3D segmentation. Higher DSC indicates better performance.

Data Source	Avg	Sac& Coc	L.Pel	R.Pel
Real (oracle)	90.19	87.55	91.62	91.41
X-CTRSNet	20.51	25.66	24.71	11.17
X2CT-GAN	45.95	43.08	50.31	44.44
ICM-*Reg*	47.59	31.98	56.82	53.97
DiffuX2CT	**74.10**	**73.63**	**73.31**	**75.35**

Table 3. Results of ablation study. V.M. denotes the view modulators.

Method	PSNR	SSIM	FID
Baseline	29.37	0.7034	17.15
w/o $\mathcal{L}_{proj}$	33.62	0.7630	16.22
w/o V.M.	32.09	0.7674	16.52
w/o l^{uv}	34.19	0.7781	14.51
ResNet	33.07	0.7593	13.77
DiffuX2CT	**34.32**	**0.7984**	**13.23**

4.5 Ablation Study

We conduct ablation studies on the LumbarV dataset to demonstrate the effectiveness of the proposed ICM and the geometry projection loss $\mathcal{L}_{proj}$. Specifically, we construct five comparison models: (1) the baseline model, which is a 3D conditional diffusion model with a simple conditioning mechanism that extracts 3D conditions by a 2D encoder with expanding and pixel-wise

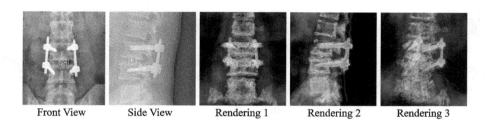

| Front View | Side View | Rendering 1 | Rendering 2 | Rendering 3 |

Fig. 7. Visualization of the reconstructed CT from real X-rays. The first and second columns are two input X-rays collected from real clinical practice.

addition operations like X2CT-GAN. (2) *DiffuX2CT* trained from scratch without $\mathcal{L}_{proj}$. (3)-(4) *DiffuX2CT* trained without the view modulators and learnable embedding $\mathbf{1}^{uv}$ respectively. (5) using a ResNetBlock to fuse the uw and vw feature planes without the view modulators. As shown in Table 3, all the proposed components contribute to the overall performance. The three models all yield good performance in perceptual metrics, illustrating the effectiveness of the 3D diffusion model in reconstructing high-quality CT images. Notably, the implicit conditioning model significantly improves the performance in terms of PSNR and SSIM, demonstrating its great effectiveness in capturing 3D structural information from 2D X-rays.

4.6 Clinical Case Study

To demonstrate the clinical utility of the proposed *DiffuX2CT*, we explore its actual clinical application in lumbar vertebra surgery. In that case, orthogonal X-rays can be easily acquired with C-arm machines, but CT images are often unavailable. This poses a great challenge to accurately locate the position and shape of implants, thereby increasing the difficulty of surgery. Fortunately, *DiffuX2CT* can potentially serve as a valuable alternative to reconstruct CT images. To verify this, we assess the performance of our *DiffuX2CT* on real-world lumbar X-ray data. Due to the absence of corresponding ground truth for the X-rays, we limit the evaluation to qualitative analysis. The reconstructed results of *DiffuX2CT* are shown in Fig. 7. Despite being trained on synthetic data, *DiffuX2CT* successfully reconstructs a realistic lumbar CT image. Especially, it effectively recovers the structure of bones and locates the position of implants, which is well aligned with real-world biplanar X-rays. This case study demonstrates that *DiffuX2CT* can potentially assist surgeons in observing the precise shapes and positions of bones and implants.

5 Conclusion

This paper presents a new 3D conditional diffusion model, DiffuX2CT, that achieves high-quality and faithful CT reconstruction from biplanar X-rays.

It is a significant breakthrough in this area. Specifically, DiffuX2CT adopts a 3D denoising model to ensure a high sample quality and global coherence across the 3D volume. More importantly, DiffuX2CT adopts a new implicit conditioning mechanism, comprised of a tri-plane decoupling generator and an implicit neural decoder, that effectively captures 3D structure information from 2D X-rays. The 3D conditions enable DiffuX2CT to achieve faithful CT reconstruction. Moreover, we construct a new lumbar dataset, named LumbarV, as a new benchmark for validating the significance and performance of CT reconstruction from biplanar X-rays. Extensive experiments on three public datasets and the LumbarV dataset demonstrate the effectiveness of our DiffuX2CT.

Limitation. The purpose of this work is not to replace the current CT reconstruction methods using hundreds and thousands of X-rays, but to offer a valuable alternative when performing CT scans is infeasible. Although DiffuX2CT can reconstruct high-quality CT images with faithful structures, *i.e.* bones, implants, it may not always accurately synthesize soft tissues, such as pulmonary nodules. One potential solution to this problem would be to introduce more X-ray views to gather additional structural information on soft tissues.

Acknowledgements. The work was supported by the National Key Research and Development Program of China (Grant No. 2023YFC3300029). This research was also supported by the Zhejiang Provincial Natural Science Foundation of China under Grant No. LD24F020007, Beijing Natural Science Foundation L223024, National Natural Science Foundation of China under Grant NO. 62076016, 62176068, and 12201024, "One Thousand Plan" projects in Jiangxi Province Jxsg2023102268, Beijing Municipal Science & Technology Commission, Administrative Commission of Zhongguancun Science Park Grant No.Z231100005923035. Taiyuan City "Double hundred Research action" 2024TYJB0127.

References

1. Abdollahi, A., Pradhan, B., Alamri, A.: VNet: an end-to-end fully convolutional neural network for road extraction from high-resolution remote sensing data. IEEE Access (2020)
2. Anciukevičius, T., Xu, Z., Fisher, M., Henderson, P., Bilen, H., Mitra, N.J., Guerrero, P.: RenderDiffusion: image diffusion for 3D reconstruction, inpainting and generation. In: arXiv:2211.09869 (2022)
3. Armato III, S.G., et al.: The lung image database consortium (LIDC) and image database resource initiative (IDRI): a completed reference database of lung nodules on CT scans. Med. Phys. **38**(2) 915–931 (2011)
4. Arnab, A., Dehghani, M., Heigold, G., Sun, C., Lučić, M., Schmid, C.: ViViT: a video vision transformer. In: ICCV (2021)
5. Barron, J.T., Mildenhall, B., Tancik, M., Hedman, P., Martin-Brualla, R., Srinivasan, P.P.: Mip-Nerf: a multiscale representation for anti-aliasing neural radiance fields. In: ICCV (2021)
6. Blattmann, A., et al.: Align your latents: high-resolution video synthesis with latent diffusion models. In: CVPR (2023)

7. Brock, A., Lim, T., Ritchie, J.M., Weston, N.: Generative and discriminative voxel modeling with convolutional neural networks. arXiv:1608.04236 (2016)
8. Cai, R., et al.: Learning gradient fields for shape generation. In: ECCV (2020)
9. Chan, E.R., et al.: Efficient geometry-aware 3D generative adversarial networks. In: CVPR (2022)
10. Chan, E.R., Monteiro, M., Kellnhofer, P., Wu, J., Wetzstein, G.: pi-GAN: periodic implicit generative adversarial networks for 3D-aware image synthesis. In: CVPR (2021)
11. Chen, Z., Zhang, H.: Learning implicit fields for generative shape modeling. In: CVPR (2019)
12. Chung, H., Ryu, D., McCann, M.T., Klasky, M.L., Ye, J.C.: Solving 3D inverse problems using pre-trained 2D diffusion models. In: CVPR (2023)
13. Chung, H., Sim, B., Ryu, D., Ye, J.C.: Improving diffusion models for inverse problems using manifold constraints. In: NeurIPS (2022)
14. Deng, Y., et al.: CTspine1k: a large-scale dataset for spinal vertebrae segmentation in computed tomography. arXiv:2105.14711 (2021)
15. Deng, Y., Yang, J., Xiang, J., Tong, X.: Gram: generative radiance manifolds for 3D-aware image generation. In: CVPR (2022)
16. Dhariwal, P., Nichol, A.: Diffusion models beat GANs on image synthesis. In: NeurIPS **34**, 8780–8794 (2021)
17. Galvin, J.M., Sims, C., Dominiak, G., Cooper, J.S.: The use of digitally reconstructed radiographs for three-dimensional treatment planning and CT-simulation. Int. J. Radiat. Oncol. Biol. Phys. **31**(4), 935–942 (1995)
18. Gao, S., et al.: Implicit diffusion models for continuous super-resolution. In: CVPR (2023)
19. Ge, R., et al.: X-CTRSNet: 3D cervical vertebra CT reconstruction and segmentation directly from 2D X-ray images. Knowl. -Based Syst. **236**, 107680 (2022)
20. Goodfellow, I., et al.: Generative adversarial nets. NeurIPS **27**(2014)
21. Harmon, S.A., et al.: Artificial intelligence for the detection of COVID-19 pneumonia on chest CT using multinational datasets. Nat. commun. 11(1), 4080 (2020)
22. Henzler, P., Rasche, V., Ropinski, T., Ritschel, T.: Single-image tomography: 3D volumes from 2D cranial x-rays. Comput. Graph. Forum **37**(2), 377–388 (2018)
23. Herman, G.T.: Fundamentals of Computerized Tomography: Image Reconstruction from Projections. Springer Science & Business Media (2009). https://doi.org/10.1007/978-1-84628-723-7
24. Ho, J., et al.: Imagen video: high definition video generation with diffusion models. arXiv:2210.02303 (2022)
25. Ho, J., Jain, A., Abbeel, P.: Denoising diffusion probabilistic models. NeurIPS **33**, 6840–6851 (2020)
26. Ho, J., Salimans, T., Gritsenko, A., Chan, W., Norouzi, M., Fleet, D.J.: Video diffusion models. NeurIPS **35**, 8633–8646 (2022)
27. Huang, Y., Taubmann, O., Huang, X., Haase, V., Lauritsch, G., Maier, A.: Scale-space anisotropic total variation for limited angle tomography. IEEE Trans. Radiat. Plasma Med. Sci. **2**(4), 307–314 (2018)
28. Jiang, L., Zhang, M., Wei, R., Liu, B., Bai, X., Zhou, F.: Reconstruction of 3D CT from a single x-ray projection view using CVAE-GAN. In: ICMIPE (2021)
29. Johnson, A.E., et al.: MIMIC-CXR, a de-identified publicly available database of chest radiographs with free-text reports. Sci. data **6**(1), 317 (2019)
30. Johnson, C.D., et al.: Accuracy of CT colonography for detection of large adenomas and cancers. New Engl. J. Med. **359**(12), 1207–1217 (2008)

31. Kasten, Y., Doktofsky, D., Kovler, I.: End-to-end convolutional neural network for 3D reconstruction of knee bones from bi-planar X-ray images. In: MICCAI (2020)
32. Khan, A., et al.: Comparing next-generation robotic technology with 3-dimensional computed tomography navigation technology for the insertion of posterior pedicle screws. World Neurosurg. **123**, e474–e481 (2019)
33. Kingma, D.P., Welling, M.: Auto-encoding variational bayes. In: ICLR (2014)
34. Kong, Z., Ping, W., Huang, J., Zhao, K., Catanzaro, B.: DiffWave: a versatile diffusion model for audio synthesis. arXiv:2009.09761 (2020)
35. Lee, S., Chung, H., Park, M., Park, J., Ryu, W.S., Ye, J.C.: Improving 3D imaging with pre-trained perpendicular 2D diffusion models. In: ICCV (2023)
36. Li, H., et al.: UV-IDM: identity-conditioned latent diffusion model for face UV-texture generation. In: CVPR (2024)
37. Li, M., Duan, Y., Zhou, J., Lu, J.: Diffusion-SDF: text-to-shape via voxelized diffusion. In: CVPR (2023)
38. Li, S., et al.: Zone: zero-shot instruction-guided local editing. In: CVPR (2024)
39. Liao, Y., Schwarz, K., Mescheder, L., Geiger, A.: Towards unsupervised learning of generative models for 3D controllable image synthesis. In: CVPR (2020)
40. Lin, C.H., et al.: Magic3D: High-resolution text-to-3D content creation. arXiv:2211.10440 (2022)
41. Liu, J., et al.: Dolce: a model-based probabilistic diffusion framework for limited-angle CT reconstruction. In: ICCV (2023)
42. Liu, J., Li, C., Ren, Y., Chen, F., Zhao, Z.: Diffsinger: singing voice synthesis via shallow diffusion mechanism. In: AAAI (2022)
43. Liu, P., et al.: Deep learning to segment pelvic bones: large-scale CT datasets and baseline models. Int. J. Comput. Assist. Radiol. Surg. **16**, 749–756 (2021)
44. Liu, X., et al.: Ladiffgan: Training GANs with diffusion supervision in latent spaces. In: CVPRW (2024)
45. Liu, Z., Ning, J., Cao, Y., Wei, Y., Zhang, Z., Lin, S., Hu, H.: Video swin transformer. In: CVPR (2022)
46. Liu, Z., Feng, Y., Black, M.J., Nowrouzezahrai, D., Paull, L., Liu, W.: MeshDiffusion: score-based generative 3D mesh modeling. arXiv:2303.08133 (2023)
47. Luo, S., Hu, W.: Diffusion probabilistic models for 3D point cloud generation. In: CVPR (2021)
48. Mildenhall, B., et al.: Nerf: Representing scenes as neural radiance fields for view synthesis. ACM, Commun. **65**(1), 99–106 (2021)
49. Mohan, K.A., et al.: TIMBIR: a method for time-space reconstruction from interlaced views. IEEE Trans. Comput. Imaging 1(2), 96–111 (2015)
50. Nichol, A., et al.: Glide: towards photorealistic image generation and editing with text-guided diffusion models. arXiv:2112.10741 (2021)
51. Nolan, T.: Head-and-neck squamous cell carcinoma patients with CT taken during pre-treatment, mid-treatment, and post-treatment (HNSCC-3DCT-RT). Cancer Imaging Archive (2022)
52. Park, J.J., Florence, P., Straub, J., Newcombe, R., Lovegrove, S.: DeepSDF: learning continuous signed distance functions for shape representation. In: CVPR (2019)
53. Peebles, W., Xie, S.: Scalable diffusion models with transformers. arXiv:2212.09748 (2022)
54. Poole, B., Jain, A., Barron, J.T., Mildenhall, B.: DreamFusion: text-to-3D using 2D diffusion. arXiv:2209.14988 (2022)
55. Ramesh, A., Dhariwal, P., Nichol, A., Chu, C., Chen, M.: Hierarchical text-conditional image generation with clip latents. arXiv:2204.06125 (2022)

56. Rombach, R., Blattmann, A., Lorenz, D., Esser, P., Ommer, B.: High-resolution image synthesis with latent diffusion models. In: CVPR (2022)
57. Ronneberger, O., Fischer, P., Brox, T.: U-net: convolutional networks for biomedical image segmentation. In: MICCAI (2015)
58. Saharia, C., et al.: Photorealistic text-to-image diffusion models with deep language understanding. NeurIPS **35**, 36479–36494(2022)
59. Saharia, C., Ho, J., Chan, W., Salimans, T., Fleet, D.J., Norouzi, M.: Image super-resolution via iterative refinement. TPAMI **45**(4), 4713–4726 (2022)
60. Schofield, R., et al.: Image reconstruction: part 1–understanding filtered back projection, noise and image acquisition. J. cardiovasc. comput. tomogr. **14**(3), 219–225 (2020)
61. Schwarz, K., Liao, Y., Niemeyer, M., Geiger, A.: Graf: generative radiance fields for 3d-aware image synthesis. NeurIPS **33**, 20154–20166 (2020)
62. Shen, L., Zhao, W., Xing, L.: Patient-specific reconstruction of volumetric computed tomography images from a single projection view via deep learning. Nat. Biomed. Eng. **3**(11), 880–888(2019)
63. Shiode, R., et al.: 2D–3D reconstruction of distal forearm bone from actual X-ray images of the wrist using convolutional neural networks. Sci. Rep. **11**(1), 15249 (2021)
64. Siasios, I.D., Pollina, J., Khan, A., Dimopoulos, V.G.: Percutaneous screw placement in the lumbar spine with a modified guidance technique based on 3D CT navigation system. J. Spine Surg. **3**(4), 657 (2017)
65. Simpson, A.L., et al.: A large annotated medical image dataset for the development and evaluation of segmentation algorithms (2019)
66. Sitzmann, V., Martel, J., Bergman, A., Lindell, D., Wetzstein, G.: Implicit neural representations with periodic activation functions. INeurIPS **33**, 7462–7473 (2020)
67. Sohl-Dickstein, J., Weiss, E., Maheswaranathan, N., Ganguli, S.: Deep unsupervised learning using nonequilibrium thermodynamics. In: ICML (2015)
68. Song, J., Meng, C., Ermon, S.: Denoising diffusion implicit models. arXiv:2010.02502 (2020)
69. Song, Y., Shen, L., Xing, L., Ermon, S.: Solving inverse problems in medical imaging with score-based generative models. In: ICLR (2022)
70. Tang, J., Wang, T., Zhang, B., Zhang, T., Yi, R., Ma, L., Chen, D.: Make-it-3D: high-fidelity 3D creation from a single image with diffusion prior. arXiv:2303.14184 (2023)
71. Tevet, G., Raab, S., Gordon, B., Shafir, Y., Cohen-Or, D., Bermano, A.H.: Human motion diffusion model. In: ICLR (2023)
72. Vahdat, A., Kreis, K., Kautz, J.: Score-based generative modeling in latent space. In: NeurIPS **34**, 11287–11302 (2021)
73. Vaswani, A., et al.: Attention is all you need. NeurIPS (2017)
74. Venkatakrishnan, S.V., Drummy, L.F., Jackson, M.A., De Graef, M., Simmons, J., Bouman, C.A.: A model based iterative reconstruction algorithm for high angle annular dark field-scanning transmission electron microscope (HAADF-STEM) tomography. TIP **22**(11), 4532–4544 (2013)
75. Venkatakrishnan, S.V., Mohan, K.A., Ziabari, A.K., Bouman, C.A.: Algorithm-driven advances for scientific CT instruments: from model-based to deep learning-based approaches. IEEE Sign. Process. Mag. **39**(1), 32–43 (2021)
76. Wang, G., Ye, J.C., Mueller, K., Fessler, J.A.: Image reconstruction is a new frontier of machine learning. TMI **37**(6), 1289–1296 (2018)
77. Wang, H., Du, X., Li, J., Yeh, R.A., Shakhnarovich, G.: Score jacobian chaining: Lifting pretrained 2D diffusion models for 3D generation. In: ICCV (2023)

78. Wang, T., et al.: Rodin: a generative model for sculpting 3D digital avatars using diffusion. In: CVPR (2023)
79. Wang, Z., Lu, C., Wang, Y., Bao, F., Li, C., Su, H., Zhu, J.: Prolificdreamer: High-fidelity and diverse text-to-3D generation with variational score distillation. arXiv:2305.16213 (2023)
80. Wang, Z., Wu, Z., Agarwal, D., Sun, J.: MedCLIP: contrastive learning from unpaired medical images and text. arXiv:2210.10163 (2022)
81. Wu, J.Z., et al.: Tune-a-video: one-shot tuning of image diffusion models for text-to-video generation. In: ICCV (2023)
82. Wu, J., Zhang, C., Xue, T., Freeman, B., Tenenbaum, J.: Learning a probabilistic latent space of object shapes via 3D generative-adversarial modeling (2016)
83. Wu, J., Mahfouz, M.R.: Reconstruction of knee anatomy from single-plane fluoroscopic X-ray based on a nonlinear statistical shape model. J. Med. Imaging **8**(1), 016001–016001(2021)
84. Xie, C., Wang, C., Zhang, B., Yang, H., Chen, D., Wen, F.: Style-based point generator with adversarial rendering for point cloud completion. In: CVPR (2021)
85. Ying, X., Guo, H., Ma, K., Wu, J., Weng, Z., Zheng, Y.: X2ct-GAN: reconstructing CT from biplanar X-rays with generative adversarial networks. In: CVPR (2019)
86. Yu, S., Sohn, K., Kim, S., Shin, J.: Video probabilistic diffusion models in projected latent space. In: CVPR (2023)
87. Zeng, B., et al.: IPDreamer: appearance-controllable 3D object generation with image prompts. arXiv:2310.05375 (2023)
88. Zeng, B., et al.: Controllable mind visual diffusion model. In: AAAI (2024)
89. Zeng, B., et al.: FNeVR: neural volume rendering for face animation. NeurIPS **35**, 22451–22462 (2022)
90. Zeng, B., et al.: Face animation with an attribute-guided diffusion model. In: CVPRW (2023)
91. Zhang, C., et al.: Xtransct: Ultra-fast volumetric CT reconstruction using two orthogonal x-ray projections via a transformer network. arXiv:2305.19621 (2023)
92. Zhang, Z., Sun, L., Yang, Z., Chen, L., Yang, Y.: Global-correlated 3D-decoupling transformer for clothed avatar reconstruction. NeurIPs **36** (2024)
93. Zhu, J., Zhuang, P.: HiFA: high-fidelity text-to-3D with advanced diffusion guidance. arXiv:2305.18766 (2023)

AdaIFL: Adaptive Image Forgery Localization via a Dynamic and Importance-Aware Transformer Network

Yuxi Li[⬤], Fuyuan Cheng[⬤], Wangbo Yu[⬤], Guangshuo Wang[⬤], Guibo Luo[✉][⬤], and Yuesheng Zhu[✉][⬤]

School of Electronic and Computer Engineering, Peking University, Shenzhen, China
yuxili@stu.pku.edu.cn, {luogb,zhuys}@pku.edu.cn

Abstract. The rapid development of image processing and manipulation techniques poses unprecedented challenges in multimedia forensics, especially in Image Forgery Localization (IFL). This paper addresses two key challenges in IFL: (1) Various forgery techniques leave distinct forensic traces. However, existing models overlook variations among forgery patterns. The diversity of forgery techniques makes it challenging for a single static detection method and network structure to be universally applicable. To address this, we propose AdaIFL, a dynamic IFL framework that customizes various expert groups for different network components, constructing multiple distinct feature subspaces. By leveraging adaptively activated experts, AdaIFL can capture discriminative features associated with forgery patterns, enhancing the model's generalization ability. (2) Many forensic traces and artifacts are located at the boundaries of the forged region. Existing models either ignore the differences in discriminative information or use edge supervision loss to force the model to focus on the region boundaries. This hard-constrained approach is prone to attention bias, causing the model to be overly sensitive to image edges or fail to finely capture all forensic traces. In this paper, we propose a feature importance-aware attention, a flexible approach that adaptively perceives the importance of different regions and aggregates region features into variable-length tokens, directing the model's attention towards more discriminative and informative regions. Extensive experiments on benchmark datasets demonstrate that AdaIFL outperforms state-of-the-art image forgery localization methods. Our code is available at https://github.com/LMIAPC/AdaIFL.

Keywords: Image Forgery Localization · Dynamic Network Architecture · Feature Importance-aware Attention

1 Introduction

With the rapid development of image editing and processing technologies, it has become more accessible for people to create realistic forged images, which may

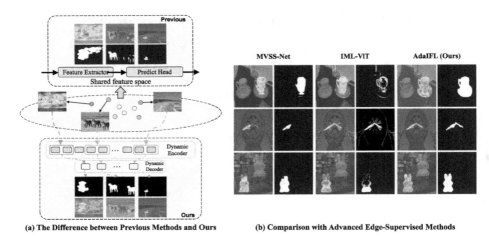

(a) The Difference between Previous Methods and Ours (b) Comparison with Advanced Edge-Supervised Methods

Fig. 1. Illustration of our contribution. Existing methods lack adaptability and flexibility in handling forged images with diverse forgery clues and patterns, leading to false alarms, missed detections, and the inability to accurately locate subtle forgery regions. In contrast, our model incorporates dynamic routing and feature importance-aware mechanisms, which can dynamically handle different forged image samples and adaptively perceive the significance of different regions, demonstrating superior performance in these aspects.

be abused to spread malicious information and pose a significant challenge to the security of media content [31]. Therefore, it is of great importance to develop an effective method to identify and locate the forged images.

In recent years, researchers have proposed many deep learning-based methods [14,25,37,39,43] to detect and localize forged regions, achieving significant progress. However, these methods still cannot achieve satisfactory results in real-life scenarios, mainly facing two challenges: (1) Fakers possess various techniques and tools to manipulate images, including object insertion, deletion, clone, and distortion, each leaving distinct artifacts and forensic traces. As illustrated in Fig. 1 (a), existing methods lack adaptability in handling forged images with diverse forgery clues and patterns. These methods usually share network structures and parameters, mapping different forged images into a common feature space without considering forgery pattern differences, resulting in limited generalization ability. (2) Artifacts and forensic traces produced by image manipulation operations are primarily located at the boundaries of forged regions. These traces are subtle and delicate, involving minor changes in lighting, texture, or the removal of small objects. This poses significant challenges for forgery localization due to sparse features, limited contextual information, and vulnerability to damage. To capture subtle artifacts and forensic traces, some methods such as IML-ViT [21] and MVSS-Net [5] use edge supervision loss to force the model to focus on the region boundaries. However, as shown in Fig. 1 (b), this hard-constrained approach easily leads to attention bias, causing the model to be overly sensitive to image edges or fail to finely capture all forensic traces. This leads to problems such as false alarms, missed detections, and the inability to accurately locate the forged regions with sharp boundaries. Therefore, it is essential to strike a more flexible and appropriate balance to prevent the model from excessively relying on boundary regions and neglecting other crucial features.

In this paper, we propose AdaIFL, a novel dynamic and importance-aware transformer network for IFL. To address the challenge of insufficient adaptability in handling forged samples, we propose a novel dynamic framework that incorporates the concept of dynamic routing mechanism into IFL. Our framework comprises a transformer-based dynamic encoder and a lightweight dynamic decoder. We customize various expert groups within different network components, constructing multiple distinct feature subspaces. Through the routing of the gating network, forged image samples can selectively activate different parts of the network, thereby digging out discriminative features associated with forgery patterns within their respective feature subspaces. This significantly enhances the model's generalization ability.

To avoid the model's over-sensitivity or neglect of boundary artifacts around forged regions, we propose feature importance-aware attention (FIA), a flexible approach that adaptively perceives the importance of different regions. One of the key components is the adaptive token aggregator (ATA), which consists of three parts: importance-aware region partitioning, aggregation scale allocation, and adaptive token aggregation. ATA aims to dynamically aggregate region features into variable-length tokens based on discriminative information of regional features, modelling forged regions of different scales and shapes. Specifically, we design a scoring network to quantify the importance of each image feature. Based on the importance score, the entire image region is partitioned into multiple sub-regions. Then a simple yet effective adaptive mechanism is used to evaluate the discriminative information, thereby determining the aggregation scale of each sub-region. In particular, the smaller aggregation scales are assigned to generate more feature tokens for regions with more discriminative regions, such as image manipulation boundaries. Finally, the clustering algorithm is employed to merge tokens guided by the aggregation scales, resulting in compact yet highly discriminative token representations. FIA directs the model's attention towards more discriminative regions, significantly enhancing the model's ability to locate various forged regions accurately.

Our main contributions are summarized as follows:

- We propose AdaIFL, a novel dynamic and importance-aware IFL framework. To the best of our knowledge, this is the first work introducing the dynamic routing mechanism to IFL, making it a pioneering contribution in this field.
- We propose a feature importance-aware attention that adaptively perceives the importance of different regions, improving the model's ability to locate various forged regions accurately.
- We conduct extensive experiments on several benchmarks and demonstrate that AdaIFL outperforms the existing state-of-the-art methods qualitatively and quantitatively.

2 Related Work

Image Forgery Localization. Existing methods typically leverage inconsistencies or discrepancies between forged and authentic regions for detection and localization. They mostly employ feature extractors to capture forgery-related information, i.e. RGB noise [3,11], high-frequency features [19,32], or edge artifacts [5,42]. [3] and [11] extract low-level artifacts from camera model fingerprints. ManTra-Net [37] utilizes both BayarConv and SRM as noise extractors simultaneously to capture rich features. CAT-Net [19] uses discrete cosine transform (DCT) coefficients to localize image manipulation. ObjectFormer [32] captures subtle forgery traces by extracting high-frequency features of the image and combining them with RGB features. TruFor [11] extracts high-level features from RGB images as well as noise-sensitive fingerprint features to locate forgeries. In addition, many methods exploit edge artifacts for forgery localization. For example, GSR-Net [42] introduces an edge detection and refinement branch, enabling the model to better identify boundary artifacts. MVSS-Net [5] employs an edge supervision branch to capture fine-grained boundary details in a shallow-to-deep manner. However, these methods use the shared feature extractor for all forged images, overlooking the differences among forgery patterns. In this paper, we propose a dynamic IFL framework that leverages adaptively activated experts to capture discriminative features associated with forgery patterns.

Mixture of Experts. Mixture of Experts (MoE) is first proposed by Jacobs et al. [15] to learn a system consisting of separated expert networks to improve model performance. Recently, Sparse Mixture of Experts has become popular in natural language processing [8,30] and computer vision [9,28,38] tasks. These networks activate a portion of them based on the input, separating the number of parameters from computational complexity. Recent works introduce MoE into vision transformers [1,26] to enhance their performance. These studies typically focus on combining the feed-forward layers of transformers with MoE or developing better routing strategies to achieve improved performance. In this paper, we apply the aforementioned dynamic routing and sparse computation techniques to the vision transformer and construct the AdaIFL network.

Feature Importance Awareness. Convolutional neural networks (CNNs) and vision transformers (ViTs) have achieved great success in computer vision tasks. They often require many computational resources to handle large-scale image data. In fact, most image features are not informative, containing a considerable amount of redundancy. Therefore, it is crucial to identify the most relevant features and prioritize them accordingly. Gated convolution [40] introduces gated units, allowing the weights at each location to be dynamically adjusted based on the inputs. Moreover, to mitigate the computational overhead of vanilla self-attention, many content-based sparse attentions have been proposed in ViTs. These methods either dynamically prune redundant tokens [27,34] based on the prediction module or directly cluster tokens [29,33,41] based on content similarity to reduce the number of tokens. In this paper, we propose a feature importance-aware attention that adaptively perceives the importance of different regions, achieving superior performance without any edge supervision.

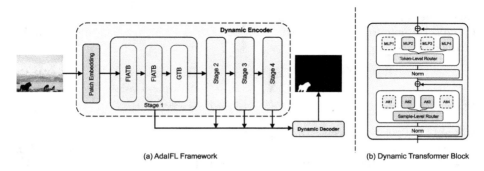

Fig. 2. (a) AdaIFL framework. AdaIFL takes the suspicious image as input and then utilizes a transformer-based dynamic encoder to extract multi-stage features. These features are passed into a lightweight dynamic decoder for multi-scale feature fusion, generating a spatial localization map to predict the forged regions. (b) Dynamic transformer block. AdaIFL framework incorporates the concept of dynamic routing mechanism into the transformer blocks (FIATB and GTB), introducing multiple specific experts within the attention and MLP layers to dig out discriminative features associated with forgery patterns. FIATB and GTB represent the Feature Importance-aware Transformer Block and Global Transformer Block, respectively.

3 Method

3.1 Framework Overview

In this paper, we propose AdaIFL, a dynamic and importance-aware image forgery localization framework. As illustrated in Fig. 2, AdaIFL introduces the dynamic routing mechanism into IFL. It tailors various expert groups for distinct network components, constructing multiple feature subspaces to specialize in learning different forgery patterns. Specifically, the framework consists of a transformer-based dynamic encoder and a lightweight dynamic decoder. The encoder involves four stages, each comprising two stacked feature importance-aware transformer blocks (FIATB) and a global transformer block (GTB). Additionally, we propose a dynamic decoder to fuse multi-stage features and predict the spatial localization map of forged regions. Its detailed structure is illustrated in Fig. 4 (a), comprising two processes: Multi-Scale Feature Fusion (MSFF) and Dynamic Decoding. In the following sections, we present details of several key components of AdaIFL.

3.2 Dynamic Transformer Block

Preliminaries: Mixture of Experts. A Mixture of Experts (MoE) layer comprises a set of expert networks, denoted as $E_1, E_2, \ldots, E_N$, along with a routing network, denoted as $\mathcal{G}$. Each expert can be implemented as an attention layer, an MLP layer, or a convolutional layer in different components of the network. The routing network $\mathcal{G}$ is responsible for determining the probability of utilizing each expert E_i and selects the top k experts as contributors to the final output. $\mathcal{G}$ can select experts based on various forms such as a single token, input sample, or task embedding:

$$\mathcal{G} = \begin{cases} G_{token}(x_i), & \text{Token-level} \\ G_{sample}(X), & \text{Sample-level} \\ G_{task}(embed(id_{task})), & \text{Task-level} \end{cases} \quad (1)$$

where G defines the specific routing strategy for gate decision and $X = \{x_i\}_{i=1}^L$ is the sequence of all tokens in the current sample. The final output of the MoE layer is a weighted sum of outputs from selected experts $E \subset \{E_1, E_2 \ldots E_N\}$:

$$y = \sum_{i \in E} \mathcal{G}(x) \cdot E_i(x) \quad (2)$$

Dynamic Transformer Block with MoE. We further incorporate the dynamic routing concept of MoE into the transformer architecture. Specifically, we introduce multiple specific experts into the attention and MLP layers. These experts are encouraged to explore unique artifacts and forensic traces associated with forgery patterns, enhancing the network's generalization capability. As shown in Fig. 2 (b), we replace the original attention layer with a set of parallel attention experts and a sample-level router, and replace the original MLP layer with a set of parallel MLP experts and a token-level router. Formally, the transformer block of the AdaIFL encoder can be formulated as

$$X'_l = \text{MoE-Att}(\text{LN}(X_{l-1})) + X_{l-1}$$
$$X_l = \text{MoE-FFN}(\text{LN}(X'_l)) + X'_l \quad (3)$$

where $\text{MoE-Att}(X) = \sum_{i \in E_{\text{Att}}} G^i_{sample}(X) \cdot E^i_{\text{Att}}(X)$, $\text{MoE-FFN}(x) = \sum_{i \in E_{\text{MLP}}} G^i_{token}(x) \cdot E^i_{\text{MLP}}(x)$, and LN denotes the layer normalization. Notably, we employ a feature importance-aware attention with MoE in FIATB, detailed in Sect. 3.3. In GTB, standard global self-attention is utilized.

3.3 Feature Importance-Aware Attention

To avoid the model's over-sensitivity or neglect of manipulation boundary region, we propose a feature importance-aware attention that adaptively perceives the importance of different regions and aggregates region features into variable-length tokens, modelling forged regions of different scales, shapes, and contents. The overall structure is shown in Fig. 3.

Importance-Aware Region Partitioning. We design a scoring network to evaluate the importance of region features for the IFL task, denoted as f_s, which is implemented as a lightweight module consisting of two MLP layers. Specifically, given an input feature $X \in \mathbb{R}^{N \times d}$, where d denotes the dimension of each feature token, and N denotes the number of feature tokens. f_s is used to quantify the importance of each feature token x_i.

$$s_i = f_s(\boldsymbol{x}_i), i = 1, \ldots, N \quad (4)$$

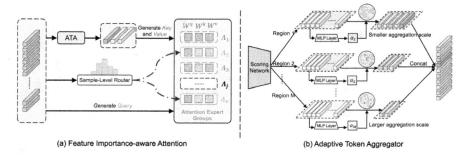

Fig. 3. (a) Illustration of feature importance-aware attention (FIA). (b) Adaptive token aggregator (ATA). From left to right, ATA involves three processes: importance-aware region partitioning, aggregation scale allocation, and adaptive token aggregation.

Furthermore, the feature tokens are sorted based on the importance scores, resulting in a sorted set of feature tokens and their respective scores, denoted as $\{x'_i\}_{i=1}^{N}$ and $\{s'_i\}_{i=1}^{N}$. Also, based on the sorted list of tokens, the entire image region is divided into m irregular sub-regions $\{R_i\}_{i=1}^{m}$ with each sub-region containing N_{R_i} tokens.

Aggregation Scale Allocation. Based on the importance of each region for forgery localization, the aggregation scales are dynamically adjusted to model forged regions of different scales and shapes. To accomplish this, we use a simple MLP layer to convert the distribution of region importance into information density factors. These factors are used to assess the discriminative information of each image region, represented as $\rho = \{\rho_1, \rho_2, \ldots, \rho_m\}$. We then apply the Softmax function to normalize the density factors and use them to generate the aggregation scales for different regions. This can be formulated as

$$\hat{\rho}_i = \frac{e^{\rho_i}}{\sum_{i=1}^{m} e^{\rho_i}}, i = 1, \ldots, m \quad (5)$$

$$c_i = N_\lambda \hat{\rho}_i, \alpha_i = \frac{N_{R_i}}{c_i} \quad (6)$$

where N_λ is the predefined total number of tokens ($N_\lambda \ll N$), c_i denotes the number of tokens allocated to sub-region R_i, and α_i denotes the aggregation scale for R_i. This adaptive approach allocates a smaller aggregation scale to more discriminative regions, generating more feature tokens. Conversely, other regions use a larger scale for coarser token aggregation.

Adaptive Token Aggregation. After allocating the corresponding aggregation scale, we need to further consider how to aggregate the tokens. Intuitively, aggregating tokens with similar semantic information can avoid excessive attention to redundant features and insufficient attention to subtle features. Inspired by [7,41], we use the clustering algorithm for token clustering and merging, resulting in more accurate and compact token representations. Specifically, we employ the DPC-KNN algorithm [7] for token clustering, which is a density

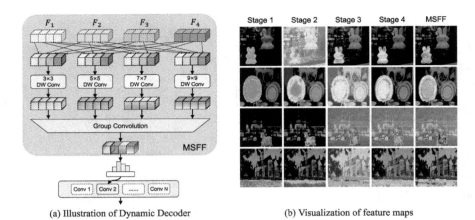

Fig. 4. (a) Illustration of dynamic decoder. It consists of multi-stage feature fusion (MSFF) and dynamic decoding. (b) From left to right are the GradCAM of the output feature maps for Stage 1, Stage 2, Stage 3, Stage 4, and MSFF. The MSFF module integrates multi-stage features, boosting the performance of forgery localization.

peak-based K-nearest neighbor clustering algorithm. Based on this algorithm, the feature tokens of each region are divided into c_i distinct clusters based on their similarity. During the aggregation process, importance scores are assigned as weights to the tokens within the same cluster, emphasizing the significance of different tokens. Thus, the token representation for each cluster is obtained:

$$\hat{x}_i = \frac{\sum_{j \in c_i} e^{s_j} x_j}{\sum_{j \in c_i} e^{s_j}} \quad (7)$$

Finally, the aggregated tokens from all regions are concatenated to obtain the final token representation $\hat{X} \in \mathbb{R}^{N_\lambda \times d}$.

Feature Importance-Aware Attention. We propose feature importance-aware attention (FIA) based on the adaptive token aggregator to avoid overfocusing on redundant features or overlooking local details. Specifically, we project the original feature tokens $X \in \mathbb{R}^{N \times d}$ as queries Q and the aggregated tokens $\hat{X} \in \mathbb{R}^{N_\lambda \times d}$ as keys K and values V. The process is defined as:

$$Q = XW^q, K = \hat{X}W^k, V = \hat{X}W^v$$
$$\text{FIA}(Q, K, V) = \text{Softmax}\left(\frac{QK^\top}{\sqrt{d}}\right) V \quad (8)$$

where $W^q, W^k, W^v \in \mathbb{R}^{d \times d}$ are the learnable matrixes.

As shown in Fig. 6, FIA directs the model's attention towards more discriminative regions, enhancing the performance of forgery localization. This is attributed to several advantages. (1) The region partitioning mechanism limits the interaction between authentic and forged regions, mitigating the feature coupling issue. Many forgery techniques aim to create semantically consistent and perceptually convincing tampered images for visual deception. Thus, direct clustering [7,41] inadvertently causes feature coupling. As shown in Fig. 6, ATA's regional partitioning mechanism mitigates the issue and reduces false alarms. (2) The adaptive aggregation mechanism can dynamically adjust the aggregation scale based on the importance of different regions, enabling the model to adapt flexibly to various scales and shapes of forged regions.

3.4 Dynamic Decoder

As shown in Fig. 4 (a), the dynamic decoder consists of two processes: multi-stage feature fusion (MSFF) and dynamic decoding, which are designed to fuse multi-stage features and achieve better localization performance. As illustrated in Fig. 4 (b), each stage focuses on capturing different levels of features. Thus, the MSFF is proposed to fully utilize the expressive capabilities of features at different stages and capture more discriminative information.

Specifically, MSFF utilizes the feature maps F_1, F_2, F_3, and F_4 from the four stages of the transformer as input and divides each feature map into four parts along the channel dimension. Then, we select one part of the features from each feature map and concatenate them to obtain four fused features. Next, depth-wise separable convolutions with different kernel sizes are applied to process these four fused features, capturing multi-scale features. Furthermore, the group convolution [18] is used to extract valuable information and filter out redundant feature representations from fused features. This can be formulated as follows:

$$Z_i = \text{Concat}\left(F_1\left[i\right], F_2\left[i\right], F_3\left[i\right], F_4\left[i\right]\right), Z_i^{'} = \text{DW}_{k_i \times k_i}\left(Z_i\right)$$
$$\hat{Z} = \text{GC}\left(\text{Concat}\left(Z_1^{'}, Z_2^{'}, Z_3^{'}, Z_4^{'}\right)\right) \tag{9}$$

where $k_i \in \{3, 5, 7, 9\}$, DW represents depth-wise separable convolution and GC represents group convolution.

Finally, we introduce a dynamic decoding module to predict the forged regions of the input image. It involves a set of parallel prediction heads $\{P_i\}_{i=1}^{n}$ and a sample router. Each prediction head P_i is implemented as a 1×1 convolution, i.e., $P_i(\hat{Z}) = W_i \hat{Z}$. In the sample router, we first perform global average pooling to generate the global embedding τ. Then, the activation probability for each head is calculated through a fully connected layer and sigmoid function, i.e., $AP\left(\hat{Z}\right) = \sigma\left(W_a \tau\right)$. The final output of the dynamic decoder is the weighted sum of the top-k prediction heads with the highest probabilities:

$$Y = \sum_{AP_i \in \text{top-k}} AP_i(\hat{Z}) \cdot P_i(\hat{Z}) \tag{10}$$

Table 1. Performance comparison of different image forgery localization methods. IOU, F1, and AUC scores are reported. The first and second rankings are shown in bold and underlined, respectively.

Method	CASIA v1			Coverage			DSO-1			NIST16			MISD			Mean		
	IOU	F1	AUC	IOU	F1	AUC	IOU	F1	AUC	IOU	F1	AUC	IOU	F1	AUC	IOU	F1	AUC
ManTraNet	.087	.543	.776	.187	.612	.849	.032	.487	.690	.119	.560	.770	.159	.567	.868	.117	.554	.791
CAT-Net v2	<u>.636</u>	.844	.873	.231	.614	.651	<u>.256</u>	<u>.632</u>	.649	.128	.553	.592	.248	.598	.628	.300	.648	.679
IF-OSN	.218	.591	.626	.202	.568	.706	.208	.602	.657	.217	.598	.720	.517	.775	.778	.272	.627	.697
PSCC-Net	.547	.799	**.938**	.280	.646	.858	.137	.552	.705	.226	.572	.745	<u>.598</u>	<u>.822</u>	<u>.943</u>	.358	.678	.838
MVSS-Net	.397	.707	.718	<u>.384</u>	<u>.699</u>	.736	.188	.570	.599	.240	.616	.642	.521	.786	.766	.346	.676	.692
HiFi-IFDL	.062	.518	.681	.077	.524	.760	.068	.509	.668	.149	.544	.713	.172	.571	.773	.106	.533	.719
TruFor	.592	.813	.930	.298	.657	<u>.871</u>	**.270**	**.634**	<u>.827</u>	.288	<u>.653</u>	<u>.834</u>	.583	.819	<u>.943</u>	<u>.406</u>	<u>.715</u>	<u>.881</u>
IML-ViT	.489	.764	.908	.271	.631	.840	.224	.598	.697	<u>.306</u>	.643	.716	.592	.818	.931	.376	.691	.818
Ours	**.661**	**.848**	**.938**	**.464**	**.745**	**.910**	.237	.615	**.838**	**.393**	**.706**	**.861**	**.606**	**.830**	**.952**	**.472**	**.749**	**.900**

3.5 Optimization

To improve the accuracy of the model in detecting forged regions at the pixel level, we utilize binary cross-entropy loss and dice loss [22].

$$\mathcal{L}_{\text{BCE}}(p,y) = \sum \left(-y_i \log p_i - (1-y_i)\log(1-p_i)\right) \tag{11}$$

$$\mathcal{L}_{\text{Dice}}(p,y) = 1 - \frac{2\sum p_i \cdot y_i}{\sum p_i^2 + \sum y_i^2} \tag{12}$$

where p_i and y_i are the prediction labels and ground truth for each pixel of the forged image, respectively. In addition, we employ metric learning loss [13] to increase the discrepancy of feature distributions between the authentic and forged regions in forged image samples following [23].

$$\mathcal{L}_q = \frac{1}{|A_i|} \sum_{k^+ \in A_i} -\log \frac{\exp(q \cdot k^+/\tau)}{\sum_{k_-} \exp(q \cdot k^-/\tau)} \tag{13}$$

where k^+ and q represent the feature embeddings of the authentic regions, k^- represents the feature embedding of the forged regions. A_i denotes the set of all k^+. Combining all of the above, our final loss function can be formulated as:

$$\mathcal{L} = \lambda_1 \cdot \mathcal{L}_q + \lambda_2 \cdot \mathcal{L}_{\text{BCE}} + \lambda_3 \cdot \mathcal{L}_{\text{Dice}} \tag{14}$$

where $\lambda_1, \lambda_2, \lambda_3$ are the parameters to balance the three terms in loss function ($\lambda_1+\lambda_2+\lambda_3=1$). In experiments, they are set as 0.5, 0.15, and 0.35, respectively.

4 Experiments

4.1 Experimental Setup

Datasets. We use the same dataset as [11,19] to train AdaIFL. These training datasets include CASIA v2 [6], IMD2020 [24], and a manipulated image dataset

Table 2. Localization performance on NIST16 dataset under various distortions. IOU and F1 scores are reported.

Distortion	PSCC-Net		MVSS-Net		TruFor		Ours	
	IOU	F1	IOU	F1	IOU	F1	IOU	F1
JREGCompress(q=100)	.217	.571	.238	.615	.285	.651	**.392**	**.707**
JREGCompress(q=50)	.224	.575	.233	.613	.272	.644	**.399**	**.710**
GammaCorrect(p=0.85)	.220	.566	.232	.611	.290	.654	**.390**	**.704**
GammaCorrect(p=1.15)	.223	.579	.241	.617	.287	.652	**.385**	**.704**
GaussianBlur(k=3)	.224	.542	.225	.614	.276	.648	**.384**	**.703**
GaussianBlur(k=15)	.146	.297	.100	.539	.251	.630	**.347**	**.685**
GaussianNoise(k=1)	.094	.538	.109	.544	.156	.575	**.185**	**.586**
GaussianNoise(k=23)	.044	.494	.057	.513	**.119**	**.552**	.110	.543

created by [19], covering various types of forgery. To comprehensively evaluate the generalization ability of AdaIFL, we conduct benchmark tests on five datasets that do not overlap with the training set, i.e., CASIA v1 [6], Coverage [35], DSO-1 [4], NIST16 [10] and MISD [16]. These datasets cover a large number of forged images, with diverse forgery types and a wide distribution of data sources.

Metrics. Following most previous work, We use pixel-level Area Under Curve (AUC), F1, and Intersection over Union (IoU) scores as the evaluation metrics, where the threshold is set to 0.5 by default.

Implementation Details. AdaIFL is implemented using PyTorch and trained in an end-to-end manner. During the training process, input images are cropped to 1024 × 1024. To prevent bias caused by imbalanced training dataset sizes, we employ the methods of [11,19] to perform equal sampling for each dataset in every training epoch. In addition, common data augmentation techniques such as flipping, scaling, blurring, and JPEG compression are employed to enhance data diversity. We use an Adam [17] optimizer with a learning rate decays from 2×10^{-4} to 1×10^{-7}.

4.2 Comparison with the State-of-the-Art Methods

We carefully choose methods with open-source code and pre-trained models for testing to ensure a fair comparison. Additionally, it is crucial to ensure that the training datasets of these models do not overlap with the test datasets. Finally, we choose eight state-of-the-art methods to conduct a comprehensive comparison in a fair manner, i.e., TruFor [11], HiFi-IFDL [12], CAT-Net v2 [19], ManTraNet [37], MVSS-Net [2], PSCC-Net [20], IF-OSN [36] and IML-ViT [21].

Quantitative Comparisons. Table 1 shows the quantitative comparison results of different models regarding IOU, F1, and AUC scores. Among all compared models, AdaIFL achieves state-of-the-art performance, with average IOU,

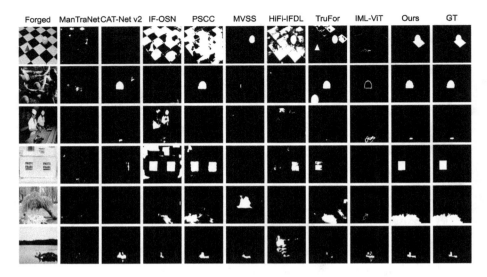

Fig. 5. Visualization of the predicted forged regions by different models. From left to right, we present the forged images, predictions of ManTraNet, CAT-Net v2, IF-OSN, PSCC-Net, MVSS-Net, HiFi-IFDL, TruFor, IML-ViT, and Ours, along with the ground truth masks. Zoom in for a better view.

F1, and AUC scores surpassing the second-best model by 6.6%, 3.4%, and 1.9%, respectively. Specifically, for F1 and IOU scores, AdaIFL achieves the best localization performance on CASIA, Coverage, NIST16, and MISD, ranking second on DSO-1. In particular, on Coverage (copy-move forgery dataset), AdaIFL's IOU and F1 scores exceed the second-best model by 8.0% and 4.6%, demonstrating our model's outstanding ability in suppressing feature coupling between forged and authentic regions. On the more challenging NIST16 dataset, our model outperforms the second-best model by 8.7% and 5.3% in IOU and F1 scores, showcasing exceptional generalization ability in dealing with various manipulation techniques and forgery patterns. Notably, using AUC as a metric may lead to overestimating the model's performance due to the dataset's highly imbalanced ratio between forged and authentic pixels. Nevertheless, our model achieves the best AUC scores on all datasets.

Qualitative Comparisons. Figure 5 shows the forgery localization results for different test images. Our method outperforms the current SOTA methods in several aspects. Firstly, our method effectively reduces false alarms and missed detections. Unlike other methods, which misidentify non-forged regions as forged or miss many forged regions, our method can accurately locate forged regions with sharp boundaries. Secondly, our method can accurately locate forged regions with various complex shapes. This is evident from the fifth and sixth rows of the test images, where other methods can only predict rough results and cannot capture detailed boundaries. In contrast, our method can accurately locate forged regions with complex shapes. Thirdly, our method demonstrates

Table 3. Ablation results on CASIA and MISD dataset using different variants of AdaIFL. IOU and F1 scores are reported.

Variants	CASIA v1		MISD	
	IOU	F1	IOU	F1
w/o FIA	.609	.817	.562	.808
w/o ATA	.617	.825	.567	.809
w/o DyD	.612	.824	.591	.821
w/o DyE	.587	.803	.569	.814
Ours	**.661**	**.848**	**.606**	**.830**

exceptional capability in accurately locating tiny and subtle forgery regions. As shown in the test images from the third and sixth rows, many methods struggle to identify these small regions, while our method accurately locates them.

Robustness Analysis. In real-world scenarios, forged images may undergo various post-processing operations. To evaluate the robustness of AdaIFL for forgery localization, we utilize common image degradation techniques, $i.e.$, Gaussian blurring, Gaussian noise, Gamma correction, and JPEG compression. The test results on NIST16, as shown in Table 2, indicate that AdaIFL demonstrates superior robustness against various degradation techniques compared to other state-of-the-art methods. Furthermore, Table 2 shows that excessive noise ($k = 23$) leads to suboptimal results in our model. It may be due to the routing process being affected by noise, resulting in suboptimal expert selection. We believe that it may be beneficial to design appropriate remedial mechanisms to ensure the accuracy of routing in various degradation scenarios.

4.3 Ablation Analysis

In this section, we analyze the effects of key components in the AdaIFL from two perspectives: feature importance awareness and dynamic network structure. Specifically, AdaIFL incorporates the idea of dynamic routing into IFL by customizing various expert groups in the encoder (DyE) and decoder (DyD), respectively. The feature importance-aware attention (FIA) adaptively perceives the importance of different regions, and one of the key components is the adaptive token aggregator (ATA), which aggregates region features into variable-length tokens, directing the model's attention towards more discriminative and informative regions. To evaluate the effectiveness of FIA, ATA, DyE, and DyD, we individually remove each component from AdaIFL and compare the test results with the full setup on CASIA and MISD datasets.

Influence of Feature Importance Awareness. As shown in Table 3, replacing FIA with the original attention results in a 5.2% decrease in IOU scores and a 3.1% decrease in F1 scores for CASIA, and a 4.4% decrease in IOU and a 2.2% decrease in F1 for MISD. After removing ATA, the IOU scores decrease by 4.4%

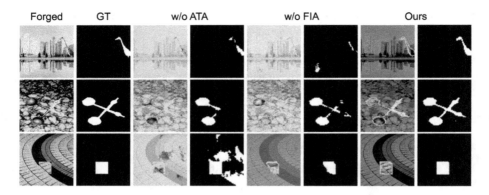

Fig. 6. Visualization of adaptive token aggregator (ATA) and feature importance-aware attention (FIA). From left to right, we display the forged images, ground truth masks, GradCAM of the feature map and prediction results without (w/o) ATA, without (w/o) FIA, and with the full setup.

on CASIA and 3.9% on MISD. Moreover, Fig. 6 shows that removing FIA or ATA leads to problems like missed detections, false alarms, and the inability to accurately locate forged regions. AdaIFL benefits from the region partitioning and adaptive aggregation mechanism, mitigating the issue of feature coupling and boosting the performance of image forgery localization. Additionally, Fig. 4 (b) shows the GradCAM of the feature maps at different stages of the transformer encoder. Guided by FIA, AdaIFL focuses on capturing diverse features at different stages. In the initial stage, the model prioritizes local details of the forged region boundary, emphasizing the fine perception of the boundary forged traces. In subsequent stages, it focuses on capturing different levels of features within the forged region. The MSFF module enhances the expressiveness of the model by fusing these features, enabling better capture of forged artifacts and forensic traces.

Influence of Dynamic Network Structure. In AdaIFL, we apply the concept of dynamic routing in both the transformer-based encoder and the lightweight decoder. As shown in Table 3, removing the dynamic expert groups significantly degrades the localization performance. Specifically, removing all dynamic expert groups from the encoder results in a decrease of 7.4% and 4.5% in IOU and F1 scores for CASIA, and a decrease of 3.7% and 1.6% in IOU and F1 scores for MISD. After removing the dynamic components from the decoder, the IOU scores for CASIA and MISD decrease by 4.9% and 1.5%, respectively. This clearly demonstrates the critical role of the dynamic routing concept in enhancing model generalization.

5 Conclusions

This paper addresses two key challenges of image forgery localization: (1) the insufficient adaptability in handling different forged images and (2) the inflexible approach to capturing boundary artifacts. To address these issues, we propose a novel dynamic and importance-aware image forgery localization framework (AdaIFL), which can dynamically handle different forged image samples and adaptively perceive the significance of different regions, enhancing the model's generalization ability. Extensive experiments demonstrate that our proposed method outperforms the state-of-the-art methods.

Acknowledgements. This work is supported by Shenzhen Science and Technology Program (No.JCYJ20230807120800001), and 2023 Shenzhen sustainable supporting funds for colleges and universities (No.20231121165240001). The authors sincerely appreciate the computing environment supported by the China Unicom Shenzhen Intelligent Computing Center.

References

1. Chen, T., Zhang, Z., Jaiswal, A., Liu, S., Wang, Z.: Sparse MoE as the new dropout: scaling dense and self-slimmable transformers. arXiv preprint arXiv:2303.01610 (2023)
2. Chen, X., Dong, C., Ji, J., Cao, J., Li, X.: Image manipulation detection by multi-view multi-scale supervision. In: Proceedings of the IEEE/CVF International Conference on Computer Vision, pp. 14185–14193 (2021)
3. Cozzolino, D., Verdoliva, L.: Noiseprint: a CNN-based camera model fingerprint. IEEE Trans. Inf. Forensics Secur. **15**, 144–159 (2019)
4. De Carvalho, T.J., Riess, C., Angelopoulou, E., Pedrini, H., de Rezende Rocha, A.: Exposing digital image forgeries by illumination color classification. IEEE Trans. Inf. Forensics Secur. **8**(7), 1182–1194 (2013)
5. Dong, C., Chen, X., Hu, R., Cao, J., Li, X.: MVSS-net: multi-view multi-scale supervised networks for image manipulation detection. IEEE Trans. Pattern Anal. Mach. Intell. **45**(3), 3539–3553 (2022)
6. Dong, J., Wang, W., Tan, T.: CASIA image tampering detection evaluation database. In: 2013 IEEE China Summit and International Conference on Signal and Information Processing, pp. 422–426. IEEE (2013)
7. Du, M., Ding, S., Jia, H.: Study on density peaks clustering based on k-nearest neighbors and principal component analysis. Knowl.-Based Syst. **99**, 135–145 (2016)
8. Du, N., et al.: Glam: efficient scaling of language models with mixture-of-experts. In: International Conference on Machine Learning, pp. 5547–5569. PMLR (2022)
9. Gross, S., Ranzato, M., Szlam, A.: Hard mixtures of experts for large scale weakly supervised vision. In: Proceedings of the IEEE Conference on Computer Vision and Pattern Recognition, pp. 6865–6873 (2017)
10. Guan, H., et al.: MFC datasets: large-scale benchmark datasets for media forensic challenge evaluation. In: 2019 IEEE Winter Applications of Computer Vision Workshops (WACVW), pp. 63–72. IEEE (2019)
11. Guillaro, F., Cozzolino, D., Sud, A., Dufour, N., Verdoliva, L.: TruFor: leveraging all-round clues for trustworthy image forgery detection and localization. In: Proceedings of the IEEE/CVF Conference on Computer Vision and Pattern Recognition, pp. 20606–20615 (2023)

12. Guo, X., Liu, X., Ren, Z., Grosz, S., Masi, I., Liu, X.: Hierarchical fine-grained image forgery detection and localization. In: Proceedings of the IEEE/CVF Conference on Computer Vision and Pattern Recognition, pp. 3155–3165 (2023)
13. He, K., Fan, H., Wu, Y., Xie, S., Girshick, R.: Momentum contrast for unsupervised visual representation learning. In: Proceedings of the IEEE/CVF Conference on Computer Vision and Pattern Recognition, pp. 9729–9738 (2020)
14. Hu, X., Zhang, Z., Jiang, Z., Chaudhuri, S., Yang, Z., Nevatia, R.: Span: spatial pyramid attention network for image manipulation localization. In: Computer Vision–ECCV 2020: 16th European Conference, Glasgow, UK, August 23–28, 2020, Proceedings, Part XXI 16, pp. 312–328. Springer (2020).https://doi.org/10.1007/978-3-030-58589-1_19
15. Jacobs, R.A., Jordan, M.I., Nowlan, S.J., Hinton, G.E.: Adaptive mixtures of local experts. Neural Comput. **3**(1), 79–87 (1991)
16. Kadam, K.D., Ahirrao, S., Kotecha, K.: Multiple image splicing dataset (MISD): a dataset for multiple splicing. Data **6**(10), 102 (2021)
17. Kingma, D.P., Ba, J.: Adam: a method for stochastic optimization. arXiv preprint arXiv:1412.6980 (2014)
18. Krizhevsky, A., Sutskever, I., Hinton, G.E.: Imagenet classification with deep convolutional neural networks. Adv. neural inf. process. syst. **25** (2012)
19. Kwon, M.J., Nam, S.H., Yu, I.J., Lee, H.K., Kim, C.: Learning JPEG compression artifacts for image manipulation detection and localization. Int. J. Comput. Vision **130**(8), 1875–1895 (2022)
20. Liu, X., Liu, Y., Chen, J., Liu, X.: PSCC-Net: Progressive spatio-channel correlation network for image manipulation detection and localization. IEEE Trans. Circuits Syst. Video Technol. **32**(11), 7505–7517 (2022)
21. Ma, X., Du, B., Liu, X., Hammadi, A.Y.A., Zhou, J.: IML-ViT: image manipulation localization by vision transformer. arXiv preprint arXiv:2307.14863 (2023)
22. Milletari, F., Navab, N., Ahmadi, S.A.: V-net: Fully convolutional neural networks for volumetric medical image segmentation. In: 2016 Fourth International Conference on 3D Vision (3DV), pp. 565–571. IEEE (2016)
23. Niloy, F.F., Bhaumik, K.K., Woo, S.S.: CFL-Net: image forgery localization using contrastive learning. In: Proceedings of the IEEE/CVF Winter Conference on Applications of Computer Vision, pp. 4642–4651 (2023)
24. Novozamsky, A., Mahdian, B., Saic, S.: Imd2020: a large-scale annotated dataset tailored for detecting manipulated images. In: Proceedings of the IEEE/CVF Winter Conference on Applications of Computer Vision Workshops, pp. 71–80 (2020)
25. Park, J., Cho, D., Ahn, W., Lee, H.K.: Double JPEG detection in mixed JPEG quality factors using deep convolutional neural network. In: Proceedings of the European Conference on Computer Vision (ECCV), pp. 636–652 (2018)
26. Rajbhandari, S., et al.: Deepspeed-MoE: advancing mixture-of-experts inference and training to power next-generation AI scale. In: International Conference on Machine Learning, pp. 18332–18346. PMLR (2022)
27. Rao, Y., Zhao, W., Liu, B., Lu, J., Zhou, J., Hsieh, C.J.: DynamicViT: efficient vision transformers with dynamic token sparsification. Adv. Neural. Inf. Process. Syst. **34**, 13937–13949 (2021)
28. Riquelme, C., et al.: Scaling vision with sparse mixture of experts. Adv. Neural. Inf. Process. Syst. **34**, 8583–8595 (2021)
29. Roy, A., Saffar, M., Vaswani, A., Grangier, D.: Efficient content-based sparse attention with routing transformers. Trans. Assoc. Comput. Linguist. **9**, 53–68 (2021)

30. Shazeer, N., Mirhoseini, A., Maziarz, K., Davis, A., Le, Q., Hinton, G., Dean, J.: Outrageously large neural networks: the sparsely-gated mixture-of-experts layer. arXiv preprint arXiv:1701.06538 (2017)
31. Verdoliva, L.: Media forensics and deepFakes: an overview. IEEE J. Sel. Top. Signal Process. **14**(5), 910–932 (2020)
32. Wang, J., Wu, Z., Chen, J., Han, X., Shrivastava, A., Lim, S.N., Jiang, Y.G.: Objectformer for image manipulation detection and localization. In: Proceedings of the IEEE/CVF Conference on Computer Vision and Pattern Recognition, pp. 2364–2373 (2022)
33. Wang, P., et al.: KVT: k-NN attention for boosting vision transformers. In: European Conference on Computer Vision, pp. 285–302. Springer (2022). https://doi.org/10.1007/978-3-031-20053-3_17
34. Wang, T., Yuan, L., Chen, Y., Feng, J., Yan, S.: PnP-DETR: towards efficient visual analysis with transformers. In: Proceedings of the IEEE/CVF International Conference on Computer Vision, pp. 4661–4670 (2021)
35. Wen, B., Zhu, Y., Subramanian, R., Ng, T.T., Shen, X., Winkler, S.: Coverage-a novel database for copy-move forgery detection. In: 2016 IEEE International Conference on Image Processing (ICIP), pp. 161–165. IEEE (2016)
36. Wu, H., Zhou, J., Tian, J., Liu, J.: Robust image forgery detection over online social network shared images. In: Proceedings of the IEEE/CVF Conference on Computer Vision and Pattern Recognition, pp. 13440–13449 (2022)
37. Wu, Y., AbdAlmageed, W., Natarajan, P.: Mantra-net: manipulation tracing network for detection and localization of image forgeries with anomalous features. In: Proceedings of the IEEE/CVF Conference on Computer Vision and Pattern Recognition, pp. 9543–9552 (2019)
38. Yang, B., Bender, G., Le, Q.V., Ngiam, J.: Condconv: conditionally parameterized convolutions for efficient inference. Adv. Neural Inf. Process. Syst. **32** (2019)
39. Yang, C., Li, H., Lin, F., Jiang, B., Zhao, H.: Constrained r-CNN: a general image manipulation detection model. In: 2020 IEEE International conference on multimedia and expo (ICME), pp. 1–6. IEEE (2020)
40. Yu, J., Lin, Z., Yang, J., Shen, X., Lu, X., Huang, T.S.: Free-form image inpainting with gated convolution. In: Proceedings of the IEEE/CVF International Conference on Computer Vision, pp. 4471–4480 (2019)
41. Zeng, W., Jin, S., Liu, W., Qian, C., Luo, P., Ouyang, W., Wang, X.: Not all tokens are equal: Human-centric visual analysis via token clustering transformer. In: Proceedings of the IEEE/CVF Conference on Computer Vision and Pattern Recognition, pp. 11101–11111 (2022)
42. Zhou, P., et al.: Generate, segment, and refine: towards generic manipulation segmentation. In: Proceedings of the AAAI Conference on Artificial Intelligence, vol. 34, pp. 13058–13065 (2020)
43. Zhou, P., Han, X., Morariu, V.I., Davis, L.S.: Learning rich features for image manipulation detection. In: Proceedings of the IEEE Conference on Computer Vision and Pattern Recognition, pp. 1053–1061 (2018)

Author Index

A
Amjadian, Ehsan 438

B
Bai, Yutong 257
Bar, Amir 257
Byeon, Wonmin 401

C
Cai, Jiting 124
Cao, Jiajiong 181
Chai, Weilong 181
Chang, Xiaojun 162
Chen, Kai 419
Chen, Ling 162
Chen, Long 365
Chen, Runnan 54
Chen, Tao 197, 365
Chen, Zhiquan 181
Cheng, Fuyuan 477
Cheng, Xiaoqiang 90
Cho, In 72

D
Darrell, Trevor 1, 257
Ding, Shuangrui 329
Dong, Runpei 214
Doublet, Félix 349

E
Elsaddik, Abdulmotaleb 162

F
Frost, Duncan 382
Fujiyoshi, Hironobu 349

G
Gao, Jianfeng 19
Ge, Jiannan 142
Ge, Jiaxin 1

Ge, Zheng 214
Geng, Haoran 214
Globerson, Amir 257
Gong, Kaixiong 197
Green, Simon 382
Gu, Qiao 382

H
Han, Chunrui 214
He, Lijun 107
He, Xiao 90
Herzig, Roei 1
Hirakawa, Tsubasa 349
Hojel, Alberto 257
Hu, Minghui 274
Huang, Qingming 311
Huang, Shijia 19

I
Iida, Tsumugi 349

J
Jeong, Jaehwan 401
Jeong, Jongoh 36
Jiang, Xinyun 365
Jiang, Xudong 274

K
Khaki, Samir 438
Kim, Jinkyu 401
Kim, Sangpil 401
Kim, Seon Joo 72
Kim, Sieun 401
Kim, Sungwoong 401
Kong, Lingdong 54

L
Lawryshyn, Yuri A. 438
Leizhang 19
Li, Chunyuan 19

Li, Fan 107
Li, Feng 19
Li, Hong 458
Li, Hongyang 19
Li, Jinke 90
Li, Liang 311
Li, Mingjie 162
Li, Pandeng 142
Li, Xiangtai 419
Li, Yining 419
Li, Yong-Lu 124
Li, Yuxi 477
Li, Zongrui 274
Liang, Xiaodan 162
Lin, Dahua 329
Lin, Haokun 162
Liu, Liu 292
Liu, Lucy Z. 438
Liu, Mingyu 124
Liu, Runkun 458
Liu, Shilong 19
Liu, Weihang 239
Liu, Xuhui 458
Liu, Youquan 54
Lou, Xin 239
Loy, Chen Change 419
Lu, Cewu 124
Luo, Guibo 477
Lv, Zhaoyang 382

M
Ma, Chenguang 181
Ma, Kaisheng 214
Ma, Yuexin 54

O
Oh, Gyeongrok 401
Otsuki, Seitaro 349

P
Pan, Liyuan 292
Peng, Xidong 54
Plataniotis, Konstantinos N. 438

Q
Qi, Zekun 214
Qian, Rui 329
Qian, Zhen 458
Qiao, Feng 54

Qiao, Yu 197
Qiao, Zhi 458
Qiu, Liang 162

R
Ren, Tianhe 19

S
Sajedi, Ahmad 438
Shi, Baifeng 1
Shi, Botian 197
Shim, Hyunbo 72
Straub, Julian 382
Su, Li 311
Subramanian, Sanjay 1
Sugiura, Komei 349
Sun, Yujing 54
Sweeney, Chris 382

T
Tian, Qi 142
Tu, Yunbin 311

W
Wang, Changbao 181
Wang, Guangshuo 477
Wang, Tai 54
Wang, Zhen 365
Wen, Yang 90

X
Xiao, Jun 365
Xie, Hongtao 142
Xie, Lingxi 142
Xu, Yue 124

Y
Yamashita, Takayoshi 349
Yan, Chenggang 311
Yang, Hunmin 36
Yang, Jainwei 19
Yang, Le 107
Yang, Yan 292
Yi, Li 214
Yoon, Kuk-Jin 36
Yu, Jingyi 239
Yu, Wangbo 477
Yuan, Haobo 419
Yuan, Jiakang 197
Yue, Xiangyu 197

Z

Zhang, Baochang 458
Zhang, Bo 197
Zhang, Dan 90
Zhang, Hao 19
Zhang, Juan 458
Zhang, Mingyu 124
Zhang, Shaochen 214
Zhang, Xiaopeng 142
Zhang, Yongdong 142

Zhen, Xiantong 458
Zheng, Dandan 181
Zheng, Qian 274
Zheng, Xue Xian 239
Zheng, Ziwei 107
Zhou, Chong 419
Zhou, Chonghua 90
Zhu, Xinge 54
Zhu, Yuesheng 477
Zou, Xueyan 19